SCIENCE IN FOCUS

STAGE 5

Dr Silvia Rudmann
Christopher Huxley
Anne Disney
Adam Sloan
Mya Skirving
Florence Coghlan
Charlotte Donovan
Stephen Zander
Celia McNeilly
Debra Smith

LEARNING DISCOVERY

COVID-19 VIRUS

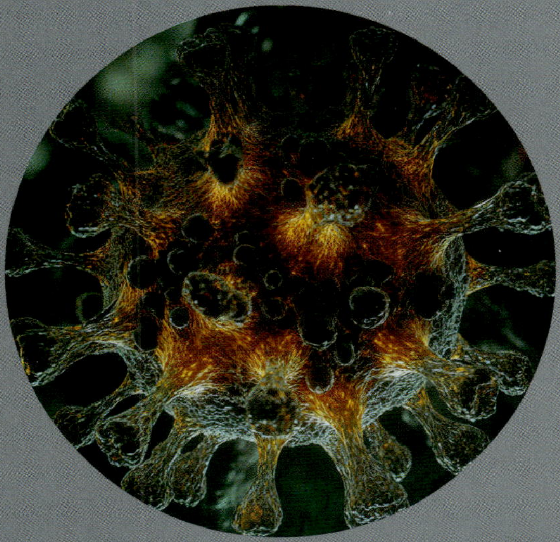

Viruses are a type of pathogen that can infect different organisms including animals, plants and even bacteria. To replicate, viruses hijack and use the machinery found inside a cell to produce more virus particles. Their ability to mutate and change rapidly, makes it hard to develop effective treatments.

COVID-19 was caused by the coronavirus SARS-CoV-2, a disease that became a global pandemic.

AF585012

Science in Focus Stage 5 NSW
1st edition
Dr Silvia Rudmann
Christopher Huxley
Anne Disney
Adam Sloan
Mya Skirving
Florence Coghlan
Charlotte Donovan
Stephen Zander
Celia McNeilly
Debra Smith

ISBN 9780170491785

Product Manager: Catherine Healy
Content Developer: Dr Trisha Downing
Editor: Jane Fitzpatrick
Proofreader: Scott Vandervalk
Series Designer: Danielle Maccarone
Series cover design: Regine Abos, Studio Regina
Series text design: Leigh Ashforth, Watershed Art & Design
Cover image: stock.adobe.com/Shanorsila
Permissions Researcher: Lumina Datamatics
Content Manager: Renee Tome
Typeset by: Lumina Datamatics

Any URLs contained in this publication were checked for currency during the production process. Note, however, that the publisher cannot vouch for the ongoing currency of URLs.

Acknowledgements
NSW Education Standards Authority

© 2025 Cengage Learning Australia Pty Limited

Copyright Notice

This Work is copyright. No part of this Work may be reproduced, stored in a retrieval system, or transmitted in any form or by any means without prior written permission of the Publisher. Except as permitted under the *Copyright Act 1968*, for example any fair dealing for the purposes of private study, research, criticism or review, subject to certain limitations. These limitations include: Restricting the copying to a maximum of one chapter or 10% of this book, whichever is greater; providing an appropriate notice and warning with the copies of the Work disseminated; taking all reasonable steps to limit access to these copies to people authorised to receive these copies; ensuring you hold the appropriate Licences issued by the Copyright Agency Limited ("CAL"), supply a remuneration notice to CAL and pay any required fees. For details of CAL licences and remuneration notices please contact CAL at Level 11, 66 Goulburn Street, Sydney NSW 2000,
Tel: (02) 9394 7600, Fax: (02) 9394 7601
Email: info@copyright.com.au
Website: www.copyright.com.au

For product information and technology assistance,
in Australia call **1300 790 853;**
in New Zealand call **0800 449 725**

For permission to use material from this text or product, please email
aust.permissions@cengage.com

National Library of Australia Cataloguing-in-Publication Data
A catalogue record for this book is available from the National Library of Australia.

Cengage Learning Australia
Level 5, 80 Dorcas Street
Southbank VIC 3006 Australia

For learning solutions, visit **cengage.com.au**

Printed in Malaysia by Papercraft
1 2 3 4 5 6 7 29 28 27 26 25

ACKNOWLEDGEMENT OF COUNTRY

Nelson acknowledges the Traditional Owners and Custodians of the lands of all Aboriginal and Torres Strait Islander Peoples of Australia. We pay respect to their Elders past and present.

We recognise the continuing connection of Aboriginal and Torres Strait Islander Peoples to the land, air and waters, and thank them for protecting these lands, waters and ecosystems since time immemorial.

Warning – Aboriginal and Torres Strait Islander Peoples are advised that this book and associated learning materials may contain images, videos or voices of deceased persons.

ABORIGINAL AND TORRES STRAIT ISLANDER PEOPLES GLOSSARY

Country/Place

Spaces mapped out that individuals or groups of Aboriginal and Torres Strait Islander Peoples of Australia occupy and regard as their own and that have varying degrees of spirituality. These spaces include lands, waters and sky.

Cultural narrative

A broad term that encompasses any cultural expression that includes (but is not limited to) knowledge and community values that are central to the identity of a particular group of Aboriginal and Torres Strait Islander Peoples.

Cultural narratives can hold information about almost anything, such as the origins of life, or can teach people about acceptable behaviour and rules, such as caring for Country. They can take the form of songs, stories, visual arts or performances. 'Cultural narrative' is a more accurate and respectful term than 'myth', 'story' or 'fable', terms that often diminish their importance.

Nation

A self-governed community of people based on a common language, culture and territory.

Peoples and Nations

We use the plural for these terms because Aboriginal and Torres Strait Islander Peoples do not belong to one nation/culture. There are many distinct Peoples and Nations. Also, some Nations consist of distinct clans or groups, so are referred to as Peoples.

Contents

Authors and contributors

Lead author

Dr Silvia Rudmann

Contributing authors

Christopher Huxley

Anne Disney

Adam Sloan

Mya Skirving

Florence Coghlan

Charlotte Donovan

Stephen Zander

Celia McNeilly

Debra Smith

Consultants

Joe Sambono
First Nations curriculum consultant

Science communication consultants

Additional science investigations

Reviewers

Eleanor Gregory, Anna Davis, Pete Byrne, Faye Paioff
Teacher reviewers

Aunty Gail Barrow, Nicole Brown, Christopher Evers, Associate Professor Melitta Hogarth, Carly Jia, Jesse King, Associate Professor Jessa Rogers, Theresa Sainty
Reviewers of the original First Nations Science Contexts pages in the *Nelson Science* series

SCIENCE IN FOCUS Learning Ecosystem

Science in Focus Stage 5 caters to all learners

Nelson has developed a detailed teaching program and scope and sequence to make it easier for teachers to plan, teach and support students using the *Science in Focus* series. Editable versions are available on Nelson MindTap.

Nelson MindTap

A flexible and easy-to-use online learning space that provides students with engaging, tailored learning experiences.

- Flexible formats: choose how you navigate using either the online eText, or offline PDFs.
- Margin links in the student book signpost multimedia student resources found on Nelson MindTap.

For students:

- Short, engaging videos with fun quizzes that bring science to life.
- Interactive activities and simulations that help you develop your science skills and knowledge.
- Content, feedback and support that you can access as you need it, which allows you to take control of your own learning.

For teachers:

- 100% modular, flexible courses let you adapt the content to your students' needs.
- Differentiated activities and assessments can be assigned directly to the student, or the whole class.
- You can monitor progress using assessment tools like Gradebook and Reports.

Security & privacy:

Nelson MindTap joined the Safer Technologies 4 Schools (ST4S) Product Badge Program in 2024. The annual ST4S assessment is part of our commitment to supporting the online security and safety of students and schools. Learn more at st4s.edu.au.

How to use this book

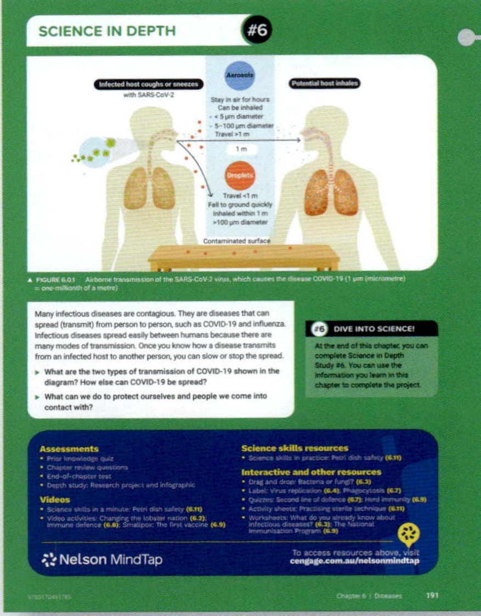

Science in depth: The opening page begins the chapter by placing the science topic into a real-life context that is both interesting and relevant to our lives.

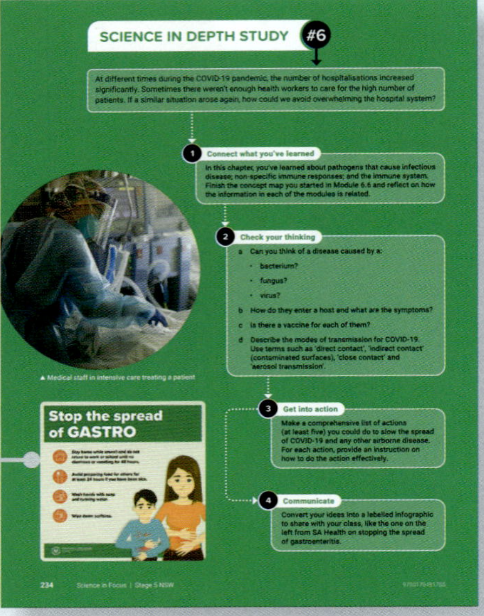

Think, do, communicate: In the Science in Depth Study, you are encouraged to reflect on and apply your learning to a set of activities, which allows you to make meaningful connections with the content and skills you have just learned.

Learning modules: Content is chunked into key concepts for effective teaching and learning.

Learning objectives: Clear, concise learning objectives give you oversight of what you are learning and set you up for success.

Key words: Key words are defined the first time they appear.

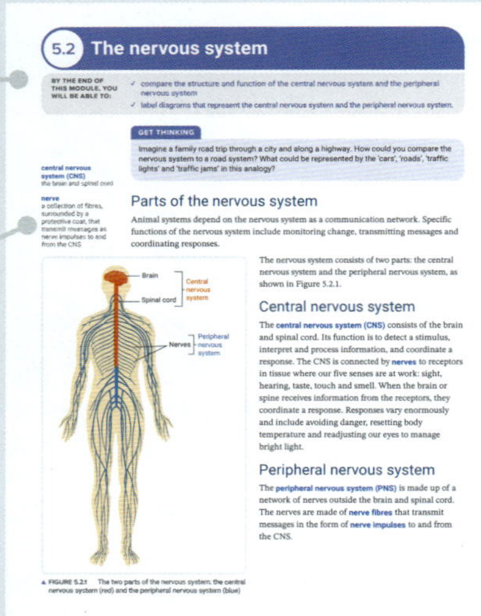

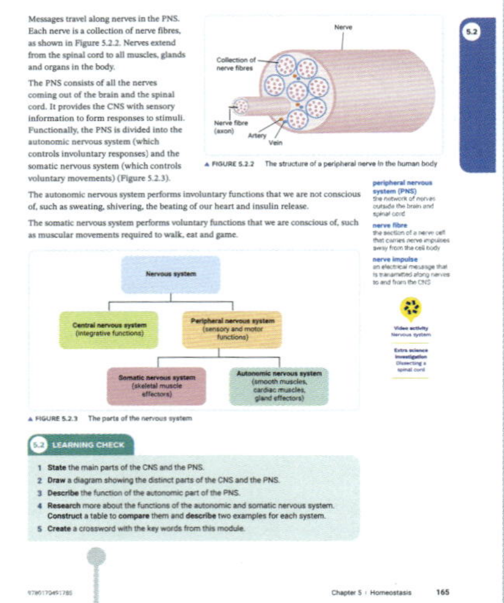

Learning check: These are engaging activities to check your understanding. Activities are presented in order of increasing complexity to help you confidently achieve the module's learning objectives. **Bolded** cognitive verbs help you clearly identify what is required of you.

Science in Context:
The NSW Science syllabus is explicitly addressed with interesting, contemporary content and activities.

Aboriginal and Torres Strait Islander Science Contexts: This content was developed in consultation with a First Nations Australian curriculum specialist. It showcases the key connections between Science and Aboriginal and Torres Strait Islander Peoples' Cultural Knowledges, with authentic, engaging and culturally appropriate science content.

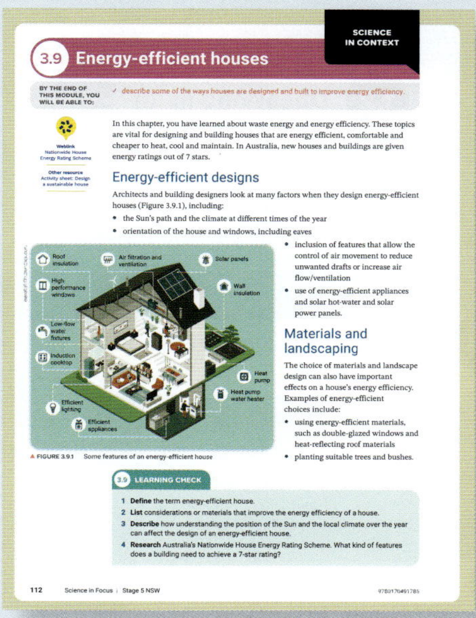

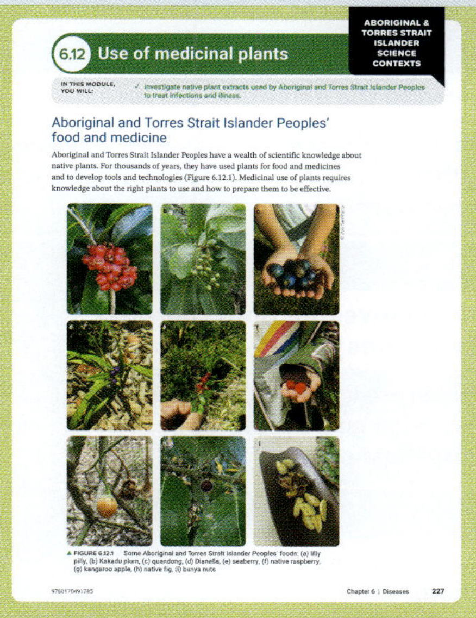

Activities: Activities are open-ended and often practical, allowing you to understand the Aboriginal and Torres Strait Islander Peoples' cultural and historical connections to science.

Science skills in focus: Each chapter focuses on specific working scientifically skills. These are explained and modelled with our 'Science skills in a minute' videos, before you put it into practice in a science investigation. The science skill is reinforced with our 'Science skills in practice' digital activities.

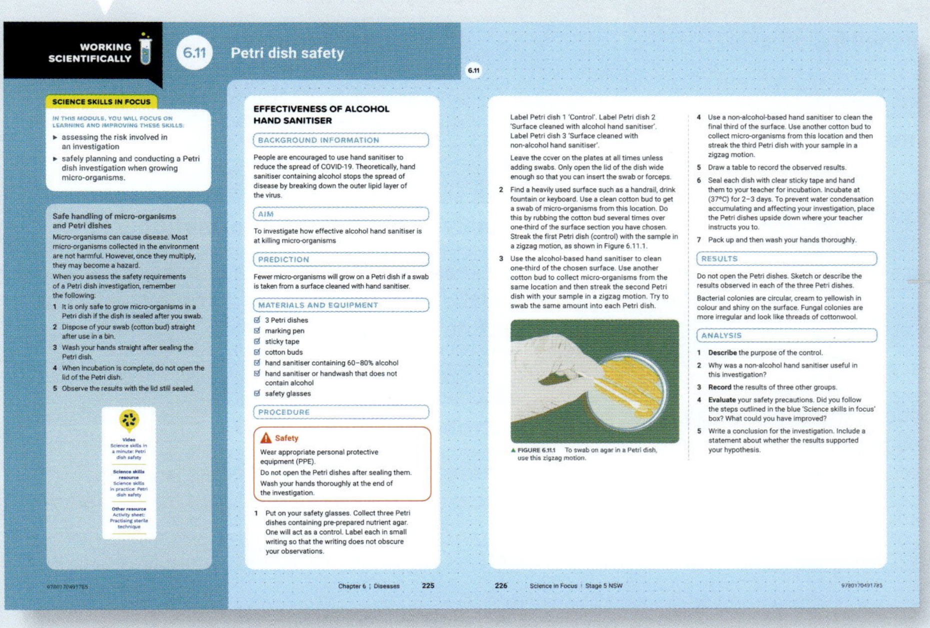

Investigations: Practise and reinforce working scientifically skills through fit-for-purpose and engaging science investigations.

Working scientifically

1.1 Reviewing the scientific process

BY THE END OF THIS MODULE, YOU WILL BE ABLE TO:

- ✓ describe how to work scientifically
- ✓ explain why the process of working scientifically may not be linear.

GET THINKING

You are in charge of decorating your garden with fairy lights for Christmas (Figure 1.1.1). How can you decide which lights are best for making your garden feel festive? What are some of the questions you will need to ask yourself so you can decide? How are you going to test the best ideas?

Hannamariah/Shutterstock.com

▲ FIGURE 1.1.1 A situation where planning and testing ideas first could be worthwhile

Video activity
What are the working scientifically processes?

Interactive resource
Crossword: Scientific processes

What is the scientific process?

In Stage 4 you learned about the processes you must follow to work scientifically. The working scientifically processes are a series of steps scientists use to conduct their investigations. The steps are:

1 Observing – using your senses and accurate equipment to make observations
2 Questioning and predicting – formulating a testable scientific question and **hypothesis** based on observations
3 Planning investigations – creating procedures to test the scientific question and hypothesis in a safe and ethical way
4 Conducting investigations – following procedures to collect data and information systematically with **accuracy** and **precision**
5 Processing data – organising data into a range of representations: tables, graphs, diagrams or models, including the use of descriptive statistics
6 Analysing data – assessing data to identify patterns, trends and relationships between variables to develop **evidence**-based arguments to draw **valid** conclusions
7 Problem-solving – selecting strategies to find solutions to scientific problems in the investigation
8 Communicating – presenting your findings and scientific arguments to different audiences.

hypothesis
a testable explanation for something based on existing knowledge; a testable statement of the predicted relationship between the independent and dependent variables

accuracy
how close a measurement is to the correct value

precision
how close repeated measurements of the same thing are to each other

evidence
verified information that supports or refutes a statement

valid
describes the extent to which an investigation tests a hypothesis

The working scientifically processes as a cycle

The scientific process is not always linear because scientists often need to reflect on their observations and findings, revise their questions and hypotheses, and repeat steps in the methods (Figure 1.1.2). This can happen when they face challenges such as unexpected results or problems with the procedure. This shows us that science is a constantly evolving process to create knowledge and solve problems.

Scientists usually start a scientific research project by observing a phenomenon and asking scientific questions. Next, scientists think creatively, plan their investigation, and set clear goals for their research. However, scientists can return to or repeat any step in the scientific process as their research progresses. This flexibility is a key part of how science works. By completing all the steps in the scientific process, scientists ensure their research is valid, **reliable**, and can be repeated by others.

reliable consistent; able to be trusted

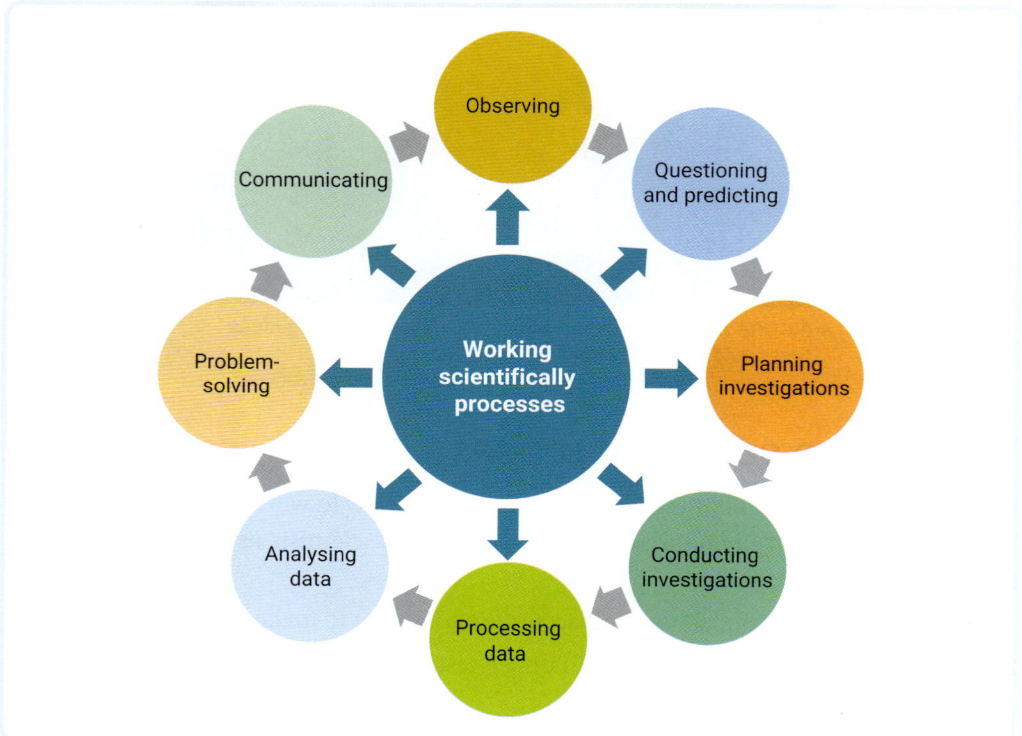

▲ **FIGURE 1.1.2** The working scientifically processes are not a linear series of steps. As the blue arrows indicate, during investigations scientists can return to or repeat any step in the process.

For example, imagine that you are investigating which shoes are best for long-distance runners.

You could start the investigation by observing different athletes and the shoes they wear, or you could start by analysing data available from shoe companies. You then need to decide what you are going to investigate; for example, the force of impact on the shoe, durability, flexibility, weight or comfort (Figure 1.1.3). Once you have determined your scientific question, you can design your investigation and test the shoes. Imagine you begin designing your test method to look at how hard the shoe's sole is and you might find this is tied to flexibility so you might go back to look further at data from shoe companies. From the companies' data, you discover that there are other factors affecting the shoe's sole and you decided to refine your hypothesis and scientific question.

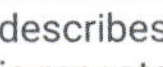

In this example, the scientific process is not linear because it involves continuous feedback loops, revisions and refinements. Each step can lead back to earlier observations, requiring new questions or hypotheses, additional experiments, or changes in method. This **iterative** approach is essential for scientific progress, as it allows for continual improvements by responding to new data and information.

iterative
describes something that is repeated multiple times, often in a cycle or process

lucky_pics/Shutterstock.com

▲ **FIGURE 1.1.3** There are many features of running shoes that could be investigated, including flexibility and resistance to force.

1.1 LEARNING CHECK

1. **Describe** the steps of the scientific process.
2. **Explain** why the scientific process is not linear. Use an example.
3. **Discuss** the importance of following the scientific process when carrying out an investigation.
4. **Create** an annotated flow chart illustrating the steps of the scientific process of an investigation to **research** the best springs for a trampoline.

1.2 Observing

BY THE END OF THIS MODULE, YOU WILL BE ABLE TO:

- ✓ describe what a scientific observation is
- ✓ describe characteristics of scientific observations
- ✓ distinguish between qualitative and quantitative observations
- ✓ explain how to make precise observations.

Video
Science skills in a minute: Making observations

Science skills resource
Science skills in practice: Observations to make classifications

GET THINKING

You are watching a baker make bread. You observe that different types of bread have different ratios and types of ingredients. The baker measures the amount of ingredients with a scale, or uses measuring spoons for small amounts of ingredients such as seeds. Why do you think the baker uses different equipment to measure different ingredients? What other observations could you make when the baker is making and baking the bread?

Scientific observations

observation
data collected through the senses (sight, smell, taste, touch or hearing) or with measuring equipment

An **observation** is a measurement or any other information collected directly from our senses or using measuring instruments. Scientific observations are usually part of the beginning of the scientific process. They provide information and data that are fundamental evidence to formulate a scientific question and hypothesis that will guide the investigation.

As shown in Figure 1.2.1, scientific observations have characteristics that distinguish them from simple, everyday observations.

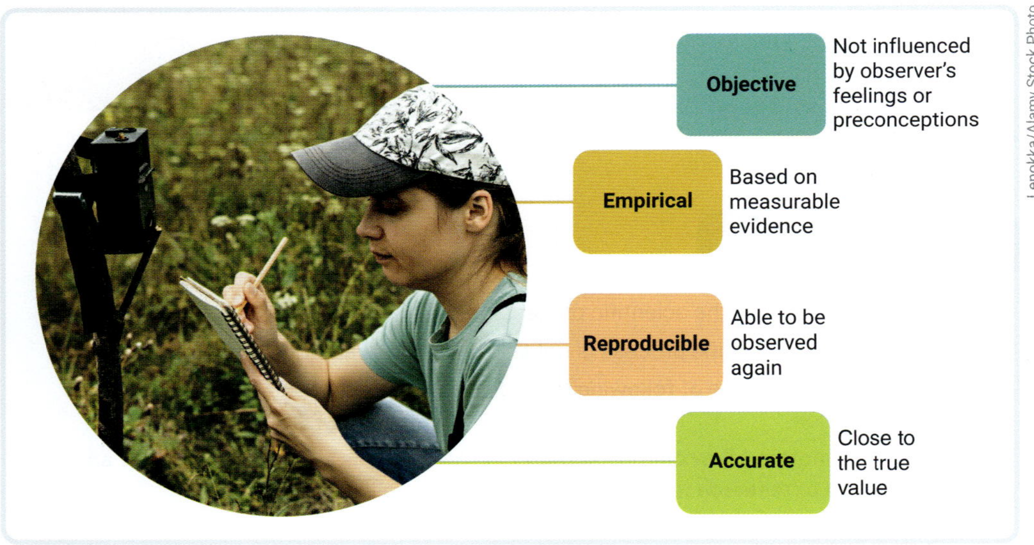

▲ **FIGURE 1.2.1** Characteristics of scientific observations

Observations can give **qualitative data** or **quantitative data** (Table 1.2.1). The type of phenomenon or object that you are observing will determine the type of observation. In science, usually a combination of both types of observation is applicable.

qualitative data
non-numerical information that relates to a quality, type, choice or opinion

quantitative data
numerical information that is counted or measured and expressed as quantities in numbers

1.2

▼ **TABLE 1.2.1** Differences between qualitative and quantitative observations

Observation type	Qualitative observation or data	Quantitative observation or data
Description	• Descriptions that do not use numbers • Descriptions based on senses: smell, taste, touch, sight and hearing	• Descriptions that involve measurements, numbers and counting
Examples of scientific observations	• Biology: behaviour of animals • Astronomy: patterns of planetary clouds and surfaces • Chemistry: changes in the colour of solution during a chemical reaction • Physics: observing a skydiver jumping from a plane	• Biology: counting the number of animals • Astronomy: measuring the intensity of light from a star • Chemistry: measuring the pH of a substance • Physics: recording the speed of a car over a certain distance

✩ **ACTIVITY**

Observations

You are walking to school one morning and you observe an unusual cloud formation in the sky (Figure 1.2.2).

1 What type of observation did you make first? **List** all the observations that you can think of.
2 How can your observations become scientific observations?
3 **Explain** how your observations could be used to start a scientific investigation.

October.___09/Shutterstock.com

▲ **FIGURE 1.2.2** Unusual cloud formation

Precision in observations

It is important to know the difference between a general observation of a phenomenon (for example, cloud formations or animals drinking water in a pond) and observations from scientific research (such as the colour of substances in a chemical reaction or temperature changes over time due to climate change).

Scientific observations are the foundation of scientific knowledge because they are generally the starting point for the initial ideas and questions that will lead to hypotheses and then an experimental design. Consequently, it is important to have precision when making observations.

Figure 1.2.3 illustrates the main steps to make precise scientific observations.

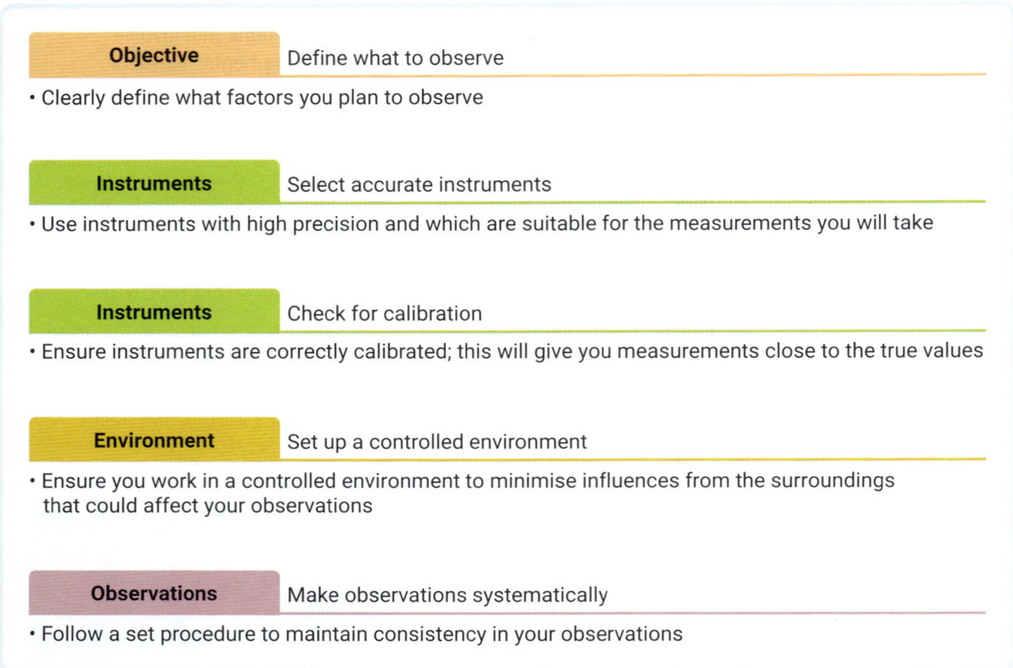

▲ **FIGURE 1.2.3** Steps to ensure precise scientific observations

1.2 LEARNING CHECK

1. **Describe** the types of scientific observations. Give examples.
2. **Outline** the characteristics of a scientific observation.
3. **Explain** the steps to ensure precise scientific observations.
4. **Explain** why scientific observations are important in research.
5. **Design** a simple investigation to find the temperatures at which a range of metals become liquid. What observations would be made?

1.3 Questioning and predicting: hypotheses

1.3

BY THE END OF THIS MODULE, YOU WILL BE ABLE TO:

- ✓ formulate scientific questions and hypotheses
- ✓ distinguish between a scientific question and a hypothesis
- ✓ predict outcomes based on observations, questions and hypotheses.

GET THINKING

You come home from school and see your paints all over the floor and your cat lying in the middle of the mess (Figure 1.3.1).

- What questions can you ask about this situation?
- What hypothesis can you make to investigate the mess?
- Are your questions and hypotheses testable?

Discuss with a classmate.

OlgaBartashevich/Shutterstock.com

▲ FIGURE 1.3.1 Can you use science to investigate this situation?

Video
Science skills in a minute: Writing hypotheses

Science skills resource
Science skills in practice: Writing hypotheses

Testable questions and hypotheses

In Stage 4, you learned about testable questions and hypotheses. To summarise, a testable question and a hypothesis each relate the variable that you are measuring (**dependent variable**) with the variable that you change in the experiment (**independent variable**).

Table 1.3.1 summarises what you learned in Stage 4 about questions and hypotheses.

dependent variable
the factor that may be affected by the independent variable; the factor that can be measured or counted

independent variable
the factor that you choose to vary in your investigation

refutable
able to be proven wrong

▼ **TABLE 1.3.1** Differences between testable questions and scientific hypotheses

	Testable questions	Scientific hypotheses
Description	• Are relevant to the investigation and based on observations • Include the observed variables • Are possible to answer by collecting data from experiments or using secondary sources • Are written as an open-ended query	• Are testable and specific to the investigation • Are **refutable** (can be proven wrong) • Link the independent and dependent variables • Are based on a prediction (not a guess) • Are statements often written as 'If... then...'
Examples	• How does temperature affect the rate of a chemical reaction? • How does the length of a pendulum's string affect its swing?	• If temperature increases, then the rate of the reaction increases. • If the string of the pendulum decreases, then the swing of the pendulum increases.

Formulating scientific questions and hypotheses

Scientific questions and hypotheses guide scientific research, frame the experimental design and consequently lead to new discoveries. Therefore, scientific questions and hypotheses need to be well formulated.

framework
a systematic structure used to plan or build a concept

You can follow a simple **framework** to formulate a scientific question and a hypothesis. Figure 1.3.2 shows a framework you can use to develop scientific questions and hypotheses.

If investigations where the hypotheses are known to be correct are the only ones undertaken, science won't be challenged and advances will be minimal.

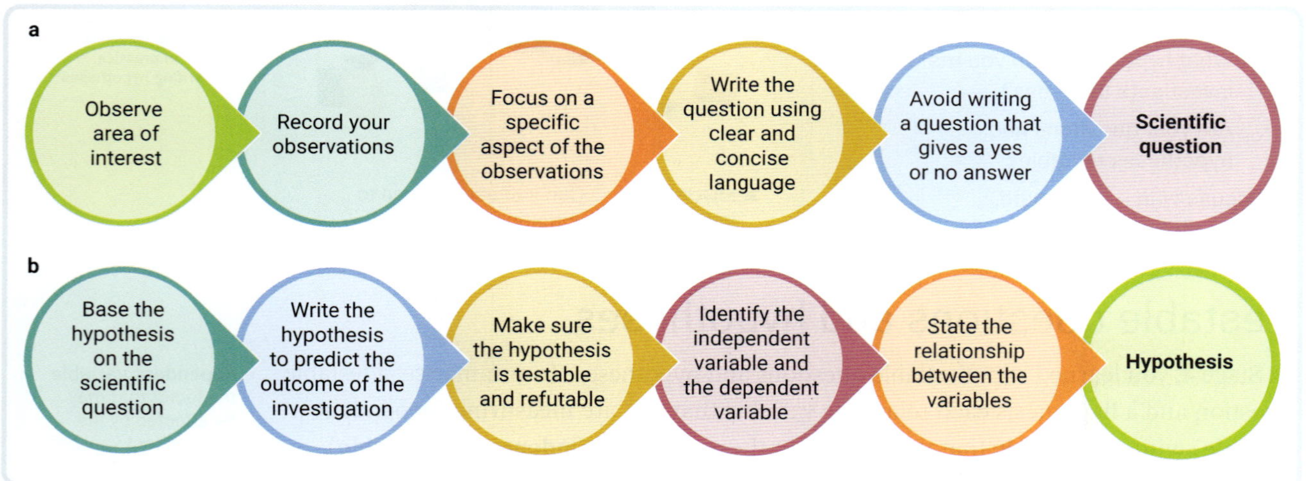

▲ FIGURE 1.3.2 Frameworks to formulate (a) a scientific question and (b) a hypothesis

literature review
a study of reports about previous related experiments and investigations

After you have formed your hypothesis, you can test it by conducting an experiment or making further observations. To check if your prediction is correct, you need to analyse the data you collected or the observations you made. You can further check your prediction by conducting a **literature review** to see what other researchers have published on the topic. Checking your prediction by reading about other investigations will improve your understanding of the phenomenon you are studying.

ACTIVITY

Formulating questions and hypotheses

For each of the following scenarios, formulate a scientific question and a hypothesis.

A Daniel works on a farm that grows tomatoes in a greenhouse under well-controlled conditions. He wants to see if having a higher light intensity will affect the growth of the tomatoes and consequently lead to production of more tomatoes.

B Scientists at a drug company are developing a new tablet. They are concerned about how soluble the tablet is in the digestive system. They have read research papers suggesting that stomach acid will affect the solubility of the tablet.

C Ecologists observed that water pollution is a big issue in many areas because it may decrease aquatic biodiversity. Secondary data shows a range of different results.

9780170491785

1.3 LEARNING CHECK

1 **Describe** the similarities and differences between a scientific question and a hypothesis.

2 **Outline** how to write a hypothesis. Give an example.

3 You have been asked to set up a simulation experiment to measure the relationship between density, mass and volume of objects.

a **Formulate** a scientific question for the experimental simulation.

b **Formulate** a hypothesis for the experimental simulation.

c **Predict** an outcome based on your question and hypothesis.

d Run this simulation using the linked PhET site (Figure 1.3.3) to test your hypothesis. Was your prediction correct? **Justify** your answer, using your observations.

Weblink
PhET: Density

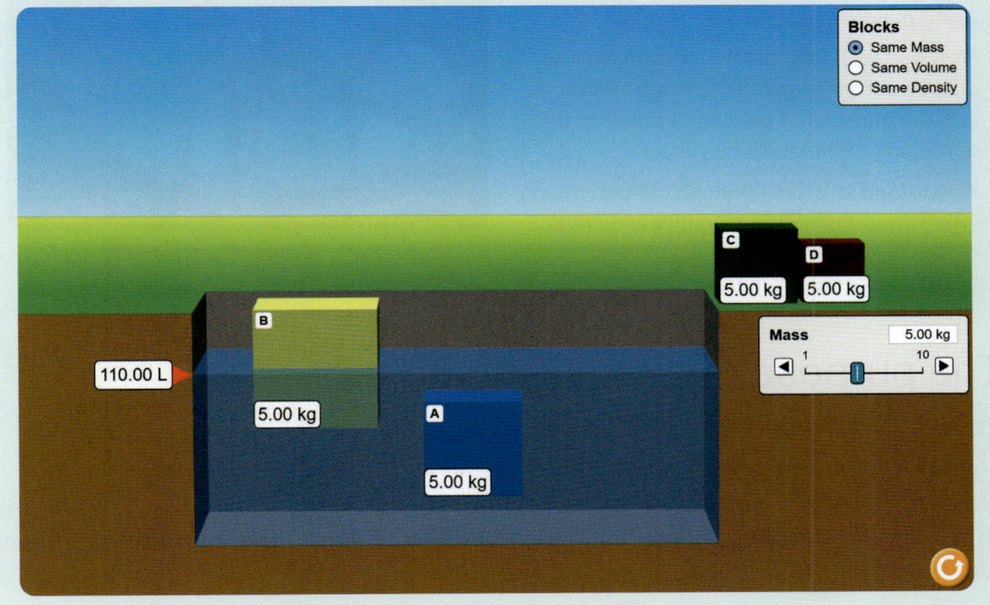

University of Colorado Boulder

▲ **FIGURE 1.3.3** A screenshot from the PhET density simulator

1.4 Planning investigations: purpose and variables

BY THE END OF THIS MODULE, YOU WILL BE ABLE TO:

- ✓ describe the purpose of an investigation
- ✓ explain how the independent variable, dependent variable and controlled variables are used in investigations
- ✓ explain the use of a control in an experiment.

Video
Science skills in a minute: Identifying variables

GET THINKING

In winter, gardeners sometimes cover some plants with clear plastic (Figure 1.4.1). Discuss with a classmate the purpose of using plastic in this way.

How can you distinguish between the purpose of using plastic to cover plants and the purpose of an investigation of this gardening technique?

How would you design an investigation about covering plants with plastic? What would be the purpose of this investigation?

S.O.E/Shutterstock.com

▲ **FIGURE 1.4.1** Covering a plant with clear plastic

The purpose of an investigation

In Stage 4, you learned that the purpose of an investigation is the reason you research a specific question. In Stage 5, you will learn that the purpose of an investigation is to gather meaningful insights to contribute to advancing scientific knowledge. The purpose guides the investigation and consequently all associated working scientifically processes. We write the purpose of the investigation as its aim.

For example, if you want to compare the fuel consumption of a hybrid car and a non-hybrid car, the purpose of your investigation will be to investigate the fuel consumption of each car. This purpose will inform how you collect data and analyse it to reach a conclusion.

Using variables in an investigation

There are three categories of variables, as shown in Figure 1.4.2.

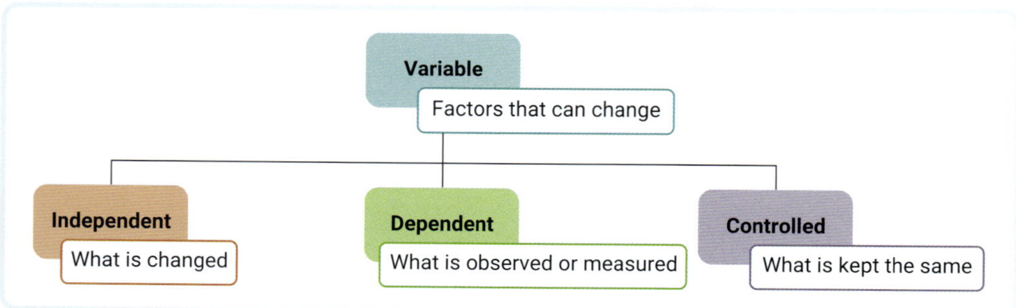

▲ **FIGURE 1.4.2** Classification of variables

9780170491785

In many investigations, scientists change one factor and observe the effects of this change on another factor. For example, imagine you are at your little sister's birthday party, where her friends are making bubbles with soapy water (Figure 1.4.3). You observe that some bubbles are bigger and others, smaller. How can you investigate that observation? You might decide to propose that the diameter of the bubble wand's loop influences the size of the bubble.

Patrick Foto/Shutterstock.com

▲ FIGURE 1.4.3 What factors could influence the size of a bubble?

The variables for this investigation are:

- independent variable – the factors that you are changing: the diameter of the loop
- dependent variable – the factor that you are measuring: the size of the bubbles
- **controlled variables** – the factors that you keep the same: soapy solution, no wind, the same person blowing the bubbles, same force of breath blowing the bubbles.

controlled variable
a factor that needs to be kept the same throughout a science investigation so that any changes in it do not influence the results

The relationship between the variables for this investigation is shown in Figure 1.4.4.

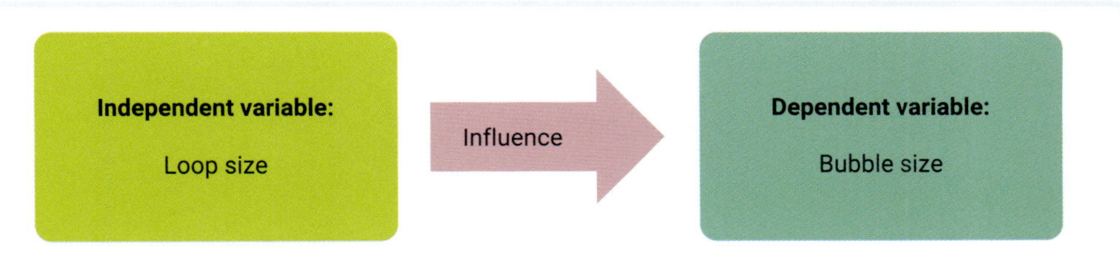

▲ FIGURE 1.4.4 The relationship between the independent variable and dependent variable

control group
a trial or group in an investigation that doesn't have the independent variable and is used as a comparison

treatment groups
all the trials or groups in an investigation that contain variations of the independent variable

In many experiments, a **control group** is needed to compare with the rest of the **treatment groups** in the experiment. The control group is identical to the one or more treatment groups, except it does not contain the independent variable. For example, an experiment to test if listening to music affects your learning could have one group in which people don't have music (the control group) and another group that listen to music while they study (the treatment group) (Figure 1.4.5). A valid experiment must have a control to show that the dependent variable remains unchanged when the independent variable is not present.

For a valid experiment that tests the hypothesis, you must clarify the variable that you are changing and the variables that you will control.

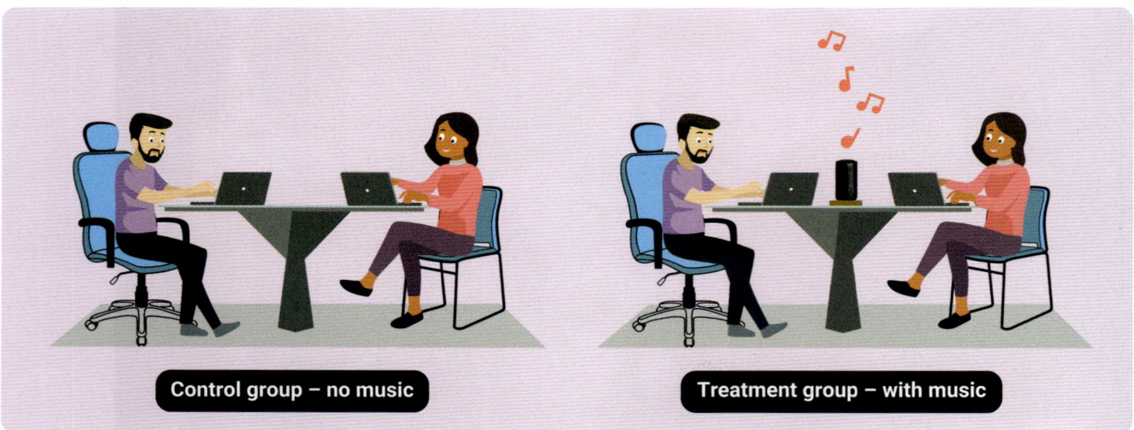

▲ **FIGURE 1.4.5** The control group is not exposed to the independent variable that is being tested.

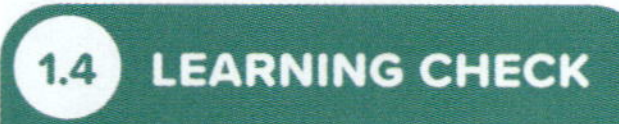

1.4 LEARNING CHECK

1. **Describe** the purpose of an investigation.
2. **Explain** the meaning and give examples of:
   a. the independent variable.
   b. the dependent variable.
   c. controlled variables.
3. **Explain** the use of a control in an experiment. Give an example.
4. **Create** a flow chart to illustrate how to test the hypothesis that eating green apples will reduce sea sickness.
5. **Demonstrate** how to use variables in an experiment to test if drinking coffee will increase the ability to recall more words in a memory test.

1.5 Planning investigations: methods

BY THE END OF THIS MODULE, YOU WILL BE ABLE TO:

✓ explain the difference between a procedure and a method
✓ explain different types of methods used in investigations
✓ explain the differences between validity and reliability.

GET THINKING

There are many different irrigation systems you can use to water your garden, such as drip systems, sprinklers, subsurface systems and spray irrigation. How will you decide which is best for your vegetable garden? What questions would help you decide?

Interactive resource
Crossword: Method and procedure

Other resources
Worksheet: Experimental design

Procedures and methods in science

The **procedure** is the set of instructions you follow when you conduct an investigation. The **method** is the steps that describe what you did to carry out your investigation.

procedure
a set of instructions to follow; written in the present tense

method
steps that were taken during an investigation, written in past tense

Methods are an important aspect of working scientifically. Methods are recorded and communicated to others so they can see what the investigation did. Once the investigation is finished, the method is published so others can use it to design their own experimental procedures.

The key differences between a procedure and a method are summarised in Figure 1.5.1.

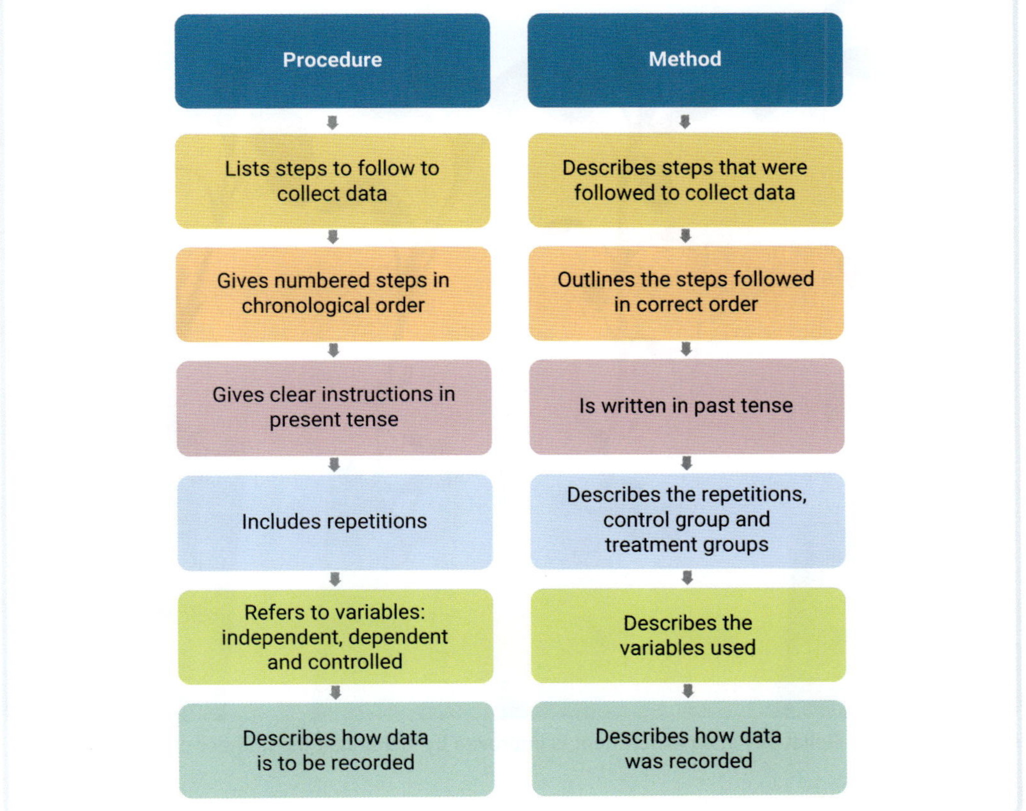

▲ **FIGURE 1.5.1** Comparison between a procedure and a method

Reliability and validity in methods

reliability
how similar the results of the same experiment are

The **reliability** of the method refers to how consistent the results are. Reliability depends on the number of repetitions in the treatment groups of the experiment (Figure 1.5.2). If an experiment is reliable, repetition of the experiment yields similar results, and the experiment is described as reproducible. This means that when other researchers repeat the experiment, they get similar results.

validity
the extent to which an investigation tests a hypothesis

The **validity** of the method refers to how well the experiment investigated the aim and purpose of the investigation and how well it tested the hypothesis.

The following questions are prompts that you can use to assess the validity and reliability of the method:

- Did the method include controlling the necessary variables?
- Was a control set up to show what happens when the independent variable wasn't present?
- Did the method include enough repetitions for reliable results?
- Was the way that data was recorded explained well?
- Did the method actually test the investigated hypothesis?

pogonici/Shutterstock.com

▲ **FIGURE 1.5.2** Reliability in an experiment is improved by increasing the number of repetitions.

1.5

Investigation methods and reliability

Choosing the right method of investigation depends on the type and purpose of a research question and the type of data needed.

Table 1.5.1 shows the different investigation methods used in science, how they are used and some features important to reliability.

For all these types of methods, you need to ensure reliability. This involves careful planning, **systematic** procedures, and meticulous and thorough record-keeping. It is also crucial to address potential **biases** and maintain consistency when you collect data.

systematic
something done in an organised way

bias
a strong preference for one thing or idea over another

▼ **TABLE 1.5.1** Different investigation methods used in science

Method	Description	Features for reliability
Controlled experiments	Involves changing one variable to determine the relationships with another variable	• Controlled variables • Repetitions of treatment groups • Control(s) • Consistent procedures
Fieldwork	Collection of data/experimental research in the field	• Well-structured, consistent procedures • Well-recorded data
Observations	Systematic watching or recording behaviours or phenomena	• Clear and detailed guidelines for making observations
Secondary data analysis	Analysis of data from other experts' published sources	• Data used from different sources has similar results • Data does not have a bias
Focus groups	Guided discussions to gather opinions	• Well-structured discussion questions
Case studies	Detailed examination of a real-world context	• Use of data from multiple sources
Longitudinal studies	Tracking the same subject over a period of time	• Consistency in method of data collection
Surveys	Asking people a series of questions to gather opinions or characteristics	• Use of validated survey instruments • Consistency in wording and format

1.5 LEARNING CHECK

1 **Describe** the difference between a procedure and a method.
2 **Describe** different types of methods used in investigations.
3 **Explain** the differences between validity and reliability in an investigation's methods.
4 **Outline** the types of investigations an ecologist could do to study the effect of plastic pollution in a local wildlife area, given they plan to obtain data by talking with knowledgeable local people.
5 **Create** a statement that explains what reliability is in investigations.

1.6 Planning investigations: safety

BY THE END OF THIS MODULE, YOU WILL BE ABLE TO:

✓ define hazards and risks in the laboratory
✓ explain the importance of safety in carrying out experiments
✓ assess risks to increase safety in the laboratory.

GET THINKING

Imagine you are helping your grandfather in his timber workshop (Figure 1.6.1). What do you see him doing to avoid injuries and keep the machines he uses safe? How do those safety measures in his workshop relate to those you need to apply in the science laboratory?

List the safety procedures that are common to a home workshop and a science laboratory.

Cavan Images/Alamy Stock Photo

▲ **FIGURE 1.6.1** Safety measures are important in many settings.

Safety in the laboratory

It is critical to ensure safety in the science laboratory so that people and the environment are not harmed. In Stage 4, you learned about laboratory safety rules and procedures for experiments. In this module, you will learn how to assess **hazards** and minimise the **risk of harm**.

hazard
something that has the potential to harm

risk of harm
the chance that damage or injury will occur

Assessing risks

Everything you and your classmates do in the laboratory involves some level of risk; for example, using the Bunsen burner, sharp objects, glassware and chemicals. You can start assessing the risks of an experimental procedure by asking questions such as:

1 What activity will you be doing in the experiment?
2 What equipment and chemicals will you use?
3 What are the potential hazards and risks of using the equipment and chemicals?
4 What are the risks of those hazards?
5 How can you reduce the risks to yourself and others?
6 How will you dispose of any chemicals safely for you and the environment?

You can use a risk assessment matrix to assess hazards and reduce the risks of harm (Table 1.6.1). The matrix combines the **likelihood** of something occurring with the severity of any potential **consequences**. Once you have considered the hazards and completed the risk assessment matrix, you need to create and follow a risk management plan to keep safe. Wear personal protective equipment (PPE) when you are completing experiments. PPE includes goggles, lab coats, gloves and enclosed leather shoes. Table 1.6.2 gives examples of some other simple risk assessments for an experiment.

likelihood
the chance something will happen

consequence
the result of a decision or action

▼ TABLE 1.6.1 An example of a risk assessment matrix

Risk	Likelihood				
	Rare	**Unlikely**	**Possible**	**Likely**	**Almost certain**
Severe: For example, potentially fatal or causing an injury or illness with permanent disability	MEDIUM	MEDIUM	HIGH	EXTREME	EXTREME
Major: For example, causing a potential lost-time injury, but not permanent disability	LOW	MEDIUM	MEDIUM	HIGH	EXTREME
Moderate: For example, causing an injury or illness requiring moderate medical treatment but not lost time	LOW	LOW	MEDIUM	MEDIUM	HIGH
Minor: For example, causing an injury potentially requiring application of first aid	LOW	LOW	LOW	MEDIUM	MEDIUM
Minimal: For example, a hazard or near-miss that requires reporting and follow-up action	LOW	LOW	LOW	LOW	LOW

▼ TABLE 1.6.2 Simple risk assessments for common experiments

Experiment or activity	Hazard	Actions to reduce the risk
Heating using the Bunsen burner	Being burned when using the Bunsen burner or working with hot equipment	• Avoid touching hot parts of the Bunsen burner • Use the yellow flame when not working • Use tongs when moving hot equipment • Leave hot equipment to cool down before putting it away
Using chemicals	Spillages when working with chemicals	• Read the hazards specified for each bottle of chemical • Do not taste or smell the spilt chemical • Tell your teacher if there is a spillage
Using beakers, test tubes or microscope slides	Cuts from broken glassware	• Avoid touching broken glass • Report to the teacher as soon as breakage happens
Using scalpels or scissors to dissect specimens	Cuts from sharp objects	• Manage sharp objects with care • Report injuries to your teacher

1.6 LEARNING CHECK

1 **Define** hazard and risk of harm.
2 **Explain** the importance of safety in carrying out experiments. Give an example.
3 You are performing an experiment to test how much of a substance dissolves as the temperature of the mixture increases. **Describe** two safety practices that you could implement when doing the experiment.
4 **Assess** the hazards in your school's science laboratory and **create** a risk assessment plan for a recent experiment.
5 **Create** a short video illustrating the safety measures that are used in your science laboratory to show at the beginning of the school year to the new Year 7 students.

1.7 Conducting investigations: equipment

BY THE END OF THIS MODULE, YOU WILL BE ABLE TO:

- ✓ identify suitable equipment to perform an investigation
- ✓ distinguish between different apparatus
- ✓ explain how to assemble apparatus for experiments
- ✓ assess the accuracy and precision of equipment.

Videos
Science skills in a minute: Lab equipment

Science skills in a minute: Dissection

GET THINKING

Imagine you accompanied someone to get their car serviced. As you enter the mechanic's workshop, you notice many different tools and equipment (Figure 1.7.1). All of these require different procedures when the mechanic uses them to assess or work on the car. How can you identify which tools and equipment are suitable for different tasks? Make a list of possible equipment in a mechanic's workshop and the purpose of each. Do you need more information on how they are used? Why?

Kristin Spalder/Shutterstock.com

▲ **FIGURE 1.7.1** A car workshop with a range of equipment

Choosing appropriate equipment

In Stage 4, you learned that many different types of equipment are used in the laboratory to perform experiments and collect measurements. Figure 1.7.2 illustrates some of the equipment that you are familiar with in the laboratory. Figure 1.7.3 shows some common set-ups using different equipment for different types of experiment and data to be collected.

Choosing the correct equipment to do an experiment increases the accuracy and precision of the data collected in the investigation. For example, to measure 15 mL of water, use a measuring cylinder, not a beaker, because the measuring cylinder is more accurate for that amount of liquid.

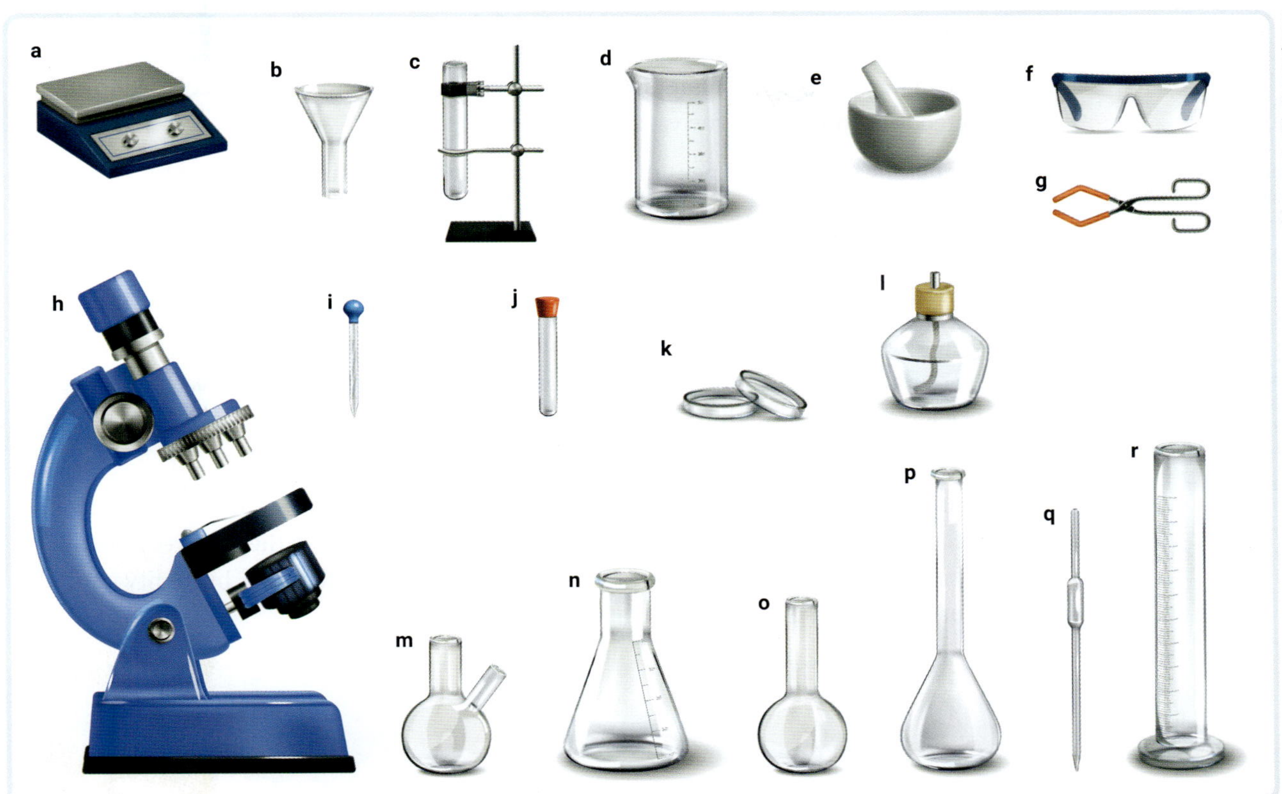

▲ **FIGURE 1.7.2** Common laboratory equipment: (a) electronic scale; (b) funnel; (c) test tube held by clamps on a retort stand; (d) beaker; (e) mortar and pestle; (f) goggles; (g) tongs; (h) microscope; (i) pipette; (j) test tube with rubber stopper; (k) Petri dishes; (l) methylated spirit burner; (m) distillation flask; (n) conical flask; (o) round flask; (p) large round flask; (q) volumetric pipette; (r) measuring cylinder

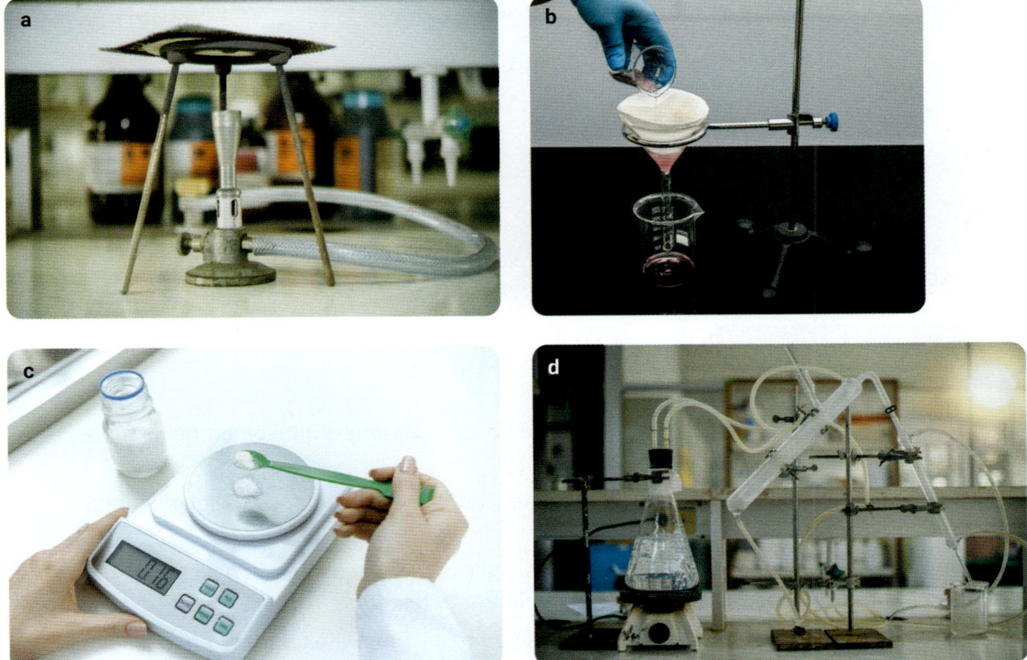

▲ **FIGURE 1.7.3** Various set-ups of science equipment: (a) heating apparatus (Bunsen burner, tripod, gauze mat); (b) filtering apparatus (retort stand, clamp, funnel, filter paper, beaker); (c) electronic scale; (d) distillation apparatus (hot plate, conical flask, condenser, beaker)

Accuracy and precision of instruments

All the instruments used in an investigation need to be accurate and precise so the data collected are close to the true values of the measurements.

accuracy (instruments)
how close the instrument measures to the true value

calibration
the process of adjusting and verifying the accuracy of a measuring instrument or device

precision (instruments)
the level of variation between measurements by an instrument

data logger
an electronic device that automatically records and stores data over time

The **accuracy** of an instrument means how well it can measure the true value of a parameter. Accuracy is verified when the instrument undergoes **calibration**.

The **precision** of an instrument refers to how much variation there is in measurements when you repeat them.

For example, an electronic scale is accurate when the weight is close to the true weight of the object that you are measuring. Its precision is high if the object's weight is the same each time you weigh it.

Data loggers (Figure 1.7.4) are electronic devices that automatically record measurements over time and store the data directly in the device or in a computer connected to it. They are digitally calibrated to the true value, so they are very precise and accurate. Data loggers measure parameters such as temperature, humidity and pressure, and log this information at set intervals of time.

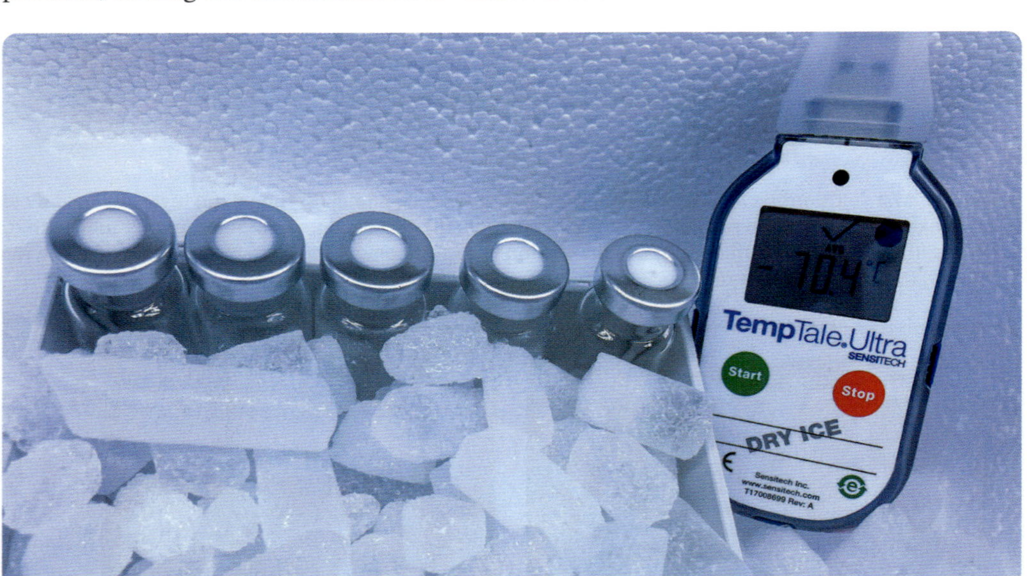

Elzbieta Krzysztof/Shutterstock.com

▲ **FIGURE 1.7.4** A data logger

1.7 LEARNING CHECK

1. a. You need to heat a beaker of water to a 65°C. **State** why a Bunsen burner is more suitable than a kitchen stove.
   b. What equipment would you select to measure 50 mL of water with high accuracy? **Explain** your choice.
2. How would you **determine** if an electronic scale is more accurate than a spring scale for weighing 100 mg of salt? **Design** an experiment to test this.
3. In a practical skills test, your teacher asks you to measure the temperature change in a chemical reaction. **Explain** the equipment you would use and why it is suitable for this purpose.

9780170491785

1.8 Conducting investigations: collecting and recording data

BY THE END OF THIS MODULE, YOU WILL BE ABLE TO:

✓ describe how to collect reliable data in an investigation

✓ use tables effectively to record data.

GET THINKING

Social media sites collect a large amount of information every second. For example, Facebook collects 2.8 gigabytes of data per minute, which includes posts, likes, comments, shares and interactions between users. Think about how that information is organised and collected. How would a company record and present a dataset that large?

Video Science skills in a minute: Measuring liquids

Quiz Measuring in Science

Recording data from investigations

In Stage 4, you learned how to collect and record data in appropriate tables. You also learned that you need to follow these conventions when creating a table:

- record the independent variable in the first column
- put the different factor(s) that changes (the dependent variable(s)) in the other columns
- put a heading with the name of the variables and the correct units at the top of each column
- (optional) include the average values in the last column of the table.

Figure 1.8.1 shows the key features of an effective table.

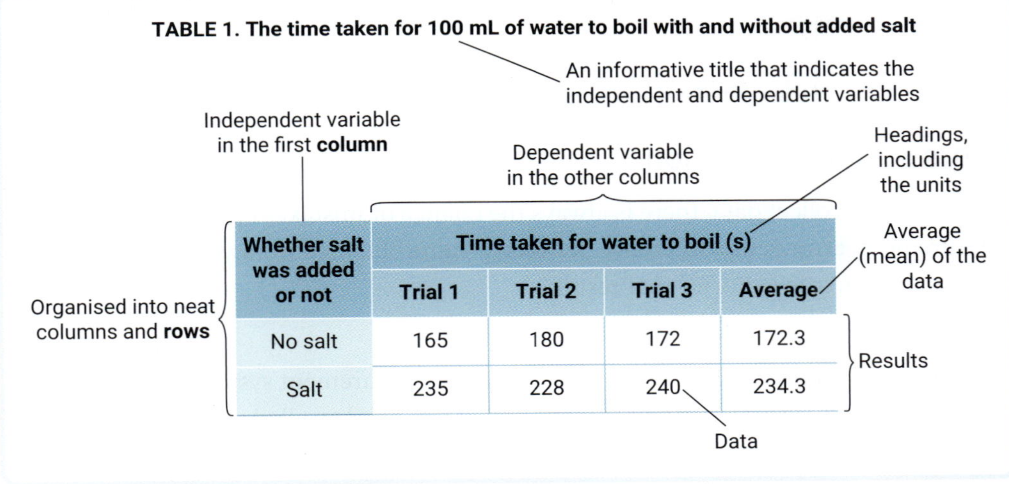

Whether salt was added or not	Time taken for water to boil (s)			
	Trial 1	Trial 2	Trial 3	Average
No salt	165	180	172	172.3
Salt	235	228	240	234.3

▲ **FIGURE 1.8.1** Features of an effective table

Collecting data from investigations

The aim of collecting data in an investigation is to provide evidence that supports or refutes the hypothesis. The collected data must represent the relationship between the independent and dependent variables mentioned in the hypothesis. Be careful to ensure your procedure collects data on these variables and not anything irrelevant. For example, if you want to test if changes in temperature affect the rate of a chemical reaction, the hypothesis could be, 'if the temperature increases then the rate of reaction will increase'. To collect data to support your hypothesis, you would use a procedure where you test the rate of a particular chemical reaction at different temperatures.

Systematic and accurate collection of data

In an experiment, the systematic and accurate collection of data contributes to the validity of the research. Systematic collection of data means you follow a structured and planned way to observe and record all of the results from the investigation. Accurate collection of data means that you make all of the observations and measurements using instruments that are calibrated and procedures that are well designed and standardised.

To increase the reliability of data, a large sample of data is needed. This is achieved by repeating the treatments of the variables or the entire experiment to ensure you are getting similar results.

Figure 1.8.2 shows how high quality data collection contributes to the validity of research.

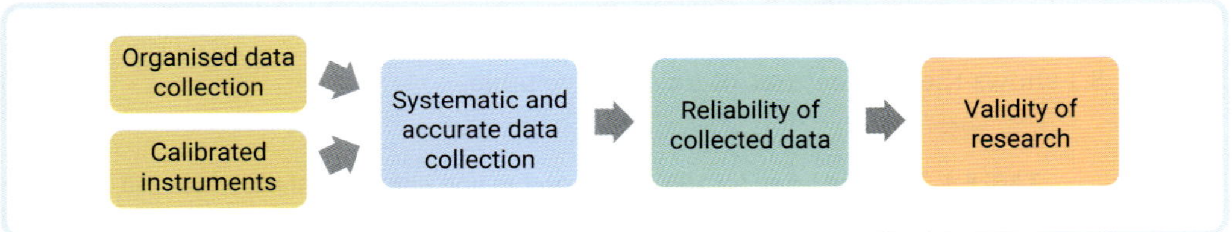

▲ **FIGURE 1.8.2** The impact of data collection on the validity of research

Errors in data collection

Whenever you collect data, there is always the chance of mistakes or errors. In scientific measurement, error is the difference between a value obtained from your measurement process and the true value for that result.

systematic error
an error that occurs in the same way with each measurement, normally due to a mistake in the procedure

random error
a small, variable error caused by slight variations in an instrument or the environment

The two main types of error are systematic and random.

- **Systematic errors** are errors due to flaws in the measurement system or experimental set-up. Once a systematic error is identified, it can be controlled or eliminated. Different types of systematic errors are shown in Figure 1.8.3.
- **Random errors** are unpredictable variations that occur in measurements due to factors that are not easily controlled. The two main types of random errors are shown in Figure 1.8.4.

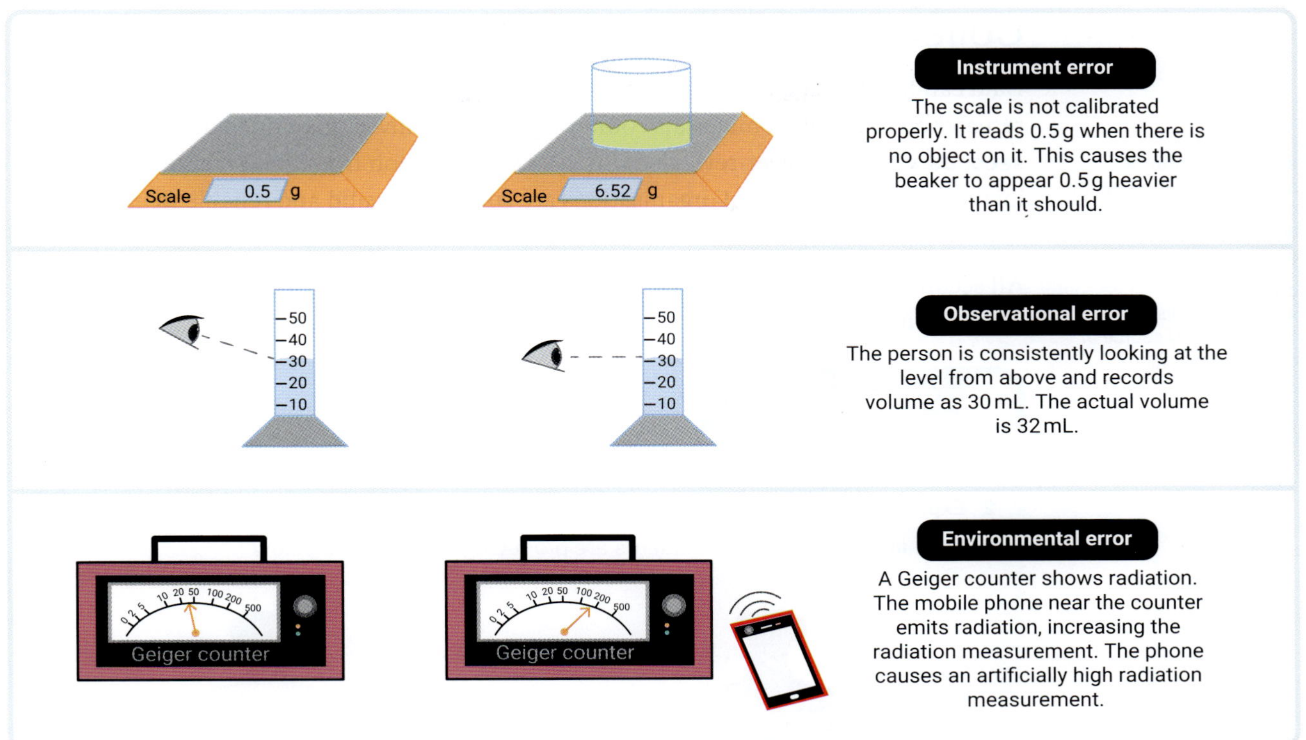

▲ **FIGURE 1.8.3** There are three types of systematic errors: instrument, observational and environmental.

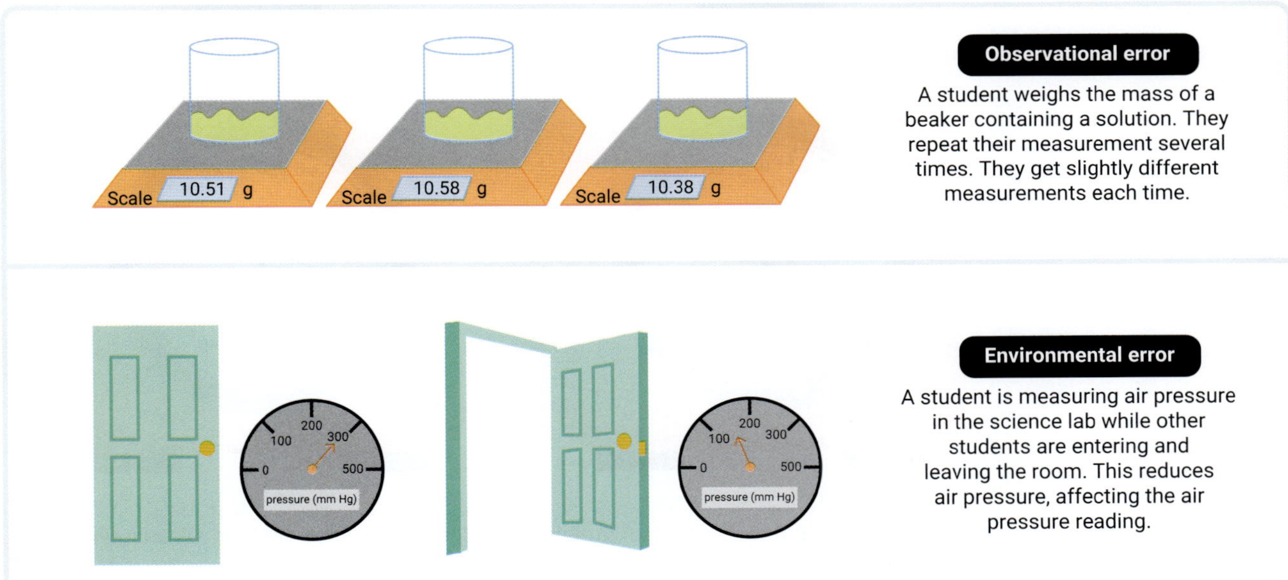

▲ **FIGURE 1.8.4** There are two types of random errors: observational and environmental.

Collecting data from secondary sources

Data can be collected from other investigations that were completed by other researchers. Published materials containing previous investigation results are called **secondary sources**. Peer-reviewed sources of information from universities, government institutions or educational organisations are considered valid and reliable to use as sources of data.

All secondary sources used in your research need to be **cited** to recognise the work of other scientists.

secondary source
a publication, information or data that has been written or collected by another person

cite
to give reference or credit to a secondary source of data or information

1.8 LEARNING CHECK

1 **Describe** the purpose of collecting data from an investigation.
2 **Explain** why data collection needs to be systematic and accurate.
3 **Explain** the type of error students are producing in the scales below.

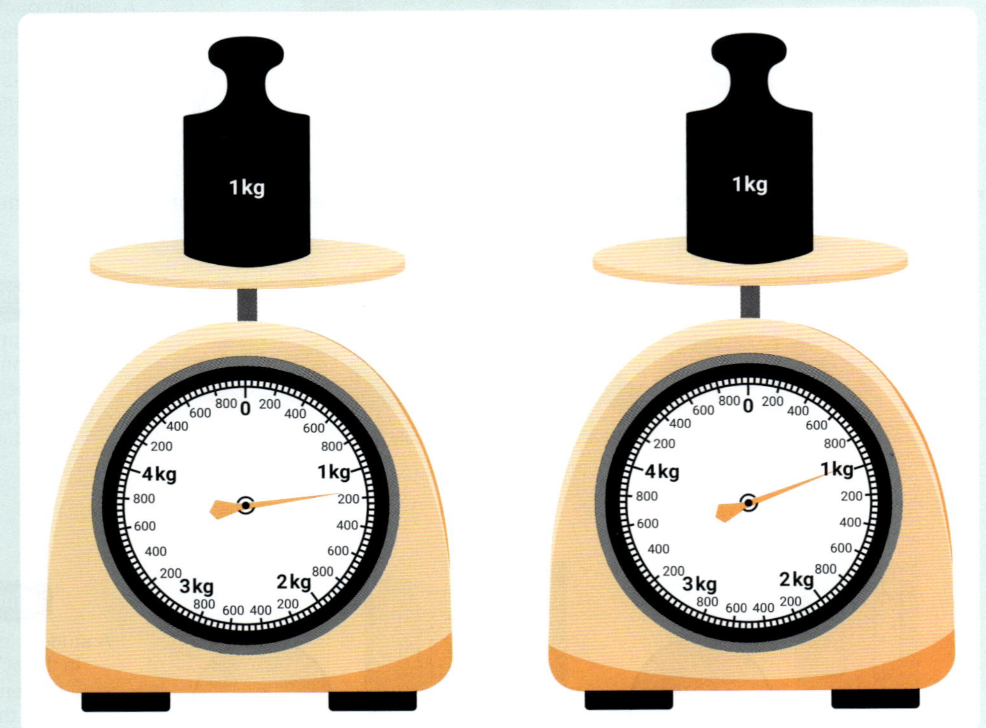

4 You are investigating the type of timber needed to build a scaffold that will sustain a 200-kg weight. What factors will you need to consider to collect valid, reliable data? **Describe** how you can reduce systematic errors.

1.9 Processing data and information

BY THE END OF THIS MODULE, YOU WILL BE ABLE TO:

- ✓ select and use different ways to organise data
- ✓ calculate simple descriptive statistics to process data
- ✓ explain how to improve the quality of data.

GET THINKING

You are part of your school's student representative council. The principal asks your council to analyse data collected from a survey about students' study habits across different year groups. The data shows that students in Year 7 study significantly less than students in Year 10. What factors could influence the data? How can you organise the data to inform the principal?

Video
Science skills in a minute: Collecting and organising data

Interactive resource
Drag and drop: Mean, median and mode

Processing data

In Stage 4, you learned that data can be presented in tables, various types of graphs, diagrams, flow charts, spreadsheets and models (Figure 1.9.1). In this module, you will learn how to select and extract information from those data representations to support or refute scientific questions and hypotheses and propose solutions to problems. Data can tell a story about the research, but if the data are not represented and processed correctly, they can be misleading.

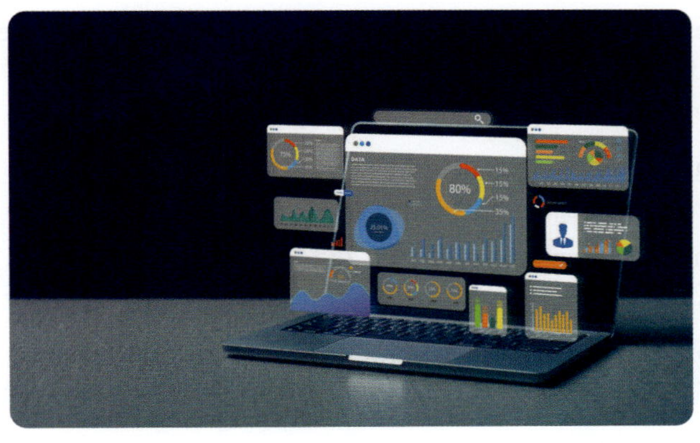

▲ **FIGURE 1.9.1** Information can be extracted from databases and scientific models.

Selecting the best data representation and understanding how to extract information from it is crucial for the analysis of the data. Correct interpretation of data provides evidence to support or refute a hypothesis.

Table 1.9.1 summarises the types of data and graphs used to represent information.

▼ **TABLE 1.9.1** Summary of types of data representation and how to extract information from them

Type of data representation	Data represented	How to extract information
Text	Observations Secondary data from other scientists	Categorise the information under specific criteria, based on the research question
Tables and spreadsheets	Numerical data – continuous or discrete data	Use descriptive statistics to summarise the data, and identify errors and outliers
Graphs	Numerical data (commonly represented in column and line graphs)	Use interpolation of the data in the graph Relate the independent and dependent variables to identify patterns and trends
Databases	Large numerical or non-numerical datasets stored in digital form	Make a selection of valid databases and apply algorithms and mathematical models
Flow charts	Processes that have steps or links between concepts	Follow the steps of the flow chart to create new annotated texts
Models	Processes (may be shown in physical or digital form)	Describe and explain the phenomena by adding more details to the model
Diagrams	Observable phenomena	Annotate and include extra resources to support the information in the diagram

Descriptive statistics

Descriptive statistics are numbers that summarise a group of scores to make data more comprehensible and understandable. This makes it easier to establish if there is a **cause-and-effect relationship** between the independent and dependent variables.

In Stage 4, you learned about basic descriptive statistics used in processing data. In Stage 5, you will learn about three broad categories of descriptive statistics:

- statistics that describe extreme values: **maximum value** and **minimum value**
- statistics that describe the centre of the dataset: **mean**, **median** and **mode**
- statistics that describe the spread or **dispersion** of values: **range**, **standard deviation** and **variance**.

Figure 1.9.2 shows how to calculate the mean, median, mode and range.

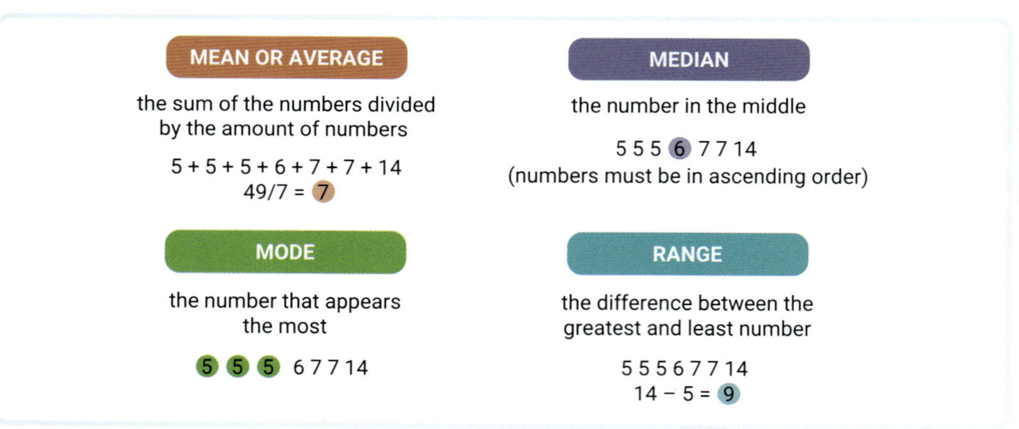

▲ **FIGURE 1.9.2** The rules for calculating the mean, median, mode and range

descriptive statistics values derived from a dataset, such as a mean or maximum value, that summarise the dataset

cause-and-effect relationship a relationship in which a change in one variable causes a change in another variable; also known as causation

maximum value the largest value in a dataset

minimum value the smallest value in a dataset

mean the calculated 'central' value of a set of numbers; an average

median the middle value in an ordered set of numbers

mode the value that occurs most frequently in a dataset

dispersion the extent to which values in a dataset are above or below the mean (average)

range a measure of spread calculated by subtracting the smallest number from the largest number in a dataset

standard deviation a calculated measure of spread representing the spread of data around the mean in a dataset

variance a measurement of the amount of spread in the values in a dataset

DATA SCIENCE

See **Module 2.6** to learn more about how to calculate and use descriptive statistics.

ACTIVITY 1.9

Calculating descriptive statistics

The science teacher has distributed the results of the mid-year examinations. The marks are listed below:

92, 77, 69, 98, 77, 85, 84, 77, 85, 87

1 Organise the data from least to greatest.
2 Calculate the mean, median and mode.

Improving the quality of data

Any data collected from an investigation may have errors or mistakes. As you learned in Module 1.8, errors can be random or systematic. Mistakes include, for example, the units not being converted correctly, typing mistakes in tables of data, having **outliers** (extreme values) in the dataset, or having data missing.

outlier
a value that differs significantly from other values in a dataset

To improve the quality of the data, there are important steps to follow.

1 Repeat the experiments to increase the reliability of the data by ensuring the measurements are similar each time.
2 Test the accuracy and precision of the instruments to ensure accurate measurements.
3 Increase the sample size to provide more data and reduce errors.
4 Clean the dataset by identifying obvious mistakes and removing them.

How data supports or refutes a hypothesis

The data you collect and process becomes the evidence to support or refute the hypothesis, answer the scientific question and make a conclusion about the purpose of the investigation. The data from an investigation may be compared with data from secondary sources, obtained in previous experiments by other scientists (published in the peer-reviewed literature), to provide confidence that the results are valid.

All scientific hypotheses should be refutable, which means they can be proven wrong. If the hypothesis is refuted by the data, this leads to a new set of ideas and questions to test again. In this way, science progresses to new discoveries.

ACTIVITY

Data that supports or refutes the hypothesis

An ecologist is working in the field to test the hypothesis that a high level of soil moisture increases the number of snails observed (Figure 1.9.3). As a trial, they set up a 10-m transect in the middle of the bush and measure the soil moisture with a probe. They also count the number of snails every metre. They obtained the data shown in Table 1.9.2.

AndreyShutov/Shutterstock.com

▲ **FIGURE 1.9.3** One variable in this investigation is the number of snails observed per metre of transect.

▼ TABLE 1.9.2 Number of snails observed per metre of transect

Transect distance	Soil moisture (%)	Number of snails
0	10	0
1	30	0
2	24	10
3	55	3
4	23	1
5	70	5
6	75	7
7	89	2
8	24	8
9	55	6
10	65	5

Processing data

1 Create a graph of the data to relate the soil moisture and the number of snails.

2 Describe the pattern and trend in the graph.

Analysis

1 Was the hypothesis supported or refuted by the data?

2 What type of problems may the ecologist encounter in the collection of data?

3 How could you improve this experiment?

1.9 LEARNING CHECK

1 **Describe** ways to organise data when making observations and measurements in the field.

2 **Calculate** the mean, median and mode for the following dataset, collected in an experiment measuring air pressure in bike tyres:

69, 77, 77, 77, 84, 85, 85, 87, 92, 98 (unit of measurement: PSI, pounds per square inch)

3 **Create** a simple experiment that includes collecting data from your class to **calculate** descriptive statistics. For example, measure height, or hand length from the wrist to the top of the middle finger.

4 **Explain** what factors you need to consider to reduce systematic and random errors if you are testing how the acidity of a substance affects the dissolving rate of a solid.

5 Students are testing the amount of microplastic in the sand banks of the local creek. **Explain** how to improve the quality of data in the investigation.

1.10 Analysing data and information

BY THE END OF THIS MODULE, YOU WILL BE ABLE TO:

- ✓ evaluate patterns and trends in data
- ✓ assess validity and reliability in data and secondary sources
- ✓ describe the sources of uncertainty
- ✓ develop evidence-based arguments from conclusions.

GET THINKING

You are a keen photographer learning new skills, and you decide to upgrade your camera. There are hundreds of different types of cameras. What characteristics (variables) are you going to analyse before you make your decision: value for money, camera features, brand or something else? Will you focus on the claims by the photography companies or will you base your decision on independent reviews or other evidence?

Quiz
Analysing data

Purpose of data analysis

We analyse the data in investigations by examining and interpreting it, so that we can make sense of the gathered information. We can draw meaningful conclusions from a thorough analysis and these will **verify** or refute the hypothesis. Careful data analysis supports us to make decisions that are based on **empirical evidence** (Figure 1.10.1)

verify
to demonstrate an argument is true by providing evidence

empirical evidence
evidence collected through scientific observation or experimentation

▲ **FIGURE 1.10.1** Key steps in analysing data to draw conclusions

Patterns, trends and inconsistencies in data

In Stage 4, you learned that a trend is a relationship between the independent and dependent variable. As the independent variable changes, the dependent variable changes in a certain way. You also learned to identify the types of patterns when data follows a relationship.

In Stage 5, we are going to further analyse data patterns and trends, and describe inconsistencies in data. Figure 1.10.2 is a scatter plot graph showing the exam results from a Year 9 Science class in relation to the number of hours studied for the test.

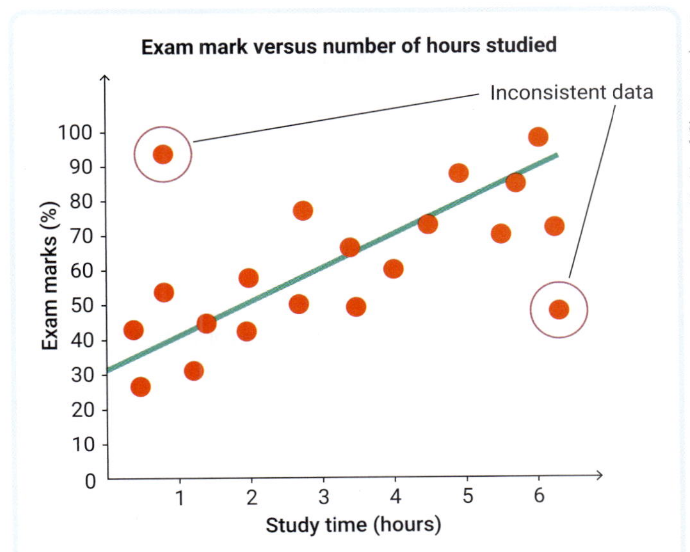

▲ **FIGURE 1.10.2** Exam marks achieved by Year 9 students graphed against the number of hours they studied

9780170491785

The first step in analysing the data is to observe patterns and trends. In this example, the data follows an increasing trend. This means that as the independent variable increases, the dependent variable also increases. A conclusion from the example data is that the more hours that students studied, the better results they achieved in the exam. Another observation from the graph is that there are two outliers (circled on the graph), values that look different and may be incorrect. This is because they do not follow the pattern of the other data points. The outliers are described as **inconsistencies in the data**.

inconsistency in the data
a data point that does not follow the pattern of other data from the same experiment or across repeated experiments

If there are inconsistencies in the data, you need to check the original data for errors or mistakes in its recording and processing. Table 1.10.1 shows some causes, consequences and ways to process inconsistencies in data.

▼ **TABLE 1.10.1** Causes, consequences and ways to process inconsistencies in data

Causes	Consequences	Processes
• Random errors • Systematic errors • Methods not testing the hypothesis • Mistakes in recording	• Inaccurate analysis • Inaccurate conclusions • Waste of time and resources	• Use appropriate data cleaning • Check accuracy and precision in measurements • Check calculations • Draw appropriate graphs

Relationships between variables

correlation
a trend in data in which one variable changes consistently as the other variable changes

positive correlation
a correlation where one variable increases as the other variable increases

negative correlation
a correlation where one variable decreases as the other variable increases

In investigations that have an independent and dependent variable, we analyse the relationship between variables to see if there is evidence to support or refute the hypothesis. The relationship can be analysed by looking for correlations. **Correlations** are statistical relationships between two variables. A **positive correlation** means that as one variable changes (for example, increases) the other changes in the same way. A **negative correlation** means that as one variable changes (for example, increases) the other changes in the opposite way (decreases). When there is no relationship between the variables, there is no correlation (Figure 1.10.3).

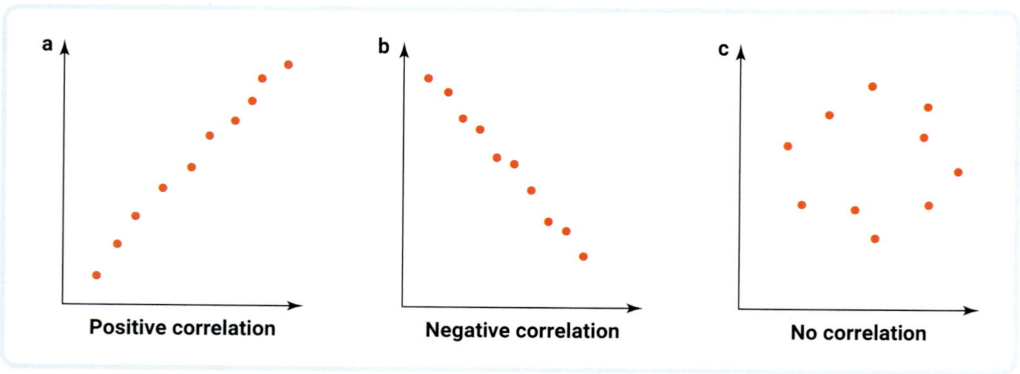

▲ **FIGURE 1.10.3** Linear correlation can be **(a)** positive or **(b)** negative. **(c)** Two variables may show no correlation.

Box plots, also known as box-and-whisker plots, are a type of graph that summarises the distribution of a dataset. They can represent the relationship of variables and show the distribution of the data, median and mean (Figure 1.10.4). The 'box' shows where the bulk of the data sits. Figure 1.10.5 shows the exam results data from the Year 9 classes in a box plot.

box plot
a graphical representation of the distribution of data

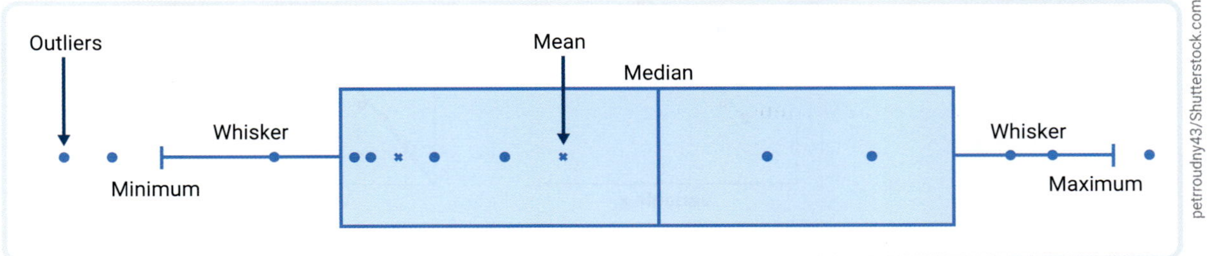

▲ **FIGURE 1.10.4** A box plot shows the spread of values in a dataset and how they compare to the median.

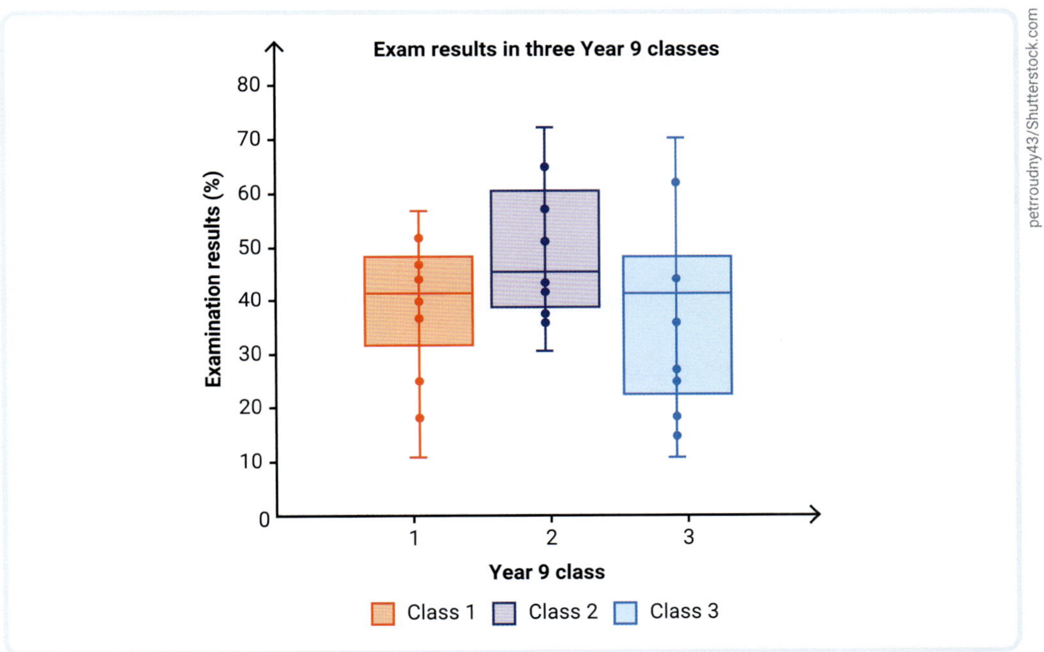

▲ **FIGURE 1.10.5** Box plots representing the distribution of the data, median and mean of the exam results for Year 9 classes

Graphed data can be used for:

- **interpolation** – using one variable to predict a value for the other variable *within* the range of the graph
- **extrapolation** – using one variable to predict a value for the other variable *outside* the range of the graph (Figure 1.10.6).

It is important to take care when doing extrapolations because we don't know that data outside the range of the graph will follow the same trend as known data.

interpolation
the estimation of an unknown value within the range of known data for a relationship

extrapolation
the estimation of an unknown value by extending a trend beyond the known values

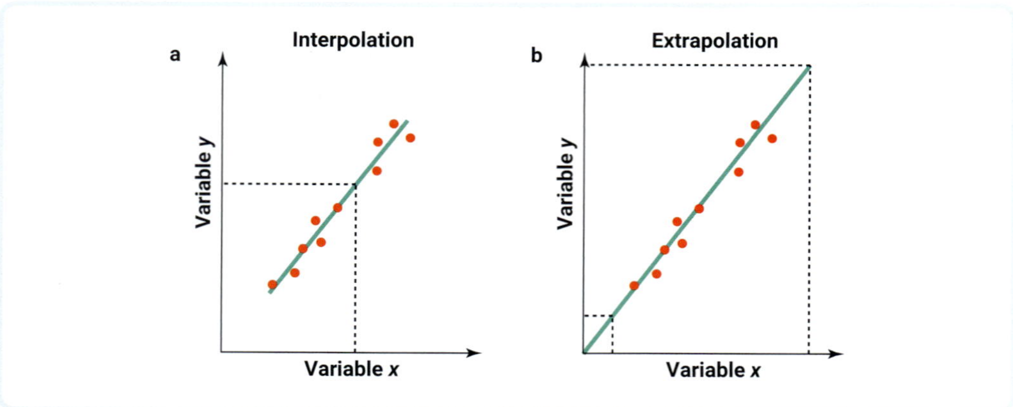

▲ FIGURE 1.10.6 (a) Interpolation means using values of *x* to predict values of *y* within the range of known data; (b) extrapolation means using values of *x* to predict values of *y* outside the range of known data.

Validity and reliability of data

To test the hypothesis of an investigation properly, data must be valid and reliable, whether it is collected first-hand data or from secondary sources. If any data collected from the investigation is unreliable or invalid, the investigation cannot be trusted (Figure 1.10.7).

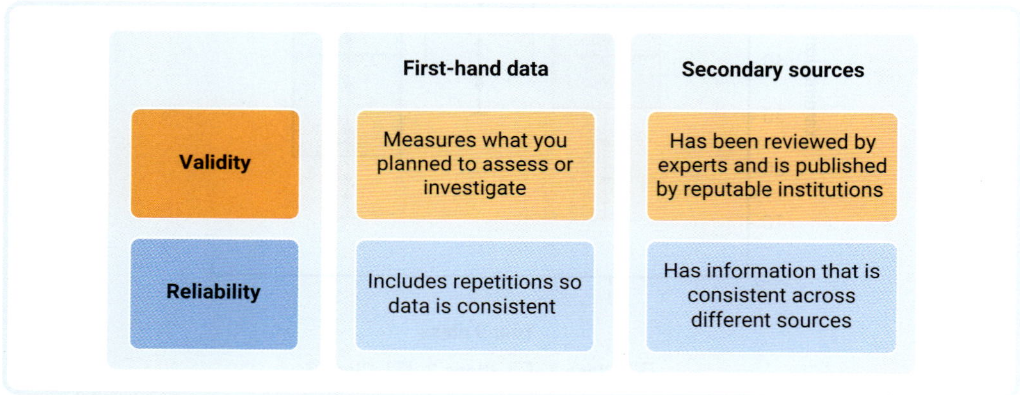

▲ FIGURE 1.10.7 How to assess the validity and reliability of data from first-hand and secondary sources

Conclusions from data

uncertainty
the range of possible values or outcomes that may be due to limitations in measurement, variability or incomplete data

The analysis of results becomes the evidence from which you can draw conclusions that support or refute the hypothesis. But how sure can you be of your conclusions? Data from investigations always has some **uncertainty**, even if all our calculations and analyses are correct (Figure 1.10.8). For example, if we investigate how many people prefer tiramisu to ice cream for dessert, we might find 20 per cent like tiramisu and 43 per cent like ice cream, and 37 per cent are unsure. The conclusion that ice cream is more popular than tiramisu has uncertainty because, for example, the results may come from just a small sample of the population. To increase confidence in a conclusion, the data need to be reliable and similar in repeated experiments.

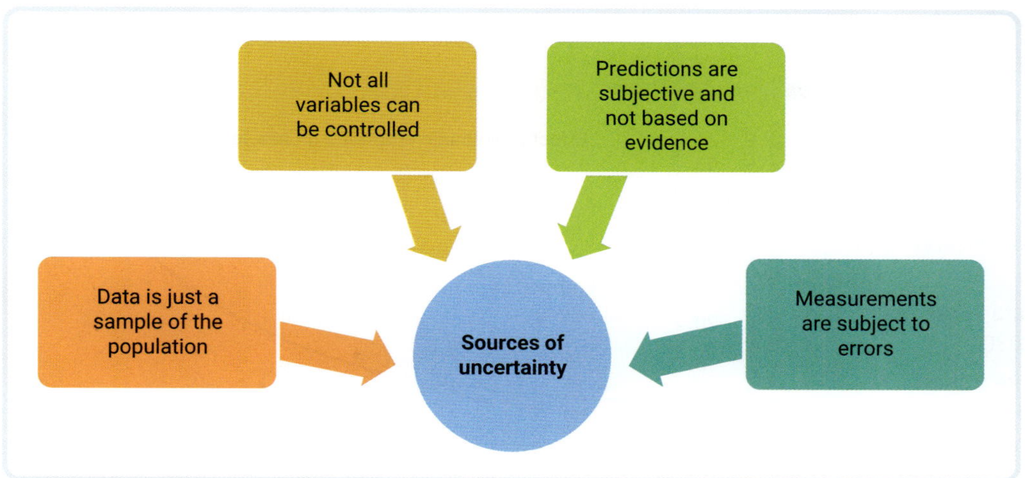

▲ **FIGURE 1.10.8** Sources of uncertainty in data

Evidence-based arguments

An evidence-based argument is a **scientific claim** supported by reliable and relevant data such as research findings, statistics or observations. The goal of an evidence-based argument is to persuade others by providing a strong foundation of proof rather than relying on opinion or speculation.

scientific claim
a statement based on evidence from a scientific investigation

There are three components in an evidence-based argument: claim, evidence and **rationale** (Figure 1.10.9).

rationale
the grounds that explain the evidence that supports the claims of the investigation

Evidence-based arguments allow us to evaluate and update scientific knowledge. Scientists have the opportunity to challenge past knowledge when they publish new findings. This is the way science advances.

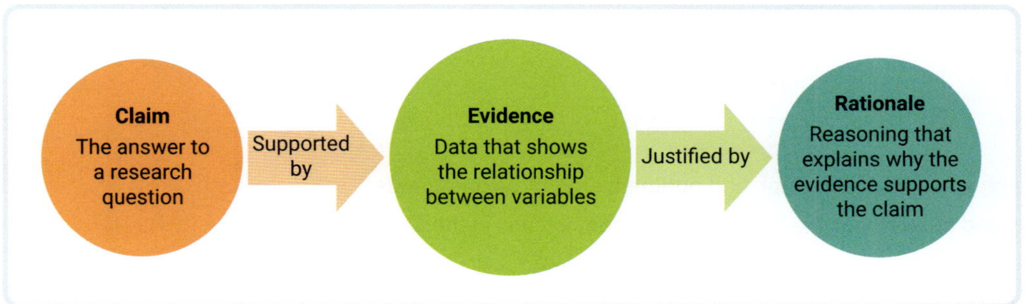

▲ **FIGURE 1.10.9** The evidence-based argument process. You can remember this using the acronym CER.

1.10 LEARNING CHECK

1 **Observe** the graph in Figure 1.10.10, which is a simulation that shows the effect of volcanic eruptions on sea-level rise.
   a What trend can you **observe** from the graph?
   b Extrapolate the data to **predict** sea-level rise in 2030.
   c **Construct** and write a short conclusion based on the evidence in the graph.

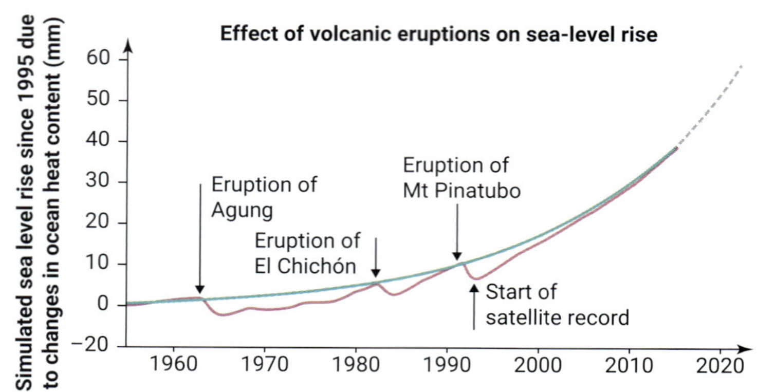

▲ **FIGURE 1.10.10** Simulated sea-level rise with (pink line) and without (green line) the effect of volcanic activity. Following an eruption, Earth cools and the ocean's heat content drops, lowering sea levels.

2 **Explain** the meaning of validity and reliability in first-hand data.

3 The dataset in Table 1.10.2 was collected to test the relationship between the levels of nitrogen in the soil and plant growth.

▼ **TABLE 1.10.2** Level of nitrogen in the soil and plant growth

Levels of nitrogen in soil (mg/kg)	Plant growth (cm)				
	Rep 1	Rep 2	Rep 3	Rep 4	Rep 5
0	5.1	7.2	5.3	4.0	6.2
20	7.0	9.0	8.0	7.5	8.8
40	10.9	11.2	11.8	11.5	12.0
60	15.2	14.8	14.9	15.1	15.4
80	15.8	16.1	17.2	16.8	15.9
100	11.1	10.9	10.8	9.9	10.6

   a Plot the results using an appropriate graph and **outline** the patterns and trends.
   b **Explain** the validity and reliability of the data and any inconsistency in the data.
   c **Construct** and write an evidence-based argument from the data.

4 A scientist is researching the effects of plastic pollution on the birth rate of emperor penguins in Antarctica. They found that as the amount of plastic pollution increases, the birth rate decreases. How certain can the scientist be about the result? **Justify** your answer.

5 A sunscreen company claims that the higher the level of SPF in sunscreen, the higher the level of protection from skin cancer.
   What data would scientists need to **analyse** to provide an evidence-based argument to support this statement?

9780170491785

1.11 Problem-solving

BY THE END OF THIS MODULE, YOU WILL BE ABLE TO:

- ✓ describe the types of problems in science
- ✓ select different problem-solving strategies
- ✓ assess problem-solving strategies
- ✓ use cause-and-effect relationships to make predictions.

GET THINKING

On 11 April 1970, Apollo 13 was launched from Kennedy Space Centre in Florida, USA. Two days into the mission, the oxygen tank in the service module ruptured, disabling the life support system. The crew needed to solve problems in order to survive and return to Earth. Compare the magnitude of solving the problems faced by the Apollo 13 crew with solving a problem in an experiment in the science lab. What criteria should you use to solve problems in both these situations? Discuss with a group of classmates.

Christian Kohler/Shutterstock.com

▲ **FIGURE 1.11.1** The Apollo 13 command module

Scientific problems

There are two types of problems in science. One is the scientific problem that you can answer using the working scientifically processes – for example, how can we recycle rubber products using a chemical process? The other type of problem is an investigation problem that arises during an experiment – for example, problems with instruments, data collection or control of variables. Figure 1.11.2 summarises the common types of problems in science.

The strategies you use will depend on the type of problem that you are solving. For example, if you want to test a claim, your problem-solving will be based on the design of the investigation as a whole. However, if your problem is that the data collected seems invalid, you will need to check such things as the methods and the calibration of instruments. Table 1.11.1 summarises problem-solving strategies.

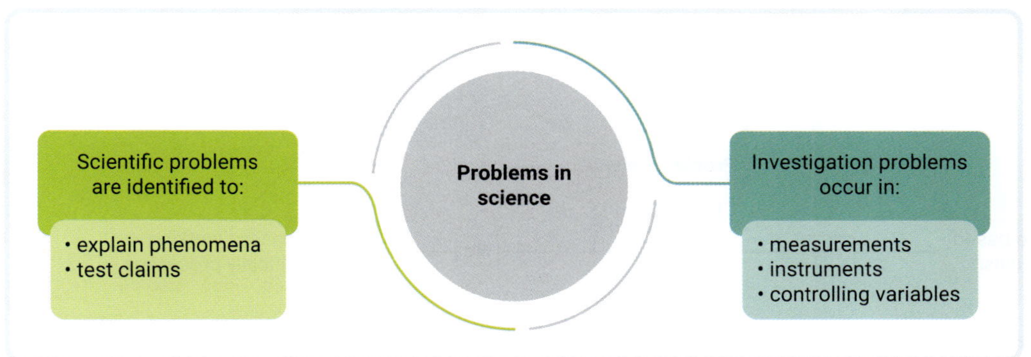

▲ **FIGURE 1.11.2** Problems in science can be scientific problems (a problem that has a hypothesis to be tested) or problems in the investigation (a problem in the process).

▼ TABLE 1.11.1 Strategies to solve common problems encountered in science investigations

Common problem	Problem-solving strategy
Anomalies in collected data and measurements	• Check apparatus and devices and calibrate if necessary • Check for mistakes in data entry • Check if units need to be converted
Differences in observations between repetitions	• Check controlled variables were all kept the same across repetitions
Data collected did not support hypothesis and did not answer the scientific question	• Check the procedures used in the investigation • Check how data was collected and recorded • Check the data analysis and type of graph used
Results do not match with previous literature	• Conduct further literature research • Check how data was collected and recorded • Check the data analysis and type of graph • Identify reasons data may differ

Once you apply a problem-solving strategy, the next step is to assess the solution. You may ask yourself:

- Did I solve the problem with this strategy?
- Do I need to start the investigation again?

See the decision flow chart below to guide your assessment of solutions to the problem (Figure 1.11.3).

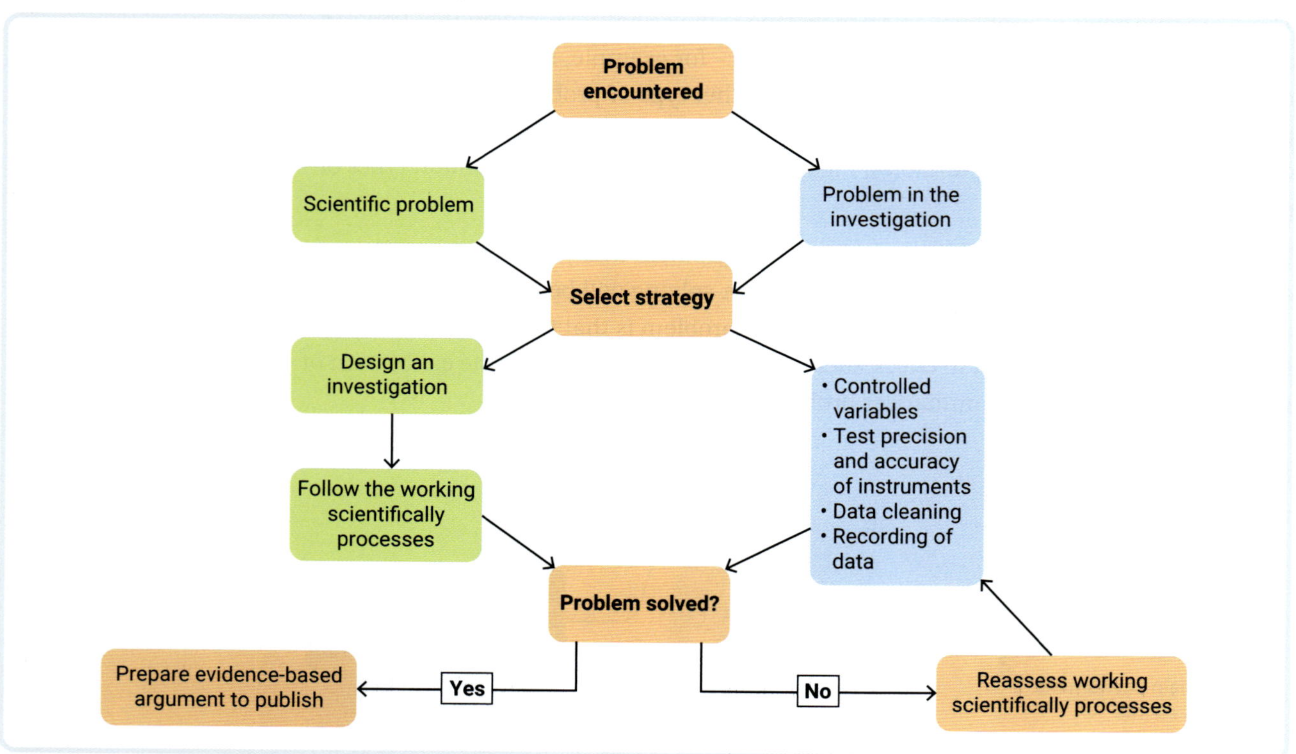

▲ FIGURE 1.11.3 Steps to assess your solution to the encountered problems

9780170491785

Cause-and-effect relationships using models

In Stage 4, you learned that models can explain phenomena or complex concepts. Models are used to make predictions that answer scientific problems. For example, models can predict how oceans may rise due to global warming.

Models help solve problems by:

- enabling forecasting – predicting future trends based on extrapolation of recorded data or by simulating the effects of natural phenomena (e.g. weather patterns, volcanic eruptions)
- increasing understanding – explaining underlying processes (disease spread, ecological relationships)
- informing decisions – helping determine the best course of action (e.g. public health campaigns, use of resources).

DATA SCIENCE 

Learn more about cause and correlation in **Module 2.7**.

1.11 LEARNING CHECK

1 **Identify** the type of each problem in the list below: is it a scientific problem or an investigation problem?
   a Results are not consistent across repetitions.
   b Microplastics were found in the stomach of dead birds on the beach.
   c Erosion is destroying the beach around northern NSW.
   d The readings from a data logger thermometer seem out of range.
   e A column graph is not showing some of the data.

2 Tom is working on the design of an investigation and he is confused about the type of instrument to use to measure the acidity of a substance. **Propose** strategies for solving this problem that you would recommend to Tom.

3 **Explain** what you need to do if you encounter a problem during an investigation.

4 **Choose** a model (for example, climate models, plate tectonic digital models) and **assess** how well it can be used to make predictions.

1.12 Communicating

BY THE END OF THIS MODULE, YOU WILL BE ABLE TO:

- ✓ communicate scientific evidence-based arguments
- ✓ follow scientific conventions in writing texts
- ✓ explain the importance of scientific texts.

Quiz
Science communication

GET THINKING

Imagine you need to present to the school committee an evidence-based argument to run a fundraising campaign for a recycling project led by students. What evidence do you need to write your argument? How does a written argument differ from one that you would make verbally?

Written texts in science

In Stage 4, you learned the steps to write scientific texts based on the purpose of the investigation and the targeted audience (Figure 1.12.1).

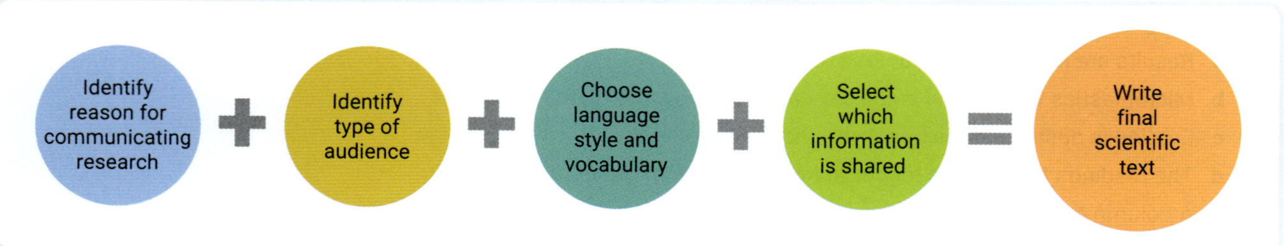

▲ **FIGURE 1.12.1** The steps in planning a written scientific text

In Stage 5, you will learn how to write a scientific argument based on evidence that you collected in your investigation. Writing a scientific argument requires a structured approach to presenting evidence and supporting the claim of your investigation with relevant data and reasoning.

A written scientific argument follows this structure:

1. State the claim – begin with a clear, concise statement of the hypothesis and the claim that you are making.
2. Provide background information – this is the context for your argument, which explains the importance of your claim to the audience and explains the relevant background from previously published research.
3. Present the evidence – concisely show the data (evidence) from the investigation by using appropriate graphs; include the analysis of the data and show descriptive statistics that help show that the data is valid and reliable.
4. Discuss the evidence – present the significance of the evidence and how it connects to your argument, including the strengths of your evidence and the limitations of your research (which may lead to future research).

9780170491785

5 Draw conclusions – summarise how the evidence collectively supports your claim by coming back to the statement of the main argument, how that relates to the evidence presented, and the significance of your findings.

6 Cite all your sources – list any secondary sources you used in your background information, as extra images or as models.

An evidence-based argument must be clear, flow logically and follow the structure of a scientific text.

DATA SCIENCE 

Learn more about investigating scientific claims in **Module 2.2**

The importance of well-written scientific text

Clear and well-written scientific text is important when sharing or publishing scientific knowledge (Figure 1.12.2). Good communication that is verified by reliable secondary sources helps advance science and makes it easier for others to trust the research.

It is important to think about your audience when writing scientific text. Will it be read by other scientists, or local community groups that need to use the information? Your audience should inform your writing style and choice of language, as well as the format of the written text.

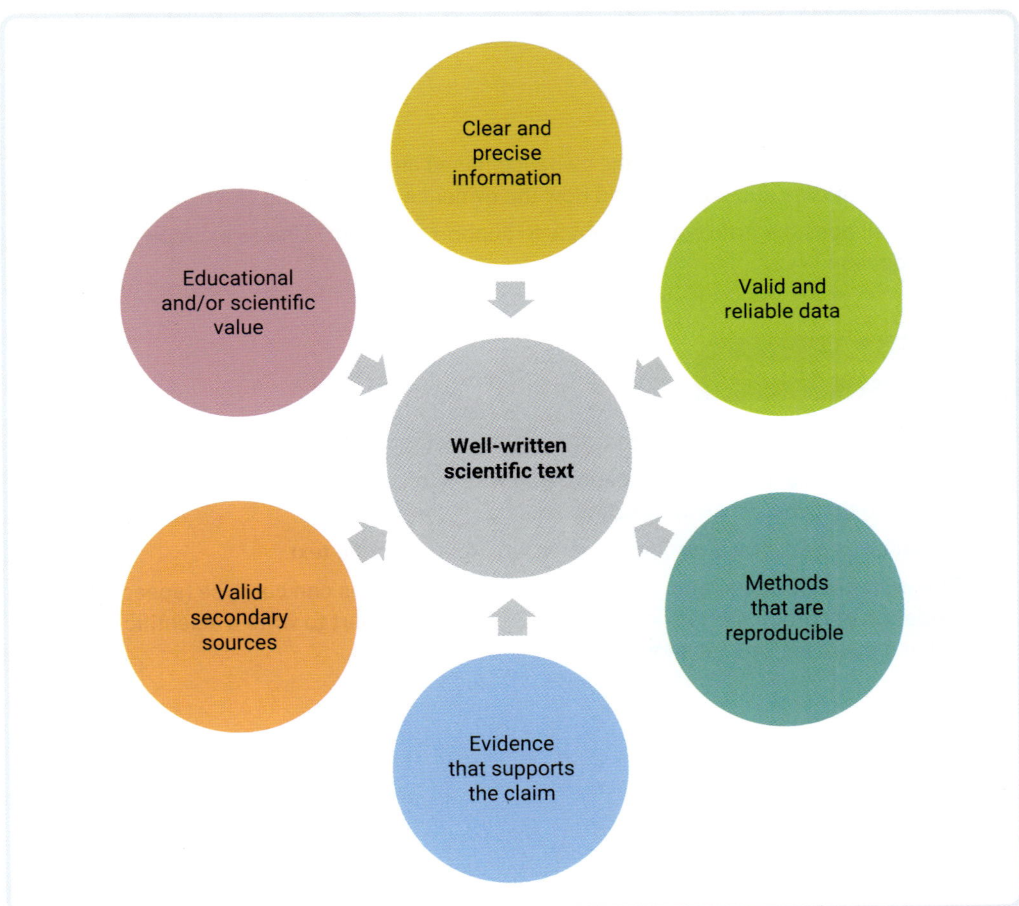

▲ **FIGURE 1.12.2** Features of well-written scientific texts

✩ ACTIVITY

Evaluate a written text

A student has created a short draft of an evidence-based argument.

The case for protecting the northern hairy-nosed wombat

The northern hairy-nosed wombat (*Lasiorhinus krefftii*), one of Australia's most endangered species, requires immediate and strong conservation efforts to prevent its extinction.

Only 300 northern hairy-nosed wombats remain in the wild (Wimmer & Taylor, 2020). The species has had severe population declines due to habitat loss, disease and environmental changes (Walker et al., 2019).

The unique role that the wombat plays as a major herbivore in its ecosystem justifies making its conservation a priority. Failure to act could result in irreversible loss of biodiversity and ecosystem function (Johnson, 2018).

References

Wimmer S and Taylor K (2020) 'Abundance of hairy-nosed wombats in the wild', *Journal of Ecology* 1:25–30.

Walker L, Smith S, Johansen T and Singh R (2019) 'Effect of environmental factors and diseases on the hairy-nosed wombat, *Lasiorhinus krefftii*', *Animal Ecology* 4:63–78.

Johnson, G (11 April 2018) 'Hairy-nosed wombat news', *Ecology Community Blog*, accessed 22 January 2024.

Analysis

1 After reading the draft, **assess** the evidence-based argument in terms of validity, reliability of data and sources, and how the evidence supports the claim.

2 How well does the article use evidence to support the claim? Use an example to **justify** your opinion.

1.12 LEARNING CHECK

1 **Develop** a claim about a topic that you are interested in.

2 **Describe** the main elements to consider in writing an evidence-based argument about the claim that you chose in Question 1.

3 **Explain** why evidence is important when writing a scientific text.

4 The evidence from research shows that gold nanoparticles can catalyse (speed up) the oxidation of carbon monoxide. **Create** a planning flow chart to write a scientific text to publish the research that shows this in a scientific journal.

1 REVIEW

REMEMBERING

1 **Describe** the steps in the scientific process.

2 **Define** scientific observation and give an example.

3 **Describe** how to write a hypothesis.

4 **List** five examples of hazards in the laboratory.

5 **Define** systematic error.

UNDERSTANDING

6 **Explain** the characteristics of scientific observations.

7 An experiment was left unattended in a Year 10 Science laboratory. The Bunsen burner was on with a beaker on a tripod above, boiling an unknown substance. **Explain** the risks that this situation created to students and the laboratory.

8 **Explain** the importance of repetition in an experiment.

9 **Describe** the difference between validity and reliability.

10 If an electronic scale is not measuring weight properly, what type of factors are going to be affected in the collection of data: accuracy, precision, validity or reliability? More than one answer is possible.

APPLYING

11 An animal behaviour scientist is collecting evidence to support the following hypothesis:

'If domestic cats are exposed to a variety of interactive toys, then their overall activity levels will increase compared to when they are exposed to no toys'.

- a **Identify** the dependent and independent variables.
- b **Identify** the controlled variables.
- c **Develop** a short procedure to test this hypothesis.

12 Sue said to the teacher, 'As soon as the door of the laboratory opened, the measurements seemed to change.' Could this statement be true? **Determine** the type of error in these measurements.

13 Data representation is important in processing data. **Explain** how data is extracted from graphs and models.

14 The following data were collected to test the rate of photosynthesis as light intensity increases.

Light intensity (lux)	Rate of photosynthesis (oxygen produced, mL/min)		
	Trial 1	Trial 2	Trial 3
100	2.5	2.4	2.5
200	4.0	4.1	4.0
300	5.4	5.6	5.5
400	5.9	6.0	6.0
500	6.2	6.2	6.2
600	5.9	6.1	6.1
700	5.8	7.1	5.9
800	5.6	5.5	5.7

- a **Plot** the data in an appropriate graph.
- b **Determine** the pattern and trend in the data.
- c **Calculate** the mean of the data.
- d **Create** a box plot graph and **explain** the distribution of the data.
- e What trend could be observed in the rate of photosynthesis at 1000 lux? **Explain** if this extrapolation is valid.
- f **Assess** the validity and reliability of the data in this investigation.
- g **Construct** a valid conclusion from this investigation.

EVALUATING

15 **Discuss** the sources of uncertainty in data.

16 In a digital model that measures the reflection of light in optic fibres, physicists encountered a problem within the investigation. The internal reflection in the fibre was not correct. **Justify** two strategies that the scientists may apply to solve the problem.

17 **Evaluate** the importance of well-written scientific text.

CREATING

18 **Plan** and conduct an investigation to test the question 'Do people who play an instrument have better coordination than people who play sport?' **Create** a depth study report to **investigate** and share your findings.

19 **Evaluate** the strength of the evidence presented in a claim that human activities are the primary driver of recent climate change. Consider the types of evidence typically used to support this claim and **discuss** how they contribute to the argument's rationale. **Construct** an evidence-based argument to support this claim.

9780170491785

DATA SCIENCE 2

SYLLABUS OUTCOMES

A STUDENT:

- assesses the use of scientific knowledge and data in evidence-based decisions and when verifying the legitimacy of claims SC5-DA2-01
- analyses data from investigations to identify trends, patterns and relationships, and draws conclusions SC5-WS-06
- selects suitable problem-solving strategies and evaluates proposed solutions to identified problems SC5-WS-07
- communicates scientific arguments with evidence, using scientific language and terminology in a range of communication forms SC5-WS-08

© 2023 NSW Education Standards Authority

THE CHAPTER RELATED TO THIS FOCUS AREA IS:

- **CHAPTER 2 – DATA SCIENCE**

ZinetroN/Shutterstock.com

2 Data science

9780170491785

SCIENCE IN DEPTH

Horizon International Images/Alamy Stock Photo

▲ **FIGURE 2.0.1** The 2019–20 bushfires burned more than 5 million hectares in New South Wales.

Bushfires occur regularly in Australia, but they are changing. Major bushfires are becoming more common as the climate warms. This makes it harder for people to predict and manage bushfires. Researchers at the University of New South Wales are using data science methods to explore large environmental datasets and have developed a model to better predict major bushfires.

- **What are major bushfires?**
- **What types of data can be used to predict bushfires?**
- **How do data scientists analyse something like a bushfire with multiple causes?**
- **How can predictive models reduce the impact of major bushfires on our communities?**

DIVE INTO SCIENCE!

At the end of this chapter, you can complete Science in Depth Study #2. You can use the information you learn in this chapter to complete the project.

Assessments
- Prior knowledge quiz
- Chapter review questions
- End-of-chapter test
- Depth study: Presentation

Videos
- Science skills in a minute: Writing a discussion **(2.2)**; Organising data **(2.6)**; Organising data into charts **(2.6)**; Representing data in different forms **(2.6)**
- Maths in science videos: Distorted graphs **(2.2)**; Analysing different graphs **(2.6)**; Mean, media and mode **(2.6)**; Correlation and causation **(2.7)**
- Video activities: Tackling scientific misinformation **(2.3)**; Confirmation bias **(2.4)**

Science skills resources
- Science skills in practice: Discussions in science reports **(2.2)**

Interactive and other resources
- Crossword: Datasets **(2.5)**
- Quizzes: Questions, hypotheses and predictions **(2.2)**; Secondary data **(2.4)**
- Worksheets: Univariate analysis **(2.6)**; Bivariate analysis **(2.6)**

Nelson MindTap

To access resources above, visit **cengage.com.au/nelsonmindtap**

2.1 Investigating questions

BY THE END OF THIS MODULE, YOU WILL BE ABLE TO:

- ✓ describe the features of investigable and non-investigable questions
- ✓ explain how available resources can determine if a question is investigable
- ✓ evaluate the validity and reliability of online sources.

GET THINKING

Consider light passing through a glass prism (Figure 2.1.1). Make a list of questions about what you see. Which questions can be answered by a scientific investigation? Can you look up some of the answers? Are there questions that cannot be answered scientifically?

▲ FIGURE 2.1.1 Sunlight interacting with a glass prism

Atlantist Studio/Shutterstock.com

Investigable questions

investigable question a question that can be answered using a fair test investigation

An **investigable question** is a type of **testable question**. Rather than a question that can be answered using secondary sources, an investigable question requires an investigation to answer it. We conduct scientific investigations to answer such questions. Scientific knowledge is created when questions are answered using scientific processes. Many investigations involve controlled experiments and matching observations and measurements to predictions arising from models or theories.

fair test an investigation that is conducted correctly to answer a scientific question

An investigable question can be answered using a **fair test**. You learned about fair tests in Stage 4. A fair test may involve a controlled experiment and the collection of primary data or secondary data. Scientists use different types of investigable questions to address a range of purposes (Table 2.1.1).

▼ TABLE 2.1.1 Types of investigable questions

Question type	Purpose	Examples
Descriptive	To create a qualitative and/or quantitative description of something	• When do cattle drink water? • What diversity of plants and animals exists in this area after a fire?
Relational	To identify associations between variables. This includes grouping, ranking or correlations	• Which type of material is the most absorbent? • How can the grains of sand be classified into groups?
Cause-and-effect	Working out if a variable causes or influences one or more other variables	• Does air humidity affect frost thickness? • How do predator numbers affect prey reproduction rates?

9780170491785

Non-investigable questions

Non-investigable questions are those we can't easily answer with a single scientific investigation. They do not involve collecting data or evidence to answer the question. Sometimes, an answer is easily found by looking it up. Often, the answer to non-investigable questions is based on the research of different scientists over many years, published in journals. This information is then published again in textbooks and on websites.

non-investigable question
a question not easily answered by scientific investigation

Some questions are too hard to answer with investigations given the time and resources that they would take. Imagine trying by yourself to answer the question 'Why is the sky blue?' (Figure 2.1.2). Where would you start? The nature of light, how the atmosphere affects light and how our eyes work are not straightforward to investigate ourselves.

Andromeda stock/Shutterstock.com

▲ **FIGURE 2.1.2** The colours in sunsets, blue skies and the ocean are all caused by complex physical processes that are beyond the reach of an individual investigation.

What makes a good investigable question?

A good investigable question has four key properties. It is:

- **specific**
- measurable
- controllable
- **achievable** (Table 2.1.2).

By checking for each of these properties in our investigations, we make sure they are **valid** and **reliable**.

specific
clearly identified or defined

achievable
capable of being completed successfully

valid
describes the extent to which an investigation tests a hypothesis

reliable
consistent; able to be trusted

▼ **TABLE 2.1.2** Properties of good investigable questions

Properties	Key considerations
Specific	• Does the question contain a **dependent variable** and an **independent variable**? • Are the variables of interest easy to identify and define?
Measurable	• Can the dependent and independent variables be measured? • How will you measure the dependent and independent variables? • When will the measurements be made? • What units of measurement will you use?
Controllable	• Which variables should you control? • How should you control them? • Are there any variables that can't be controlled?
Achievable	• Is there enough time to complete the investigation? • Do you have the resources you need to conduct the investigation? • Who can help you conduct the investigation?

dependent variable
the factor that may be affected by the independent variable; the factor that can be measured or counted

independent variable
the factor that you choose to vary in your investigation

The importance of resources

resource
something used to complete a task or achieve a goal

A **resource** is something used to complete a task or achieve a goal. The apparatus and tools we use and the time available for the task are all resources.

Access to appropriate instruments is important. Some of the equipment you will use in Stage 5 is shown in Figure 2.1.3.

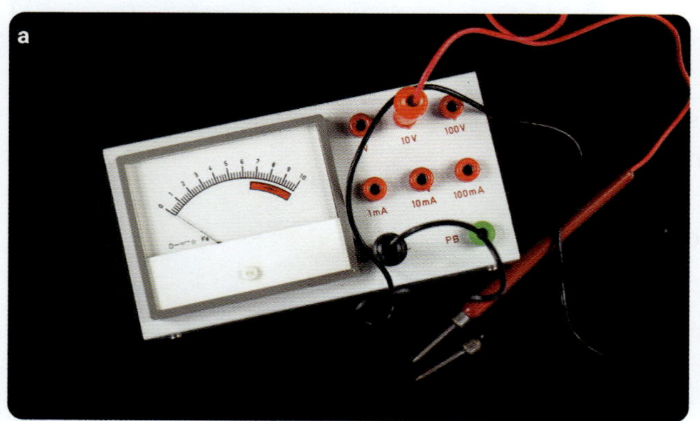

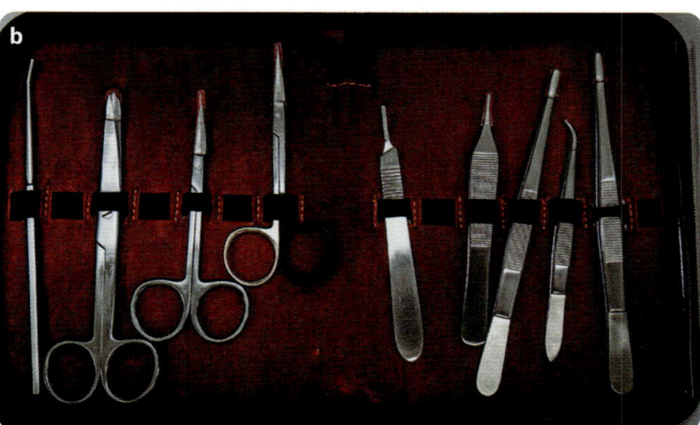

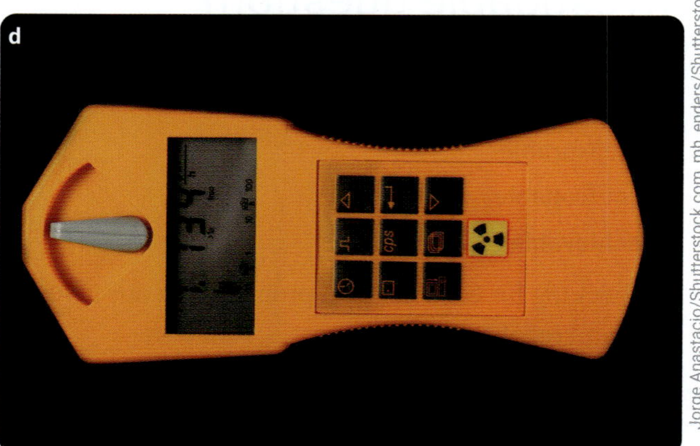

Jorge Anastacio/Shutterstock.com, mh_enders/Shutterstock.com, Noiel/Shutterstock.com, xpixel/Shutterstock.com

▲ **FIGURE 2.1.3** Four instruments you may use in Stage 5 science: **(a)** a meter used for measuring electricity, **(b)** dissecting tools for studying anatomy, **(c)** pH paper for measuring acidity, and **(d)** a Geiger counter for measuring radiation

Verifying online information

Background research helps create good investigable questions. It can tell us what science already knows about the subject. But how do we check that what we read is valid and reliable? It's likely you will often use online resources for background research, and they can vary widely in how reliable they are. Table 2.1.3 outlines some questions you can use to verify online sources.

9780170491785

▼ TABLE 2.1.3 Questions to ask when verifying online sources

Issue	Questions
Who	• Who wrote the information? • Who is the author? • Is the person qualified to write on the subject? • What else have they had published? • Are they a scientist, a science journalist, or someone else?
What	• What does the information add to my research? • Is the information related to the topic I'm researching? • Does the information add to my knowledge? • Does it support other sources? • Is the information too simple or complex for my needs?
Where	• Where was the information published? • Was the information published on a credible site? • Does the site describe how items are selected for publication? • Is the purpose of the site to inform, sell, entertain or persuade?
When	• When was the information published? • How recently was the information updated or revised? • If it refers to other sources, are those references recent? • Are there more up-to-date sources available?
Why	• Why did the author write the report? • Is there any information that suggests the authors could have a **bias**?
How	• How was the information gathered? • How was the data in the source originally gathered? • Is the same information also in a scientific journal or a published textbook?

bias
a strong preference for one thing or idea over another

2.1 LEARNING CHECK

1 **Define** investigable question and provide an example.
2 **Outline** the features of non-investigable questions.
3 **Explain** why appropriate resources are needed to make an investigation achievable.
4 **Describe** two features of an online source that would make it reliable.
5 **Identify** the following questions as investigable or non-investigable. **Justify** your answers.
  a Does phosphorus increase the number of algae in pond water?
  b Why do magnets stick to nails?
  c Is creating static electricity easier on a dry or a humid day?
  d What is the mass of a proton?
  e Does sugar dissolve in methylated spirits?

2.2 Supporting claims

BY THE END OF THIS MODULE, YOU WILL BE ABLE TO:

- ✓ identify claims that can be scientifically tested with an investigation or series of investigations
- ✓ explain the evidence and reasoning supporting a claim using data from investigations
- ✓ construct a written scientific argument using a range of evidence to support a claim.

Maths in science video
Distorted graphs

Quiz
Questions, hypotheses and predictions

GET THINKING

A fact about bandicoots (Figure 2.2.1) is that their diet includes truffle-like fungi that grow underground. What makes a scientific statement a fact? If you have evidence to support a hypothesis, is it enough to simply report these things? Draw a labelled diagram to show how evidence is used to support a hypothesis.

Susan Flashman/Shutterstock.com

▲ **FIGURE 2.2.1** A southern brown bandicoot (*Isoodon obesulus*)

Questions, claims and hypotheses

claim
a statement asserting something is true, real or a fact

fact
a piece of information that is supported by verified evidence

Science creates reliable knowledge by testing claims. A **claim** is a statement declaring something is true or a fact, but it is not necessarily true. Every day we make claims that are opinions based on limited or untested evidence. A **fact** is something that is supported by verified evidence.

A scientific claim is a particular type of claim that is based on existing scientific knowledge and is testable. A person may claim that heavy things fall faster than light things and strongly believe it, but is it true? For example, someone might claim that the speed of a falling feather depends on its shape (Figure 2.2.2). Such a claim can be tested with an experiment in which variables, such as object shape and air resistance, are controlled. The outcome of the experiment will support the claim or show it to be false (refute it).

Siwakorn1933/Shutterstock.com

▲ **FIGURE 2.2.2** Does the feather's shape affect how fast it falls?

Identifying claims that can be scientifically tested

hypothesis
a testable explanation for something based on existing knowledge; a testable statement of the predicted relationship between the independent and dependent variables

null hypothesis
the hypothesis that a relationship between variables being studied does not exist

A **hypothesis** is a predicted explanation or answer to an investigable question. Figure 2.2.3 shows that a good hypothesis is based on evidence from initial observations and background research of the subject. It also shows that a hypothesis will affect decisions about measurement and instruments. Sometimes, the hypothesis will state that a relationship between the variables being studied does not exist. Such a hypothesis is called a **null hypothesis** (Table 2.2.1).

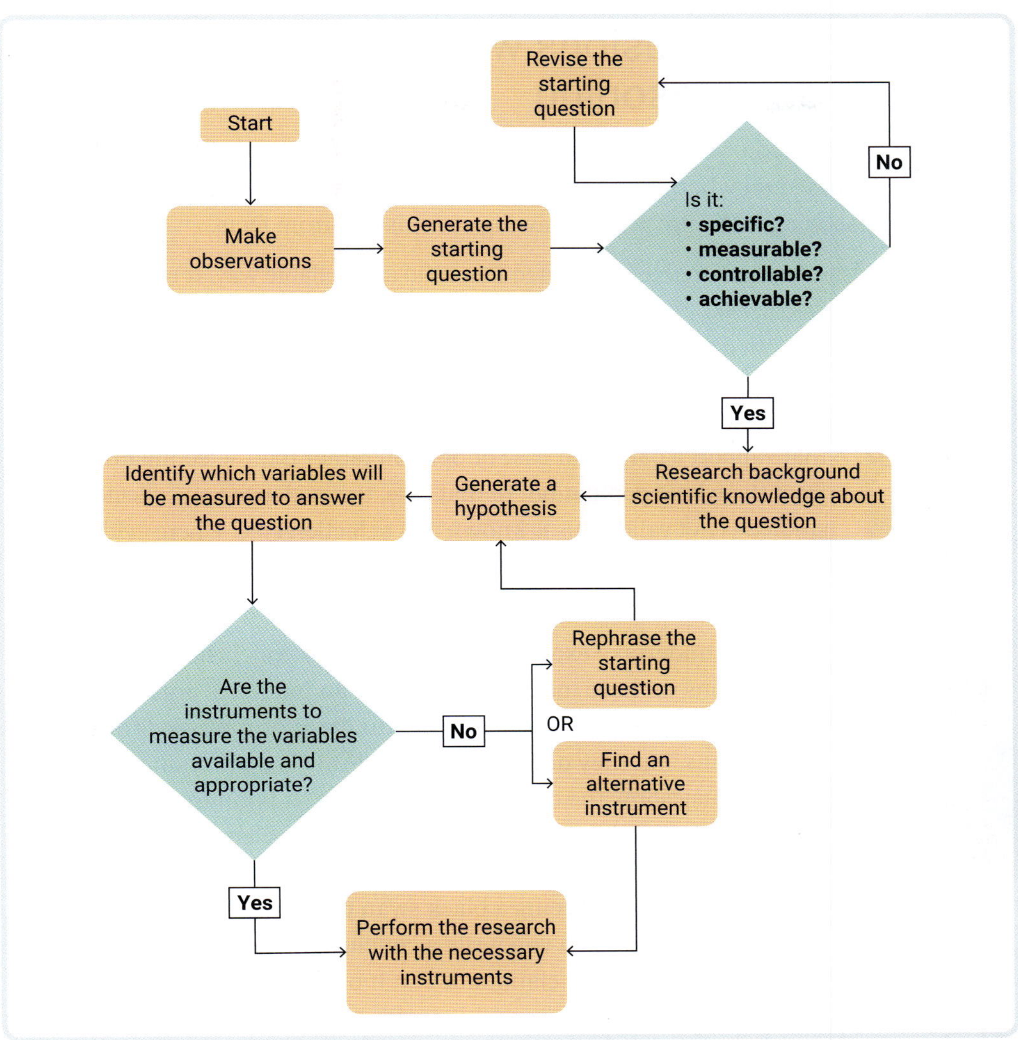

▲ **FIGURE 2.2.3** A flow chart of an investigation, from the initial observations to the final experiment

▼ **TABLE 2.2.1** Examples of hypotheses and null hypotheses

Investigable question	Hypothesis	Null hypothesis
Does the strength of an acid affect how long it takes limestone to dissolve?	If the strength of hydrochloric acid is increased then the time for the limestone to dissolve will decrease.	There is no relationship between the strength of hydrochloric acid and the time for limestone to dissolve.
Using a constant force, how does the acceleration of a rolling ball change as the mass of the ball increases?	A constant force will create increasing acceleration on a rolling ball as the mass of the ball decreases.	For a given force, there is no relationship between the acceleration of a rolling ball and the mass of the ball.

Valentyn Volkov/Shutterstock.com

▲ **FIGURE 2.2.4** Can you develop a hypothesis about temperature and dissolving using the observation that tablets seem to dissolve faster in warm water?

A good hypothesis identifies the variables you will measure in an investigation. Think about the hypothesis that if the temperature is increased, the time it takes for a tablet to dissolve will decrease. The temperature is controlled (Figure 2.2.4), so it is the independent variable. The time for the tablet to dissolve is the dependent variable.

Supporting conclusions with evidence and reasoning

reasoning the process of drawing conclusions from facts

selection bias selecting subjects that do not accurately represent the population they are drawn from

To support a conclusion, you need **evidence** and reasoning. Evidence from our investigations, based on what we measure and observe, is referred to as **empirical evidence**. **Reasoning** is the process of drawing conclusions from facts.

Evidence needs to be measured and collected in a valid and reliable way. It may be drawn from measurements or a set of consistent observations. For example, imagine you want to gather evidence about a population. If you measure something carefully many times using a large random sample, the data you gather will likely represent what is being studied. However, suppose you mostly select individuals that support a particular case or opinion rather than selecting randomly from the whole population. In that case, the evidence you collected and studied is not valid. This is known as a **selection bias**, where evidence is specifically selected to support a case or opinion (for more on this see Module 2.4).

Dark_Side/Shutterstock.com, KontroledKaos/Shutterstock.com

▲ **FIGURE 2.2.5** A white swan and black swan. The black swan was the counter-example that disproved the long-held argument that all swans are white.

When we undertake experiments, we base our conclusions on reliable observations we have made. This is called an inductive argument. An inductive argument may be based on hundreds or thousands of similar observations, but they do not *prove* something is true. A single counter-example may show the conclusion to be false. A famous example of the limits of an inductive argument is about all swans being white. To people in England, they *are* all white – thousands of them – but a visit to Australia and seeing a black swan proves the statement false (Figure 2.2.5). So, when you write a conclusion, be careful to say that the evidence supports your claim, not that it proves it.

Writing good scientific arguments

Video Science skills in a minute: Writing a discussion

Science skills resource Science skills in practice: Discussions in science reports

Good scientific arguments are vital to the scientific process. When you write scientific arguments in the discussion of your results, it is important to address both the meaning of the results and the validity and reliability of the methods you used.

Module 1.10 explains a good scientific argument as consisting of three parts: a claim, evidence from the data, and a rationale (the reasons why the evidence supports the claim). For example, consider the results of an experiment to determine which of six metals or metal alloys has the highest density (Table 2.2.2).

▼ **TABLE 2.2.2** Summary results of density measurements for four metals. The means, medians and ranges are based on five measurements made for each metal.

Material	Density (g/cm³)			
	Mean	**Median**	**Range**	**Published values**
Copper	8.8	8.8	0.3	8.8
Iron	7.3	7.3	0.6	7.2
Tin	7.2	7.2	0.6	7.3
Zinc	6.8	6.3	2.6	7.1

9780170491785

▲ **FIGURE 2.2.6** The metal elements (a) copper, (b) zinc, (c) iron and (d) tin

The argument that, of these metals, copper has the greatest density is shown in Table 2.2.3.

▼ **TABLE 2.2.3** The argument for copper having the greatest density of the four metals tested

Claim	The test material with the greatest density is copper.
Evidence	In the table of summary results, copper has the greatest mean and median density. The density range for copper suggests it was measured with precision.
Rationale (reasoning)	Copper has the highest mean value and therefore has the greatest density of the four metals.

When writing an argument in your investigation report, use the claim–evidence–rationale structure to clarify the link between your claim and the evidence. Make it easy for the reader to understand how your evidence supports the claim.

2.2 LEARNING CHECK

1 **Identify** two important features of scientific claims.
2 **Describe** the roles of evidence and reasoning in supporting a claim.
3 **Explain** the role of evidence in an inductive argument.
4 **Construct** and write a scientific argument about the relationship between force and the distance an object moves, using Table 2.2.4 and the claim–evidence–rationale structure.

▼ **TABLE 2.2.4** Force applied to a mass and the distance it moves

Force (newtons)	Distance mass moves (cm)
0.05	0.04
0.10	0.08
0.15	0.12
0.20	0.16
0.25	0.20

2.3 Science and pseudoscience

BY THE END OF THIS MODULE, YOU WILL BE ABLE TO:

✓ describe the differences between science and pseudoscience
✓ describe examples of pseudoscience claims
✓ determine if an assertion of a claim or theory is pseudoscientific.

Video activity
Tackling scientific misinformation

GET THINKING

What are some claims you have heard that seem to lack a scientific basis? How can you assess a claim to see if it is scientific or not? Make a list of reasons why you should avoid unscientific health advice.

Differences between science and pseudoscience

pseudoscience
beliefs or practices that claim to be scientific but have not been supported by fair tests

Science and pseudoscience differ in important ways. In Stage 4, **pseudoscience** was defined as beliefs or practices that claim to be scientific but have not been supported by fair tests. In contrast, science is knowledge gained from scientific processes, including fair tests. The scientific process also involves a set of attitudes to knowledge and its evaluation, such as rigour and openness, that may not be shared by people who support pseudoscience (Table 2.3.1).

▼ TABLE 2.3.1 Key differences between science and pseudoscience

Feature	Science	Pseudoscience
Scepticism	Uses a critical approach to assess new information. Evidence from investigations is used to support claims.	Beliefs are not open to critical examination. Belief determines what is accepted as evidence.
Rigour	The consequences of a hypothesis are explored in detail.	Concepts and hypothesis are not usually open to review or research.
Humility	Even established theories can potentially be proved wrong.	A certainty that truths are correct. An understanding that beliefs are not open to being proved false.
Openness	Knowledge and techniques are shared with others.	Sometimes knowledge is kept secret or restricted by those who believe in the theory.
Connectedness	New knowledge must link to the broader knowledge of the scientific community.	Knowledge does not have to be consistent with scientific knowledge.
Evolving understanding	Once tested, new evidence and ideas can replace older evidence and knowledge.	Evidence supporting beliefs adds to a pseudoscience's knowledge base but does not change what is believed.

Why do people believe in pseudoscience?

homeopathy
an alternative medicine based on unsupported scientific ideas

Some pseudoscience supporters promote untested theories, and others deny scientific ideas. Untested pseudoscience theories include astrology, chromotherapy, **homeopathy**, phrenology and crystal healing. Areas that deny scientific ideas and claims include tobacco disease denial, climate change denial, scepticism about genetically modified foods and scepticism about vaccines.

9780170491785

There are strong psychological motives for why some people support pseudoscience ideas. Pseudoscience can provide a sense of belonging to a community and a sense of being in control. Peer pressure also leads some people to embrace pseudoscience beliefs. Sometimes, people believe they feel better after receiving **alternative medicine** treatments such as homeopathy (Figure 2.3.1). This can be due to the **placebo effect** of receiving a treatment, rather than the effectiveness of the treatment itself.

▲ FIGURE 2.3.1 Homeopathy includes the belief that very diluted chemicals can cure diseases.

solomonjee/Shutterstock.com

The denial of scientific knowledge in some pseudoscience areas can also be motivated by a pursuit of profit or a reluctance to change behaviour. For example, tobacco use was identified as a cause of lung cancer as early as the 1940s. By the 1960s, research findings in a range of scientific disciplines had confirmed the claim. However, the tobacco industry disputed the science well into the 1980s, attacking research findings, funding positive advertising (Figure 2.3.2) and using paid experts in court cases to deny responsibility for the effects of their products.

alternative medicine therapy or practice unsupported by scientific evidence or medical experts that aims to improve the health of people

placebo effect a positive effect due to a patient's belief in a treatment

Ralf Liebhold/Shutterstock.com

▲ FIGURE 2.3.2 Tobacco companies produced advertising implying that smoking was safe long after its harmful nature was known.

How to identify pseudoscience claims

We can use three key features to identify pseudoscientific claims:

- they often rely on confirmation bias rather than a systematic study of evidence
- they lack openness to evaluation by experts
- the scientific process has not been used in developing and testing theories.

Reliance on confirmation bias

confirmation bias intentionally selecting data to support a case or opinion

Confirmation bias is the tendency to favour information that confirms or strengthens a belief. For example, a climate-change denier might use a temporary period of cool weather as 'evidence' that the climate is not warming. The fact that the weeks on either side of the cool weather period experienced higher than usual temperatures would not be considered as evidence to judge this pseudoscience theory by the climate-change denier.

Lack of openness to expert evaluation

In scientific research, methods and evidence are shared with the wider community of scientists for assessment. Pseudoscience may involve some form of research, but generally the results and methods are not revealed.

New Age a philosophy that seeks healing and self-development through practices such as meditation, astrology and alternative medicine

For over 4000 years, alchemists sought a cure for all diseases, ways to turn lead into gold and medicine to make someone immortal (Figure 2.3.3). Alchemy is now considered an example of a pseudoscience, in part because it did not make its ideas and methods open to evaluation. Gradually, theories tested by science replaced the spiritual side of alchemy, but such ideas can still be found in some **New Age** belief systems. Some discoveries alchemists made have contributed to developing the current scientific disciplines of chemistry and medicine.

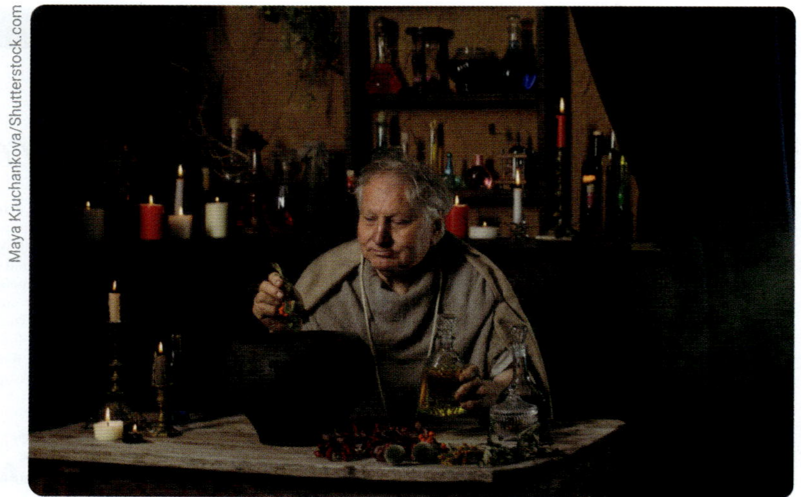
Maya Kruchankova/Shutterstock.com

▲ **FIGURE 2.3.3** Alchemy is today regarded as a pseudoscience.

Absence of the scientific process

Pseudoscience makes claims that cannot be tested or turned into investigable questions. If such questions are possible, the results of investigations do not support the claims.

While lacking good scientific practice themselves, some pseudoscience supporters criticise shortcomings in scientific investigations that do not agree with their beliefs or ignore well-conducted studies that cast doubt on their beliefs. Homeopathy is an example of a pseudoscience where this practice occurs. Homeopathy is a system that believes that the smaller the dose of a medicine, the greater its effect. It also believes that a medicine producing symptoms similar to a disease in healthy people can be used to treat that disease.

A pseudoscience case study: anti-vaccination beliefs

2.3

Vaccination has been thoroughly tested, and found safe and effective in reducing the effects of disease. As with most medical practices, a small number of people have bad effects. However, the health community considers vaccination to be beneficial and safe for most people. During the COVID-19 pandemic, misinformation about the safety of vaccination caused many people to protest and avoid vaccination (Figure 2.3.4), even though it was proven to reduce death rates and serious side effects.

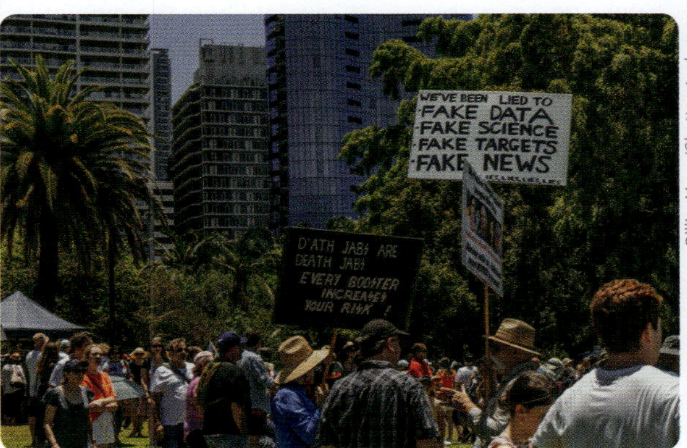

GillianVann/Shutterstock.com

▲ **FIGURE 2.3.4** Anti-vaccination beliefs can affect health outcomes during a pandemic.

Some people also resist the use of vaccinations on philosophical and religious grounds.

Some anti-vaccination ideas are misinformation created by people who seek to benefit from making people mistrust the vaccine. For example, some people sell alternative therapies, and some lawyers have sought compensation on behalf of people who have claimed a vaccine negatively affected them.

In 1998, the British medical journal *The Lancet* published a scientifically fraudulent paper that claimed there was a link between measles-mumps-rubella vaccination in children and autism. Despite scientific research finding that the study was poorly performed and that the author stood to benefit financially from the claims, media attention generated a fear of vaccinations in many people. In 2010, the paper was retracted and the author was banned from practising medicine in the UK. In 2019, a study of 660 000 children over a period of 11 years found no link between the vaccine and autism.

Pseudoscience can have serious effects on global health. For example, the World Health Organization estimates that in 2022, 136 000 deaths occurred globally due to measles, mainly in unvaccinated or incompletely vaccinated children, some of whom were unvaccinated by parental choice.

2.3 LEARNING CHECK

1 **Define** pseudoscience.
2 **Describe** three ways in which science and pseudoscience are different.
3 **List** three features of a pseudoscience claim.
4 **Identify** a pseudoscience claim and **explain** why the claim is not scientific.
5 **Describe** two reasons why people believe pseudoscience claims.

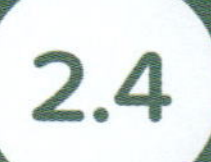

2.4 Data and scientific claims

BY THE END OF THIS MODULE, YOU WILL BE ABLE TO:

✓ describe biases that can distort the collection and analysis of data

✓ explain how the manipulation of an investigation's data and conclusions can lend support to a specific viewpoint.

Video activity
Confirmation bias

Quiz
Secondary data

GET THINKING

Have you ever left out a measurement from a table of results? How can the absence or addition of data in an investigation affect the results? Can an investigation's findings be altered during the analysis of data? How might this happen?

Bias in experimental investigations

You must always be on the lookout for bias in experimental methods, collected data and claims. A bias occurs during sampling or analysis when data is selected that favours one possible answer over others. Bias can occur at any stage of a scientific investigation. Figure 2.4.1 shows some biases that can occur at different stages of an experimental investigation. The main types of bias are also summarised in Table 2.4.1.

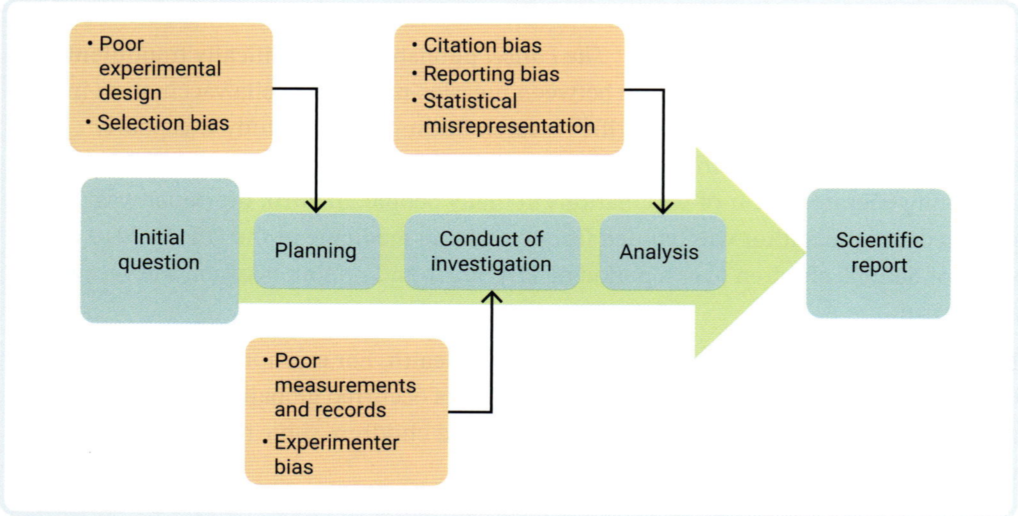

▲ **FIGURE 2.4.1** Examples of bias that can occur at different stages of a research investigation

9780170491785

▼ **TABLE 2.4.1** Types of bias in scientific investigation and reporting

Bias	Description and examples
Poor design	Failing to control variables, using small datasets and selecting instruments that lack precision all distort the meaning of results
Selection bias	Selecting groups for study that do not accurately represent the population being studied
Measurement bias	Not accurately recording data or information, possibly due to errors in data collection, bias in the data collector, inconsistent measurement methods or unreliable instruments
Reporting bias	Under-reporting unexpected or undesirable research results
Experimenter bias	Interpreting data in a way that supports pre-existing beliefs about the investigation outcomes (a form of confirmation bias)
Citation bias	Selecting references that support specific experimental outcomes and ignoring studies that do not support the experimental claims
Statistical misrepresentation	• Choosing evidence that supports a specific claim • Using deceptive charts • Treating small differences as meaningful rather than being due to natural variation • Favouring results from small sample groups • Repeatedly comparing variables until a relationship is found by chance, but presenting it as a planned investigation (also called data-dredging)

Some of the common types of bias are described in more detail below.

Selection bias

Imagine using the Australian basketball team as a sample to determine the average height of Australian males (Figure 2.4.2). In 2024–25, the average height of this group was 200 cm (2 metres). This is an example of selection bias, where subjects are selected who do not accurately represent the population they are drawn from. A good investigation plans how data is collected to fairly represent the population being sampled.

▲ **FIGURE 2.4.2** Using the Australian men's basketball team to find the average height of Australian men would be an example of selection bias.

Citation bias can be thought of as a form of selection bias. It occurs when an author selects references that support their claim but leaves out references that do not support their claim or that give different evidence or conclusions.

Statistical misrepresentation

There are many ways that statistics can be manipulated to support a bias (Table 2.4.1). Creating misleading graphs is a common form of statistical manipulation. For example, the axes of graphs can be manipulated to maximise differences (Figure 2.4.3), alter the rates of change (Figure 2.4.4) or even suggest an opposite effect to the one that occurred (Figure 2.4.5).

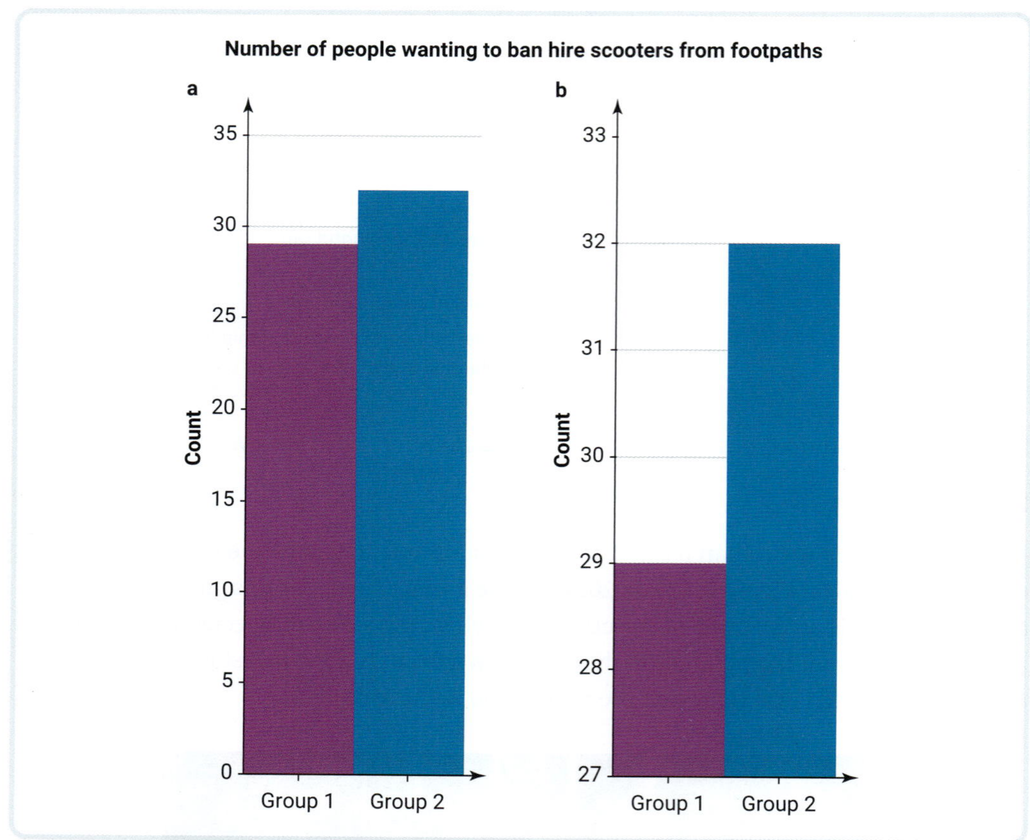

▲ **FIGURE 2.4.3** Exaggerating differences by manipulating scales. (a) Small differences can be made to look as if they are (b) large differences.

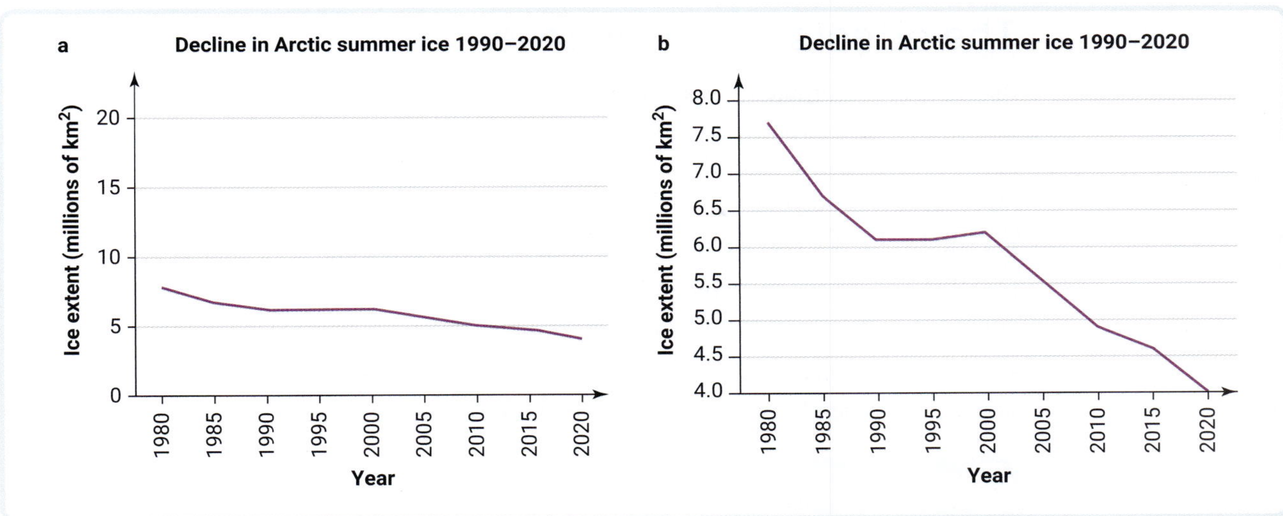

▲ **FIGURE 2.4.4** Reducing or increasing apparent change. The trend can be made to look **(a)** small or **(b)** large by stretching the axis.

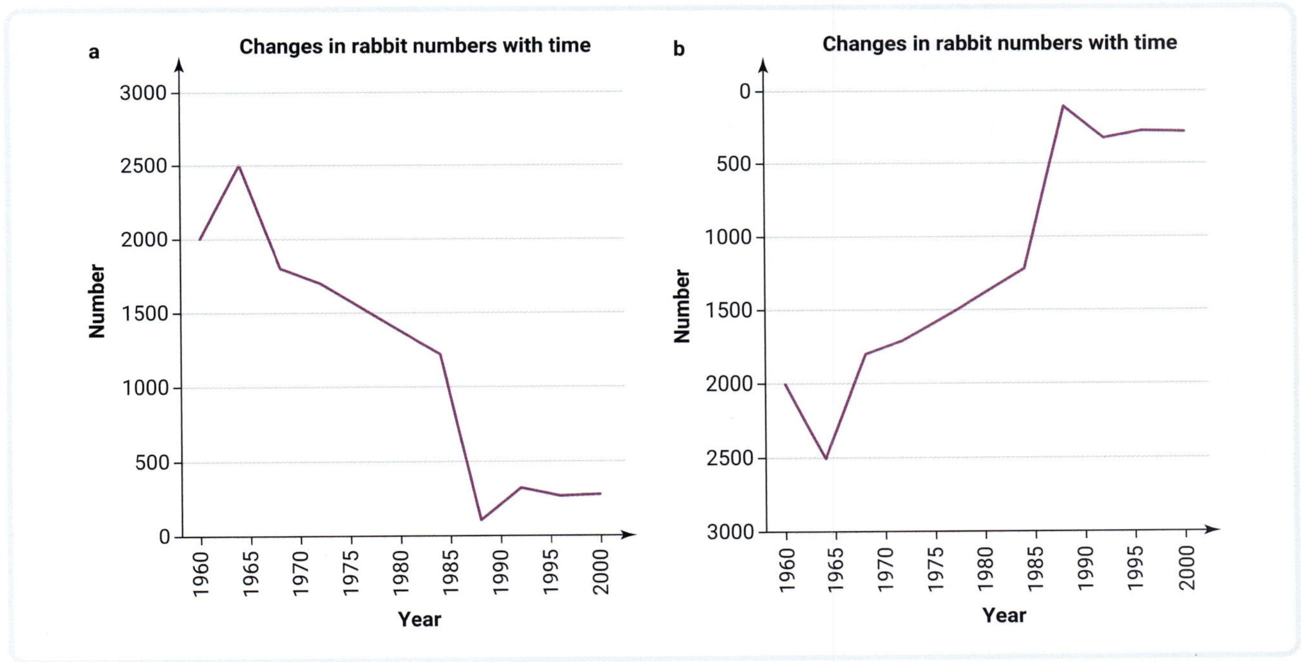

▲ **FIGURE 2.4.5** Ignoring graphing conventions makes it harder to understand the data. For example, **(a)** one scale can be **(b)** reversed to make a decrease look like an increase.

Confirmation bias

Imagine running 20 trials of an experiment but only reporting on those that support the outcome you want to occur. The reported trials may be chance events and may misrepresent the experimental results. This is an example of confirmation bias, where evidence is selected to support a specific claim. For example, some supporters of homeopathy use confirmation bias to defend their beliefs. They only select studies that support their ideas but do not acknowledge studies that report no effect from homeopathy methods.

The effects of bias in research

Bias in an experimental investigation leads to claims that support a particular viewpoint. However, that viewpoint is not likely to be a valid conclusion as some data has been given greater weight than other data or has been manipulated. As a result, the investigation is not a fair test.

Reviewers usually identify bias in published papers, but some biased papers pass peer review. These papers potentially affect how other researchers answer their own questions. Time and resources may be wasted proving the biased viewpoint is wrong.

Sometimes, the media may pick up on biased research and introduce false beliefs into society. It is much harder to correct wrong information than it is to produce it.

2.4 LEARNING CHECK

1 **Define** bias.
2 **Describe** two types of bias that can affect the analysis of experimental data.
3 **Explain** how altering the vertical axis can minimise differences in a bar graph.
4 **Explain** citation bias.
5 Is measurement bias always due to bias by the data collector? **Explain** your response.
6 How can the biased reporting of an investigation affect other scientists? **Justify** your answers.

2.5 Large datasets

BY THE END OF THIS MODULE, YOU WILL BE ABLE TO:

- ✓ describe the features and applications of large datasets
- ✓ outline how data collection and analysis are conducted with large datasets.

GET THINKING

What is the largest table of data you have analysed so far in class? How hard would it be to analyse an experiment that generated 100 000 rows of data in a spreadsheet? Make a list of things you could do that would make it easier to analyse a large dataset.

Interactive resource
Crossword: Datasets

What is a large dataset?

In Stage 4, we defined big data as large, complex datasets that are difficult to process and analyse. But what would a data scientist classify as large? To a data scientist, a **large dataset** is one that cannot be easily stored in a computer's working memory or RAM. This is a lot of data, given most personal computers today have a RAM of around 8 to 16 gigabytes. Consider a downloaded file of astronomy data, which is 4.8 megabytes. It represents a dataset of seven variables for more than 118 thousand stars (Figure 2.5.1). To us, this would be a very large dataset, but not to a data scientist.

ESA - C. Carreau

▲ **FIGURE 2.5.1** The Gaia satellite measured the position and brightness of more than 2 billion objects in the Milky Way Galaxy. Such projects produce vast amounts of data.

Features of a large dataset

A large dataset has two key features: it is large enough to be statistically reliable and it requires **computer analysis** to find and describe associations, patterns and trends.

Statistical reliability means the data produces consistent results when it is analysed using the same methods repeatedly. In contrast, when you repeat analysis using small sample sizes from a much larger population, you might calculate averages that are different from each other and from the true measurement. Increasing the sample size will likely result in averages closer to the true measurement.

large dataset
a dataset large enough to be statistically reliable and require computer analysis

computer analysis
a detailed examination of something using a computer

statistical reliability
the ability to produce consistent results when repeatedly analysed using the same methods

Use of large datasets in science

Large datasets are valuable in science because they are likely to contain patterns and correlations that will help answer questions. If it is correctly collected, more data can reduce errors, which is why large datasets are more reliable. This is particularly important when analysing highly variable data, such as weather data (Figure 2.5.2). Large language models (LLMs), such as ChatGPT, are artificial intelligence programs trained using vast amounts of text. They can generate answers to text-based questions using connections found within the datasets.

Minerva Studio/Shutterstock.com

▲ **FIGURE 2.5.2** Predicting the movement and effects of storms requires large datasets of past storm behaviour.

NASA Earth Observatory image by Michala Garrison, using Landsat data from the U.S. Geological Survey. Story by Lindsey Doermann

▲ **FIGURE 2.5.3** An image of snow-covered mountains taken with remote sensors; such an image contains a large amount of data.

Collecting and accessing large datasets

Large datasets can be collected and accessed in many ways. A laboratory data logger may generate thousands of lines of data in a few minutes during an experiment. Satellites and telescopes can collect huge amounts of data very quickly. Images are also a rich source of data, and a single image can contain a large dataset. For example, an image of snow cover can provide data that can help analysts to estimate how snowfall changes from season to season (Figure 2.5.3). The internet holds a wide range of data stored at public sites. Organisations such as the Commonwealth Scientific and Industrial Research Organisation (CSIRO) and NASA, as well as the Registry of Open Data on Amazon Web Services, have large datasets available for public use. Government websites also provide publicly available datasets.

Analysing large datasets

Large datasets require computer analysis mainly because of their size. Imagine working out the average of several thousand values on your calculator; that would take a long time! Computer software, such as spreadsheets and programming languages, provide statistical and graphical tools that speed up data processing, presentation and analysis (Figure 2.5.4).

Weblink
CSIRO educational datasets

Programming languages, such as Python and the R statistical language, are used extensively by research scientists. Languages such as C++ and JavaScript are also used for modelling and games development (Figure 2.5.5).

Chay_Tee/Shutterstock.com

▲ **FIGURE 2.5.4** Statistical software helps data analysts manage large datasets, simplifies calculating statistics and creates visualisations.

Gorodenkoff/Shutterstock.com

▲ **FIGURE 2.5.5** Programming languages used in science are also used for games development.

Computer software can also help speed up data preparation and analysis, including identifying and correcting data formatting errors or duplicate data. Some software can rapidly graph data to save time.

2.5 LEARNING CHECK

1 **Describe** the two key characteristics of a large dataset.
2 **Explain** why large datasets require computer analysis.
3 **List** three sources of large datasets.
4 **Outline** two ways that the use of computers speeds up data preparation and analysis.

2.6 Analysing datasets

BY THE END OF THIS MODULE, YOU WILL BE ABLE TO:

✓ identify the benefits of univariate analysis in describing data
✓ outline methods used in a univariate analysis
✓ describe the purpose of bivariate analysis
✓ compare the information provided by univariate and bivariate analysis.

Univariate analysis

The first stage of data analysis is called **univariate analysis** (from 'uni-' meaning one and 'variate' meaning variable). Its purpose is to describe and summarise data. Univariate analysis also helps us check for missing values and **outliers** and ensures the data is as reliable as possible. Each variable, or column of data, is analysed in turn using a set of relevant questions (Figure 2.6.1).

Videos
Science skills in a minute: Organising data
Organising data into charts
Representing data in different forms

univariate analysis an analysis that involves only one variable that describes the data

outlier a value that differs significantly from other values in a dataset

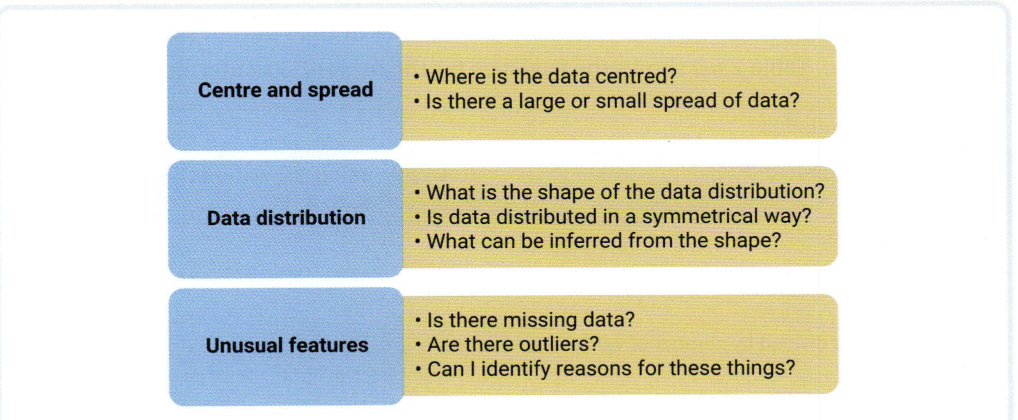

▲ **FIGURE 2.6.1** Some questions to ask when you perform univariate analysis

There are three methods used in univariate analysis:

- descriptive statistics (see Module 1.9)
- frequency/grouped data tables
- graphical analysis.

You should use all three methods when conducting univariate analysis (Figure 2.6.2).

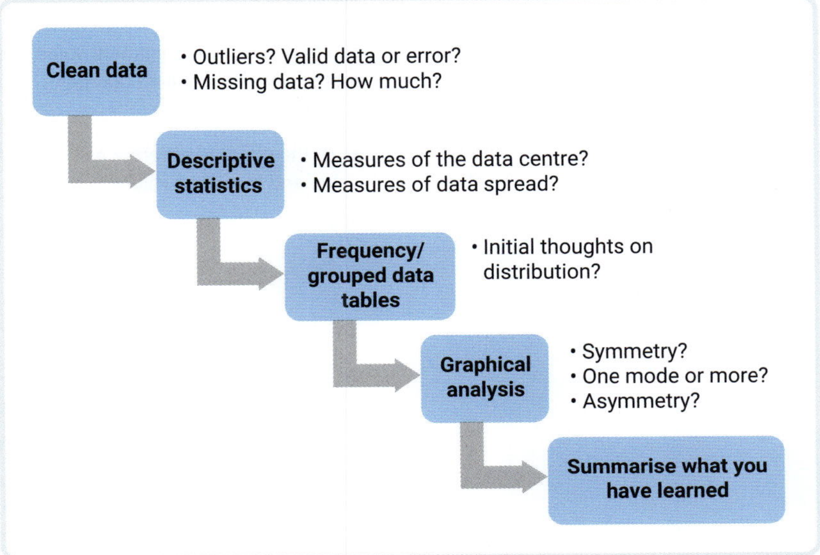

▲ **FIGURE 2.6.2** The steps of univariate analysis

Maths in science videos
Analysing different graphs

Mean, median and mode

Worksheets
Univariate analysis

Bivariate analysis

Data cleaning

Data cleaning is the process of fixing errors and other problems in datasets. There are often missing values in large datasets that need to be carefully considered and managed. For example, just deleting rows with missing data may affect the validity of your analysis. Data errors can also occur when dealing with large amounts of data. For example, some outlier data can be errors. Errors affect statistics and can affect the outcome of your analysis. When you look at outliers:

- carefully consider whether they are errors or just possible extreme values
- do not remove outliers from your analysis unless you are sure they are errors
- always report how you deal with outliers in the analysis section of your science report.

Descriptive statistics

descriptive statistics values derived from a dataset, such as a mean or maximum value, that summarise the dataset

Descriptive statistics, sometimes called summary statistics, identify the centre and variation in data. They also provide information used to create frequency tables and graphs, such as box plots. Table 2.6.1 contains a summary of the relevant statistics we can use. For example, Table 2.6.2 shows the statistics for the mass of 30 watermelons in a crate.

▼ **TABLE 2.6.1** A summary of descriptive statistics

Type	Statistic	Definition or calculation	Example, using the dataset: 2, 3, 3, 4, 5, 6, 8, 9
Extreme values	Maximum	Largest value	9
	Minimum	Smallest value	2
Centre of dataset	Mean	Sum of values divided by number of values	$\frac{(2+3+3+4+5+6+8+9)}{8} = 5$
	Median	Centre value in an ordered dataset	Lies between 4 and 5, so the median is 4.5
	Mode	Most frequently occurring value in the distribution of scores. There may be more than one mode	3
Spread of values	Range	Distance between maximum and minimum value, calculated by subtracting the smallest number from the largest number in a dataset	$9 - 2 = 7$
	Variance*	Arithmetic mean of the squared differences between each value and the mean value	Mean = 5 Square difference of each value from the mean: $(2-5)^2 = 9$ $(3-5)^2 = 4$ $(3-5)^2 = 4$ $(4-5)^2 = 1$ $(5-5)^2 = 0$ $(6-5)^2 = 1$ $(8-5)^2 = 9$ $(9-5)^2 = 16$ Sum of squared differences: $9+4+4+1+0+1+9+16 = 44$ Variance: $\frac{44}{5} = 8.8$
	Standard deviation*	The square root of the variance. A measure of the spread of scores around the mean	Using a calculator: 2.3 (rounded to one decimal place)

*You may not be required to be able to calculate this type of statistic but you should know what the statistic describes.

▼ TABLE 2.6.2 Descriptive statistics for a crate of watermelons

Statistic	Value
Number of measurements (*n*)	30
Missing data	Nil
Mean	10.1 kg
Median	9.8 kg
Mode	9.9 kg
Minimum	8.6 kg
Maximum	14.3 kg
Range	5.7 kg
Possible outlier	14.3 kg

Tables used in univariate analysis

Tables summarise data and identify patterns. A grouped frequency table presents data in groups called classes. A class contains a range of values that doesn't overlap with other classes, so a value can only be in one class. Grouping data helps to summarise large datasets.

Table 2.6.3 is a grouped frequency table showing the mass of a sample from a specific fish species. Note that 14 of the 30 masses (47%) occur in the class of values that contain the mean, 10.1 kg.

▼ TABLE 2.6.3 A grouped frequency table of fish sample masses

Mass (kg)	Frequency
1.450–1.499	1
1.500–1.549	4
1.550–1.599	4
1.600–1.649	14
1.650–1.699	5
1.700–1.749	1
1.750–1.799	0
1.800–1.849	1
Total	30

Graphs used in univariate analysis

The main graphs used in univariate analysis are histograms and box plots for **continuous data** and bar charts and pie charts for **discrete data**. Figure 2.6.3 shows a histogram and a box plot of the data from the fish mass example. In this example, individual points are plotted next to the box plot for comparison but this is not required in a box plot. These displays make it easier to analyse the data and notice the following:

- There are a relatively small number of fish with a mass greater than 1.70 kg.
- The data seems slightly weighted to the lower masses but is close to symmetrical (evenly spread around the mean), considering the small sample size.

continuous data
data from measurements that may include whole numbers and any value between them

discrete data
data where there is only a limited number of possible values

Published scientific research on this fish species shows that a mass of 1.82 kg is in the top 1 per cent of masses, so the value may not be an outlier.

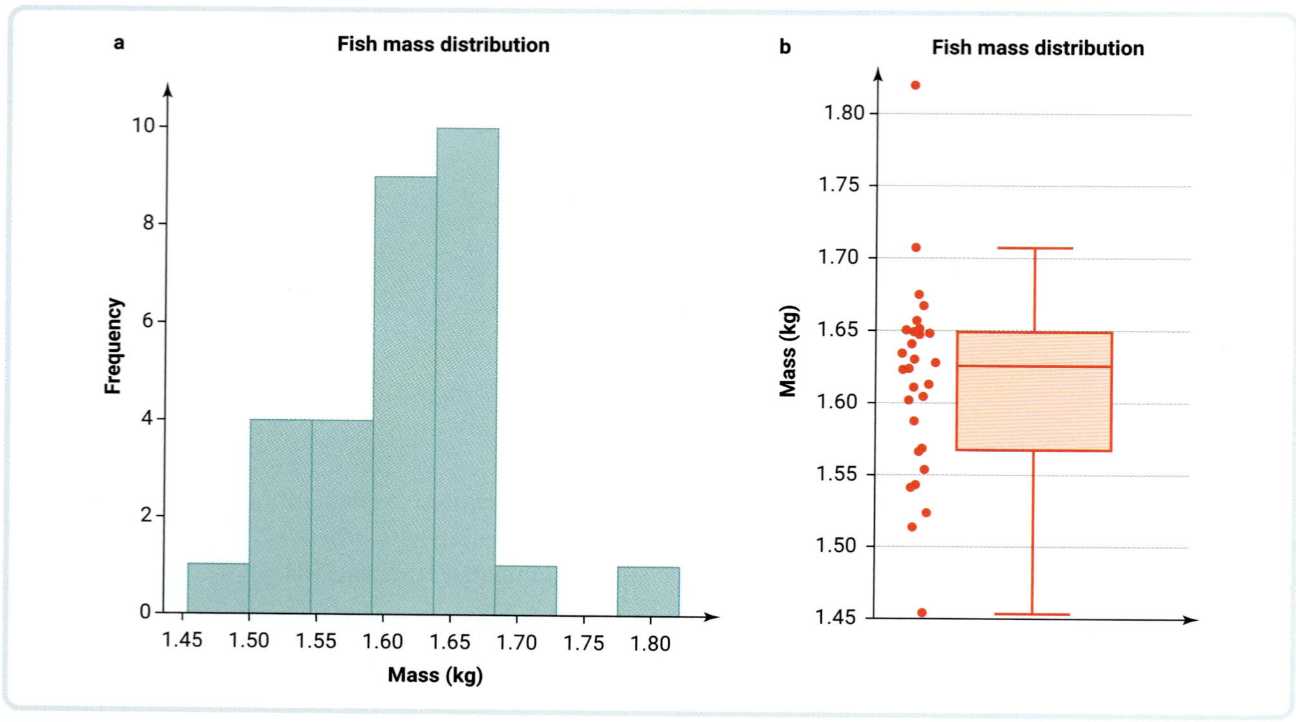

▲ FIGURE 2.6.3 (a) Histograms and (b) box plots are two methods of graphing univariate data.

Bivariate analysis

bivariate analysis
an analysis involving two variables that identifies and explains trends, correlations or groupings

correlation
a trend in data in which one variable changes consistently as the other variable changes

line of best fit
the line that best represents the trend of a set of data points

Bivariate analysis ('bi-' meaning two and 'variate' meaning variable) compares two variables and aims to identify trends, **correlations** or grouping. While univariate analysis describes data, bivariate analysis identifies correlations and relationships, and generates explanations.

You are familiar with creating line graphs with a **line of best fit**. This is one example of bivariate analysis.

Analysing correlations

Table 2.6.4 and Figure 2.6.4 are from an investigation to determine the relationship between an applied force and how far the force moves a trolley. There is a strong linear correlation between the variables. Based on the graph, a claim can be made that the distance the trolley moves is proportional to the force applied for the range of forces studied.

9780170491785

▼ TABLE 2.6.4 Results from a study of applied force and the resulting distance a trolley moves

Force applied (N)	Distance trolley moves (m)			
	Trial 1	Trial 2	Trial 3	Average
0.26	0.33	0.55	0.42	0.44
0.47	0.87	0.68	0.71	0.75
0.68	1.21	0.88	0.99	1.03
1.03	1.70	1.51	1.52	1.58
1.26	1.82	1.82	1.82	1.82
1.49	2.17	2.00	2.18	2.12

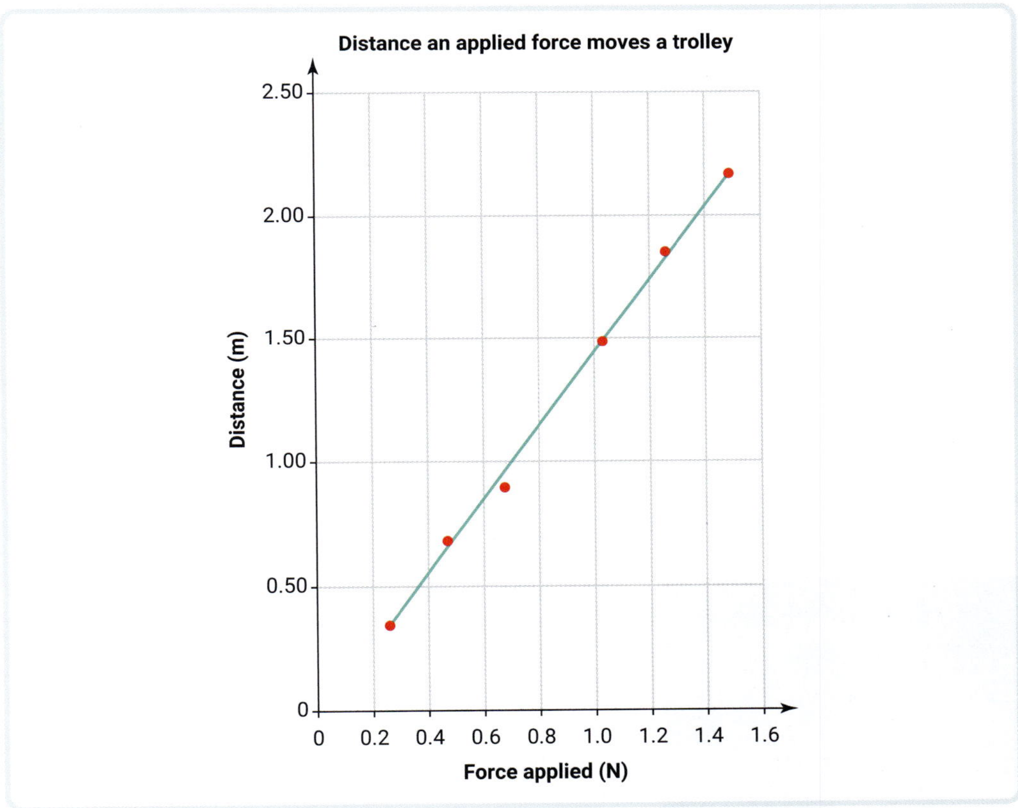

▲ FIGURE 2.6.4 A graph of applied force versus distance a trolley moves shows a strong linear correlation.

A scatter plot may show a pattern without needing a line of best fit. Figure 2.6.5 shows the relationship between the duration of an eruption and the time interval between eruptions for a famous geyser (an erupting hot spring) in the United States called 'Old Faithful' (Figure 2.6.6).

Figure 2.6.5 shows that a linear trend is present, but you may also notice the scatter plot graph reveals two clusters. The cluster in the bottom left corner of the graph shows data points for eruptions that have a short duration, and the cluster in the top right shows data points for eruptions with a longer duration.

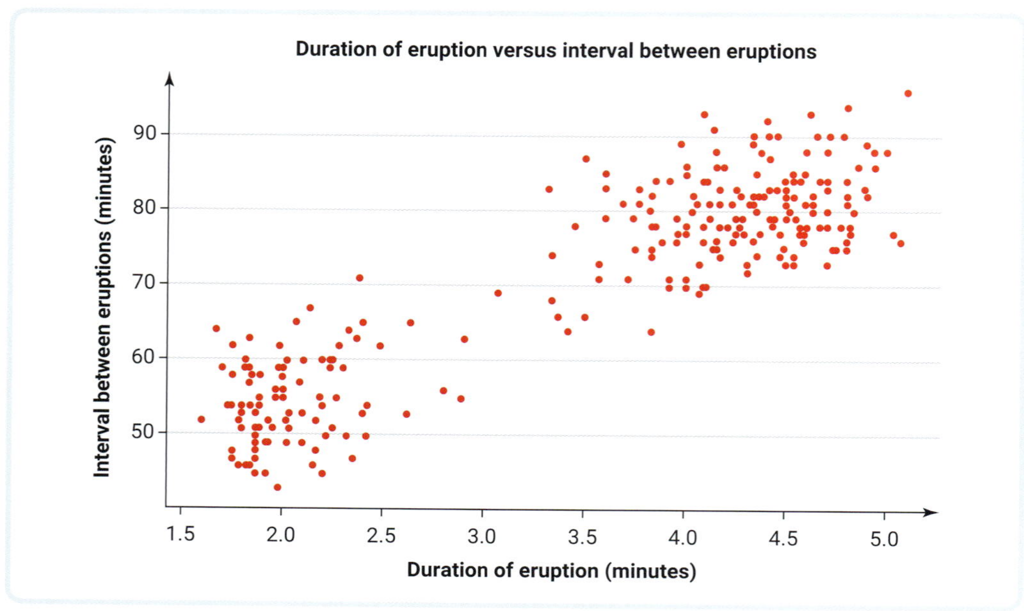

▲ **FIGURE 2.6.5** A scatter plot of the duration of eruptions versus the interval between eruptions for the 'Old Faithful' geyser shows a pattern.

▲ **FIGURE 2.6.6** The 'Old Faithful' geyser in the USA erupting. A geyser is a rare type of spring that periodically ejects hot water and steam into the air.

Mapping continuous data against discrete data

Analysing large datasets often involves mapping continuous data against discrete data. For example, the average relative sea level (continuous data) each month (discrete data) can be shown in a line graph (Figure 2.6.7). The linear trend shows a rise in sea level, but there is a lot of variation due to weather and ocean currents. When continuous data is graphed against discrete data, lines connect the points to show that the values form part of a continuous dataset.

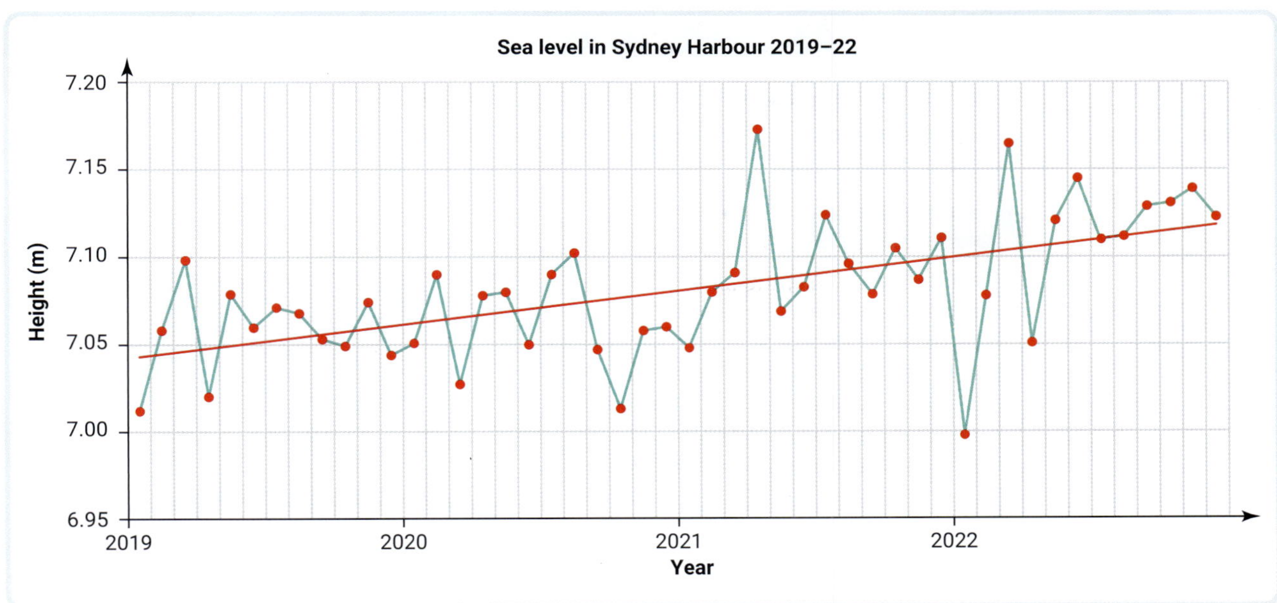

▲ FIGURE 2.6.7 Relative sea level per month is variable but can be shown as a line

Comparing univariate and bivariate analysis

Table 2.6.5 summarises the differences between univariate and bivariate analysis. Remember that both types of analysis work together. To be confident about the claims we make from bivariate analysis, we need to be confident in the quality of the data for the variables we use – which is why univariate analysis is so important.

▼ TABLE 2.6.5 A comparison of univariate and bivariate analysis methods

Characteristic	Univariate analysis	Bivariate analysis
Number of variables analysed	One	Two
Purpose of analysis	To describe	To explain
Does it deal with causes/relationships?	No	Yes
Features and methods	• Statistics • Mean, median, range, maximum, minimum • Frequency distributions • Graphs • Bar graphs, histograms, box plots	• Analysis of two variables at the same time • Independent and dependent variables • Correlations • Relationships • Explanations

2.6 LEARNING CHECK

1 **Identify** the purpose of univariate analysis.
2 **Describe** three methods used in univariate analysis.
3 **Explain** the information provided by:
  a the mean.
  b the range.
4 **Describe** the purpose of bivariate analysis.
5 **Contrast** the purpose and features of univariate and bivariate analysis.

2.7 Cause and correlation

BY THE END OF THIS MODULE, YOU WILL BE ABLE TO:

✓ identify causal and correlation relationships
✓ describe how causal relationships can be established.

Maths in science video
Correlation and causation

GET THINKING

Consider an investigation studying lifestyle and skin cancer (Figure 2.7.1). If the study finds a positive correlation between the amount of exercise and the frequency of skin cancer, does this prove that exercise causes skin cancer? How do you explain this correlation?

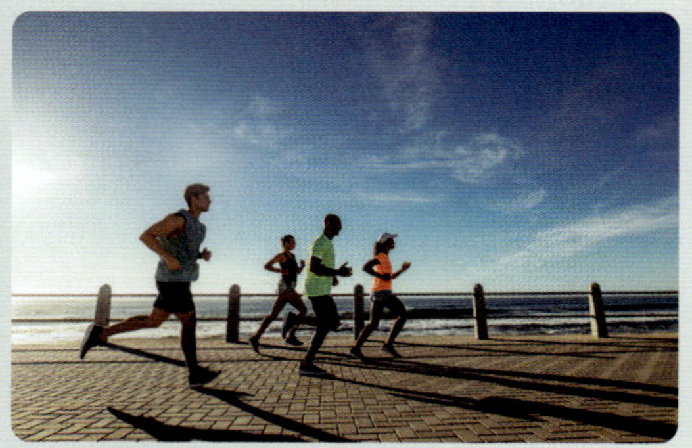

Jacob Lund/Shutterstock.com

▲ **FIGURE 2.7.1** Does exercise cause skin cancer?

cause-and-effect relationship
a relationship in which a change in one variable causes a change in another

Scientific investigations frequently aim to find the cause of something, but this can be difficult. This module examines the differences between a **cause-and-effect relationship** (also called a causal relationship) and a correlation. A causal relationship between two variables exists when a change in one variable causes the other variable to change. For example, a cause-and-effect relationship exists between the volume of a metal object and temperature. A change in temperature causes the metal object to expand or shrink.

Correlation

In Stage 4, you learned that correlation describes the closeness of the relationship between two variables (Figure 2.7.2) and how one variable changes when the other variable changes. The correlation between two variables can be strong or weak. In experiments, large random errors caused by imprecise measurements can cause a strong correlation to appear weak.

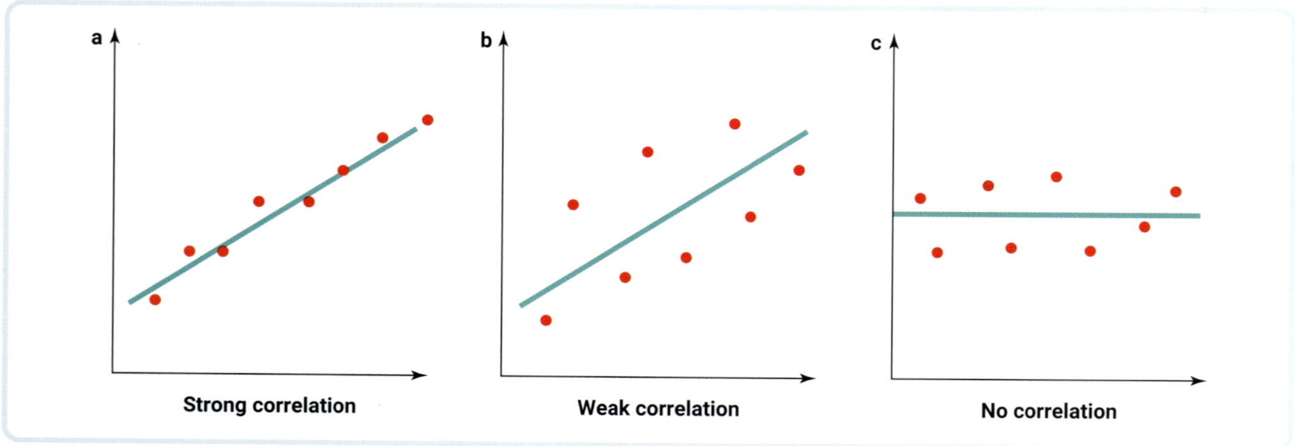

▲ **FIGURE 2.7.2** The degree of correlation can be (a) strong or (b) weak, or (c) there can be no correlation.

9780170491785

Causation

Causation is when one event causes another event to occur. Three factors are required to show that a causal relationship exists:

- a correlation exists between the variables
- the variable causing the change occurs *before* the variable that changes
- no confounding variables influence the relationship.

causation
when a change in one variable causes a change in another variable

Establishing that cause happens before effect

One of the best ways to provide evidence of a causal relationship between two variables is through an experiment that involves a treatment group (or groups) and a **control group**. The treatment applied to the treatment group(s) is the change we make to the independent variable. The control group is treated the same way as the treatment group except that it does not receive the treatment.

control group
a trial or group in an investigation that doesn't change the independent variable and is used as a comparison

We can investigate whether any changes in treatment bring about subsequent changes in the treatment group. If a correlation is established between the variables, then we know it is the effect of changes in the independent variable on the dependent variable.

The control group is important. It must be the same as the treatment group in every way except for receiving the treatment. One approach is to assign subjects or samples for the investigation to the control and treatment groups randomly (Figure 2.7.3). This balances variations in the features of the subjects or samples so the two groups are as similar as possible. For example, in plant studies, a population of plants is randomly divided into control or treatment groups.

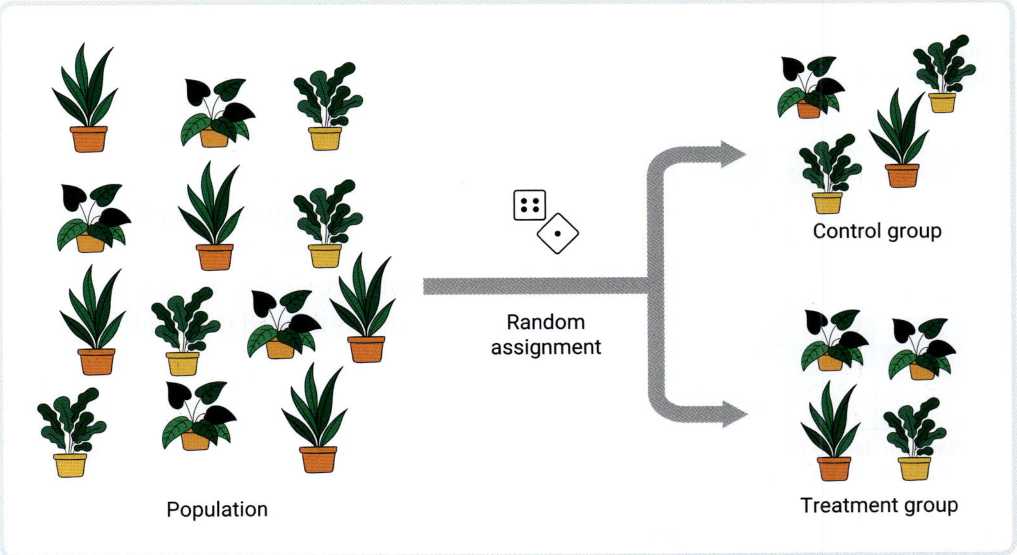

▲ **FIGURE 2.7.3** The random assignment of subjects or samples into treatment and control groups helps to ensure the experiment is valid.

confounding variable a variable that influences both the independent and dependent variable, producing a false relationship between them

spurious false, or not what it appears to be

Dealing with confounding variables

A **confounding variable** is a variable that influences both the independent variable and the dependent variable, falsely making it look like there is a relationship between them (Figure 2.7.4). For example, as ice cream sales increase in summer, so do the number of lifesaver rescues. However, one is not causing the other. They are both due to more people being at the beach as the weather warms. In this example, the confounding variable is the daily temperature. Changes in the confounding variable cause the two variables we are studying (in this example, ice cream sales and lifesaver rescues) to change together, wrongly suggesting a causal relationship might exist. Such a relationship is called a **spurious** correlation.

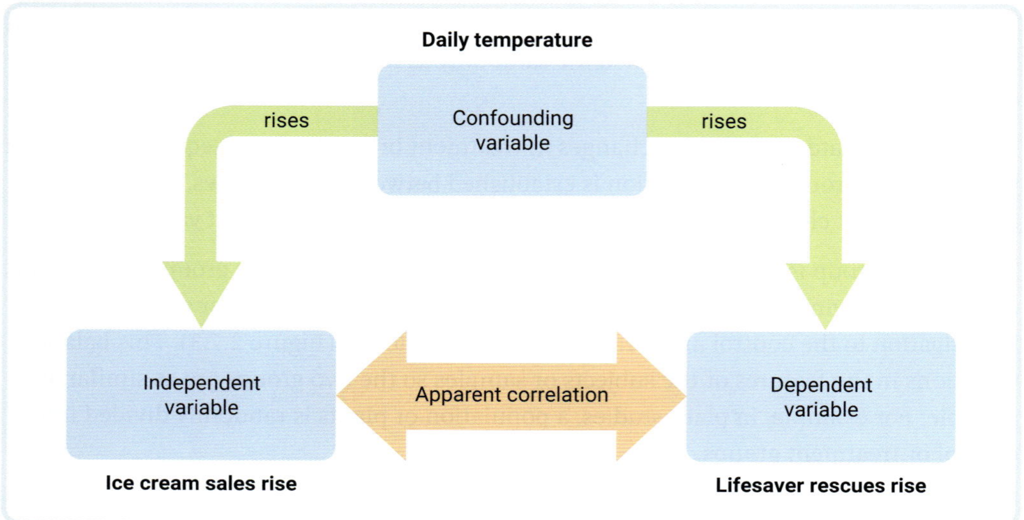

▲ **FIGURE 2.7.4** A confounding variable creates an apparent correlation between the variables being studied.

To avoid confounding variables, scientists pay careful attention to the design of an experiment. For example, in an agricultural study, the location a plant grows (such as a field) may affect the results. To minimise confounding variables, the field can be randomly divided into control and treatment areas (Figure 2.7.5).

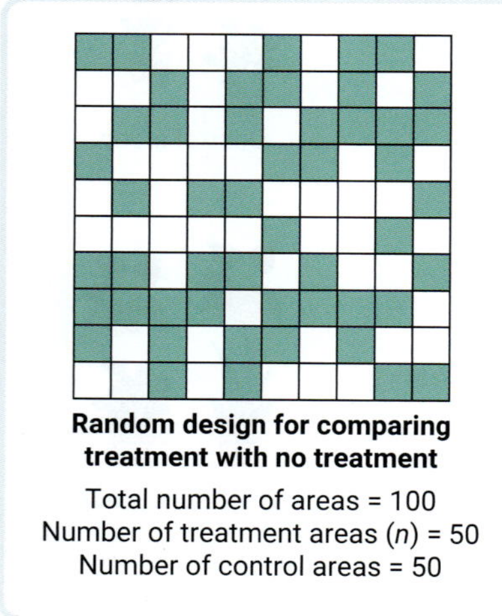

◀ **FIGURE 2.7.5** In agricultural research, the location within a field may be a confounding factor. Randomly assigning parts of the field to control and treatment areas may help make sure that location doesn't affect results.

Proving causation in medical research

Careful experimental design is essential in medical research, where many factors may influence an outcome. For example, when studying diabetes, factors such as the age, fitness, diet and smoking habits of participants may all affect the experiment's outcome. Random assignment to treatment and control groups helps to ensure a spread of factors between the groups. This means that one group will not be affected by a factor more than the other group.

A double-blind design helps to remove researcher bias as a confounding factor. In a **double-blind study**, neither the researchers nor the participants know whether they have been assigned to the control or treatment group until the results are analysed.

double-blind study
research method where neither the participants in a study nor the researchers know which groups are receiving treatments

In a trial of a new drug, a sugar pill may be provided to the control subjects instead of the drug being tested. This way, those in the control group in the trial will have no way of knowing if they are receiving the treatment or not. The sugar pill, with no medical benefit or effect, is known as a placebo, and it may help determine if a positive effect occurs due to the patient's belief in the treatment.

2.7 LEARNING CHECK

1 **Define** correlation.
2 **Identify** each statement as an example of causation or correlation.
   a In a controlled experiment, applying a nutrient to the treatment group results in a 15 per cent increase in growth compared to the control group.
   b As height above Earth increases, the weight of your body decreases.
   c Over time, an electric current flowing through a wire results in the wire becoming hot.
   d In spring, an increase in grass and an increase in rabbits occur at the same time.
3 **Explain** the difference between correlation and causation.
4 **Explain** why a controlled experiment may provide evidence for causation between a treatment and its effect.
5 **Describe** two ways an experiment may be designed to reduce the effect of confounding variables.

SCIENCE IN CONTEXT

2.8 Bioinformatics

BY THE END OF THIS MODULE, YOU WILL BE ABLE TO:

- ✓ describe some applications of bioinformatics
- ✓ explain why the increasing use of genetic data requires specialised knowledge and skills.

bioinformatics
the science of collecting and analysing complex biological data

base
a part of the structure of DNA that codes genetic information about an organism

genome
the complete set of genetic material present in an organism

exabyte
a very large amount of information, equal to a billion gigabytes

Bioinformatics and the growth of biological data

Bioinformatics is the science of collecting and analysing complex biological data (Figure 2.8.1). Its aim is to rapidly analyse and use genetic data to treat disease. In the last 40 years, the amount of biological data has skyrocketed. For example, in 1990, the Human Genome Project set out to sequence the human genome, which includes 3.2 billion pairs of **bases**. A **genome** is the complete set of genetic material present in an organism. The human genome was published in full in 2003.

Bioinformatics specialists search and analyse huge files, solve problems involving data storage and privacy, and develop artificial intelligence methods to assist in their research. They work with medical specialists and biological researchers to develop new or more efficient medical techniques.

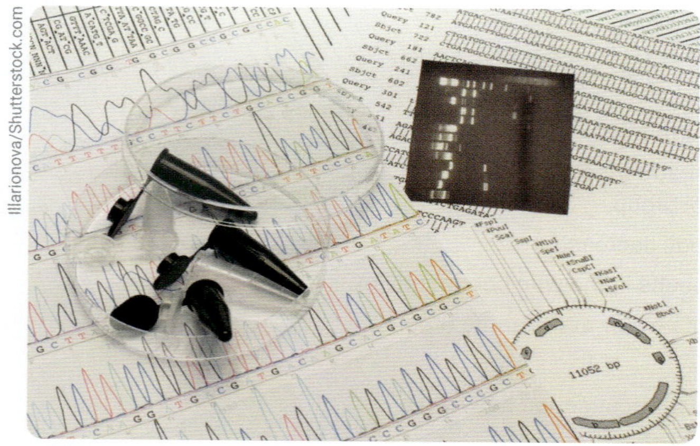
Illarionova/Shutterstock.com

▲ FIGURE 2.8.1 Bioinformatics uses biological data for medicine and biological research.

IM Imagery/Shutterstock.com

▲ FIGURE 2.8.2 The vast amounts of data used in bioinformatics and healthcare are often stored in data servers.

The amount of data generated in sequencing a genome is astonishing. Your genome would require approximately 100 gigabytes of storage. By 2025, it is estimated that 40 **exabytes** (that's 40 billion gigabytes) of storage will be needed for the genomes sequenced each year (Figure 2.8.2). In comparison, YouTube, the world's biggest online video sharing platform, creates about 2 exabytes of data annually.

Such large amounts of data require specialised skills for storage and analysis. Think about how challenging it would be to find a specific piece of information among such vast quantities of data and the cost of storage. However, as more people have their genomes sequenced, the cost to manage this data is likely to decrease due to efficiencies of scale and improvements in technology.

Applications of bioinformatics

There are many practical uses for bioinformatics. For example, many common human diseases (such as heart disease, diabetes and cancer) have a strong genetic component. Genome-based research creates new methods of diagnosis and more effective methods of treatment. It also allows rare genetic diseases to be detected in a way unavailable two decades ago.

9780170491785

Genomic information is increasingly used in cancer treatment. For example, a genetic test can show the probability that a patient will react severely to certain chemotherapy drugs. **Precision cancer medicine** uses a patient's genetic make-up and knowledge of their lifestyle and how their body works to develop individualised treatments. While effective, such treatments are expensive.

precision cancer medicine
a branch of medicine that uses a patient's genetic make-up and knowledge of their body and lifestyle to develop individualised cancer treatments

Bioinformatics also helps in the study of viruses that cause disease; for example, allowing researchers to:

- identify new viruses
- study how virus and host interact
- improve understanding of epidemics and pandemics, such as the COVID-19 pandemic (Figure 2.8.3).

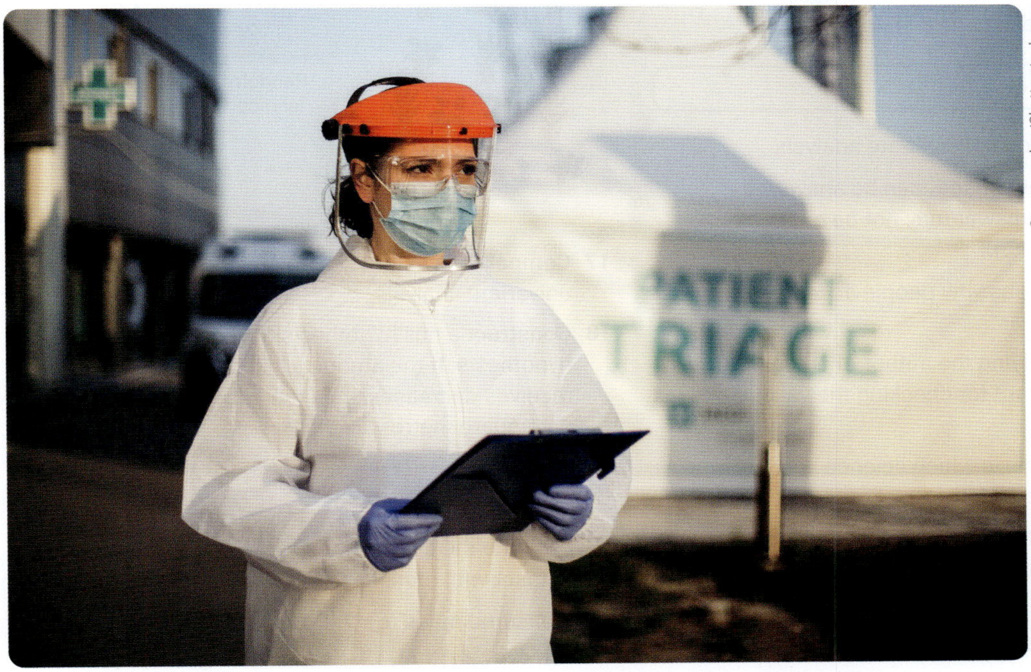

Cryptographer/Shutterstock.com

▲ **FIGURE 2.8.3** Bioinformatics provides data to help front-line medical staff in a pandemic understand how disease spreads, how diverse its strains are and the risks it presents to patients.

2.8 LEARNING CHECK

1 **Describe** bioinformatics.
2 **Explain** why data produced by bioinformatics requires large amounts of storage.
3 **Describe** two uses of bioinformatics.
4 **Explain** why specialised knowledge and skills are needed to **analyse** and store biological data.

2 REVIEW

REMEMBERING

1 **Define** the term investigable question.

2 **List** the three parts of a good scientific argument.

3 **Define** bias and give an example.

4 **State** two types of questions used in a univariate analysis.

5 **List** three things required to show a cause-and-effect relationship between two variables.

UNDERSTANDING

6 **Describe** the purpose of bivariate analysis.

7 **Explain** why a question that can be answered by looking up the answer may not be an investigable question.

8 **Describe** two reasons why people accept pseudoscience ideas.

9 **Explain** why identifying the author of an online resource is important for checking the validity of the information.

10 **Describe** three ways in which science and pseudoscience are different.

11 **Construct** a table to **compare** the purpose and number of variables used in univariate and bivariate analysis.

12 **Identify** each statement as an example of causation or correlation.

- **a** As latitude increases, the weight of a kilogram increases.
- **b** In a controlled experiment, plants treated with fertiliser increased their dry mass 8 per cent more than the control.
- **c** The longer a current flows through a lamp, the more heat is generated by the lamp.

13 **Describe** the work of a bioinformatics specialist.

APPLYING

14 **Explain** why altering the vertical scale on a graph can distort the trend shown.

15 **Explain** why increasing the sample size in an investigation increases statistical reliability.

16 The figure below represents a sample selection from a large population. **Explain** why randomly choosing the sample from the population makes an investigation more reliable.

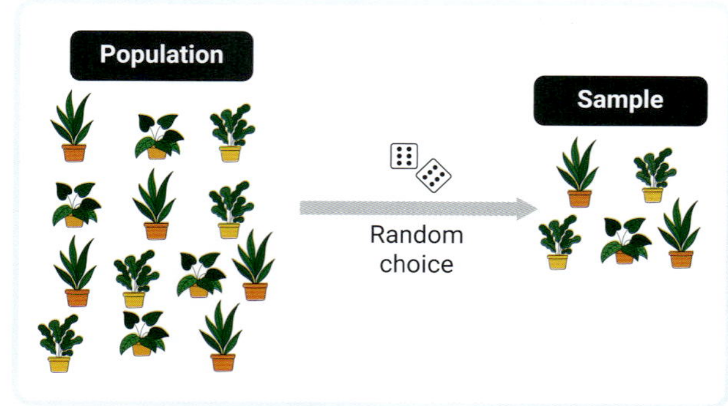

17 **Compare** how statistics and graphs represent data in univariate analysis.

18 **Construct** a null hypothesis from the following hypothesis: If the amount of light a garden bed receives decreases, then the number of weeds per square metre will decrease.

ANALYSING

19 **Evaluate** the centre and spread of the following dataset with statistics.

Measured mass (g): 2.4, 2.5, 2.1, 1.6, 2.8, 2.5, 3.0, 2.6, 2.4

20 **Compare** and **describe** the relationship between the variables in the following table using a line graph.

Variable X (g)	Variable Y (°C)
0.2	19.8
0.3	20.0
0.4	20.2
0.6	20.5
0.8	20.7
1.0	20.8

21 **Examine** and **describe** how outliers affect the accuracy and position of a line of best fit.

9780170491785

EVALUATING

22 **Assess** how attitudes to knowledge and its evaluation influence the reliability of scientific knowledge.

23 **Compare** univariate analysis and bivariate analysis.

24 The figure below shows box plots for six samples, each of 150 measurements from a population. **Evaluate** how a large sample size contributes to consistency across multiple samples.

25 **Explain** and **justify** the importance of careful experimental design in establishing a causal relationship between two variables. Give examples of two types of problems that can occur.

CREATING

26 **Design** an investigation into the relationship between heart rate and height in a class of students. In your answer, **identify**:

- **a** the hypothesis you will test.
- **b** how you will deal with confounding factors.
- **c** how the data will be collected.
- **d** the analysis procedure you will use.

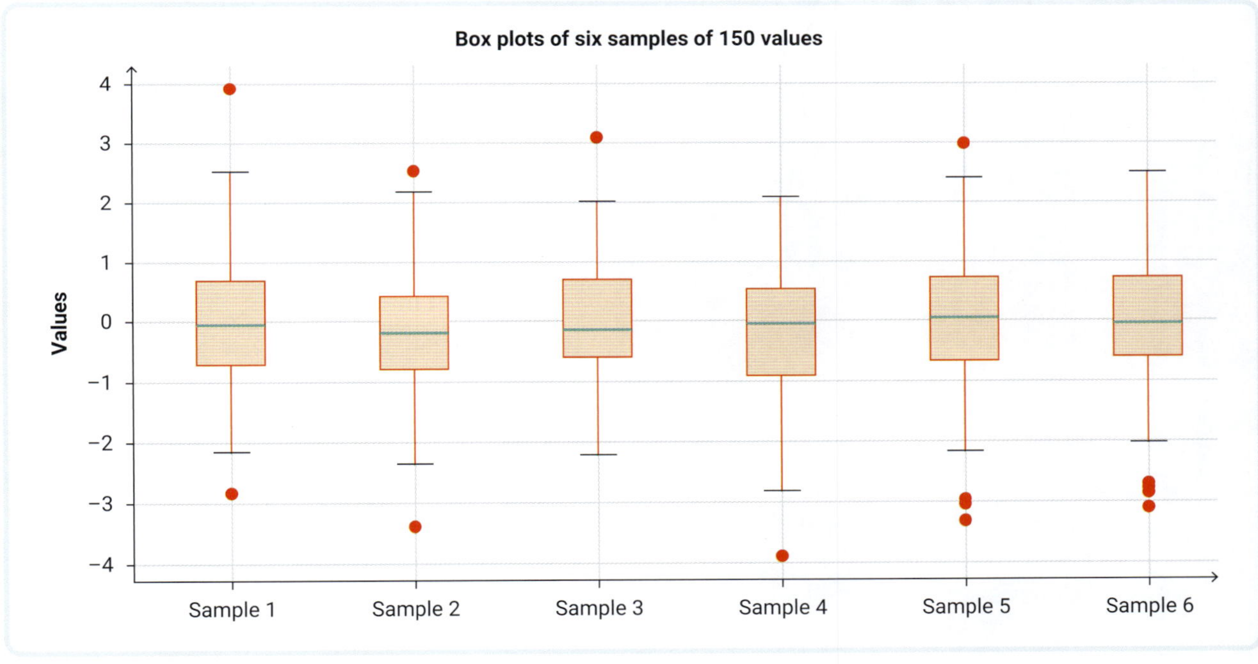

SCIENCE IN DEPTH STUDY

1 Connect what you've learned

In this chapter, you've learned about using data to answer investigative questions and how to create scientific arguments using gathered evidence. Create a flow chart for an investigation. Begin by identifying an investigative question about extreme bushfires and end with a scientific argument about what the data shows.

2 Check your thinking

Formulate an investigable question using the questions from the Science in Depth #2 at the start of the chapter. Where will the data come from to answer your question? How much data will you need to make your conclusions reliable?

3 Get into action

Gather information relevant to your question about major bushfires using 3–4 reliable scientific sources. Can you apply the analysis techniques from the chapter to the data? How do you judge the validity and reliability of your investigation?

4 Communicate

Use the knowledge and understanding of major bushfires you have gained to create a presentation using a scientific argument about what you have found and present it to your class.

ENERGY

SYLLABUS OUTCOMES

A STUDENT:

- evaluates current and alternative energy use based on ethical and sustainability considerations SC5-EGY-01
- selects and uses scientific tools and instruments for accurate observations SC5-WS-01
- follows a planned procedure to undertake safe, ethical, valid and reliable investigations SC5-WS-04
- selects suitable problem-solving strategies and evaluates proposed solutions to identified problems SC5-WS-07

© 2023 NSW Education Standards Authority

CHAPTERS RELATED TO THIS FOCUS AREA ARE:

- CHAPTER 3 – CONSERVATION OF ENERGY
- CHAPTER 4 – ELECTRICAL ENERGY

winyuu/Shutterstock.com

3 Conservation of energy

SCIENCE IN DEPTH 

NurPhoto/NurPhoto/Getty Images

▲ **FIGURE 3.0.1** The design of Formula 1 cars includes features that allow efficient energy transformations so they can accelerate and travel rapidly.

Formula 1 racing teams aim to have their vehicles travel a particular distance in the shortest possible time. Formula 1 races have strict rules about engine size and design, and the type of fuel that can be used. As a result, racing teams are always looking to maximise the efficiency of the energy transformations that occur in race cars. This is the transformation of the chemical potential energy of the fuel into the kinetic energy of the moving vehicle. The team that can do this the most efficiently is more likely to win the race.

The most significant challenge in maximising efficiency is the effect of air resistance. Air resistance is a frictional force that opposes the motion of the vehicle, which slows it down.

- **What features of the design of the vehicle body are aimed at maximising efficiency by minimising air resistance? List at least five features.**
- **How do you think that the Formula 1 racing teams test the efficiency of their designs?**

 **DIVE INTO SCIENCE!**

At the end of this chapter, you will complete Science in Depth Study #3. You can use the information you learn in this chapter to complete the project.

Assessments
- Prior knowledge quiz
- Chapter review questions
- End-of-chapter test
- Depth study: Poster

Videos
- Science skills in a minute: Measurement errors **(3.4)**; Evaluating secondary evidence **(3.7)**
- Video activities: Energy efficiency ratings **(3.3)**; Improving energy efficiency at home **(3.6)**

Science skills resources
- Science skills in practice: Measurement errors **(3.4)**; Evaluating secondary evidence **(3.7)**
- Extra science investigations: Transforming gravitational energy **(3.1)**; Transferring and transforming energy **(3.2)**; Energy efficiency **(3.3)**

Interactive and other resources
- Drag and drop: Calculating energy efficiency **(3.1)**
- Label: Sankey diagrams **(3.2)**
- Activity sheets: Cotton reel spinner **(3.1)**; Design a sustainable house **(3.9)**
- Worksheets: Gravitational potential energy **(3.1)**; Energy transformations **(3.2)**; Energy flow in devices **(3.2)**; Analysing an energy system **(3.2)**; Energy revision **(3.R)**

Nelson MindTap

To access resources above, visit **cengage.com.au/nelsonmindtap**

3.1 Law of conservation of energy

BY THE END OF THIS MODULE, YOU WILL BE ABLE TO:

- ✓ recall and explain the law of conservation of energy and understand how it applies to energy transfers and transformations
- ✓ model energy transfers and transformations using energy flow diagrams.

Extra science investigation
Transforming gravitational energy

GET THINKING

Make a list of the possible *sources* of energy in a school science laboratory. Make a second list of all the different ways energy is *used* in a school science laboratory.

Types of energy

Energy is important in everything that we do. It exists in many different forms, some of which are shown in Table 3.1.1. The general definition for energy is the ability to do work. We can think of it as the ability or capacity to make something happen. We measure energy in joules (J), kilojoules (kJ) or megajoules (MJ).

▼ **TABLE 3.1.1** Some different forms of energy

Form of energy	Definition	Example of work performed
Kinetic energy	The energy of something that has mass and is moving	Air particles in wind contain kinetic energy because they are moving and can turn turbines.
Thermal energy	The energy associated with the temperature and kinetic energy of particles	A hot stovetop full of thermal energy may produce steam that moves away from the stove.
Sound energy	The energy associated with the vibrations of matter caused by sound waves	A sound makes your eardrums vibrate.
Light and radiation energy	The energy carried by electromagnetic waves	Light from the Sun is converted into electrical energy in a solar panel.
Electrical energy	The energy carried by moving charges (electrons)	Electrical energy flowing to a television can produce light through the screen.
Gravitational potential energy	The energy stored within something as a result of its position when under the influence of a gravitational force	Objects that are dropped from heights cause damage.
Elastic potential energy	The energy stored within something when it is stretched or compressed away from its natural resting position	A stretched rubber band flings an object across a room when released.
Chemical potential energy	The energy stored within a chemical, which can be released when the bonds between atoms change during a chemical reaction	When wood is burned, chemical energy is released as heat and light.
Nuclear potential energy	The energy stored within an atom's nucleus, which can be released during a nuclear reaction	At nuclear power plants, nuclear reactions release energy from radioactive nuclei.

Understanding conservation of energy

law of conservation of energy
when energy is transferred or transformed, the total amount of energy remains the same

isolated system
a system in which no energy or matter is exchanged with the surroundings

When modelling and understanding where energy comes from and how it moves around, an important fundamental law of physics is the **law of conservation of energy**. This law states that *energy cannot be created or destroyed in an* ***isolated system***. An isolated system is one that does not allow matter or energy to enter or leave. This means that energy can only be transferred or transformed. Therefore, when there is an:

- energy increase within the system, the energy must have come from somewhere or something else outside of the system

9780170491785

- energy decrease within the system, the energy must have gone to somewhere or something else outside of the system.

Imagine a person diving from a high platform into a pool (Figure 3.1.1). At the top of the platform, the diver has gravitational potential energy but no kinetic energy because they are not moving. At the bottom of the dive when the diver meets the water, the diver has less gravitational potential energy, but they have a lot of kinetic energy because they are moving very fast. The gravitational potential energy has transformed into kinetic energy as the diver falls.

According to the law of conservation of energy, the total amount of energy that exists always stays the same.

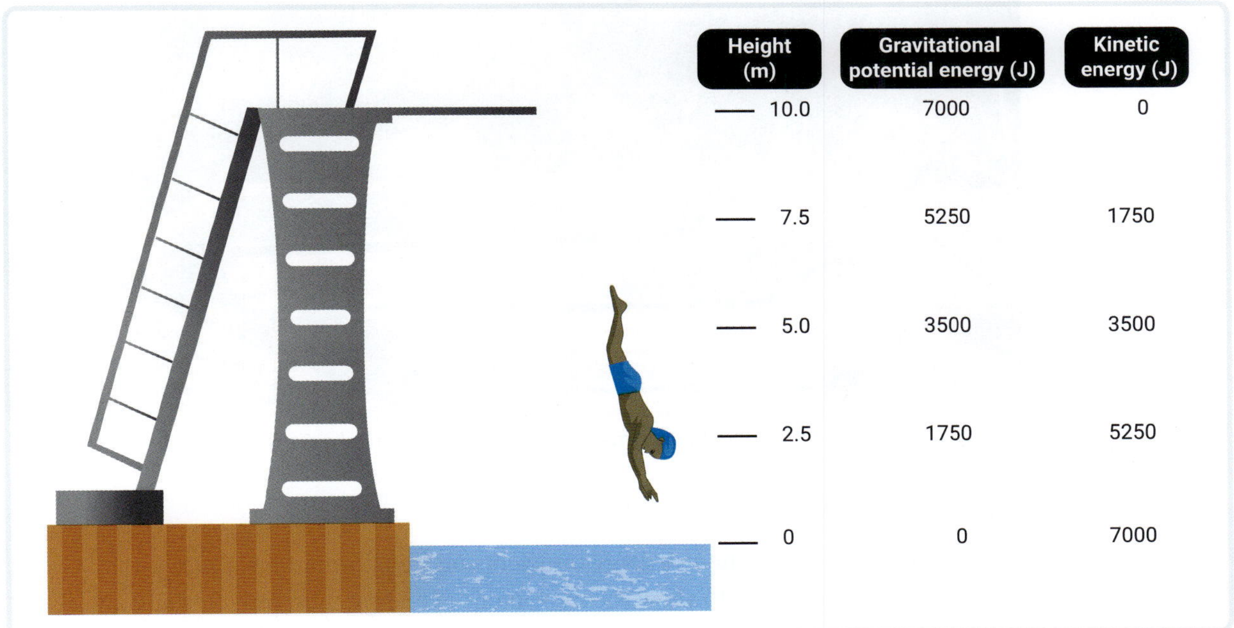

▲ **FIGURE 3.1.1** A diver loses gravitational potential energy and gains kinetic energy.

Energy can be moved around in two ways: energy transfer and energy transformation.

Energy transfer

An energy transfer is when a particular form of energy is passed from particle to particle, object to object or space to space without the type of energy changing. For example:

- kinetic energy in a cricket bat transferring to kinetic energy in a cricket ball when the ball is hit (Figure 3.1.2)
- thermal energy transferring from a hot iron to a shirt when ironing
- sound energy transferring from a speaker through air particles to your ears.

poltu shyamal/Shutterstock.com

▲ **FIGURE 3.1.2** A moving cricket bat transfers kinetic energy when it hits a cricket ball.

Other resources
Activity sheet: Cotton reel spinner

Worksheet: Gravitational potential energy

Energy transformation

An energy transformation happens when energy changes from one form into another. An example of an energy transformation is when a diver's gravitational potential energy changes to kinetic energy as they fall into a pool. Some other examples of energy transformations are:

- electrical energy turning into light and thermal energy in a light bulb
- light energy changing into electrical energy in a solar panel (Figure 3.1.3)
- chemical energy from fuel transforming into kinetic and thermal energy in a combustion engine.

Daria Nipot/Shutterstock.com

▲ **FIGURE 3.1.3** A solar panel transforms light energy into electrical energy.

Energy flow diagrams

energy flow diagram
a visual representation of energy movement using arrows and boxes

An **energy flow diagram** shows energy transfers and transformations. Energy flow diagrams use arrows to show the movement of energy, with transfers and transformations often distinguished by different colours. An energy flow diagram for a parent pushing a child on a swing is shown in Figure 3.1.4.

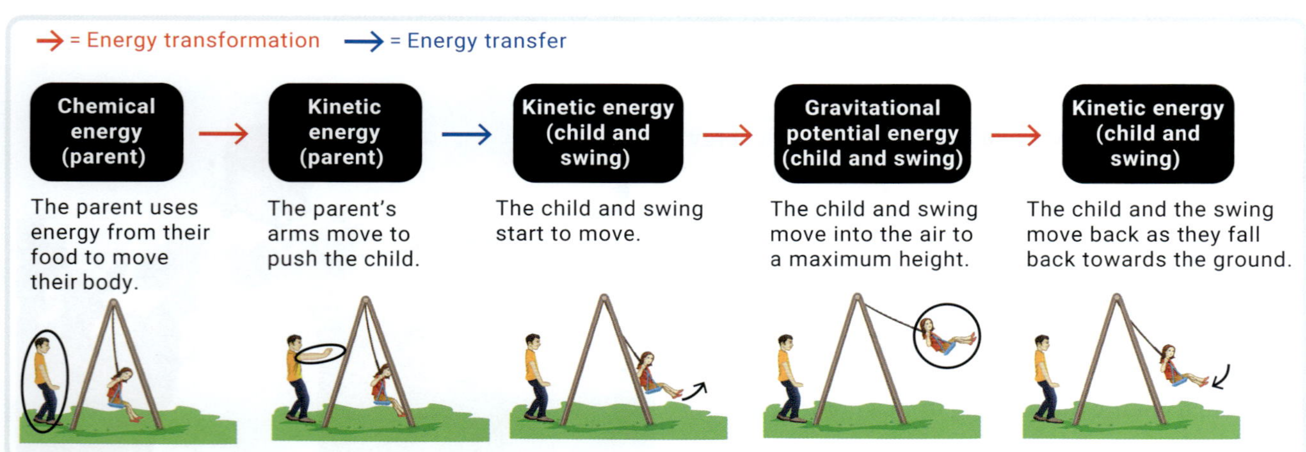

▲ **FIGURE 3.1.4** An energy flow diagram for a parent pushing a child on a swing

9780170491785

Figure 3.1.5 shows a simple energy flow diagram for the transformations of energy in the production of electricity in a hydroelectric power plant.

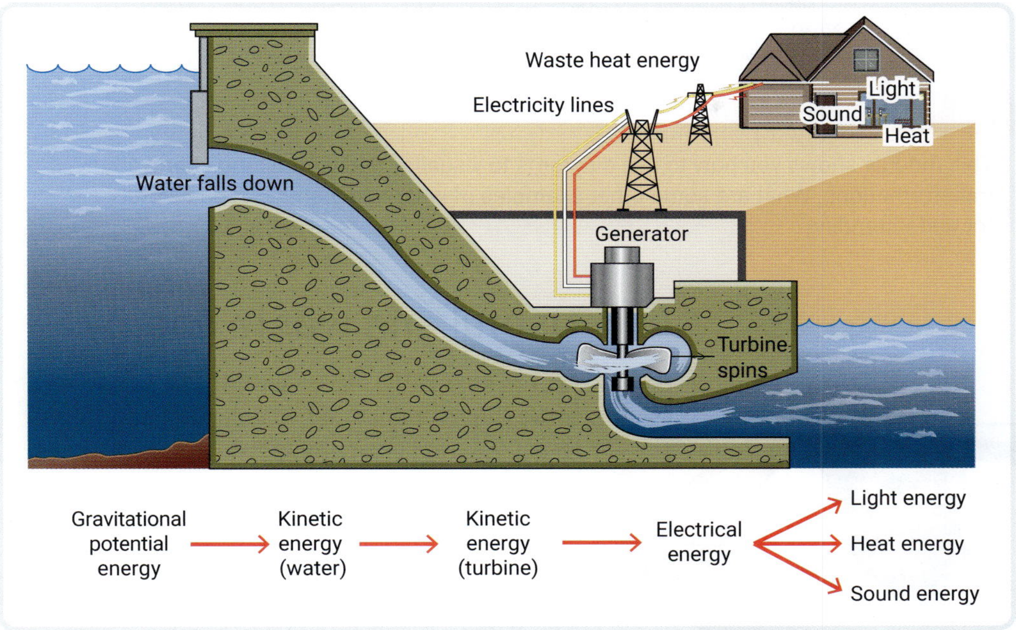

▲ **FIGURE 3.1.5** In a hydroelectric power plant, the gravitational potential energy of water stored at height is transformed into kinetic energy and then electrical energy.

3.1 LEARNING CHECK

1 Copy and **complete** the following table. For each example, **identify** the type of energy input and output and then **classify** whether this is an example of an energy transfer or an energy transformation. The first example has been done for you.

Situation	Energy input type	Energy output type	Transfer or transformation?
Burning wood in a campfire	Chemical potential energy (in wood)	Thermal and light energy (flames)	Transformation
Riding a seesaw			
Watching television			
Hitting a golf ball			
Doing a bomb dive in a pool			
Heating a pie in an oven			
Sling-shotting a rock			
Boiling water on a stove			

2 **Distinguish** between an energy transfer and an energy transformation.

3 **Construct** an energy flow diagram for using an electric stove to fry an egg.

4 As a car brakes to a stop, it starts with a maximum amount of kinetic energy and ends with none. Considering the law of conservation of energy, **explain** how this happens and the energy transformations and transfers that take place.

3.2 Visualising energy transformations

BY THE END OF THIS MODULE, YOU WILL BE ABLE TO:

✓ construct and interpret Sankey diagrams to model different energy transfers and transformations.

Interactive resource
Label: Sankey diagrams

GET THINKING

This module shows you how to use diagrams to visualise and understand energy transformations. What are some other examples of where diagrams can be helpful for understanding?

Useful and waste energy

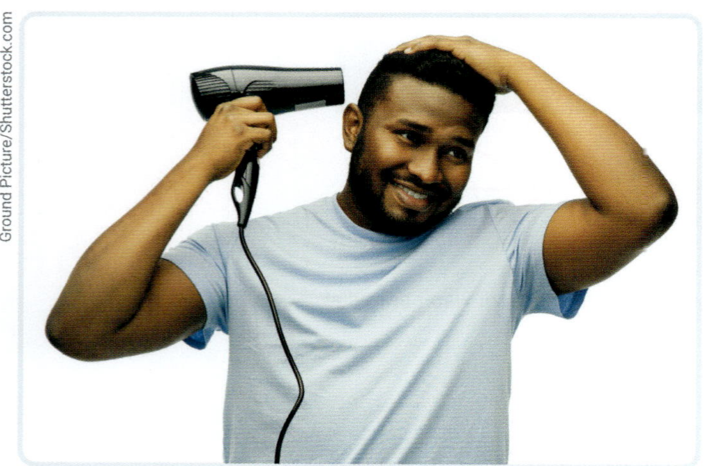

Ground Picture/Shutterstock.com

▲ FIGURE 3.2.1 A hairdryer produces waste sound energy when it blows hot air.

We use energy transfers and transformations to make things happen. When an energy transfer or transformation is intentional, it produces the type of energy you need or want. For example, when you throw a ball, you want the ball to have kinetic energy when you release it. This energy is referred to as **useful output energy**. The useful output energy for a light bulb is light and the useful output energy for a radiator is heat. In all energy transfers and transformations, there are also some unwanted energy transfers and transformations that cannot be avoided. The energy produced by these is known as **waste output energy**. This waste energy is often heat, such as a light bulb getting warm, or sound, such as a hairdryer producing noise (Figure 3.2.1).

useful output energy the output energy of a process or action that is intended and useful

waste output energy the output energy of a process or an action that is not intended and not useful for the main purpose of the process or action

Some other examples of useful and waste energy are given in Table 3.2.1.

▼ TABLE 3.2.1 Some examples of useful and waste energy in different situations

Process	Input energy	Useful output energy	Waste output energy
Throwing a ball	Chemical potential energy	Kinetic energy	Thermal energy of thrower and air; sound of ball moving
Using a light bulb	Electrical energy	Light energy	Thermal energy of warm bulb
Boiling water in a kettle	Electrical energy	Thermal energy of water	Thermal energy of steam
Watching television	Electrical energy	Light and sound energy	Thermal energy of television
Wind turbine	Kinetic energy of wind	Kinetic energy of turbine	Thermal energy of air; sound of moving blades

Sankey diagrams

A **Sankey diagram** is an energy flow diagram that:

- shows measurements of more than one output energy during an energy transfer or transformation
- identifies each output energy as useful or waste
- quantifies energy amounts and shows proportions of energy that flow in different directions.

Sankey diagram
a type of flow chart that uses arrows of various sizes to indicate the amount of energy transferring or transforming in a system

Sankey diagrams show flows of energy as arrows, where the thickness of the arrow indicates how much energy there is. When energy is transferred or transformed, the arrow splits into multiple arrows. Useful energy transfers and transformations are shown as arrows going left to right. Waste energy flows are directed downwards. Different energy types are sometimes shown in different colours.

To satisfy the law of conservation of energy, the total thickness of output arrows must always add up to the total thickness of the input arrows.

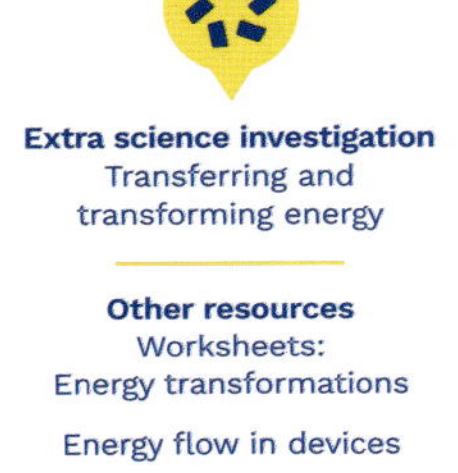

Extra science investigation
Transferring and transforming energy

Other resources
Worksheets:
Energy transformations

Energy flow in devices

Analysing an energy system

Figure 3.2.2 shows a Sankey diagram for a light bulb. The input energy is shown as 200 J of electrical energy. Because light is the useful output energy, the transformation with light as an output is represented horizontally. Thermal energy is a waste energy output of this energy transformation and so is drawn pointing downwards. The useful light energy is 150 J, which is three-quarters or 75 per cent of the input energy. Therefore, the light energy arrow is three-quarters the size of the input electrical energy arrow. The waste thermal output energy is 50 J, which is one-quarter or 25 per cent of the input energy. Therefore, the downwards arrow is one-quarter of the size of the input energy arrow.

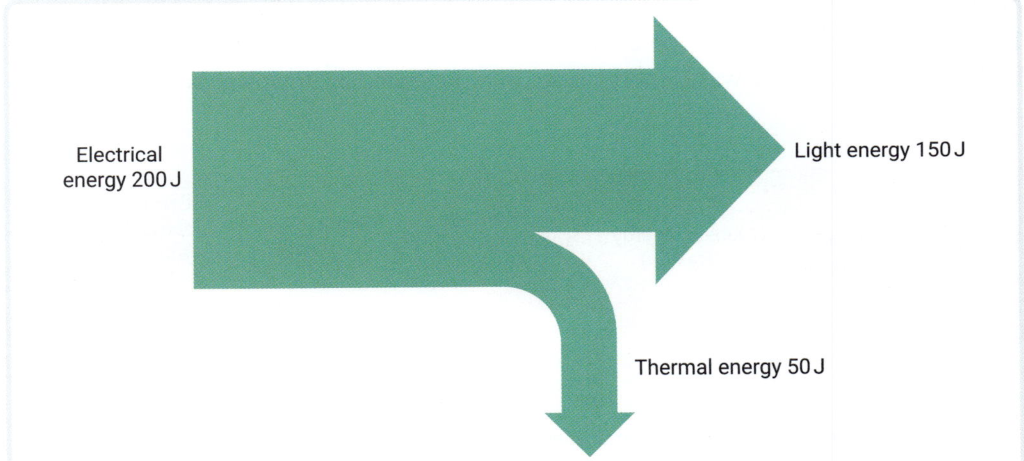

▲ **FIGURE 3.2.2** A Sankey diagram for a compact fluorescent lamp (CFL) that uses 200 J of electrical energy

If you watch a bouncing ball, you'll notice the height of the ball's bounce decreases with each bounce, as shown in Figure 3.2.3. This occurs because, with each bounce sequence, energy is transferred and energy is transformed and, therefore, waste energy is produced.

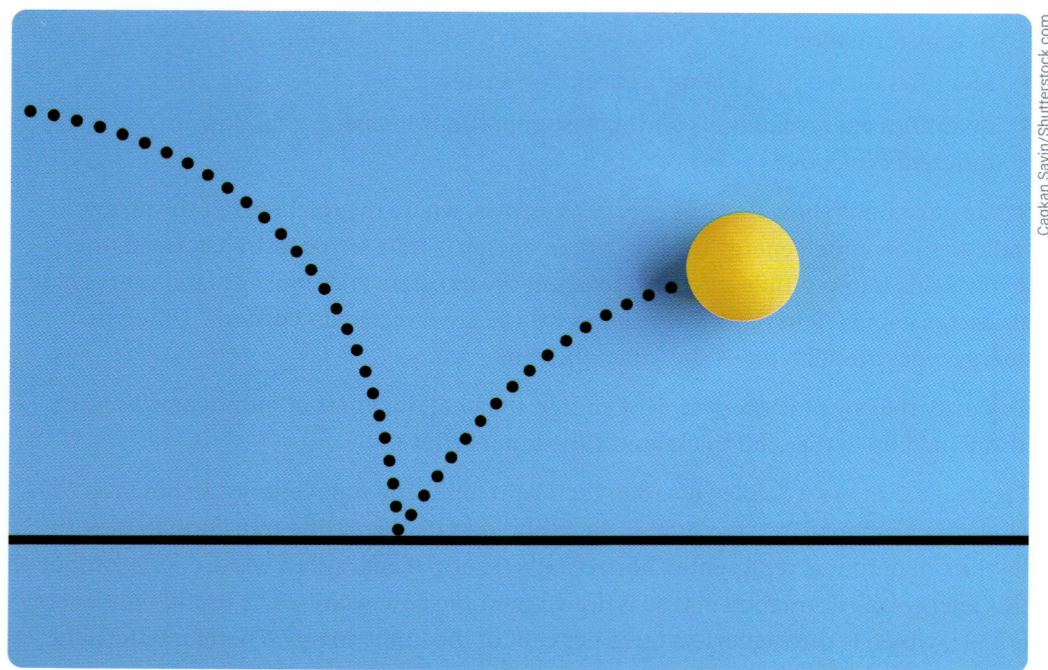

▲ **FIGURE 3.2.3** The height of a ball's bounce becomes lower with each bounce as energy is transferred and transformed.

A Sankey diagram for a ball bouncing is shown in Figure 3.2.4. This diagram shows multiple energy transfers and transformations in the process rather than a single one. The different types of energy (potential, kinetic, elastic and thermal energy) are shown in different colours. The waste energy is again represented by downwards arrows.

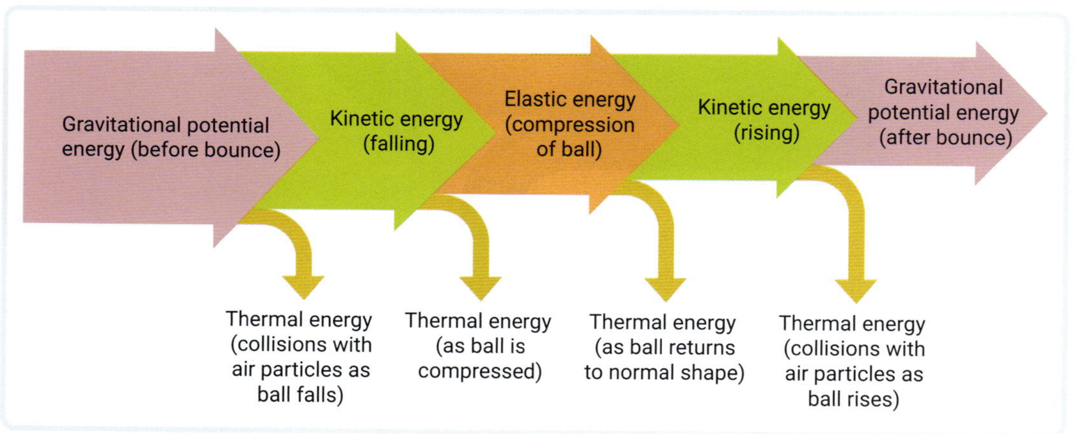

▲ **FIGURE 3.2.4** A Sankey diagram for a ball bouncing

You'll see that the thickness of the arrows decreases from left to right. The decrease in the ball's energy, which we observe in the decreased height of each bounce, is visually represented by the reduction in the width of the arrows.

9780170491785

3.2 LEARNING CHECK

1 **Identify** the useful output energy and a possible waste output energy for a:
   a washing machine.
   b toaster.
   c blender.

▲ FIGURE 3.2.5 (a) A front-loading washing machine, (b) a toaster and (c) a jug-style benchtop blender – each have useful output energy and waste output energy.

2 **Construct** a Sankey diagram to show the energy transformation for a kettle that uses 80 kJ and produces 60 kJ of thermal energy.

3 A hairdryer converts 600 J of energy into 300 J of kinetic energy, 200 J of thermal energy and 100 J of sound energy.
   a **Identify** which type of energy is the wasted energy.
   b **Construct** a Sankey diagram to represent the energy transformations.

4 For the Sankey diagram in Figure 3.2.6, **identify** the:
   a amount of waste energy.
   b amount of useful output energy.
   c useful output energy as a fraction of the total energy input.

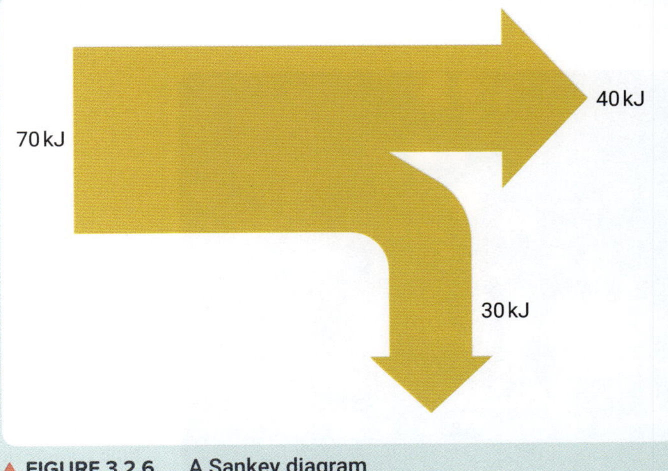

▲ FIGURE 3.2.6 A Sankey diagram

3.3 Calculating energy efficiency

BY THE END OF THIS MODULE, YOU WILL BE ABLE TO:

- ✓ define the concept of energy efficiency and calculate energy efficiency for different processes
- ✓ use examples to explain why it is important to calculate energy efficiency.

GET THINKING

What does the term 'efficiency' mean to you? How do you think it might be relevant to ideas about energy and energy transformations? Write down some ideas before you start this module.

Energy efficiency

energy efficiency a measure of how much input energy is converted to useful output energy, often stated as a percentage

Energy efficiency is a measure of how much input energy is converted into useful energy. Processes that are highly efficient convert a large amount of input energy into useful output energy and have small amounts of waste energy. Low-efficiency devices or processes produce a lot of waste energy.

Energy efficiency is usually measured as a percentage and is found by calculating the percentage of the input energy that is converted into useful output energy:

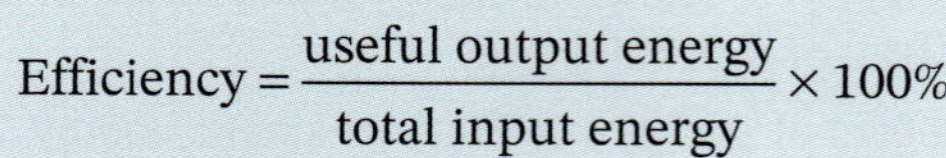

$$\text{Efficiency} = \frac{\text{useful output energy}}{\text{total input energy}} \times 100\%$$

Video activity
Energy efficiency ratings

Interactive resource
Drag and drop: Calculating energy efficiency

Extra science investigation
Energy efficiency

How to calculate energy efficiency

The following worked examples show how to calculate the energy efficiency of a light bulb (Worked example 3.3.1) and an electric stovetop (Worked example 3.3.2).

WORKED EXAMPLE 3.3.1

Calculate the energy efficiency of a light bulb that has a total input energy of 200 J and a useful output energy of 20 J.

Champiofoto/Shutterstock.com

FIGURE 3.3.1 A light bulb transforms electrical energy to light energy.

9780170491785

THINKING PROCESS	WORKING
Step 1: Identify the known and unknown variables.	Total input energy = 200 J Useful output energy = 20 J Efficiency = ?
Step 2: Identify the appropriate relationship.	$\text{Efficiency} = \frac{\text{useful output energy}}{\text{total input energy}} \times 100\%$
Step 3: Substitute known values.	$\text{Efficiency} = \frac{20}{200} \times 100\%$
Step 4: Rearrange, using algebra if necessary, and solve.	Efficiency = 10% The light bulb is 10% energy efficient.

WORKED EXAMPLE 3.3.2

When preparing dinner, Charlie heats up a frying pan on an electric stove, which uses 800 J of electrical energy. Calculate the efficiency of heating the pan on the stove if:

- **400 J of thermal energy is absorbed by the pan**
- **100 J of sound energy is produced**
- **300 J of thermal energy is lost to the air and other parts of the stove top.**

New Africa/Shutterstock.com

FIGURE 3.3.2 Electric stoves transform electrical energy to thermal energy to cook food.

THINKING PROCESS	WORKING
Step 1: Identify the known and unknown variables.	Total input energy = 800 J Useful output energy = 400 J Efficiency = ?
Step 2: Identify the appropriate relationship.	$\text{Efficiency} = \frac{\text{useful output energy}}{\text{total input energy}} \times 100\%$
Step 3: Substitute known values.	$\text{Efficiency} = \frac{400}{800} \times 100\%$
Step 4: Rearrange, using algebra if necessary, and solve.	Efficiency = 50% The stove to pan heating process is only 50% energy efficient.

Calculating useful output and waste energy

If the energy efficiency of a device or process is given with the total input energy, you can calculate the useful output energy and waste output energy (Worked example 3.3.3) from the following equations:

$$\text{Useful output energy} = \frac{\text{efficiency} \times \text{total input energy}}{100}$$

$$\text{Total input energy} = \text{useful output energy} + \text{waste output energy}$$

WORKED EXAMPLE 3.3.3

A kettle in the school staff room uses 70 kJ of electrical energy and is only 80 per cent efficient. Calculate how much waste energy is produced by the kettle.

To calculate this, we need to break this example down into two parts.

PART A: CALCULATE THE USEFUL OUTPUT ENERGY

THINKING PROCESS	WORKING
Step 1: Identify the known and unknown variables.	Total input energy = 70 kJ = 70 kJ × $\frac{1000\text{ J}}{1\text{ kJ}}$ = 70 000 J Efficiency = 80% Useful output energy = ?
Step 2: Identify the appropriate relationship.	Useful output energy = $\frac{\text{efficiency} \times \text{total input energy}}{100}$
Step 3: Substitute known values.	Useful output energy = $\frac{80 \times 70\,000}{100}$
Step 4: Rearrange, using algebra if necessary, and solve.	Useful output energy = 56 000 J

PART B: CALCULATE THE WASTE OUTPUT ENERGY

THINKING PROCESS	WORKING
Step 1: Identify the known and unknown variables.	Useful output energy = 56 000 J Total input energy = 70 000 J Waste output energy = ?
Step 2: Identify the appropriate relationship.	Total input energy = useful output energy + waste output energy
Step 3: Substitute known values.	70 000 = 56 000 + waste output energy
Step 4: Rearrange, using algebra if necessary, and solve.	Waste output energy = 70 000 − 56 000 Waste output energy = 14 000 J × $\frac{1\text{ kJ}}{1000\text{ J}}$ = 14 kJ There is 14 kJ of waste energy.

Comparing efficiency

Incandescent light bulbs used to be the most common types of light bulbs, but have been mostly phased out because of their low energy efficiency. Compact fluorescent lights (CFLs) and light-emitting diodes (LEDs) are now common. They have higher efficiency, which means they use less electricity to produce the same amount of light (Figure 3.3.3). People choose more efficient light bulbs to save electricity, to save money and for environmental reasons such as reducing carbon emissions.

a

b

c

▲ **FIGURE 3.3.3** Different types of light bulbs have different energy efficiencies. **(a)** An incandescent light bulb is 10–20 per cent energy efficient. **(b)** A compact fluorescent light bulb is 70–85 per cent energy efficient. **(c)** A light-emitting diode is 80–90 per cent energy efficient.

The Australian Energy Rating Label scheme allows people to compare the energy efficiencies of different appliances. You will learn more about this scheme in Chapter 4.

3.3 LEARNING CHECK

1 **Calculate** the efficiency of a torch that converts 100 J of electrical energy and produces 40 J of waste thermal energy.

2 During an all-day movie marathon, a remote control uses 1500 J of electrical energy from the batteries and is only 40 per cent efficient. **Calculate** the:
   a useful output energy.
   b waste output energy.

3 Look at the three types of heater below. **Imagine** you have to **analyse** the energy efficiency of each. What information would you need? **Explain** why it is important to be able to **calculate** energy efficiency.

a

Electric oil column heater

b

Infrared heater

c

Electric fan heater

WORKING SCIENTIFICALLY

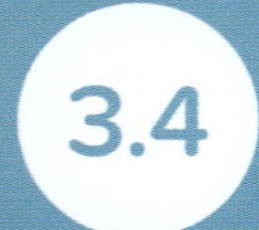

3.4 Measurement errors

SCIENCE SKILLS IN FOCUS

IN THIS MODULE, YOU WILL FOCUS ON LEARNING AND IMPROVING THESE SKILLS:

- identifying and reducing the effect of different types of measurement errors
- measuring bounce height and efficiency.

Random and systematic errors

We can categorise the errors we make in science investigations into two broad categories: random errors and systematic errors. Random errors are usually small errors, caused by slight variations in an instrument or the environment, that do not produce the same error every time. Like the name suggests, random errors are usually due to chance.

Systematic errors are often due to something going wrong in the procedure. Systematic errors can be easy to spot because the errors occur in a predictable way.

When we take measurements, we can accidentally introduce both these types of errors into our data collection processes.

Examples of measurement errors include:

- random: fluctuations in air temperature, air pressure and humidity affecting the elasticity of a rubber band; an error in reading the volume of liquid in a measuring cylinder; or a small crack in the ground affecting the height of a ball's bounce
- systematic: a calibration error in an electronic balance; hot weather the temperature of experimental equipment to rise; incorrectly rounding down or up of a measurement; or a small amount of a substance left on the electronic scales while samples are weighed.

We can reduce the effect of random errors in our datasets by conducting multiple trials in our investigations and increasing the size of our datasets. We can reduce the possibility of systematic errors by ensuring we follow the procedure consistently, using calibrated instruments, evaluating the procedure and sharing our data with others to get feedback on possible errors.

MEASURING THE EFFICIENCY OF A BALL'S BOUNCE

AIM

To measure and compare the efficiency of a ball bounce for two balls made of different materials

MATERIALS AND EQUIPMENT

- ☑ 2 different balls that will bounce (e.g. ping pong ball, tennis ball, bouncy ball)
- ☑ metre ruler
- ☑ putty-like adhesive
- ☑ video recording device
- ☑ electronic scales

PROCEDURE

1. List all the possible sources of random and systematic errors that could occur in this investigation.
2. Hold a metre ruler against a flat wall so that it measures from the ground up.

9780170491785

3 Secure the ruler in place with the adhesive. Alternatively, you can hold the ruler in place against the wall.
4 Weigh each ball on the electronic scales and record their masses.
5 Hold the first ball so that the bottom of the ball is at the 1.0 m mark.
6 Do a test bounce to confirm approximately where the ball will bounce to when dropped from this height.
7 Set up the video recording device so that it will record the area level with where the ball will bounce up to.
8 Hold the ball again so that the bottom of the ball is at the 1.0 m mark.
9 Start recording and drop the ball. Allow it to bounce once before stopping it and the recording.
10 Use the recording to measure the highest point reached by the bottom of the ball.
11 Record this height as the bounce height for the 1.0 m drop height.
12 Repeat steps 8–11 for two more trials.
13 Calculate an average bounce height for the 1.0 m drop height.
14 Repeat steps 6–12 for drop heights of 0.8 m, 0.6 m, 0.4 m and 0.2 m.
15 Repeat steps 4–14 for the other ball.

RESULTS

Record your results in a table like Table 3.4.1.

1 Use your table to calculate the input energy for each drop height, using the formula:

$$E_{Pdrop} = mgh_{drop}$$

where E_{Pdrop} is the gravitational potential energy at the drop height (J), m is the mass of the ball (kg), g is the gravitational acceleration value of 9.8 m/s^2, and h_{drop} is the drop height of the ball (m).

For example, for a ball that weighs 0.250 kg and a drop height of 0.8 m:

$h_{drop} = 0.8$ m
$g = 9.8$ m/s^2
$m = 0.25$ kg

$$E_{Pdrop} = mgh_{drop} = 0.25 \times 9.8 \times 0.8 = 1.96 \text{ J}$$

▼ TABLE 3.4.1 Experimental results for the bounce height of each ball at different drop heights

Ball type	Mass (kg)	Drop height (m)	Bounce height (m)			
			Trial 1	Trial 2	Trial 3	Average
		1.0				
		0.8				
		0.6				
		0.4				
		0.2				
		1.0				
		0.8				
		0.6				
		0.4				
		0.2				

▼ **TABLE 3.4.2** Processed data table for ball bounce efficiency

Ball type	Drop height (m)	Input energy (J)	Useful output energy (J)	Efficiency (%)
	1.0			
	0.8			
	0.6			
	0.4			
	0.2			
				Average
	1.0			
	0.8			
	0.6			
	0.4			
	0.2			
				Average

2 Calculate the useful output energy for each drop height using the formula:

$$E_{Pbounce} = mgh_{bounce}$$

where $E_{Pbounce}$ is the gravitational potential energy at the bounce height (J), m is the mass of the ball (kg), g is the gravitational acceleration value of 9.8 m/s^2, and h_{bounce} is the bounce height of the ball (m).

3 Calculate the efficiency of the ball bounce for each drop height and record this in Table 3.4.2.

4 Calculate the average efficiency of a ball bounce for each type of ball and record this in Table 3.4.2.

ANALYSIS

1 **Construct** a scatter graph of bounce height (m) (y-axis) versus drop height (m) (x-axis) for both balls. Include both sets of data on the one plot and include a legend to **identify** the different balls.

2 **Construct** a column graph to **compare** the average efficiency of the balls.

3 **Identify** the trend shown in the scatter plot from step 1.

4 **Compare** the average bounce efficiencies of the balls.

5 **Discuss** the variability of the trials used to **calculate** the average bounce height.

6 **Explain** how random measurement errors may have affected the data recorded.

7 Consider the list of possible errors you created at the start of the procedure. **Identify** other experimental errors (random and systematic) from this experiment and **describe** how they may have affected the data collected.

8 Suggest two ways in which this **experiment** could be improved.

CONCLUSION

Draw a conclusion that directly responds to the aim of the experiment.

9780170491785

3.5 Thermal waste energy

BY THE END OF THIS MODULE, YOU WILL BE ABLE TO:

✓ recognise thermal energy as a common form of waste energy and give examples of how this occurs.

GET THINKING

Think about one meal that you know how to make. From start to finish, identify all the waste materials or energy that might be produced during the process of making your meal. Are there any ways you could avoid producing this waste?

Friction and waste energy

When a ball rolls to a stop along the ground, it can seem as though the kinetic energy is leaving the ball and disappearing. However, the law of conservation of energy says that this is not the case. During many energy transfers or transformations, thermal energy can be produced as waste output energy. This is due to the interactions between particles. As the ball rolls, the interaction between the ball's surface and the ground makes particles vibrate, which increases their temperature and produces thermal energy (Figure 3.5.1). This production of thermal energy is often referred to as **frictional heat loss** because the force of **friction** is being overcome when particles rub past each other.

frictional heat loss
waste thermal energy produced as a result of the vibration of particles when moving past each other in a process where thermal energy production is not intended or useful

friction
a resistance force that results when two surfaces rub against one another

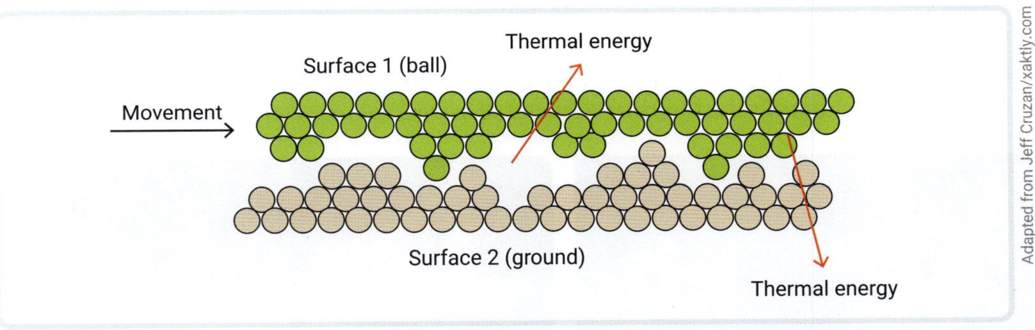

▲ **FIGURE 3.5.1** Thermal energy is produced when particles on the ball's surface and the ground rub against each other and vibrate.

Another example of this is when you rub your hands together. The kinetic energy from the movement of your hands is transformed to thermal energy as the particles in your skin vibrate when rubbed together. Consider the diver shown in Figure 3.5.2. The energy transformation from gravitational potential energy to kinetic energy includes some thermal energy as the air particles vibrate when the diver moves through the air. Thermal energy is produced as the diver and air particles vibrate against one another. We often hear this described as 'air resistance'. This means that the diver isn't going quite as fast when they hit the water as they might have been if there were no frictional heat losses along the way.

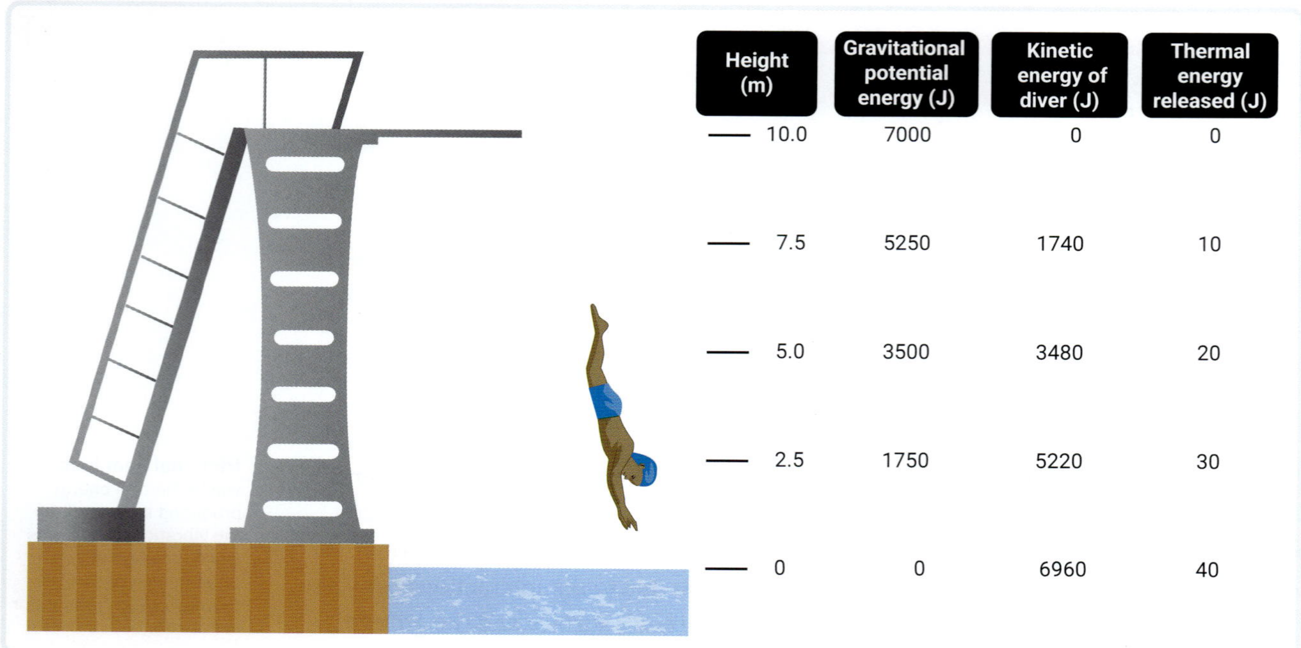

Height (m)	Gravitational potential energy (J)	Kinetic energy of diver (J)	Thermal energy released (J)
10.0	7000	0	0
7.5	5250	1740	10
5.0	3500	3480	20
2.5	1750	5220	30
0	0	6960	40

▲ **FIGURE 3.5.2** A diver loses gravitational potential energy and gains kinetic energy while producing some thermal energy as waste.

When meteors or other objects enter the atmosphere, they travel so fast through the air particles that the frictional heat losses produced can be so great that they burst into flames. Often, they burn up completely before they hit the ground. We see this happening as shooting stars in the night sky (Figure 3.5.3).

Belish/Shutterstock.com

aapp/Shutterstock.com

▲ **FIGURE 3.5.3** Fast-moving meteors burn up in Earth's atmosphere and appear as shooting stars.

Many electrical devices or appliances produce thermal energy that we don't need or use when transforming electrical energy into other forms such as light, sound and kinetic energy (Figure 3.5.4). This is why light bulbs, speakers, chargers, televisions, mobile phones and other devices often get warm.

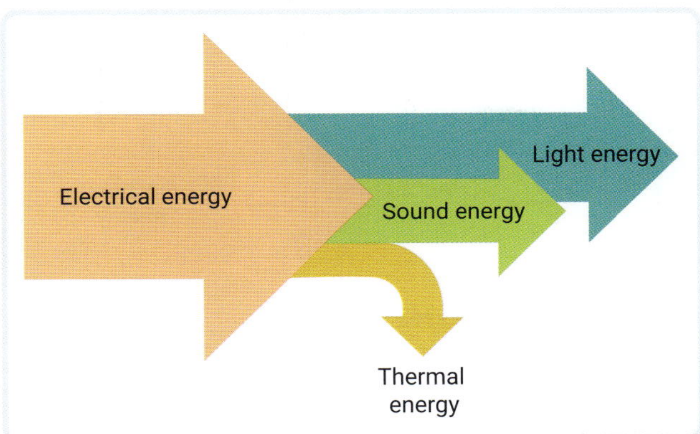

FIGURE 3.5.4 A television converts electrical energy to sound and light energy with some waste thermal energy produced.

When the purpose of an energy transformation is to produce thermal energy, sometimes thermal energy can still be considered a waste output energy when it is transferred to other objects or particles. For example, boiling water on the stove involves losing thermal energy to the stovetop, the pot and the air surrounding the pot (Figure 3.5.5). Not all input electrical energy is being transformed into the thermal energy of the water.

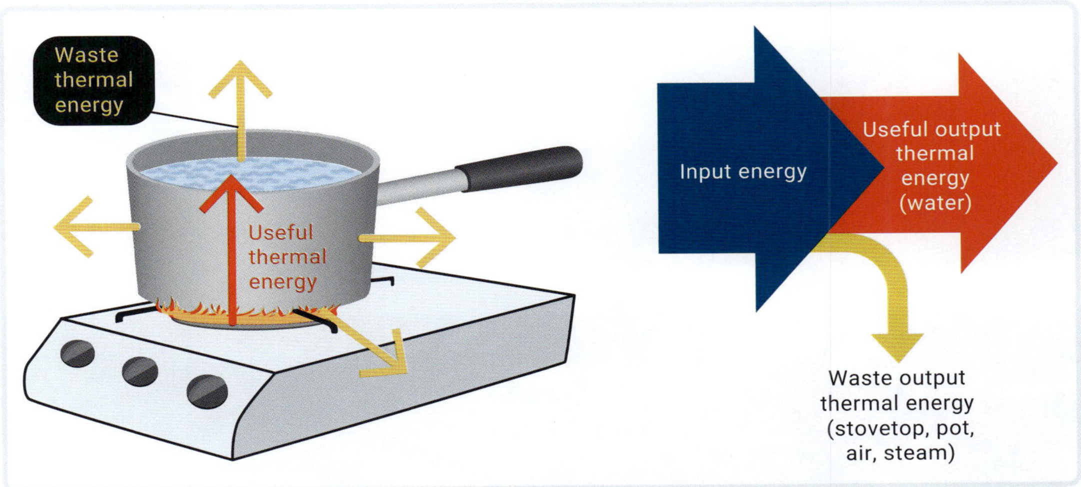

FIGURE 3.5.5 There is waste thermal energy produced when boiling water in a pot.

Thermal energy losses in electricity generation

Electricity generation in a power station involves the following steps (Figure 3.5.6).

1. Chemical potential energy (usually coal) is converted to thermal energy during a chemical reaction (combustion).
2. Thermal energy heats water to steam. The steam contains both thermal energy and kinetic energy.
3. Kinetic energy of steam is transferred to kinetic energy of a turbine, which then spins.
4. The spinning turbine produces electrical energy (electricity), using a generator.

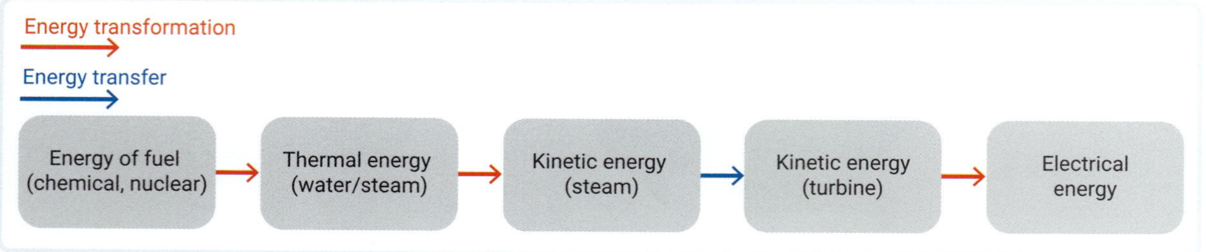

▲ **FIGURE 3.5.6** The energy flow diagram for most electricity generation processes

During the energy transfers and transformations between the fuel and the kinetic energy of the turbine, there are many waste energy losses, including:

- thermal energy to the heating of components of power station
- thermal energy in steam that escapes or leaves the system after passing the turbine (Figure 3.5.7)
- thermal energy in other exhaust gases that are produced during the reactions of the fuel
- sound energy produced during the movement of mechanical parts
- thermal energy of moving parts due to friction.

Clare Louise Jackson/Shutterstock.com

▲ **FIGURE 3.5.7** A lot of thermal energy is lost in electricity production when hot steam is released to the atmosphere.

There is a lot of waste energy when using a fuel to produce electricity. As a result, power generation facilities (power plants) use specially designed processes and equipment to reduce energy losses and increase efficiency.

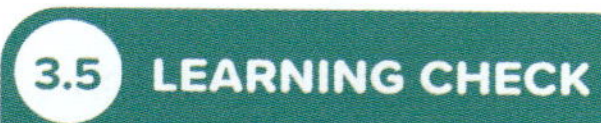

3.5 LEARNING CHECK

1 **Research** the processes for electricity generation by hydroelectric, coal-fired and nuclear power stations. **Construct** simple energy flow diagrams for these processes.

2 **Identify** three examples or situations where frictional heat losses occur.

3.6 Reducing waste energy

BY THE END OF THIS MODULE, YOU WILL BE ABLE TO:

- ✓ describe the different forms of waste energy
- ✓ explain how energy efficiency can be increased and give examples of ways waste energy can be used for other purposes.

GET THINKING

Look around your school or home. Can you see any objects, materials or technologies that have been designed to reduce waste energy?

Video activity
Improving energy efficiency at home

Other types of waste energy

In Module 3.5, we looked at thermal waste energy. There are also other types of waste energy.

Sound waste energy

Sound energy is a common form of waste energy. There are many energy transformations that are noisy because sound, which is produced by moving parts and vibrations, is released into the air. Some common examples are:

- sound made by vehicles when travelling (engine noises, tyres on the road)
- sound produced by the moving parts of machinery
- sound made by a food blender.

Highways often have large noise barriers that absorb sound or reflect it back to the road.

Electromagnetic radiation waste energy

When light is the intended output of an energy transformation, such as within a light bulb, torch or flame, there are usually other forms of electromagnetic radiation produced. For example, a light bulb can also produce small amounts of ultraviolet (UV) radiation and heat as infrared radiation (Figure 3.6.1).

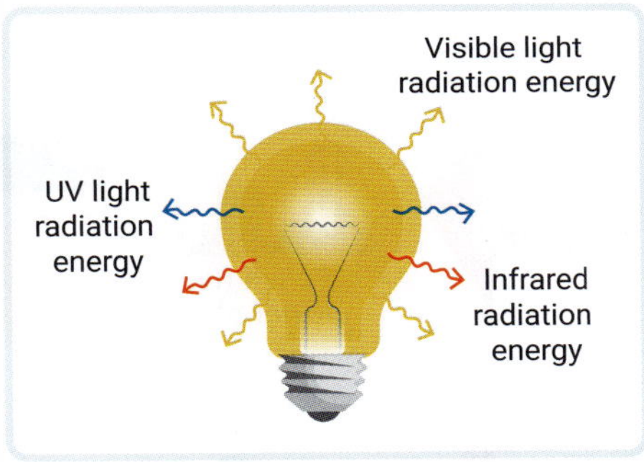

▲ **FIGURE 3.6.1** Waste energy from light globes may include other types of electromagnetic radiation.

Increasing energy efficiency

There are three common ways to increase the efficiency of an energy transfer or transformation (Figure 3.6.2).

- Redesign the energy transfer or transformation process with new technologies that have higher conversions of input energy to useful output energy, such as the development of LED light bulbs.
- Reduce the unwanted energy transfers so that less total input energy is required; for example, add insulation to a building (Figure 3.6.3).

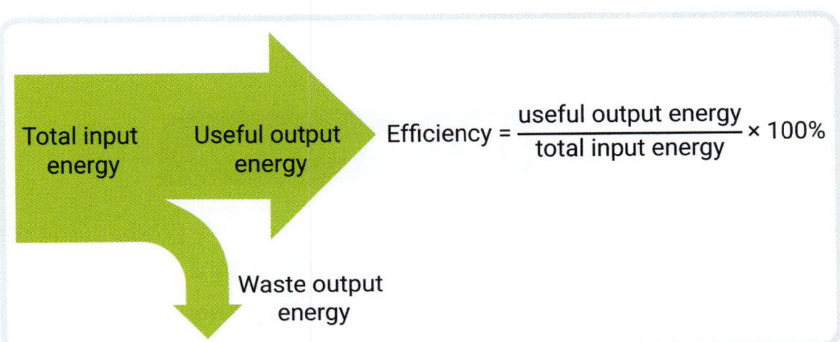

▲ **FIGURE 3.6.2** The energy efficiency of a process can be increased at different stages of the process: by redesigning the process to increase the energy conversion, by reducing waste, and by reusing the waste output energy.

insulation material used to minimise heat transfer

cogeneration the process of using waste thermal energy in power generation for another purpose

- Reuse the waste output energy; for example, convert the kinetic energy of steam leaving a kettle to sound energy that indicates the water is boiled.

Redesigning energy transfer and transformation processes often takes a long time and requires large amounts of funding. Reducing and reusing waste energy are generally more practical and affordable ways to increase energy efficiency.

Reducing thermal waste energy

Bilanol/Shutterstock.com

▲ **FIGURE 3.6.3** Insulation in the walls of buildings reduces the transfer of heat in and out of rooms.

In a power plant, water is heated and the resulting steam moves fast enough to turn a turbine. As the turbine spins, equipment becomes very hot. Heat transfers to the surrounding air and a lot of thermal energy is lost as the hot steam escapes.

Minimising heat transfer can reduce the thermal energy losses. This can be done by:

- insulating with materials that reduce heat conduction to equipment or air
- minimising the surface area of containers from which heat loss may occur
- lining containers with reflective surfaces to reduce radiative heat loss.

Houses have **insulation** inside walls, and coverings and double-glazing on windows to reduce heat gain and loss. This increases the efficiency of cooling and heating systems, saving energy and money, and reducing emissions.

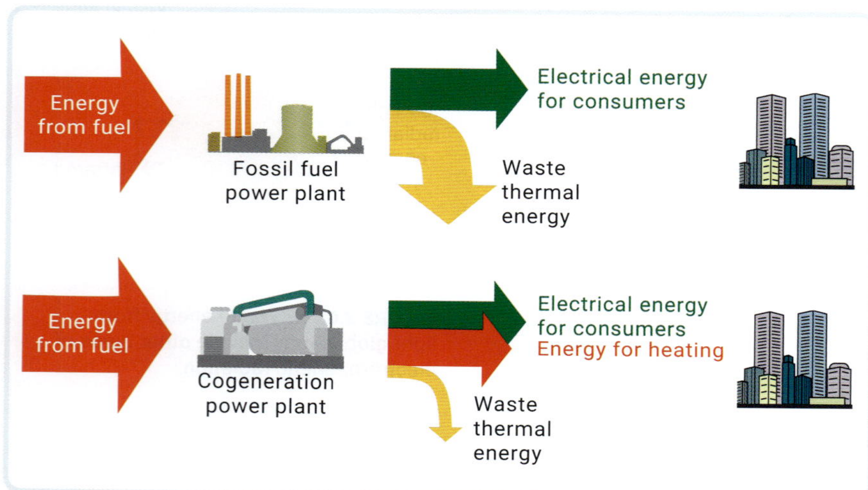

▲ **FIGURE 3.6.4** Cogeneration can increase efficiency and reduce fuel consumption by turning waste thermal energy into useful energy.

Reusing thermal waste energy

Some power stations reuse waste thermal energy. **Cogeneration** or combined heat and power generation is the process of using the waste thermal energy for another purpose. Typically, waste steam is redirected to heat nearby buildings or is used by factories in other industrial processes (Figure 3.6.4). Therefore, this process saves energy because less thermal energy is wasted.

3.6 LEARNING CHECK

1. Consider an electricity production process that supplies 300 MJ of electrical energy and wastes 1000 MJ of thermal energy.
   a. **Calculate** the energy efficiency of the process.
   b. If 800 MJ of thermal waste energy could be reused and classified as useful output energy during cogeneration, **calculate** the new energy efficiency of the process.
2. **Justify** why a coal plant fitted with a cogeneration system might be a more environmentally friendly choice than one without.

9780170491785

WORKING SCIENTIFICALLY

3.7 Evaluating secondary evidence

SCIENCE SKILLS IN FOCUS

IN THIS MODULE, YOU WILL FOCUS ON LEARNING AND IMPROVING THESE SKILLS:

- identifying strengths and limitations of primary and secondary evidence and their sources
- defining reliability and validity
- evaluating the reliability and validity of primary and secondary evidence
- drawing conclusions from evidence based on interpretation and evaluation.

Reliability and validity of secondary evidence

Research or evidence that you use or collect from other researchers is known as **secondary evidence**. This is because someone other than you has conducted the experiment, observation or research. When you use secondary evidence, it is important to think about the reliability and validity of the evidence and its source.

Reliability is the extent to which the evidence can be trusted or considered reliable for use. Reliability is usually based on where the evidence has come from or the method used to collect it. For example, scientific evidence published in a prestigious journal is likely to be more reliable than the observations from Bob next door!

The **validity** of the evidence judges how well the evidence answers the specific research question or aim. For example, if a researcher wants to know how plastic pollution is affecting the health of polar bears, but only collects data about penguins, any conclusion they make about their research wouldn't be considered valid. This is because their data does not directly show the relationship between plastic pollution and polar bears.

Some attributes of sources and evidence that can support and question reliability and validity are outlined in Table 3.7.1.

Evaluating the reliability and validity of the evidence and the source will help you decide whether the conclusions drawn from the evidence are appropriate or not. If you identify limitations with secondary evidence, you may still be able to use it, but you should consider the limitations and discuss them when drawing your own conclusions based on the research or information.

TABLE 3.7.1 Strengths and limitations of different secondary evidence and sources

Strengths	Limitations
Secondary evidence has strong reliability if it is: • written by a reputable university or government department • supported or backed up by the research of other academics or leaders in the field • recent, up-to-date research.	Secondary evidence has limited reliability if it: • is written by a corporate or commercial organisation that may have financial or political reasons for publishing it • has an anonymous author • is out of date or not the most recent research in the field.
Secondary evidence has strong validity if the: • data or ideas are clear and directly related to the research question or aim • methods for research and data collection are clear and scientifically appropriate.	Secondary evidence has limited validity if: • its ideas are only vaguely connected to the research topic • the methods of data collection are unclear.

Video
Science skills in a minute: Evaluating secondary evidence

Science skills resource
Science skills in practice: Evaluating secondary evidence

EVALUATING EVIDENCE TO CHOOSE AN ENERGY-EFFICIENT LIGHTING SOLUTION

AIM

To choose the best type of light bulb for lighting a bedroom based on critical evaluation of available research and evidence

PROCEDURE

Evidence has been collected from a variety of sources. Four pieces of evidence are presented in Table 3.7.2 in the results section.

RESULTS

▼ **TABLE 3.7.2** Different types of evidence about different light bulbs and their efficiency

<table>
<tr><th colspan="2">Evidence</th><th>Source</th><th>Information about the source</th></tr>
<tr><td>Quote</td><td>'When I switched all the bulbs in our house to halogen bulbs, we saved a lot on our electricity bill, and in terms of light in the house–no difference!'</td><td>Mark, your friend's dad</td><td><ul><li>Mark is an engineer.</li><li>Mark is a very nice man.</li><li>Mark is a close friend and always gives good advice.</li></ul></td></tr>
<tr><td>Quantitative data</td><td><table><tr><th>Light bulb</th><th>Power required for 500 lumens of light (W)</th></tr><tr><td>LED</td><td>5–8</td></tr><tr><td>CFL</td><td>7–9</td></tr><tr><td>Halogen</td><td>28</td></tr><tr><td>Incandescent</td><td>40</td></tr></table>© Commonwealth of Australia 2019/ energyrating.gov.au (CC BY 3.0 AU)</td><td rowspan="2">Australian Government website</td><td rowspan="2"><ul><li>The initiative is led by the Australian Government, state and territory governments, and the New Zealand Government.</li><li>It is managed by the Greenhouse Energy Minimum Standards Regulator and the Energy Efficiency Advisory Team made up of government officials. These officials undertake activities to improve the energy efficiency of appliances and equipment, including energy rating labelling, setting minimum energy performance standards, and education and training programs.</li></ul></td></tr>
<tr><td>Graph</td><td>© Commonwealth of Australia 2019/ energyrating.gov.au (CC BY 3.0 AU)</td></tr>
<tr><td>Quantitative data</td><td>Se_vector/Shutterstock.com</td><td>Aussie Hardware company website</td><td><ul><li>Aussie Hardware is a leading Australian retailer of home and lifestyle products.</li><li>The company also owns other commercial brands and companies.</li><li>Digital branding for the company includes online posts for advice, inspiration and product reviews.</li></ul></td></tr>
</table>

9780170491785

ANALYSIS

1 **Analyse** the pieces of evidence separately to **identify** what each piece can tell us about the best choice for an energy-efficient light bulb.

2 **Construct** a table such as Table 3.7.3 to **identify** strengths and limitations associated with each piece of evidence.

▼ TABLE 3.7.3

Evidence	Strengths	Limitations

3 Rank each piece of evidence in order of least to most reliable. **Justify** your choice using the information in Table 3.7.1.

4 Rank each piece of evidence in order of least to most valid. **Justify** your choice using information in Table 3.7.1.

CONCLUSION

Draw a conclusion about which type of light bulb you would choose to use in your bedroom to save energy. **Justify** your choice by referring to your responses in the analysis and evaluation section.

3.8 Heat transfer and conservation in clothing and bedding

ABORIGINAL & TORRES STRAIT ISLANDER SCIENCE CONTEXTS

IN THIS MODULE, YOU WILL:

✓ investigate the materials used by Aboriginal Peoples to manufacture clothing and bedding to suit climactic conditions.

© Joe Sambono

▲ **FIGURE 3.8.1** A contemporary possum skin cloak from south-western Victoria

© Joe Sambono

▲ **FIGURE 3.8.2** Possum fur fibres are hollow, which means they trap a layer of air, making the fur an excellent insulator.

Clothing worn by Aboriginal Peoples in cold climates

Many parts of Australia get cold, especially alpine areas, Tasmania and southern parts of mainland Australia. Clothing prevents heat loss from the body. Aboriginal Peoples have long made cloaks from animal furs to control heat loss. In cool climates, clothing was made from the furs of a variety of animals, including wallabies, kangaroos, possums, platypuses and quolls (Figure 3.8.1). The furs are traditionally worn with the fur next to the body, trapping a layer of air as thermal insulation. The Noongar Peoples (south Western Australia) have long manufactured kangaroo skin cloaks called buka that traditionally were worn with the fur facing inwards for warmth. Sometimes, furs were rubbed with animal fat to provide a further layer of insulation and waterproofing.

Possum skin cloaks are a culturally important item of clothing, often showing designs that depict connections to Country, kinships, family groups or significant stories. Possum fur is highly unusual because the fibres have a hollow structure. This traps air within the fibre to provide insulation (Figure 3.8.2). Although a possum pelt is smaller than that of many other native mammals such as kangaroos, possum skin provides superior insulation from cold.

Many Aboriginal and Torres Strait Islander Peoples of Australia express their cultural identity through continuing important cultural practices. For example, some Aboriginal Peoples continue to manufacture possum skin cloaks. The Gunditjmara and Yorta Yorta Peoples of Victoria are renowned for the quality of their contemporary cloaks.

Bedding made by Aboriginal Peoples

Mattresses and pillows were manufactured from a variety of natural materials. The Bundjulung Peoples (northern New South Wales) used dried grasses to fill mattresses.

9780170491785

The Ngarrindjeri Peoples (South Australia) used dried seaweed to make mattresses. The Dyirbal Peoples (northern Queensland) manufactured blankets from sheets of bark from the banana fig. The fibre from the seeds of native kapok trees (Figure 3.8.3) was widely used as stuffing for pillows. On the Cape York Peninsula, the pods of the large-leaved mangrove were used to provide a soft bedding material. The filling in mattresses, pillows and blankets provides comfort and insulates against weather conditions.

Image Professionals GmbH/Alamy Stock Photo

▲ **FIGURE 3.8.3** Fibre from the seeds of the native kapok tree is used for stuffing for pillows.

The efficiency of materials in preventing heat loss

☆ ACTIVITY

Materials and equipment

- 3 small bottles or jars
- hot water
- 3 thermometers
- insulating materials – natural materials such as feathers, wool, fabric, cottonwool, tissues, newspaper or other material you have chosen
- 3 plastic containers/tubs
- measuring cylinder
- jug

Procedure

1. Place a bottle or jar in a plastic container.
2. Fill the surrounding space with one type of insulation material.
3. Repeat steps 1 and 2 for the second insulation material.
4. Place the third bottle or jar in the third container but do not add insulating material.
5. Measure 100 mL of hot water and add it to the first bottle.
6. Immediately **record** the temperature of the water in the bottle and leave the thermometer in the bottle or jar.
7. Repeat steps 5 and 6 for the other two bottles/jars.
8. Observe the thermometers until one bottle/jar reaches room temperature.
9. When one bottle/jar reaches room temperature, observe and **record** the temperatures in all bottles/jars.

Analysis

1. What do your results show?
2. Which material provided the best thermal insulation?
3. Did you conduct a fair test? For example, how did you control the amount of material you used for insulation?
4. How does this model Aboriginal and Torres Strait Islander Peoples' knowledge and selection of materials for insulation?

SCIENCE IN CONTEXT

3.9 Energy-efficient houses

BY THE END OF THIS MODULE, YOU WILL BE ABLE TO:

✓ describe some of the ways houses are designed and built to improve energy efficiency.

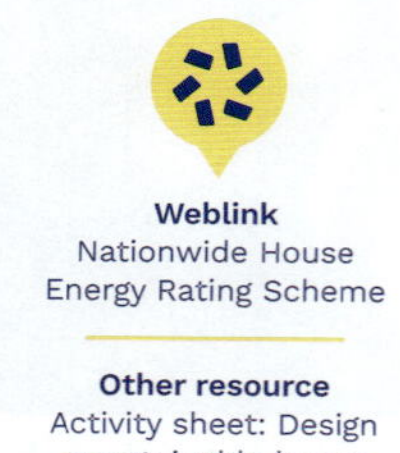

Weblink
Nationwide House Energy Rating Scheme

Other resource
Activity sheet: Design a sustainable house

In this chapter, you have learned about waste energy and energy efficiency. These topics are vital for designing and building houses that are energy efficient, comfortable and cheaper to heat, cool and maintain. In Australia, new houses and buildings are given energy ratings out of 7 stars.

Energy-efficient designs

Architects and building designers look at many factors when they design energy-efficient houses (Figure 3.9.1), including:

- the Sun's path and the climate at different times of the year
- orientation of the house and windows, including eaves
- inclusion of features that allow the control of air movement to reduce unwanted drafts or increase air flow/ventilation
- use of energy-efficient appliances and solar hot-water and solar power panels.

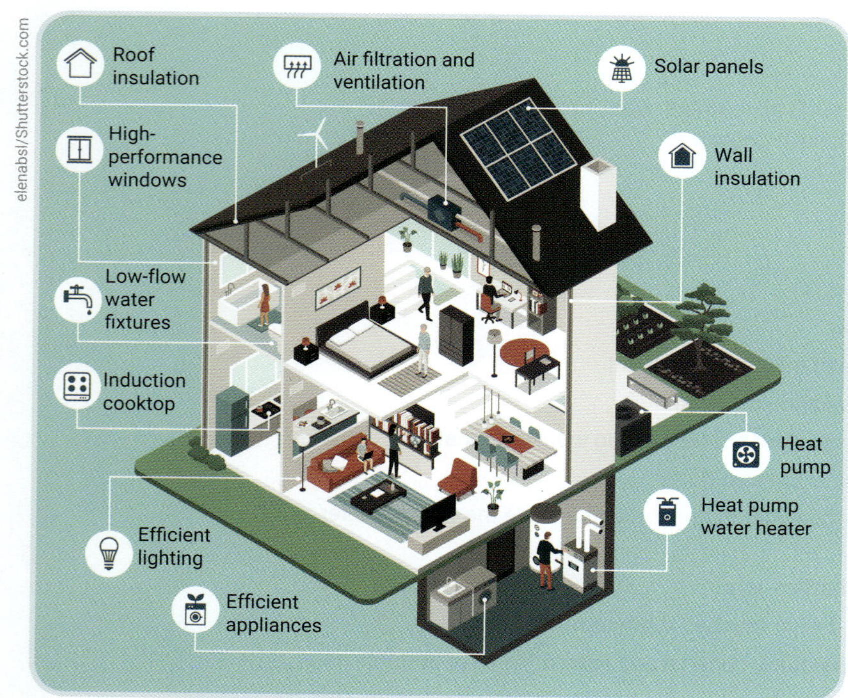

▲ FIGURE 3.9.1 Some features of an energy-efficient house

Materials and landscaping

The choice of materials and landscape design can also have important effects on a house's energy efficiency. Examples of energy-efficient choices include:

- using energy-efficient materials, such as double-glazed windows and heat-reflecting roof materials
- planting suitable trees and bushes.

3.9 LEARNING CHECK

1 **Define** the term energy-efficient house.
2 **List** considerations or materials that improve the energy efficiency of a house.
3 **Describe** how understanding the position of the Sun and the local climate over the year can affect the design of an energy-efficient house.
4 **Research** Australia's Nationwide House Energy Rating Scheme. What kind of features does a building need to achieve a 7-star rating?

9780170491785

3 REVIEW

REMEMBERING

1 **Recall** the law of conservation of energy.

2 **Recall** the formula for calculating energy efficiency.

3 **Describe** the process of cogeneration.

4 **Define**:

a waste energy output.

b useful energy output.

c Sankey diagram.

UNDERSTANDING

5 **Define** the following terms and give an example for each.

a Energy transfer

b Energy transformation

6 **Describe** the concept of energy efficiency and what it measures.

7 **Explain** how output energy is classified as useful or waste.

APPLYING

8 A ball thrown into the air starts with 450 J of kinetic energy. If it loses 15 J through frictional heat losses on its way up, determine how much gravitational potential energy it has at its maximum height before returning to the ground.

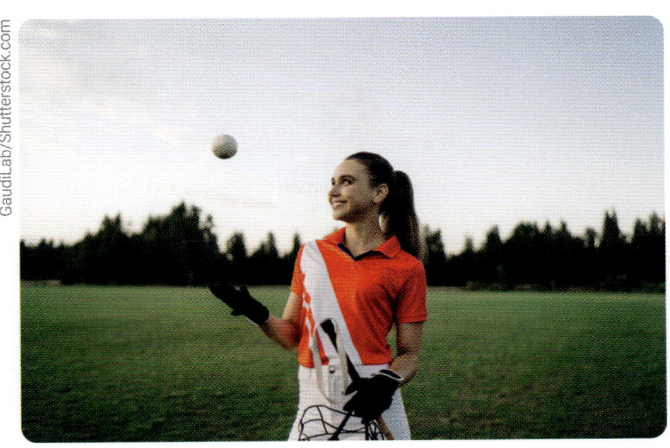

GaudiLab/Shutterstock.com

9 **Explain** how a sound barrier next to a freeway is relevant to wasted energy.

10 **Calculate** the efficiency of a process that has 200 J of input energy and 40 J of useful output energy.

11 Ceiling fans convert electrical energy into kinetic energy to move air. **Calculate** the kinetic energy produced per second from a ceiling fan that uses 400 J/s and is 80 per cent efficient.

Siraphol/Getty Images

12 **Calculate** how much waste energy is produced by a fridge that uses 700 000 J of electrical energy and is only 60 per cent efficient.

13 **Explain** why bicycle brakes get hot.

14 **Compare** the advantages and disadvantages of possum and kangaroo furs as materials used to make cloaks.

ANALYSING

15 If you walked to the shops, you might need about 200 kJ of chemical energy from your food. If you drove a car there, the chemical energy from the petrol needed would be much more. **Explain** the difference in terms of useful output energy and waste output energy.

16 The following diagram is a new Zoned Energy rating system used for air-conditioners. **Interpret** the main message from this diagram and **identify** one key limitation of the system. **Explain** why this limitation exists.

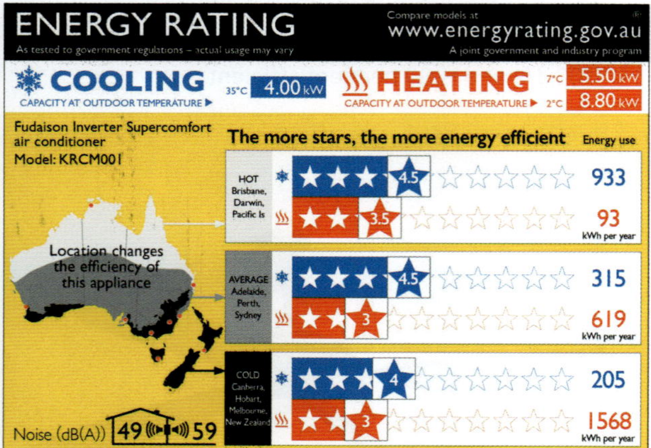

17 When a car is moving, it transforms chemical energy into kinetic energy. A lot of waste heat is produced. **Construct** a Sankey diagram to illustrate this. The Sankey diagram does not need to be to scale because not enough information has been given to do so.

EVALUATING

18 **Explain** why energy efficiency is important when choosing appliances.

19 Consider two power plants: one that has cogeneration design features and one that does not. **Construct** a table to evaluate the advantages and disadvantages of each design.

20 **Compare** the energy transformations when using a hairdryer and when hitting a ball with a bat. Use Sankey diagrams to **identify** similarities and differences in these two processes.

CREATING

21 **Construct** energy flow diagrams for two hot water systems: one that heats water via wind-generated electricity and one that heats water via solar-generated electricity.

22 **Construct** a Sankey diagram to scale for the following situation.

A slingshot transforms elastic potential energy into kinetic energy to send objects flying through the air. A slingshot holding a small pebble is stretched backwards and stores 40 J of elastic potential energy. The pebble is projected upwards when the slingshot is released, transferring 38 J of kinetic energy into the pebble. During its flight, the conversion of kinetic energy to gravitational potential energy has 90 per cent efficiency.

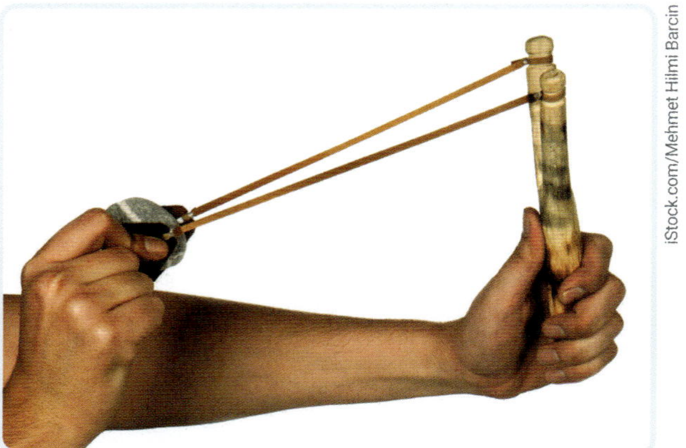
iStock.com/Mehmet Hilmi Barcin

Other resources
Worksheet:
Energy revision

9780170491785

SCIENCE IN DEPTH STUDY

1 Connect what you've learned

In this chapter, you've learned about the law of conservation of energy, energy efficiency, waste energy and many applications of these ideas. Create a mind map to show how the main ideas you have learned about are connected.

2 Check your thinking

Identify the waste energy transformations as a Formula 1 racing vehicle moves along the track. Ensure you are specific about:

- where on the vehicle the transformation happens
- what form the waste energy has
- where the energy goes in each case.

▲ An image of a car design produced using 3D software, with lines of light showing its aerodynamics (how the air passes over the car)

3 Get into action

Imagine you are head of a Formula 1 racing team and have to test the efficiency of a new design feature. You have access to:

- a racing track
- a powerful computer to run software that models how air flows over different car designs (see top image)
- sophisticated wind-tunnel technology, which also provides the means to test and visualise a car's air resistance (see bottom image).

Using this technology, the movement of air over the surface of the car can be analysed. Designers use this information to make cars more aerodynamic by reducing air resistance (friction with air). Make an action plan for the steps you will need to take to test your new feature in a cost-effective way.

▲ An image produced using engineering research software that models how air flows over a car body in a wind-tunnel

4 Communicate

You are planning to build a commercial wind-tunnel facility and need to promote it to prospective clients who build trucks, small planes and cars. Create a poster that would highlight the reasons why using wind-tunnel testing is an important step in the design of efficient vehicles.

Electrical energy

4.1 Sources of electricity (p. 118)
A range of different forms of energy can be transformed into electrical energy.

4.2 Electrical energy production (p. 122)
Different electrical energy sources have different strengths and weaknesses.

4.3 Electrical circuits (p. 126)
Electrical circuits have some requirements to operate, such as a conductive path, a power source and a load. Circuits can have a range of other components, such as a switch, light bulbs and a battery.

4.4 Series and parallel circuits (p. 128)
There are two ways to design an electrical circuit.

4.5 WORKING SCIENTIFICALLY: **Building electrical circuits** (p. 130)
Comparing series and parallel circuits

4.6 Drawing electrical circuit diagrams (p. 133)
Circuit diagrams are drawn to represent a range of electrical circuits using conventions and symbols.

4.7 Measuring quantities in electrical circuits (p. 137)
Quantities can be measured in electrical circuits in different ways.

4.8 WORKING SCIENTIFICALLY: **Measuring with apparatus** (p. 142)
Comparing brightness, power and efficiency in LED and filament bulbs, and comparing resistance in devices

4.9 Global energy use and future needs (p. 146)
Energy requirements and our use of energy sources have impacts on the planet and are gradually changing.

4.10 SCIENCE IN CONTEXT: **Solar energy in Australia** (p. 152)
The use of solar energy in Australia is growing.

9780170491785

SCIENCE IN DEPTH

▲ **FIGURE 4.0.1** A wind turbine transforms the kinetic energy of moving air into electrical energy.

Electrical energy has improved people's lives all around the world. It has made huge improvements possible in health, communication, personal comfort, transport, leisure and safety. Consider the impact of streetlights, refrigeration, air-conditioning, computers, power tools, water pumps and lifts: these are just a small fraction of the ways in which electrical energy has improved our lives.

The way electricity is generated has changed over time. For example, solar energy now offers an exciting alternative to power generated from fossil fuels. However, the use of solar power in Australia is both an opportunity and a challenge.

- **Think about the things you have done so far today. In what ways has electricity been important to each of them?**
- **Imagine your town or city without any electricity. How does electricity change the way your town or city operates?**
- **How might this widespread use of electricity by individuals, businesses and governments be bad?**
- **How can renewable energy sources, such as solar and wind energy, change the way we live?**

DIVE INTO SCIENCE!

At the end of this chapter, you can complete Science in Depth Study #4. You can use the information you learn in this chapter to complete the project.

Assessments

- Prior knowledge quiz
- Chapter review questions
- End-of-chapter test
- Depth study: Science investigation

Videos

- Science skills in a minute: Risk assessments **(4.5)**
- Video activities: How batteries work **(4.1)**; How we measure electricity **(4.7)**; Cost of building Australia's future energy needs **(4.9)**

Science skills resources

- Science skills in practice: Risk assessments **(4.5)**
- Extra science investigations: Investigating electrical circuits **(4.7)**

Interactive and other resources

- Quiz: Series versus parallel circuits **(4.4)**
- Worksheet: Drawing electrical circuits **(4.6)**

To access resources above, visit **cengage.com.au/nelsonmindtap**

4.1 Sources of electricity

BY THE END OF THIS MODULE, YOU WILL BE ABLE TO:

- ✓ describe the various types of energy that can be transformed into electrical energy
- ✓ explain the three processes involved in the transformations that produce the electricity we use.

Video activity
How batteries work

GET THINKING

We will often switch on or plug in an electrical device, such as a phone, without thinking about where the electricity comes from that makes the device work. What if electricity was available only some of the time? How would this affect the way you use electrical devices and appliances?

In Chapter 3, you learned how to use flow charts to represent energy transfers and transformations. But where does the electrical energy we use in our homes, businesses, industries and schools come from? Electrical energy can be produced via energy transformations from any of three types of devices: batteries, (photovoltaic) solar cells and generators.

Batteries

batteries
devices that transform chemical energy into electrical energy

Batteries are devices that transform chemical potential energy (or chemical energy) into electrical energy. They do this through chemical reactions that occur within the battery.

Batteries come in a range of shapes and sizes depending on the amount of energy that needs to be transformed and the type of transformation required (Figure 4.1.1).

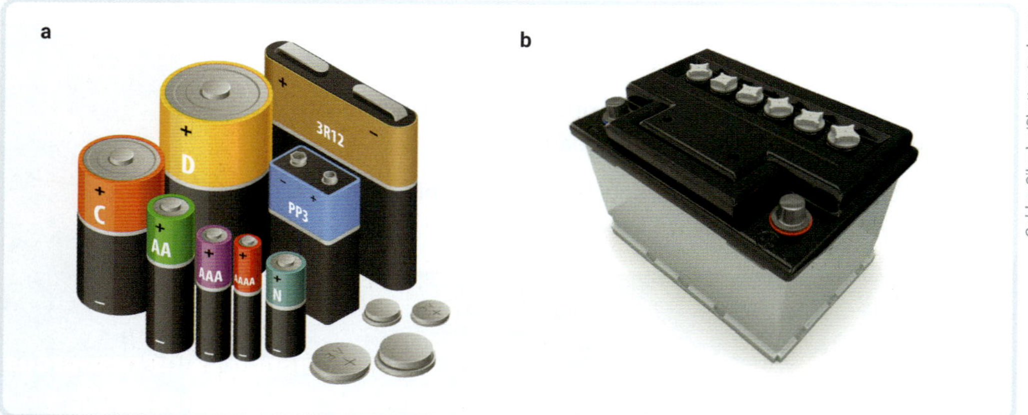

Golden Sikorka/Shutterstock.com, Maxx-Studio/Shutterstock.com

▲ **FIGURE 4.1.1** (a) Household batteries come in different sizes and shapes; (b) a car battery

9780170491785

Chemical reactions inside the battery result in a flow of electricity. A battery is designed to direct this flow of electricity through a component or device where an energy transformation can occur (Figures 4.1.2 and 4.1.3).

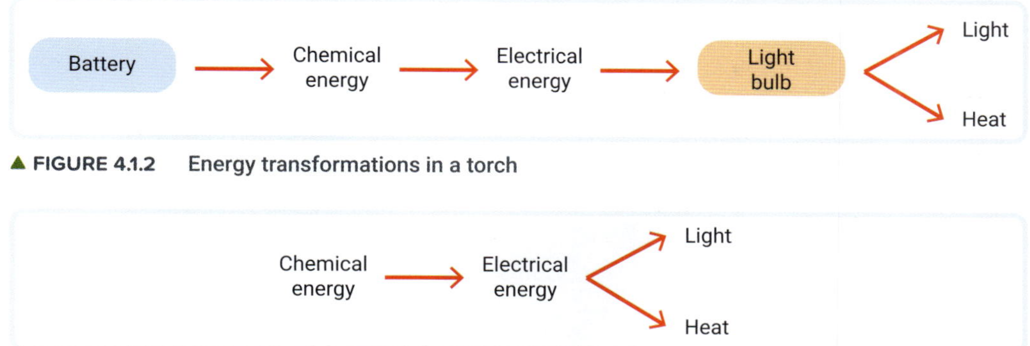

▲ **FIGURE 4.1.2** Energy transformations in a torch

Chemical energy → Electrical energy → Light / Heat

▲ **FIGURE 4.1.3** A flow chart for chemical energy being transformed into light and heat

Solar cells

Solar cells, or photovoltaic cells, are devices that transform light energy into electrical energy.

solar cell
a device that transforms light energy into electrical energy

Solar cells can vary in size from those that power small devices, such as calculators and garden path lamps, to large-scale industrial solar installations that can provide electricity for a whole town (Figure 4.1.4).

SB2010 studio/Shutterstock.com, Jenson/Shutterstock.com

▲ **FIGURE 4.1.4** (a) A hand-held calculator powered by a solar cell; (b) a large-scale solar array

silicon
a common element, used in many electronic devices; the main component of solar cells

The energy transformation that occurs in a solar cell is caused by the interaction between two thin layers of slightly different **silicon** that occurs when light strikes them. This results in a flow of electricity, which can be used to power devices (Figure 4.1.5).

Radzas2008/Shutterstock.com

▲ **FIGURE 4.1.5** A solar panel transforms light energy into electrical energy.

Generators

generator
a device that transforms kinetic energy into electrical energy

Generators are devices that transform the kinetic energy of a spinning object (usually a turbine) into electrical energy.

Generators can be used in anything from small devices, such as hand-held torches, up to large-scale industrial generator installations, which can provide electricity to entire cities (Figure 4.1.6). Large-scale generators are often called power plants.

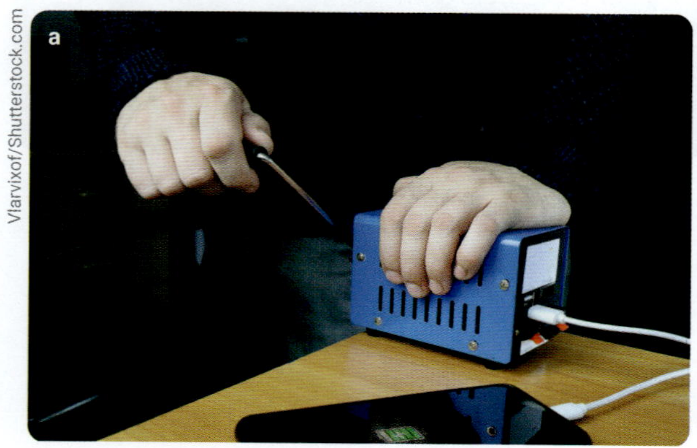

Vlarvixof/Shutterstock.com

Daria Nipot/Shutterstock.com

▲ **FIGURE 4.1.6** **(a)** A small, hand-powered generator; **(b)** a large-scale electrical generator, also known as a power plant

9780170491785

In a generator, energy transformation occurs when conducting wires (usually made of copper) move within a magnetic field (Figure 4.1.7). This movement can be caused by water, air or steam passing through a turbine, or even someone pedalling a bicycle. People have used this principle of electricity generation for the past 150 years. Module 4.2 describes large-scale electricity generation in more depth.

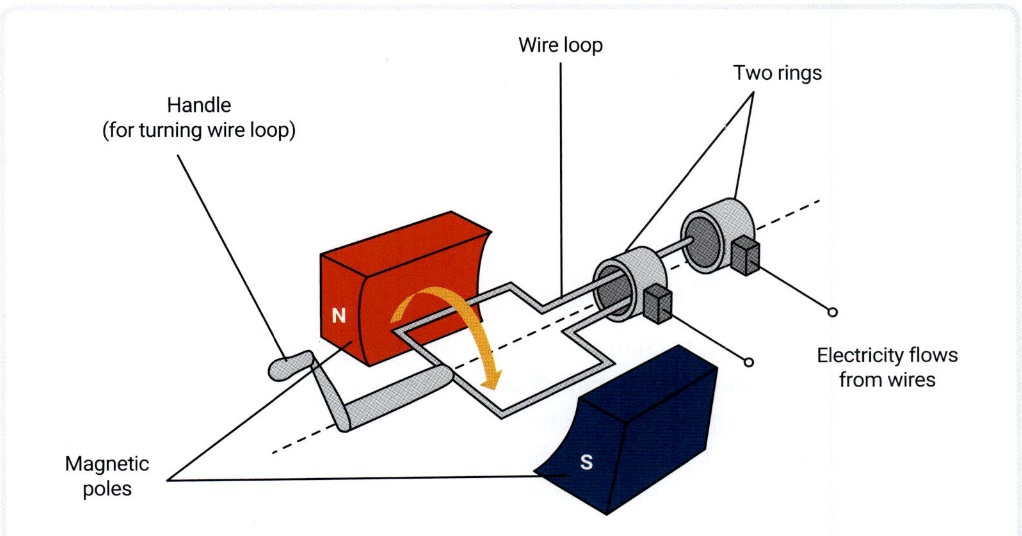

▲ **FIGURE 4.1.7** The components of an electrical generator

4.1 LEARNING CHECK

1. **List** all the devices you can think of in your home that use batteries as a source of electrical energy.
2. **Draw** an energy flow chart to represent the energy transformation when a battery is used to power a torch.
3. Garden lights are commonly powered by a battery and a small solar panel. **Explain** the energy transformations involved.
4. Most towns and cities are powered by an electric generator (or generators) that are located far from populated areas. **Justify** why.

4.2 Electrical energy production

BY THE END OF THIS MODULE, YOU WILL BE ABLE TO:

- ✓ explain how energy can be transformed through energy systems
- ✓ compare and contrast different sources of energy used to produce electricity.

GET THINKING

We rely on energy transformations in all kinds of situations. For example, electricity production relies on energy transformations. Think about an electric scooter. It can only move if the chemical energy stored in its battery is transformed into electrical energy, which is then transformed into kinetic energy. As you complete this module, imagine what it would be like if energy transformations were not possible in your daily life.

Transforming energy

energy transformation
the changing of one type of energy into another type of energy

Transforming energy means to change or convert one type of energy into another type of energy. We rely on **energy transformations** in all aspects of our life. One important example of energy transformation you learned about in Stage 4 is when green plants transform light energy into chemical energy (a type of potential energy) by photosynthesis. Our bodies use the chemical energy stored in the food we eat to keep us alive by transforming it into other types of energy.

Many devices and appliances transfer and transform energy. In many instances, there will be both energy transfer and energy transformations occurring, especially in complex energy systems. Table 4.2.1 provides some examples of energy transformations.

▼ **TABLE 4.2.1** Simple examples of energy transformations

Example	Energy transformation(s) involved
Burning a log 	• Chemical energy stored in the wood transforms into heat and light energy.
Turning on a torch 	• Chemical energy stored in the battery transforms into electrical energy and then into light energy in the bulb. • When left on, the bulb will get hot, showing that some electrical energy has transformed into heat. • When an electric current flows, some heat is generated by the friction of the charges moving through the wires.
Knocking a glass off a table 	• Gravitational potential energy is transformed into kinetic energy as the glass falls. • Some kinetic energy is transformed into sound energy when the glass hits the floor and shatters.

Transforming energy into electrical energy

Electrical energy is one of the most common types of energy we depend on every day. We use electricity in our homes, at work or at school, in industry and in many other places. Today's society relies heavily on the large scale transformation of energy into electrical energy.

Non-renewable energy sources

Non-renewable energy sources include fossil fuels such as coal, oil and natural gas. Some countries also use nuclear power. Coal has been an important energy source for electricity generation in Australia for many years. Using fossil fuels to produce electrical energy in power stations relies on transforming their chemical potential energy into heat by burning the fuel. The heat is then used to change the state of liquid water into high-pressure steam. The steam turns a turbine, which drives a generator, so that the kinetic energy of the turbine is transformed into electrical energy (Figure 4.2.1).

non-renewable energy source
a source of energy that is finite in nature; that is, used at a faster rate than it can be produced

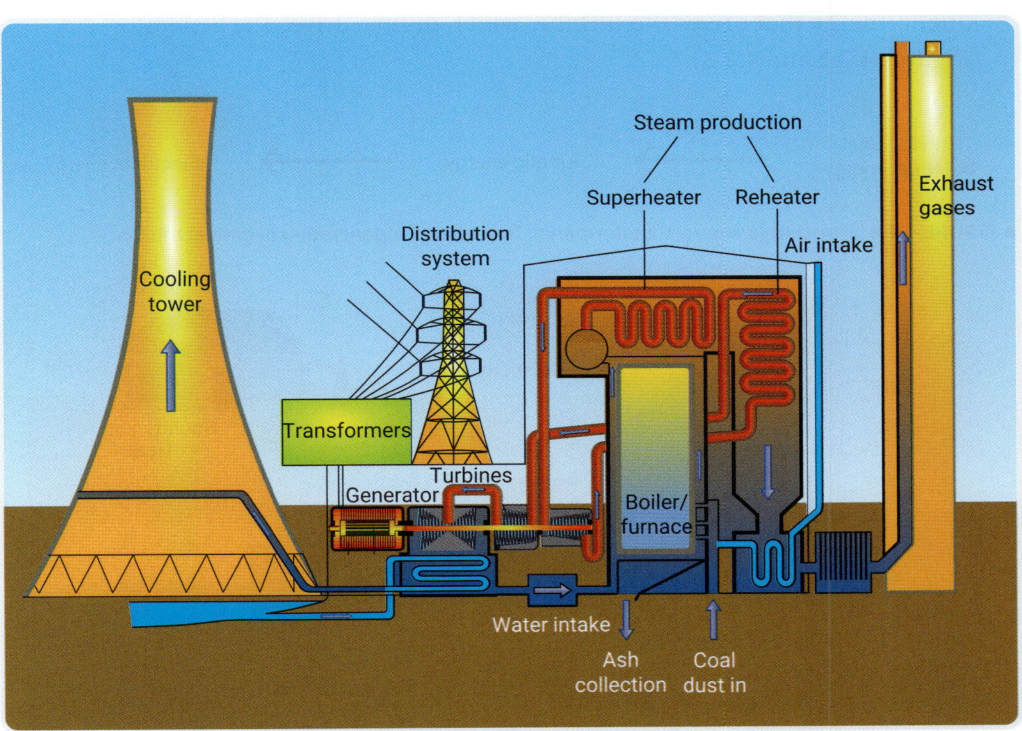

▲ FIGURE 4.2.1 A coal-fired power station

There is now a significant push from the community to shift away from coal and other fossil fuels, and to instead use energy sources that produce fewer **greenhouse gases** and reduce the impact on climate change.

greenhouse gases
gases in the atmosphere that can trap heat and affect global surface temperatures and other aspects of the climate

Removing fossil fuels from underground also leads to environmental damage and risks, from the huge pits dug to access coal to coal dust, oil spills and gas explosions.

Nuclear power also has significant environmental impacts caused by mining and the need to dispose of nuclear waste material. There is also a small but high-impact risk of pollution from the malfunction of a nuclear power plant (as occurred at Fukushima, Japan, in the aftermath of a tsunami in 2011).

Renewable energy sources

renewable energy source
a source of energy that can be produced at a faster rate than it can be used

biofuels
fuels that are made from biological sources that can replace fossil fuels

hydroelectricity
electrical energy produced by transforming the gravitational energy of falling water into kinetic energy to drive a turbine

Renewable energy sources are sources of energy that can be produced at a faster rate than they are used; that is, they can be 'renewed'.

- **Biofuels**, like fossil fuels, are burned to release heat. This heat is used to turn turbines, which drive generators. The advantage of biofuels is that the fuel is made from recycled biological material that would otherwise be waste material. The energy captured in these materials is used as a fuel rather than ending up being incinerated or breaking down in the environment.
- In the production of **hydroelectricity**, gravitational potential energy is used. Water drops from a height, transforming the gravitational potential energy into kinetic energy that drives the turbine, which in turn drives the generator (Figures 4.2.2 and 4.2.3).

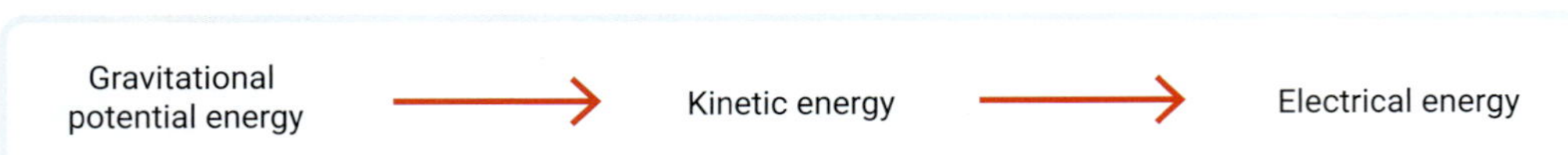

▲ FIGURE 4.2.2 A simple energy transformation flow chart for generating hydroelectricity

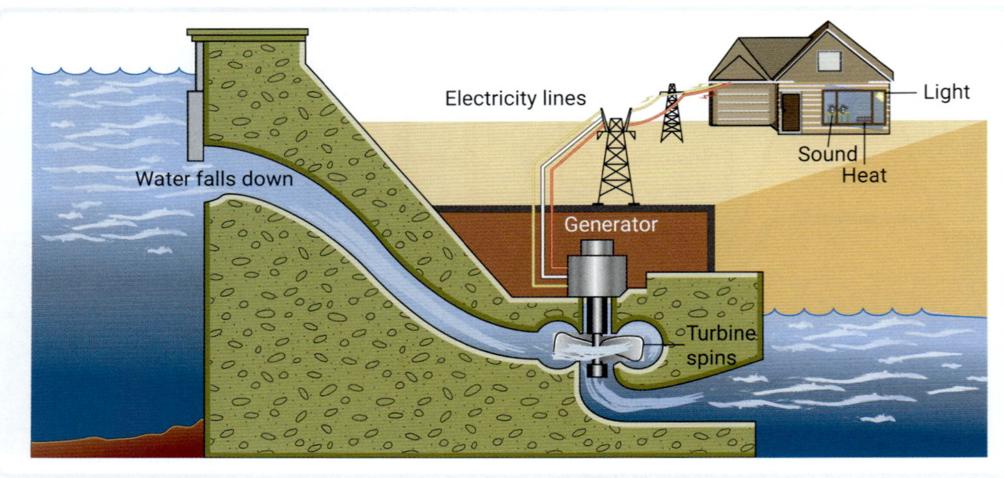

▲ FIGURE 4.2.3 Transforming gravitational potential energy into electrical energy using a hydroelectric power station

tidal energy
electricity generated from the ebb and flow of the tides

- **Tidal energy** can be used to generate electricity. The kinetic energy of moving water is used to drive a turbine and produce electricity. Some systems use the difference in the heights of the tides to act like the dams in hydroelectric power stations. The water is forced through pipes to rotate turbines.

- **Geothermal energy** is the heat that is trapped in rocks close to heat sources deep within Earth's crust. Geothermal energy is used in New Zealand as a source of electricity. Water, usually pumped into these regions via pipes, is heated and comes back to the surface as steam, under pressure, in separate pipes. The significant kinetic energy of the steam drives a turbine to transform the kinetic energy into electrical energy.
- **Wind power** generators transform the kinetic energy of the wind to drive a turbine to produce electricity.
- **Solar power** relies on converting light energy from the Sun directly into electrical energy in solar cells.

geothermal energy heat that is trapped in rocks close to heat sources deep within Earth's crust

wind power electricity generated by harnessing the kinetic energy of wind to drive a turbine

solar power electricity generated directly from sunlight, using solar cells

Wind and solar power often use large battery systems to store power. When power is available, electrical energy from these sources charges the batteries, converting electrical energy into chemical potential energy. This energy can be converted back into electrical energy when demand is high or supply of energy from wind or sunlight is low.

4.2 LEARNING CHECK

1. **Define** a non-renewable source of electrical energy.
2. **Outline** why environmental organisations oppose the use of non-renewable sources of electrical energy.
3. **Explain** how energy can be transformed from one form into another.
4. Generators rely on a substance of some type to cause the turbines to turn. **Compare** the substances that turn turbines in each energy source (ignore solar power).
5. **Justify** the choice made by governments across Australia to increase the use of renewable sources of electrical energy.

4.3 Electrical circuits

BY THE END OF THIS MODULE, YOU WILL BE ABLE TO:

✓ identify the requirements of a simple electrical circuit
✓ draw symbols to represent common components of electrical circuits in circuit drawings.

GET THINKING

A simple circuit, like the one in a torch, must have certain components to work. What are those components? Even simple circuits can have a number of wires and components. This makes it is hard to see what is going on and almost impossible to describe it to someone. How could describing a circuit be simplified?

electrical circuit
a complete loop that can conduct a continuous flow of electricity so that electrical potential energy can be transferred

complete loop
an unbroken, continuous path for electricity

electrically conductive material
a substance that enables electricity to flow through it

power pack
an adjustable power supply used in the science laboratory that plugs into a power point

load
any component in a circuit that transforms electrical energy

component
any part of an electrical circuit

switch
a component that can complete or disconnect the conducting path of electricity in an electrical circuit

resistor
a component that regulates the flow of electricity in an electrical circuit

ammeter
an instrument used to measure current (the rate of electricity flow in a circuit)

voltmeter
an instrument used to measure voltage (the energy transformed when electricity passes through a component in a circuit)

Electrical energy travels through an **electrical circuit** to reach a device that then transforms the electrical energy into other forms of energy. An electrical circuit needs certain features to function.

Requirements of an electrical circuit

There are many different types of electrical circuits, but they all have some important features in common. To be able to function, an electrical circuit must have:

- a conductive path – a **complete loop** made of an **electrically conductive material**
- a power source – a battery or **power pack** that provides electrical power
- a **load** – an electrical **component** that can transform electrical energy into other forms of energy. A circuit has to have a load or too much current will flow. Examples of loads include a light globe (electrical energy to light energy) and the element (heating device) in a kettle (electrical energy to heat).

An example of a simple circuit is shown in Figure 4.3.1.

Components in an electrical circuit

Electrical circuits can be simple or complicated, with a wide range or number of components. Common components include a:

- **switch**
- light bulb
- battery
- **resistor**
- **ammeter**
- **voltmeter**.

▲ **FIGURE 4.3.1** Some common components in a simple circuit

For components to work, they must be connected in one or more conductive loops in the electrical circuit.

9780170491785

An introduction to circuit diagrams

The easiest way to represent an electrical circuit is in a **circuit diagram**. This is a simplified diagram that represents the different components of an electrical circuit and how they are connected. A circuit diagram is a much easier way to communicate the important details about an electrical circuit than a photo or a description.

circuit diagram
a simple diagram of an electrical circuit that uses standard symbols

A circuit diagram is not drawn to scale. This means that it may not show you the true size of the components relative to each other or their actual position in the circuit. A circuit diagram commonly uses symbols to represent components. The symbols for some common components in an electrical circuit are shown in Figure 4.3.2. These symbols are universal, which means anyone can understand and use them to build electrical circuits. Module 4.6 explains in detail how to draw circuit diagrams.

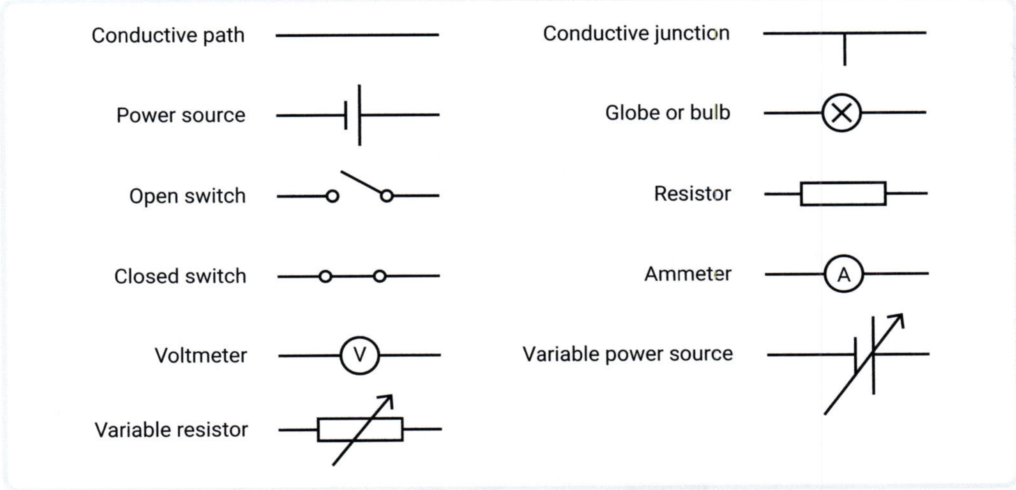

▲ **FIGURE 4.3.2** Some of the symbols used in circuit diagrams to represent common components of electrical circuits

An example of a simple circuit diagram is shown in Figure 4.3.3. This circuit contains:

- a closed switch
- two light globes
- one battery.

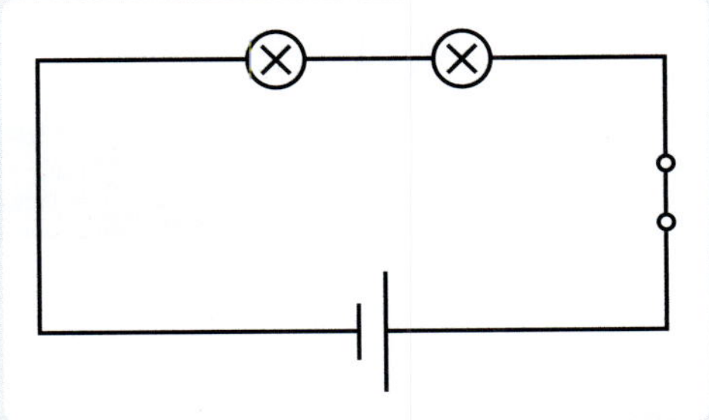

▲ **FIGURE 4.3.3** A simple circuit diagram

4.3 LEARNING CHECK

1 **List** the requirements of a simple electrical circuit.
2 **Identify** three components that could be included in an electrical circuit.
3 Why doesn't a circuit work when the switch is open?
4 **Explain** why symbols are used for components of electrical circuits.
5 **Predict** the meaning of an arrow on some circuit symbols.

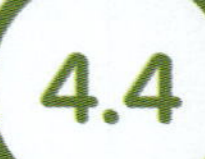

4.4 Series and parallel circuits

BY THE END OF THIS MODULE, YOU WILL BE ABLE TO:

- ✓ describe the defining features of series and parallel circuits
- ✓ explain the differences in how series and parallel circuits function.

Quiz
Series verses parallel circuits

GET THINKING

You have probably noticed that when someone turns off the kitchen light, the other lights in the house stay on; or that when someone in the next room turns off the television, the microwave can keep running. How does the wiring of a house make this possible?

Electrical charges can flow through conductive paths in a series circuit or a parallel circuit. There are also more complicated circuits that combine these two types.

Series circuits

series circuit
an electrical circuit where wires and components are connected end-to-end

In a **series circuit**, all components are connected end-to-end, forming a single path for electricity to flow through. Since series circuits have only one path, all of the electricity must flow through each component of the series circuit, one after another.

If there is a broken or disconnected component in a series circuit, then no current can move through the circuit at all. Figure 4.4.1 shows an example of simple series circuit.

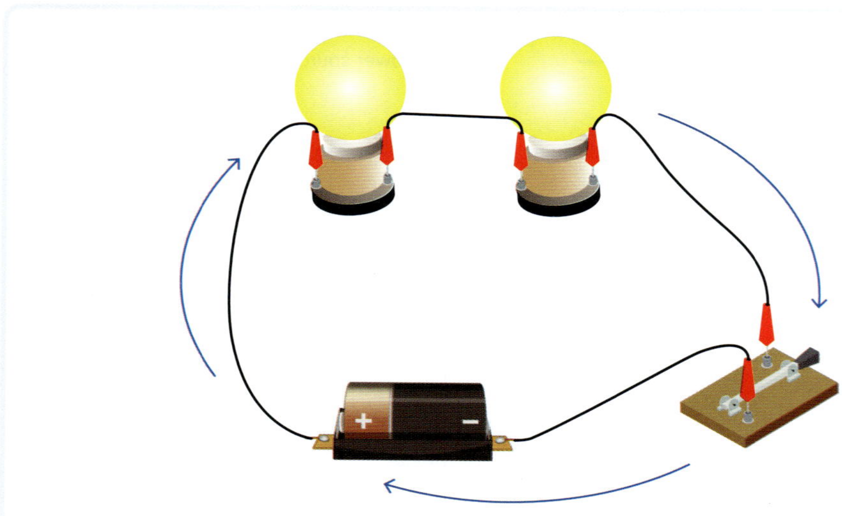

▲ **FIGURE 4.4.1** A simple series circuit featuring two light bulbs, a battery and a switch. The blue arrows show the direction of the electrical current, flowing from the positive terminal (+) of the battery towards the negative terminal (–).

9780170491785

Parallel circuits

In a **parallel circuit**, the electrical current has two or more paths to move through.

parallel circuit
an electrical circuit where the current can travel along two or more paths

If there is a broken or disconnected component in a parallel circuit, then no current can move through that part of the circuit. However, the electrical current can still flow in the other parts of the circuit that are not affected by the disconnected component. Figure 4.4.2 shows an example of a parallel circuit.

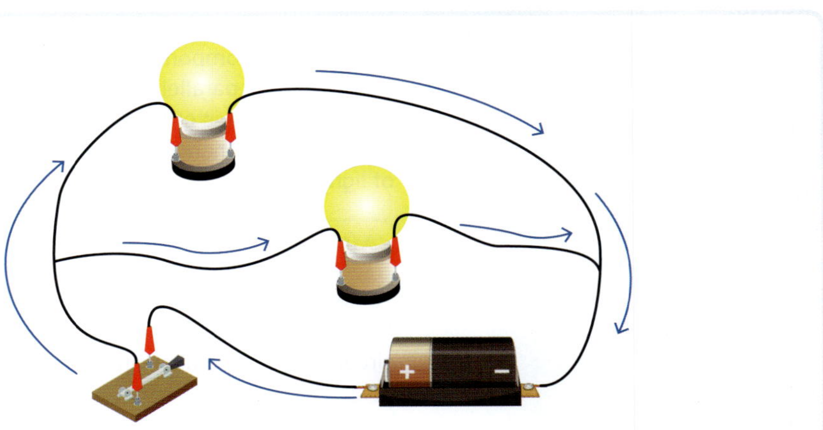

▲ **FIGURE 4.4.2** A simple parallel circuit with two light bulbs, a switch and a battery. The electrical current flows from the positive terminal of the battery along both paths towards the negative terminal of the battery.

4.4 LEARNING CHECK

1 **Outline** the key difference between a series circuit and a parallel circuit.
2 **Compare** what happens if one bulb is removed from each of the circuits illustrated in Figures 4.4.1 and 4.4.2.
3 **Describe** a change that would need to be made to the circuit shown in Figure 4.4.2 so that each of the bulbs could be turned off individually.
4 Long chains of fairy lights can vary significantly in price. Often, when used over a long period, a few bulbs may stop working. In cheaper versions, sometimes all the bulbs may stop lighting up at the same time. **Explain** why this happens, using series and parallel circuits in your answer.
5 **Explain** why it is better to electrically connect a house using parallel circuits rather than series circuits.

WORKING SCIENTIFICALLY

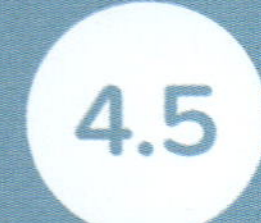

4.5 Building electrical circuits

SCIENCE SKILLS IN FOCUS

IN THIS MODULE, YOU WILL FOCUS ON LEARNING AND IMPROVING THIS SKILL:

- using diagrams to construct different types of electrical circuits.

Here are some important rules to follow when constructing electrical circuits.

1 Look closely at the circuit diagrams *before* beginning the investigation

2 Carry out a risk assessment *before* beginning the investigation by asking yourself these questions:

   a Is there anything you will be doing or using in the investigation that could cause harm? For example, will you work with power sources, hot objects or glass objects?

   b What are the hazards? For example, is there risk of electrocution, burns, cuts?

   c What could you do to reduce the risk of harm? For example, you might aim to not touch hot objects, and to ensure your hands are dry.

3 Gather all components you need before building the circuit.

4 Build your circuit by following the circuit diagram.

5 Check the circuit you have built against the circuit diagram before beginning the investigation.

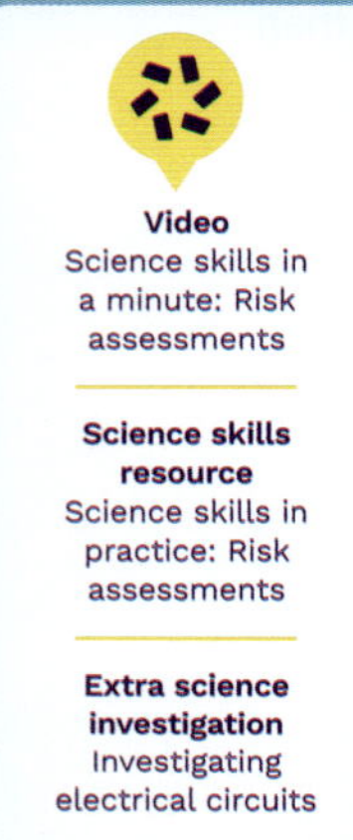

COMPARING SERIES AND PARALLEL CIRCUITS

BACKGROUND INFORMATION

The way that components are added to a circuit (in series or in parallel) will determine what happens when the circuit is changed. We can qualitatively observe these changes by comparing the brightness of light bulbs.

AIM

To determine how the brightness of lights bulbs changes when components are added to a series circuit and to a parallel circuit

MATERIALS AND EQUIPMENT

- ☑ power source (battery or power pack)
- ☑ switch
- ☑ 2 light bulbs
- ☑ 12 connecting wires

 **Safety**

When constructing, adjusting or packing away an electrical circuit, always turn the power off at the main switch and unplug the power source. Only once you are ready to collect data should the power source be plugged in and the main switch turned on.

Only use the circuit switch to connect the circuit for as long as you need to record the data. Do not handle circuit components after use; they can get hot and may need time to cool down.

Ensure your hands are dry when touching the circuit.

PROCEDURE

1 Write a risk assessment for this investigation, using the questions in the blue 'Science skills in focus' box on this page to help you.

2 Write a hypothesis for the experiment.

9780170491785

3 Construct an electrical circuit with one bulb and one switch, as shown in Figure 4.5.1.

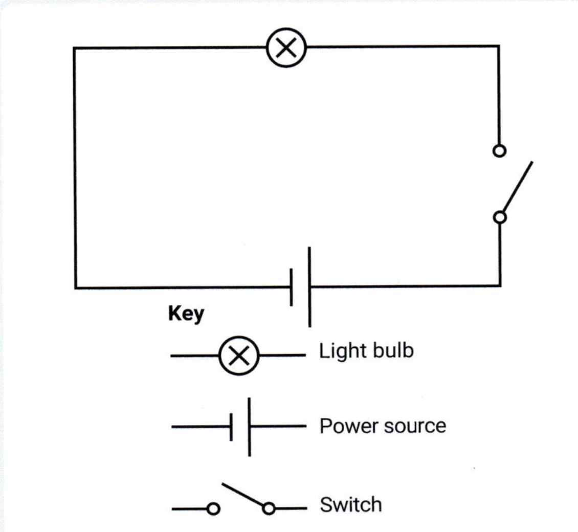

▲ **FIGURE 4.5.1** Circuit with one bulb

4 Close the switch. Observe the brightness of the bulb. Take a photo and record your observation for one bulb in a table like Table 4.5.1.

5 Construct an electrical circuit with two bulbs in a series circuit, as shown in Figure 4.5.2.

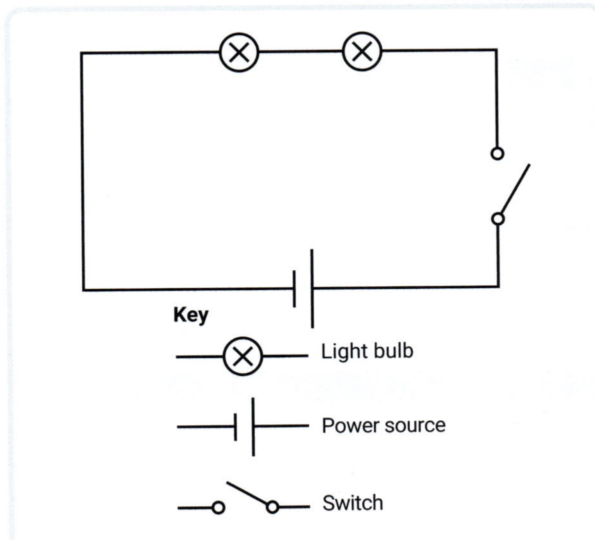

▲ **FIGURE 4.5.2** Circuit with two bulbs in series

6 Close the switch. Observe the brightness of both bulbs. Take a photo and record in your table your observation of whether each bulb is brighter, dimmer or the same as in the one-bulb circuit.

7 Construct an electrical circuit with two bulbs in parallel, as shown in Figure 4.5.3.

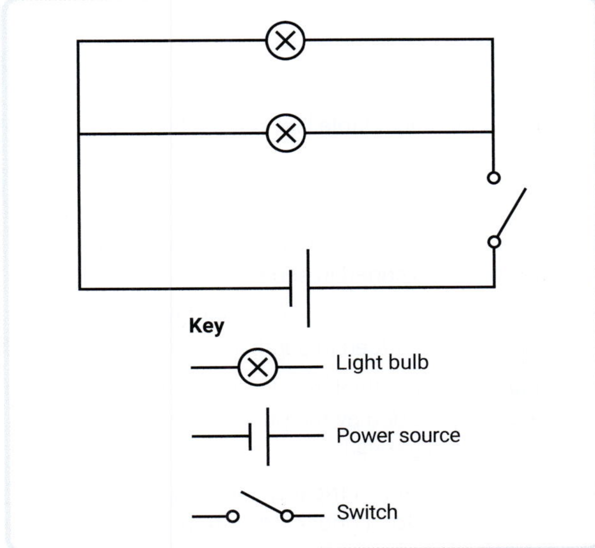

▲ **FIGURE 4.5.3** Circuit with two bulbs in parallel

8 Close the switch. Observe the brightness of both bulbs. Take a photo and record in your table your observation of whether each bulb is brighter, dimmer or the same as in the one-bulb circuit.

▼ TABLE 4.5.1 Observations of bulb brightness in different circuits

Circuit design	Bulb 1 – Observation	Bulb 2 – Observation	Brightness comparison
One bulb			
Two bulbs in series			
Two bulbs in parallel			

RESULTS

Record your data in a table like Table 4.5.1.

ANALYSIS

1 What do you notice about the brightness of bulb 1 when bulb 2 is added in series?
2 What do you notice about the brightness of bulb 1 when bulb 2 is added in parallel?
3 **Explain** what you think has happened to the flow of electricity provided by the power source when bulb 2 was added in parallel.
4 What method of electrical circuit **design** do you think would be best for a house? **Justify** your answer.

CONCLUSION

Write a conclusion for this experiment. Remember to state if your results support your hypothesis.

FURTHER ANALYSIS

You can use a simulation (computer model) such as PhET to further investigate and understand processes that occur in a range of electrical circuits (Figure 4.5.4). Try repeating the procedure for this investigation using the simulation provided in the link.

You can 'build' circuits and make a wide range of changes quickly and easily through the simulation. You can make observations of the moving charges and can take measurements, which you will learn more about in Module 4.8.

Weblink
PhET circuit construction kit

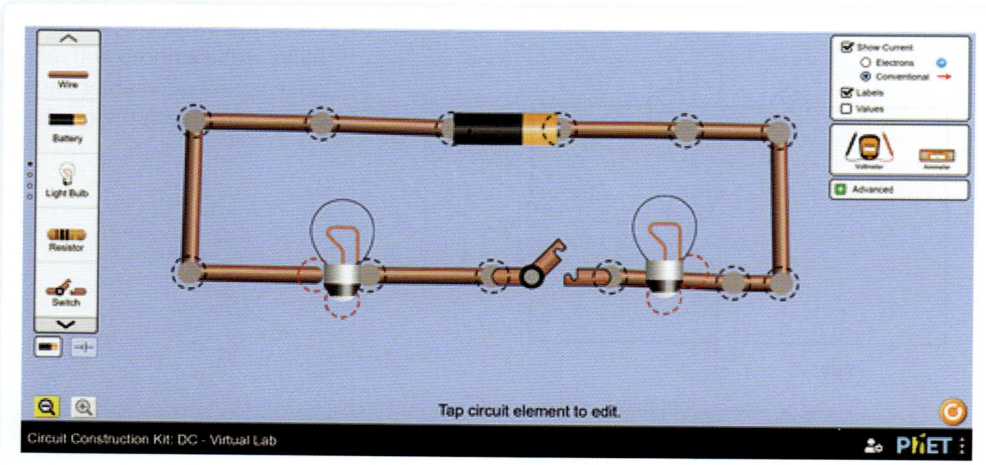

▲ FIGURE 4.5.4 A screenshot from the PhET virtual circuit construction kit

9780170491785

4.6 Drawing electrical circuit diagrams

BY THE END OF THIS MODULE, YOU WILL BE ABLE TO:

- ✓ use and construct diagrams to represent data
- ✓ use diagrams to communicate information.

GET THINKING

We can communicate scientific concepts in many different ways. For electrical circuits, we usually use circuit diagrams instead of descriptions. Can you think of other examples of when it easier to communicate scientific or technical information using diagrams instead of words?

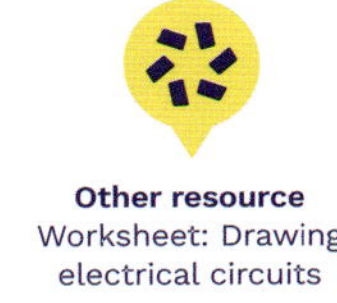

Other resource
Worksheet: Drawing electrical circuits

Rules of circuit diagrams

There are some simple rules you should follow when drawing circuit diagrams. Creating a clear diagram will make it easier for other people to use it.

1. Use a sharp pencil and ruler to draw the diagram.
2. Represent the conductive path (wire) with a straight line. Represent direction changes in the wire with right angles. Circuit diagrams should, therefore, be shown as a rectangle.
3. Don't draw wires crossing.
4. Use the right symbols to represent components in the circuit (Figure 4.6.1).

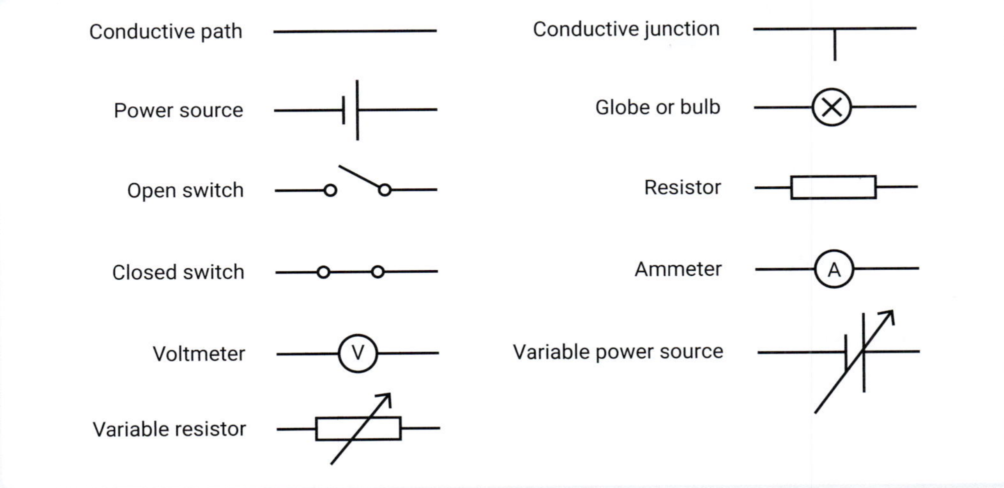

▲ **FIGURE 4.6.1** Common symbols used in circuit diagrams

5. If there is just one component on one side of the rectangle, place it in the centre of the side. If there is more than one component on one side of the rectangle, draw the symbols for the components evenly distributed.
6. Ensure the symbols are drawn in the same order as the components are connected in the circuit.
7. The circuit should form a closed loop or loops. This means you should be able to trace a path around the circuit without any breaks (unless an open switch is represented).

Examples of circuit diagrams

In Module 4.4, two simple circuits were shown as drawings. Figures 4.6.2 and 4.6.3 show how these circuits can also be represented as circuit diagrams.

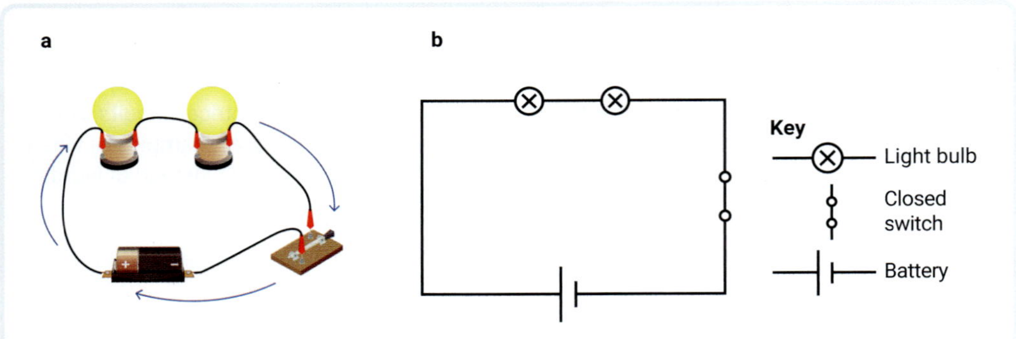

▲ **FIGURE 4.6.2** A simple series circuit, containing a power source, two light bulbs and a switch, shown as **(a)** a drawing and **(b)** a circuit diagram.

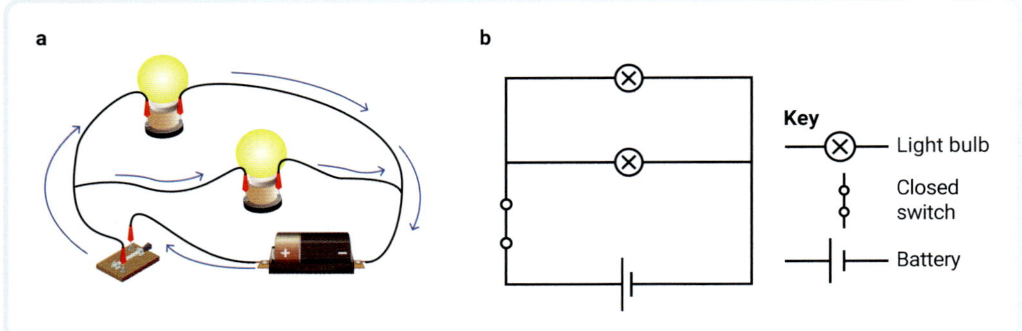

▲ **FIGURE 4.6.3** A simple parallel circuit, containing two light bulbs, a switch and a battery, shown as **(a)** a drawing and **(b)** a circuit diagram

Some further examples of circuit diagrams show circuits with:

- a power source, switch and two resistors (each represented by a rectangle) in series (Figure 4.6.4)
- a power source and two resistors in parallel, where each resistor has a switch on its branch (Figure 4.6.5).

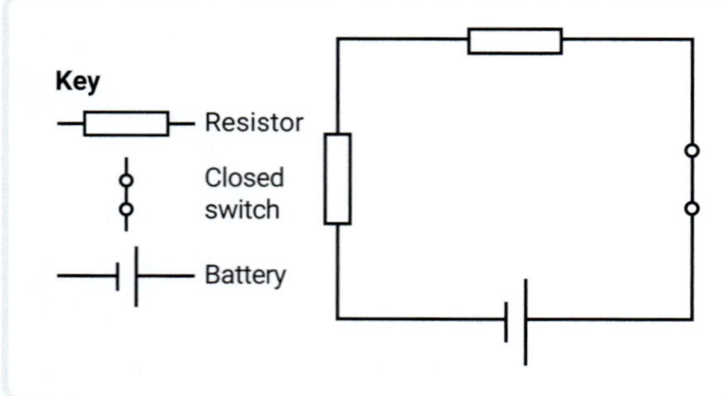

▲ **FIGURE 4.6.4** A circuit diagram of a simple electrical circuit with a power source, two resistors and a switch, connected in series

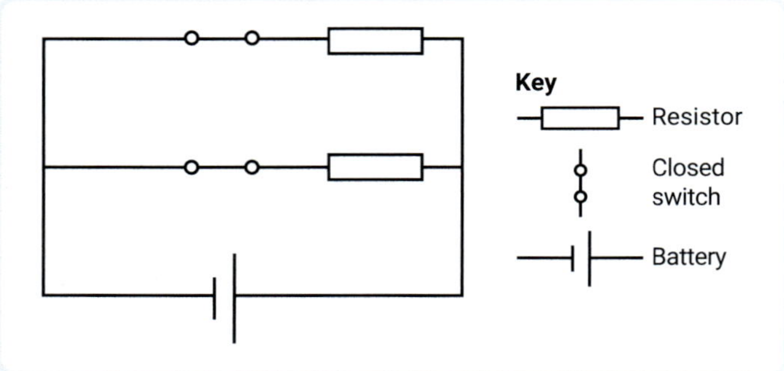

▲ **FIGURE 4.6.5** A circuit diagram of a simple electrical circuit with a power source and two resistors with a switch each, connected in parallel

4.6 LEARNING CHECK

1 **Draw** the following circuits:
  a a series circuit with three resistors and a switch.
  b a parallel circuit with three resistors and one switch that can disconnect them all.

2 **Match** the provided diagram with the appropriate description.

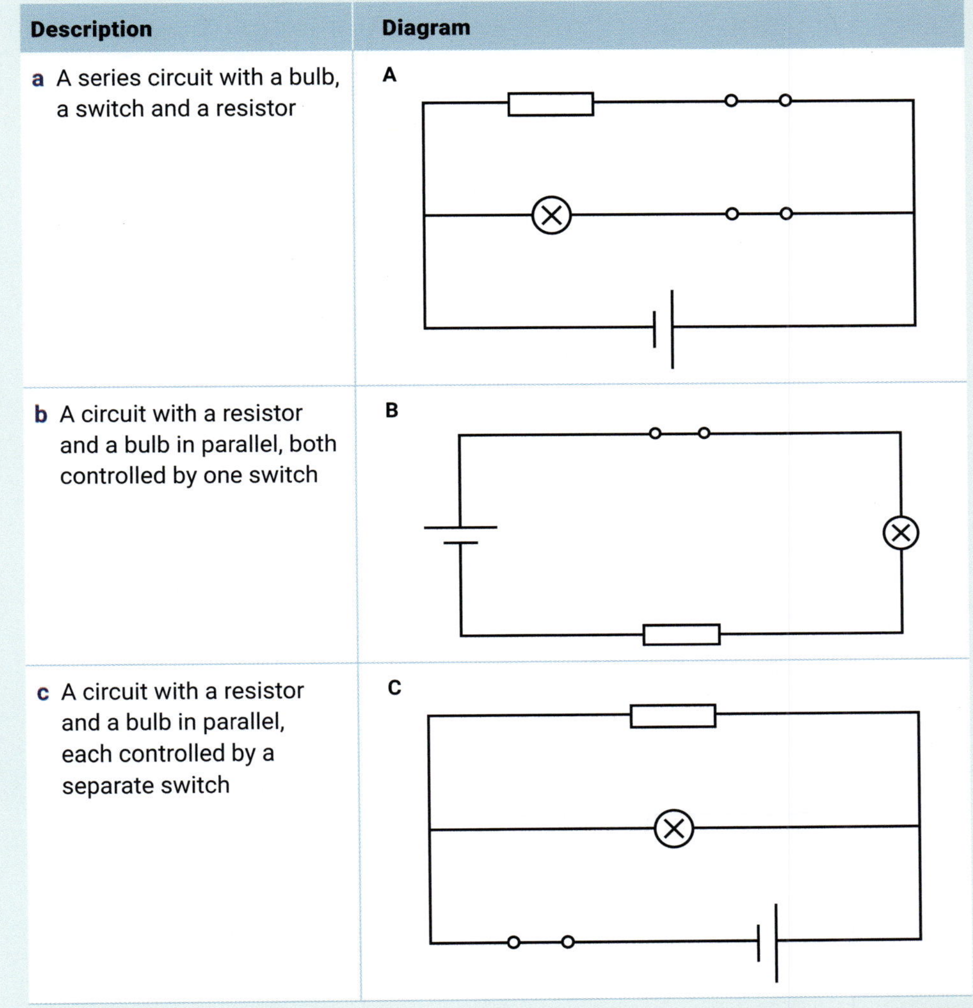

Description	Diagram
a A series circuit with a bulb, a switch and a resistor	A
b A circuit with a resistor and a bulb in parallel, both controlled by one switch	B
c A circuit with a resistor and a bulb in parallel, each controlled by a separate switch	C

3 **Identify** the errors in each of the three circuit diagrams below and **explain** why these features are wrong.

a

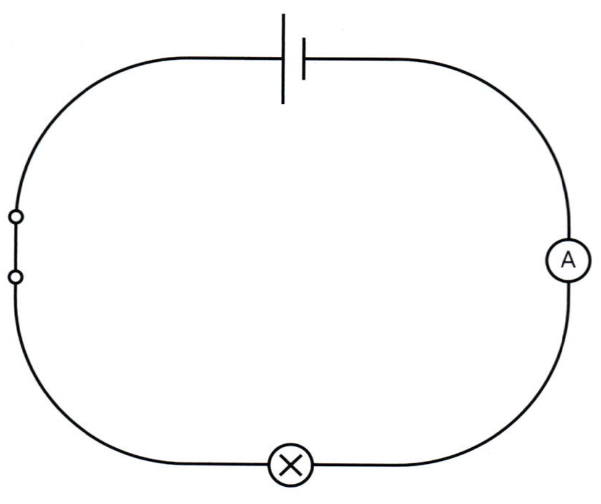

b

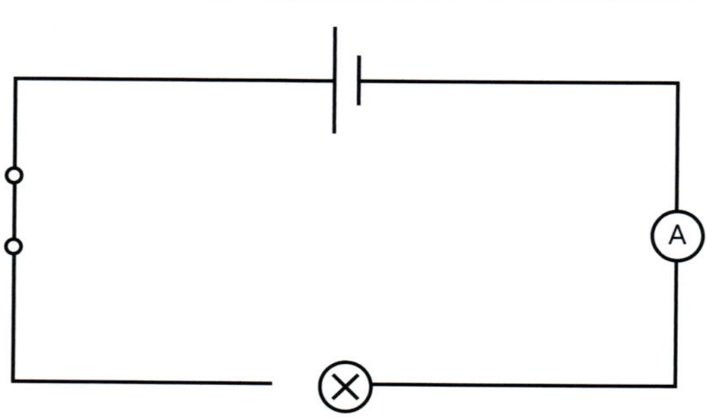

c

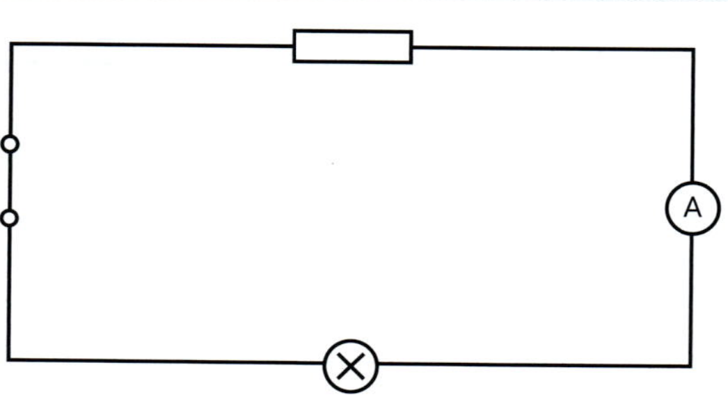

9780170491785

4.7 Measuring quantities in electrical circuits

BY THE END OF THIS MODULE, YOU WILL BE ABLE TO:

- ✓ describe the current, voltage and resistance of an electrical circuit
- ✓ explain how these quantities are measured.

GET THINKING

Have you ever tried to use a torch and discovered that the light was too dim? What would you need to measure beforehand to find out if the battery would be able to make the torch light properly? How would you measure it?

Video activity
How we measure electricity

Current

current
the amount of charge that passes a point in an electrical circuit in one second

Current is defined as the flow of electricity. It is caused by the movement of electrons. You can think of the current in an electrical circuit as being like the flow of water through a waterway. Current is represented in equations by the symbol I. (Note that, by convention, current is shown in diagrams as flowing from the positive terminal to the negative terminal, even though electrons actually flow in the opposite direction.)

Current is measured as the amount of electrical charge that moves past a point in the circuit in 1 second. An electric current occurs when electrons move in one direction. If electrons move in random directions, there is no electric current (Figure 4.7.1).

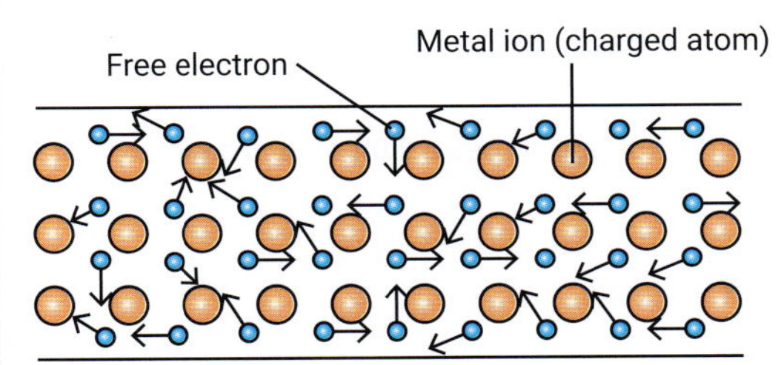

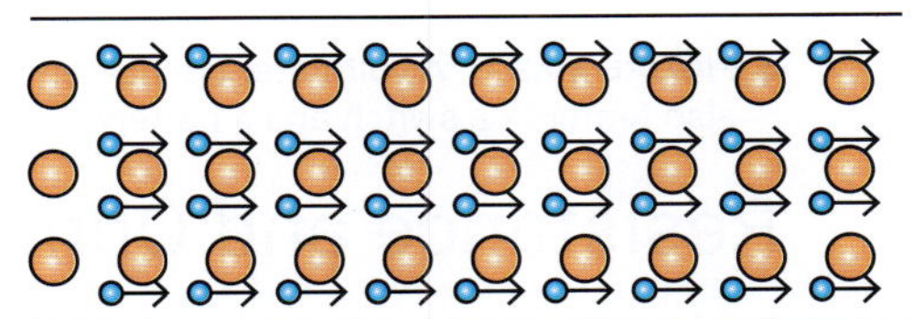

▲ **FIGURE 4.7.1** A model of current, showing the movement of electrons in a conductor

You can measure current with a device called an ammeter. An ammeter measures current in units of amperes, shortened to A. The circuit diagram symbol for an ammeter is a circle with a capital A inside. Figure 4.7.2 shows an electrical circuit with an ammeter. An ammeter is connected in series to measure the current flowing in a circuit.

◀ **FIGURE 4.7.2** An ammeter can be connected to measure the current flowing through a series circuit.

Voltage

voltage
a measure of the difference in electrical energy between two points in a circuit

When a charge moves between two points in an electrical circuit, some of its electrical energy may be transformed into another type of energy, such as light. Alternatively, the charge may gain energy when it moves through a power source and has electrical energy added to it. **Voltage** is a measure of the amount of electrical energy lost or gained by charges as they move between two points in an electrical circuit. Voltage is represented by the symbol V when used in equations. A voltmeter is a device that measures voltage. It is shown in circuit diagrams as a circle with a capital V inside. Voltage is measured in units called volts (also shown as V). Figure 4.7.3 shows an electrical circuit with a voltmeter. A voltmeter is connected in parallel to measure the voltage across a component.

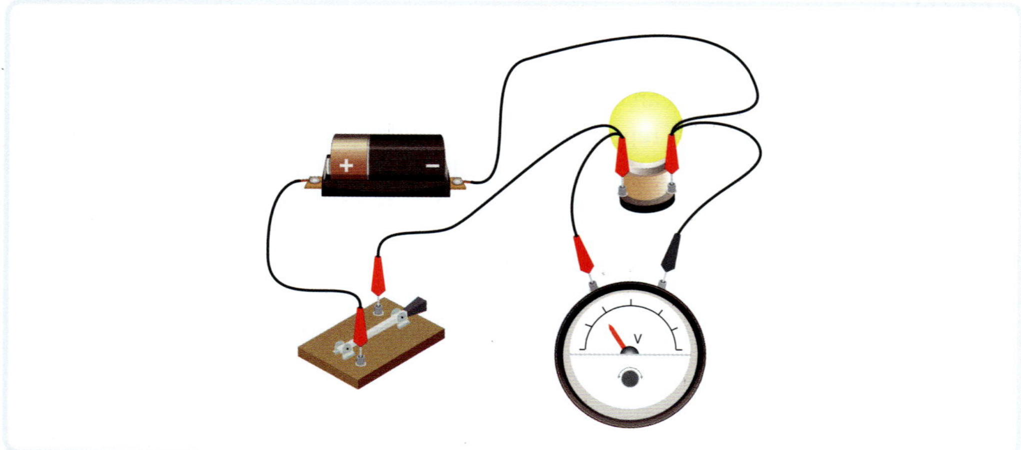

▲ **FIGURE 4.7.3** A voltmeter connected to measure the voltage across a light bulb in a circuit that also features a switch and a battery

Resistance and Ohm's law

Resistance is a measure of how much a circuit component opposes the flow of current through it. All circuit components, even conductive wires, have some resistance. Resistance is represented in equations by the symbol R. The resistance of a circuit component is defined as the ratio of the voltage across a component to the electric current that flows through the component. This is expressed with the equation known as Ohm's law:

$$R = \frac{V}{I}$$

This equation is often written as $V = IR$.

If a component has high resistance, then a large amount of electrical energy will be lost by electrons moving through it. This happens even if the number of electrons per second moving through it – the current – is small.

9780170491785

4.7

In contrast, electrons flowing through a component with low resistance will lose a small amount of electrical energy, even when the current is high. Components such as filament light bulbs and voltmeters have high resistance. Components such as ammeters have low resistance.

Resistance is measured in units called ohms (symbol Ω) and is represented in equations by *R*.

WORKED EXAMPLE 4.7.1

An ammeter measures the current flowing through a light bulb to be 0.2 A (*I*). A voltmeter connected to either side of the bulb measures the bulb's voltage to be 230 V (*V*). What is the resistance (*R*) of the bulb?

THINKING PROCESS	WORKING
Step 1: Identify the known and unknown variables.	$V = 230$ V $I = 0.2$ A
Step 2: Identify the appropriate relationship.	$R = \frac{V}{I}$
Step 3: Substitute known values.	$R = \frac{230}{0.2}$
Step 4: Rearrange, using algebra if necessary, and solve.	$R = 1150\ \Omega$

WORKED EXAMPLE 4.7.2

What current (*I*) would flow through an electric motor with a resistance (*R*) of 920 Ω when it is connected to a 230 V power source (*V*)?

THINKING PROCESS	WORKING
Step 1: Identify the known and unknown variables.	$V = 230$ V $R = 920\ \Omega$
Step 2: Identify the appropriate relationship.	$R = \frac{V}{I}$
Step 3: Substitute known values.	$920 = \frac{230}{I}$
Step 4: Rearrange, using algebra if necessary, and solve.	$I = \frac{230}{920} = 0.25$ A

Energy transformations in a circuit

You have learned that an electrical circuit consists of a power source, a load (electrical component that transforms electrical energy) and a complete conductive path.

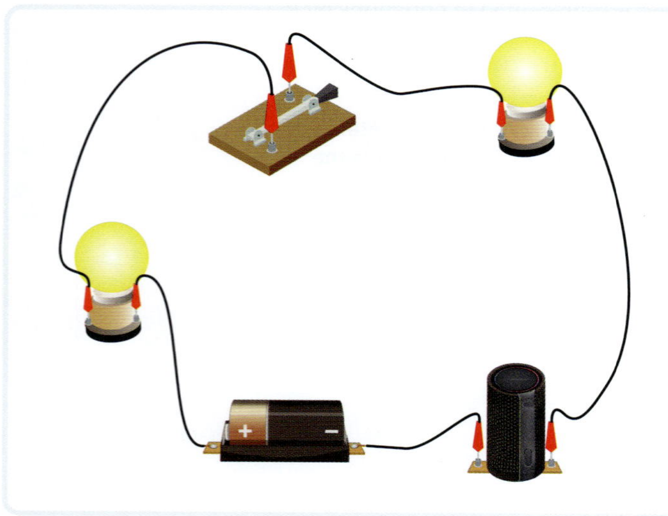

▲ **FIGURE 4.7.4** A speaker in a series circuit with two light bulbs, a battery and a switch

The power source provides electrical energy to the electrons in the conducting material (wire) and the components. The components in the circuit transform some of this electrical energy into other forms of energy, such as light, heat and sound. As a result, the electrical energy of electrons decreases as they move through the components. In contrast, when the electrons move through the power source, their electrical energy increases.

Electrical components are added to electrical circuits because they transform energy to do useful work. An electric motor is put in an electrical circuit to transform electrical energy into kinetic energy; for example, to make a lift move or to power an electric car. A speaker is put into an electrical circuit to transform electrical energy into sound energy to play music (Figure 4.7.4). An oven is put into an electrical circuit to transform electrical energy into heat energy to cook food.

Measuring energy transformations

Most electrical components and devices transform some electrical energy into unwanted heat or sound. For example, a kettle makes a noise as it boils water, and your phone or laptop charger gets hot as it charges your device.

Measuring the light, heat or sound energy a device or component emits is challenging. Generally, we make measurements of electrical energy transformations using information about the electrical energy put into the device rather than the device's output.

Earlier in this module you learned that electrical current (I) is a measure of how much charge passes a point in the circuit per second and that voltage (V) is a measure of how much energy is lost or gained by charges as they move through a component. It follows that the product of current and voltage can be used to determine how much electrical energy has been transformed per second (also described as the rate of electrical energy transformation) by a device or component. This measure is called power (P). The unit for power is the watt (W). The equation $P = V \times I$ explains this relationship between current, voltage and power. We cannot directly measure the electrical energy transformed by a device or component or even by a whole circuit, but we can determine it by measuring current and voltage and using this equation.

9780170491785

4.7

WORKED EXAMPLE 4.7.3

An ammeter measures the current (*I*) flowing through a light bulb to be 0.2 A. A voltmeter connected to either side of the bulb measures the bulb's voltage *V* to be 230 V. What is the power *P* of the bulb?

THINKING PROCESS	WORKING
Step 1: Identify the known and unknown variables.	$V = 230$ V $A = 0.2$ A
Step 2: Identify the appropriate relationship.	$P = V \times I$
Step 3: Substitute known values.	$P = 230 \times 0.2$
Step 4: Rearrange, using algebra if necessary, and sclve.	$P = 46$ W

Electrical devices in your home have a power rating that is often displayed somewhere on the device (like in Figure 4.7.5) or is published online. For example, the power rating of a kettle or microwave is usually 1000 W, while the power rating of a calculator is around 0.00002 W.

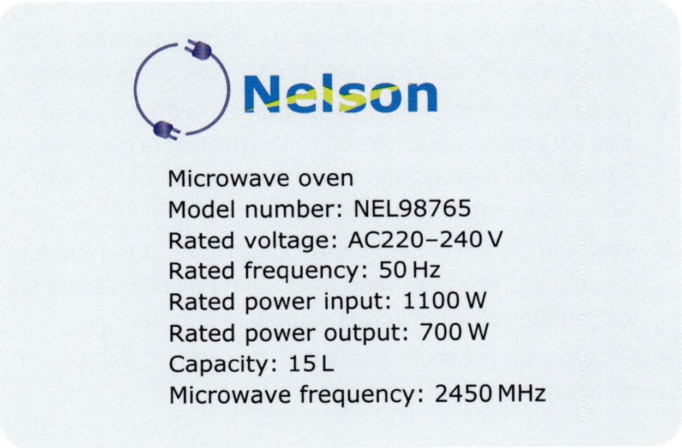

Microwave oven
Model number: NEL98765
Rated voltage: AC220–240 V
Rated frequency: 50 Hz
Rated power input: 1100 W
Rated power output: 700 W
Capacity: 15 L
Microwave frequency: 2450 MHz

FIGURE 4.7.5 The electrical specifications label of a microwave oven shows the power input and other details.

4.7 LEARNING CHECK

1. **Create** a table with the following terms:
   - electrical current
   - voltage
   - resistance.

   In the table, **define** each term and **explain** how it is measured.
2. **Calculate** the resistance of an electric toothbrush motor that has 0.05 A current flowing through it when connected to a 6 V battery.
3. **Calculate** the power of a bicycle light that uses a 2 V battery and has a 0.02 A current flowing through it.

WORKING SCIENTIFICALLY

4.8 Measuring with apparatus

SCIENCE SKILLS IN FOCUS

IN THIS MODULE, YOU WILL FOCUS ON LEARNING AND IMPROVING THESE SKILLS:

- using appropriate measuring equipment to collect current and voltage data
- following a method to undertake an effective investigation with a range of electrical circuits.

A voltmeter connected in parallel with (to either side of) a component or device in a circuit can measure the voltage of that component. An ammeter connected in series (next to) a component or device in a circuit can measure the current flowing through that component. Voltage and current data can be used to calculate the component's resistance and power and compare parts of a circuit.

Voltage (V) = current (I) × resistance (R) $V = I \times R$

Power (P) = voltage (V) × current (I) $P = V \times I$

How to use an ammeter

1 Place the ammeter in the electrical circuit in series with the component for which you are collecting data.
2 Close the switch. If the needle moves to the negative side (of an analogue meter) or the display is negative (on a digital meter) then swap the location of the two leads connected to the ammeter.
3 Keep the switch closed long enough to record a value for current. You may need to estimate a value based on the middle value if the current varies slightly over time.
4 Remember to always record your measurements with the correct unit of measurement (amps or A).

How to use a voltmeter

1 Place the voltmeter in the electrical circuit in parallel with the component for which you are collecting data. This means you need to place an additional lead from each side of the component and connect them to each side of the voltmeter.
2 Close the switch. If the needle moves to the negative side (of an analogue meter) or the display is negative (on a digital meter) then swap the location of the two leads connected to the voltmeter.
3 Keep the switch closed long enough to record a value for voltage. You may need to estimate a value based on the middle value if it varies slightly over time.
4 Record your measurements with the correct unit of measurement (volts or V).

INVESTIGATION 1: COMPARING BRIGHTNESS, POWER AND EFFICIENCY IN LED AND FILAMENT BULBS

BACKGROUND INFORMATION

Over recent years, LED lighting has become increasingly popular for use in homes and businesses, traffic lights and streetlights, and in everyday appliances and tools, such as torches and bicycle lights. It is popular because it can transform more electrical energy into light and less into heat.

AIM

To compare the energy consumption of a filament lamp and an LED bulb

MATERIALS AND EQUIPMENT

- ☑ power source (power pack)
- ☑ switch
- ☑ 12 V filament bulb
- ☑ 12 V LED bulb
- ☑ ammeter
- ☑ voltmeter
- ☑ 4 connecting wires
- ☑ light meter (optional)

9780170491785

⚠ Safety

When constructing, adjusting or packing away an electrical circuit, always turn the power off at the main switch and unplug the power source. Only plug in the power source and turn on the main switch when you are ready to collect data.

Only use the circuit switch to connect the circuit for as long as you need to record the data.

Do not handle circuit components after use; they can get hot and may need time to cool down.

Ensure your hands are dry during the investigation.

PROCEDURE

1 Write a hypothesis for the experiment.

2 Construct an electrical circuit with the filament bulb, as shown in Figure 4.8.1.

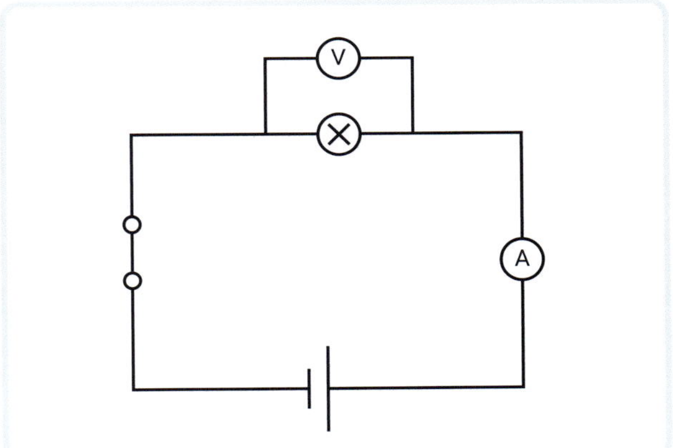

▲ **FIGURE 4.8.1** An electrical circuit to measure the electrical energy used by a lamp

3 Close the switch and record data for the filament bulb in a table like Table 4.8.1. Either **describe** the brightness by qualitative judgement (dull/bright/very bright) or measure it using a light meter.

4 Calculate a value for the power of the light bulb and record the value in the table.

5 Construct an electrical circuit with an LED bulb replacing the filament bulb. You may need to wait before handling the filament bulb, as it may be hot.

6 Close the switch and record the data for the LED bulb in your table. Again, either record the brightness as a qualitative judgement or measure it using a light meter.

7 Calculate a value for the power of the LED bulb and enter the value in the table.

RESULTS

Record your data in a table like Table 4.8.1.

ANALYSIS

1 Did the LED or filament bulb feel very warm when you packed away the equipment?

2 What does the bulb brightness and warmth tell you about the energy transformations of the two bulbs?

3 What aspect of the experiment was or would be improved by using a light meter rather than qualitative judgement when recording data about the brightness of the bulbs?

CONCLUSION

Write a conclusion for this experiment. Remember to state if your results support your hypothesis.

▼ **TABLE 4.8.1** Electrical data for a filament bulb and an LED bulb

	Voltage (V)	Current (A)	Power (W) $P = V \times I$	Brightness (qualitative judgement/light meter value)
Filament bulb				
LED bulb				

INVESTIGATION 2: COMPARING RESISTANCE IN DEVICES

BACKGROUND INFORMATION

How well a material conducts electricity depends on how its atoms are arranged and how many electrons are free to move. All materials have some resistance to the flow of an electrical current – even copper, which is used in electrical wiring. The term 'conductor' is used to describe materials that offer low resistance, and the term 'insulator' is used to describe materials that provide high resistance.

Insulating coating on wires (Figure 4.8.2) is vital to electrical safety because it enables wires in electrical circuits to be used in devices and buildings, and to be handled without the risk of an electric shock.

While the terms 'insulator' and 'conductor' suggest that materials can be classified into these two groups, in reality materials fall along a sliding scale between highly conductive and highly insulating. Plastic, wood, rubber and air are good electrical insulators. Metals are good conductors, but some are better than others. For example, silver and copper are more conductive than iron and lead.

VladyslaV Travel photo/Shutterstock.com

▲ **FIGURE 4.8.2** Insulated electrical wires made of copper

AIM

To investigate the conductive ability of different materials

MATERIALS AND EQUIPMENT

- ☑ power source (battery or power pack)
- ☑ switch
- ☑ 3 different materials to be tested (e.g. pencil lead, wooden ruler, aluminium foil, piece of lead)
- ☑ light bulb
- ☑ ammeter
- ☑ voltmeter
- ☑ 4 connecting wires

⚠ Safety

When constructing, adjusting or packing away an electrical circuit, always turn the power off at the main switch and unplug the power source. Only plug in the power source and turn on the main switch when you are ready to collect data.

Only use the circuit switch to connect the circuit for as long as you need to record the data.

Do not handle circuit components after use; they can get hot and may need time to cool down.

Ensure your hands are dry during the investigation.

PROCEDURE

1 Write a hypothesis for the experiment.

2 Construct an electrical circuit with Material A, as shown in Figure 4.8.3.

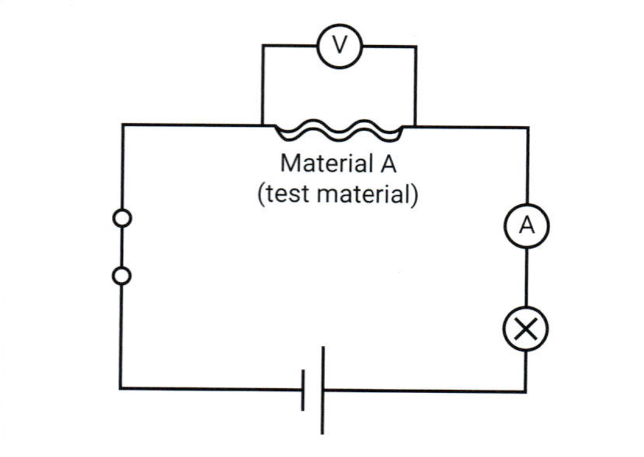

▲ **FIGURE 4.8.3** An electrical circuit for testing the conductivity of a material

9780170491785

3 Close the switch and record data for Material A in a table like Table 4.8.2.

4 Calculate a value for the resistance of Material A and enter the value in the table.

5 Construct an electrical circuit with Material B replacing Material A. You may need to wait before handling the Material A sample, as it may be hot.

6 Close the switch and record data in the table for Material B.

7 Calculate a value for the resistance of Material B and enter the value in the table.

8 Construct an electrical circuit with Material C replacing Material B. You may need to wait before handling the Material B sample, as it may be hot.

9 Close the switch and record data in the table below for Material C. Calculate a value for the resistance of Material C and enter the value in the table.

10 You may need to wait before handling the Material C sample when packing up your equipment, as it may be hot.

RESULTS

Record your data in a table like Table 4.8.2.

▼ **TABLE 4.8.2** Electrical data for three materials

	Voltage (V)	**Current (A)**	**Resistance (Ω)** $R = \frac{V}{I}$
Material A			
Material B			
Material C			

ANALYSIS

1 Place the three materials in order from the material with the lowest resistance to the one with the highest resistance.

2 What aspects of the experiment could be changed to make the investigation more reliable (that is, to produce more consistent results)?

3 What aspects of the experiment could be changed to make the investigation more valid (that is, to improve how well it tests our hypothesis)?

4 What was the purpose of the light bulb in the investigation?

CONCLUSION

Write a conclusion for this experiment. Remember to state if your results support your hypothesis.

4.9 Global energy use and future needs

BY THE END OF THIS MODULE, YOU WILL BE ABLE TO:

- ✓ explain the need for a change in the world's energy demands and sources
- ✓ describe how energy labels help consumers make informed energy choices
- ✓ describe ways in which consumers are changing Australia's energy usage with personal choices.

Video activity
Cost of building Australia's future energy needs

GET THINKING

Think about how much electrical energy you and your family use. Do you know the source of your electricity? How do your family's energy demands change during the day, week and year? How do you think your energy demands could change in the future?

We need energy every day in almost every aspect of our lives. Electrical energy is vital to activities in homes, businesses and government services.

The world's energy consumption is increasing as population rises and people gain more access to electrical appliances and devices. Producing the energy to meet these needs significantly affects the local and global environment. Electrical energy is generated using different methods, each with different impacts on the environment. Some impacts include removal of resources from the ground, land-clearing, emission of greenhouse gases, production of nuclear waste, and the flooding of waterways and valleys.

To maintain Earth's environmental health, we need a plan that helps us meet future global energy needs while minimising harm. This plan must include reducing energy consumption by improving current energy use, and developing alternative energy sources that have less environmental impact. Australia's National Energy Productivity Plan was released in 2015 by the ministers for energy across Australia's states and territories (Figure 4.9.1).

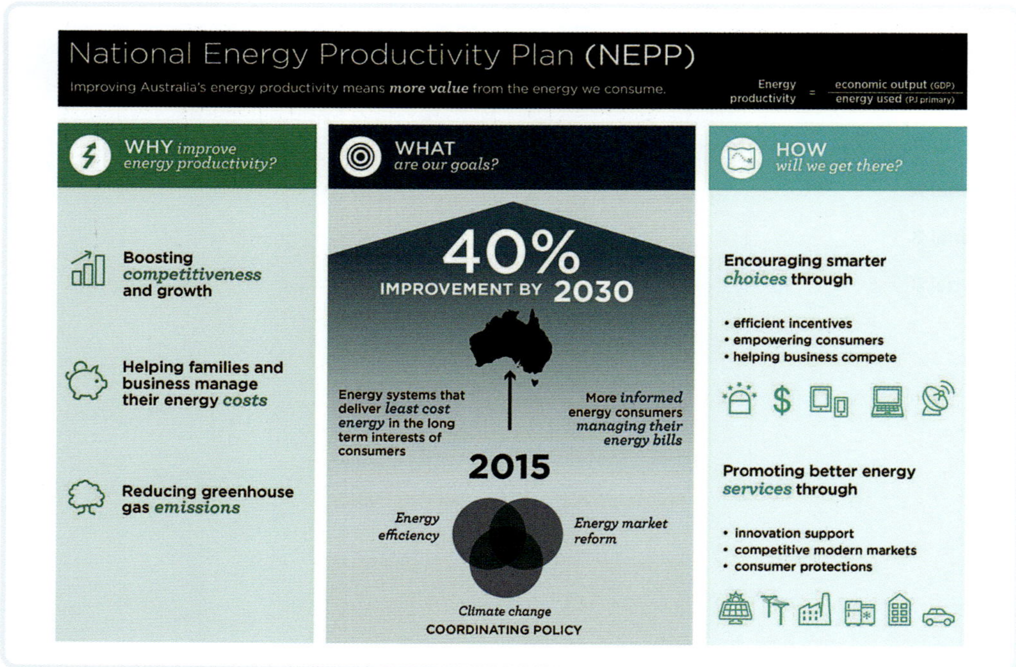

▲ **FIGURE 4.9.1** Australia's National Energy Productivity Plan

9780170491785

Improving how we use energy

Making more efficient use of energy will be vital to maintaining our lifestyles if we want to reduce environmental impacts in the future. Governments can create policies and laws designed to reduce energy demand by businesses, industries and individuals. Choices made by individuals and households will also help. Sometimes, governments provide an incentive (often financial) or the means for people to make choices that reduce their electricity demands. One such Australian government policy is the energy rating system.

Energy rating of electrical appliances

Electrical appliances bought in Australia must display an energy rating label. Energy rating labels are regulated by the Australian Government's Australian Energy Regulator and the Energy Rating Label scheme has operated for over 35 years. Energy rating labels can be used to compare the energy efficiency and, therefore, running costs of various appliances based on assumptions about how often and for how long an 'average' household will run the appliance.

The labels help people make informed choices about the appliances they buy, by providing two important pieces of information:

- the star rating of the appliance
- the appliance's energy consumption (Figure 4.9.2).

The star rating refers to energy efficiency; that is, how much energy an appliance uses compared to similar products. More stars means the device is more efficient than other models with a similar size and features.

Most products are given between 1 and 6 stars. As technology continues to improve, so does energy efficiency. That is why some super-efficient appliances are now available in shops and online with an extra row of stars, extending the star rating system to 10 stars.

FIGURE 4.9.2 Example of an energy rating label

When comparing star ratings of different appliances, it is important to remember that the appliances have to be of similar size and have similar features. For example, a 10 kg washing machine and a 6 kg machine may both get a 5-star rating, but the 10 kg washing machine will use more energy.

Energy consumption in your house is generally measured in kilowatt-hours (kWh). This is the unit of measurement you will see on electricity bills. We can use the energy consumption rating to work out the estimated running cost of an appliance, with the formula:

$$\text{Annual running cost} = \text{total energy consumption (kWh)} \times \text{electricity tariff (dollars/kWh)}$$

electricity tariff
the price a consumer pays for the energy they use

For example, if a refrigerator has an energy consumption of 458 kWh and your **electricity tariff** is \$0.307 per kWh, then the estimated annual running cost of your refrigerator is \$140.61 (458 kWh × \$0.307).

Weblink
Australian Energy Regulator

People can find their electricity tariff in the service calculation section of their electricity bill. The Australian Energy Regulator website provides more information about electricity tariffs.

Energy trends and predictions

Data about recent energy trends is needed to help predict how energy demand may change in future. This includes information about sources of energy and energy demand. Some data about Australian use of energy can be seen in Figures 4.9.3 and 4.9.4. Predicting when and where demand will change can help governments plan to reduce waste and choose the most appropriate sources of energy.

DATA SCIENCE

Learn more about the importance of analysing large datasets in **Modules 2.5** and **2.6**.

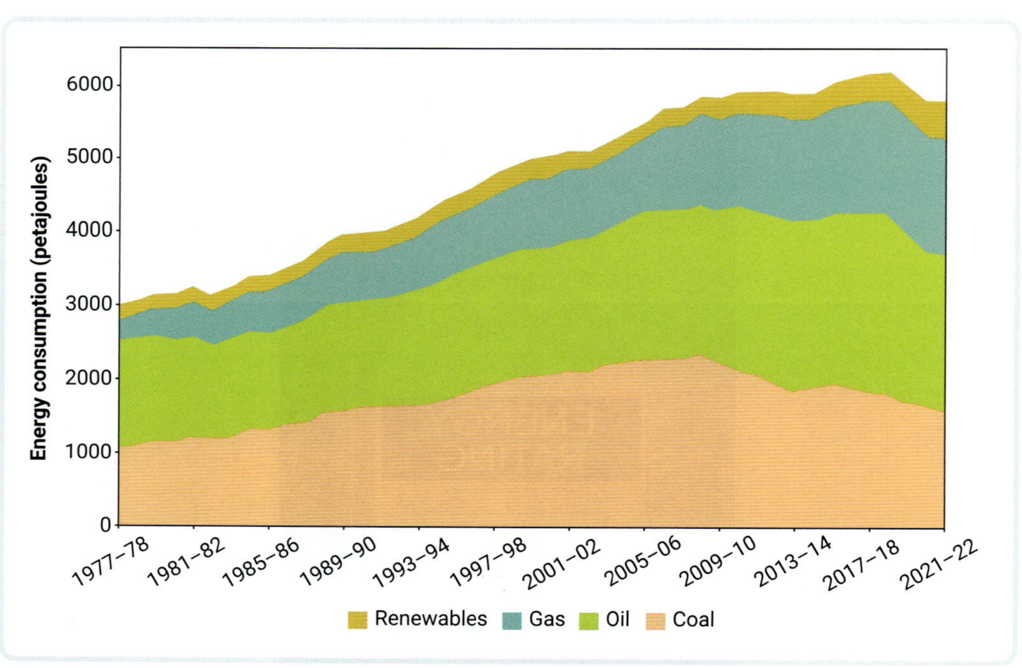

▲ **FIGURE 4.9.3** Australia's energy consumption by fuel type. One petajoule is 10^{15} joules.

9780170491785

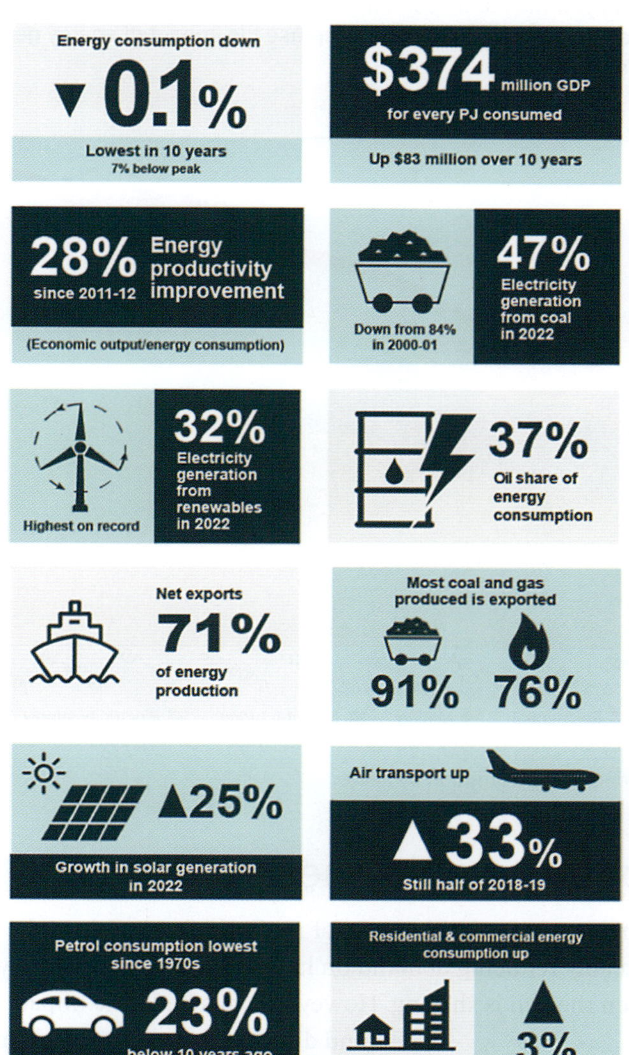

- Australia's energy consumption fell slightly in 2021–22, a third successive decline
- Over two-thirds Australian energy production is exported, including most coal and gas
- Renewable electricity generation at record levels, almost one third of all electricity
- Transport energy use increased with air transport recovery, road use still down
- The imported share of refined petroleum product consumption rose to record levels

Australian Energy Update 2023, Commonwealth of Australia 2, licensed under CC BY 4.0. https://www.energy.gov.au/sites/default/files/Australian%20Energy%20Update%202023_0.pdf

▲ **FIGURE 4.9.4** A summary of Australia's energy statistics in 2021–22

Data like this is analysed by government and private organisations to identify opportunities to build new renewable energy projects.

ACTIVITY

Design an energy policy

Identify two interesting aspects of Figure 4.9.5. **Imagine** you are the Australian Minister for the Environment and you wish to do something about one particular aspect of energy use. Outline a policy that you would introduce to decrease the impact of energy generation on the environment in this area.

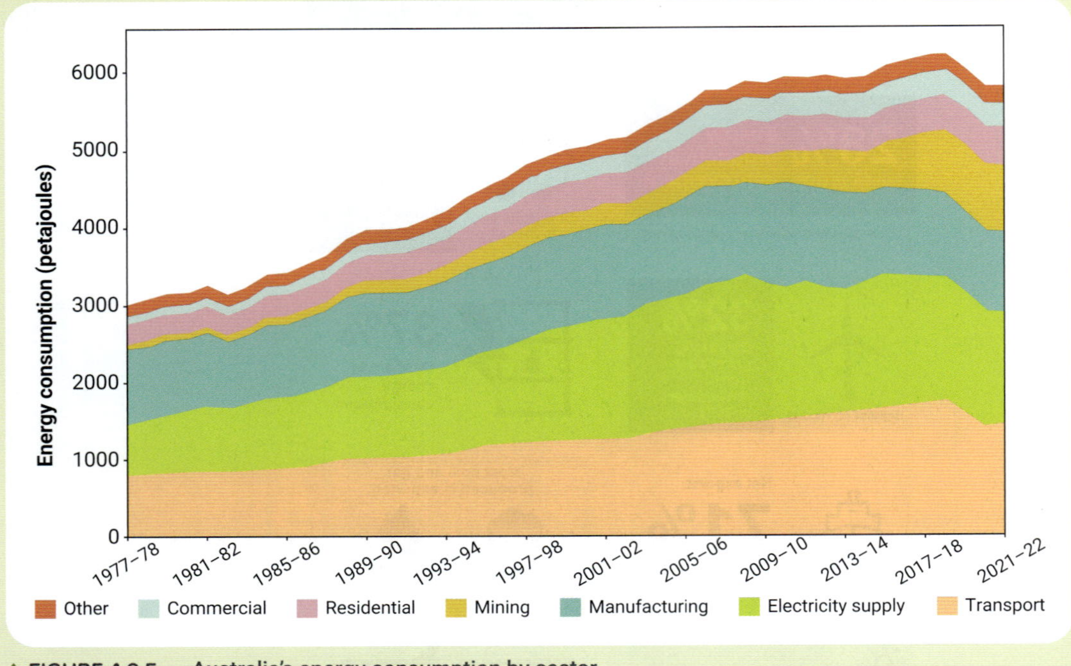

▲ **FIGURE 4.9.5** Australia's energy consumption by sector

Demand for renewable energy

Australia has the greatest household uptake of rooftop solar panels in the world. This has the benefit of noticeably reducing demand on large-scale electrical energy production during the day, when the sun is shining. However, it has had little impact on the peak demand during the evening, when many people come home and turn on their appliances, such as air-conditioners, heaters, ovens and lighting. Large battery units are being designed to store solar power for use at such times. As they become cheaper to produce, more homes and businesses in Australia are expected to consider installing solar power (Figure 4.9.6). Module 4.10 describes solar power generation in more detail.

petovarga/Shutterstock.com

▲ **FIGURE 4.9.6** A household solar charging and usage system

As discussed in Module 4.2, large-scale battery systems may also enable wider use of solar and wind power at an energy-grid level.

Electrical vehicles (EVs) are powered by substantial battery units. The number of people choosing to buy an EV is increasing rapidly in Australia. If this increase continues, it may have a greater positive impact, potentially helping lower Australia's greenhouse gas emissions. This will happen if the electrical power used to charge EVs comes from renewables rather than fossils fuels.

Research activities

ACTIVITY

From waste heat to useful energy

Researchers at the University of New South Wales have demonstrated in an experiment that it is possible to generate electricity from the infrared waves radiated away from Earth at night, although in much smaller amounts than when sunlight falls on solar panels. Researchers hope this form of electricity could one day be developed into a viable alternative energy source.

Research this discovery and, using the output power they have managed to produce, determine how many square metres of infrared panels would be required to provide 0.4 kWh in power (enough for an average household overnight, from 11 pm to 7 am).

Cyclist powers toaster

Watch the clip 'Cyclist powers toaster' at the weblink provided.

Are you surprised by how much energy is needed to toast a single piece of bread?

Research the power rating of an average oven and, using the information from the video, determine how many cyclists it would take to bake a loaf of bread.

Weblink
Cyclist powers toaster

Energy on Eigg

Research the Isle of Eigg on the west coast of Scotland to answer the following questions.

1 What methods do the locals use to generate electrical energy?
2 How can these methods usually provide enough electrical energy, even though Scottish weather can be inconsistent?
3 What techniques do the locals use to ensure the demand for electrical energy is never likely to overload the energy system?

Energy efficiency plan

Create a plan to reduce energy demand in your home or school. The plan should consider energy demand and supply during both day and night, and all relevant energy users. Once the plan is complete, estimate the reduction in energy demand it is likely to achieve.

4.9 LEARNING CHECK

1 **Identify** one government policy that was created to decrease energy demand.
2 **Examine** the graph of Australian energy consumption by fuel type in Figure 4.9.3. **Identify** the trend in coal consumption since 2009–10 and the trend in gas consumption since 2001.
3 Study the Australian Energy Statistics 2021–22 snapshot (Figure 4.9.4) and **describe** one notable increase and one notable decrease in energy use.
4 **Examine** the graph of Australian energy consumption by sector (Figure 4.9.5) and **determine** the percentage of total energy consumed by the mining and manufacturing sectors combined.
5 Elaine drives an EV to work each day in a building with solar panels. **Explain** how Elaine could minimise her electricity bill at home without adding to the infrastructure of her home.

SCIENCE IN CONTEXT

4.10 Solar energy in Australia

BY THE END OF THIS MODULE, YOU WILL BE ABLE TO:

✓ describe the use of solar energy in Australia and examine the reasons for adoption of solar panels by individuals, industries and communities.

What is solar energy?

Solar energy is the process of converting the electromagnetic radiation from the Sun into electricity or thermal energy. Solar cells or photovoltaic (PV) cells transform light energy to electrical energy. Solar cells are usually grouped together in modules, which are further grouped together to form panels or systems (Figure 4.10.1). Solar panels are placed on roofs or in open fields on solar farms where they can transform large amounts of light energy into electricity. Solar energy production is a renewable energy source because it does not involve a fuel that is in limited supply. Using solar panels does not generate greenhouse gases.

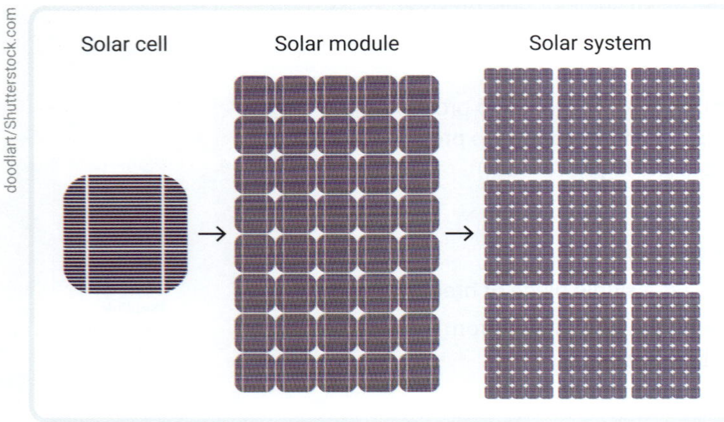

doodlart/Shutterstock.com

▲ FIGURE 4.10.1 Solar cells are grouped in modules and within panels or systems for placement on roofs or in solar farms.

Australian researchers have led the way in the development of solar cell technologies. The passivated emitter and rear cell (PERC) solar cell technology was invented by Australian researchers at the University of New South Wales in 1983. This technology remains the basis for more than 85 per cent of all new solar panel modules used globally.

Solar use in Australia

Solar energy is the fastest growing electricity generation process in Australia. In 2023, solar cells generated approximately 6 per cent of Australia's total electricity.

Adam Calaitzis/Shutterstock.com

▲ FIGURE 4.10.2 In 2023, rooftop solar panels generating more than 930 megawatts of power were installed on New South Wales homes.

Globally, Australia has the highest uptake of residential solar panel use, with one in four homes running a solar panel system (Figure 4.10.2). Homes use solar power to generate electricity and to heat water.

Large-scale solar farms are used for producing hydrogen fuel and energy for large-scale heating and cooling systems (Figure 4.10.3). They also provide electricity for industrial processes and regions. As of May 2024, there were 10 solar farms in operation in New South Wales and another nine farms under construction.

9780170491785

The switch to solar

It is no surprise that solar energy is a popular choice in Australia, given we are one of the sunniest countries on the planet. Some of the many reasons people are switching to solar are:

- technological advancement resulting in cheaper and more efficient solar cells
- government investment in large-scale renewable energy production
- financial incentives (discounts and rebates) for people and small businesses to install solar systems
- 'feed-in tariff' schemes where residents and small businesses can sell their additional electricity back to the electricity supply and earn money
- personal choice to minimise environmental impacts.

Jacqui Martin/Shutterstock.com

▲ **FIGURE 4.10.3** Solar farms are used in Australia for large-scale electricity production.

4.10 LEARNING CHECK

1 Interview people you know who have had solar panels installed on their home or at their workplace.
   a **Summarise** their reasons for installing solar panels.
   b **List** the reasons why other people do not have solar panels installed.
2 **Identify** three reasons why there have been so many solar panels installed in Australia. For each reason, **explain** why and how it has resulted in more installations.
3 **Create** a flyer that could be dropped in letterboxes around your neighbourhood to encourage people to install solar panels.
4 Look at the two cars in Figure 4.10.4.
   a **Compare** how each car is powered.
   b **Evaluate** which car you think is the better option.
   c **Discuss** the advantages and disadvantages of powering a car with solar energy.

▲ **FIGURE 4.10.4** Two different ways of powering cars

4 REVIEW

REMEMBERING

1 **List** the three ways electrical energy can be produced.

2 **Identify** the following electrical energy sources as either renewable or non-renewable: tidal energy, natural gas, biofuels, wind power and coal.

3 **Label** the components **A–F** in the circuit diagram below.

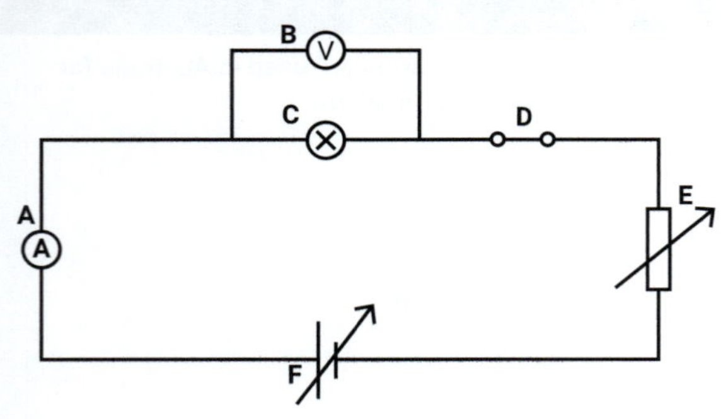

4 **Define** the terms current and voltage.

5 **Identify** which of the following circuits is in

a series.

b parallel.

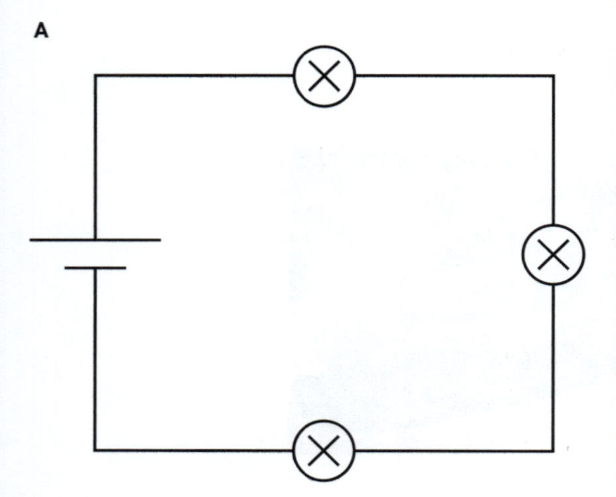

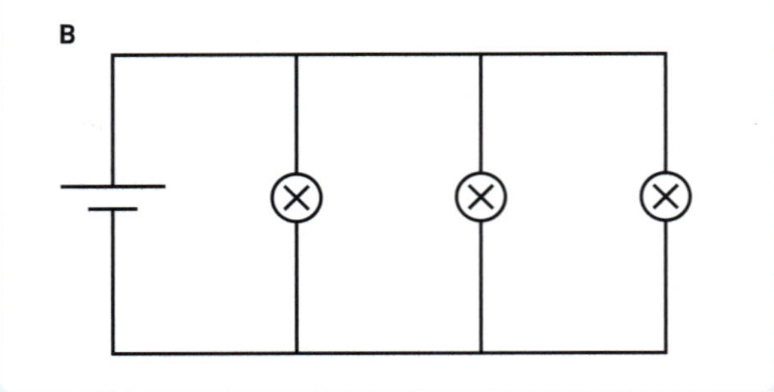

UNDERSTANDING

6 **Describe** the process that occurs in an electrical generator.

7 **Explain** the advantages and disadvantages of hydroelectric power generation.

8 **Describe** the method of connecting a voltmeter to an electrical circuit.

9 **Explain** the advantages of using a circuit diagram rather than a sentence to **describe** a circuit.

10 a **Calculate** the current that would need to flow through a bulb with 10 W of power when connected to a 120 V power source.

b **Determine** the resistance of the bulb.

9780170491785

APPLYING

11 **Predict** the potential renewable energy sources of an island in the Pacific Ocean near the equator that is geologically active and mountainous and gets plenty of sunshine and rain.

12 **Explain** why parallel circuits should be used in the electrical wiring of households.

13 **Determine** what happens to the power output of a light bulb in a simple circuit when the voltage is kept constant and the resistance is doubled.

14 **Identify** the most appropriate form of renewable energy for a desert community. **Justify** your choice.

ANALYSING

15 Australia is aiming to achieve a target of 82 per cent renewable energy by 2030. **Justify** why investing in battery and energy technologies is an important part of that target.

16 **Explain** the lack of geothermal energy in Australia.

17 Consider how an ammeter is connected in a circuit. **Propose** a reason that ammeters have very low resistance.

18 **Explain** why global energy demand has increased over the last 20 years.

EVALUATING

19 Conduct some research and then **evaluate** the effectiveness of financial incentives and policies used by the NSW government to promote the use of less energy. A good starting point is the NSW Climate and Energy Action website.

Weblink
NSW Climate and Energy Action

20 **Discuss** the issues limiting the uptake of battery units by Australian households with solar panels.

CREATING

21 Robert is designing the lighting for a room and wants three lights that can be turned on and off separately. **Draw** a circuit diagram for an electrical circuit for Robert's lights that would allow this.

22 **Create** a poster to help primary school students understand the need to use less electrical energy. Include some simple ideas on how they can do this.

SCIENCE IN DEPTH STUDY

1 Connect what you've learned

In this chapter you've learned about how:

- to use ammeters and voltmeters to measure quantities of current and voltage in electrical circuits
- to draw circuit diagrams and how to connect circuits in series and parallel
- devices can transform other forms of energy into electrical energy, such as a (photovoltaic) solar cell transforming light energy into electrical energy.

Create a mind map to show how the main ideas you've learned are connected.

2 Check your thinking

Conduct some research to find out how the current and voltage of batteries connected in series differs from the current and voltage of batteries connected in parallel. How would you use an ammeter and a voltmeter to test this?

3 Get into action

Consider how connecting solar panels in series might have a different outcome from connecting them in parallel. Design an investigation to determine the difference. For this investigation, complete the following:

- List the equipment you need to investigate circuits using three solar cells.
- Indicate how the investigation could be designed in a way to maximise its reliability.
- Indicate how the investigation could be designed in a way to maximise its validity.
- Draw an electrical circuit diagram of the two experimental set-ups with the three solar cells. Use a symbol of a circle with 'SC' written inside it to represent a solar cell.

4 Communicate

Write a valid and reliable procedure that you could use to compare the electrical output of an array of solar cells connected in series with an array of solar cells connected in parallel.

9780170491785

DISEASE

SYLLABUS OUTCOMES

A STUDENT:

- explains how an understanding of the causes of disease can be used to prevent and manage the spread of disease SC5-DIS-01
- analyses data from investigations to identify trends, patterns and relationships, and draws conclusions SC5-WS-06
- communicates scientific arguments with evidence, using scientific language and terminology in a range of communication forms SC5-WS-08

© 2023 NSW Education Standards Authority

CHAPTERS RELATED TO THIS FOCUS AREA ARE:

- CHAPTER 5 – HOMEOSTASIS
- CHAPTER 6 – DISEASES

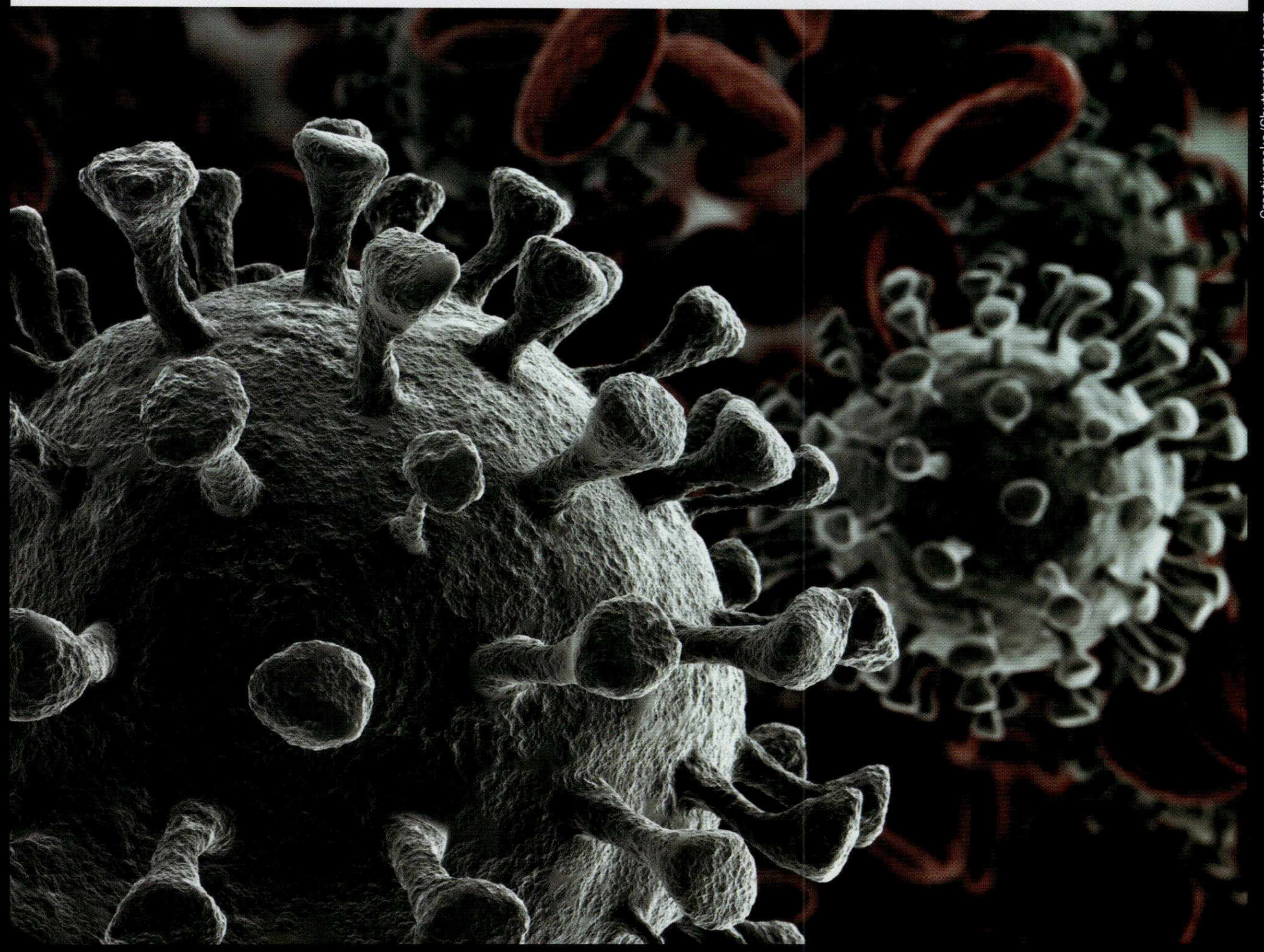

Creativeneko/Shutterstock.com

5 Homeostasis

9780170491785

SCIENCE IN DEPTH 

▲ FIGURE 5.0.1 Sweating is one way our bodies maintain a set body temperature.

Do you get really hot in summer or during exercise? In 2024, people living in Wilcannia, in far west New South Wales, experienced a 44.9°C day, while the weather station at Smithville in the state's far north-west recorded a temperature of 46.3°C. Air-conditioning and sweating are two different ways of cooling down. Air-conditioning is expensive, and generally not good for the environment. Sweating is our body's natural mechanism for cooling down; however, excessive sweating can be life-threatening due to dehydration.

▶ **What alternative, effective strategies can we choose instead of air-conditioning to help us stay cool on a hot day?**

 **DIVE INTO SCIENCE!**

At the end of this chapter, you can complete Science in Depth Study #5. You can use the information you learn in this chapter to complete the project.

Assessments

- Prior knowledge quiz
- Chapter review questions
- End-of-chapter test
- Depth study: Mind map and infographic

Videos

- Science skills in a minute: Using a microscope **(5.6)**; Identifying trends **(5.9)**
- Video activities: Nervous system **(5.2)**; The brain **(5.4)**; Endocrine system **(5.5)**; What is diabetes? **(5.10)**

Science skills resources

- Science skills in practice: Using a microscope to examine cells **(5.6)**; Identifying trends **(5.9)**
- Extra science investigations: Negative feedback loops **(5.1)**; Dissecting a spinal cord **(5.2)**; Dissecting a brain **(5.4)**

Interactive and other resources

- Drag and drop: Different types of neurons **(5.3)**; Maintaining body temperature **(5.7)**
- Label: Parts of the brain **(5.4)**; Negative feedback loops **(5.8)**
- Worksheets: Colour in a neuron **(5.3)**; Endocrine glands **(5.5)**; Multiple sclerosis **(5.5)**

To access resources above, visit **cengage.com.au/nelsonmindtap**

5.1 Homeostasis

BY THE END OF THIS MODULE, YOU WILL BE ABLE TO:

- ✓ define homeostasis, positive feedback mechanism, negative feedback mechanism, stimulus, receptor, coordinating centre, effector and response
- ✓ draw a model of a negative feedback mechanism and describe each step
- ✓ explain the function of a negative feedback mechanism in relation to homeostasis.

Extra science investigation
Negative feedback loops

GET THINKING

If you observe the eyes of a cat in different conditions, you will notice that their pupils change. Why does this change happen? Do you think this is a voluntary or involuntary response?

Homeostasis is a balancing act

On hot days, setting your air-conditioner to a specific temperature keeps your house at a constant, comfortable temperature. When the temperature of the room goes above or below the set temperature, the air-conditioner's sensors detect the change, and the air-conditioner switches on or off. Similarly, when your body's temperature rises above or below 37°C, processes inside your body 'switch on' or 'switch off' to regulate your body's temperature and keep it stable. The balancing act of maintaining a stable internal environment is one part of **homeostasis**. All organisms need homeostasis to survive changes in the external environment.

homeostasis
the maintenance of a constant internal environment necessary for survival

tolerance range
the range of a particular condition inside the body that an organism can survive

Tolerance range and homeostasis

Your body responds to changes in internal or external conditions, such as temperature, blood glucose levels, water and salts, and uses different processes to keep internal factors within the specific ranges you can tolerate. This known as a **tolerance range**. If the conditions inside your body change too far from these ranges, you can become sick or even die (Figure 5.1.1).

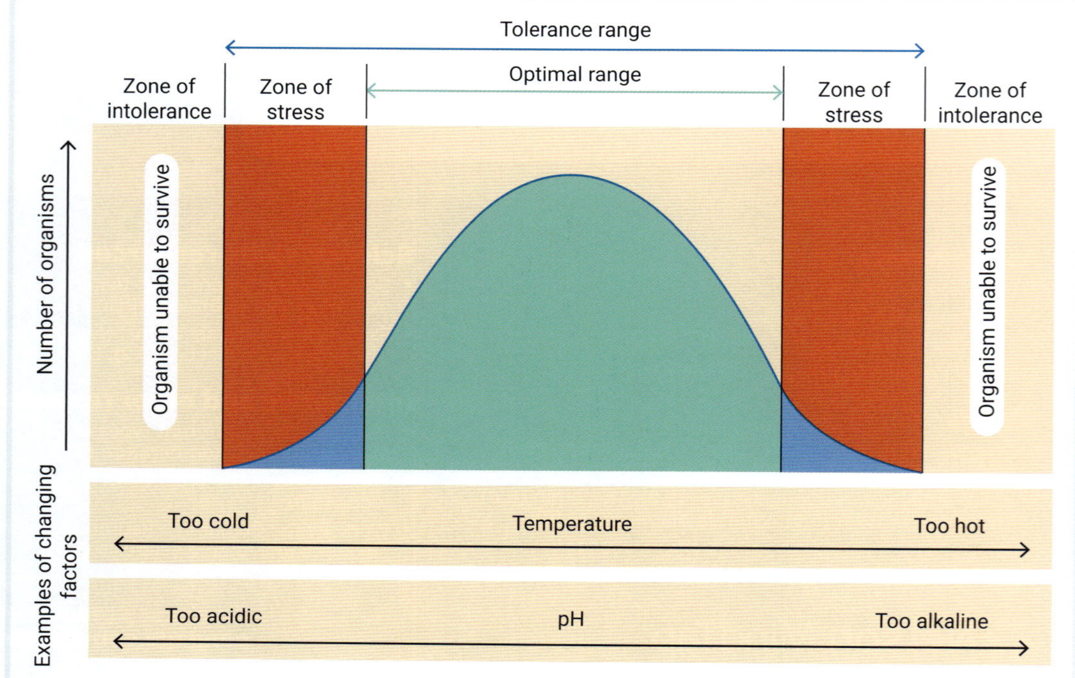

▶ **FIGURE 5.1.1** For survival, internal factors need to be maintained within a tolerance range. The body works best when these factors are within optimal range.

9780170491785

Our bodies function best when factors are within an optimal range. Beyond this range, our bodies become stressed, but we can usually survive unless the factor deviates into the zone of intolerance. Different organisms have different ranges within which they can survive.

Stimulus–response models

Scientists have developed a model called the stimulus–response model to help us understand the mechanisms that work to maintain our body's constant internal environment (Figure 5.1.2).

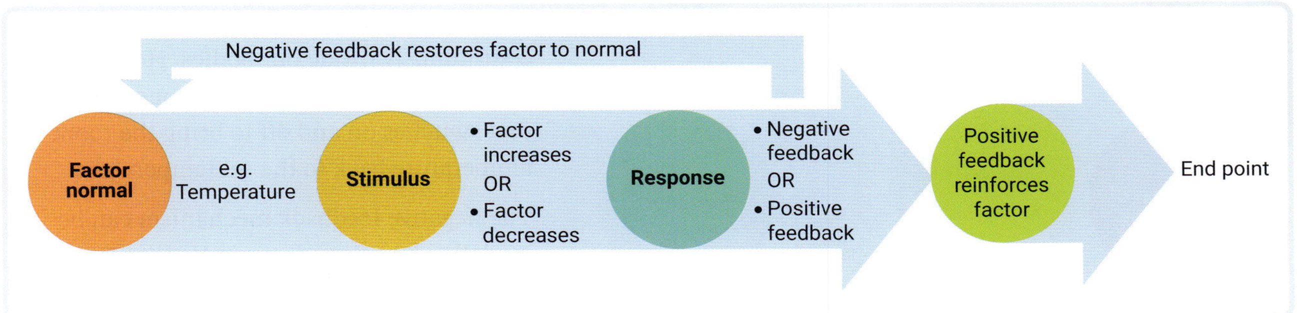

▲ **FIGURE 5.1.2** The stimulus–response model

Stimulus: detecting change

A **stimulus** is a mechanism that causes a change in the environment (inside or outside the body) that can be detected by one of the body's **receptors**. For example, an increase or decrease in environmental temperature (e.g. a cold wind) is a stimulus. A receptor is any part of the body (cell, tissue or organ) that can receive information from a stimulus. For example, thermoreceptors in the skin and brain are receptors for temperature. A receptor detects a stimulus when conditions go above or below its normal narrow range. So, when your body is cold in winter or hot in summer, receptors will detect these changes.

Response: reacting to change

A **coordinating centre** receives information from the receptor, processes the information and coordinates a **response**. For example, the hypothalamus is the part of the brain that acts as the coordinating centre for temperature regulation. The coordinating centre sends signals to an **effector** to carry out a response. An effector is any muscle or gland that does the work to return the body's condition to normal. For example, when your body is too cold, the hypothalamus sends a message to muscles attached to your hair follicles, telling them to contract. This makes your hairs stand up and conserves heat. You may know this as 'goosebumps'.

Returning to normal range

A response is the action the body takes after a stimulus, usually to return conditions inside the body to the normal range. When the response reduces the condition or factor back to its optimal range, the mechanism is known as **negative feedback**. When the response makes the condition or factor move further away from normal, the mechanism is called **positive feedback**.

stimulus
a change in a factor above or below the optimal range

receptor
a specialised cell that detects a stimulus; may be internal or external

coordinating centre
an organ or tissue that receives and processes information from receptor cells and coordinates a response

response
the action a body takes due to detection of a stimulus, usually to return conditions inside the body to the normal range

effector
a muscle or gland that receives a message from the coordinating centre and carries out a response

negative feedback
a mechanism in which a response reduces a condition or factor back to its optimal range

positive feedback
a mechanism in which a response reinforces the condition or factor until an end point

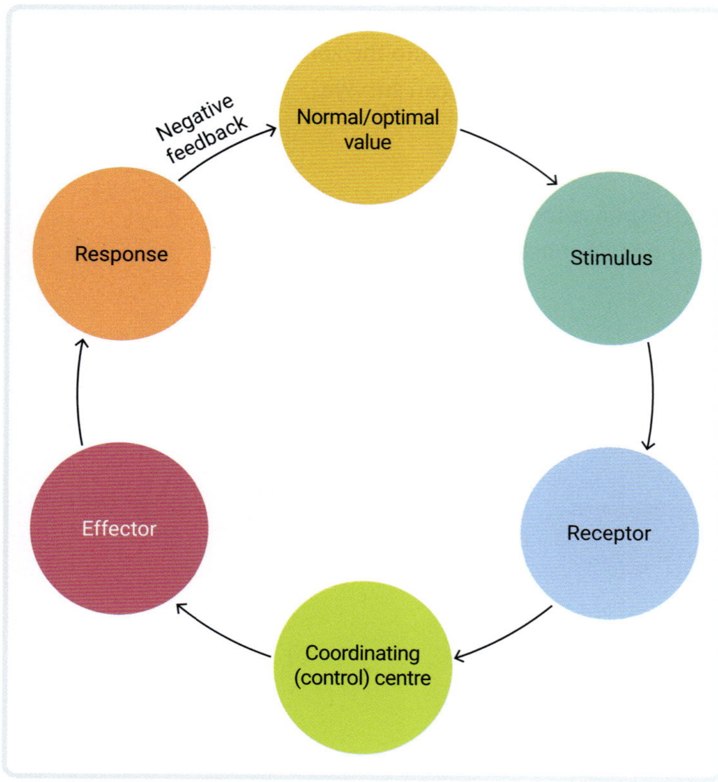

▲ **FIGURE 5.1.3** A negative feedback model

Negative feedback

When the body detects a stimulus, it will act to restore normal conditions. The effects of the stimulus can be reduced through a negative feedback mechanism.

We use the negative feedback model to understand the steps involved in homeostasis. The model represents homeostasis as a circuit, which suggests this process occurs continuously inside our bodies. However, in reality it is a bit like the air-conditioner that switches on and off to bring the room temperature back to the set temperature.

The negative feedback mechanism can be modelled by a flow diagram (Figure 5.1.3).

Positive feedback

If a response forces a stimulus to increase instead of returning to normal, the mechanism is referred to as positive feedback. Positive feedback is rare. One example is blood clotting. When skin is cut, special white blood cell fragments, known as platelets, release clotting factors. This causes more platelets to be transported to the wound to help form a clot, preventing blood loss and forming a scab. Positive feedback continues until a certain point has been reached, such as the finished formation of a blood clot.

ACTIVITY

▲ **FIGURE 5.1.4** The ice bucket challenge

Ice bucket challenge

Have you done the ice bucket challenge? Many celebrities fundraise for charity. The challenge involves pouring a bucket of ice water over your head, as shown in Figure 5.1.4.

Analysis

1. Look at the stimulus–response model in Figure 5.1.2. What part of the model does the cold water represent?
2. What is the effect of the cold water on the internal temperature?
3. How does the body respond to this change to return to 'balance'?
4. Do you think that the body's internal temperature remains constant at all times?
5. **Create** a flow chart like the stimulus–response model to **explain** how the body uses homeostatic mechanisms during the ice bucket challenge.

9780170491785

5.1 LEARNING CHECK

1 **Define** homeostasis.
2 **Describe** negative feedback.
3 **Describe** two examples to illustrate the stimulus–response model.
4 **Compare** and **explain** the principles of negative and positive feedback mechanisms by copying and completing the following table.

Principle	Negative feedback	Positive feedback
Role in homeostasis and the survival of an organism		
Deviation of factor (e.g. temperature) from normal		
Frequency (frequent or rare)		

5 Look at the homeostasis feedback mechanisms used to maintain constant temperature inside the body, shown in Figure 5.1.5.
   a **Identify** and **explain** the role of sensors, the control centre and effectors in the feedback mechanism.
   b You are snowboarding and your jacket is not warm enough. **Create** your own annotated version of Figure 5.1.5 to show how your body will regulate its temperature while snowboarding.

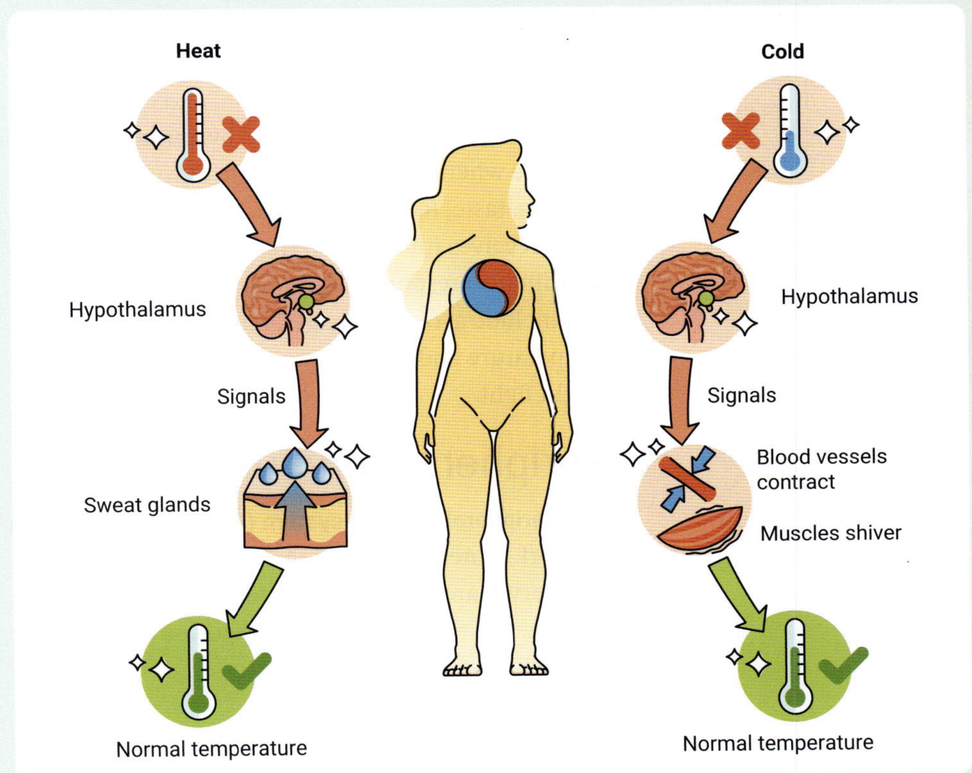

▲ **FIGURE 5.1.5** Homeostatic feedback mechanisms for body temperature

5.2 The nervous system

BY THE END OF THIS MODULE, YOU WILL BE ABLE TO:

- ✓ compare the structure and function of the central nervous system and the peripheral nervous system
- ✓ label diagrams that represent the central nervous system and the peripheral nervous system.

GET THINKING

Imagine a family road trip through a city and along a highway. How could you compare the nervous system to a road system? What could be represented by the 'cars', 'roads', 'traffic lights' and 'traffic jams' in this analogy?

central nervous system (CNS)
the brain and spinal cord

nerve
a collection of fibres, surrounded by a protective coat, that transmit messages as nerve impulses to and from the CNS

Parts of the nervous system

Animal systems depend on the nervous system as a communication network. Specific functions of the nervous system include monitoring change, transmitting messages and coordinating responses.

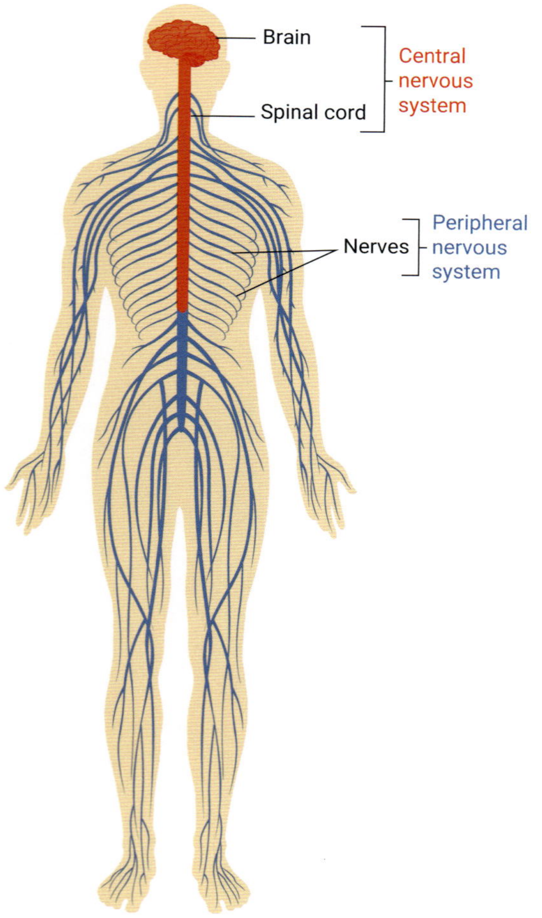

The nervous system consists of two parts: the central nervous system and the peripheral nervous system, as shown in Figure 5.2.1.

Brain
Spinal cord
Central nervous system
Nerves
Peripheral nervous system

▲ FIGURE 5.2.1 The two parts of the nervous system: the central nervous system (red) and the peripheral nervous system (blue)

Central nervous system

The **central nervous system (CNS)** consists of the brain and spinal cord. Its function is to detect a stimulus, interpret and process information, and coordinate a response. The CNS is connected by **nerves** to receptors in tissue where our five senses are at work: sight, hearing, taste, touch and smell. When the brain or spine receives information from the receptors, they coordinate a response. Responses vary enormously and include avoiding danger, resetting body temperature and readjusting our eyes to manage bright light.

Peripheral nervous system

The **peripheral nervous system (PNS)** is made up of a network of nerves outside the brain and spinal cord. The nerves are made of **nerve fibres** that transmit messages in the form of **nerve impulses** to and from the CNS.

9780170491785

Messages travel along nerves in the PNS. Each nerve is a collection of nerve fibres, as shown in Figure 5.2.2. Nerves extend from the spinal cord to all muscles, glands and organs in the body.

The PNS consists of all the nerves coming out of the brain and the spinal cord. It provides the CNS with sensory information to form responses to stimuli. Functionally, the PNS is divided into the autonomic nervous system (which controls involuntary responses) and the somatic nervous system (which controls voluntary movements) (Figure 5.2.3).

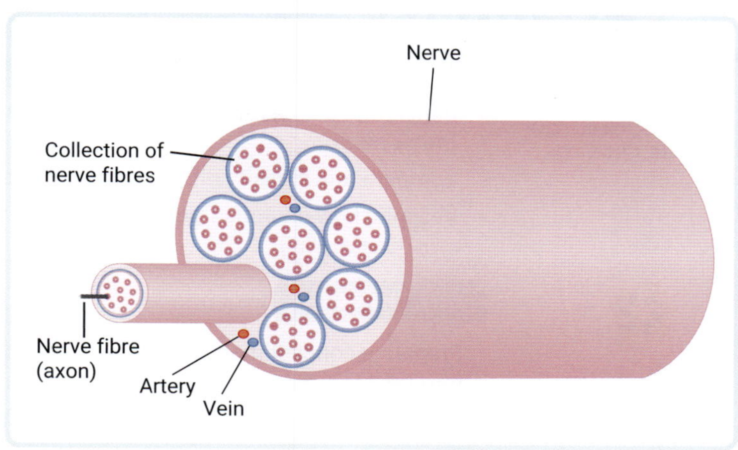

▲ **FIGURE 5.2.2** The structure of a peripheral nerve in the human body

The autonomic nervous system performs involuntary functions that we are not conscious of, such as sweating, shivering, the beating of our heart and insulin release.

The somatic nervous system performs voluntary functions that we are conscious of, such as muscular movements required to walk, eat and game.

peripheral nervous system (PNS)
the network of nerves outside the brain and spinal cord

nerve fibre
the section of a nerve cell that carries nerve impulses away from the cell body

nerve impulse
an electrical message that is transmitted along nerves to and from the CNS

Video activity
Nervous system

Extra science investigation
Dissecting a spinal cord

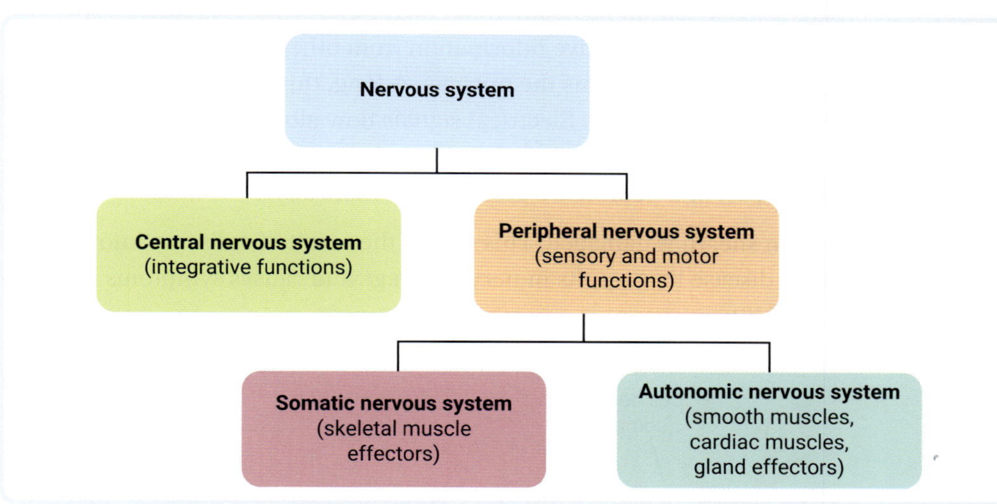

▲ **FIGURE 5.2.3** The parts of the nervous system

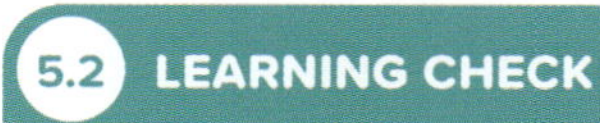

5.2 LEARNING CHECK

1 **State** the main parts of the CNS and the PNS.
2 **Draw** a diagram showing the distinct parts of the CNS and the PNS.
3 **Describe** the function of the autonomic part of the PNS.
4 **Research** more about the functions of the autonomic and somatic nervous system. **Construct** a table to **compare** them and **describe** two examples for each system.
5 **Create** a crossword with the key words from this module.

5.3 Neurons

BY THE END OF THIS MODULE, YOU WILL BE ABLE TO:

- ✓ define and draw labelled diagrams of a neuron, a sensory neuron, an interneuron and a motor neuron
- ✓ compare and classify different types of neurons
- ✓ describe how neurons transmit messages and describe the function of a neurotransmitter.

Interactive resource
Drag and drop: Different types of neurons

Other resource
Worksheet: Colour in a neuron

GET THINKING

Prepare a concept map to summarise the content in Modules 5.1 and 5.2. As you work through Modules 5.3–5.5, add new information to your concept map. You may wish to include diagrams to help your understanding.

Neurons

Information is carried around the human body at mesmerising speeds, up to 120 metres per second ($m\,s^{-1}$). The structures that carry the information are highly specialised nerve cells called **neurons**. There are an estimated 100 billion neurons in the brain and spinal cord combined, and many more specialised neurons in the peripheral nervous system.

As shown in Figure 5.3.1, a neuron has branching, finger-like structures called **dendrites** around much of its cell body that receive information from other cells in the form of chemical signals. The **cell body** contains the nucleus. A long, thin fibre, called an **axon**, extends from one side of the cell body. Electrical signals flow along the axon towards the next neuron. The **myelin sheath** is a fatty, insulating layer around each axon. The functions of each part of a neuron are described in Table 5.3.1.

Sometimes the myelin sheath is damaged. This can be the result of a disease known as multiple sclerosis – a disease that results in nerve damage and causes symptoms such as weakness, fatigue and blurry vision.

neuron
a specialised cell that can transmit nerve impulses; also known as a nerve cell

dendrite
a branching network at the end of a neuron that receives information from other neurons

cell body
the part of a neuron that contains the cytoplasm, including the nucleus

axon
a long, thin fibre that carries electrical impulses from the cell body of a neuron towards the next neuron

myelin sheath
a protective coat around an axon that increases the speed of nerve impulses

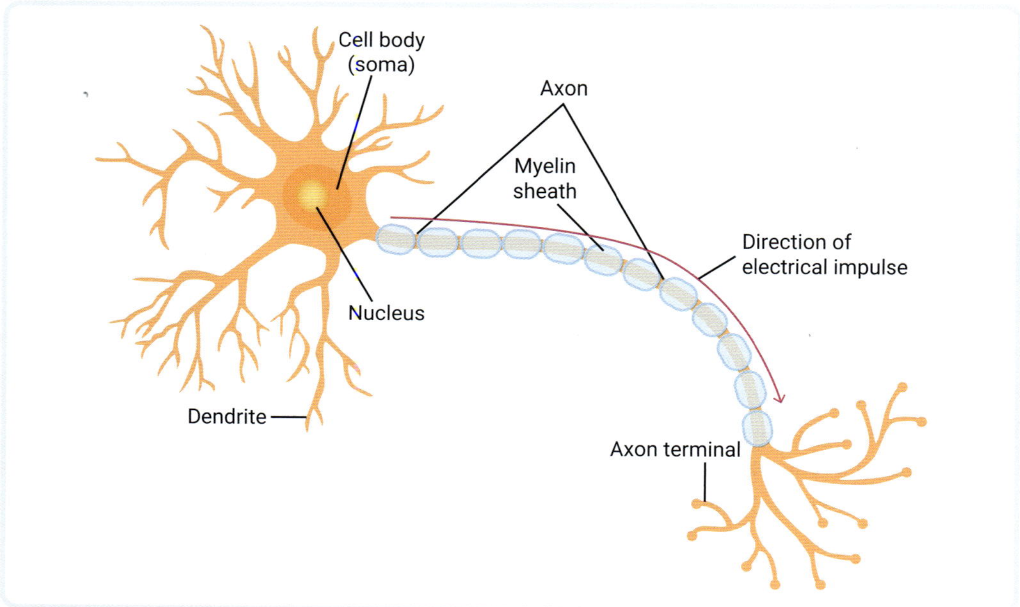

▲ **FIGURE 5.3.1** The basic structure of a neuron and the direction of the nerve impulse

9780170491785

▼ **TABLE 5.3.1** The basic neuron structures and their functions

Structure	Functions
Dendrite	Receives information from other neurons in the form of chemical signals
Cell body	Contains the nucleus
Axon	Conducts (carries) electrical impulses from the cell body, or another cell's dendrites, to the end of the neuron (axon terminals)
Myelin sheath	Increases the speed of nerve impulses Provides an insulating (high electrical resistance) fatty layer that protects the nerve fibres from other electrical signals

Neuron communication

A neuron conducts an electrical signal along its axon, as shown in Figure 5.3.1. In some cases, the signal needs to transform into a chemical signal to flow across the gap (**synapse**) between most neurons. If there were no gaps, the network of neurons would conduct the electrical impulses continuously and uncontrollably. Imagine your home without light switches. You would not have any control over switching the lights off and on.

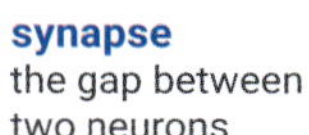

synapse
the gap between two neurons

Synapses provide some communication control. When an electrical impulse arrives at the axon terminal, it activates the release of a **neurotransmitter** (Figure 5.3.2). The neurotransmitter delivers the message across the synapse to the dendrite of the next neuron, or a muscle or gland cell. At the synapse, the impulse can be modified to be made bigger or smaller depending on the needs of the body. Once the message has been delivered across the synapse, the message is converted back into an electrical impulse.

neurotransmitter
a chemical signal that delivers a message across a synapse, to be converted back to an electrical impulse

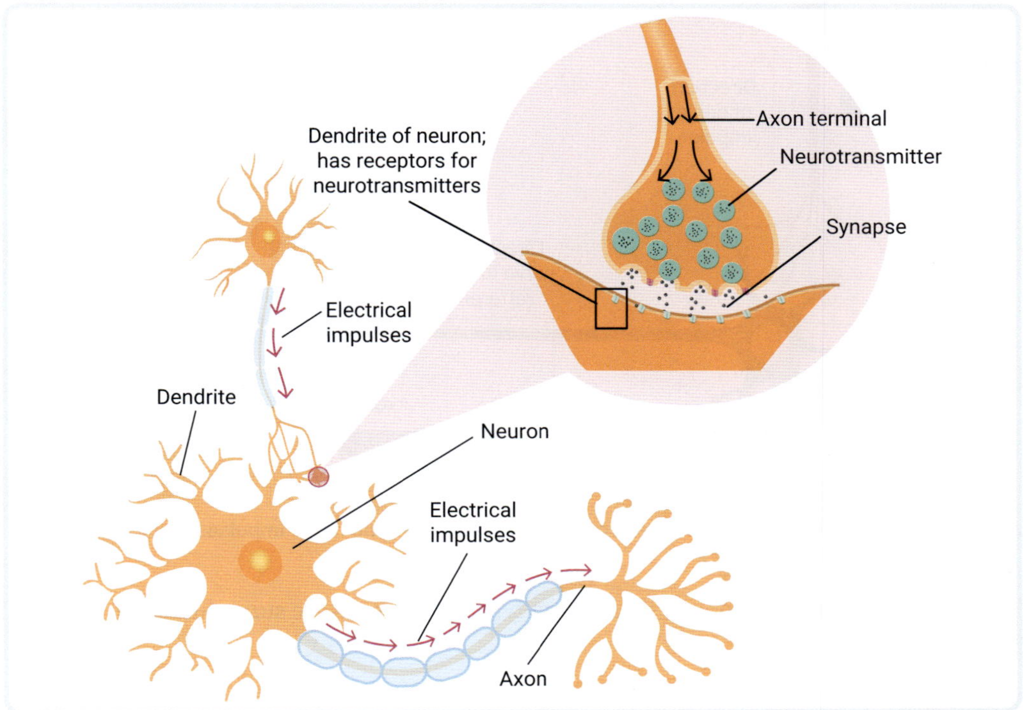

▲ **FIGURE 5.3.2** Neurotransmitters pass messages across a synapse from the axon terminal of one neuron to the receptors on the dendrite of the next neuron.

Three types of neurons

Figure 5.3.3 shows the structure and function of the three main types of neurons – sensory neurons, motor neurons and interneurons. The function of each type is possible because of their unique structure. They can be classified according to the direction in which they send nerve impulses.

- **Sensory neurons** have sensory structures that detect changes in their environment. They send an electrical impulse with this information towards the CNS (**afferent direction**).
- **Interneurons** are shorter neurons that have many branching dendrites to communicate a lot of information between sensory neurons and motor neurons.
- **Motor neurons** have long axons reaching effector muscles or glands with information sent from the CNS (**efferent direction**).

sensory neuron
a neuron that sends electrical impulses from receptors to the CNS

afferent direction
from the sensory receptor towards the central nervous system

interneuron
a short neuron that sends nerve impulses between sensory and motor neurons within the CNS

motor neuron
a neuron that sends electrical impulses from the CNS to effectors

efferent direction
from the central nervous system to the effector (muscle or gland)

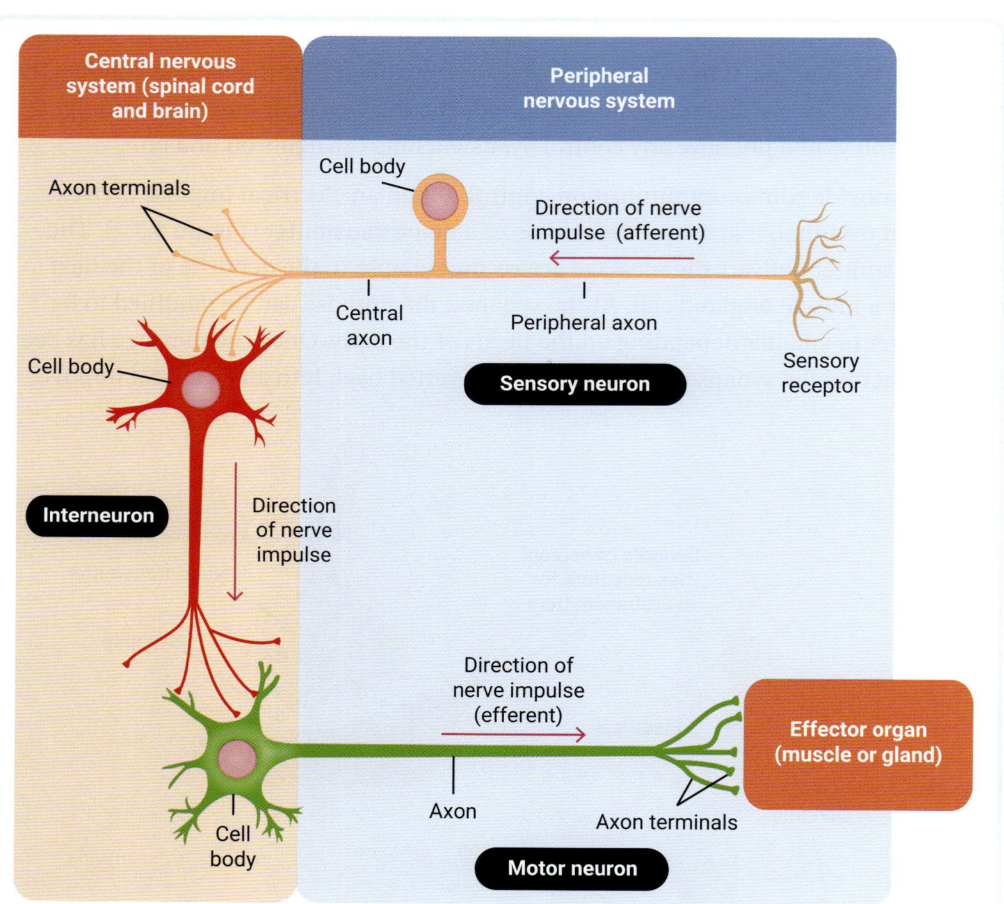

© Helder T. Moreira/Cengage

▲ **FIGURE 5.3.3** The three types of neurons work together to send nerve impulses towards and away from the CNS.

The structures and functions of these three types of neurons are summarised in Table 5.3.2.

TABLE 5.3.2 The structure and function of the three main types of neurons

	Sensory neuron	Interneuron	Motor neuron
Structure	One relatively long axon, with sensory receptors at one end, and dendrite branches at the other end, and the cell body positioned between the two	Relatively short axons with cell body and numerous branching dendrites	Many branching dendrites at one end of a large cell body with a long axon ending with axon terminals
Location	Sensory organs such as eyes and skin in PNS	CNS, within brain and spinal cord	Dendrites in CNS; axon terminals in PNS
Function	To detect changes and transmit nerve impulses towards spinal cord or brain	To transmit nerve impulses from a sensory neuron or other interneurons to a motor neuron or other interneurons	To transmit nerve impulses from brain and spinal cord to a muscle or gland to carry out a response
Direction of impulse	From sensory receptors to CNS (afferent direction)	Between sensory and motor neurons	From CNS to an effector muscle or a gland (efferent direction)

5.3 LEARNING CHECK

1 Visit the Queensland Brain Institute website to view various stained microscope images of neurons. **Observe** and **describe** a typical neuron.

2 **Define** neurotransmitter.

3 Three common neurotransmitters are acetylcholine, serotonin and dopamine. **Choose** one to **research**. Write five sentences about its function.

4 **Explain** why the axon part of a neuron is covered in myelin.

5 **Investigate** the speed of your nerve impulses (reaction time). Ask a partner to hold a ruler between your thumb and first finger (don't grip the ruler). Line up 0 cm at your finger. Your partner drops the ruler while you try to catch it. **Record** the length on the ruler where you caught it (Figure 5.3.4). Swap and repeat. The person with the shortest length may have the fastest impulses!

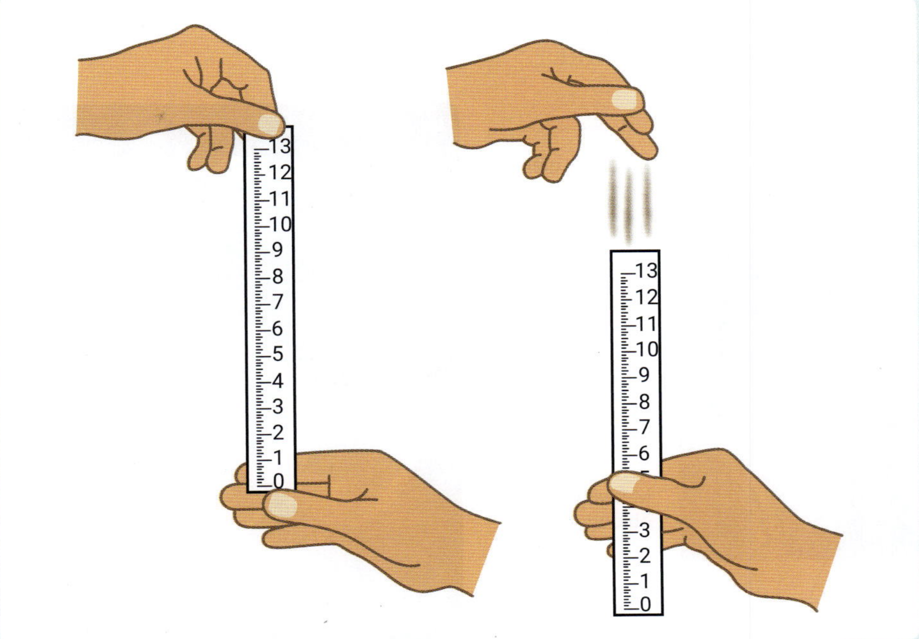

▲ **FIGURE 5.3.4** Testing reaction time with a ruler

Weblink
Queensland Brain Institute

5.4 The brain

BY THE END OF THIS MODULE, YOU WILL BE ABLE TO:

- ✓ describe the structures and functions of the brain and its parts: cerebrum, cerebellum, pituitary gland, hypothalamus and brain stem
- ✓ explain how changes to the brain can affect body functions.

Video activity
The brain

Interactive resource
Label: Parts of the brain

Extra science investigation
Dissecting a brain

GET THINKING

What do you think happens inside your brain when are you listening to music or reading? How many areas of your brain do you think are working together right now? After you read this module, reflect on which part of the brain fascinates you the most.

Brain facts

What memories do you have of last school holidays? Where are those memories filed away? They are stored in a very complex organ, the **brain**. Situated within the protective bones of the cranium (skull), the brain is about 1.4 kg of living tissue that does your thinking, learning, feeling and remembering. The brain is the coordinating centre of the CNS.

brain
a very complex organ; the coordinating centre of the CNS

The brain contains grey matter (mostly neuron cell bodies) and white matter (mostly bundles of axons found in the interior part of the brain). Nerve impulses between neurons are generated when the brain is performing its major functions such as learning new things, feeling emotions, monitoring and analysing sensory information, and coordinating a response.

hemisphere
one of the two symmetrical halves of the brain

The brain has a symmetrical structure: it has a right **hemisphere** and a left hemisphere. Most of the knowledge we have today about the brain was gained through technological advancements in brain imaging such as magnetic resonance imaging (MRI). The left hemisphere controls motor function on the right side of the body and the right hemisphere of the brain controls motor function on the left side of the body. The image in Figure 5.4.1a was made possible because of MRI.

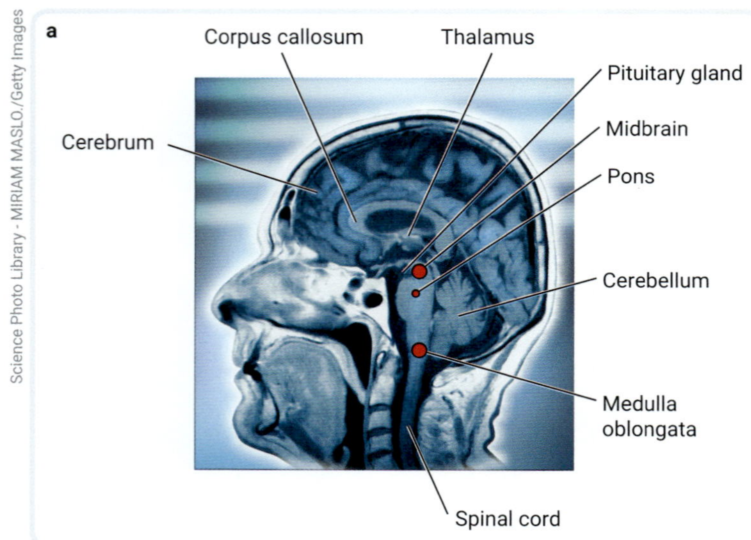

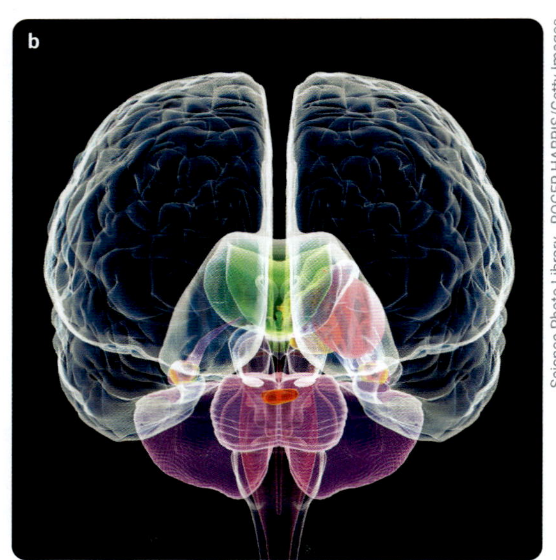

▲ **FIGURE 5.4.1** **(a)** An MRI image of one hemisphere of the brain. The brain stem consists of the midbrain, pons and medulla oblongata (marked in red). **(b)** The symmetrical structure of the brain: the hemispheres.

External brain anatomy

The largest part of the brain, the **cerebrum**, is above the **cerebellum** and **brain stem**. Major functions of the cerebrum include controlling movement, learning, emotion, memory and perception. The wrinkly, grey, outer layer is the cerebral cortex, the information processing centre of the brain. Scientists studying the brain have found that four sections of the cerebrum, called lobes, carry out distinct functions, as shown in Figure 5.4.2.

The two hemispheres communicate via the **corpus callosum**, which is a collection of axons that extend across the two hemispheres. The corpus callosum is near the centre of the brain. It allows comparison and combination of the sensory inputs from the left and right sides of the body.

The cerebellum is tucked under the cerebrum and behind the brain stem. When the cerebellum is damaged or old, people become uncoordinated and can lose their balance. This indicates that some of the functions of the cerebellum include coordination when moving, and balance; for example, when walking.

The brain stem consists of the midbrain, the pons and the medulla oblongata.

- The midbrain receives sensory information, such as sight and sound, and transfers it to the cerebrum.
- The pons transmits messages between the PNS and midbrain.
- The medulla controls automatic (and involuntary) activities such as breathing, heart rate, digestion and vomiting.

cerebrum
the largest part of the brain; controls movement, learning, emotion, memory and perception

cerebellum
a smaller section at the back of the brain; controls movement and balance

brain stem
consists of the midbrain, the pons and the medulla oblongata; controls automatic and involuntary activities such as breathing, heart rate, digestion and vomiting

corpus callosum
a collection of axons that extend across the two hemispheres of the brain that compares and combines the sensory inputs from the left and right sides of the body

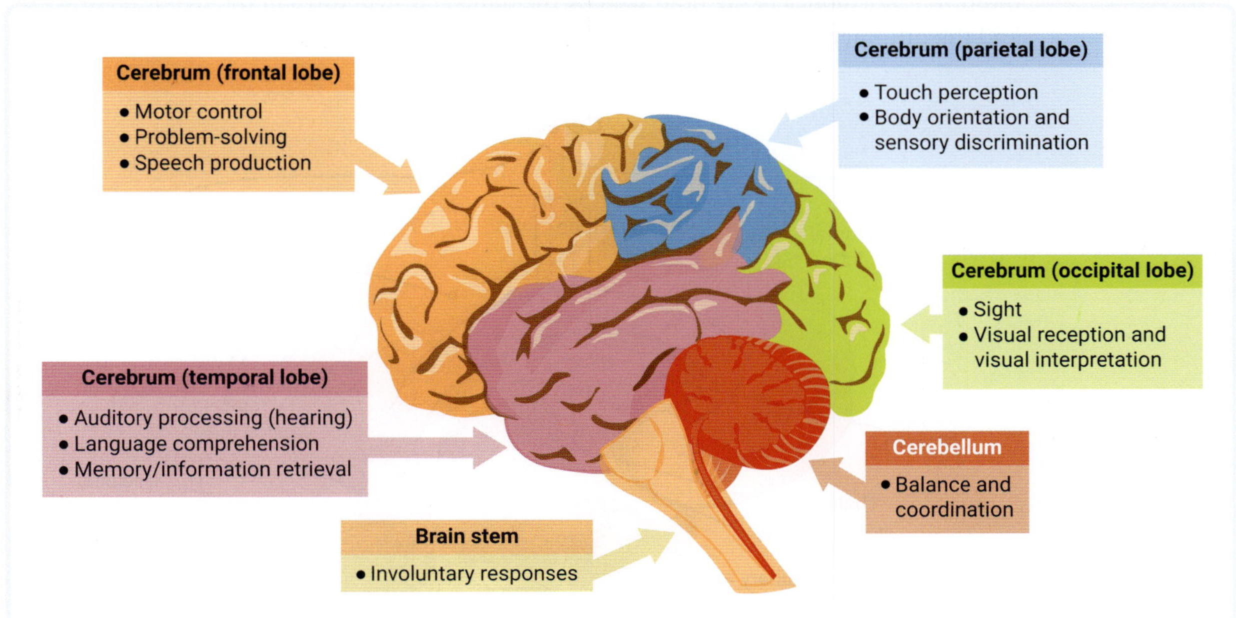

▲ **FIGURE 5.4.2** The external anatomy of the brain

Internal brain anatomy

hypothalamus
an almond-sized structure of the brain; plays a major role in homeostasis

pituitary gland
a gland that produces and releases several hormones; is connected to, and is controlled by, the hypothalamus

homeostatic
related to homeostasis

The **hypothalamus** works with the **pituitary gland** to play a major role in homeostasis. The hypothalamus (*hypo* = low) is located below the cerebrum and in front (anterior) of the brain stem (Figure 5.4.3). Although it is only the size of an almond, it has important **homeostatic** and other functions. Major functions of the hypothalamus include:

- detecting changes in temperature and water levels within the body
- coordinating negative feedback responses
- regulating hunger and thirst
- controlling the fight-or-flight response
- producing hormones
- controlling the pituitary gland.

secrete
produce and release a substance (often a hormone or enzyme)

Feeling thirsty? The hypothalamus produces antidiuretic hormone (ADH) and stores it in the pituitary gland. Your pituitary gland can release ADH when you need to conserve water and one of its effects is to make you thirsty. The pituitary gland can **secrete** (produce and release) many other hormones, such as growth hormones, which are essential for homeostasis and sexual development. The pituitary gland sits just under and is connected to the hypothalamus.

If a person's brain is injured (e.g. by a sudden force to the head, called a traumatic brain injury) the hypothalamus and the pituitary gland may not function properly. If this happens, it can reduce the person's growth or weight gain, or they may have unregulated internal water levels and body temperature. These symptoms are caused by less of various important hormones being produced by the damaged structures in the hypothalamus and the pituitary gland.

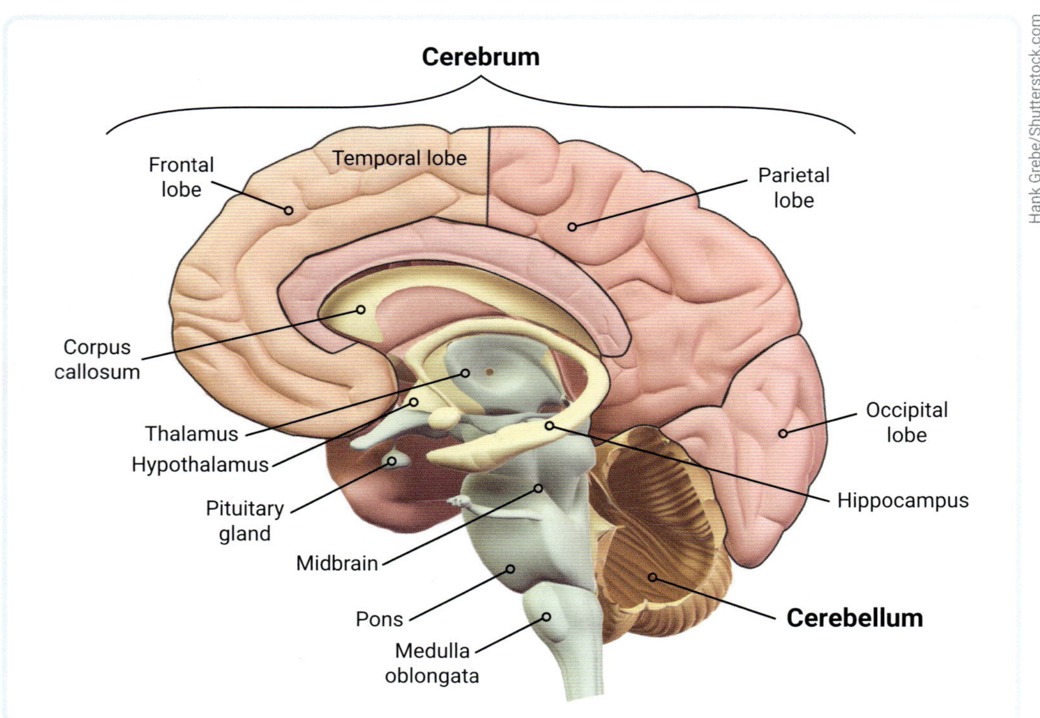

Hank Grebe/Shutterstock.com

▲ **FIGURE 5.4.3** The internal anatomy of the brain

5.4 LEARNING CHECK

1 **Identify** the part of the brain that controls several homeostasis processes in the body.

2 **Describe** the structure of the brain in two or three sentences.

3 **Summarise** the functions of different parts of the brain by copying and completing the following table.

Part of the brain	Function
Corpus callosum	
Brain stem	
Pituitary gland	
Cerebellum	
Hypothalamus	

4 An F1 driver recently suffered a traumatic brain injury during a car accident. One of his symptoms is that he is excessively thirsty all the time. What information about the brain could you share with the driver that may help him understand the cause of this disruption to water homeostasis? **Research** traumatic brain injury to find other symptoms.

5 Copy the following diagram and **label** the main parts of the brain's external anatomy. In each box, add an example of its function (e.g. cerebellum: balance when riding a skateboard).

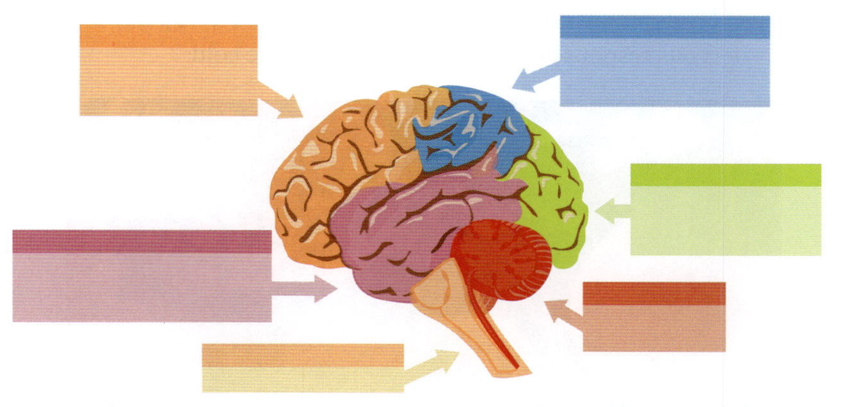

6 **Observe** the MRI image in Figure 5.4.4. It shows a damaged area in the top of the left hemisphere. What part of the brain is affected? **Discuss** what body functions could be impaired by the damage.

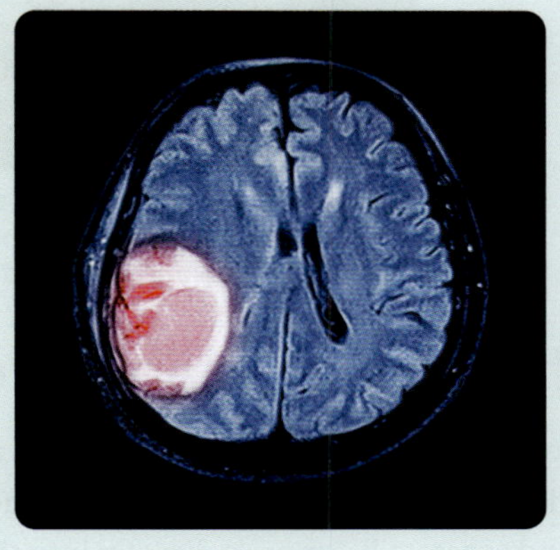

Puwadol Jaturawutthichai/Shutterstock.com

▶ **FIGURE 5.4.4** An MRI of a brain, showing a damaged area (highlighted red)

5.5 The endocrine system

BY THE END OF THIS MODULE, YOU WILL BE ABLE TO:

- ✓ describe the endocrine system, including the main glands and their major hormones and functions
- ✓ describe, explain and compare the functions of different types of hormones
- ✓ compare the endocrine and nervous systems and demonstrate their interdependence.

Video activity
Endocrine system

Other resources
Worksheets:
Endocrine glands

Multiple sclerosis

GET THINKING

Return to the concept map you started at the beginning of Module 5.3. Do you now have more information to add to the map? Scan this module to preview which vocabulary and concepts you could add.

Endocrine system

endocrine system
a network of glands that secrete hormones into the bloodstream to be transported to target cells

gland
a tissue that releases hormones

hormone
a chemical messenger

The **endocrine system** consists of a network of **glands** that secrete chemical messengers – **hormones** – into the bloodstream (Figure 5.5.1). In the bloodstream, hormones are transported to receptors, which receive the information. The messages instruct organs to perform a variety of responses needed for homeostasis, sexual development and survival.

Hormones travel in the blood from a gland to a targeted receptor on a cell. Generally, hormones provide a slower communication system than electrical messages (nerve impulses), but they can have a longer effect. For example, hormones that are responsible for reproductive development cause a slow gradual response compared with the instantaneous nervous response of shivering when you are cold.

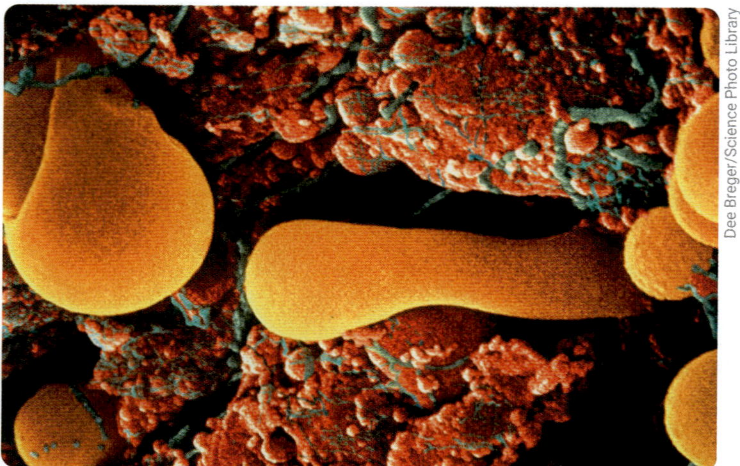

Dee Breger/Science Photo Library

▲ **FIGURE 5.5.1** A coloured scanning electron micrograph of the secretion of thyroid hormones from thyroid tissue. Hormones enter the capillary network (blue) and circulate throughout the bloodstream.

The main human endocrine glands

Are you still growing? During puberty, and for a few years afterwards, you may experience a series of growth spurts due to growth hormones released by the pituitary gland. Endocrine glands secrete hormones directly into the bloodstream. They regulate many human body

9780170491785

functions, including growth.

Figure 5.5.2 shows the location of the main endocrine glands: the hypothalamus, pituitary, thyroid, parathyroid, pancreas, thymus, gonads (testes in males and ovaries in females), pineal and adrenal glands.

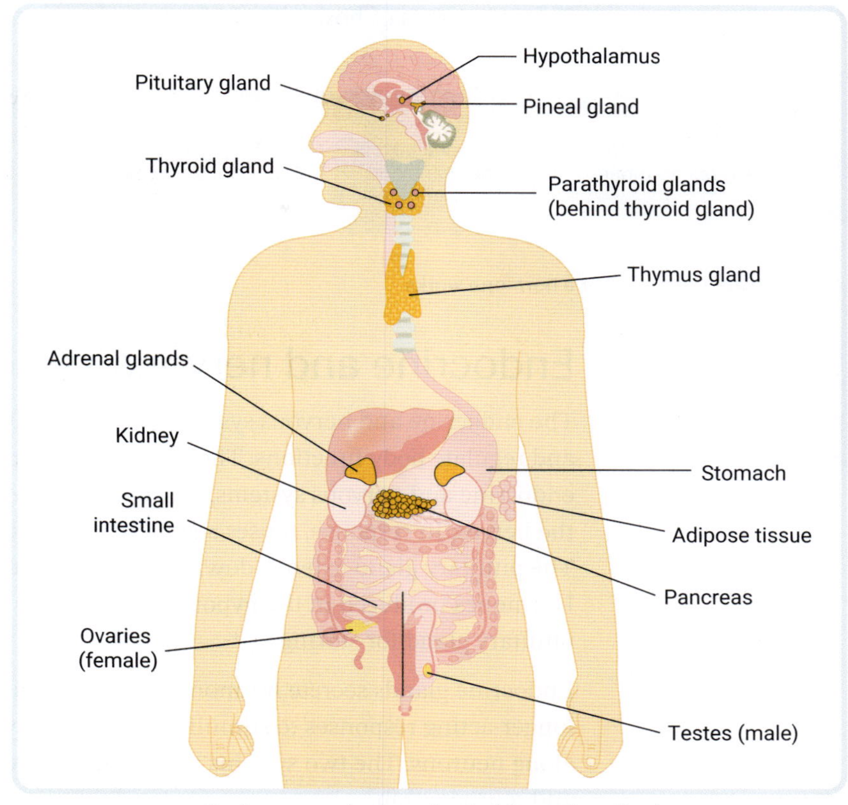

▲ **FIGURE 5.5.2** The human endocrine glands (shown in yellow)

Hormones

Hormones secreted by the hypothalamus, pituitary, thyroid, parathyroid, pancreas and adrenal glands are involved in negative feedback mechanisms. Hormones can be classified according to their chemical class, as:

- water-soluble proteins (long chains of amino acids)
- peptides (short chains of amino acids)
- amines (hormones made from a single amino acid)
- fat-soluble steroids (made from the lipid cholesterol).

The chemical class helps us predict their action when they arrive at a target cell. Some examples of hormones and their functions are listed in Table 5.5.1.

▼ **TABLE 5.5.1** Some of the human endocrine glands, major hormones and their functions

Gland	Hormone example	Hormone class	Hormone function
Pituitary gland	Antidiuretic hormone (ADH)	Peptide	Controls water regulation by increasing reabsorption of water in kidneys
	Growth hormone	Peptide	Stimulates growth
Thyroid gland	Thyroxine	Protein	Controls metabolic rate
Pancreas	Insulin	Protein	Lowers blood glucose level
	Glucagon	Protein	Raises blood glucose level
Ovaries	Progesterone	Steroid	Regulates the menstrual cycle; prepares and supports the process of pregnancy
Testicles	Testosterone	Steroid	Develops and maintains male sexual characteristics
Pineal gland	Melatonin	Amine	Regulates the biological 'sleep–wake' rhythm

The main steps of how a hormone works are summarised in Figure 5.5.3.

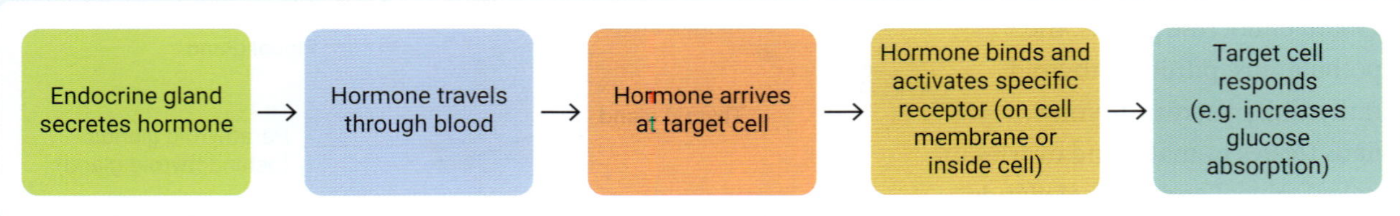

▲ **FIGURE 5.5.3** How a hormone works

Endocrine and nervous systems working together

The endocrine and nervous systems act separately and together to communicate about and regulate body functions. The hypothalamus plays a major role in coordinating the endocrine and nervous systems. The hypothalamus receives sensory information from the nerves. In response, it uses a combination of endocrine and nervous messaging. The pituitary gland extends down from the hypothalamus and is joined to it by axons. Hormones produced in the hypothalamus can be transported through the axons to the pituitary gland for storing.

Endocrine glands secrete hormones that travel through the blood to coordinate slower, longer-acting responses to stimuli. The nervous system uses fast electrical nerve impulses along neurons. The two systems can work together to achieve homeostasis, development and reproduction.

5.5 LEARNING CHECK

1 **Describe** how a hormone works, step by step.

2 Copy and complete the table of glands and hormones by adding the missing information. Extend the table by adding three more glands and their hormone examples and functions.

Gland	Hormone example	Hormone function
Pituitary gland		Controls water regulation by increasing reabsorption of water in kidneys
	Growth hormone	Stimulates growth
	Insulin	
	Glucagon	

3 **Create** a table with one similarity and two differences between hormones and nerve impulses.

4 **Explain** the separate and combined roles of the endocrine system and nervous system.

5.6 Using a microscope to examine cells

SCIENCE SKILLS IN FOCUS

IN THIS MODULE, YOU WILL FOCUS ON LEARNING AND IMPROVING THESE SKILLS:

- selecting and using appropriate microscope equipment, including digital technologies, to record and communicate structures involved in homeostasis
- analysing and connecting a variety of data and information to identify and explain relationships.

Using microscopes

Microscopes can magnify objects up to 400 times their actual size, which is why we use microscopes to see some cells. Using microscopes is an important science skill. When using a microscope, remember the following:

- Carry the microscope with two hands.
- Click the lowest objective lens into place first. After focusing images on a low magnification, turn the revolving nose to click higher objective lenses into place.
- If your school owns a digital microscope, you can use it to capture and record images or videos on a computer. You can use your laptop if it is compatible.
- To calculate total magnification, multiply the objective lens magnification by the eyepiece lens magnification. Look for the values on the microscope parts.

Locating cells under a microscope is difficult. Use a combination of viewing specimens under the microscope and looking at photos of stained specimens.

Draw and label a section of what you see, such as Figure 5.6.1. Figure 5.6.2 shows the labelled parts of the microscope.

Legger/Dreamstime LLC

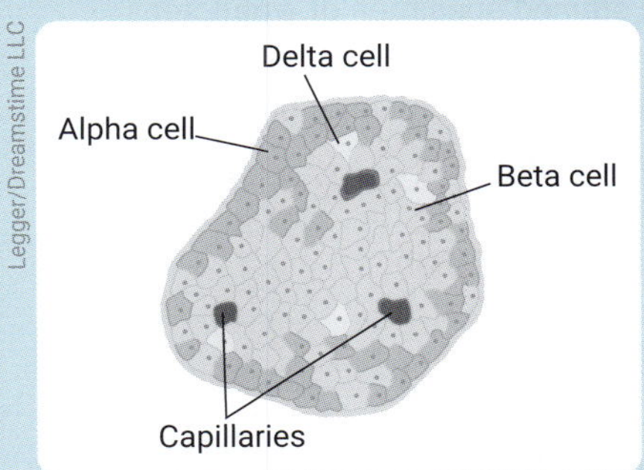

▲ **FIGURE 5.6.1** A sketch of an islet of Langerhans

Video
Science skills in a minute: Using a microscope

Science skills resource
Science skills in practice: Using a microscope to examine cells

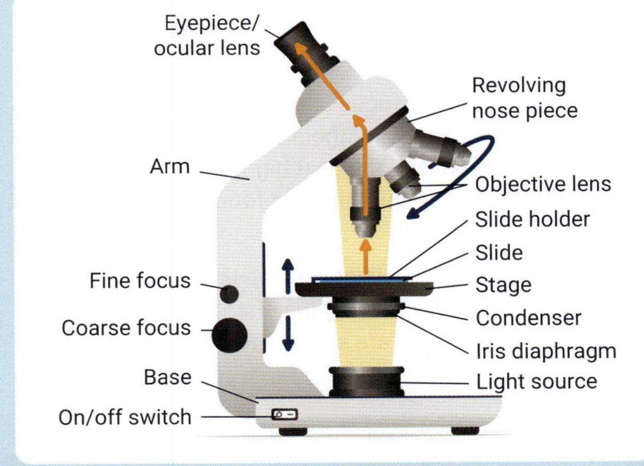

▲ **FIGURE 5.6.2** The parts of a microscope

MICROSCOPE INVESTIGATION OF PANCREAS AND BRAIN CELLS

BACKGROUND INFORMATION

The pancreas is an endocrine gland that secretes hormones for blood glucose regulation. The hormone-producing cells are in clumps called islets of Langerhans. The brain controls all the body's nervous system functions, such as breathing, voluntary movement and sensory processes. Neurons are cells that transmit information between the brain and the rest of the body.

AIM

To conduct a microscope investigation of the islets of Langerhans (the site of alpha and beta cells) and the structure of brain cells (neurons)

MATERIALS AND EQUIPMENT

- ☑ prepared and stained specimen of a pancreas section showing an islet of Langerhans
- ☑ prepared and stained specimen of brain cells
- ☑ access to the internet
- ☑ light microscope
- ☑ digital microscope (optional)

PROCEDURE

1. Set up your microscope by turning on the light and positioning the stage close to the objective lens.
2. Turn the revolving nosepiece until the lens with the lowest magnification clicks into place.
3. Place a prepared slide of an islet of Langerhans on the stage and use the stage clips to secure it in place.
4. Turn the coarse focus knob slowly until the image comes into focus. Without any further adjustments, turn the lens with the next highest magnification into position. Adjust the fine focus knob until the image is in focus.
5. Draw a sketch of what you see under low magnification, showing the distribution of islet cells. If the parts are stained, identify and label them on your diagram.
6. Focus a section of the specimen under high magnification. Draw a diagram of one islet, labelling what you can. Include a title and the total magnification next to each of your diagrams. You may see something like Figure 5.6.3a.
7. Repeat steps 1–6 with the brain cells (neurons) specimen.
8. Conduct an internet search of at least 10 different microscope photos of islets of Langerhans. Draw one more sketch, one-third of a page in size, labelling the islet of Langerhans, and alpha and beta cells.
9. Search online for images of brain cells. Draw the cells and label axons, dendrites and cell body. An example of a suitable image is shown in Figure 5.6.3b.

ANALYSIS

1. **Describe** the location of alpha and beta cells.
2. **Describe** the difference between the pancreas cells and the brain cells in terms of the structures you can see.
3. **Describe** what you could see under low power compared with what you could see under high power for both specimens, pancreas and brain cells.
4. **Evaluate** the pros and cons of using a light microscope compared with using an electron microscope to look at these cells.

CONCLUSION

Write a scientific conclusion for this investigation.

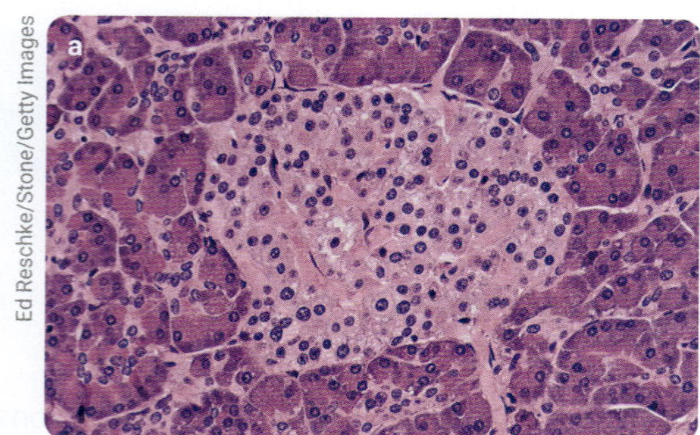

Ed Reschke/Stone/Getty Images

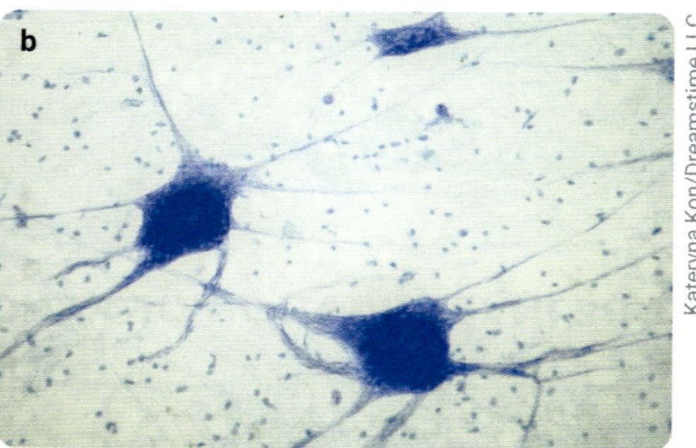

Kateryna Kon/Dreamstime LLC

▲ **FIGURE 5.6.3** (a) A light micrograph of pancreas cells. An islet of Langerhans shows as a patch of pale cells. (b) A light micrograph of human brain tissue, showing neurons with cell bodies and axons

9780170491785

5.7 Temperature regulation

BY THE END OF THIS MODULE, YOU WILL BE ABLE TO:

- ✓ describe several mechanisms that help control body temperature
- ✓ explain how and why body temperature is controlled
- ✓ draw a labelled negative feedback loop for temperature regulation.

GET THINKING

Working in groups of three, discuss what happens when you exercise, stand outside on a hot day or dive into a cold swimming pool. How long does it takes for your body to stop feeling hot or cold? Write some dot points with your ideas.

Interactive resource
Drag and drop: Maintaining body temperature

Temperature variation

Humans are exposed to extreme variations in temperature. In 2022, the temperature in Australia ranged from −11.7°C to 50.7°C.

Brook Mitchell/Stringer/ Getty Images News/Getty Images

AAP Images/STEVEN SAPHORE

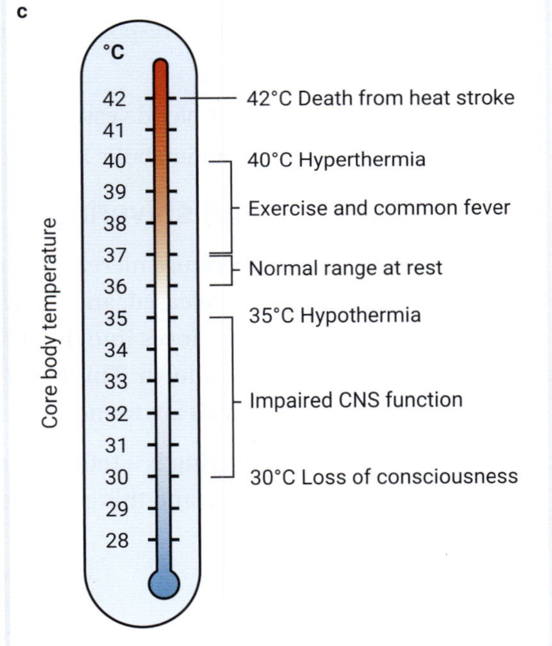

▲ **FIGURE 5.7.1** Australia experiences extreme temperature variations from **(a)** hot to **(b)** cold. **(c)** Even small temperature deviations in the body's core temperature can have significant impacts.

Regardless of these changes in the external environment, our internal body temperature must stay close to 37°C. Small deviations away from this temperature can cause a range of medical issues (Figure 5.7.1c). Just 5°C above our normal 37°C can cause death from heat stroke. This means that humans have a relatively small tolerance range and a narrow temperature setting for optimal activity. How do we regulate our temperature to stay within such a small tolerance range?

Mechanisms of temperature regulation

The human body has several mechanisms for regulating internal body temperature:

- **Behavioural mechanisms** are simple actions you can take to change your temperature, such as moving into cold water to cool down or moving into a sunny spot or putting on a coat to warm up.
- **Structural mechanisms** are built-in physical features that help to regulate body temperature, such as a thick, **insulating** layer of hair on your head to keep your head warm.
- **Physiological mechanisms** are internal mechanisms that your body does automatically. They are controlled by both your nervous system and your endocrine system, and coordinated by the hypothalamus. We will focus on these for the rest of the chapter.

insulating reducing the transfer/loss of heat

Mechanisms that are responses to when your body temperature rises above 37°C include:

- sweating
- dilation of blood vessels
- a decrease in metabolic activity.

Mechanisms that are responses to when your body temperature drops below 37°C include:

- shivering
- constriction of blood vessels
- an increase in metabolic activity.

Sweating and shivering

When the temperature increases, it may cause your sweat glands to open. Sweat made of water and salt is released, and the water draws heat from your body as it evaporates. As the water vapour moves into the surrounding air, it takes the heat energy away with it, cooling your body down. This effect is known as **evaporative cooling**. Figure 5.7.2 shows where sweat glands are found in the skin.

evaporative cooling the cooling effect that occurs when water evaporates

In contrast, shivering is a reflex action activated when you are very cold. Muscles rapidly contract and relax, and this generates heat.

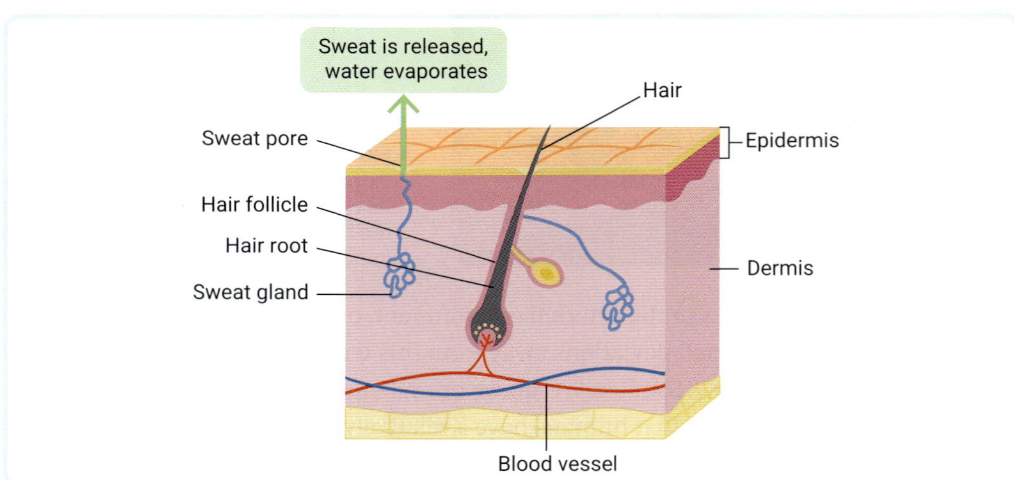

▲ **FIGURE 5.7.2** Sweat glands in the skin release sweat. When the water from sweat evaporates, it has a cooling effect.

Regulation by negative feedback

In Module 5.1, you learned about negative feedback loops. A negative feedback loop can be drawn to represent the key steps in maintaining an optimal internal body temperature. The hypothalamus has a set optimal temperature of 37°C. It responds to changes in the internal or external temperature by coordinating a response that uses both the nervous system and the endocrine system. The systems send messages to effector muscles or glands to carry out a response that counteracts the stimulus. You can see examples of these negative feedback loops in Figure 5.7.3.

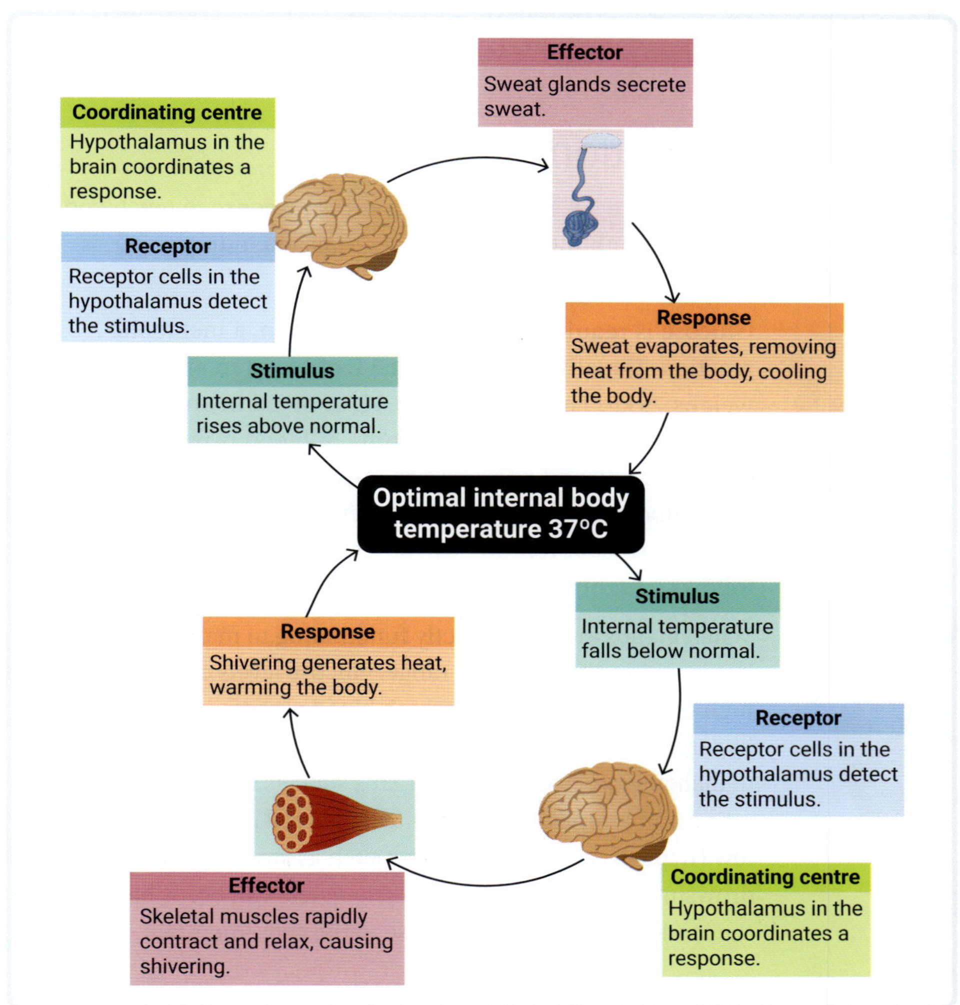

FIGURE 5.7.3 A model of two negative feedback loops for an increase or decrease in internal body temperature

5.7 LEARNING CHECK

1. **Define** shivering.
2. **Explain** why internal body temperature in humans needs to be regulated.
3. **Create** a diagram of a negative feedback loop for an increase in internal body temperature above 37°C, using dilation of blood vessels as the response mechanism.
4. **Create** a diagram of a negative feedback loop for a decrease in internal body temperature below 37°C, using constriction of blood vessels as the response mechanism.
5. **Create** a crossword of 5–10 words for this module. Hand it to a partner to test it out.

5.8 Blood glucose regulation

BY THE END OF THIS MODULE, YOU WILL BE ABLE TO:

- ✓ describe several mechanisms that assist in the control of blood glucose levels
- ✓ explain how and why blood glucose is controlled
- ✓ draw a labelled negative feedback loop for blood glucose regulation.

Interactive resource
Label: Negative feedback loops

GET THINKING

You may know someone who has diabetes, a common condition that affects many Australians. What do you know about the condition? How do you think it relates to homeostasis? Write down your answers and then revisit them when you've completed the module.

Blood glucose variation

blood glucose
the amount of glucose in blood

Have you ever been 'hangry' (Figure 5.8.1)? Hunger is a stress response to low blood glucose. When we are hungry, our inhibition to other emotions, including anger, is lowered. **Blood glucose** is the amount of glucose in blood.

iStock.com/-101PHOTO-

▲ **FIGURE 5.8.1** 'Hangry' = hungry + angry

Glucose is required for cellular respiration, a metabolic reaction. Usually there is about 5 L of blood in an adult who weighs 75 kg. There is closer to 4 L of blood in an adolescent. The level of glucose in this volume of blood is regulated by the endocrine and nervous systems. The range of glucose in adult blood is 3.3–7 g. Glucose levels rise after every meal by 1–2 g. Glucose levels are lowest just after waking, which usually follows a night of fasting.

When glands are not correctly functioning to maintain levels within the tolerance range, glucose levels can rise very high or fall low. Persistently high blood glucose is known as hyperglycaemia. This may indicate low **insulin** production due to diabetes, and can damage organs such as the kidneys, nerves, retina and arteries. In contrast, when blood glucose is persistently low (hypoglycaemia), symptoms such as sweating, shaking, drowsiness, seizures and unconsciousness can occur because of the inadequate supply of glucose to the cells and brain.

Mechanisms of blood glucose regulation

insulin
a hormone secreted by the pancreas; controls how much glucose is in the blood

glucagon
a hormone secreted by the pancreas; breaks down glycogen into glucose

Our bodies have several mechanisms to regulate blood glucose levels. Behavioural mechanisms include eating a healthy diet. We also have physiological mechanisms (internal processes) that are triggered by our nervous and endocrine systems.

Regulation of blood glucose is largely controlled by the endocrine system. The pancreas is a gland that detects blood glucose levels and secretes two types of hormones that can decrease or increase blood glucose: insulin and **glucagon**.

Stimulus: blood glucose too high

Your body secretes varying amounts of insulin to keep blood glucose within a normal range. Insulin is secreted from beta cells in the pancreas. Insulin causes glucose to leave the bloodstream and enter the cells. It does this by binding to a target cell, which activates the transport of glucose across the cell membrane. Once inside cells, glucose can be used in cellular respiration to release energy. If your blood glucose rises, this is a stimulus that is detected by the pancreas. In response, more insulin is secreted. Less insulin is secreted when your blood glucose drops below normal, as shown in Figure 5.8.2.

Stimulus: blood glucose too low

When your blood glucose is low, another response happens: the secretion of glucagon from alpha cells in the pancreas. Glucagon breaks down stored **glycogen** to glucose. Glucagon travels to the liver and attaches to liver cells. Liver cells convert glycogen into glucose molecules and release them into the bloodstream, increasing your blood glucose levels (Figure 5.8.2). A saying to help you remember the hormone that is released when glucose levels are low is: 'When the **gluc**ose is **gon**e, release **glucagon**'.

glycogen
a store of glucose in the liver and muscles

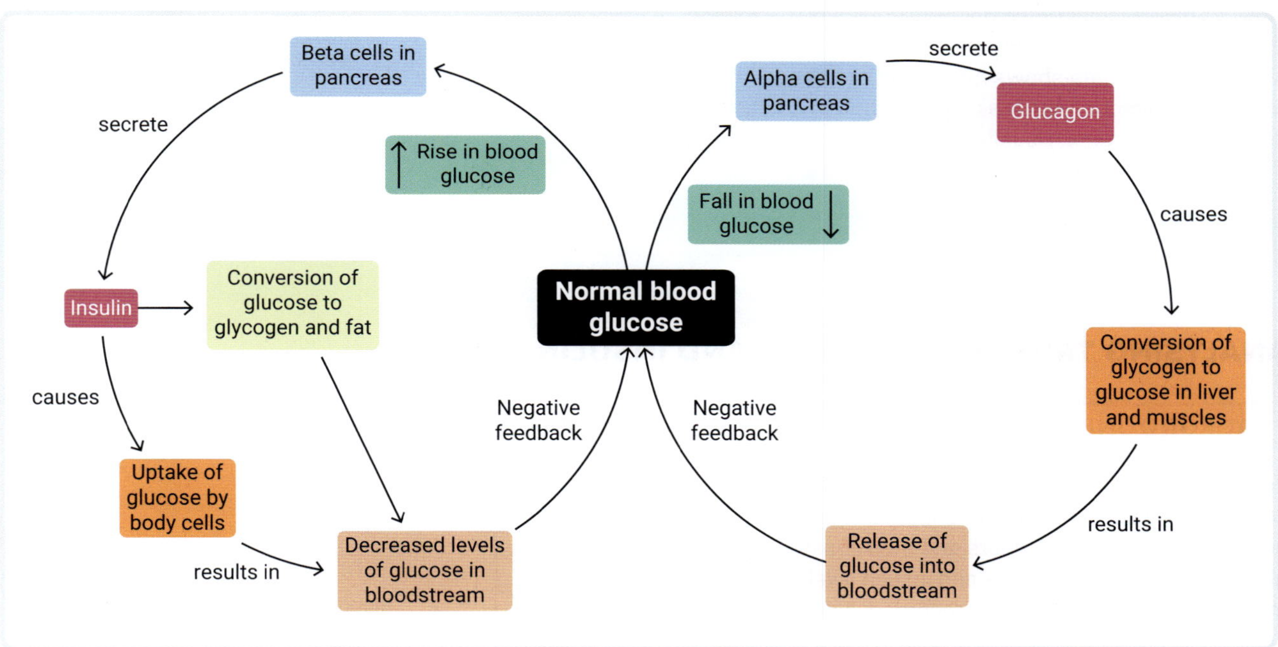

▲ **FIGURE 5.8.2** The negative feedback loops of glucose regulation

5.8 LEARNING CHECK

1 **State** the hormone released from:
   a alpha cells.
   b beta cells.
2 **Define** blood glucose.
3 You eat three pieces of delicious birthday cake at your friend's party. **Draw** an annotated negative feedback loop mechanism to illustrate how your body will regulate the sugar you have eaten.
4 **Analyse** Figure 5.8.2. Start at one point in the 'double' loop and write one sentence per stage to **summarise** the two complete cycles blood glucose negative feedback.

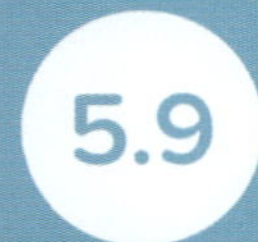

5.9 Identifying trends between variables

SCIENCE SKILLS IN FOCUS

IN THIS MODULE, YOU WILL FOCUS ON LEARNING AND IMPROVING THIS SKILL:

- analysing data to identify and explain relationships between variables.

During practical investigations, scientists collect, organise and process data. The next step is to analyse or interpret the data to form conclusions.

When they analyse data, scientists usually look to identify three main things:

- patterns
- trends
- relationships.

Scientists look at data to see if something repeats over time. **Patterns** in scientific data give us information about the past and may help us make predictions. An example of a pattern is the cyclic rise and fall of glucose levels in the blood.

A **trend** is when data shows movement or change in a particular direction, such as a population of koalas decreasing over time. Usually, scientists need a great deal of data to be confident that it is showing a trend.

A **relationship** is a trend in which there is mathematical correlation between two or more variables; for example, the more it rains, the higher the water level in a rain gauge. Another example could be a reduction in the consumption of junk food resulting in an improvement in blood cholesterol levels.

Video
Science skills in a minute: Identifying trends

Science skills resource
Science skills in practice: Identifying trends

ANALYSING TRENDS IN GLUCOSE AND INSULIN DATA

BACKGROUND INFORMATION

You will be examining the graphed results of a study conducted by specialist doctors. In the study, healthy participants fasted for 5 hours and then consumed glucose. Their blood glucose and insulin levels were monitored, and averages were recorded, over 5 hours.

AIM

To analyse trends in glucose and insulin data

PROCEDURE

Analyse the two curves in Figure 5.9.1.

ANALYSIS

1 **Describe** the blood glucose concentration between 0 and 5 hours.

2 **Explain** why insulin fluctuates between 2 and 4 hours.

3 **Explain** the relationship between blood glucose and insulin levels.

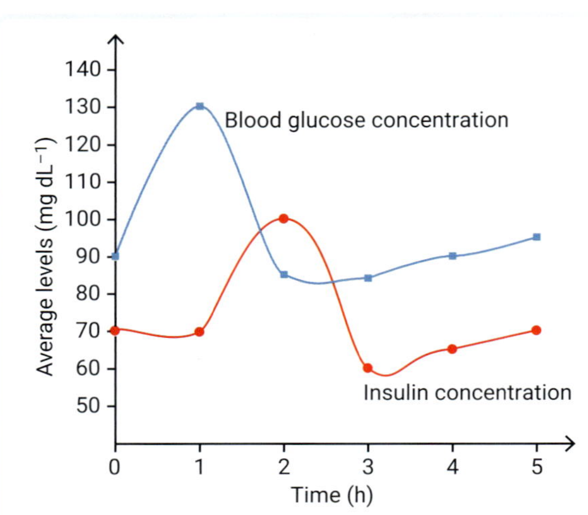

▲ **FIGURE 5.9.1** Mean (average) concentration of insulin and blood glucose 0–5 hours after glucose consumption

SCIENCE IN CONTEXT

5.10 Use of technology to control blood glucose

BY THE END OF THIS MODULE, YOU WILL BE ABLE TO:

- ✓ describe diabetes monitoring and the use of data to make action plans
- ✓ investigate and evaluate how technologies and engineering enable advances in diabetes monitoring.

Insulin – a vital hormone

Video activity
What is diabetes?

Insulin is a vital hormone because it is the only hormone that can lower blood glucose levels. Insulin is produced by the beta cells in the islets of Langerhans in the pancreas. In a healthy person, insulin is secreted in response to rising blood glucose levels. The job of insulin is to attach to receptors on target cells to activate the transfer of glucose from the blood into cells for use in cellular respiration.

What is diabetes?

About 1.8 million Australians have diabetes. People with diabetes cannot regulate their blood glucose levels because their bodies do not produce enough insulin. When they eat food containing glucose, the glucose stays in the blood instead of entering cells to make energy. This makes people with diabetes feel lethargic, dizzy and hungry. Long-term symptoms can be life-threatening.

There are three main types of diabetes: type 1, type 2 and gestational diabetes. Diabetes can be managed either through strict diets or by regularly injecting insulin (Figure 5.10.1).

Keeping blood glucose levels in a healthy range prevents short-term and long-term complications. Diabetes technology has developed over time, making the monitoring and treatment processes more accurate and less invasive.

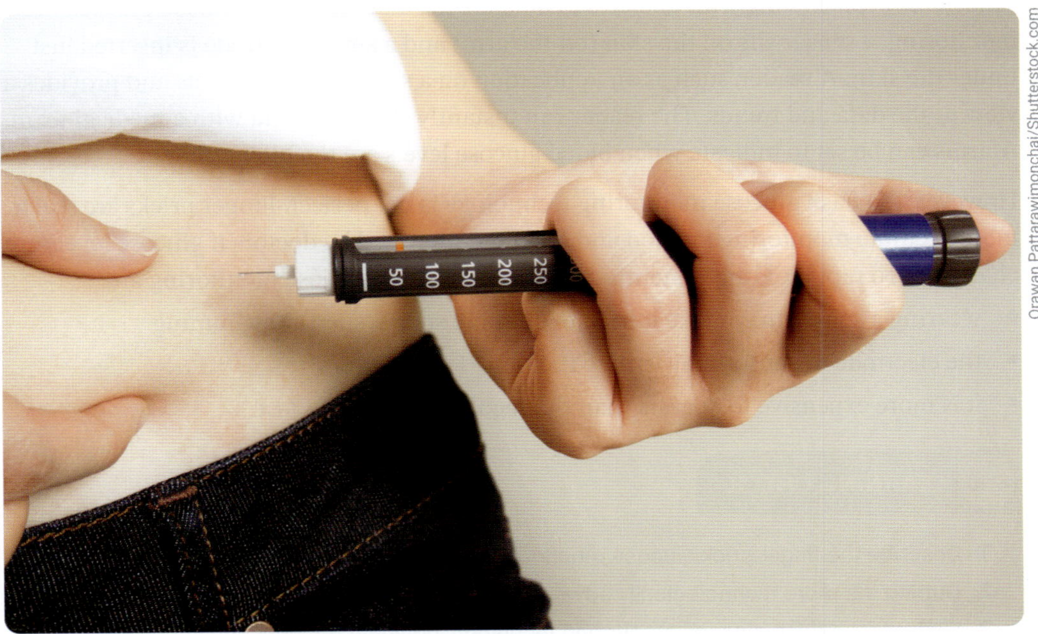

Orawan Pattarawimonchai/Shutterstock.com

▲ **FIGURE 5.10.1** People with type 1 diabetes usually have to inject themselves with insulin several times a day to manage their blood glucose levels.

Blood glucose monitoring

A person with diabetes can use a blood glucose monitor to measure their blood glucose levels. They use a small, needle-like device to prick their finger. The person squeezes their finger so a small sample of blood appears. They transfer the blood onto a test strip and, within 10 seconds, they can read a result and record it in a journal. If their blood glucose value is too far away from normal, the person may need to inject insulin or have something to eat.

New technology and engineering now allow people to use the blood prick test three times a day or to monitor glucose continuously using a sensor inserted under the skin, in a process called flash glucose monitoring (Figure 5.10.2).

Proxima Studio/Shutterstock.com

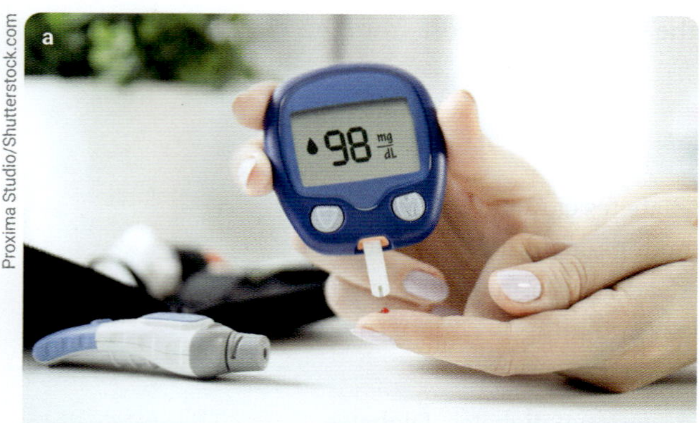

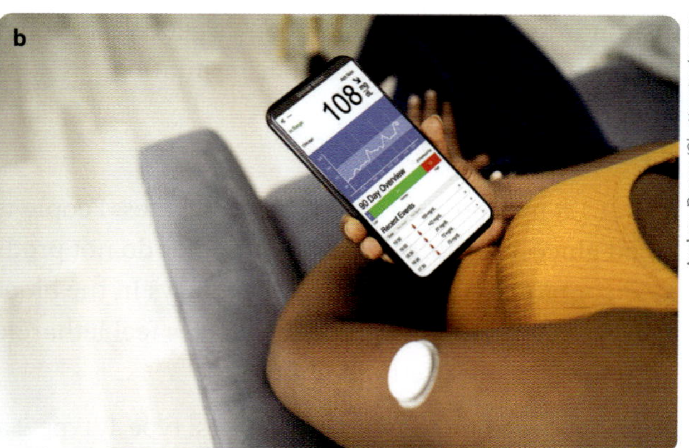

Andrey_Popov/Shutterstock.com

▲ **FIGURE 5.10.2** (a) Finger-prick testing: a person punctures their finger to draw a drop of blood. (b) Flash glucose monitoring involves continuous monitoring with a sensor and compatible smartphone.

One benefit of flash glucose monitoring is that it collects continuous data, which can give more information about changes in the person's blood glucose levels over time. This helps people living with diabetes to manage the disease better. In flash glucose monitoring, a sensor sits on the skin (on the arm) and a small electrode is inserted just under the skin. A reader, such as an application on a smartphone, collects and provides a view of the data and trends. The reader can instantly share the data with a medical team. Alarms can be set to indicate when blood glucose levels are too high or low.

However, the flash glucose monitoring technology is not perfect. Finger pricks are still a more accurate monitoring method and are often preferred by doctors if glucose levels are changing too rapidly. Other potential drawbacks of flash glucose monitoring are:

- wearing a sensor may make it hard to do some sports, such as swimming
- the electrode or sensor may irritate the skin
- it is more expensive to operate than the regular finger-prick test.

5.10 LEARNING CHECK

1 **Describe** the finger-prick glucose monitoring technique.
2 **Outline** the more technologically advanced flash glucose monitoring technique.
3 **Evaluate** the flash glucose monitoring technique, outlining the pros and cons.

9780170491785

5 REVIEW

REMEMBERING

1 **Describe** a stimulus–response model.

2 Copy and complete the following table by identifying or describing the functions of the parts of a negative feedback model.

Part	Function
Stimulus	
	A specialised cell that detects the stimulus. The receptor may be internal or external.
Coordinating (control) centre	
	A muscle or gland that receives a message from the coordinating (control) centre and carries out a response.
Response	

3 Copy and complete the following table by identifying the three types of neuron according to their function.

Function	Type of neuron
Transmits nerve impulses from a sensory neuron to a motor neuron	
Detects changes and transmits nerve impulses towards the spinal cord or brain	
Transmits nerve impulses from the spinal cord towards a muscle or gland to carry out a response	

4 **List** the three parts of the brain stem.

5 **Identify** the following statements as true or false:

a The cerebellum controls involuntary responses.

b The brain stem controls involuntary responses.

c The cerebrum controls muscle movements.

UNDERSTANDING

6 **Draw** a generalised model of a negative feedback loop.

7 **Explain** how the message in an electrical signal can get across a synapse.

8 **Compare** positive feedback and negative feedback and include an example of each.

9 **State** the internal body temperature tolerance range and optimal temperature for humans.

10 **Describe** the disease diabetes.

APPLYING

11 Read the following scenario and **identify** the type of feedback mechanism involved. **Justify** your answer by using the stimulus–response model.

After a run along the beach on a dry summer's day, Dave was feeling thirsty. The thirst centre in the hypothalamus part of his brain detected low water levels in his blood, so it sent a hormone to his kidneys to increase permeability (an increase in the ease of water flow), allowing water to return to his blood. This caused a decrease in the amount of water becoming part of his urine and moving into his bladder. When Dave went to the toilet, he found he only excreted a tiny amount of urine.

12 **Explain** how the structure of a sensory neuron allows for its function.

13 Using an example, **explain** the main steps of how a hormone works.

14 **Create** a diagram of a negative feedback loop for a decrease in internal body temperature below 37°C, using metabolic heat production as the mechanism. Use the negative feedback model in Figure 5.1.3 as a guide.

15 **Explain** how sweating works in regulating body temperature.

16 **Explain** the roles of insulin and glucagon in regulating blood glucose after a meal is consumed.

17 **Explain** why people who cannot produce sufficient insulin feel tired and lethargic.

ANALYSING

18 **Analyse** the following MRI image and locate and **label** eight parts of the brain.

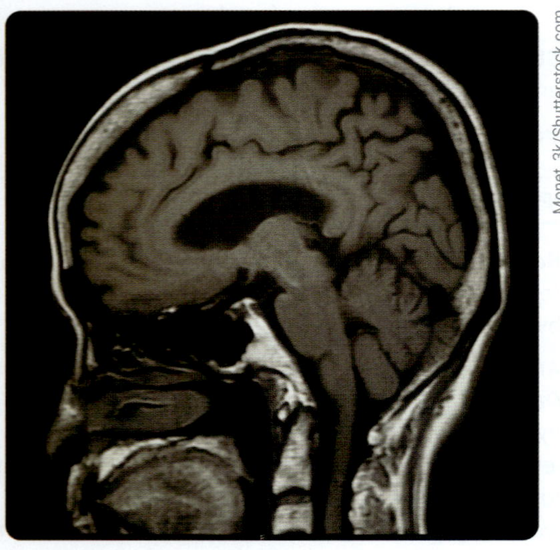
Monet_3k/Shutterstock.com

19 The following diagram depicts two homeostasis processes.

a **Identify** the two processes.

b **State** which type of feedback loop is shown in each case.

c **Write** out the steps shown in more detail, using the labels as subheadings.

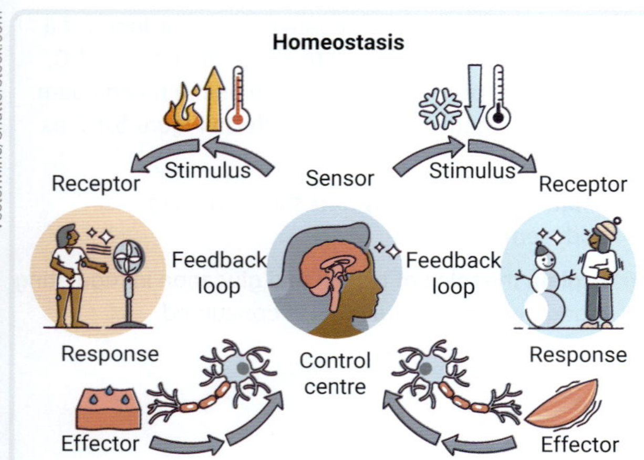

VectorMine/Shutterstock.com

EVALUATING

20 Look at the image of Robert Wadlow. He grew to be 2.52 m tall.

History and Art Collection/Alamy Stock Photo

a How tall are you in comparison?

b **State** the hormone that may have caused the excessive growth.

c **Evaluate** whether there was too much or too little of the hormone secreted.

d **Name** the gland the hormone was excreted from.

CREATING

21 Produce a short, engaging presentation for Year 7 students that explains what a negative feedback loop is, and how it is important in keeping us healthy. Include:

- what the term 'homeostasis' means
- information on the terms stimulus, sensor, receptor, response and effector
- an image of a negative feedback loop
- examples of different negative feedback loops
- reasons why negative feedback loops are important for keeping our body in a steady state.

22 **Create** a model in the form of a diagram or poster that demonstrates the flow of an electrical impulse along the three types of neurons between the PNS and the CNS.

9780170491785

SCIENCE IN DEPTH STUDY #5

1 Connect what you've learned

In this chapter, you've learned about homeostasis and temperature regulation. Finish the concept map that you started in Module 5.3 and reflect on how the concepts in each of the modules are related.

2 Check your thinking

Think back to when you last experienced a really hot day.

- Name one mechanism your body uses to regulate temperature.
- Can the body regulate its own temperature when the external temperature is very high?
- At which times of the day did you rely on an air-conditioner to stay cool?
- Can you explain why our negative feedback systems are not effective in really hot weather?
- What are some smart ideas that can assist your negative feedback processes to keep you within human temperature tolerance range? Ideas could relate to cold drinks, clothing, houses and fitness.

3 Get into action

Make a list of your ideas for keeping cool and state how each idea helps a negative feedback process. For example, if you chose eating ice, you could explain that some internal heat will be used to melt the ice, cooling the body down.

4 Communicate

Convert your ideas into a labelled infographic to share with your class.

6 Diseases

SCIENCE IN DEPTH

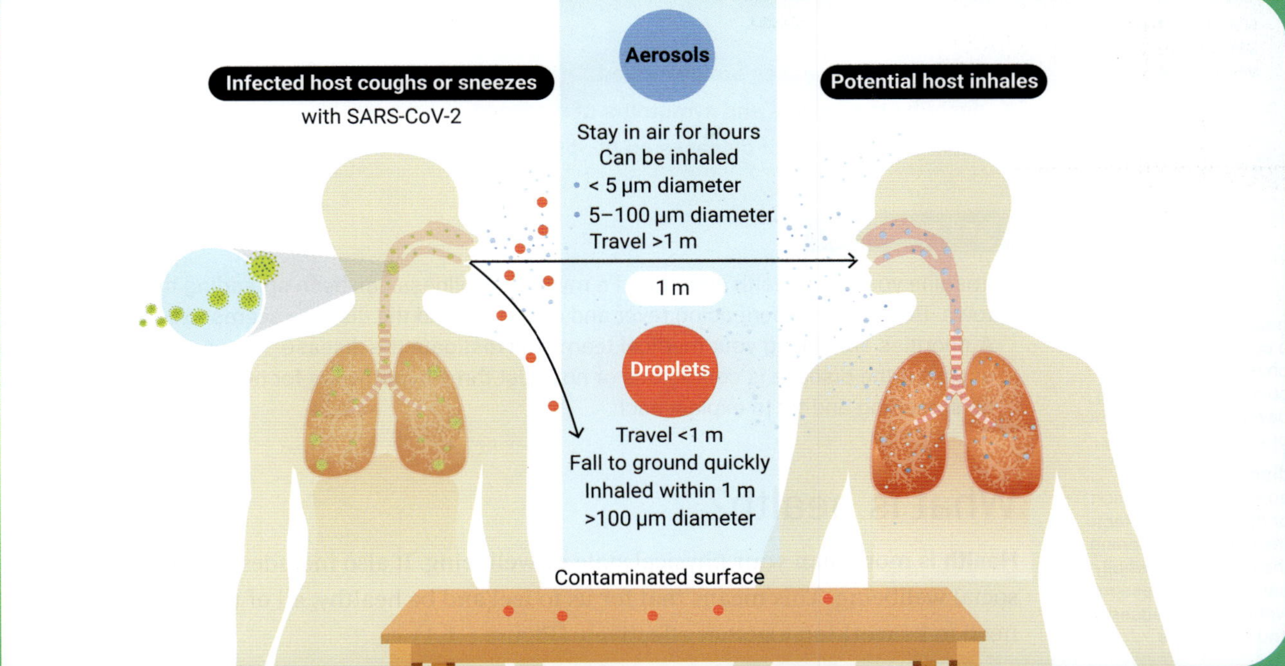

▲ **FIGURE 6.0.1** Airborne transmission of the SARS-CoV-2 virus, which causes the disease COVID-19 (1 µm (micrometre) = one-millionth of a metre)

Many infectious diseases are contagious. They are diseases that can spread (transmit) from person to person, such as COVID-19 and influenza. Infectious diseases spread easily between humans because there are many modes of transmission. Once you know how a disease transmits from an infected host to another person, you can slow or stop the spread.

- **What are the two types of transmission of COVID-19 shown in the diagram? How else can COVID-19 be spread?**
- **What can we do to protect ourselves and people we come into contact with?**

#6 DIVE INTO SCIENCE!

At the end of this chapter, you can complete Science in Depth Study #6. You can use the information you learn in this chapter to complete the project.

Assessments

- Prior knowledge quiz
- Chapter review questions
- End-of-chapter test
- Depth study: Research project and infographic

Videos

- Science skills in a minute: Petri dish safety **(6.11)**
- Video activities: Changing the lobster nation **(6.2)**; Immune defence **(6.8)**; Smallpox: The first vaccine **(6.9)**

Science skills resources

- Science skills in practice: Petri dish safety **(6.11)**

Interactive and other resources

- Drag and drop: Bacteria or fungi? **(6.3)**
- Label: Virus replication **(6.4)**; Phagocytosis **(6.7)**
- Quizzes: Second line of defence **(6.7)**; Herd immunity **(6.9)**
- Activity sheets: Practising sterile technique **(6.11)**
- Worksheets: What do you already know about infectious diseases? **(6.3)**; The National Immunisation Program **(6.9)**

To access resources above, visit **cengage.com.au/nelsonmindtap**

6.1 What is disease?

BY THE END OF THIS MODULE, YOU WILL BE ABLE TO:

- ✓ define disease
- ✓ identify how diseases are described
- ✓ describe the causes and symptoms of diseases
- ✓ assess the impact of diseases on people.

GET THINKING

Imagine you are a health expert and a mysterious illness has been spreading through your town. People are experiencing fever and coughing, and the disease seems to be contagious. Authorities are asking your medical team to investigate the disease. How are you going to assess if the disease is contagious or not? List three hypotheses for how the disease might be spreading and try to explain each.

health
a condition of total physical, mental, and social wellness, going beyond just the absence of illness or disability

disease
an abnormal condition or disorder that affects the normal functioning of the body, either wholly or partially, and is associated with specific causes and symptoms

What is health?

Health is more than your physical state or wellbeing. It also includes your mental and social wellbeing. This means that for us to feel and be healthy, all of the body systems need to be working together effectively (Figure 6.1.1)

Physical
- body systems working effectively
- good nutrition and exercise
- good sleeping patterns

Mental
- managing stress well
- positive mindset
- psychological resilience

Social
- sense of belonging
- positive relationships
- effective interactions with others

▲ **FIGURE 6.1.1** Factors that are part of good health

Defining disease

Think about how you feel when you are sick. You might feel pain, have no energy, or be unable to sleep properly. Feeling sick is often the result of having a **disease**. A disease is a condition or **disorder** that affects the normal functioning of the body in some way. Diseases are associated with specific causes and **symptoms**.

Causes of diseases

There are many factors that can cause diseases, such as:

- **pathogens** (bacteria, viruses, fungi, protists)
- genetic and autoimmune conditions
- the environment (exposure to pollutants)
- lifestyle (smoking, drinking, nutritional choices) (Figure 6.1.2).

disorder
a group of symptoms that disrupts your normal body functions but does not have a known cause

symptom
something a person experiences that may indicate the existence of disease

pathogen
an organism that causes disease in its host

Pathogens
- bacteria
- fungi
- viruses
- protists

Genetic/autoimmune factors
- inherited diseases
- mutations

Environment
- pollution
- radiation
- toxins
- dust

Lifestyle
- smoking
- drinking alcohol
- poor diet
- substance abuse

▲ **FIGURE 6.1.2** The main causes of diseases

Types of disease

Diseases can be described and categorised in various ways based on their causes, how long they last and how they affect the body. Understanding these descriptions helps us to diagnose, treat and prevent different health conditions. Table 6.1.1 describes the main types of diseases. Many diseases are described in a combination of ways; for example, lung cancer may be the result of a person being exposed to pollution in their workplace (**occupational**). At the same time, it is a non-infectious disease because it is not transmissible to others.

occupational related to an occupation or profession

▼ **TABLE 6.1.1** Types of disease

Disease category	Description	Examples
Infectious	Caused by a pathogen and transmissible from person to person	Influenza, COVID-19, malaria
Non-infectious	Not caused by a pathogen and not transmissible from person to person	Cancer, heart disease, diabetes
Chronic	Develops over a long period of time	Asthma, rheumatoid arthritis
Acute	Has rapid onset and short duration	Appendicitis, acute bronchitis
Degenerative	Causes gradual deterioration of cells, tissues and organs	Alzheimer's disease, Parkinson's disease
Autoimmune	Occurs when the immune system attacks the own's body tissues or organs	Lupus, multiple sclerosis, type 1 diabetes
Deficiency	Caused by a lack of essential nutrients	Scurvy (due to lack of vitamin C), rickets (due to lack of vitamin D), anaemia (caused by low iron levels)
Occupational	Related to the person's occupation	Asbestosis, carpal tunnel syndrome, noise-induced hearing loss, cancer, silicosis
Psychological disorder	Affects mental health and behaviour	Depression, anxiety, bipolar disorder

Impacts of diseases

Diseases negatively affect a person's body, mind and social wellbeing. For example, a person who has cancer can experience serious symptoms and pain, have to attend ongoing treatment, and will often feel anxious about their **prognosis**. Their illness will also affect their friends' and family's mental and social wellbeing.

prognosis predictions made by health professional about how a disease will progress

Diseases also have an economic impact. The cost of disease is not only carried by the ill person and their family, but also can have a massive effect on governments because they run the hospitals, clinics and health system that must prevent and treat disease. You will learn more about prevention and control of disease in Module 6.9.

6.1 LEARNING CHECK

1 **Identify** three components of health.
2 **Describe** what is meant by 'social wellbeing' in the context of health.
3 **Explain** the difference between an acute disease and a chronic disease, providing an example of each.
4 **Explain** why the absence of disease does not necessarily mean a person is in good health.
5 **Discuss** how lifestyle choices can impact both physical and mental aspects of health, providing examples.

6.2 Non-infectious diseases

BY THE END OF THIS MODULE, YOU WILL BE ABLE TO:

✓ define non-infectious diseases
✓ describe the types and causes of non-infectious diseases
✓ analyse data for a common non-infectious disease.

Video activity
Changing the lobster nation

GET THINKING

Working in groups of three of four, debate this statement: 'The rise in non-infectious diseases is a direct consequence of modern lifestyle choices.' Do you agree or disagree with this statement? Why? Consider the role of diet, physical activity, stress and environmental factors in your debate. Write a short argument for the affirmative or the negative team.

What are non-infectious diseases?

Non-infectious diseases are those conditions that are not caused by a pathogen (bacteria, virus, fungus, protist) and, as a result, cannot be transmitted from one person to another (Figure 6.2.1). For this reason, non-infectious diseases are also called non-communicable diseases.

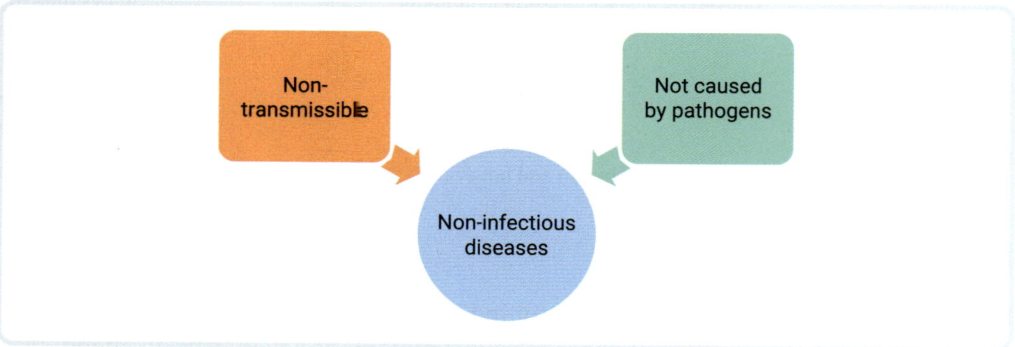

▲ **FIGURE 6.2.1** Characteristics of non-infectious diseases

Risk factors

predisposition
a likelihood of developing a particular disease or condition

The chance of having a non-infectious disease can be increased by the following factors:

- genetic **predisposition**: some people carry genes that cause or contribute to inherited diseases
- environmental: exposure to pollutants, radiation, toxins
- lifestyle choices: poor diet, smoking, alcohol consumption, lack of exercise.

Often, a non-infectious disease is caused by a combination of more than one factor. For example, liver diseases, such as cirrhosis, are often related both to the excessive consumption of alcohol (environmental) and to a genetic predisposition to liver diseases (genetics).

9780170491785

Incidence of non-infectious disease

Disease **incidence** means how many new cases appear in a population over a period of time. This is often expressed as the number of new cases per 1000 or 100 000 people per year. It is calculated using the formula:

incidence
the number of new disease cases appearing in a population over a period of time

$$\text{Incidence} = \frac{\text{number of new cases}}{\text{total population in the area of study}} \times 1000$$

To reduce the incidence of non-infectious diseases in a population, health experts must use a comprehensive approach that considers risks factors and promotes healthier lifestyles choices (Figure 6.2.2). For example, the long-running Quit campaign is designed to reduce the incidence in Australia of diseases related to smoking. This campaign encourages people to quit smoking by providing resources, support and information about the consequences of smoking. The campaign has achieved its aims by making people aware of the effects of smoking on health. It has reduced the rate of smoking in the Australian population.

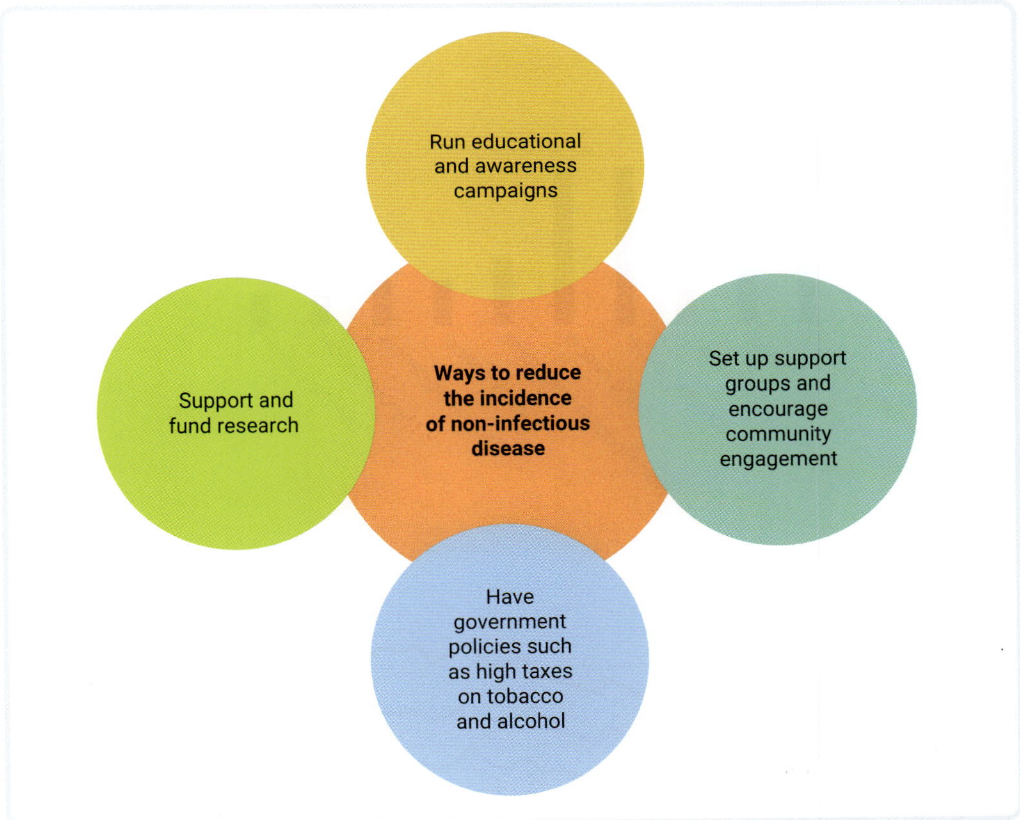

▲ **FIGURE 6.2.2** Reducing the incidence of non-infectious diseases requires a combination of different strategies.

☆ ACTIVITY 1

Incidence of melanoma in Australia

Melanoma is the third most diagnosed cancer in Australia, with 18 200 people diagnosed in 2023 alone. The incidence of melanoma is increasing. By analysing the incidence of the disease, we can gain ideas about how to reduce its incidence.

Procedure

1 Look at the graph in Figure 6.2.3 showing new cases of various cancers in Australia. Using the incidence formula, calculate the incidence of melanoma and prostate cancer in Australia per 100 000. In 2023, the total population of Australia was 26 966 789.

2 Look closely at Figure 6.2.4. Identify the trend of incidence of melanoma in males and females.

Analysis

1 What is the incidence of melanoma and prostate cancer in 2023?

2 **Describe** the accuracy of the data in Figure 6.2.3. Is it a record of all cases, or is it an estimate?

3 There is a gap between the incidence of melanoma trend between males and females. **Discuss** this difference. What do you think could be the cause?

4 **Write** an overall analysis of the data as a conclusion.

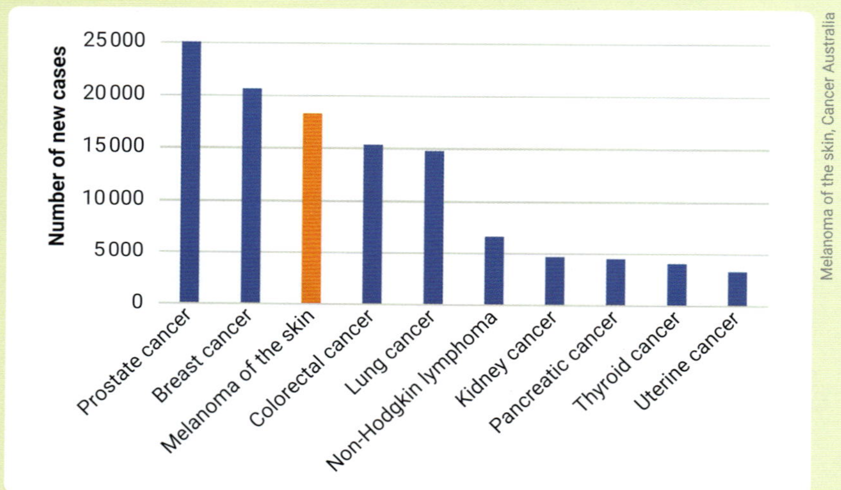

▶ **FIGURE 6.2.3** Estimated new cancer cases in Australia in 2023

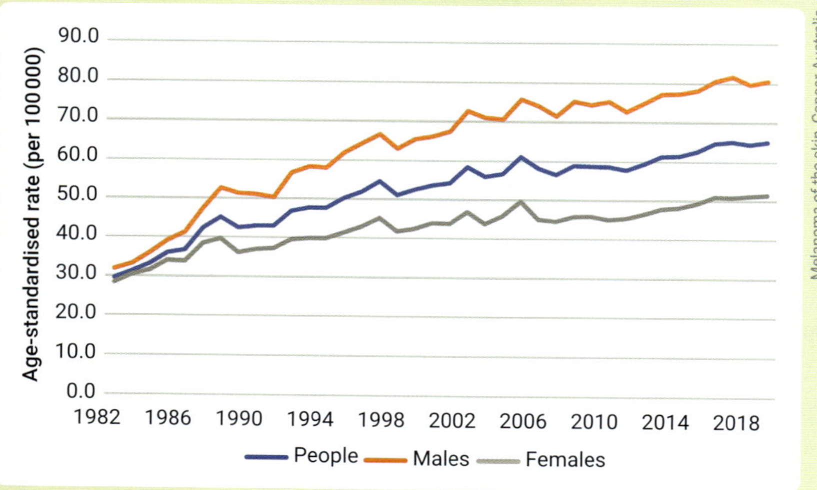

▶ **FIGURE 6.2.4** Age-standardised incidence rates for melanoma by sex between 1982 and 2019. (Age standardisation is a statistical method used to make it easier to compare rates of disease between populations.)

9780170491785

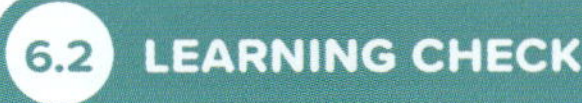

6.2 LEARNING CHECK

1 **Define** non-infectious disease and give two examples.
2 **Describe** the main characteristics of non-infectious diseases.
3 Using an example, **describe** two factors that contribute to the development of a non-infectious disease.
4 Table 6.2.1 shows the numbers of new cases internationally for some major non-infectious diseases.

▼ TABLE 6.2.1 Number of new cases per year internationally of three major non-infectious diseases

Non-infectious disease	New cases (millions of people per year)
Cardiovascular diseases	17.9
Cancers (all types)	4.1
Diabetes	2.0

a Considering that the current world population is 8.2 billion people, **calculate** the incidence of these diseases per 10 000 people.
b Add an extra column in the table with the results of the incidence calculations and **construct** an appropriate graph to represent this data.
c **Compare** the pattern of the data for new cases versus the incidence you calculated in part **a. Describe** the differences.
d **Choose** one of the diseases from the table and **design** a flyer to educate people about it, with the aim of reducing the incidence of that disease.

6.3 Infectious diseases

BY THE END OF THIS MODULE, YOU WILL BE ABLE TO:

- ✓ define infectious disease, pathogen, host and transmission
- ✓ describe three types of pathogens: bacteria, fungi and protists
- ✓ explain how pathogens are transmitted.

Interactive resource
Drag and drop: Bacteria or fungi?

Other resource
Worksheet: What do you already know about infectious diseases?

GET THINKING

Check this list of infectious diseases for ones you are familiar with. Have you suffered from any of them? Are there any that you haven't heard of?

- Influenza
- Conjunctivitis
- Whooping cough
- Ross River virus infection
- COVID-19
- Measles
- Chickenpox

What is an infectious disease?

Infectious diseases are diseases that can be passed on – they are transmissible (or communicable). Any agent that causes an infectious disease is known as a pathogen. Pathogens include micro-organisms, such as bacteria, fungi and protists, and non-living agents, such as viruses. We will look at viruses in more detail in Module 6.4. Note that not all micro-organisms are pathogens.

host
an organism infected with a pathogen

transmission
the transfer of a pathogen from one organism or reservoir to a new host

reservoir
a source of infection such as a habitat or population of organisms where a pathogen can replicate and survive for long periods

A **host** is an organism infected with a pathogen. An infection occurs when a pathogen enters a host, establishes itself and replicates. **Transmission** is when a pathogen passes to a new host. This can happen from an infected host or from a **reservoir**, which is another habitat or population of organisms where the pathogen can replicate and survive for long periods. If transmission is direct from an infected host to a new host, the disease is described as contagious.

If a pathogen replicates but there are no symptoms, the infected host is described as asymptomatic but contagious (able to spread the disease to another person).

Types of pathogens

Three types of micro-organisms can cause infectious disease – bacteria, fungi and protists. They each have unique structures that help identify them. Scientists study a pathogen's structure and life cycle to make informed decisions about how to treat disease and manage outbreaks.

Bacteria

Bacteria are unicellular, which means each bacterium consists of only one cell. An average bacterium is a microscopic 1–10 micrometres (μm) in length. Like all cells, bacteria have a membrane that surrounds the cytosol, and a cell wall. The cytosol is the semi-liquid substance inside all cells. Bacteria are prokaryotes. A prokaryote has no membrane-bound organelles such as mitochondria or nucleus. This means the chromosome is found in the cytosol. Some bacteria possess a tail-like flagellum that helps them move.

Tuberculosis (TB) is an example of a disease caused by bacteria (Figure 6.3.1). In 2022, TB killed approximately 1.3 million people. TB is thought to be the world's oldest **pandemic**. Symptoms of TB include coughing (with mucus or blood), chest pains and fever. Infection occurs by inhaling airborne droplets containing the bacteria. The bacteria reproduce inside host cells, resulting in the symptoms. Reproduction involves a single cell dividing into two identical daughter cells. Active TB can cause serious illness and death.

pandemic
a disease that has spread rapidly across the world, having grown from an epidemic

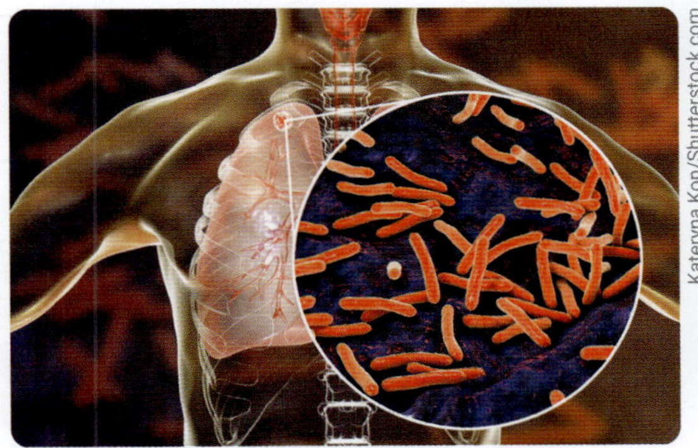

Kateryna Kon/Shutterstock.com

▲ **FIGURE 6.3.1** Tuberculosis is a disease caused by a bacterial pathogen.

Fungi

You may have seen the white fungus that grows on bread that looks like white, fuzzy cotton. The diverse fungal world includes large organisms, such as mushrooms, as well as very small forms that were only revealed with the invention of the microscope. Fungi have a eukaryotic cell structure with membrane-bound organelles, including mitochondria and a nucleus. They also have a cell wall made of a fibrous substance called chitin.

Moist, warm parts of a body provide ideal conditions for a fungal disease called tinea (athlete's foot) (Figure 6.3.2). Symptoms include an itchy, scaly rash, and peeling and inflamed skin. This contagious disease can be transmitted by direct contact with infected skin or contaminated surfaces such as the floor or shoes.

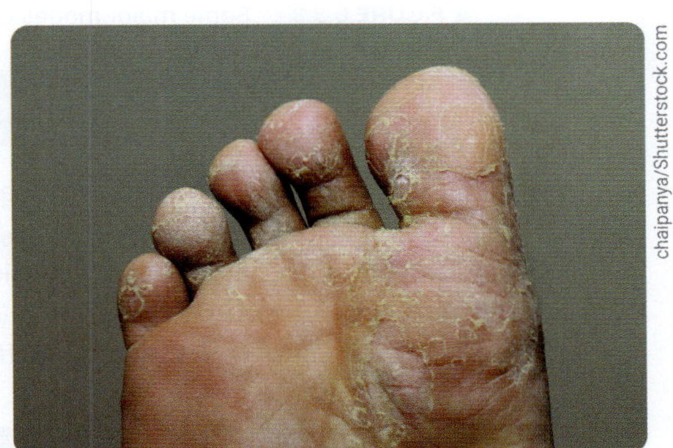

chaipanya/Shutterstock.com

▲ **FIGURE 6.3.2** Tinea is a disease caused by a fungal pathogen. The warm, moist environment between the toes provides ideal conditions for reproduction.

Protists

Usually microscopic, protists can look very different from each other, and can be animal- or plant-like. They are eukaryotic organisms. Most protists are single celled and usually live in water.

The disease malaria is caused by the protist *Plasmodium*.

Methods of pathogen transmission

Pathogens infect a host by transferring from a source. A source could be an infected person or a reservoir. Respiratory (lung) diseases such as tuberculosis, influenza and COVID-19 can be transmitted when a person inhales contaminated airborne droplets. If another living thing transports and assists the entry of a pathogen into a host, from another host or reservoir, it is known as a **vector**. A common example of a vector animal is a mosquito. Some diseases can be transmitted by several means. For example, COVID-19 can be transmitted by direct contact (touch) or through the air via droplets or **aerosols**.

vector
an organism that transmits a pathogen from an animal or a plant to another animal or plant

aerosol
fine droplets of saliva or mucus containing pathogens

Some major modes of pathogen transmission are summarised in Figure 6.3.3.

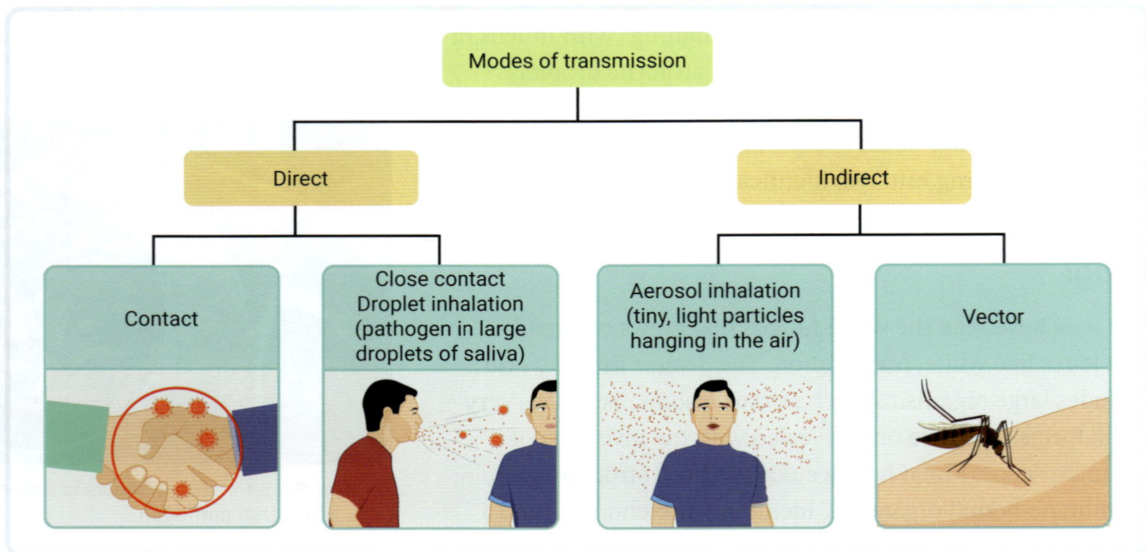

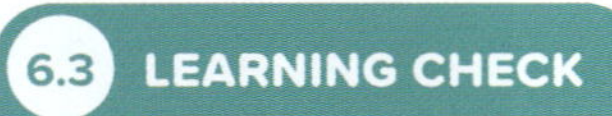

▲ **FIGURE 6.3.3** Some major modes of transmission of pathogens

6.3 LEARNING CHECK

1 **Define** the terms pathogen and micro-organism.

2 **Describe** the disease tuberculosis (TB), including the type of pathogen, symptoms and mode of transmission.

3 Study Figure 6.3.4, which shows a series of photos taken during an experiment on the effectiveness of different face masks in protecting against airborne pathogens.

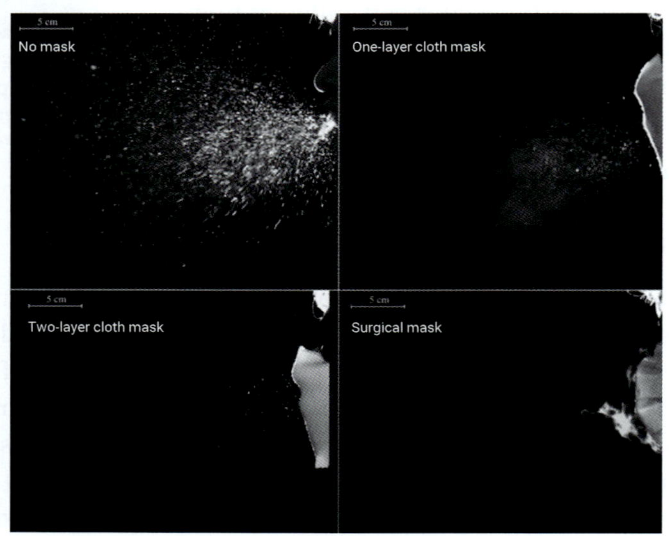

▲ **FIGURE 6.3.4** An experiment testing the effectiveness of different types of masks when sneezing

State the dependent and independent variables. Write a statement that summarises the results. **State** one controlled variable. **State** a variable you think was not controlled that may affect the validity.

6.4 Viruses

6.4

BY THE END OF THIS MODULE, YOU WILL BE ABLE TO:

- ✓ describe a virus and give examples
- ✓ explain how viruses replicate
- ✓ compare viruses with bacteria.

GET THINKING

Viruses are very different types of pathogens from bacteria and fungi. Do you know how they are different? In 2 minutes, write down all the things you already know about viruses.

Interactive resource
Label: Virus replication

What are viruses?

Viruses are non-living pathogens. We can only see viruses if we use an electron microscope because they are so small, around 30–300 nanometres long (1 nanometre (nm) = one-billionth of a metre). The size and shape of viruses vary greatly, as shown in Figure 6.4.1.

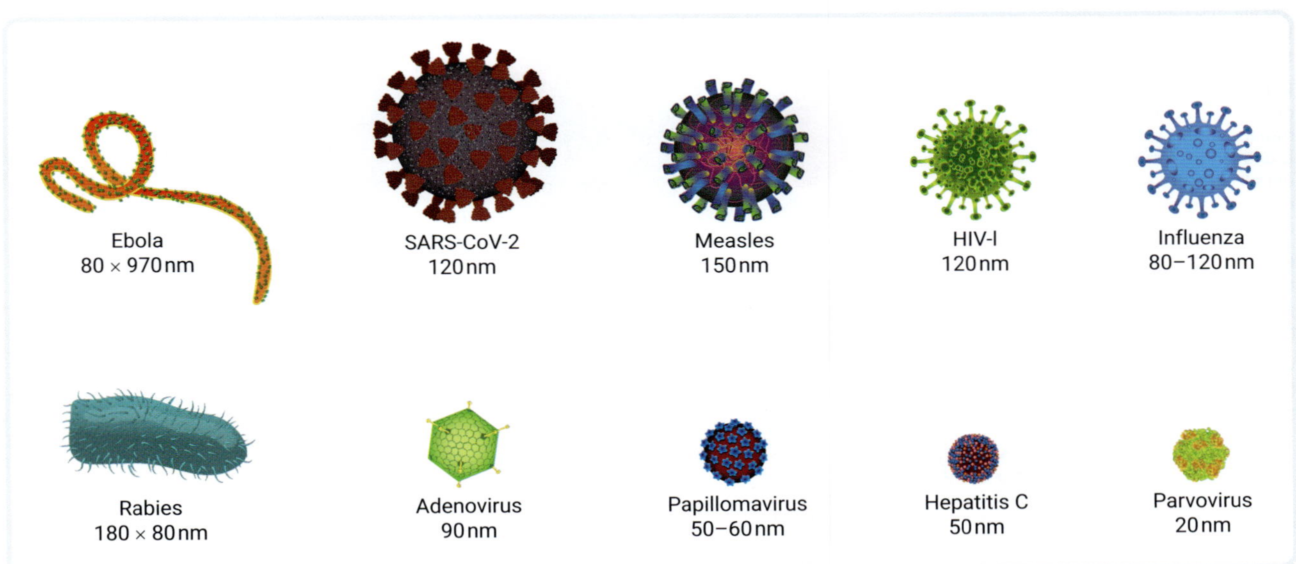

▲ **FIGURE 6.4.1** Viruses are very diverse. These are some of the most common human viruses with their relative size.

There are thought to be more than 200 different types of viruses that can infect humans and cause disease.

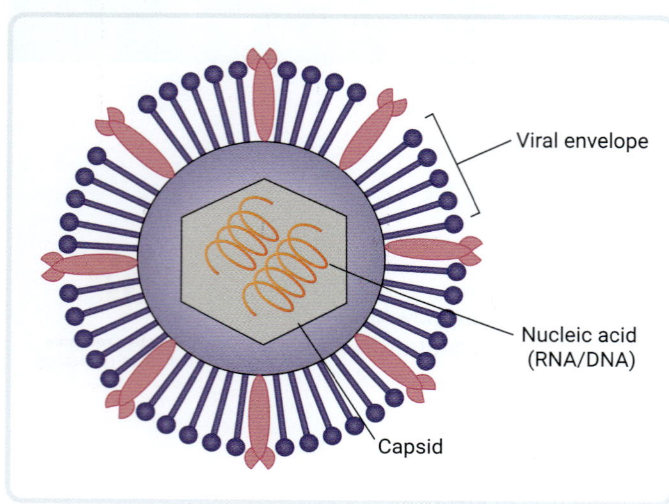

▲ **FIGURE 6.4.2** The general structure of viruses

Viruses consist of genetic material (nucleic acid), either **DNA** or **RNA**, covered by a protective layer of protein called a **capsid**. The basic structure of a virus is shown in Figure 6.4.2. In some viruses, the capsid is protected by another layer called a **viral envelope**.

Viruses can only replicate (produce new viruses) inside a host cell. When a virus infects an organism, it recognises receptors on the surface of a host cell and attaches to them. It then enters through the cell membrane and inserts its nucleic acid (DNA or RNA) into the host cell. Once inside, the virus's nucleic acid takes control and directs the host cell to make many copies of the virus parts. The parts are assembled into new viruses and released by the cell. This can cause the cell to die. The host usually experiences symptoms when a virus is replicating. Figure 6.4.3 represents the main steps of how a virus replicates inside the host cell.

DNA
deoxyribonucleic acid, the molecule that makes up the genetic material inside the nucleus of cells and in some viruses

RNA
ribonucleic acid, a molecule similar to DNA found in the cytoplasm of cells; the genetic material found in many viruses

capsid
the protein coating that protects the genetic material inside a virus

viral envelope
the outer layer of many viruses

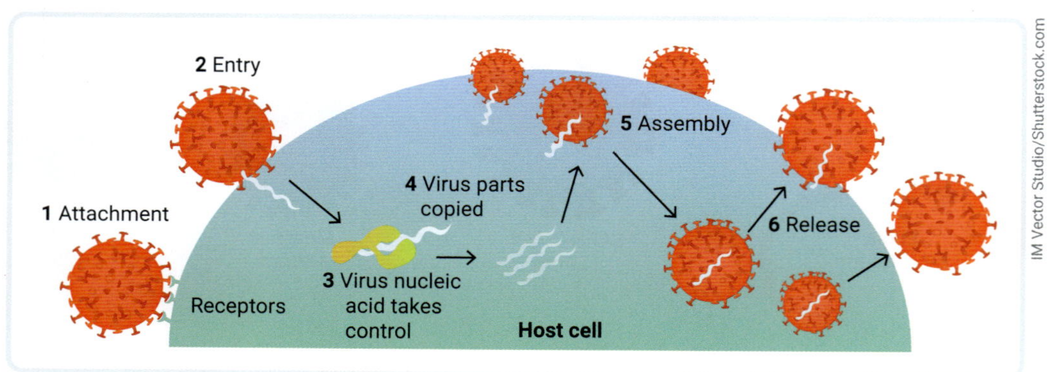

▲ **FIGURE 6.4.3** How a virus replicates inside a host cell

COVID-19

COVID-19 is a viral disease that has spread globally. The World Health Organization (WHO) declared the disease a pandemic on 11 March 2020, three months after it was first reported. COVID-19 is caused by severe acute respiratory syndrome coronavirus 2 (SARS-CoV-2) (Figure 6.4.4).

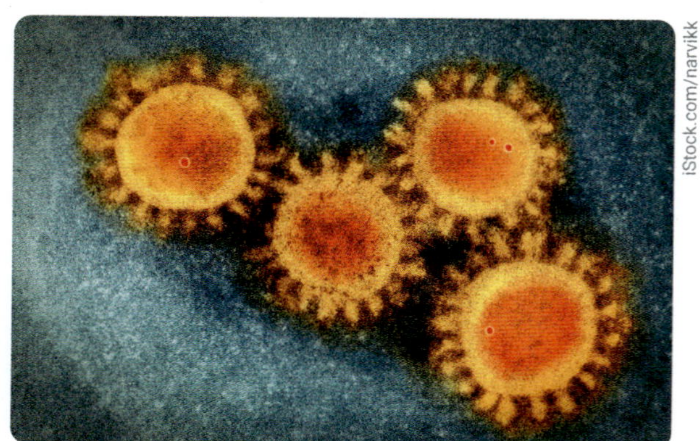

▶ **FIGURE 6.4.4** An electron microscope photo (micrograph) of SARS-CoV-2 virus particles

The virus has had many variants since 2019. Variants are produced when some of the RNA mutates. A mutation is a permanent change in the nucleic acid that codes for the production of virus parts. WHO names the variants after letters of the Greek alphabet, such as alpha, beta, gamma, delta and omicron.

The most common symptoms of COVID-19 are:

- fever
- tiredness
- cough
- loss of taste or smell
- headache
- shortness of breath
- sore throat
- muscle and joint pain.

Most infected people develop mild to moderate illness. Older people, or those with compromised immune systems, may develop severe disease symptoms, such as difficulty breathing and chest pain. It can take 5–14 days from when someone is infected to develop symptoms. This is called the incubation period. During this period, the infected person is contagious. This is the reason why, for much of the pandemic, infected people and their close contacts were often asked to quarantine (isolate themselves) or wear masks.

Ross River virus

Ross River virus infection is a disease caused by a virus, delivered to humans by mosquitoes (vectors) when they feed on our blood. Symptoms include fever, chills, headaches, swollen joints and muscle pain. A rash may occur for 7–10 days (Figure 6.4.5).

nechaevkon/Shutterstock.com

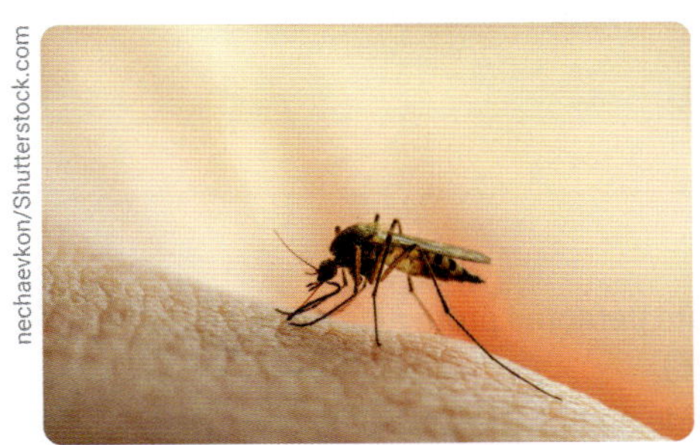

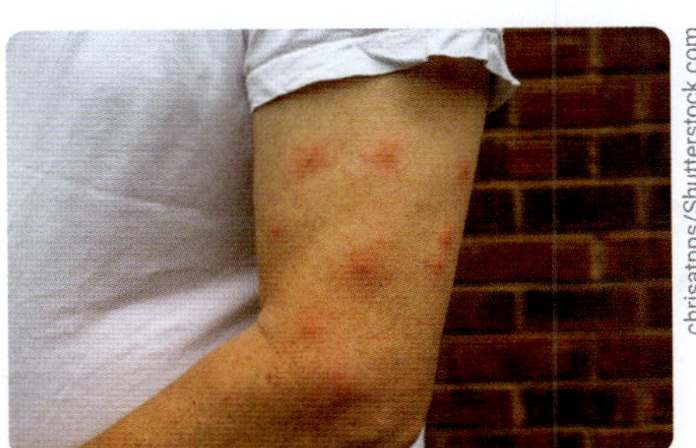

chrisatpps/Shutterstock.com

▲ **FIGURE 6.4.5** Ross River virus infection is transmitted to humans by mosquitoes (vectors).

6.4 LEARNING CHECK

1 **Describe** how viruses replicate and spread.

2 Write two or three sentences to **explain** why viruses are classified as non-living.

3 **Choose** one virus and one bacterial pathogen to **research**.

a Draw a diagram of each and **label** its parts.

b Copy and complete the table to **compare** them.

	Bacteria example	Virus example
Size and shape		
Living or non-living		
How it reproduces		
How it is transmitted		

WORKING SCIENTIFICALLY

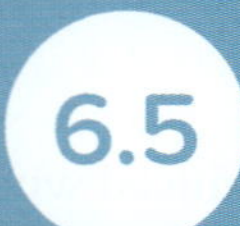

6.5 Using models to communicate scientific concepts

SCIENCE SKILLS IN FOCUS

IN THIS MODULE, YOU WILL FOCUS ON LEARNING AND IMPROVING THESE SKILLS:

- modelling the transmission of an infectious disease
- assessing the effectiveness of a model of infectious disease transmission
- communicating scientific concepts with models.

Scientists use scientific models to understand processes, make predictions and communicate their ideas.

In this investigation you will analyse and communicate how well the model represents the spread of a disease. You will need to answer the following questions to assess your model:

- How well did the model explain or represent the spread of the disease?
- Was the model adequate to represent the spread of the disease?
- What were its limitations?
- What needs to be improved in the model to better represent the spread of the disease?
- How can you use this model to predict the spread of other diseases?

By analysing these aspects of the model, you can describe how the disease could spread and make predictions about outbreaks.

MODELLING TRANSMISSION OF AN INFECTIOUS DISEASE

AIM

To evaluate a model for the transmission of an infectious disease

MATERIALS AND EQUIPMENT

- ☑ class set of cups containing 50 mL water
- ☑ cup containing 50 mL of dilute (0.1 M) sodium hydroxide solution
- ☑ dropper bottle with phenolphthalein indicator solution
- ☑ class set of plastic droppers
- ☑ safety glasses

PROCEDURE

Note: Your teacher will keep the identity of the cup with sodium hydroxide in it a secret, but also keep track of who collects it.

 **Safety**

Wear appropriate personal protective equipment (PPE).

Sodium hydroxide and phenolphthalein can irritate eyes and skin. Wash with water if contact is made.

1 Wear safety glasses. Collect a clear plastic cup with 50 mL of liquid (either water or sodium hydroxide solution) already poured in. All cups should contain the same volume of liquid.

2 One cup will contain sodium hydroxide (NaOH) solution. It is a clear colourless liquid that looks like water. It also has a high pH (very basic substance). The student who collects this cup will be 'patient zero'. All the other cups will contain water.

3 Collect a dropper.

4 When your teacher asks you to, walk slowly around the classroom until the teacher says 'stop'. Use the dropper to add 10 mL of your liquid into the cup of another student. Repeat this two more times until you have added to the cups of three other students.

By doing this, you are modelling the spread of infection through body fluids such as droplets of mucus or saliva, which can contain pathogens.

5 Record who you shared your liquid with, in order.

6 A volunteer student can use a dropper bottle with phenolphthalein indicator solution to add a few drops to each student's cup, to test who has been infected. Phenolphthalein is an acid–base indicator. If the colourless liquid turns pink (Figure 6.5.1), the person is 'infected'.

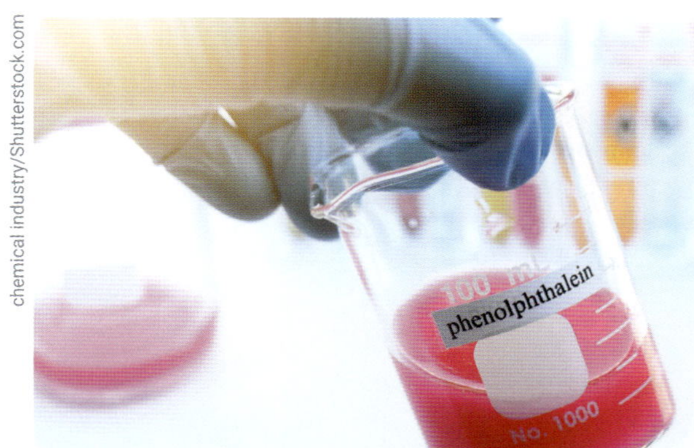

chemical industry/Shutterstock.com

▲ **FIGURE 6.5.1** Phenolphthalein indicator turns pink in basic solutions.

7 A table like the one below can be drawn on the board for all class members to contribute to. Add all names of students (givers) in the class in the first column, and then highlight the 'infected' names to process the data.

Name of giver (potential source of infection)	1st receiver	2nd receiver	3rd receiver

8 Discuss and determine who could be patient zero (the original source of the 'infection').

ANALYSIS

1 **Assess** this model of how a disease spreads, using the questions listed in the bullet list at the start of the investigation.

2 **Discuss** how patient zero was identified in the COVID-19 pandemic. Do you think that doctors used a similar method to the one in this model?

6.6 First line of defence against disease

BY THE END OF THIS MODULE, YOU WILL BE ABLE TO:

✓ describe the first line of defence against disease

✓ explain how the skin, stomach acid, cilia and secretions prevent infectious disease.

GET THINKING

Prepare a concept map, similar to Figure 6.6.1, to summarise the content in the modules you have studied so far. Add to it after you finish each module. You may wish to include diagrams within the map to help your understanding.

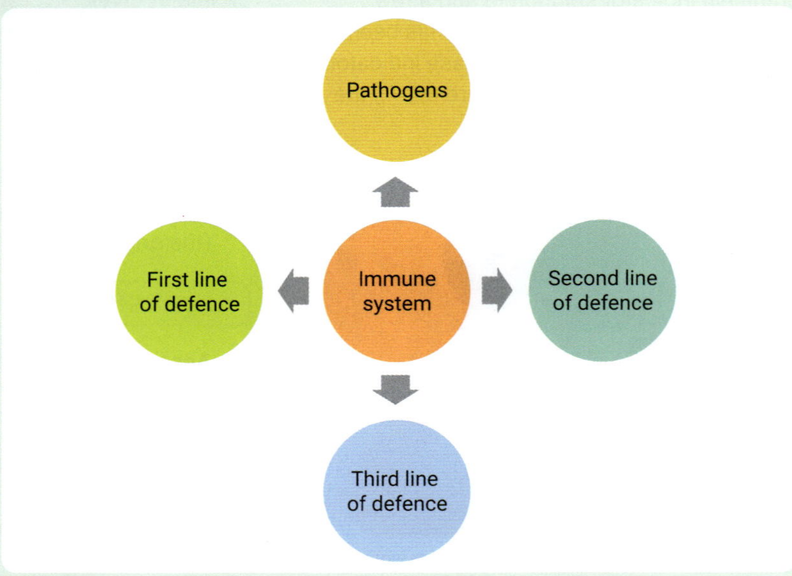

▲ FIGURE 6.6.1 Create a concept map to link what you have learned so far.

The immune system's three lines of defence

The human immune system has three main lines of defence against infectious disease. The first and second lines of defence are non-specific, which means they defend against *any* foreign bodies or invading pathogens. We will look at the second line of defence in more detail in Module 6.7.

In Module 6.8, you will learn about a third line of defence – the adaptive and specific immune response. This is when highly specialised cells and antibodies defend the body against specific pathogens.

cilia
the hair-like extensions of mucous membranes that pushes pathogens up and out of the body

mucus
a thick, sticky liquid in our airways that captures pathogens

lysozyme
an enzyme that breaks down pathogens

The first line of defence: external prevention

The first line of defence in the immune system is about prevention. This initial defence has two main types of barriers:

- **physical barriers:** the skin, **cilia** and **mucus** (mucus traps particles and cilia move them out)
- **chemical barriers:** stomach acid and secretions that contain **lysozymes** (saliva, tears, reproductive tract secretions, sweat).

These barriers are non-specific because they use the same methods to help protect the body from any type of pathogen invasion (Figure 6.6.2).

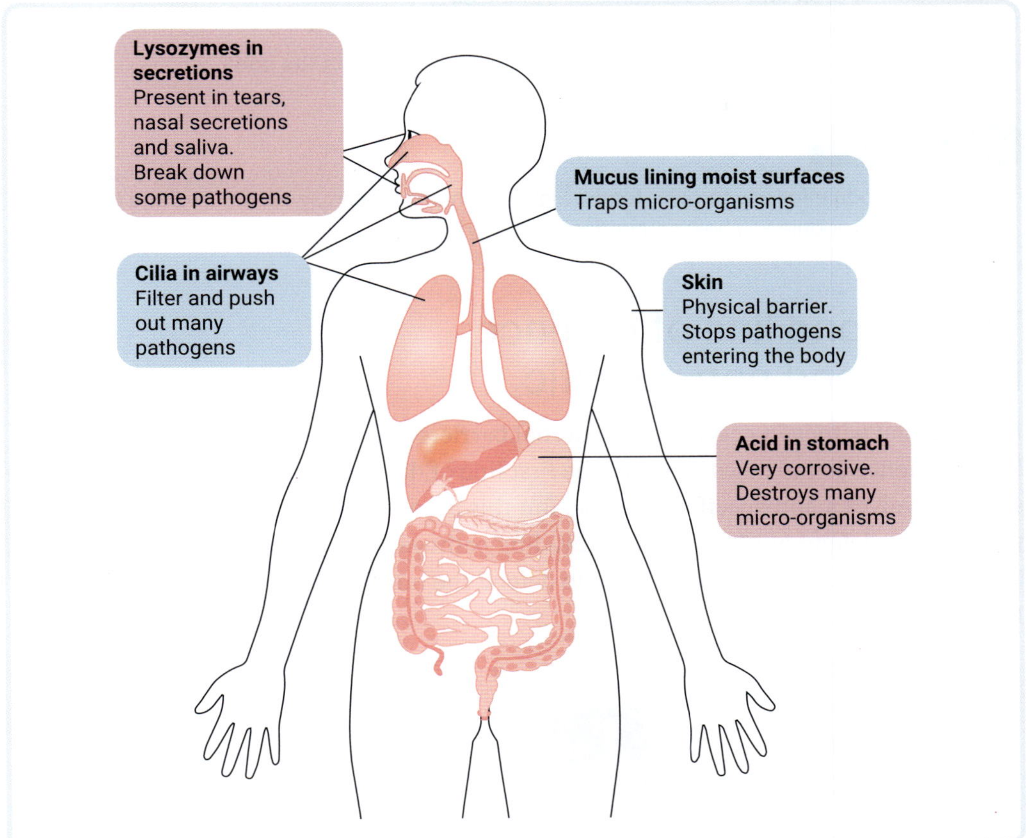

▲ **FIGURE 6.6.2** The first line of defence of the immune system has physical (blue) and chemical (pink) parts.

In the context of disease defences, the surfaces of organs such as the lungs, intestines and stomach are all considered external to the body. This means that all parts of the first line of defence are external.

6.6 LEARNING CHECK

1 **Describe** the main barriers in the first line of defence of the immune system.
2 **Draw** a human body silhouette and **label** your drawing with the list you wrote for Question 1. Include the functions of those parts in your diagram.
3 There are four main reflexes of the human body that assist in the first line of defence: coughing, sneezing, vomiting and diarrhoea. **Choose** one and **research** it to find out what function it serves in your body. Write 3–5 sentences about its function.

6.7 Second line of defence against disease

BY THE END OF THIS MODULE, YOU WILL BE ABLE TO:

- ✓ describe and draw labelled diagrams of a phagocyte, a lysosome, phagocytosis, inflammation and fever
- ✓ list the symptoms of inflammation and explain how it protects the body.

Quiz
Second line of defence

Interactive resource
Label: Phagocytosis

GET THINKING

Scan this module for any familiar terms. Have you had a wound swell, turn red and become painful? This module will help you understand why.

The second line of defence: internal and non-specific

If a potential pathogen gets past the external barriers and enters the body, the second line of defence commences. Just like the first line of defence, this second type of immune system involves a non-specific response. But this time, it is internal processes and cells that work to defend our body against invading pathogens. The processes of **phagocytosis**, **inflammation** and **fever** work together to defend the human body, as summarised in Figure 6.7.1.

phagocytosis
a process in which blood cells called phagocytes engulf and destroy a pathogen

inflammation
a response triggered by damaged cells or a pathogen in the body; characterised by redness, swelling, heat and pain

fever
a non-specific response to infection where the body temperature exceeds the normal 37°C

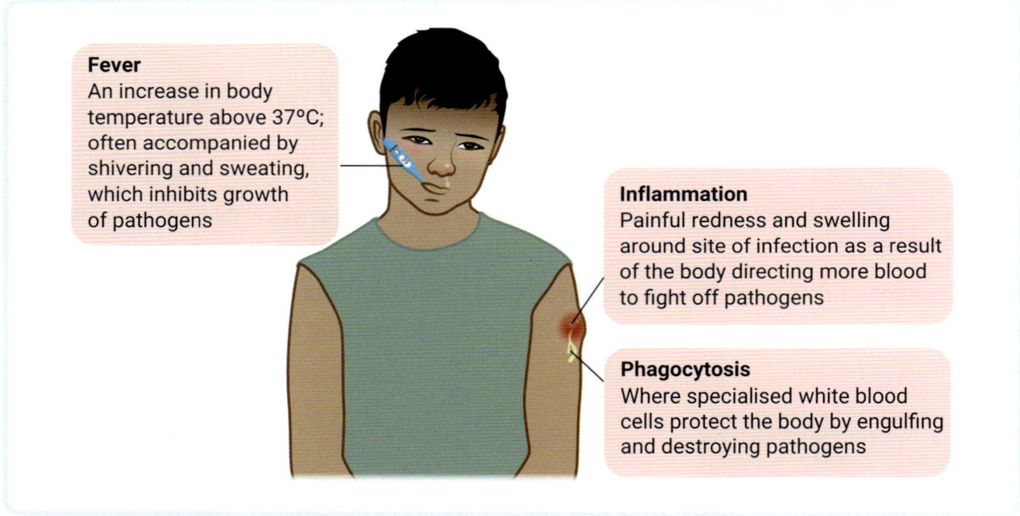

▲ **FIGURE 6.7.1** Processes of the internal, second line of defence of the immune system

Phagocytes and phagocytosis

We have many different types of **white blood cells** as part of our immune system. Specialised white blood cells called **phagocytes** are like personal 'biological weapons' that can engulf (surround and take in) and destroy pathogens in the process of phagocytosis.

white blood cell
a specialised cell in the blood that defends the body against disease

phagocyte
a specialised white blood cell that can engulf and destroy pathogens

There are different types of phagocytes in our blood, each with slightly different functions. However, they are all large white blood cells capable of identifying 'non-self' cells and changing their shape to surround invading substances. As a phagocyte engulfs the cell, **lysosome** organelles digest (break down) the pathogen using enzymes. Figure 6.7.2 shows the main stages of phagocytosis.

lysosome
a cell organelle that contains digestive enzymes

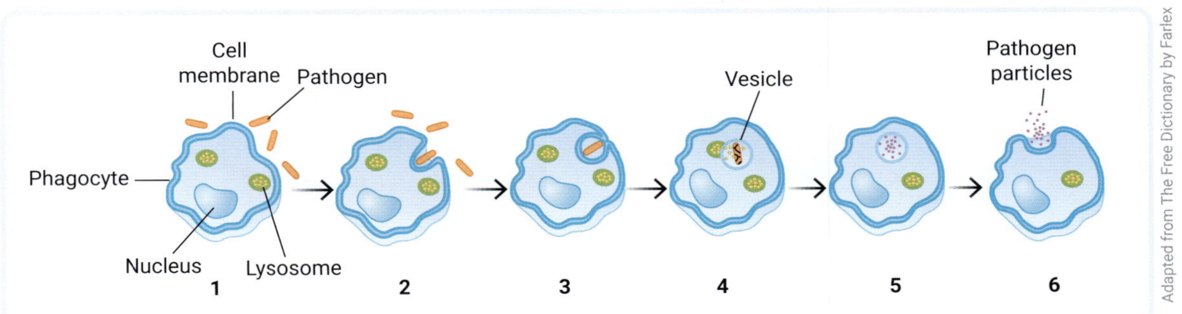

▲ **FIGURE 6.7.2** The process of phagocytosis

The process of phagocytosis consists of the following steps:

1. A phagocyte recognises a potential pathogen (e.g. a bacterium).
2. The phagocyte changes its shape to surround the pathogen.
3. The cell membrane engulfs the pathogen enclosing it in a membrane and forming a vesicle.
4. The vesicle containing the pathogen fuses with a lysosome, which injects a mix of enzymes to digest the pathogen.
5. The pathogen breaks down into small particles.
6. The particles are released from the phagocyte.

Inflammation

Inflammation usually hurts, but it is a defence system that helps protect the body from pathogens. Inflammation is triggered by damaged cells in the body or by a pathogen that got past the barriers of the immune system's first line of defence (Figure 6.7.3). The inflammatory response sends for immune cells, such as phagocytes, to remove dead cells and begin repairing tissue.

The signs of inflammation are:

- redness
- swelling
- heat
- pain.

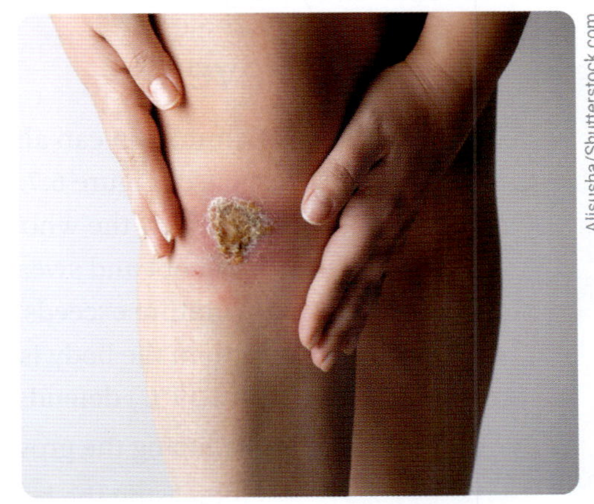

▲ **FIGURE 6.7.3** This scab is showing signs of infection.

Figure 6.7.4 shows the main stages of inflammation.

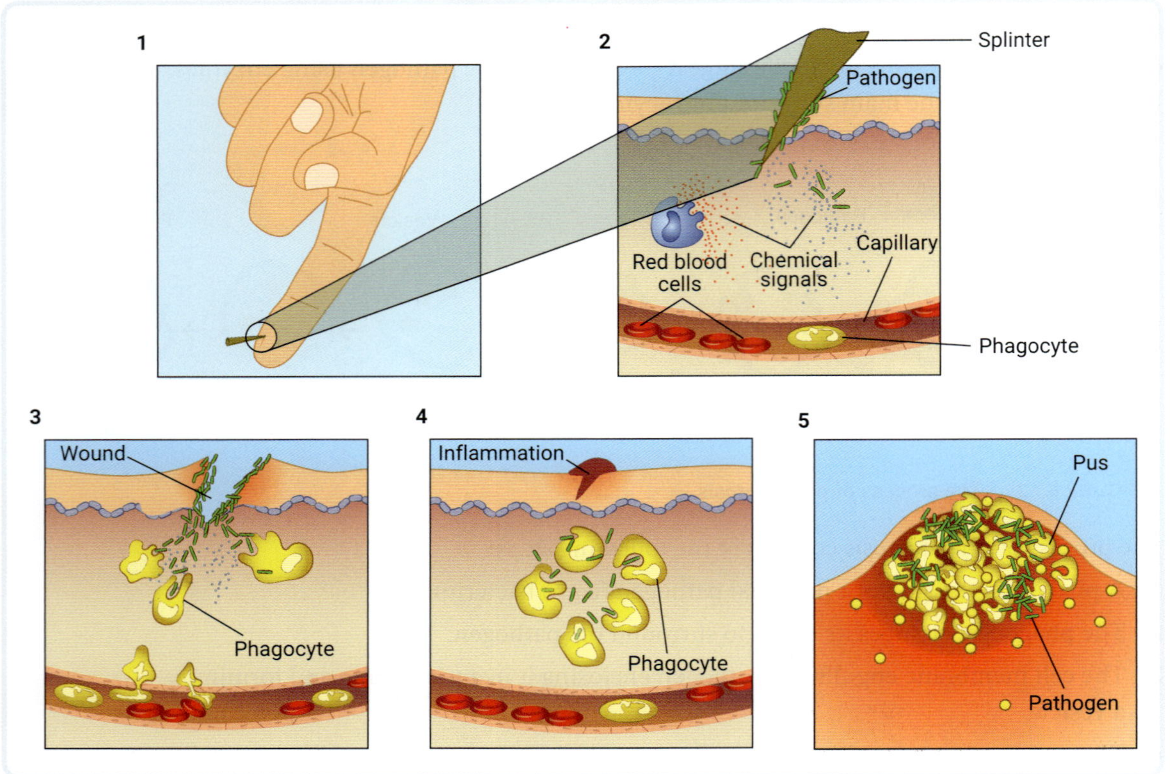

▲ **FIGURE 6.7.4** The process of inflammation

The process of inflammation consists of the following steps:

1 The body suffers a wound, which is an entry point for pathogens.

2 Cells near the wound release chemical signals. These cause blood vessels to dilate and become more permeable, which increases blood flow near the wound. This results in redness, heat, swelling and pain.

3–4 Phagocytes arrive at the site of the wound to destroy the pathogens.

5 Pus (a mix of used phagocytes, dead cells and tissue fluids) may form but will eventually be removed by phagocytes.

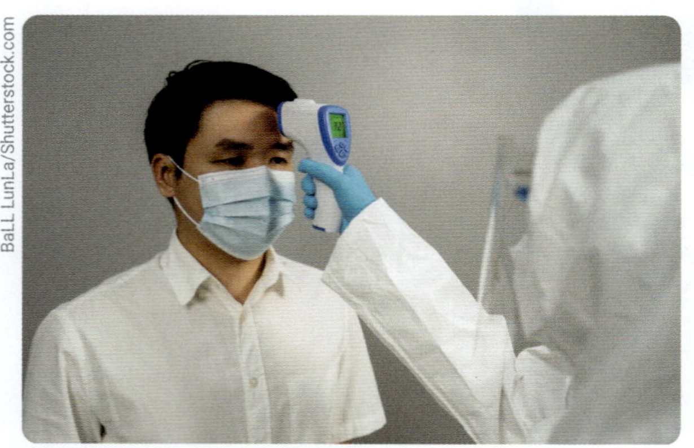
BaLL LunLa/Shutterstock.com

▲ **FIGURE 6.7.5** A temperature above 39.0°C is considered a high fever.

Fever

Inflammation can also lead to fever, another non-specific response (Figure 6.7.5). Instead of a localised response, fever affects the whole body. Symptoms include feeling unwell, hot and sweaty, and shivering. When internal temperature exceeds the normal 37°C in response to an infection, the body is experiencing fever. A high fever (about 39°C) defends the body by:

- reducing the growth rate of the pathogen
- increasing the body's metabolic activity, which makes the immune cells respond faster.

Using a digital thermometer

☆ ACTIVITY

6.7

Test the effectiveness of measuring body temperature by the 'under the arm' method with a digital thermometer. You will need to ask your teacher if you can use a thermometer. Conduct multiple trials on three people and calculate an average body temperature for each person. Disinfect the thermometer before using it on different people.

6.7 LEARNING CHECK

1 **Define** phagocyte.

2 **Describe** the steps of phagocytosis.

3 **Draw** an annotated diagram of the steps of the inflammation response.

4 **Compare** (**discuss** the similarities and differences between) the first and second lines of defence in terms of:
   - the processes and parts of the body involved
   - whether they are specific or non-specific.

   You can do this by using a Venn diagram, as shown below.

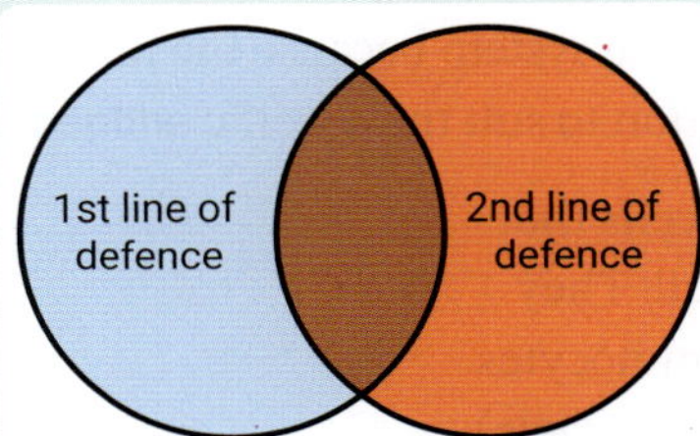

5 Write a paragraph to **explain** why processes in the second line of defence are described as 'non-specific'. Use examples to **justify** your reasoning.

6 Lysosomes are found in all animal cells. Figure 6.7.6 shows an animal cell with lysosomes.

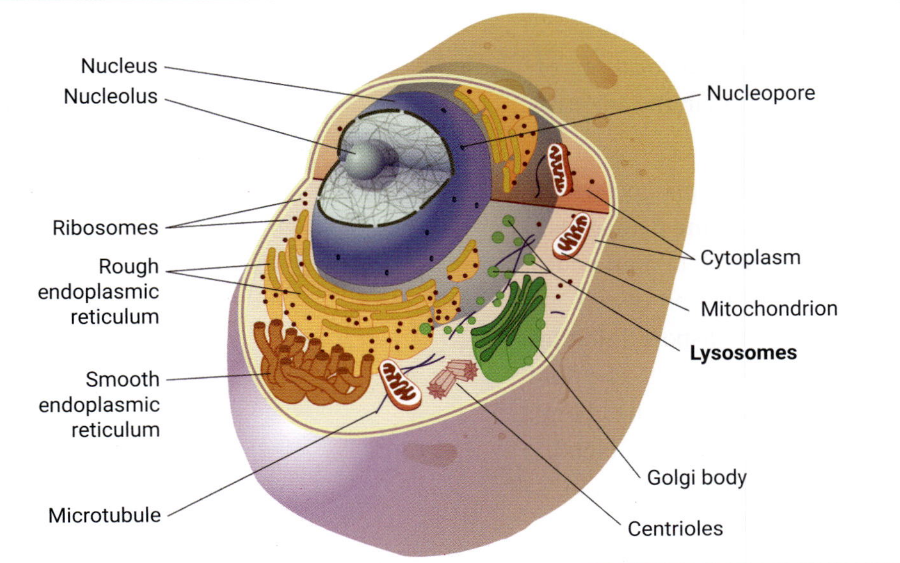

▲ **FIGURE 6.7.6** An animal cell with lysosomes

What is the function of the lysosomes? What type of human cell uses lysosomes in the second line of defence?

6.8 Third line of defence: adaptive immune system

BY THE END OF THIS MODULE, YOU WILL BE ABLE TO:

- ✓ classify and describe the function of the major white blood cells of the immune system
- ✓ identify where the immune cells are produced in the body
- ✓ define antigen, antibody, lymphocyte, T cell, B cell and memory cell
- ✓ describe the different parts of the adaptive immune system and how they work together.

Video activity
Immune defence

GET THINKING

Have you had a cold recently? How long did it last: 7 to 10 days? This is approximately how long it generally takes for someone's third line of defence to launch a specific response and destroy the pathogens.

The third line of defence: adaptive and specific

antibody
a Y-shaped protein that binds to one type of antigen

antigen
a substance on a pathogen that stimulates the adaptive immune response

lymphocyte
a type of white blood cell involved in adaptive immune responses

B cell
a type of white blood cell involved in the humoral (antibody)-mediated immune response

T cell
a type of white blood cell involved in the cell-mediated immune response

The third line of defence to disease is the adaptive immune system. It responds directly to a specific invading pathogen by producing an 'army' of immune cells and **antibodies**. It is known as the adaptive immune system because the immune cells and antibodies need to adapt in structure and function to suit the specific **antigen** present on the pathogen.

Lymphocytes are immune cells that are a type of white blood cell. They have two main types: B-lymphocytes (**B cells**) and T-lymphocytes (**T cells**). B cells are produced and mature in the bone marrow (B for bone marrow). T cells are produced in the bone marrow too, but they mature in the thymus (so T for thymus). Figure 6.8.1 shows the locations of the organs involved in the adaptive immune system.

The spleen contains phagocytes that fight any invading pathogens that are in the blood. Lymph nodes are connected by lymph vessels that absorb and transport fluid that leaks out of blood vessels. The fluid inside lymph vessels is called lymph. Pathogens can travel out of blood vessels into lymph and collect in the lymph nodes, where a variety of immune cells (lymphocytes and phagocytes) can target them.

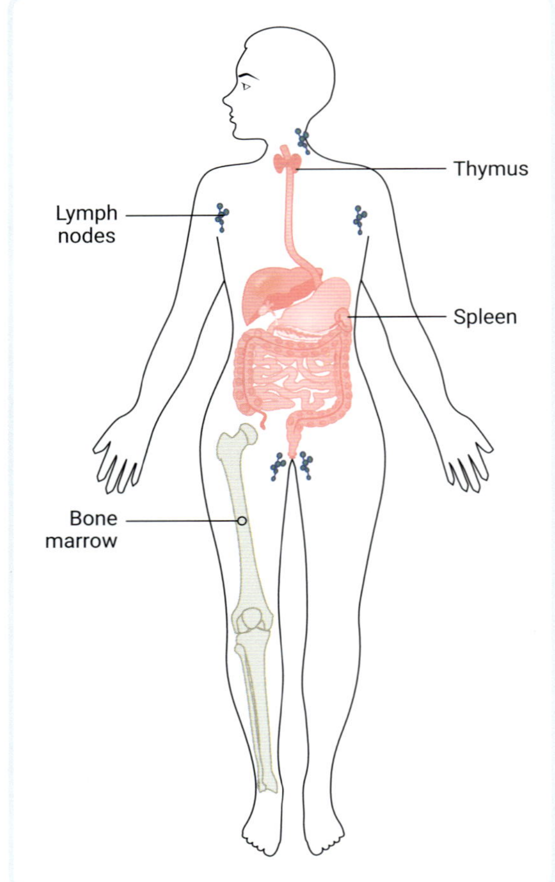

▲ **FIGURE 6.8.1** Organs involved in the adaptive (specific) immune system

Antigens and antigen-presenting cells

When pathogens get past the barriers and processes of the first and second lines of defence, they can be targeted by the adaptive immune system. How does the adaptive immune system know which cells to destroy? And how do the immune cells and antibodies know what specific structures to develop?

The answer to both questions is 'antigens'. All cells, including pathogens, have antigens (usually proteins) on their surface (Figure 6.8.2). Viruses and toxins contain antigens too. Antigens stimulate an immune response if identified by the immune system. There are specialised white blood cells that can identify 'non-self' antigens, which are antigens that do not belong in the body. These specialised white blood cells are called antigen-presenting cells (APCs). APCs, such as phagocytes, identify, capture and present antigens to T cells and B cells. T cells and B cells are stimulated to adapt their structures to be ready to defend against the specific pathogen they have met. Some B cells produce antibodies, which are Y-shaped proteins that recognise and bind to a specific type of antigen on the surface of a pathogen.

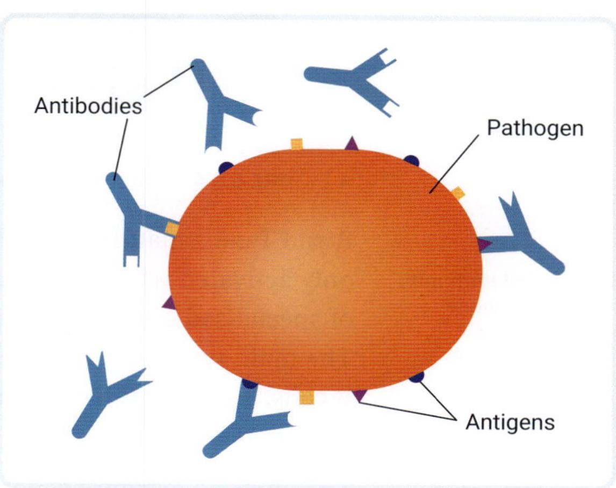

▲ **FIGURE 6.8.2** Antibodies recognise and bind to specific antigens.

Figure 6.8.3 shows the relative sizes of a virus, a bacterium and two types of blood cell.

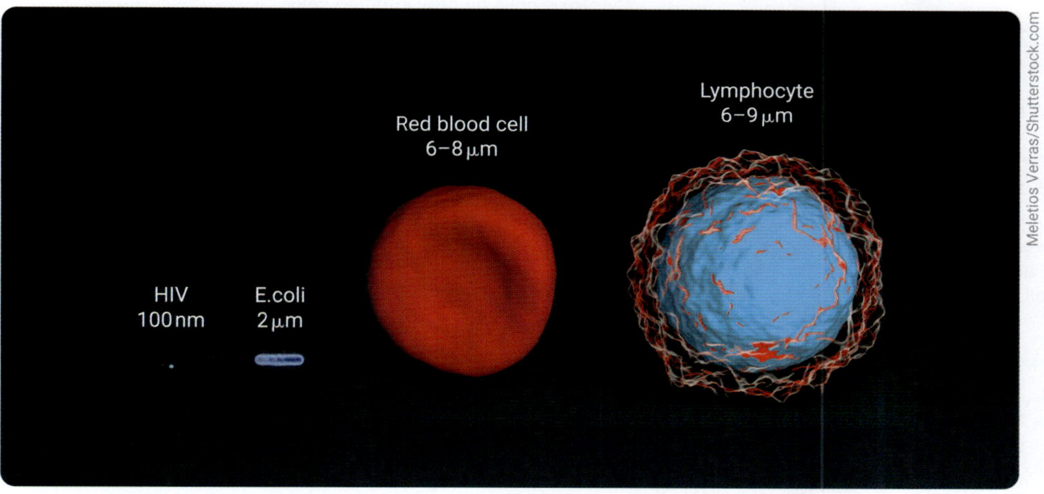

Meletios Verras/Shutterstock.com

▲ **FIGURE 6.8.3** A lymphocyte compared with a blood cell, a bacterial cell and a virus

A coordinated response

Have you watched a movie where a medical team carries out surgery? Each specialist has a different role during the surgery. One applies anaesthetic, while another cuts and sews, but all work together to achieve the same purpose – to heal the patient. Our B cells and T cells are a team of specialised cells with different roles to achieve the same purpose – to destroy the pathogen and remember the antigen in case it invades again.

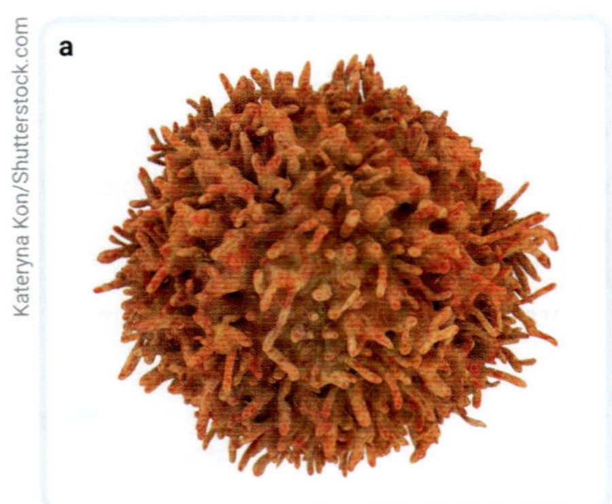

Kateryna Kon/Shutterstock.com

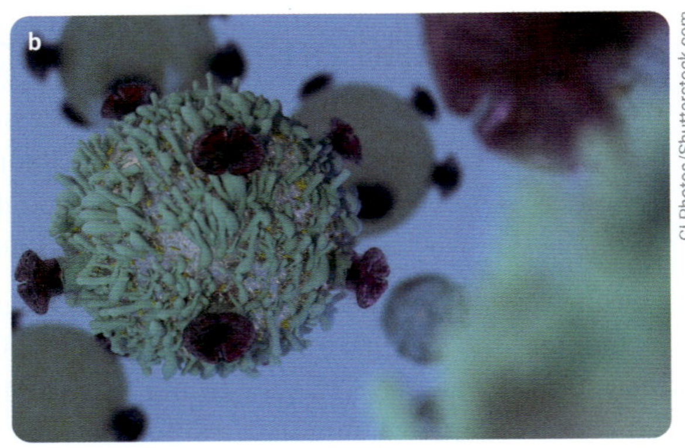

CI Photos/Shutterstock.com

▲ **FIGURE 6.8.4** (a) B cell and (b) T cell

cell-mediated immunity the adaptive immune response prompted by T cells on infected host cells

humoral immunity the adaptive immune response prompted by B cells, and antibodies specific to the antigen, on pathogens found outside of host cells

B and T cells mature into different types of specialised cells (Figure 6.8.4). They are only activated when they recognise specific antigens on a pathogen. The main types of specialised T cells are helper T cells, cytotoxic (killer) T cells and memory T cells. The main types of B cells are plasma B cells (which produce antibodies) and memory B cells. T cells and B cells work together to provide a coordinated immune response (Figure 6.8.5).

The immune response generated by T cells is classified as **cell-mediated immunity**. The immune response generated by B cells is classified as **humoral immunity**. Table 6.8.1 summarises the main types and functions of B cells and T cells.

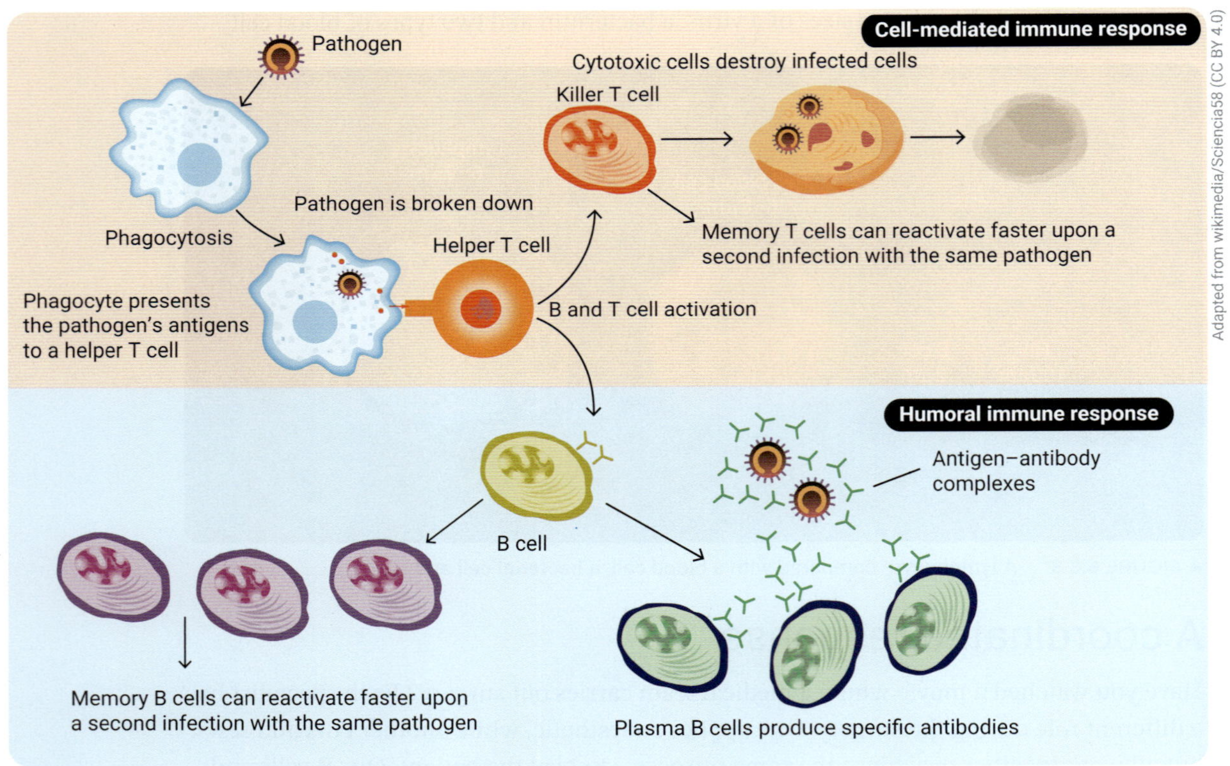

Adapted from wikimedia/Sciencia58 (CC BY 4.0)

▲ **FIGURE 6.8.5** The coordinated cell-mediated and humoral immune responses. This diagram shows the main functions of T cells and B cells.

▼ **TABLE 6.8.1** Types and functions of the main types of T cells and B cells

Type of cell	How are they activated?	Function and type of response
Helper T cell	Immature helper T cells mature when presented with an antigen	• Signal immature B cells to become plasma B cells • Cell-mediated immunity
Plasma B cell	Immature B cells mature into plasma B cells and multiply when signalled by helper T cells	• Make lots of specific antibodies that act on the antigen • Antibodies attach to the antigen on the surface of a pathogen forming the antigen–antibody complex and neutralising the effect of the pathogen • Humoral immunity
Cytotoxic (killer) T cell	Antibodies act like a marker for cytotoxic (killer) T cells to destroy the pathogen	• Release a toxic chemical when it encounters an infected host cell • The chemical destroys the damaged cell and the pathogen inside • Cell-mediated immunity
Memory T cell and B cell	Some of the specialised T and B cells form 'memories' of a specific pathogen	• Generate a faster and more intense immune response if a host is exposed again to the same pathogen • B cells: humoral immunity • T cells: cell-mediated immunity

6.8 LEARNING CHECK

1 **Define**:
  a antigen.
  b antibody.
  c T cell.
  d B cell.
  e memory cell.

2 **Describe** the function of white blood cells.

3 **Identify** the correct cell type (B cell or T cell) for the following functions:
  a Humoral response
  b Produces antibodies
  c Marks pathogens ready for destruction

4 **Draw** a diagram of your immune system and **label** the main barriers, bone marrow, thymus, lymph nodes and spleen. Then **construct** a table to **record** the parts and their functions.

5 **Analyse** Figure 6.8.5. Write out the steps of the coordinated cell-mediated and humoral immune responses.

6.9 Artificial immunity

BY THE END OF THIS MODULE, YOU WILL BE ABLE TO:

- ✓ define and compare natural immunity and artificial immunity
- ✓ describe how vaccination stimulates the adaptive immune response
- ✓ describe the principles of herd immunity.

Video activity
Smallpox: The first vaccine

Quiz
Herd immunity

Other resource
Worksheet: The National Immunisation Program

GET THINKING

Imagine you have to convince someone that vaccines are important. Scan the information in this module and list as many facts as you can to support your argument.

Natural immunity

Natural immunity (adaptive immunity) can be achieved by the production of antibodies after the invasion of a pathogen. During the coordinated response, antibodies and **memory cells** are actively produced, specific to one antigen. This ensures the body is ready for a quicker and stronger immune response in the case of reinfection with the same pathogen. This usually provides long-term protection from reinfection.

natural immunity
the protection obtained from the production of antibodies after exposure to an infectious disease

memory cell
a mature B cell or T cell that 'remembers' information about a specific antigen

artificial immunity
the protection obtained from the production of antibodies after receiving a vaccine or antibodies

vaccine
a substance that stimulates a response in the immune system to produce antibodies

vaccination
administration of a vaccine to provide protection from a specific disease

Artificial immunity from antibodies or vaccination

One way you can gain **artificial immunity** is to receive a sample containing antibodies either produced in someone who recovered from the same disease or a sample containing antibodies in the laboratory. It provides immediate protection against the pathogen. For example, a treatment for COVID-19 is the antibody drug sotrovimab. The drug works by binding to the protein spikes on the virus, which prevents it from replicating. This drug is used early in the course of an infection to reduce the severity of the disease. This is a short-term treatment because the new host does not produce memory cells.

You can gain longer-term artificial immunity by having a **vaccine**. A vaccine is a preparation of a weakened (attenuated) or inactivated (dead) micro-organism or virus or part of one that is used to artificially induce immunity to a specific disease (Figure 6.9.1).

When you get vaccinated, your immune system responds as if it had been naturally exposed to a pathogen. The vaccine stimulates the production of mature B cells and T cells, antibodies and memory cells specific to the antigen. **Vaccination** is usually administered by injection, but some vaccines are given by mouth as a liquid or tablet.

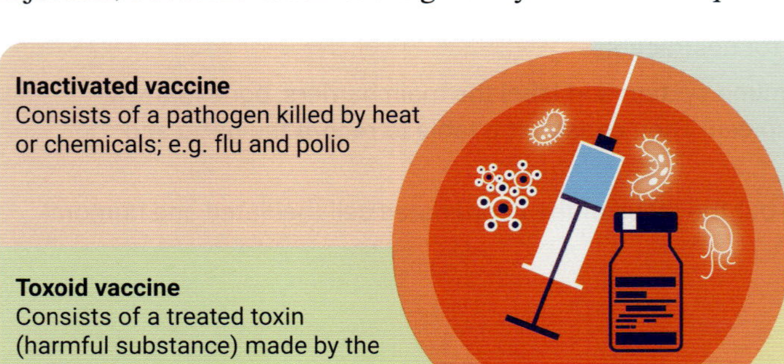

▲ **FIGURE 6.9.1** The four main types of vaccines

Artificial immunity is generally safer than natural immunity because the symptoms that follow the vaccine are usually not as severe as those of the infection of a pathogen. One of the greatest vaccine success stories is the eradication (elimination) of smallpox. Smallpox is a highly infectious and deadly disease that killed an estimated 300 million people in the 20th century alone (Figure 6.9.2). It is the only human disease that has been eradicated.

Immunisation for smallpox was first documented in 1796 when Dr Edward Jenner inoculated a boy with a mild cowpox, and then a few months later with the smallpox virus. The boy did not develop the disease. It was not until the 1970s that the World Health Organization coordinated a global vaccination program for smallpox, which eventually resulted in its eradication. The last recorded case was in 1977.

In Australia, the national immunisation program provides vaccines for 13 different diseases for your age group and younger. You may have received a dose (booster shot) in Year 7 for human papillomavirus, diphtheria, tetanus and whooping cough.

Science History Images/Alamy Stock Photo

▲ **FIGURE 6.9.2** Smallpox before it was eradicated. The boy on the right and the girl on the left had previously been vaccinated.

Boosters

The live attenuated (weakened) vaccines, such as those used for measles and mumps, can provide a lifetime of protection after one or two doses. Other vaccines, such as tetanus, need to be given in additional doses, or **boosters**, because, over time, antibodies and memory B and T cells decline.

booster
an additional dose of a vaccine, designed to provide a higher level of protection when antibodies have decreased

COVID-19 vaccines

COVID-19 emerged in 2019. Its rapid spread and severe impact on people meant that urgent protection was needed. As soon as the RNA of the virus was sequenced, it was shared globally and collaboratively, allowing for more than 250 vaccines to be developed.

Vaccines need to go through a series of trials before they can be approved by the Australian Government (Figure 6.9.3). Scientists:

- conduct laboratory research (exploratory phase)
- test for safety and effectiveness on animals (such as ferrets or mice) (preclinical trials)
- conduct human clinical trials to ensure the vaccine is safe and effective.

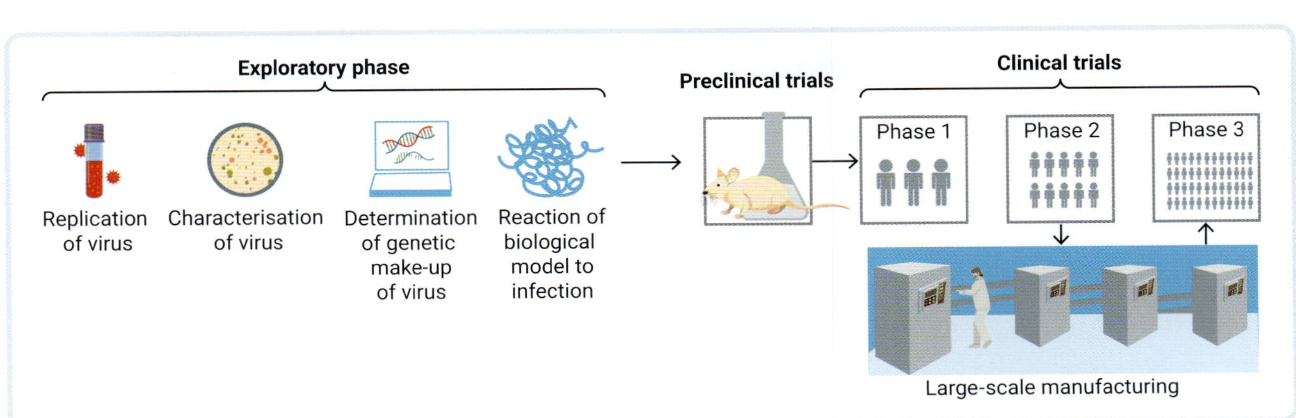

▲ **FIGURE 6.9.3** A vaccine needs to go through a series of steps before it can be approved.

Herd immunity

herd immunity
a level of protection gained when a certain percentage of a community is immune to an infectious disease, which breaks the chain of transmission

herd immunity threshold
the level of population immunity needed to stop the spread of a specific disease

Herd immunity is a level of protection gained when a certain percentage of people in a community is immune to an infectious disease (Figure 6.9.4). This breaks the chain of transmission and effectively stops the spread of disease. Different diseases have different levels of **herd immunity threshold** to stop their spread, depending on how transmissible they are. For example, the vaccination threshold to stop the spread of measles is 95 per cent of the population, whereas polio has a threshold of 80 per cent. Natural immunity can contribute to herd immunity; however, artificial immunisation by vaccination is a safer method.

Can we achieve herd immunity for COVID-19? Very contagious diseases like COVID have a higher threshold of immunisation to achieve herd immunity. Additionally, herd immunity only works if a person can maintain their level of immunity. With COVID-19, we know people can still become infected even after a vaccine, and there is also evidence that people can get reinfected with different variants of COVID-19. This is because natural and artificial immunity wanes over time, or the vaccine is ineffective against new variants.

So why should we bother with COVID-19 vaccinations? A study by the US Centers for Disease Control and Prevention (CDC), conducted in 2021, showed that unvaccinated people are more than twice as likely to be reinfected with COVID-19 than those who are vaccinated. In early 2022, there was approximately a one in 1000 chance of being admitted to intensive care if you were double vaccinated, but a 15 in 1000 chance if you were unvaccinated. For most people, vaccination reduces the severity of COVID-19.

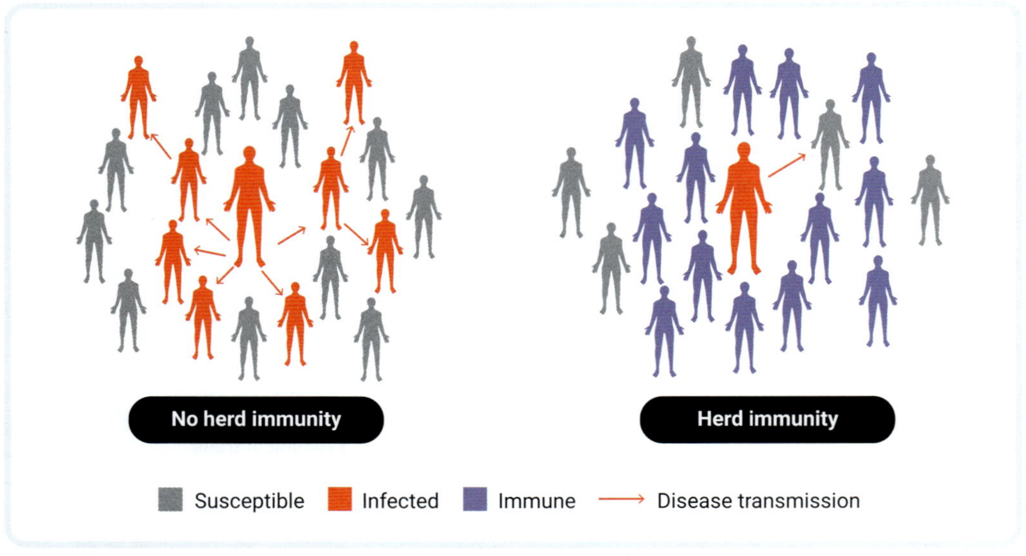

▲ **FIGURE 6.9.4** Herd immunity reduces the transmission of disease.

9780170491785

6.9 LEARNING CHECK

1 **Define** vaccine.

2 **Analyse** Figure 6.9.5. **Describe** the trend shown in this graph between vaccination and deaths and intensive care unit (ICU) admissions.

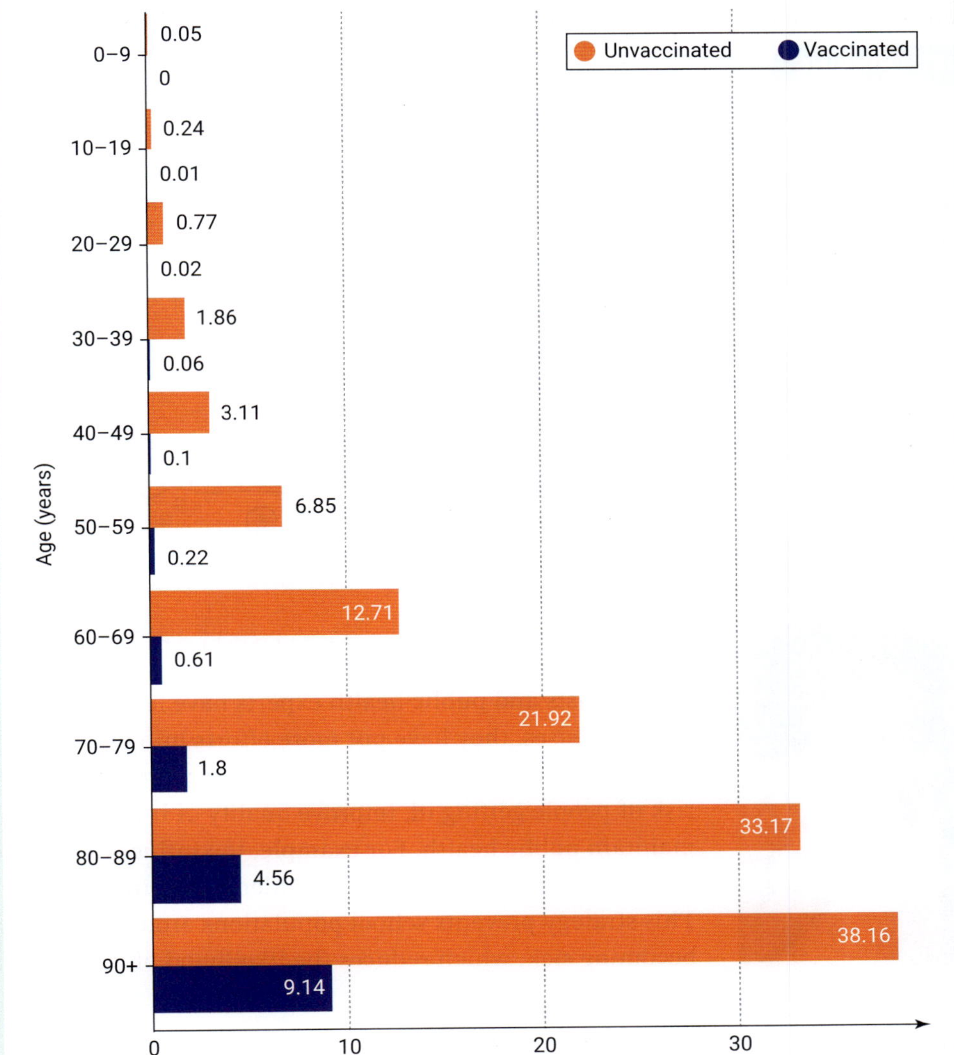

Copyright Guardian News & Media Ltd 2022, based on New South Wales health department surveillance reports

▲ **FIGURE 6.9.5** The proportion of COVID-19 cases with a severe outcome (death or ICU admission) by age and vaccination status from June 2021 to 1 January 2022

3 Copy and complete the following table to **compare** natural immunity and artificial immunity through vaccination.

Characteristic	Natural immunity	Artificial immunity (vaccination)
Is it specific or non-specific to a pathogen?		
Are antibodies made in response?		
Do memory cells form?		

6.10 Preventing disease

BY THE END OF THIS MODULE, YOU WILL BE ABLE TO:

- ✓ describe disease prevention
- ✓ outline the categories of disease prevention
- ✓ outline strategies to prevent diseases
- ✓ explain strategies to prevent and control the spread of diseases.

GET THINKING

You come back to school on a Monday, and you see this poster placed around your school (Figure 6.10.1). Discuss with your school friends the purpose of this poster. What is the idea behind the slogan 'personal hygiene keeps you healthy'?

▶ **FIGURE 6.10.1** A public health poster

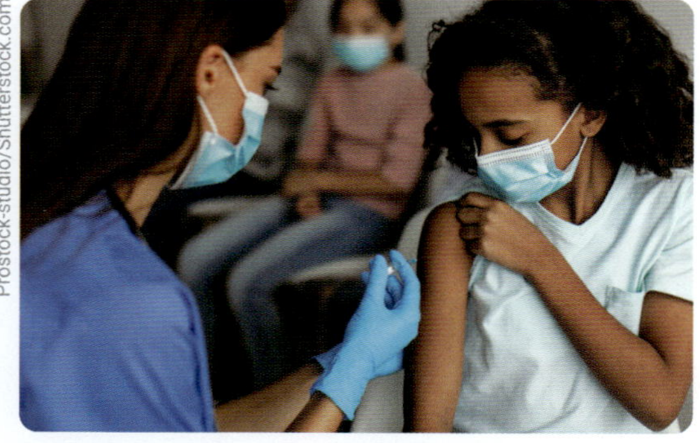

▲ **FIGURE 6.10.2** Regular influenza vaccinations in schools can reduce the spread of disease through other populations.

What is disease prevention?

As scientists and public health experts have learned more about diseases, they have put more efforts into prevention of disease. Effective preventive measures can reduce the risk of people getting ill, improve quality of life, and help maintain public health. For example, the influenza vaccine is given for free to teachers and students (Figure 6.10.2). This strategy prevents school populations around Australia becoming sick, which reduces the likelihood of outbreaks in the broader community.

Strategies to prevent diseases

Disease prevention includes making plans and taking actions to prevent illnesses happening and to reduce how severe disease is when it occurs. It has three levels:

- **primary prevention**: actions to prevent a disease before it occurs; for example, vaccinations, promoting healthy lifestyle choices
- **secondary prevention**: early detection and treatment of diseases to prevent them getting worse; for example, regular screenings, check-ups
- **tertiary prevention**: management and reduction of the impact of an ongoing disease, including rehabilitation and support to improve quality of life.

The importance of prevention

Preventing diseases is often more effective and less costly than treating them after they occur. It reduces healthcare costs, minimises suffering and promotes healthier communities. Disease prevention is the responsibility of everyone in the community, from the individual to government and other organisations (Table 6.10.1). Individuals can reduce disease by eating healthily, having regular physical activity, and avoiding harmful substances, such as alcohol and tobacco. The government and other non-government organisations use educational campaigns to inform people about how to prevent specific diseases. The government also runs vaccination and screening programs.

▼ **TABLE 6.10.1** Disease prevention strategies

Responsibility	Strategies	Example	Prevention of
Individual	Good diet and nutrition	Eating a balanced diet of fibres, proteins and carbohydrates	Obesity Nutritional deficiencies
	Physical activity	Walking regularly, attending gym sessions, swimming, running	Obesity Bone-related diseases Cardiovascular diseases
	Avoid harmful substances	Stop smoking and alcohol consumption Avoid recreational drugs	Lung cancer Kidney and liver diseases
	Managing stress and social wellbeing	Meditation Community engagement	Mental-health-related diseases
Government/ non-government organisations	Educational campaigns	QUIT (smoking) Slip, Slop, Slap, Seek, Slide (skin cancer) COVID-19 hygiene campaign Television advertisements for other diseases (e.g. bowel cancer testing)	Create awareness of diseases
	Vaccination programs	National vaccination programs	Create herd immunity
	Screening programs	Breast cancer Bowel cancer Cervical cancer	Cancer
	Mental health support	headspace Beyond Blue Dementia Australia	Mental-health-related diseases

☆ ACTIVITY

Analysing infectious disease data

Procedure

Look at the world map in Figure 6.10.3. It represents the percentage of respiratory specimens (samples from infected people) that tested positive for influenza by transmission zone (particular regions of the world). The pie charts use colour to show the percentage of various virus subtype in each zone and the colours on the map show the percentage of influenza cases for the zone.

Analysis

1 What is the most common virus subtype across the world?

2 What areas of the world had the lower percentages of influenza?

3 What areas of the world had the higher percentages of influenza?

4 **Infer** possible reasons that some areas of the world have higher percentages of influenza than others.

5 **Formulate** a hypothesis about the distribution of influenza subtypes that could be tested by further analysis of the world map.

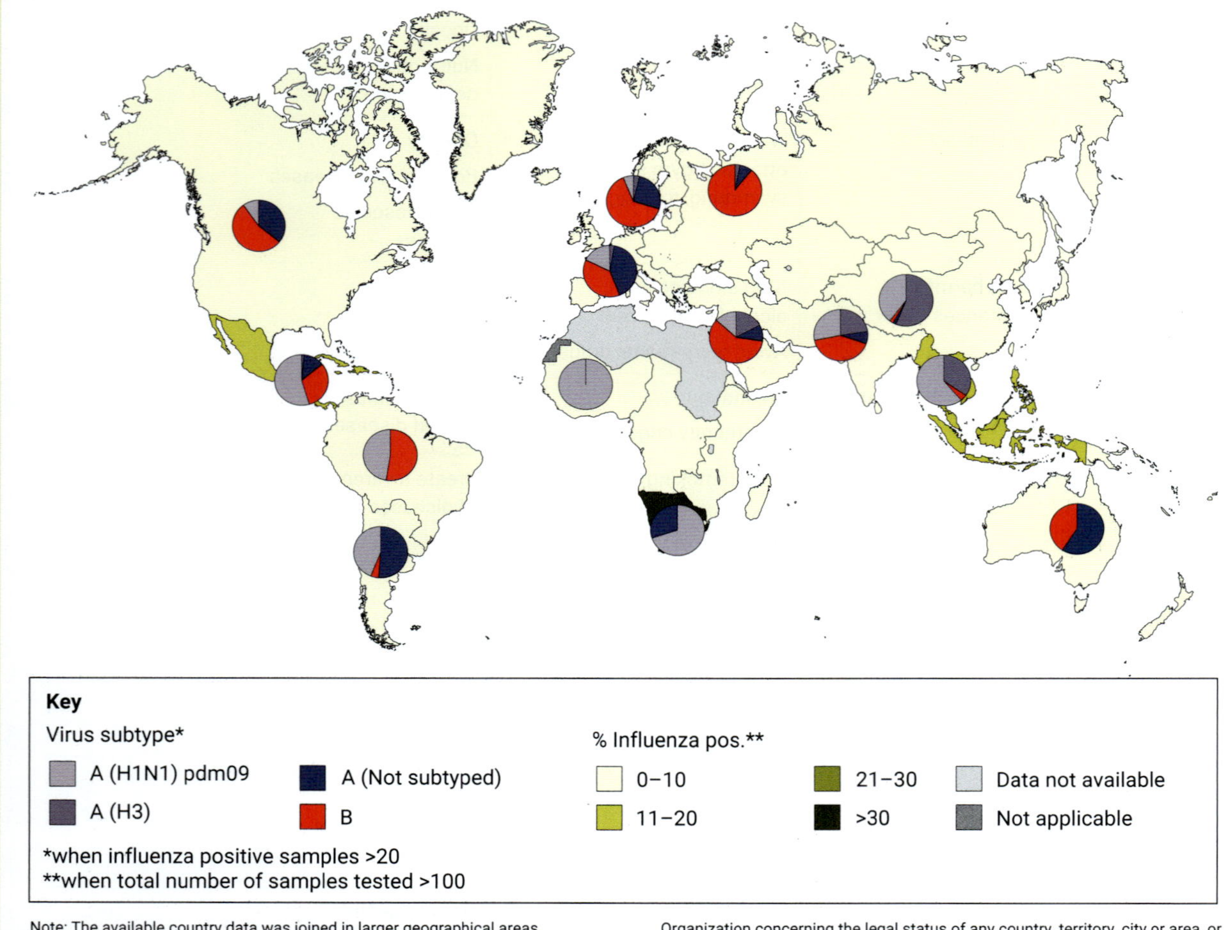

Note: The available country data was joined in larger geographical areas with similar influenza transmission patterns to be able to give an overview (https://www.who.int/publications/m/item/influenza_transmission_zones). The displayed data reflect reports of the week from 4 June 2023 to 17 June 2023, or up to two weeks before if no sufficient data were available for that area.

The boundaries and names shown and the designations used on this map so not imply the expression of any opinion whatsoever on the part of the World Health Organization concerning the legal status of any country, territory, city or area, or of its authorities, or concerning the delimitation of its frontiers or boundaries. Dotted and dashed lines on maps represent approximate border lines for which there may not yet be full agreement.

Data source: Global Influenza Surveillance and Response System (GIRS), FluNet (www.who.int/tools/flunet). Copyright WHO 2023. All rights reserved.

▲ **FIGURE 6.10.3** The percentage of respiratory specimens that tested positive for influenza, by influenza transmission zone

Influenza Update N° 448, World Health Organization, June 26, 2023. https://www.who.int/publications/m/item/influenza-update-n-448

9780170491785

Analysing non-infectious disease data

Type 2 diabetes is a non-infectious disease in which the body loses the capacity to produce enough insulin in the pancreas. Some people have a genetic make-up that gives them a strong predisposition to this disease.

Procedure

Look carefully at Figures 6.10.4 and 6.10.5. These graphs use data gathered by the Australian Institute of Health and Welfare.

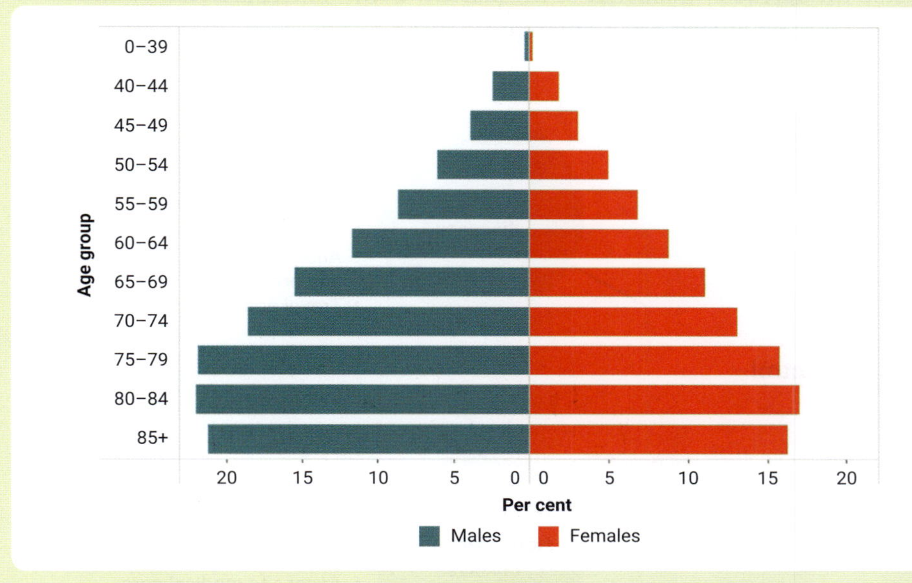

FIGURE 6.10.4 Prevalence of type 2 diabetes by age and sex in Australia

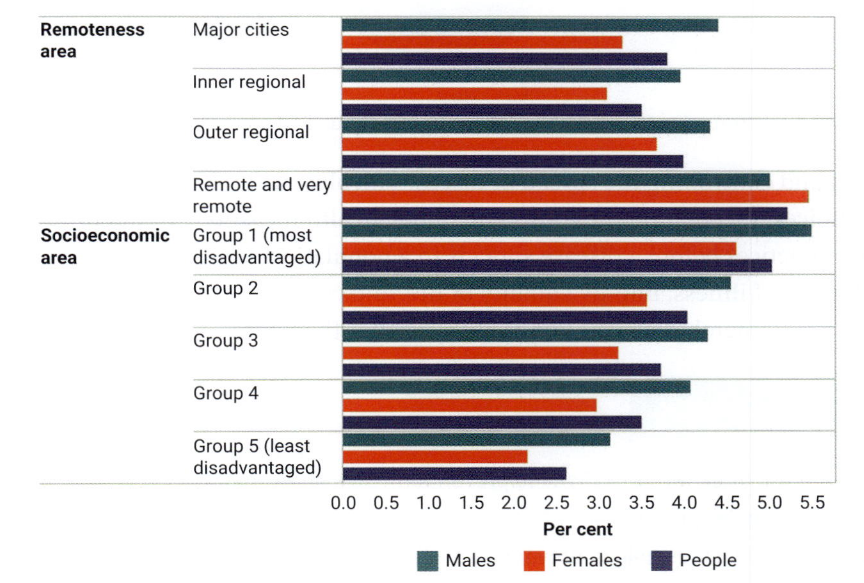

FIGURE 6.10.5 Prevalence of type 2 diabetes by population group and sex, 2021

Analysis

1 Which sex has a higher prevalence of type 2 diabetes? Why do you think this is?
2 **Describe** the prevalence in remote areas and cities. Infer a reason for the difference.
3 **Discuss** the difference in prevalence between Groups 1 and 5 in Figure 6.10.5.
4 **Create** an inquiry question about type 2 diabetes based on this activity's data.

How to control the spread of disease

The prevention of infectious diseases and non-infectious diseases is different due to the different ways these diseases spread or occur in a population.

Sometimes, despite having many different prevention measurements in place, infectious diseases can still spread in a population, or the incidence of a non-infectious diseases can continue to increase. When that happens, control methods need to be used to slow or stop the spread or incidence of the disease. Control measures can be similar to prevention measures but are often different (Figure 6.10.6).

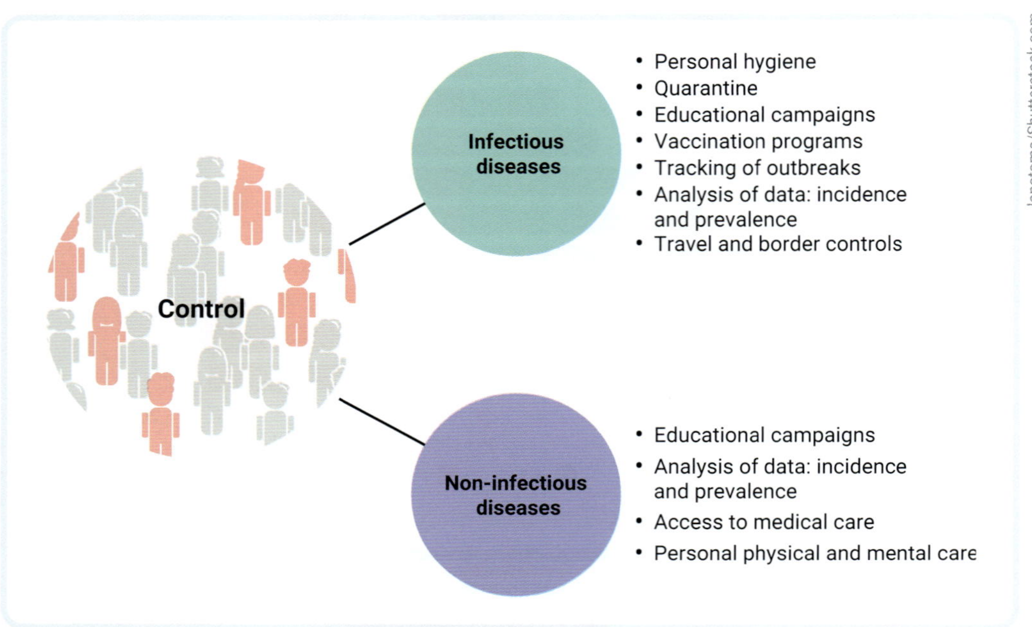

▲ **FIGURE 6.10.6** Strategies to control the spread of infectious and non-infectious diseases

Preventing and controlling the spread of diseases involves a combination of healthy lifestyle choices, vaccination programs and public health measures. By adopting preventive practices before a disease spreads in a population, we can significantly reduce the risk of illness, improve quality of life, and contribute to the wellbeing of communities. Remember, prevention is always better than cure.

6.10 LEARNING CHECK

1 What are the three levels of disease prevention? Give an example of each.
2 How does regular physical activity contribute to disease prevention?
3 **Explain** how vaccines help in preventing diseases in a population. **Outline** an example.
4 **State** two public health initiatives that contribute to disease prevention.
5 Imagine you are the chief public health adviser of the government. An outbreak of an infectious disease is happening in one area of the community. **Outline** four strategies that you would plan to present to the government to prevent and control the spread of the disease.

WORKING SCIENTIFICALLY

6.11 Petri dish safety

SCIENCE SKILLS IN FOCUS

IN THIS MODULE, YOU WILL FOCUS ON LEARNING AND IMPROVING THESE SKILLS:

- assessing the risk involved in an investigation
- safely planning and conducting a Petri dish investigation when growing micro-organisms.

Safe handling of micro-organisms and Petri dishes

Micro-organisms can cause disease. Most micro-organisms collected in the environment are not harmful. However, once they multiply, they may become a hazard.

When you assess the safety requirements of a Petri dish investigation, remember the following:

1. It is only safe to grow micro-organisms in a Petri dish if the dish is sealed after you swab.
2. Dispose of your swab (cotton bud) straight after use in a bin.
3. Wash your hands straight after sealing the Petri dish.
4. When incubation is complete, do not open the lid of the Petri dish.
5. Observe the results with the lid still sealed.

Video
Science skills in a minute: Petri dish safety

Science skills resource
Science skills in practice: Petri dish safety

Other resource
Activity sheet: Practising sterile technique

EFFECTIVENESS OF ALCOHOL HAND SANITISER

BACKGROUND INFORMATION

People are encouraged to use hand sanitiser to reduce the spread of COVID-19. Theoretically, hand sanitiser containing alcohol stops the spread of disease by breaking down the outer lipid layer of the virus.

AIM

To investigate how effective alcohol hand sanitiser is at killing micro-organisms

PREDICTION

Fewer micro-organisms will grow on a Petri dish if a swab is taken from a surface cleaned with hand sanitiser.

MATERIALS AND EQUIPMENT

- ☑ 3 Petri dishes
- ☑ marking pen
- ☑ sticky tape
- ☑ cotton buds
- ☑ hand sanitiser containing 60–80% alcohol
- ☑ hand sanitiser or handwash that does not contain alcohol
- ☑ safety glasses

PROCEDURE

 **Safety**

Wear appropriate personal protective equipment (PPE).

Do not open the Petri dishes after sealing them.

Wash your hands thoroughly at the end of the investigation.

1. Put on your safety glasses. Collect three Petri dishes containing pre-prepared nutrient agar. One will act as a control. Label each in small writing so that the writing does not obscure your observations.

Label Petri dish 1 'Control'. Label Petri dish 2 'Surface cleaned with alcohol hand sanitiser'. Label Petri dish 3 'Surface cleaned with non-alcohol hand sanitiser'.

Leave the cover on the plates at all times unless adding swabs. Only open the lid of the dish wide enough so that you can insert the swab or forceps.

2 Find a heavily used surface such as a handrail, drink fountain or keyboard. Use a clean cotton bud to get a swab of micro-organisms from this location. Do this by rubbing the cotton bud several times over one-third of the surface section you have chosen. Streak the first Petri dish (control) with the sample in a zigzag motion, as shown in Figure 6.11.1.

3 Use the alcohol-based hand sanitiser to clean one-third of the chosen surface. Use another cotton bud to collect micro-organisms from the same location and then streak the second Petri dish with your sample in a zigzag motion. Try to swab the same amount into each Petri dish.

▲ **FIGURE 6.11.1** To swab on agar in a Petri dish, use this zigzag motion.

4 Use a non-alcohol-based hand sanitiser to clean the final third of the surface. Use another cotton bud to collect micro-organisms from this location and then streak the third Petri dish with your sample in a zigzag motion.

5 Draw a table to record the observed results.

6 Seal each dish with clear sticky tape and hand them to your teacher for incubation. Incubate at (37°C) for 2–3 days. To prevent water condensation accumulating and affecting your investigation, place the Petri dishes upside down where your teacher instructs you to.

7 Pack up and then wash your hands thoroughly.

RESULTS

Do not open the Petri dishes. Sketch or describe the results observed in each of the three Petri dishes.

Bacterial colonies are circular, cream to yellowish in colour and shiny on the surface. Fungal colonies are more irregular and look like threads of cottonwool.

ANALYSIS

1 **Describe** the purpose of the control.

2 Why was a non-alcohol hand sanitiser useful in this investigation?

3 **Record** the results of three other groups.

4 **Evaluate** your safety precautions. Did you follow the steps outlined in the blue 'Science skills in focus' box? What could you have improved?

5 Write a conclusion for the investigation. Include a statement about whether the results supported your hypothesis.

9780170491785

6.12 Use of medicinal plants

ABORIGINAL & TORRES STRAIT ISLANDER SCIENCE CONTEXTS

IN THIS MODULE, YOU WILL:

✓ investigate native plant extracts used by Aboriginal and Torres Strait Islander Peoples to treat infections and illness.

Aboriginal and Torres Strait Islander Peoples' food and medicine

Aboriginal and Torres Strait Islander Peoples have a wealth of scientific knowledge about native plants. For thousands of years, they have used plants for food and medicines and to develop tools and technologies (Figure 6.12.1). Medicinal use of plants requires knowledge about the right plants to use and how to prepare them to be effective.

© Joe Sambano

▲ **FIGURE 6.12.1** Some Aboriginal and Torres Strait Islander Peoples' foods: (a) lilly pilly, (b) Kakadu plum, (c) quandong, (d) Dianella, (e) seaberry, (f) native raspberry, (g) kangaroo apple, (h) native fig, (i) bunya nuts

Joe Sambono

▲ **FIGURE 6.12.2** Tea tree is used by the Yaegl People to treat the symptoms of coughs and colds.

Native plants to treat coughs and cold

When you have a cold, you might use a medicated rub to soothe coughs and breathe more easily. Usually, the rub is covered with a warm cloth to help the medicines in the rub vaporise. Aboriginal Peoples have been using medicated treatments for coughs and colds in this way for thousands of years. Prior to European colonisation, the Wiradjuri People (central New South Wales) built steam pits lined with eucalyptus leaves to treat coughs and colds. The eucalyptus vapour was released by heating the leaves over a fire with possum rugs placed on top to create a steam pit. The Yaegl People (Coffs Harbour region, New South Wales) heat the leaves of tea tree (*Melaleuca alternifolia*) to relieve the symptoms of coughs and colds (Figure 6.12.2). Today, many common products that are used to treat coughs and colds contain eucalyptus or melaleuca oil.

Native plants to treat skin infections

There are many examples of plant extracts being used to treat or prevent skin infections. The Bundjalung Peoples (northern New South Wales/south-east Queensland) have long crushed tea tree leaves and applied them to skin wounds and infections, traditionally held in place with layers of paper bark (Figure 6.12.3).

Joe Sambono

© Joe Sambono

▲ **FIGURE 6.12.3** (a) Paper bark is prepared as a type of bandage to (b) hold treatments, such as crushed tea leaves, in place on the body.

When European colonists first travelled to Bunjalung Nation, they were affected by skin infections caused by high rainfall, injuries and insect bites. Supplies of Western treatments were scarce. European colonists observed the Bunjalung Peoples' use of tea tree and adopted these methods to control infections. In the 1920s, scientist Dr AR Penfold investigated the antiseptic properties of tea tree. His study reaffirmed the knowledge of the Bunjalung Peoples: that the leaves contained powerful antiseptics that were more than 10 times stronger than other disinfectants of the time. The active components of tea tree have antimicrobial, antifungal, antiviral and anti-inflammatory properties. Tea tree grows mostly along the coastal areas of south-east Queensland and northern New South Wales.

Aboriginal and Torres Strait Islander Peoples in other parts of Australia use different plants to treat infections. The Peoples of the Torres Strait Islands have traditionally used coconut plant extracts to treat wound infections. Traditional Owners of the Northern Territory continue to use traditional medicine such as emu bush leaves (*Eremophila* species) to treat sores and cuts.

These examples give a glimpse of Aboriginal and Torres Strait Islander Peoples' medicinal knowledge and use of native plants that continue today.

☆ ACTIVITY

Antimicrobial effect of products containing native plant extracts used by Aboriginal and Torres Strait Islander Peoples

Follow your teacher's instructions. Be aware that some handwashes contain ingredients that can cause allergic or sensitivity reactions. Do not use these products if you have skin allergies or skin sensitivities. If you notice a reaction during this investigation, immediately wash your hands and seek medical advice.

Make sure your agar plates are securely taped once they have been swabbed.

Materials and equipment

- ☑ 2 nutrient agar plates
- ☑ handwash containing tea tree or eucalyptus extract
- ☑ sterile swabs
- ☑ sterile water
- ☑ tape
- ☑ incubator

 **Safety**

Wear appropriate personal protective equipment (PPE).

Do not open the Petri dishes after sealing them.

Wash your hands thoroughly at the end of the investigation.

▷

Procedure

1 Dip a sterile swab into sterile water to moisten it.

2 Run the swab from your wrist at the base of your thumb up and over each of your fingers, down to the base of your wrist on the other side of your hand.

3 Roll the swab gently over the surface of one of the agar plates to cover the entire surface.

4 Tape the agar plate to seal it.

5 Use a handwash containing tea tree or eucalyptus extract to clean your hands. Make sure you rub the handwash between your fingers (Figure 6.12.4).

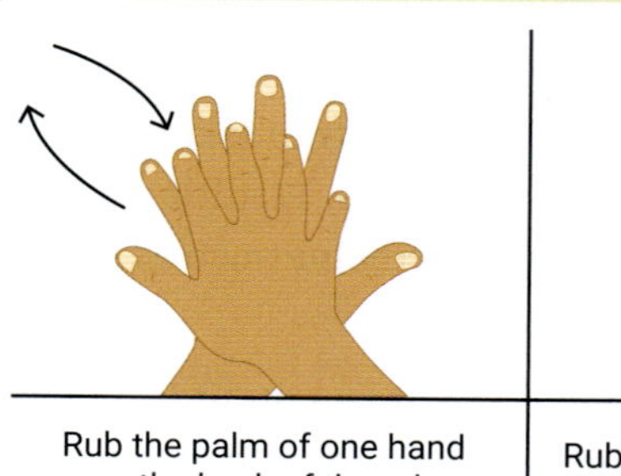

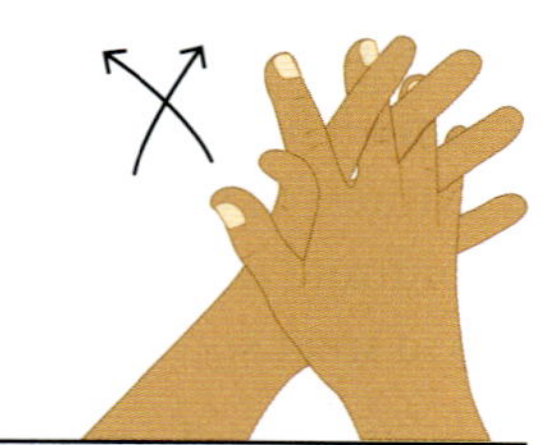

▲ **FIGURE 6.12.4** How to wash your hands with handwash

6 Rinse off the handwash and wait for 5 minutes for your hands to air dry.

7 Repeat the hand swab in the same way as before, using a clean swab and rolling it onto the second agar plate.

8 Seal the plates with tape.

9 Incubate the plates at 37°C for 24–48 hours.

10 Count the number of colonies on the agar plates before and after treatment with the handwash.

11 Record your results in a table.

12 Calculate the percentage difference in the number of bacteria:

$$\text{Difference in number of microbial colonies (\%)} = \frac{\text{number of colonies before handwash} - \text{number of colonies after handwash}}{100}$$

Analysis

1 **Analyse** the results and draw a conclusion about the effect of the handwash on skin micro-organisms.

2 How do your results **compare** with those of other students in the class? Was one handwash more effective than the other?

3 Was this a fair test? Why or why not? What did you do to try to make this a fair test?

4 **Propose** another investigation in which the plant extract could be tested more specifically for its antimicrobial activity.

9780170491785

SCIENCE IN CONTEXT

6.13 Technological advances in disease treatment

BY THE END OF THIS MODULE, YOU WILL BE ABLE TO:

✓ outline how advances in technology address disease or disorder treatment.

GET THINKING

If you had an unlimited amount of money to develop a technology to help treat a disease or disorder, what would you choose? Describe how the technology would help to address the symptoms or treatment of the disease.

Spray-on skin technology

Burn injuries can be severe and challenging to treat. Because the body's first line of defence – the skin – is so damaged, it is easy for infection to enter the body. People with burns often need long recovery periods and have to go through complex medical procedures. Fortunately, in the 1990s Australian researchers made a groundbreaking development in treating burns by developing spray-on skin technology.

In 1993, Professor Fiona Wood and Marie Stoner began working with a method of growing skin tissue in the laboratory to treat burns, called CEA (cultured epithelial autograft). From this, they next developed a 'spray-on' approach to treatment.

Spray-on skin is a repair technique that involves taking a small piece of healthy skin (a biopsy) from a burns victim and using it to grow new skin cells in the laboratory (Figure 6.13.1). The cells are put into an enzyme 'cocktail' to form a suspension (they are suspended in solution). The cells are then rinsed and the suspension is filtered, then sprayed onto the patient's wound, over the whole burned area. This method provides an even spread of live cells on the wound.

This type of technology has a big advantage over skin grafts because the person's wound heals faster, and they have less pain, less chance of infection and less scarring. Professor Wood patented the technique in 1993 and today it is used worldwide under the registered name ReCell.

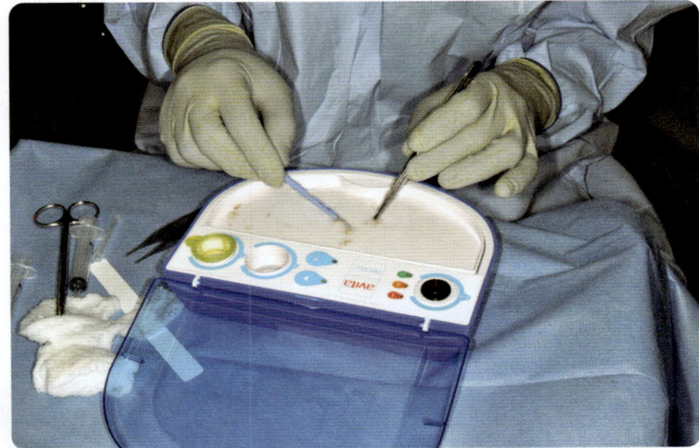

▲ **FIGURE 6.13.1** A step in the preparation of ReCell® cells for spray-on skin technology

6.13 LEARNING CHECK

1. What is spray-on skin technology?
2. **Outline** the steps to create the spray-on skin for a burn patient.
3. **Describe** the advantages of spray-on skin technology over skin grafts.
4. **Research** the career of Professor Fiona Wood and **create** a timeline of her investigations and achievements.

6 REVIEW

REMEMBERING

1 **List** three different types of pathogens.

2 **Describe** how scientists classify a disease as infectious.

3 To which line of defence does inflammation belong?

UNDERSTANDING

4 **Identify** the type of biological molecule, and **state** the function of, a lysozyme found in tears.

APPLYING

5 Copy and complete the following table to **identify** the mode of transmission of each disease as direct contact, airborne or vector.

Disease	Information	Mode of transmission
Tuberculosis	Airborne droplets can be inhaled within 1 m.	
Malaria	An *Anopheles* mosquito has a blood feed and simultaneously injects the pathogen into the blood of a host.	
Smallpox		

6 **Explain** what antigens are and where they are found on pathogens.

7 **Explain** why a booster vaccine is required for some vaccination programs.

8 Copy and complete the following table to **classify** the different types of non-specific immune defence systems.

Non-specific immune system	System parts	Line of defence
External	1 Skin (secretions/barrier) 2 3 4 5	First
Internal	1 Inflammation 2 3	Second

9 Using the term 'antimicrobial properties', **explain** why the Bundjalung Peoples of northern New South Wales/south-east Queensland traditionally use tea tree leaves to treat wounds.

10 When humans suffer a skin wound, they may experience inflammation before healing, as shown in the diagram below.

List the signs of inflammation at the site of a wound.

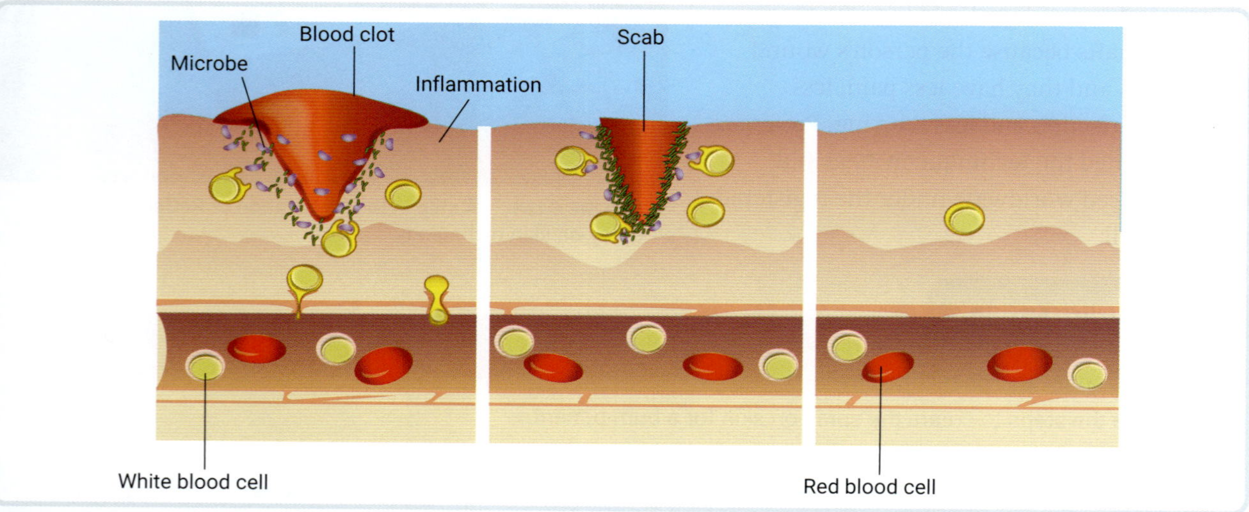

11 Several parts of the body have mucous membranes. The following image shows one of these areas, the lungs.

State the function of mucus in the human body.

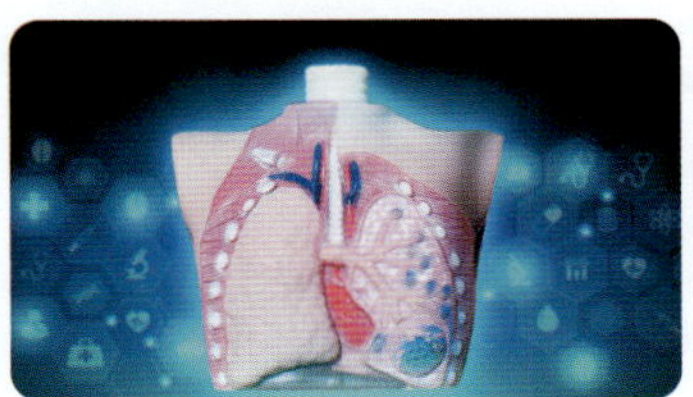

Antonio Marca/ Shutterstock.com

12 **Compare** natural immunity with artificial immunity by copying and completing the following table.

	Natural	Artificial
Is exposure to antigen deliberate or non-deliberate?		
Do antibodies develop?		
Does long-term immunity result?		

13 **Draw** a diagram showing the steps of phagocytosis.

Label your diagram with short statements to **explain** what happens at each stage.

ANALYSING

14 **Explain** the meaning of the term 'non-specific' in the context of the immune system.

15 **Explain** the impact of a break, or tear, in the skin on the body's defence against disease.

16 **Compare** one infectious and one non-infectious disease in terms of transmission and prevention.

17 A study was conducted at the University of New South Wales on how often medical students touch their face. The study found that:

- the 26 students touched their face an average of 23 times per hour
- just under half of all touches (1024 touches out of a total of 2345) were to areas that contain mucous membranes (eyes, nose or mouth).

a **Explain** why this behaviour might increase the spread of disease.

b **Outline** a message you think should be given to society about hand hygiene.

18 The image below represents herd immunity to a virus. The red figure represents an infected person. The grey figures represent unvaccinated people or those vulnerable to infection. The blue figures represent people with either natural or artificial immunity. Imagine that each infected person is capable of infecting another two people. **Explain** what would happen if:

a the scenario remained the same.

b the number of people with immunity was halved.

c everyone had immunity.

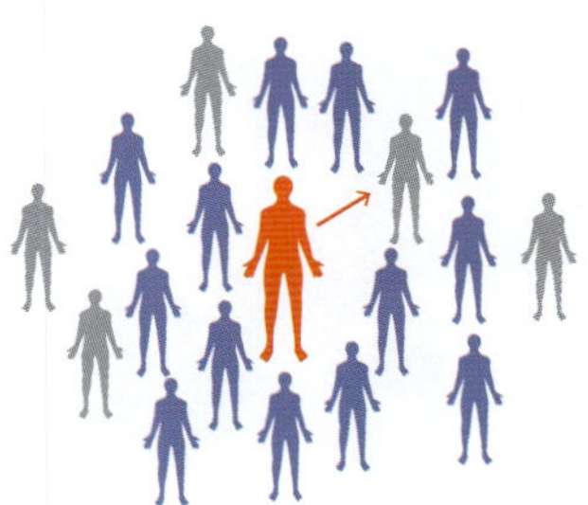

EVALUATING

19 **Evaluate** the secondary prevention strategies applied during COVID-19.

20 In 2020, scientists conducted experiments to determine how long SARS-CoV-2 (COVID-19) could remain infectious on a surface. One study found sunlight could 'kill' (disrupt) the virus within 14 minutes. By mid-2021, scientists had stopped studying contaminated surfaces and had turned to research and experiments on the effectiveness of wearing masks to slow the spread of the virus.

Explain why scientists may have changed their research focus. **Justify** your reasoning.

CREATING

21 **Create** a table summarising the structures of a bacterium, a fungus and a virus. Include three features for each type of pathogen.

22 **Design** your own poster to be used in the classroom to **explain** the three lines of defence.

SCIENCE IN DEPTH STUDY

At different times during the COVID-19 pandemic, the number of hospitalisations increased significantly. Sometimes there weren't enough health workers to care for the high number of patients. If a similar situation arose again, how could we avoid overwhelming the hospital system?

1 Connect what you've learned

In this chapter, you've learned about pathogens that cause infectious disease; non-specific immune responses; and the immune system. Finish the concept map you started in Module 6.6 and reflect on how the information in each of the modules is related.

▲ Medical staff in intensive care treating a patient

2 Check your thinking

a Can you think of a disease caused by a:
- bacterium?
- fungus?
- virus?

b How do they enter a host and what are the symptoms?

c Is there a vaccine for each of them?

d Describe the modes of transmission for COVID-19. Use terms such as 'direct contact', 'indirect contact' (contaminated surfaces), 'close contact' and 'aerosol transmission'.

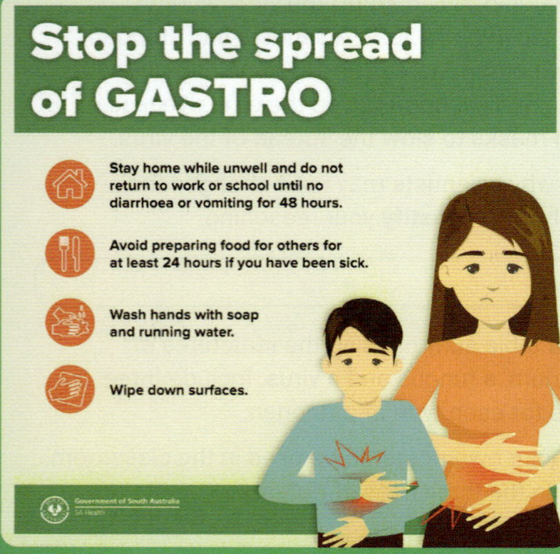

3 Get into action

Make a comprehensive list of actions (at least five) you could do to slow the spread of COVID-19 and any other airborne disease. For each action, provide an instruction on how to do the action effectively.

4 Communicate

Convert your ideas into a labelled infographic to share with your class, like the one on the left from SA Health on stopping the spread of gastroenteritis.

MATERIALS

SYLLABUS OUTCOMES

A STUDENT:

- assesses the uses of materials based on their physical and chemical properties SC5-MAT-01
- designs safe, ethical, valid and reliable investigations SC5-WS-03
- selects suitable problem-solving strategies and evaluates proposed solutions to identified problems SC5-WS-07
- communicates scientific arguments with evidence, using scientific language and terminology in a range of communication forms SC5-WS-08

© 2023 NSW Education Standards Authority

CHAPTERS RELATED TO THIS FOCUS AREA ARE:

- CHAPTER 7 – BONDING
- CHAPTER 8 – ORGANIC COMPOUNDS

Trygve Finkelsen/Shutterstock.com

Bonding

SCIENCE IN DEPTH

▲ **FIGURE 7.0.1** Some products made from metals mined in Australia

Every day, you use products that are made from minerals and resources that are extracted and produced in Australia. Each of the three objects in Figure 7.0.1 are made from metals we mine here. What features of each metal make them useful in the products shown?

None of the metals shown are found in the ground as pure metals. Activities like refining need to occur to extract and produce useful materials from ore (rocks containing minerals or metals).

- **What minerals and resources do we mine in Australia?**
- **Do you know of any mining that occurs near where you live?**
- **How does a rock containing a metal get treated to extract the pure metal?**
- **Are there any environmental issues involved with mining and refining minerals and resources?**

DIVE INTO SCIENCE!

At the end of this chapter, you can complete Science in Depth Study #7. You can use the information you learn in this chapter to complete the project.

Assessments
- Prior knowledge quiz
- Chapter review questions
- End-of-chapter test
- Depth study: Project and scientific report

Videos
- Science skills in a minute: Justifying conclusions **(7.5)**; Presenting data in tables **(7.11)**
- Video activities: Atom structure: Energy shells **(7.3)**; Mendeleev and the periodic table **(7.4)**; Introduction to chemical bonding **(7.6)**

Science skills resources
- Science skills in practice: Justifying conclusions **(7.5)**; Data tables **(7.11)**
- Extra science investigations: The physical properties of metals and non-metals **(7.12)**

Interactive and other resources
- Simulation: Build an atom **(7.2)**
- Crossword: Bonding **(7.14)**
- Drag and drop: Physical versus chemical properties **(7.1)**; Types of bonding **(7.6)**; Metal properties **(7.12)**
- Label: Electrons and shells **(7.3)**; Predicting ion charge **(7.7)**
- Quizzes: Ionic compounds **(7.10)**; Metal atom bonds **(7.12)**
- Worksheets: Atomic structure and properties **(7.2)**; Using the Bohr model **(7.3)**; Electron configuration and the periodic table **(7.4)**; Practising ionic formulas **(7.9)**

To access resources above, visit **cengage.com.au/nelsonmindtap**

7.1 Properties and uses of Australian minerals and resources

BY THE END OF THIS MODULE, YOU WILL BE ABLE TO:

- ✓ identify minerals and resources extracted in Australia
- ✓ compare chemical and physical properties of different materials
- ✓ explain the relationship between a substance's properties and its use.

Interactive resource
Drag and drop: Physical versus chemical properties

GET THINKING

Australian mines produce a wide variety of metals, minerals and resources, including coal, aluminium, iron, uranium, nickel and lithium. How many of these have you heard of? See if you can identify a use for each of the metals, minerals or resources listed here. Are there any other similar products you know that are produced or mined in Australia?

Australian minerals and resources

mineral
a naturally occurring inorganic solid with a neatly ordered crystal structure and characteristic composition

Australia is rich in **mineral** deposits. Australia produces large amounts of iron ore, gold, uranium, zinc, lithium, coal and lead. Figure 7.1.1 shows the locations of Australia's mineral resources.

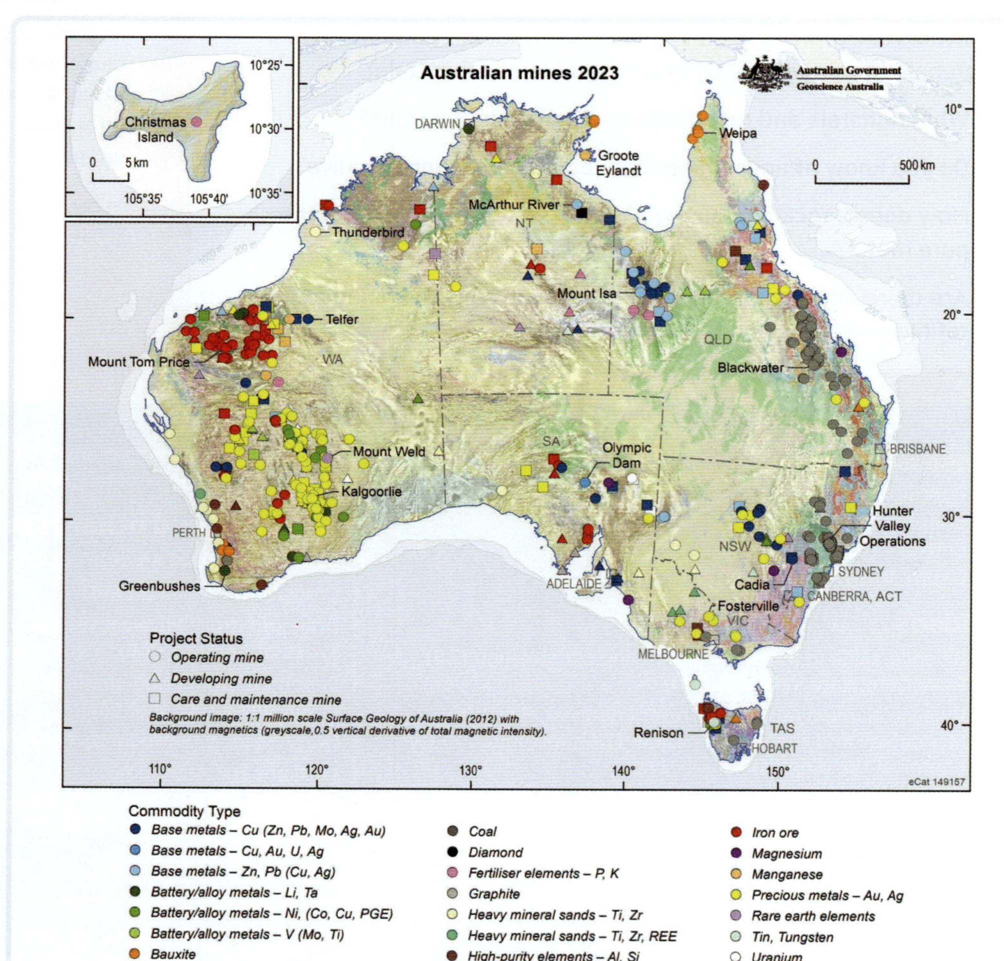

© Commonwealth of Australia (Geoscience Australia) 2021

▶ **FIGURE 7.1.1** Australia has many sites where minerals, metals and other resources are extracted from the ground.

9780170491785

A mineral is a substance with a specific chemical composition from which useful materials like metals can be extracted. An **ore** is a rock that contains minerals that we can use to produce a specific resource.

ore
a rock that contains one or more minerals containing valuable substances

An example of an ore mined in Australia is bauxite, a mixture of rocks that contains minerals rich in aluminium. After mining, refineries extract and process the aluminium minerals from bauxite to produce aluminium oxide and then aluminium metal. The stages of this process are seen in Figure 7.1.2.

▲ FIGURE 7.1.2 (a) Bauxite ore is processed to produce (b) aluminium oxide that is then refined into (c) aluminium metal.

Substances and their properties

Substances are often classified according to the properties they have in common. Scientists divide properties into two categories: physical and chemical.

Physical properties

Physical properties are features of a substance that can be observed or examined without changing the composition of a substance. Examples are colour, hardness, density, and melting and boiling points (Figure 7.1.3). You have already investigated many of these in your Stage 4 studies.

physical properties
the properties of a substance that can be observed or examined without changing its composition

Chemical properties

The features of a substance that determine how it reacts with other substances are called **chemical properties**. You cannot observe the chemical properties of a substance the way you can the physical properties. But if two substances have similar chemical properties, you will be able to tell by observing how they each react with the same substance. If they react in a similar way, they have similar chemical properties.

chemical properties
the properties of a substance that determine how it reacts when combined with other substances

Chemical properties include:

- **flammability** (the ability to catch fire)
- **toxicity** (how poisonous a substance is)
- how much **corrosion** the substance undergoes
- sensitivity to light (which, for example, causes newspaper to turn yellow over time).

flammability
the ability of a substance to catch fire

toxicity
a measure of how poisonous a substance is

corrosion
the breakdown of a metallic substance due to chemical reactions with substances in the environment, such as oxygen or water

Consider this example. Paper will easily catch fire, but the metal mercury does not. However, paper is unlikely to harm you, whereas mercury is highly toxic and causes brain and nerve damage. Paper and mercury have different chemical properties. Paper is flammable and non-toxic, whereas mercury is not flammable but is toxic.

Figure 7.1.3 lists some physical and chemical properties.

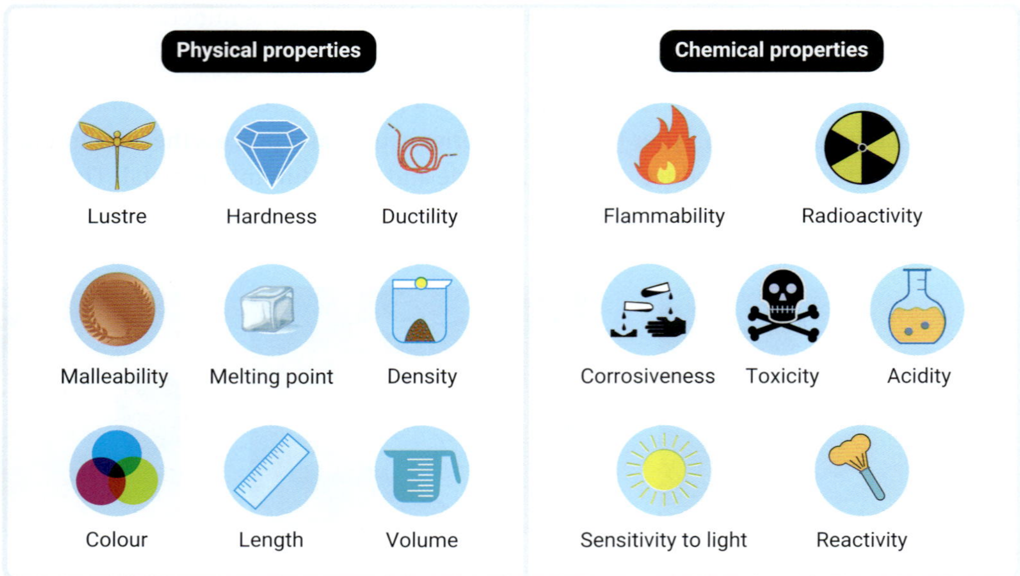

▲ **FIGURE 7.1.3** Some examples of physical and chemical properties

imageBROKER.com GmbH & Co. KG/Alamy Stock Photo

▲ **FIGURE 7.1.4** The properties of aluminium make it well-suited to being used for wires to span long distances between power poles.

Properties and uses of substances

The properties of a substance determine what it is used for. Copper wiring is used in electrical wiring in homes because it has excellent electrical conductivity. However, it is not commonly used in long distance power lines. Instead, a form of aluminium, like that used for soft-drink cans, is used. Aluminium has properties that are better suited for this purpose. It conducts electricity almost as well as copper, but it is half the density and costs less to manufacture (Figure 7.1.4).

ACTIVITY 7.1

Useful properties of Australian minerals and resources

1 Copy the table below.

Material	Item	Chemical or physical property that makes it suitable for its purpose
Aluminium	Can	Lightweight (low density), not easily corroded
Iron	Fence	
Gold	Jewellery	
Nickel	Batteries	
Coal	Fuel	

Identify the chemical and/or physical properties of each material that makes it suitable for its purpose. The first item has some features already identified. You may need to research some materials.

2 Sometimes other factors can be important when deciding on the best material for a purpose. For example, silver is a better electrical conductor than copper and aluminium. Why don't we use silver for electrical wiring in houses and in transmission lines?

3 **List** some factors other than chemical and physical properties that may influence why a particular material is used.

7.1 LEARNING CHECK

1 **List** three physical properties and three chemical properties.

2 **Explain** the difference between an ore and a mineral.

3 Lead is used for fishing sinkers and weights. Aluminium is used in soft-drink cans and aeroplane panels.

   a Based on their uses, **predict** and **compare** the physical properties of lead and aluminium.

   b Do some research and **compare** the chemical properties of each metal.

4 There are four main types of iron ores that are mined. These are haematite, magnetite, titanomagnetite and ironstone. **Predict** how the ores would be similar and different to each other and then do some research to check your predictions.

7.2 Review of atomic structure

BY THE END OF THIS MODULE, YOU WILL BE ABLE TO:

- ✓ describe the subatomic particles in atoms
- ✓ describe the structure of an atom in terms of its subatomic particles.

Interactive resource
Simulation: Build an atom

Other resource
Worksheet: Atomic structure and properties

GET THINKING

In chemistry, numbers and chemical symbols are often used to describe information about chemicals and atoms. Find examples of numbers and symbols being used in this module and suggest what the numbers might mean.

Subatomic particles

You may recall from Stage 4 that **atoms** are the smallest particles of matter that form a chemical **element**. Atoms have a central **nucleus** that contains two types of **subatomic particles**: **protons** and **neutrons**. Protons are positively charged, and neutrons have no charge. Negatively charged **electrons** are found in electron shells or **energy shells** around the nucleus.

Protons and neutrons are about the same size. Electrons are much smaller, approximately $\frac{1}{1840}$ the size of a proton or neutron (Table 7.2.1).

atom
the smallest part of an element that gives the element its chemical properties

element
a pure substance made up of only one type of atom; it cannot be broken down into a simpler substance

nucleus
the dense centre of an atom; it contains protons and neutrons, so it is positively charged

subatomic particle
a particle inside an atom, such as a proton, a neutron or an electron

proton
a positively charged particle in the nucleus of an atom

neutron
an uncharged particle in the nucleus of an atom

electron
a negatively charged particle that moves in space around the nucleus of an atom

energy shell
a level around a nucleus containing electrons of the same energy

isotopes
atoms of an element with the same number of protons but a different number of neutrons

▼ **TABLE 7.2.1** Subatomic particles

Subatomic particle	Symbol	Location in the atom	Charge	Relative mass
Proton	p	Nucleus	+1	1
Neutron	n	Nucleus	0	1
Electron	e	In shells around the nucleus	−1	$\frac{1}{1840}$

Atoms, elements and isotopes

There are over 100 different elements. These include many substances you should be familiar with, such as hydrogen, magnesium, copper and oxygen. An element contains atoms with the same number of protons. Different elements have atoms with different numbers of protons. For example, all carbon atoms have six protons, and all oxygen atoms have eight protons.

An element has atoms with the same number of protons, but it can have forms with different numbers of neutrons. These are called **isotopes**.

In Figure 7.2.1 you can see three types of carbon atoms that we refer to as carbon isotopes. Each carbon isotope has six protons; however, the number of neutrons in the nucleus differs.

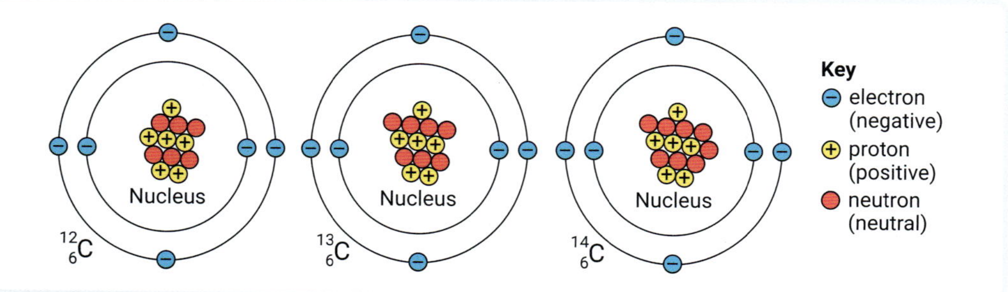

▶ **FIGURE 7.2.1** The structure of carbon atoms

9780170491785

Atomic number and mass number

The number of protons in an element is represented by its **atomic number (Z)**. From Table 7.2.2, you can see that carbon has an atomic number of six. As the number of protons is constant in an element, the atomic number for that element is always the same. Every element has a unique atomic number.

atomic number (Z) the number of protons in a nucleus; the same for every atom of the same element

▼ **TABLE 7.2.2** Isotopes of carbon

Isotope	Number of protons	Number of neutrons	Total number of particles in the nucleus
Carbon-12	6	6	12
Carbon-13	6	7	13
Carbon-14	6	8	14

The atomic number also represents the number of electrons in an atom. All atoms are electrically neutral. They have no overall charge. The number of positive protons is balanced by the number of negative electrons. This means the number of protons and electrons in an atom is equal.

mass number (A) the total number of protons and neutrons in the nucleus of an atom

The **mass number (A)** is the total number of protons and neutrons in the nucleus. The mass number is used with the element name to describe the isotopes. For example, a carbon atom with six protons and six neutrons has a mass number of 12. This isotope is carbon-12.

The atomic number and mass number can be placed next to the element symbol (Figure 7.2.2).

Mass number —— 12
Atomic number —— 6
$^{12}_{6}C$

▲ **FIGURE 7.2.2** Notation showing the mass number and atomic number for carbon-12

Modelling subatomic particles in isotopes

☆ ACTIVITY

Use three different colours of plasticine to create small spheres. Each colour will represent a different type of subatomic particle. Lithium is an element with atomic number 3. Make models of lithium-6 and lithium-7 that show all subatomic particles. Put the electrons in a circle around the nucleus for now (placement of electrons is covered in the next module).

a **Create** a table showing the number of each subatomic particle in each isotope.

b How are the two isotopes similar ? How are they different?

7.2 LEARNING CHECK

1 **Define** mass number, atomic number and isotope.

2 **Identify** the number of protons, neutrons and electrons for the following.

a $^{23}_{11}Na$ **b** $^{39}_{19}K$ **c** $^{14}_{6}C$ **d** $^{35}_{17}Cl$

3 Subatomic particles can have a similar or different size, charge or location.

a **Identify** a similarity between any two of the subatomic particles.

b **Identify** a difference between any two of the subatomic particles.

4 A student claims that two elements are isotopes. The two elements have the same number of neutrons and electrons but different numbers of protons. **Explain** whether the student's claim is correct.

7.3 Bohr's model of the atom

BY THE END OF THIS MODULE, YOU WILL BE ABLE TO:

✓ describe the Bohr model
✓ write electron configurations for the first 20 elements.

Video activity
Atom structure: Energy shells

Interactive resource
Label: Electrons and shells

excited
the state of an atom or electron when it absorbs energy

GET THINKING

This module describes the Bohr model of the atom. We use models to represent complex scientific ideas or objects we cannot observe directly. What scientific models do you know from your studies so far? How can models be helpful to your learning?

The Bohr model

The Bohr model of the atom, shown in Figure 7.3.1, is also called the planetary atomic model. It is based on a series of assumptions:

- Electrons have energy and exist in circular orbits around the nucleus in energy shells.
- All electrons in the same energy shell have the same amount of energy.
- Electrons in shells that are further away from the nucleus have more energy.

Bohr proposed that electrons in atoms are arranged in energy shells and they stay in those shells unless the atom is heated and they become **excited**. When the atom is heated, an electron absorbs this energy and moves to a higher energy level, further away from the nucleus. This electron is then unstable.

Unstable electrons release energy by returning to their original energy shell. The energy is released as light. The amount of energy released is related to how many energy shells the electron moved up and down (Figure 7.3.2).

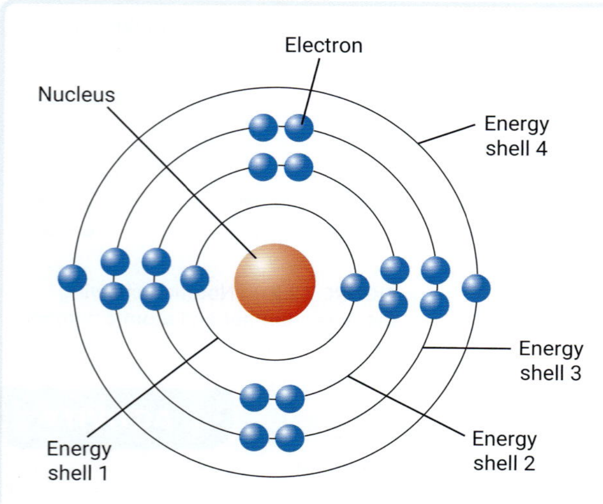

▲ **FIGURE 7.3.1** The Bohr model of the calcium atom

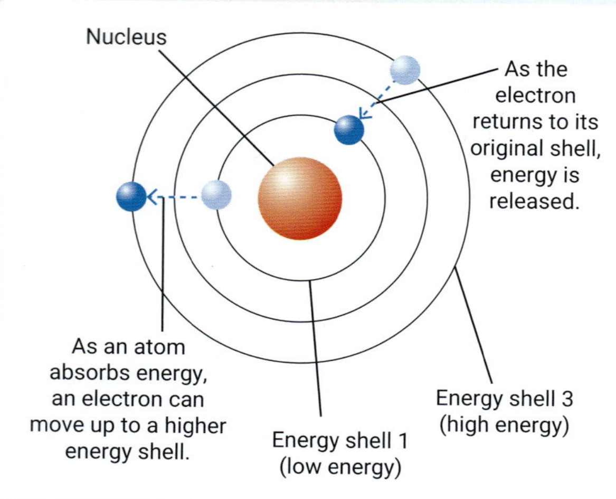

▲ **FIGURE 7.3.2** Electrons can move to higher energy shells (away from the nucleus) when an atom absorbs energy. Electrons release energy when they move down energy shells (towards the nucleus).

Electron configuration

The arrangement of electrons in energy shells is called the **electron configuration**. The electron configuration of any atom can be determined by knowing the set of rules that apply:

- Rule 1: Each shell has a maximum number of electrons given by the formula $2n^2$.
- Rule 2: Electrons fill the lower shells first.
- Rule 3: Electrons fill each shell to a certain number, then start to fill the next shell.

9780170491785

You can find the maximum number of electrons that can fit into a shell by applying the formula $2n^2$, where n represents the number of the energy shell (Table 7.3.1).

▼ **TABLE 7.3.1** The maximum number of electrons in each energy shell

Shell number (n)	Maximum number of electrons
1	$2 \times 1^2 = 2$
2	$2 \times 2^2 = 8$
3	$2 \times 3^2 = 18$
4	$2 \times 4^2 = 32$
5	$2 \times 5^2 = 50$

For the first 20 elements, they fill in the specific order shown by a–d in Figure 7.3.3 and listed here:

- first shell ($n = 1$) fills with two electrons
- second shell ($n = 2$) fills with eight electrons
- third shell ($n = 3$) fills with eight electrons (note that this shell can hold 18 but fills with only eight electrons at this stage)
- fourth shell ($n = 4$) with two electrons (this is the shell with the highest energy).

electron configuration the arrangement of electrons in energy shells in an atom

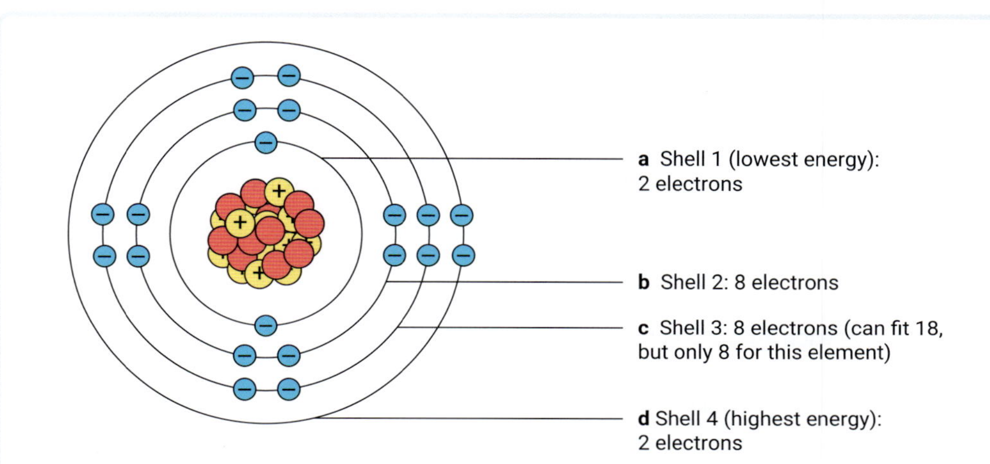

▲ **FIGURE 7.3.3** The rules show that atoms fill their electron shells in the order shown.

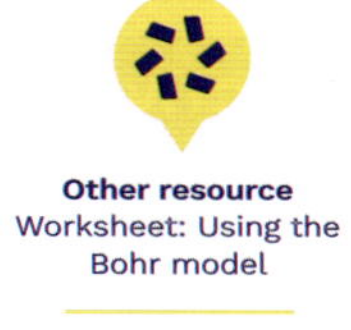

Other resource
Worksheet: Using the Bohr model

Let's look at the specific example of carbon with six electrons (Figure 7.3.4). Following the rules above:

- two electrons go into the first shell. It is now full and there are four electrons remaining
- four electrons go into the second shell. There are no electrons remaining.

The electron configuration is represented in Figure 7.3.5.

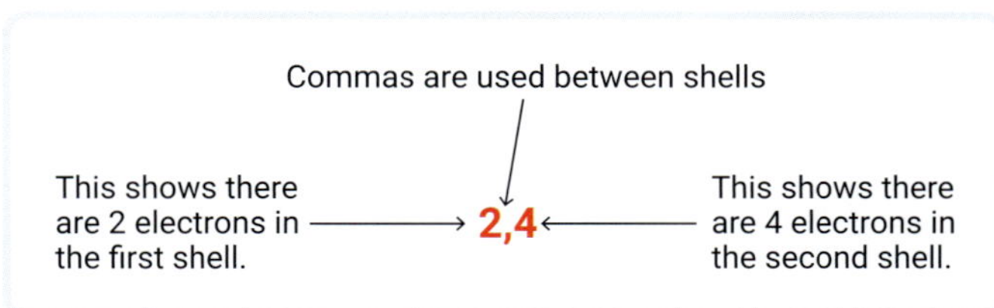

▲ **FIGURE 7.3.5** The electron configuration of carbon

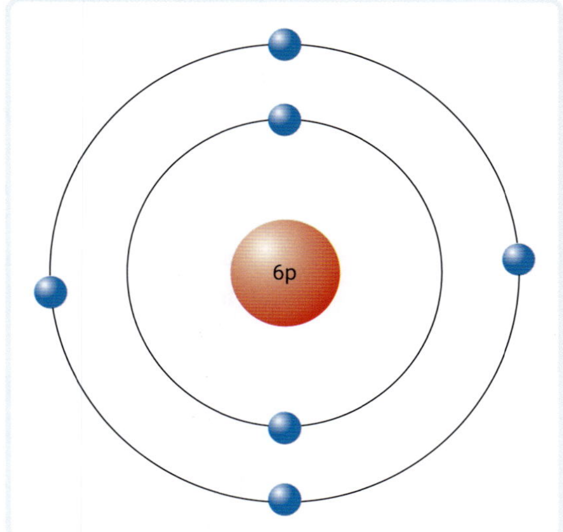

▲ **FIGURE 7.3.4** The Bohr model of the carbon atom

The electron configurations of the first 10 elements in the periodic table are shown in Table 7.3.2.

▼ **TABLE 7.3.2** The electron configurations of the first 10 elements

Element	Number of electrons	Electron configuration
Hydrogen	1	1
Helium	2	2
Lithium	3	2,1
Beryllium	4	2,2
Boron	5	2,3
Carbon	6	2,4
Nitrogen	7	2,5
Oxygen	8	2,6
Fluorine	9	2,7
Neon	10	2,8

Configurations written in this way have links to the position of the atom in the periodic table and provide more complex information about the way electrons are arranged in atoms.

☆ **ACTIVITY**

Modelling electron configuration

You can do this activity by drawing models or by using plasticine. In this activity, you are going to model each atom's electron configuration until you can extend Table 7.3.2 to include all the first 20 elements. Copy Table 7.3.2 and use your copy to record your answers.

- Use a small circle or small sphere to represent the nucleus.
- Draw a circle around the nucleus to represent the first energy shell.
- Place one electron in the shell (by drawing a dot or using a small plasticine sphere). This represents hydrogen.
- Place a second electron in the first shell. What element is this?
- Draw a second circle with a larger circumference around the first one. What does this represent?
- Add electrons one by one until you have eight electrons in this circle. After each electron is added, **identify** the element, and write the electron configuration in your table.
- Continue the modelling process with a third energy shell, and then a fourth, until you have built up the elements 11–20 to add to your table.
  - **a** How does this modelling help you understand electron configuration?
  - **b** How is this model inaccurate?

7.3 LEARNING CHECK

1 Write the electron configuration of:
  a fluorine (9 electrons).
  b potassium (19 electrons).

2 Use your table to **identify** the elements represented by:
  a 2,3.
  b 2,8,4.

3 **Describe** the Bohr model of the atom.

4 **Explain** how you can **predict** the number of electrons each shell of an atom is capable of holding.

7.4 Atomic properties and the periodic table

BY THE END OF THIS MODULE YOU WILL BE ABLE TO:

- ✓ explain the organisation of the periodic table in terms of atomic structure
- ✓ describe electronegativity and ionisation energy as properties of atoms
- ✓ use the periodic table to compare the electronegativity and ionisation energy of different elements.

GET THINKING

Many words in chemistry are common words we use every day that have specific meanings in science. In this module, the word 'group' refers to a particular set of elements on the periodic table. Are there other scientific words you know like this that might cause confusion if used incorrectly in science?

Video activity
Mendeleev and the periodic table

Other resource
Worksheet: Electron configuration and the periodic table

In Stage 4 you learned about the structure of the periodic table. Chemists use the periodic table to predict the properties of atoms, and how atoms will behave when they undergo chemical bonding and in chemical reactions.

How atoms combine together is called **chemical bonding**. When atoms bond, the key thing to know is what the electrons in each atom will do. In chemical bonds, electrons can be:

- shared
- transferred
- donated.

chemical bonding
how atoms combine together to form larger structures

The properties of atoms called **electronegativity** and **ionisation energy** help chemists predict what different atoms will do when they bond. Understanding bonding and electron movement can also help chemists predict the **chemical reactions** that will take place. Chemical reactions will be examined further in Chapter 13.

electronegativity
the ability of an atom to attract shared electrons in a bond

ionisation energy
the energy required to remove an electron from the valence shell of an atom

chemical reaction
a process that occurs when a substance changes to produce a new substance

Review of the periodic table

The current understanding of the elements is represented by the **periodic table** in Figure 7.4.1. Each square contains an element symbol (e.g. H), the element name (e.g. hydrogen) and the atomic number (e.g. 1). If you look along each row and read it like a line of text in a book, you will see the elements are arranged in order of increasing atomic number.

The vertical columns are called **groups**. There are 18 groups in total, and the group numbers are on the top of each column. The horizontal rows are called **periods**. There are seven periods, and the period numbers are next to each row on the left.

periodic table
a method of arranging elements by increasing atomic number

group
a vertical column on the periodic table

period
a horizontal row on the periodic table

Electron arrangement and the periodic table

In Module 7.3, you learned how to determine electron configuration. The electron configuration of an element is directly related to the position of the element on the periodic table (Figure 7.4.2)

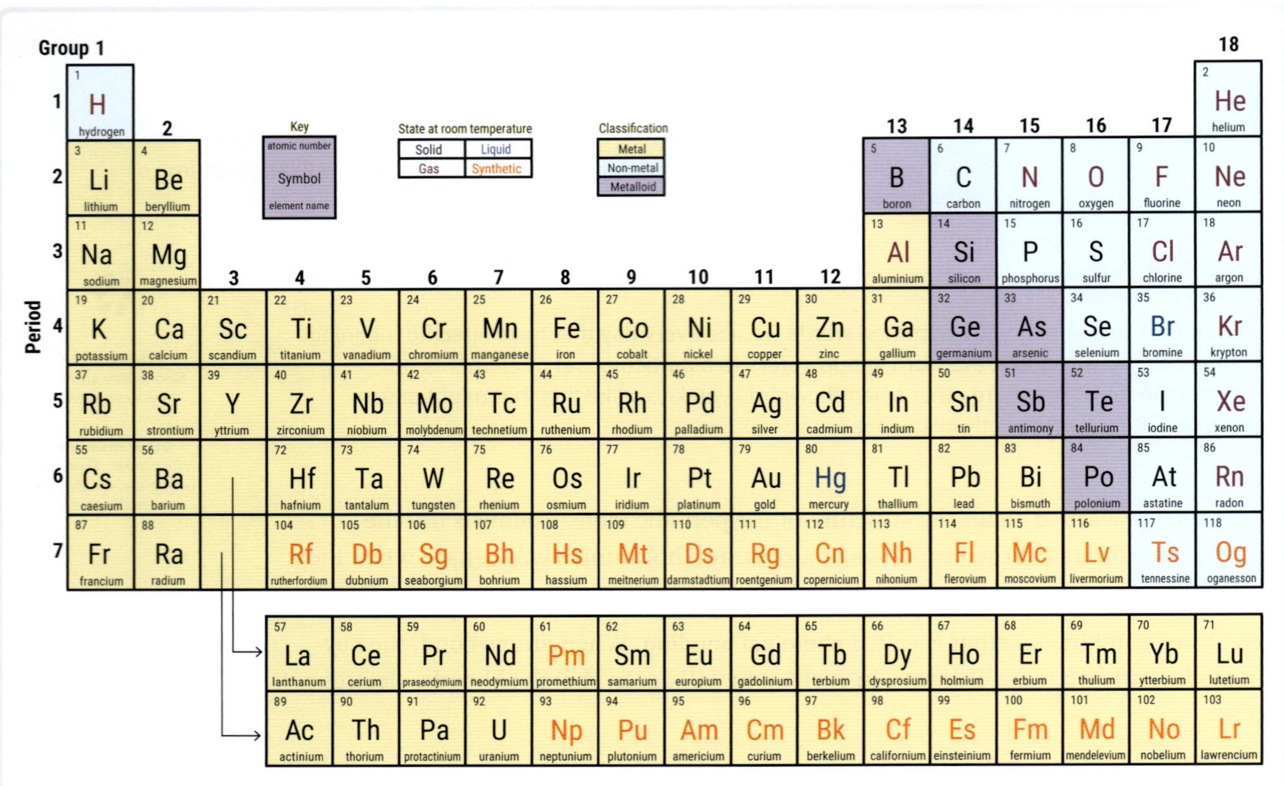

▲ **FIGURE 7.4.1** The periodic table (current as of November 2024)

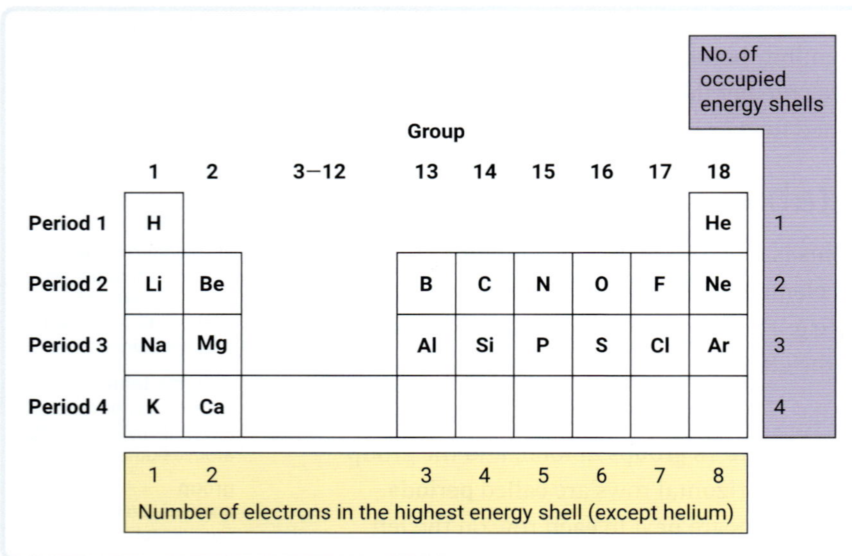

▲ **FIGURE 7.4.2** Part of the periodic table, showing the number of electrons in the highest energy shell below each group

If we look at group 1, we see the following electron configurations:

- H = 1
- Li = 2,1
- Na = 2,8,1
- K = 2,8,8,1

All elements in group 1 have one electron in their highest energy shell.

You can repeat this process for elements in group 2. They all have two electrons in their highest energy shell. This pattern repeats across groups 1, 2 and 13–18. The number of electrons in the highest shell can be seen listed along the base of the periodic table in Figure 7.4.2. The electrons in the highest energy shell are known as **valence electrons**.

valence electron
an electron in the highest energy shell of an atom

Elements that have the same number of valence electrons are in the same group on the periodic table and behave in similar ways during bonding and in chemical reactions.

Properties of atoms: electronegativity

Electronegativity is the ability of an atom to attract shared electrons when forming a chemical bond. Elements that are more electronegative attract electrons more strongly than elements of low electronegativity. For example, Figure 7.4.3 shows a molecule of hydrogen chloride, produced when a hydrogen atom and a chlorine atom share electrons to form a structure called a molecule. The pair of electrons between the hydrogen and chlorine atoms is not shared evenly. The chlorine atom is more electronegative than the hydrogen atom, so it attracts the electron pair more strongly.

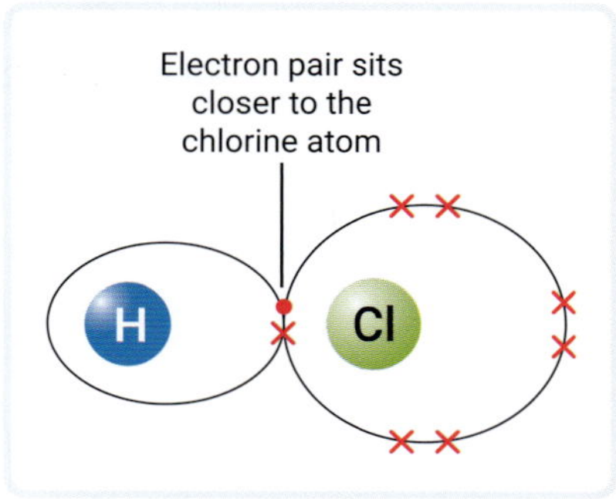

▲ **FIGURE 7.4.3** The chlorine atom attracts the shared pair of electrons in hydrogen chloride more than the hydrogen atom.

Figure 7.4.4 shows the electronegativity values for selected elements. You might notice the table looks like the periodic table, showing elements from groups 1, 2 and 13–18. There are several features to note:

- Group 18 elements all have an electronegativity of zero. This is because they do not form bonds with other atoms.
- Going across the periodic table from group 1 to group 17, the electronegativity of atoms increases.
- Going down the periodic table from period 1 to period 6, the electronegativity of atoms decreases.

	Group 1	2	13	14	15	16	17	18
Period 1	**H** 2.20							**He** 0
Period 2	**Li** 0.98	**Be** 1.57	**B** 2.04	**C** 2.55	**N** 3.04	**O** 3.44	**F** 3.98	**Ne** 0
Period 3	**Na** 0.93	**Mg** 1.31	**Al** 1.61	**Si** 1.90	**P** 2.19	**S** 2.58	**Cl** 3.16	**Ar** 0
Period 4	**K** 0.82	**Ca** 1.00			**As** 2.18	**Se** 2.55	**Br** 2.96	**Kr** 0
Period 5	**Rb** 0.82	**Sr** 0.95					**I** 2.66	**Xe** 0
Period 6	**Cs** 0.79	**Ba** 0.89						**Rn** 0

▲ **FIGURE 7.4.4** Electronegativity values allow elements to be compared.

The metals in groups 1 and 2 have very low electronegativity values. This means metals do not attract electrons strongly. Non-metals have high electronegativity values and strongly attract electrons.

Elements with low electronegativity (metals) do not tend to share electrons when they bond. Elements with high electronegativity tend to gain or share electrons when they bond.

Properties of atoms: ionisation energy

Ionisation energy is the energy required to remove an electron from the valence shell of an atom (Figure 7.4.5).

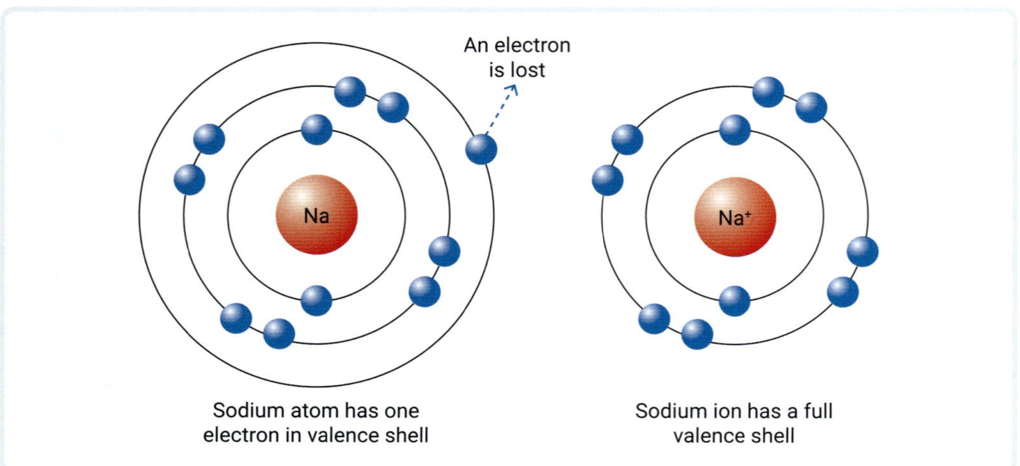

▲ **FIGURE 7.4.5** A sodium atom losing an electron from its valence shell. The energy needed for this process to occur is called ionisation energy.

The energy required to remove an electron from different elements is shown in Figure 7.4.6. Ionisation energies for metals are lower than for non-metals. This means that metals lose electrons easily, because it does not require much energy to remove them. Non-metals do not tend to lose electrons, because it requires a lot of energy for this to happen.

	Group 1	Group 2	Group 13	Group 14	Group 15	Group 16	Group 17	Group 18
Period 1	**H** 1320							**He** 2380
Period 2	**Li** 526	**Be** 905	**B** 810	**C** 1090	**N** 1410	**O** 1320	**F** 1690	**Ne** 2090
Period 3	**Na** 504	**Mg** 740	**Al** 580	**Si** 790	**P** 1020	**S** 1000	**Cl** 1260	**Ar** 1526
Period 4	**K** 425	**Ca** 600			**As** 953	**Se** 950	**Br** 1150	**Kr** 1360
Period 5	**Rb** 410	**Sr** 560					**I** 1020	**Xe** 1180
Period 6	**Cs** 380	**Ba** 510						**Rn** 1040

▲ **FIGURE 7.4.6** The ionisation energies for selected elements (measured in **kJ/mol**)

kJ/mol
kilojoules per mol; the unit of the amount of energy a substance has

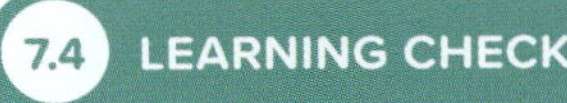

7.4 LEARNING CHECK

1 **Identify** the group of elements with an electronegativity value of zero.

2 **Explain** the difference between electronegativity and ionisation energy.

3 Use the electronic configurations of the group 17 elements to **describe** why they all have 7 electrons in their valence shell.

4 **Compare** the trends in electronegativity and ionisation energy for the period 2 elements (lithium to fluorine).

5 **Explain** whether the following atoms are more likely to share electrons or lose electrons during bonding.

  a Lithium

  b Fluorine

  c Magnesium

  d Oxygen

WORKING SCIENTIFICALLY

7.5 Justifying conclusions

SCIENCE SKILLS IN FOCUS

IN THIS MODULE, YOU WILL FOCUS ON LEARNING AND IMPROVING THESE SKILLS:

- using evidence from experiments and data to form justified conclusions
- examining trends of electronegativity and ionisation energy on the periodic table.

Using evidence to form justified conclusions

Scientists develop laws and theories that help us to predict and explain phenomena. Scientists perform experiments, gather data and look for patterns. When they see patterns, they will often propose a new hypothesis. Scientists perform new experiments to confirm or reject the hypothesis. The most important part of this is using the experimental evidence to form a conclusion.

A simple example is if you were to try to dissolve salt in water. Does temperature affect the amount of salt that can dissolve? A student proposed that heating water would increase the amount of salt that would dissolve. Some students conducted an experiment to verify this (Table 7.5.1).

Using experimental evidence to justify your conclusion is an important skill.

A poor conclusion for this experiment would be 'The mass of salt that dissolves increases as temperature increases'.

This conclusion does not provide any detail about the reasons this decision was made. Conclusions can be a little longer (or may be explained in more detail in a discussion section). A better conclusion that uses the data might be:

'Student 3's data was discarded in this experiment because it was clearly outlier data (it didn't follow the same pattern as the rest of the data). It seemed there was a malfunctioning hot plate, and the water was not heating up. From the data for students 1, 2 and 4, it seems as if increasing the temperature has influenced the mass of salt able to be dissolved. At the lowest temperature (30°C), an average of 36.4 g salt dissolved. At the highest temperature (90°C), an average of 38.8 g salt dissolved. This shows that as the temperature increased, the mass of salt dissolved also increased. This was consistent across all three students. Thus, the hypothesis is supported.'

Conclusions should:

- be more than one sentence long
- refer to the initial aim or hypothesis being tested
- use data from the experiment – the highest and lowest data points usually show the trend clearly
- summarise findings – do not repeat each piece of data from the experiment
- link your findings to the science – explain why your results occurred.

TABLE 7.5.1 Dissolving salt in water

Temperature (°C)	Maximum mass salt dissolved (g)			
	Student 1	Student 2	Student 3	Student 4
30	36.5	36.2	36.5	36.5
50	37.2	37.4	36.5	37.1
70	37.9	37.8	36.6	37.8
90	38.8	38.8	36.6	38.7

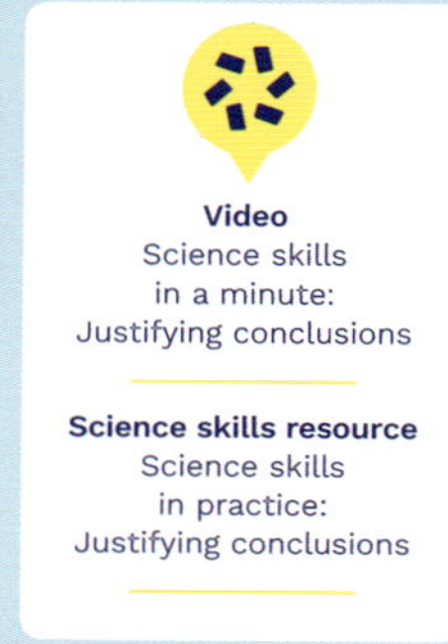

Video
Science skills in a minute: Justifying conclusions

Science skills resource
Science skills in practice: Justifying conclusions

9780170491785

PERIODIC TABLE TRENDS

AIM

To use provided data to investigate how position in a group and/or period affects the electronegativity and ionisation energy of an element

PROCEDURE

1 Using Excel (or grid paper), plot a graph of atomic number versus the electronegativity for the period 2 and period 3 elements (Tables 7.5.2 and 7.5.3).
   - Plot the atomic number on the horizontal axis (scale should be from 0–18).
   - Plot the electronegativity on the vertical axis (scale should be from 0–4).
   - Plot each point for the period 2 elements, then join the plotted points with ruled straight lines.
   - Plot each point for the period 3 elements, then join these plotted points with ruled straight lines.

2 Create a second graph with the same information but having ionisation energy on the vertical axis (use a scale of 500–2200 kJ/mol).

▼ **TABLE 7.5.2** Electronegativity and ionisation energy values for period 2 elements

Period 2 element	Atomic number	Electronegativity value	Ionisation energy value (kJ/mol)
Lithium	3	0.98	526
Beryllium	4	1.57	905
Boron	5	2.04	810
Carbon	6	2.55	1090
Nitrogen	7	3.04	1410
Oxygen	8	3.44	1320
Fluorine	9	3.98	1690
Neon	10	0	2090

▼ **TABLE 7.5.3** Electronegativity and ionisation energy values for period 3 elements

Period 3 element	Atomic number	Electronegativity value	Ionisation energy value (kJ/mol)
Sodium	11	0.93	504
Magnesium	12	1.31	740
Aluminium	13	1.61	580
Silicon	14	1.90	790
Phosphorus	15	2.19	1020
Sulfur	16	2.58	1000
Chlorine	17	3.16	1260
Argon	18	0	1526

ANALYSIS

1 **Explain** why you think the period 1 elements were not used for this investigation.

2 Which elements could be excluded from this graph set? **Justify** your choice.

3 For graph 1 (atomic number versus electronegativity):
   a **Describe** any similarities between the trends for the period 2 and period 3 elements.
   b **Describe** any differences between the trends for the period 2 and period 3 elements.
   c How do you think the period 4 element trend would **compare** to the ones you have plotted here?

4 For graph 2 (atomic number versus ionisation energy):
   a Is there anything strange about the trend seen in both the period 2 and period 3 elements?
   b **Compare** the trend lines of the period 2 and period 3 elements.
   c **Explain** how this graph could be used to **predict** which elements are likely to lose electrons.

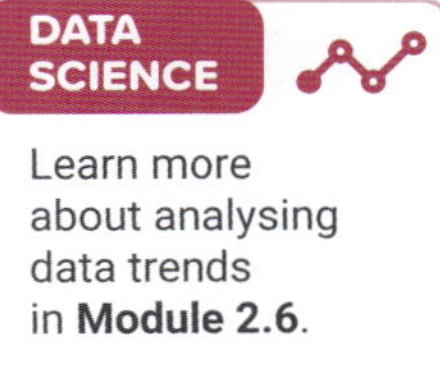

DATA SCIENCE

Learn more about analysing data trends in **Module 2.6**.

7.6 Bonding and stable atoms

BY THE END OF THIS MODULE, YOU WILL BE ABLE TO:

- ✓ explain how atoms achieve a stable electron configuration
- ✓ explain why atoms form bonds and identify the different types of bonding
- ✓ describe how atomic structure relates to the properties of elements.

Video activity
Introduction to chemical bonding

Interactive resource
Drag and drop: Types of bonding

GET THINKING

Chemistry uses terms such as 'compound', 'mixture', 'element' and 'atom'. In this chapter, you will come across a lot of new terms that have very specific meanings. Create a word list in your class notes so you can revise definitions easily.

Why are single atoms rare?

Most elements do not exist as single atoms. In fact, only the noble gases and mercury vapour can exist as single atoms. All other elements are unstable when they are individual atoms. Why is this?

In Module 7.4, you learned that the electron configuration plays a large role in the chemical behaviour of elements. The reason most atoms are unstable is related to the number and arrangement of their electrons.

Stable atoms have a full outer energy shell. This means they have eight electrons in the outer shell. You may recall that the only group on the periodic table that has this arrangement is the noble gases (Figure 7.4.1, page 248). All other elements have between one and seven electrons in the outer shell. This makes them unstable.

Atoms other than noble gases undergo changes to achieve the same stable electron configuration as a noble gas. Figure 7.6.1 shows one way that atoms can combine to make stable structures.

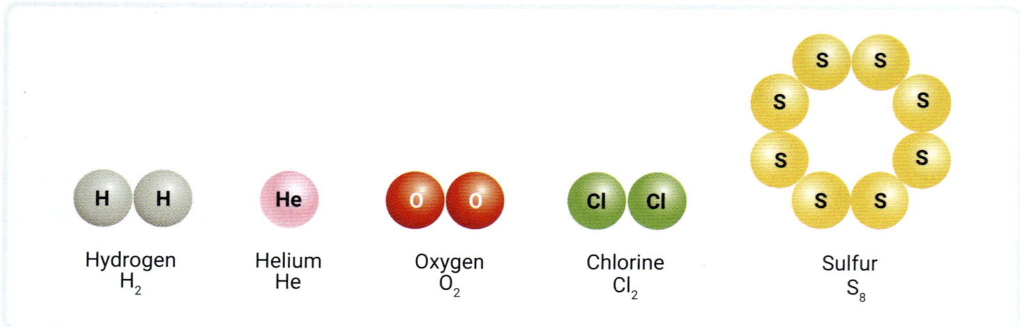

▲ **FIGURE 7.6.1** Only the noble gases and mercury vapour exist as single atoms. Other elements form molecules and compounds to achieve a stable electron configuration.

9780170491785

Gaining or losing electrons to become stable

Figure 7.6.2a shows a sodium atom with electron configuration 2,8,1. Figure 7.6.2b shows the stable configuration of the neon atom, 2,8. You can see that if sodium lost its one valence electron, then it would have the same electron configuration as neon. Figure 7.6.2c shows a sodium atom that has lost one electron. This is now a sodium ion. We will look at **ion** formation in detail in Module 7.7.

ion
a charged particle formed when an atom loses or gains valence electrons

The chlorine atom has electron configuration 2,8,7. If it gained one more electron, it would have the same configuration as argon, 2,8,8.

Thus, atoms can achieve a stable electron configuration by gaining or losing electrons.

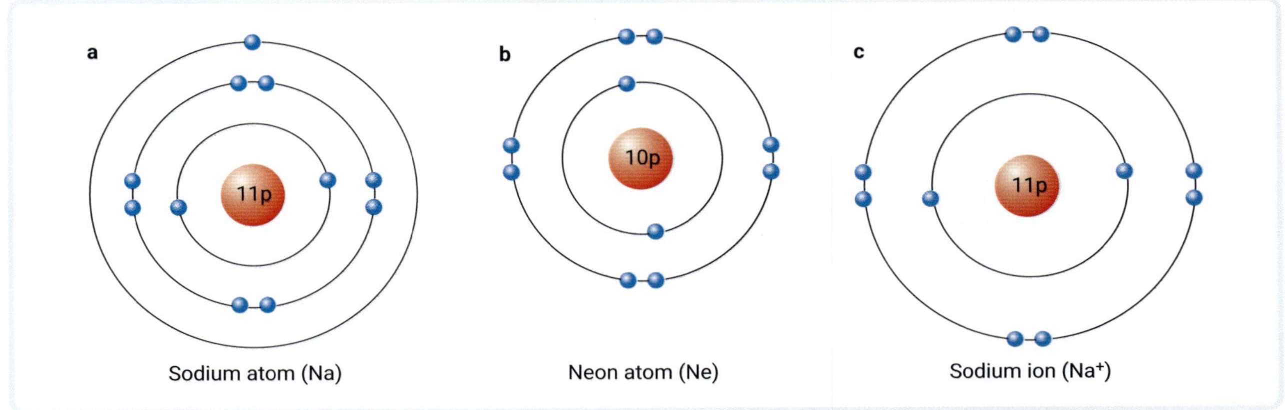

▲ **FIGURE 7.6.2** A sodium atom **(a)** has a stable electron configuration like neon **(b)** when it loses an electron to become a sodium ion **(c)**.

Sharing electrons to become stable

valency
the number of electrons an atom loses, gains or shares to form a stable electron configuration

Some atoms share electrons to achieve a stable electron configuration. Figure 7.6.3 shows two chlorine atoms with electron configuration 2,8,7. Each of the atoms needs only one more electron to achieve a stable 2,8,8 configuration. Each chlorine atom shares one electron with the other chlorine atom. This way they have their own electrons, plus the extra one they need to achieve a stable configuration.

Valency is a term used to describe how many electrons an atom needs to gain, lose or share to become stable. For example, in Figure 7.6.3, the chlorine has a valency of 1 because it is sharing one electron to become stable. In Figure 7.6.2, the sodium has a valency of 1 as it is losing one electron to become stable.

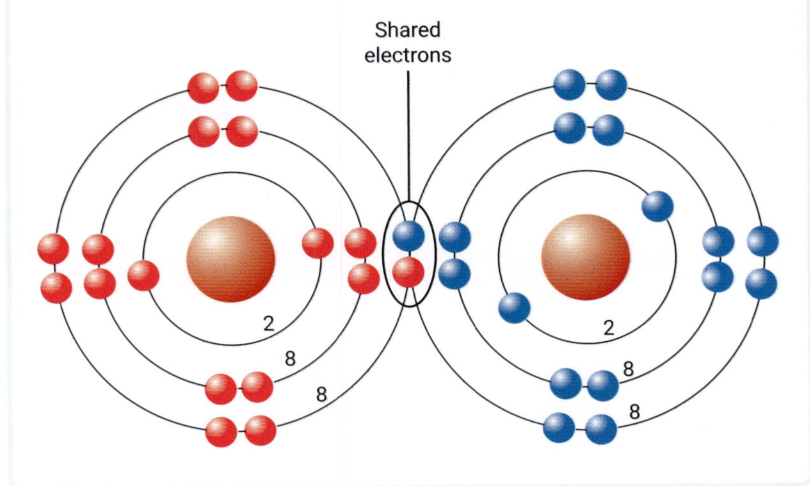

▲ **FIGURE 7.6.3** Two chlorine atoms sharing electrons to achieve a 2,8,8 electron configuration

chemical bonding
the joining of atoms via the transfer, loss or sharing of electrons to achieve a stable electron configuration

compound
two or more different elements joined by a chemical bond

Types of bonding

In Module 7.4, bonding was defined as 'how atoms combine together to form larger structures'. We can now define **chemical bonding** with reference to electrons. Chemical bonding occurs when atoms lose or gain electrons, or when they share electrons. Bonding involves atoms joining together to have a stable electron configuration.

Bonding can occur in elements. This happens when atoms of the same element join to form structures. Examples of element structures can be seen in Figure 7.6.1.

Bonding also creates **compounds** – where two or more different elements join to form substances, such as carbon dioxide (CO_2) or water (H_2O).

Three types of bonding can occur (Table 7.6.1). Each involves a different method of achieving a stable electron configuration. We will cover each type of bonding in more detail in the rest of this chapter.

▼ **TABLE 7.6.1** The types of bonding

Type of bonding	Method of achieving stable electron configuration	Examples of substances with this bonding
Metallic bonding	Electrons are lost from metal atoms and become free-moving in the metal structure.	Copper, iron, magnesium
Ionic bonding	Electrons are lost from some atoms and gained by others, so the atoms form ions.	Sodium chloride, calcium oxide
Covalent bonding	Electrons are shared between atoms.	Water, oxygen, carbon dioxide

7.6 LEARNING CHECK

1 **Explain** why only one group of the periodic table exists as single atoms.
2 Write the electron configuration of calcium and **propose** how it might achieve a stable electron configuration.
3 Use examples to **describe** two methods by which atoms achieve a stable electron configuration.
4 By describing similarities and differences between the types of bonding, **compare**:
   a metallic bonding and covalent bonding.
   b ionic bonding and metallic bonding.

7.7 Forming ions

BY THE END OF THIS MODULE, YOU WILL BE ABLE TO:

- ✓ explain how atoms form ions by gaining or losing electrons
- ✓ explain why atoms of elements in the same group form ions with the same charge.

GET THINKING

Look at Figure 7.7.3. It shows how you can use the periodic table to make conclusions about elements based on their groups rather than learning about each individual element. Why would this make it easier to learn chemistry? If you couldn't make general conclusions like this, what would be your alternative?

Interactive resource
Label: Predicting ion charge

What is an ion?

An ion is a charged particle that forms when an atom loses or gains valence electrons. Atoms are electrically neutral because the number of positive protons equals the number of negative electrons. When the number of electrons is changed, this balance no longer exists, and an ion is formed.

The structures of a sodium atom and a sodium ion are shown in Figure 7.7.1. The sodium atom has 11 protons and 11 electrons, which gives it zero net charge. The one valence electron makes it unstable, so it loses this electron to another atom to form a stable 2,8 electron configuration. The sodium ion has 11 protons, but only 10 electrons. It has a 1+ charge overall because it now has one more positive proton than it does negative electrons.

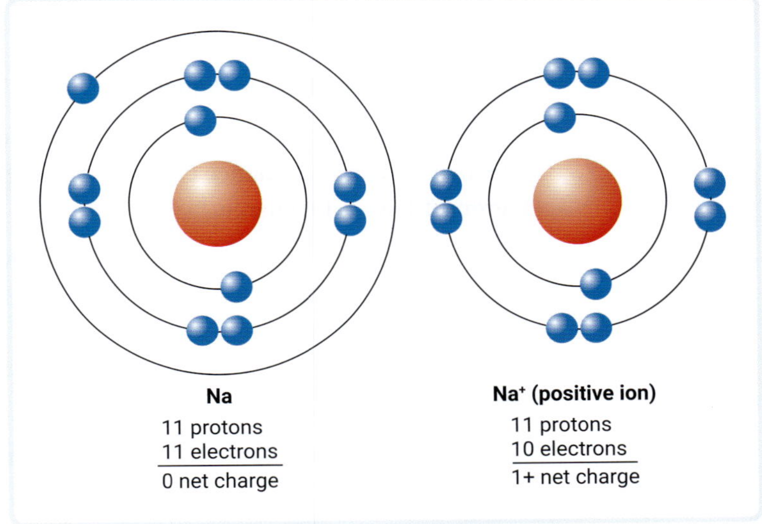

▲ **FIGURE 7.7.1** A sodium atom and a sodium ion

Positive and negative ions

Positive ions are called **cations**. Cations form when atoms lose electrons. This results in the ion having more protons than electrons and a net positive charge.

Negative ions are called **anions**. Anions form when atoms gain electrons. This results in the ion having more electrons than protons and a net negative charge.

cation
an ion with a positive charge

anion
an ion with a negative charge

So, what determines whether an electron is lost or gained from an atom? Like most chemical processes, it's all about what requires the least energy. Atoms will gain or lose as few electrons as possible to achieve a stable electron configuration, because each electron lost or gained requires the same amount of energy.

For example, oxygen has the electron configuration 2,6. An oxygen atom can either gain two electrons or lose six electrons. It is energetically easier for the atom to gain two electrons than lose six, so the oxygen atom will form a negative anion.

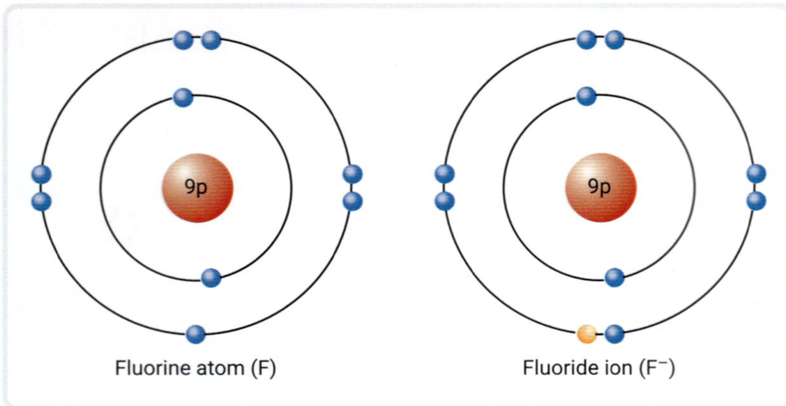

▲ **FIGURE 7.7.2** The formation of a negative fluoride ion by a fluorine atom gaining an electron

Atoms with one, two or three valence electrons will lose electrons and form positive cations.

Atoms with five, six or seven valence electrons will gain electrons and form negative anions. For example, Figure 7.7.2 shows the formation of a negative fluoride ion by gaining an electron.

Predicting the charge of an ion

The periodic table can be used to predict the charge of an ion for elements in groups 1, 2 and 13–17.

We know that the elements in group 1 all have one valence electron. We also know that elements with one valence electron form a 1+ ion. This means the ions of all elements in group 1 have a 1+ charge.

We can use the same logic for group 2 (two valence electrons, charge of 2+) and for group 3 (three valence electrons, charge of 3+). Groups 1, 2 and 3 contain metals. This means that metal ions are always positive.

Most elements in group 4 do not form ions. Elements such as carbon and silicon do not form ions. The non-metal elements in groups 15, 16 and 17 form ions of charge 3–, 2– and 1–, respectively. This information is summarised in Figure 7.7.3.

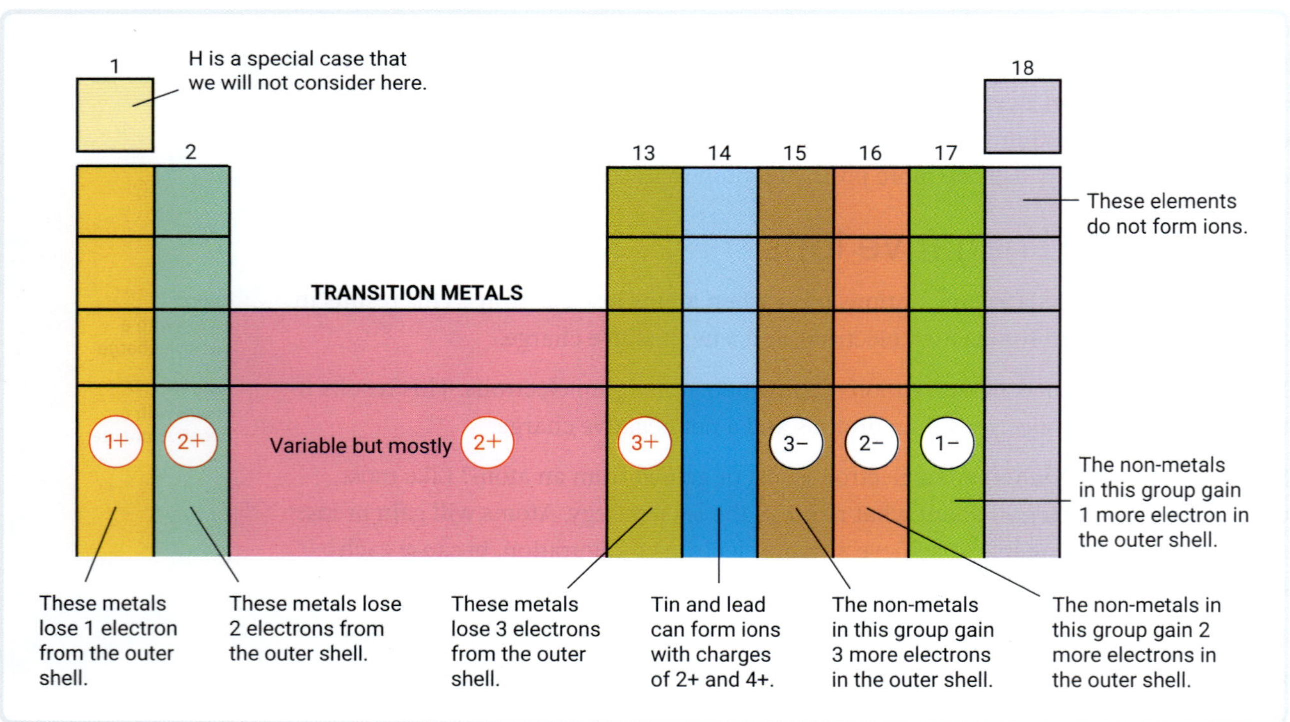

▲ **FIGURE 7.7.3** Using the periodic table to predict the charge on the ion of an element

9780170491785

ACTIVITY

7.7

Ion quiz

1 Pair up with a classmate. For this game, you can use a copy of the periodic table that only has the names of the elements and groups written on it.

2 One student is to call out the names of elements at random from groups 1, 2 and 13–17 for 60 seconds (set a timer). The second student is to **identify** the charge on the ion. Do not move onto the next element until you have the charge correct.

3 Swap roles. The winner of the game is the person who gets the most correct ions in 60 seconds.

7.7 LEARNING CHECK

1 **Describe** the similarities and differences between an anion and a cation.

2 Using Figure 7.7.1 as a model, **justify** the charge on the:
   a sulfide ion.
   b aluminium ion.
   c calcium ion.
   d oxide ion.

3 **Identify** two elements that do not form ions.

4 **Explain** why the magnesium atom does not form a 6− ion.

5 **Describe** how you can use the periodic table to **predict** the charge on ions in some groups.

7.8 Bonding between metals and non-metals

BY THE END OF THIS MODULE, YOU WILL BE ABLE TO:

- ✓ describe the structure and formation of ionic compounds
- ✓ describe the changes that occur in ionic compounds when they become molten or aqueous.

GET THINKING

Models can be used to represent things that are too small to see, like atoms. How do models help to explain a concept or allow you to visualise how something works?

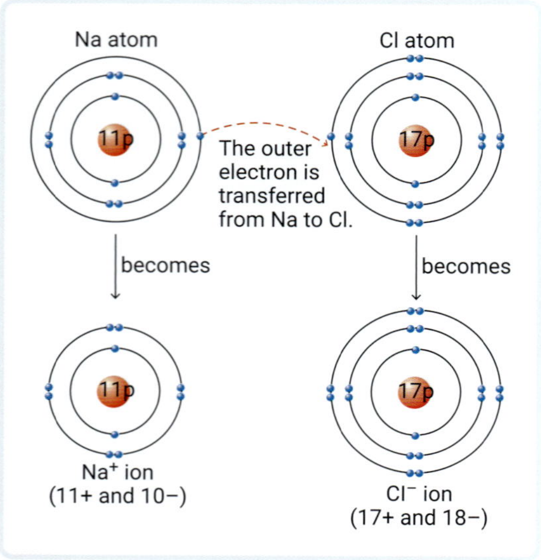

Electron transfer diagrams

When an ionic bond forms, electrons are transferred between a metal and a non-metal. Metals lose electrons to form positive ions (cations). The electrons are transferred to non-metal atoms that take the electrons to form negative ions (anions). The ionic bond is the electrostatic force of attraction between metal cations and non-metal anions. All ionic substances are compounds. They always contain two or more elements.

Figure 7.8.1 shows a sodium atom transferring its one valence electron to a chlorine atom. Both atoms form ions and achieve a stable electron configuration. The sodium ion (Na^+) is positive, and the chloride ion (Cl^-) ion is negative.

Na atom

Cl atom

11p

17p

The outer electron is transferred from Na to Cl.

becomes

becomes

11p

17p

Na^+ ion (11+ and 10−)

Cl^- ion (17+ and 18−)

▲ **FIGURE 7.8.1** Electron transfer during the formation of sodium chloride

Structure of ionic solids

An ionic solid is a continuous lattice of ions, as seen in Figure 7.8.2. The sodium chloride lattice is made from alternating positive sodium ions and negative chloride ions. Electrostatic forces hold the oppositely charged ions together in a strong ionic bond.

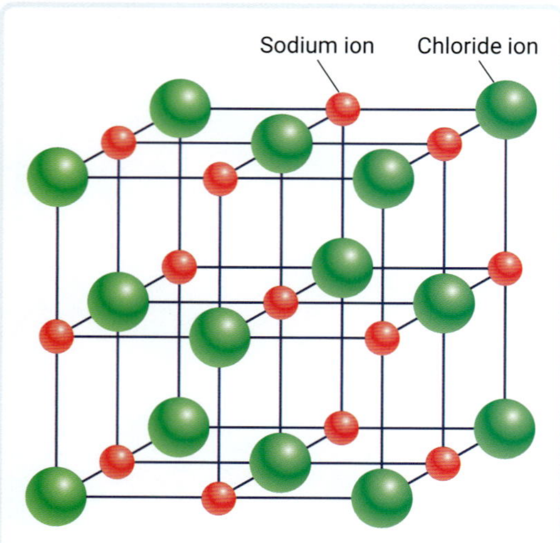

▲ **FIGURE 7.8.2** Ionic solids such as sodium chloride are a lattice of alternating positive and negative ions.

Structure of molten and dissolved ionic substances

When enough heat is applied to an ionic solid, it may melt. Melting points of ionic compounds are generally high, so you will not often see a molten ionic substance, but they do exist. When an ionic solid melts, the ionic bonds between the positive and negative ions weakens and breaks. This means the ions are now free to move and no longer fixed in position (Figure 7.8.3).

9780170491785

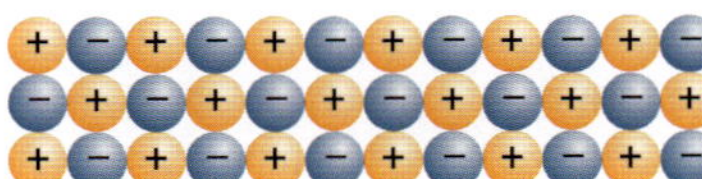

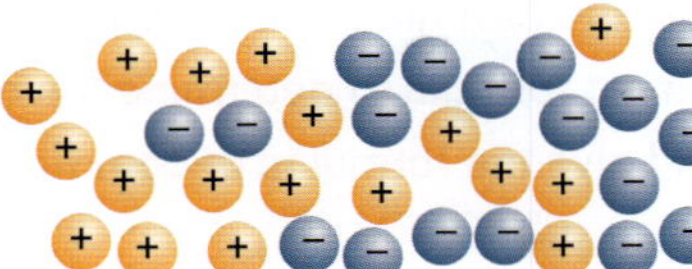

▲ **FIGURE 7.8.3** Ionic solids have ions fixed in position, while molten ionic substances and ionic substances in solution have ions that are free to move.

When an ionic substance is placed into water, it may or may not dissolve. Some ionic substances are soluble in water, like sodium chloride (table salt). Some ionic substances are not soluble in water, like calcium carbonate (limestone).

If an ionic substance is soluble, then the ions separate from the fixed lattice and are free to move. This is similar to how they move in a molten ionic substance, except that the ions are surrounded by water molecules. This is called as an ionic solution. It can also be called an aqueous solution because it contains a solute dissolved in water.

7.8 LEARNING CHECK

1 **Draw** electron transfer diagrams for the following pairs of elements.
   - **a** Magnesium and oxygen
   - **b** Lithium and fluorine
   - **c** Calcium and chlorine (Hint: You will need two chlorine atoms in your diagram – the chlorine atoms take one electron each from the calcium.)
   - **d** Lithium and sulfur
   - **e** Aluminium and oxygen

2 **Describe** the ionic bond.

3 **Explain** what happens to the ionic lattice structure when it dissolves in water.

4 Use the concepts of ionisation energy and electronegativity to **explain** why metals donate electrons, but non-metals accept electrons.

5 **Compare** the structures of solid, molten and aqueous ionic substances.

7.9 Formulas of ionic compounds

BY THE END OF THIS MODULE, YOU WILL BE ABLE TO:

- ✓ determine the formula of an ionic compound, given its name
- ✓ use the ionic formula to determine the charge on an ion
- ✓ name and give the formulas of common polyatomic ions.

Other resource
Worksheet: Practising ionic formulas

GET THINKING

Have a look through this module. Are there any ideas in here that are familiar to you from mathematics? Many ideas and ways of thinking in science are mathematically based. Try to apply your mathematical knowledge to this module to improve your understanding.

Ionic formulas

ionic formula
the chemical representation of an ionic substance showing the number and type of atoms present

Ionic compounds are represented by **ionic formulas** that show us two pieces of information:

1 the chemical elements that are present; for example, NaCl shows a compound with sodium and chlorine
2 the ratio of ions in the compound; for example, Na_2O shows that there are two atoms of sodium to one atom of oxygen in this compound.

Ionic compounds are electrically neutral. This means they have no overall charge. The positive ion charge (or charges) balances the negative ion charge (or charges). You can see how this balance gives us the correct ionic formulas in the examples below.

Ionic compounds with two elements

Simple ionic compounds contain a single metal element and a single non-metal element. To write the correct formula, we need to:

1 determine the charge on each ion
2 determine how many of each atom are required to balance the overall charge
3 write the correct balanced formula.

Example 1: sodium chloride

Sodium is in group 1, so its ion has a 1+ charge: Na^+.

Chlorine is in group 7, so its ion has a 1– charge: Cl^-.

With one of each ion, the overall charge is zero, so the formula is NaCl.

As there are no numbers in the formula, this shows one of each type of ion.

Example 2: calcium bromide

Calcium is in group 2, so its ion has a 2+ charge: Ca^{2+}.

Bromine is in group 17, so its ion has a 1– charge: Br^-.

Two Br^- are needed to balance the Ca^{2+}, so the formula is $CaBr_2$.

Subscript numbers are used to indicate if more than one ion is required to balance the formula – in this case, the 2 for the bromide ion.

9780170491785

Example 3: aluminium oxide

Aluminium is in group 13, so it has a 3+ charge: Al^{3+}.

Oxygen is in group 16, so it has a 2− charge: O^{2-}.

To balance this formula, we need to find the lowest common multiple of two and three. The lowest common multiple of two and three is six. If we have two Al^{3+} and three O^{2-}, then there is a total of 6+ and 6− charges and the overall charge is zero. Therefore, the formula is Al_2O_3.

What is a polyatomic ion?

A polyatomic ion is an ion that contains two or more elements. The ion has an overall charge that can be negative or positive. Some common polyatomic ions are shown in Table 7.9.1 and in Figure 7.9.1. The charge doesn't belong to any one atom in the polyatomic ion. It belongs to the ion overall.

▼ **TABLE 7.9.1** Common polyatomic ions

Name	Formula
Ammonium ion	NH_4^+
Hydroxide ion	OH^-
Carbonate ion	CO_3^{2-}
Nitrate ion	NO_3^-
Sulfate ion	SO_4^{2-}
Phosphate ion	PO_4^{3-}

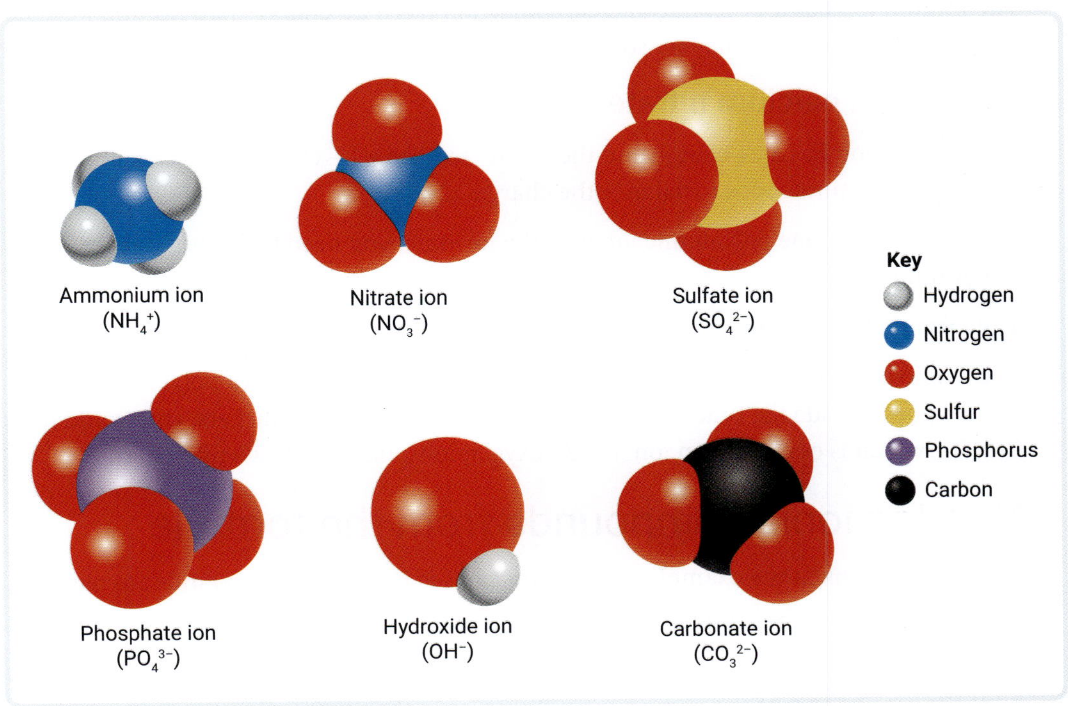

▲ **FIGURE 7.9.1** Common polyatomic ions

Transition metal ions

The transition metals occur in the centre of the periodic table. They include the metals in groups 3–12. Transition element chemistry is complicated, and you will cover parts of this in your senior studies.

You may recall that you cannot predict the charge of a transition metal ion from where it is on the periodic table. Generally, transition metal elements form 2+ ions. However, some of the elements can form ions of different charges. For example, iron can form Fe^+, Fe^{2+} and Fe^{3+} ions. Copper commonly forms Cu^+ and Cu^{2+} ions. Scientists use roman numerals in brackets as part of the name to show the charge on the ion (Table 7.9.2).

▼ **TABLE 7.9.2** Writing the names of the different iron and copper ions

Ion name	Formula
Iron(I) ion	Fe^+
Iron(II) ion	Fe^{2+}
Iron(III) ion	Fe^{3+}
Copper(I) ion	Cu^+
Copper(II) ion	Cu^{2+}

Ionic compounds with polyatomic ions

Polyatomic ions act as a single ion when you write a formula.

Example 1: sodium hydroxide

The sodium ion has a 1+ charge and hydroxide ion has a 1− charge. One of each ion is required, so the formula is NaOH.

Example 2: lithium carbonate

The lithium ion has a 1+ charge and carbonate ion has a 2− charge. Two lithium ions are required for each carbonate ion, so the formula is Li_2CO_3.

Example 3: calcium nitrate

The calcium ion has a 2+ charge and the nitrate ion has a 1− charge. One calcium ion needs two nitrate ions to balance the charge.

When more than one polyatomic ion is in the formula, you need to use brackets in the formula.

The formula for calcium nitrate is $Ca(NO_3)_2$.

The whole polyatomic ion goes inside the brackets and the number of polyatomic ions needed goes outside the bracket as a subscript. If you didn't use brackets, you would have $CaNO_{32}$, which is one nitrogen ion and 32 oxygen ions, instead of two nitrate ions.

Naming ionic compounds from the formula

If you have a look at the examples above, you can see how ionic compounds are named.

- The first part of the name is the metal element. It keeps the name of the metal.
- The second part of the name is the non-metal element or elements.
- If the second part of the name is a single element, then the ending changes to '-ide'. For example, 'chlorine' changes to 'chloride'.
- If the second part is a polyatomic ion, it keeps the name of the polyatomic ion. For example, 'potassium nitrate' includes the polyatomic ion 'nitrate'.

Determining the charge on an ion from the formula

If you know one of the ions in an ionic formula, you can determine the charge on the other ion. Consider zinc chloride ($ZnCl_2$). Zinc is a transition metal, so we cannot predict its charge from the periodic table, but we know the chloride ion has a 1− charge. As there are two chloride ions, the zinc must have a 2+ charge to balance the compounds. Therefore, the zinc ion is Zn^{2+}.

7.9 LEARNING CHECK

1 **Define** polyatomic ion.

2 **Explain** the difference between the copper(I) ion and the copper(II) ion.

3 **Name** the following ionic compounds.

- **a** KCl
- **b** Li_2O
- **c** Na_2CO_3
- **d** $CuSO_4$
- **e** $Mg(NO_3)_2$
- **f** CaS
- **g** $Zn_3(PO_4)_2$
- **h** $FeCl_3$

4 Write the formula of each of the following ionic compounds.

- **a** Sodium bromide
- **b** Magnesium oxide
- **c** Copper(II) hydroxide
- **d** Potassium phosphide
- **e** Iron(II) chloride
- **f** Lithium nitrate
- **g** Nickel(I) carbonate
- **h** Iron(III) hydroxide
- **i** Aluminium sulfide
- **j** Sodium phosphate

5 In the following formulas, X represents an ion with an unknown charge. Use the formula to **determine** the charge on X in each compound.

- **a** XCl
- **b** X_2O
- **c** $X(OH)_3$
- **d** CaX
- **e** MgX_2

7.10 Properties of ionic compounds

BY THE END OF THIS MODULE, YOU WILL BE ABLE TO:

✓ use the structure of ionic substances to explain their properties.

Quiz Ionic compounds

GET THINKING

Remember in Stage 4, when you looked at the properties of metals and non-metals. How are the properties linked to the structure? Brainstorm at least one property of metals and non-metals and explain how their structure causes this property.

Ionic compounds are brittle solids with high melting points. They do not conduct electricity as a solid but are conductive when molten or dissolved in water.

Melting point

To melt an ionic substance, the ionic bond needs to be weakened or broken. As the bond is very strong, a large amount of energy is required to weaken or break it. This means that ionic substances have high melting points. Melting points of ionic substances can vary from a few hundred degrees to more than 1000°C.

Brittle nature

Ionic solids are brittle. If you strike them with force, they will crack or shatter. The reason for this is shown in Figure 7.10.1. When a force is applied to the ionic lattice, part of the lattice shifts. As a result, positive charges line up with each other, and negative charges line up with each other. Like charges repel, causing the structure to crack and shatter.

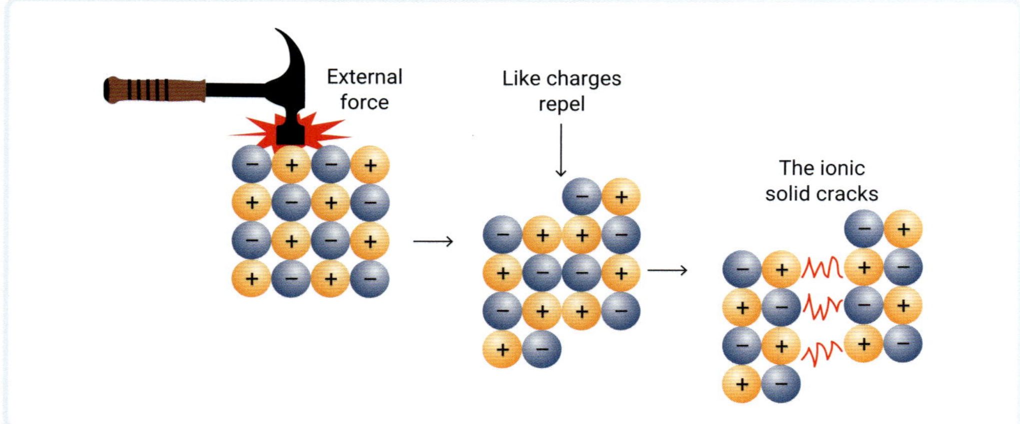

▲ FIGURE 7.10.1 Ionic solids are brittle and shatter when force is applied.

Conduction of electricity

A substance will conduct electricity if it has charged particles that are free to move. As the charged particles move, they carry charge through the substance. A solid ionic lattice has the particles fixed in place. As a result, an ionic solid will not conduct electricity because the particles are not free to move.

If an ionic solid melts (is converted to a molten form) or is dissolved in water (is converted to an aqueous form), the ions become free-moving. Therefore, both molten and dissolved ionic compounds can conduct electricity, as shown in Figure 7.10.2.

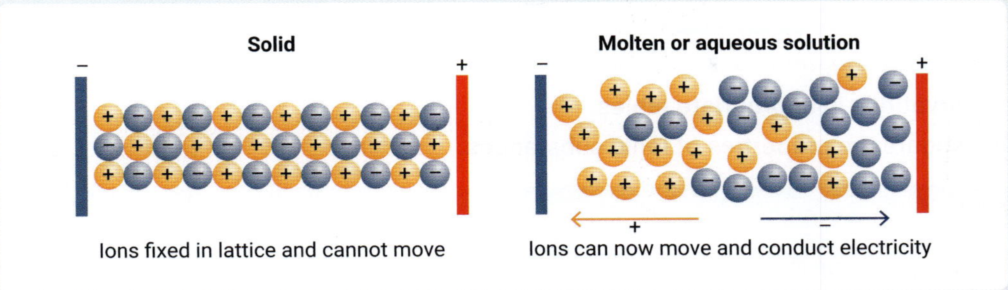

▲ **FIGURE 7.10.2** Ionic compounds do not conduct electricity as solids, only in molten or aqueous form.

7.10 LEARNING CHECK

1. Sodium chloride is an ionic substance that is soluble in water. **Predict** and **explain** how its electrical conductivity will change when solid sodium chloride is converted to aqueous form.
2. **Draw** a labelled diagram to show why ionic substances are brittle.
3. Ionic substances have high melting points that cover a range from a few hundred degrees to more than 1000°C. **Propose** an explanation for these observations.

WORKING SCIENTIFICALLY

7.11 Presenting data in tables

SCIENCE SKILLS IN FOCUS

IN THIS MODULE, YOU WILL FOCUS ON LEARNING AND IMPROVING THESE SKILLS:

- presenting data and observations in appropriate tables
- conducting tests to investigate the properties of ionic substances.

Tables are used to record and organise experimental results so they are easy to read and provide a clear, concise summary of what you found. Results tables can include quantitative measurements; for example, density, mass or length. They can also include qualitative observations; for example, colour changes, whether something is brittle, or whether something is opaque or transparent.

Below you will find some general tips for creating results tables, but just as each experiment is different, each results table will have a different structure.

- Plan your table before you start – consider how many rows and columns you will need. Generally, a results table will have more rows than columns, otherwise text can become difficult to read.
- Ensure each row and column heading fully describes what is in the table. Headings like 'Trial 1' are not helpful as they do not describe the data. Try to use what you are measuring or observing; for example, 'Mass of substance' or 'Volume of liquid'.
- Ensure your results table has a title, uses ruled lines and has units included in the column headings if you are taking quantitative measurements.

Video
Science skills in a minute: Presenting data in tables

Science skills resource
Science skills in practice: Data tables

PROPERTIES OF IONIC SUBSTANCES

AIM

To investigate some of the physical properties of ionic substances, including solubility and electrical conductivity

Safety

Take care when handling an electrical circuit. Always turn the power off at the main switch and unplug the power source when not in use.

Only use the circuit switch to connect the circuit for as long as you need to record the data.

Ensure your hands are dry.

Do not handle circuit components straight after use; they can get hot and may need time to cool down.

MATERIALS AND EQUIPMENT

- samples of solid ionic substances, including sodium chloride, calcium carbonate, copper sulfate and potassium nitrate
- 4 × 250 mL beakers
- 4 watch glasses
- 400 mL deionised water
- 4 spatulas – one for each ionic solid
- simple electric circuit, as seen in Figure 7.11.1

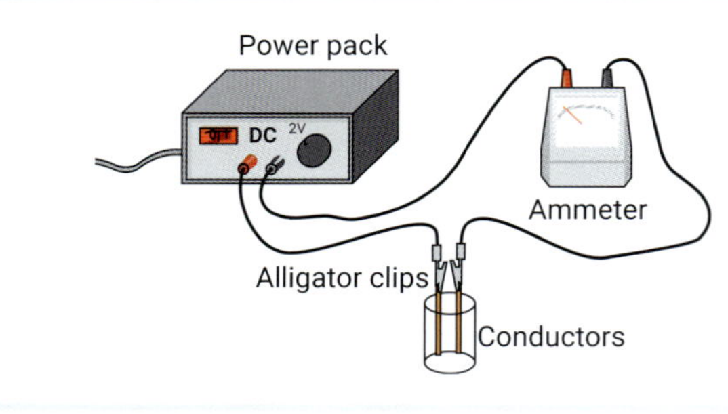

▲ **FIGURE 7.11.1** Equipment for testing conductivity

PROCEDURE

You will make the same observations and conduct the same tests for each of the four substances. Make notes on your observations for each of the tests and observations.

GENERAL OBSERVATIONS

1 Place about half a spatula of one ionic substance on a clean, dry watch glass.

2 Write down observations about the appearance of the ionic substance. This might include colour, shape of the solid particles, or whether it is shiny or dull.

TEST FOR CONDUCTIVITY OF IONIC SOLID

1 Create a small pile of your ionic solid so that all the solid particles are touching each other.

2 Hold your conductors in the electric circuit to the pile, ensuring the two conductors do not touch each other.

3 If the ammeter records a value, then the ionic solid conducts electricity. Test the substance then record your observation.

SOLUBILITY AND TEST FOR CONDUCTIVITY OF SOLUTION

1 Place 100 mL of deionised water into a 250 mL beaker.

2 Place the ionic solid from your watch glass into the deionised water and stir thoroughly for at least a minute.

3 Determine if the ionic substance is soluble. Record your observations of this process.

4 If the ionic substance is soluble, use the conductors and the electric circuit to test for ability to conduct electricity. Record your observation.

RESULTS

Repeat these tests for all ionic substances. Record your observations in a suitable table. A table you might use is shown in Table 7.11.1, or you can create your own. Remember to follow the tips from the blue Science skills in focus box.

ANALYSIS

1 Why do you think you were not asked to do a test for melting point?

2 Were all the substances soluble? What observations did you make of the substances that did not dissolve?

3 Were there any issues with the conductivity tests? **Describe** any difficulties you had while performing this experiment.

4 Are there any conclusions you can make about the conductivity of ionic solids and solutions?

5 Why is it harder to form a conclusion about the solubility of ionic substances?

CONCLUSION

Write a final conclusion about the relationship between the structure of ionic substances and their conductivity.

▼ **TABLE 7.11.1** Testing the properties of ionic compounds

Ionic substance	General observations	Conductivity as a solid	Solubility	Conductivity as a solution

7.12 Bonding in metals

BY THE END OF THIS MODULE, YOU WILL BE ABLE TO:

✓ describe how metal atoms bond

✓ use the structure of metals to explain their physical properties.

Quiz
Metal atom bonds

Interactive resource
Drag and drop:
Metal properties

Extra science investigation
The physical properties of metals and non-metals

GET THINKING

In chemistry, we often talk about the properties of substances. There are physical properties and chemical properties. What do you think the difference is? Properties can include hardness, boiling point, reactivity, colour and density. Is it easy to identify which properties are physical and which are chemical? Why, or why not?

Uses of metals

Metallic elements dominate the periodic table, making up nearly 75 per cent of all elements. Metals are widely used in society for construction, musical instruments and electrical wiring. The transport industry relies on metals for the structural components of cars, trains and planes as well as railway lines. In your kitchen, you will find metal storage, food cans, cutlery, saucepans and other cooking utensils.

metallic lattice
the organised structure formed with rows of metal cations surrounded by delocalised electrons

delocalised electrons
electrons in a metallic lattice that do not belong to any particular atom

Metals are widely used because of their physical properties. Physical properties relate to observable and/or measurable properties of an object, such as density or boiling point.

The structure of metals

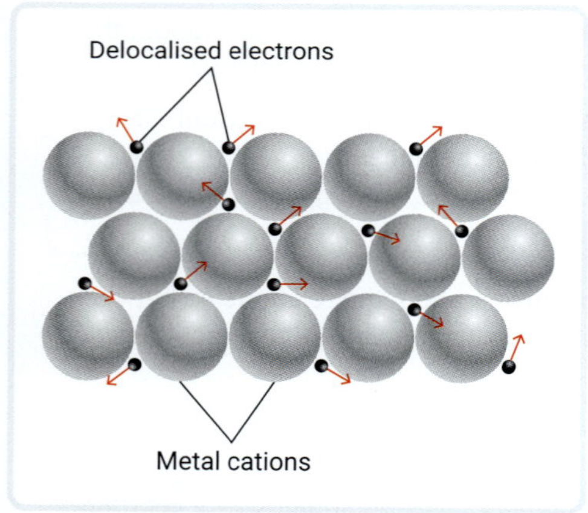

▲ **FIGURE 7.12.1** The internal structure of metals

Physical properties are determined by the structure of the substance. Metal atoms lose electrons to achieve a stable electron configuration. For example, magnesium has two valence electrons, so loses two electrons to form the Mg^{2+} ion.

The metal cations arrange themselves in an ordered structure called a **metallic lattice** (Figure 7.12.1). The electrons that were lost by the metal atoms are called **delocalised electrons**. They do not belong to any atom but are free to move around the lattice formed by the metal ions.

The lattice is held together by the force of attraction between the positive metal ions and the negative electrons. This force of attraction is the **metallic bond**.

metallic bond
the force of attraction between metal cations and delocalised electrons

lustrous
shiny when cut or polished

malleable
able to be beaten into different shapes

ductile
able to be stretched into a wire

Metallic properties

Metals share a common set of physical properties. All properties can be explained by the structure of the metallic lattice.

Metals mostly have a medium–high boiling point and are hard and dense. Metals are **lustrous** (shiny), good conductors of heat and electricity, **malleable** (their shape can be changed) and **ductile** (they can be stretched into a wire), with the exception of mercury, which is the only non-solid metal (Figure 7.12.2).

9780170491785

Hardness and density

The metallic bond is a very strong **electrostatic bond**. This binds the metal cations and delocalised electrons into a very close-packed structure.

The close-packed lattice can absorb a lot of force without distorting, snapping or breaking. This makes metals hard. The close-packed nature of the particles also means that a large amount of mass is present in a given volume. This makes metals denser than many other substances.

▲ FIGURE 7.12.2 Mercury is the only non-solid metal at room temperature.

Melting point

To melt a metal, the electrostatic attraction between the cations and electrons needs to be weakened or broken. The strong metallic bond means a large amount of energy is needed to disrupt the force of attraction. This means that metals have medium–high melting points.

Different metals have different melting points. Cadmium has a relatively low melting point for a metal, only 321°C. Aluminium has a melting point of 660°C, whereas nickel has a melting point of 1453°C.

Engineers and designers choose metals for different purposes based on their melting point. Tungsten has a melting point of 3400°C. It is used where the metal used must have a high melting point, such as in industrial heating elements that convert electricity to heat.

▲ FIGURE 7.12.3 Metals that are cut or polished show lustre.

Lustre

Metals are described as having lustre when they are polished or freshly cut (Figure 7.12.3). Metals that have been exposed to air for a long time often develop a coating on the surface, making them dull. If you sand this layer away or cut the metal, it will be shiny.

The shine from metals comes when light reflects off the delocalised electrons moving through the metallic lattice structure, as seen in Figure 7.12.4.

electrostatic bond
the force of attraction between positive and negative particles

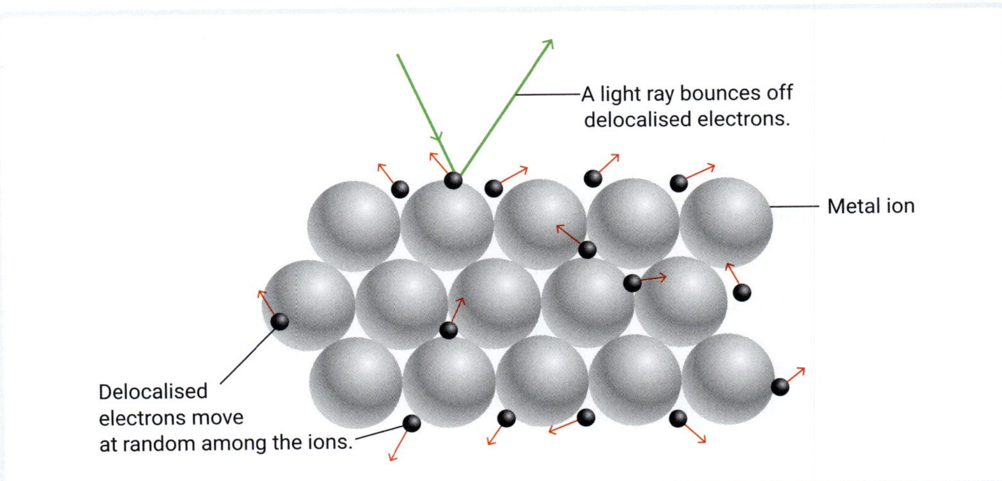

◀ FIGURE 7.12.4 Metals are shiny because the delocalised electrons reflect light.

Conduction of heat and electricity

Metals conduct both heat and electricity well. If you apply electricity or heat to one end of a piece of metal, it will move through to the other end. If you have ever left a metal spoon in a saucepan while it was heating, you would have quickly learned this when you went to pick it up again! This is why wooden spoons are used for stirring, as they do not conduct heat very well.

Metals are used in a lot of electrical appliances to allow electricity to flow. Other materials, such as plastics, are non-conductive, so they do not allow electricity to flow. Metals are used for contact points, internal wiring or anywhere you need electricity to flow through a circuit.

Both properties can be explained by the movement of the delocalised electrons. As they move through the lattice structure, the electrons easily carry electrical energy and heat energy through the metal (Figure 7.12.5).

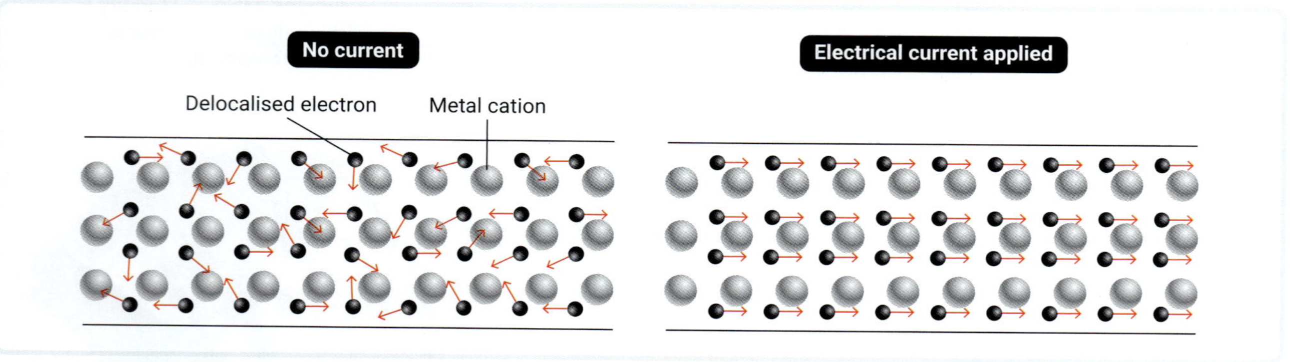

▲ **FIGURE 7.12.5** The movement of electric current through a metal lattice. When a current is applied, the electrons are attracted to the positive terminal.

Not all metals conduct heat and electricity with the same efficiency. One of the best conductors is copper. Copper wiring is most often used for electrical wiring in household circuits and electrical appliances.

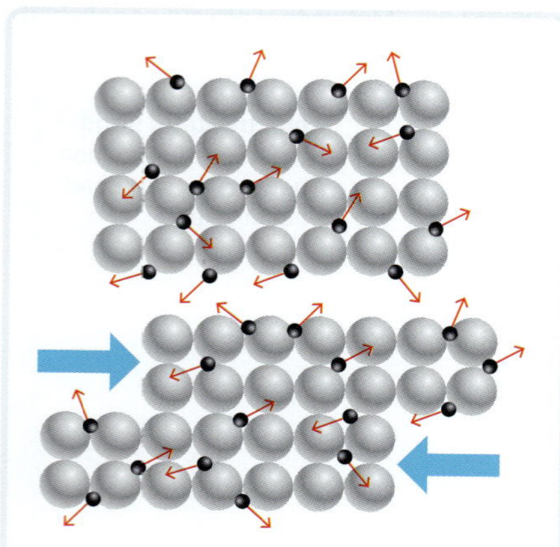

▲ **FIGURE 7.12.6** Metals are malleable and ductile because the cations and delocalised electrons can rearrange themselves to form a different shape.

Malleability and ductility

All metals, except mercury, are malleable and ductile. A malleable metal can be beaten into new shapes. A ductile metal can be stretched or drawn into a wire. These two properties make metals very useful because they can be converted into shapes such as containers or furniture. They can also be used to form wires. Copper wiring carried phone and internet signals for many years, until replaced by fibre optic cables.

Figure 7.12.6 shows what happens when force is applied to a metallic lattice. Instead of breaking, the metal cations and delocalised electrons simply rearrange themselves into the new shape. This does not damage the lattice in any way and it retains its strength and other properties, just in a different shape!

9780170491785

Modelling a metallic lattice

☆ ACTIVITY

7.12

1 You will need a flat container (such as a takeaway food container) of small polystyrene beads. This is going to act as a model for showing how a metallic lattice gives specific properties.

2 **Construct** a table like the one shown in Table 7.12.1.

▼ **TABLE 7.12.1** Modelling a metallic lattice

Property	Ways the model is good	Ways the model is poor

3 Using the headings throughout Module 7.12, select three properties and put them into the three rows in your table. For each property you selected, **describe** how the model:

a shows the features of the property.

b could be confusing.

Hint: An example is the property of lustre. The container of polystyrene beads would show the light hitting the lattice structure (good), but it could be confusing, as it does not reflect light the way the delocalised electrons would in a real metal (poor).

7.12 LEARNING CHECK

1 **Draw** the structure of a metallic lattice and **identify** the key features of the structure.

2 **Explain** why most metals are dense.

3 **Explain** two reasons many electrical circuit components are made from metal.

A.B.G./Shutterstock.com

▲ **FIGURE 7.12.7** An electrical circuit

WORKING SCIENTIFICALLY

7.13 Observing physical properties

SCIENCE SKILLS IN FOCUS

IN THIS MODULE, YOU WILL FOCUS ON LEARNING AND IMPROVING THIS SKILL:

- describing observations about the physical properties of metals.

PROPERTIES OF METALS

AIM

To investigate the physical properties of metals, including lustre, hardness, electrical conductivity and malleability

 **Safety**

Take care when handling an electrical circuit. Always turn the power off at the main switch and unplug the power source when not in use.

Only use the circuit switch to connect the circuit for as long as you need to record the data.

Ensure your hands are dry.

Do not handle circuit components straight after use; they can get hot and may need time to cool down.

MATERIALS AND EQUIPMENT

- ☑ samples of metals such as magnesium, aluminium, iron, zinc, copper and tin
- ☑ hammer and wooden board (for the malleability test)
- ☑ quartz crystal or iron nail (for the hardness test)
- ☑ sandpaper (for the lustre test)
- ☑ simple electrical circuit, as seen in Figure 7.13.1 (for the electrical conductivity test)

PROCEDURE

You will conduct four tests, repeating each step for all the metal samples. Take photographs or videos and make notes on your observations for each metal.

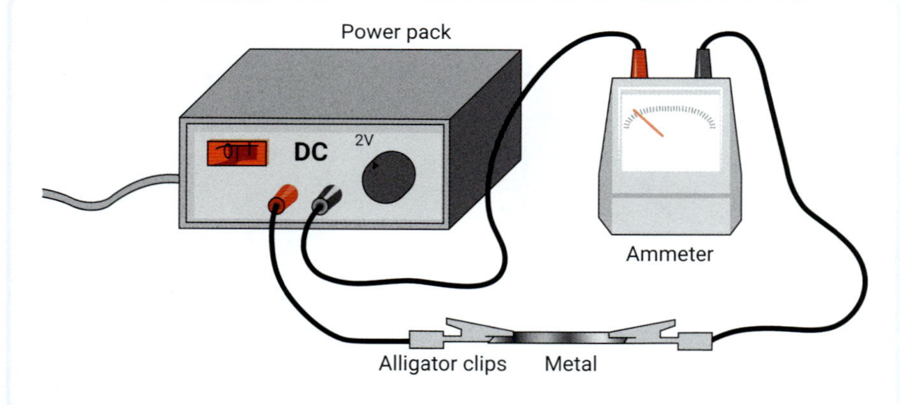

▲ **FIGURE 7.13.1** The equipment needed for testing conductivity
Note: You can substitute the ammeter for a light globe, which will light up if the substance conducts electricity.

LUSTRE TEST

1. Use sandpaper to clean each metal for about 30–45 seconds.
2. Record the metal colour and whether it is shiny or not.

HARDNESS TEST

1. Use an iron nail or quartz crystal (or both) to try to scratch each metal.
2. Record whether each metal could be scratched and whether it was easy or difficult to scratch.

ELECTRICAL CONDUCTIVITY TEST

1. Set up the equipment as shown in Figure 7.13.1, placing each metal in turn into the circuit.
2. Record the ammeter reading or brightness of the light globe. If there is no reading on the ammeter or the globe does not light up, the substance does not conduct electricity.

MALLEABILITY TEST

1. Taking care to avoid your fingers and crush damage, try to hammer the metal into a new shape.
2. Record the level of difficulty it took to change the shape of the metal – easy, medium or difficult.

RESULTS

Create a table(s) to record your results. You may wish to use the format shown in Table 7.13.1 or design your own.

As an extension, you may wish to rank each metal for each observation. For example, if you have four metals, rank their lustre from 1 to 4, with 1 being the shiniest and 4 being the least shiny.

ANALYSIS

1. Did any of the metals not conduct electricity? Sometimes if a light globe is used in these experiments, it does not light up for all metals. **Propose** a reason for this observation.
2. Rank the metals in order of hardness. Suggest a reason for the variability in results based on the structure of metals.
3. Metal used for structures such as bridges and buildings needs to be hard and not very malleable, as it must withstand massive forces to do its job. Which of the metals you tested would be most suitable for structural uses?
4. **Describe** problems that arose during this experiment that might affect your confidence when making a conclusion about the properties of metals.

CONCLUSION

Based on this experiment, make a justified conclusion about the properties of metals.

▼ **TABLE 7.13.1** Testing the properties of metals

Metal	Lustre observations	Hardness observations	Electrical conductivity observations	Malleability observations
Metal 1 name				
Metal 2 name				
Metal 3 name				
Metal 4 name				

7.14 Bonding in non-metals

BY THE END OF THIS MODULE, YOU WILL BE ABLE TO:

✓ describe how non-metal atoms bond to form covalent molecules and covalent network structures.

Interactive resource
Crossword: Bonding

GET THINKING

Pair up with a classmate and make a list of all the new chemistry words you can remember from this chapter. Have you added them all to your word list? Look for new words in this module to add to your word list.

Sharing electrons

Non-metal atoms occur mostly in groups 14–18 of the periodic table. You already know that group 14 elements such as carbon and silicon do not form ions, so how do they get a stable electron configuration?

Non-metal atoms such as chlorine can bond with other non-metal atoms such as oxygen. If you look at the electron configuration of chlorine (2,8,7) and oxygen (2,6), you will see they both need to gain electrons.

When a non-metal bonds with another non-metal, forming ions is not an option for achieving a stable electron configuration because there is no atom to give them electrons. Instead, non-metal atoms bond with other non-metal atoms by sharing valence electrons.

This means that an electron belongs to two atoms at the same time. Let's look at a simple example. Figure 7.14.1 shows two hydrogen atoms. Hydrogen atoms have one electron in their valence shell. They need to gain one more electron to achieve the stable configuration of the helium atom. Each hydrogen atom shares its one electron with the other hydrogen atom. The shared electrons belong to both atoms, so each has the same configuration as the stable helium.

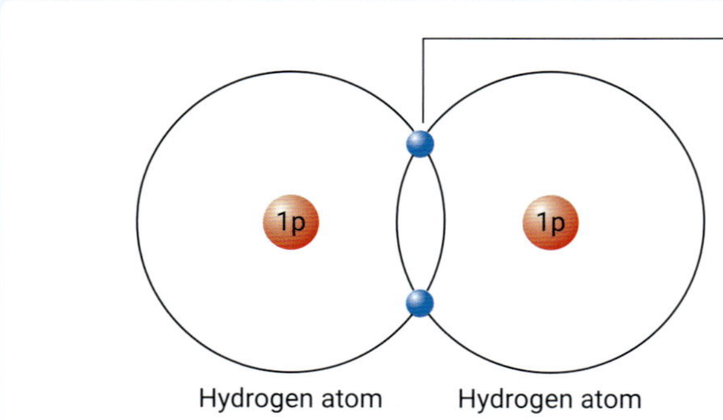

▲ **FIGURE 7.14.1** Hydrogen atoms share electrons to achieve a stable configuration.

Figure 7.14.2 shows the electrostatic force of attraction between the shared pair of negative electrons and the positive nucleus of each atom. This force of attraction is called a **covalent bond**. It is very strong and holds the two atoms together very tightly.

covalent bond
an electrostatic force of attraction between a shared pair of electrons and the nuclei of the atoms sharing the electrons

7.14

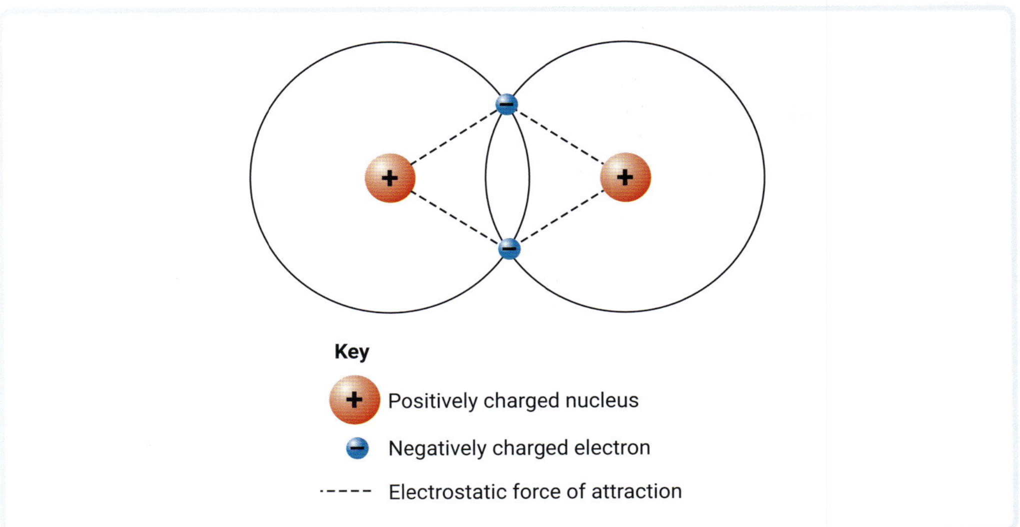

▲ **FIGURE 7.14.2** A covalent bond is an electrostatic force of attraction.

In some cases, atoms need to share more than one of their electrons. Figure 7.14.3 shows two oxygen atoms. Oxygen has the electron configuration 2,6, so it needs two more electrons to form a stable 2,8 configuration. Each oxygen atom shares two of its valence electrons with the other oxygen atom. The two shared pairs form a double covalent bond.

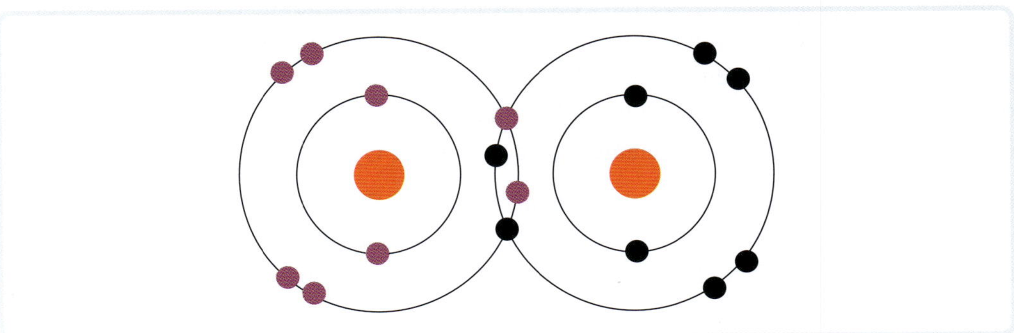

▲ **FIGURE 7.14.3** Oxygen forms a double covalent bond.

Figure 7.14.4 shows the seven valence electrons of chlorine and one valence electron of hydrogen. Because each atom needs one more electron to fill its valence shell, they contribute one electron each to the shared pair.

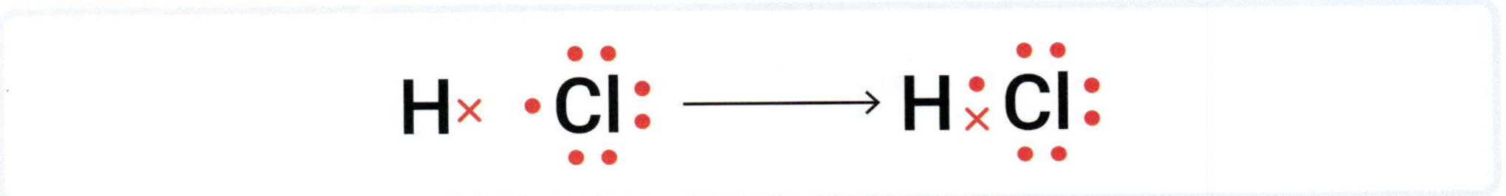

▲ **FIGURE 7.14.4** An electron dot diagram for HCl

structural formula
a two-dimensional representation of a molecule that shows bonds between atoms drawn as lines

There are a number of different ways to represent covalent bonds. Figure 7.14.5 shows (from left to right) an electron dot diagram, a **structural formula** and a physical model. Electron dot diagrams show the arrangement of valence electrons in covalent structures, with electrons being represented by dots or crosses.

H : H

Electron dot formula
The symbol H represents the nucleus of the hydrogen in this case.

H—H

Structural formula
Each pair of electrons in an electron dot formula is represented by a short line.

Physical model
This shows how atoms have partially merged.

▲ **FIGURE 7.14.5** Ways of representing covalent bonds (shared electron pairs).

A structural formula shows each covalent bond, or shared electron pair, as a single line. For the oxygen example in Figure 7.14.3, a double line would be used because there are two pairs of electrons (O=O).

A physical model is a simple representation of the shape of the resulting atom arrangement.

Forming covalent molecules

covalent molecule
a distinct structure formed when two or more non-metal atoms join through covalent bonding

When most non-metals join by covalent bonding, they form a structure known as a **covalent molecule**. A molecule is two or more atoms joined together by covalent bonds to make a distinct structure.

Unlike the lattice structures you have seen for metals and ionic substances, molecules have a fixed number of atoms in them. The hydrogen molecule formed in Figure 7.14.1 only ever has two hydrogen atoms and is represented by H_2 to show the two atoms. The oxygen molecule in Figure 7.14.3 is made from two atoms of oxygen (O_2). Molecules containing the same atoms are elements (such as H_2 or O_2).

Molecules can have different atoms in them, as seen in Figure 7.14.6. Water is a molecule where an oxygen atom forms covalent bonds with each of two hydrogen atoms. The molecule is represented by H_2O. Molecules with different atoms are compounds.

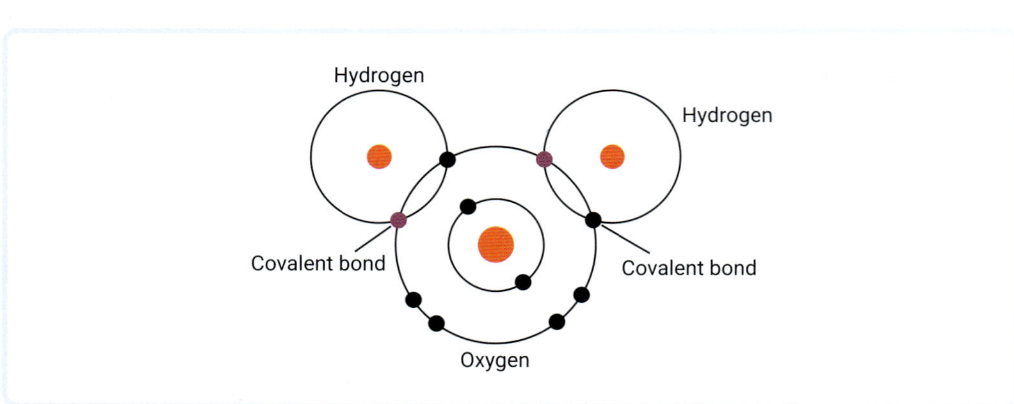

▲ **FIGURE 7.14.6** Water is an example of a covalent compound.

Forming covalent networks

The elements carbon and silicon can form a different type of covalent structure called a **covalent network**. Carbon and silicon both have four valence electrons, and this allows a symmetrically shaped lattice to form. Each carbon or silicon atom forms four bonds with other carbon or silicon atoms to form large lattices, as seen in Figure 7.14.7 and Figure 7.14.8.

covalent network
a structure formed when non-metal atoms join in a covalently bonded lattice

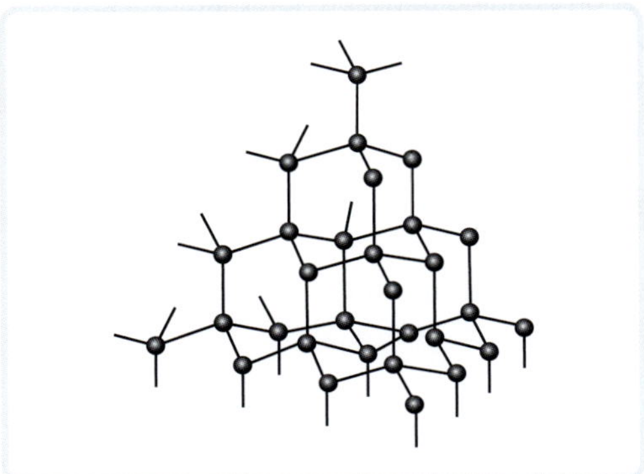

▲ **FIGURE 7.14.7** The structure of diamond; each carbon atom is bonded to four other carbon atoms in a continuous lattice.

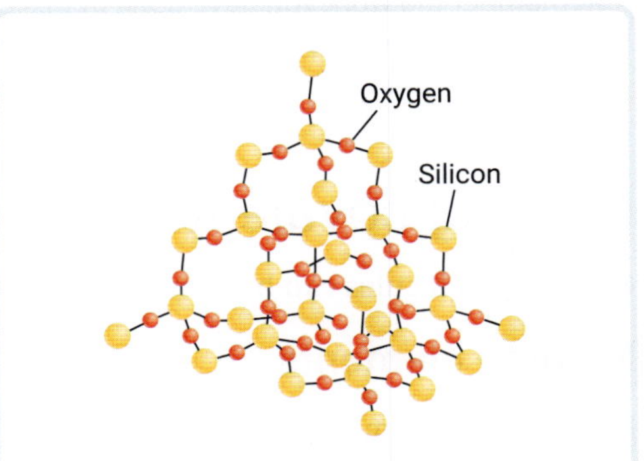

▲ **FIGURE 7.14.8** Silicon bonds with oxygen to form silicon dioxide in a continuous lattice.

7.14 LEARNING CHECK

1 Using fluorine atoms bonding together as an example, **explain** why non-metal atoms sometimes share electrons.

2 **Explain** why chlorine atoms form a single covalent bond, while oxygen atoms form a double covalent bond.

3 **Predict** how many pairs of electrons would be shared between two nitrogen atoms.

4 **Compare** the structure of a covalent molecule and a covalent lattice.

5 **Explain** why oxygen (O_2) and water (H_2O) are both molecules, but only oxygen is an element.

7.15 Formulas of covalent molecules

BY THE END OF THIS MODULE, YOU WILL BE ABLE TO:

- ✓ determine the formula of a covalent compound based on its name
- ✓ use the formula of a covalent compound to write its name.

GET THINKING

Remember the work you have done in this chapter on bonding between non-metals? How do covalent substances bond? What happens to the electrons in a covalent bond? What is the name of the structure that forms between two or more non-metal atoms?

Naming covalent molecules

Covalent molecules can be elements such as H_2 or O_2, or compounds such as H_2O or NO_2.

Naming elements

Molecules of elements can have a range of formulas, including H_2, O_2, P_5 or S_8. No matter how many atoms in the molecule, the element is usually just given the name of the atom.

However, there are some exceptions to this rule. For example, the element oxygen exists as O_2 (oxygen) and O_3 (trioxygen – common name 'ozone'). These alternative forms of the elements are called **allotropes**.

allotrope
an alternative form of an element with different physical and chemical properties

Naming compounds

The naming of compounds is best explained by looking at the molecules containing nitrogen and oxygen. There are a number of these compounds, including NO, NO_2, N_2O and N_2O_4.

Generally:

- the first element keeps its name, as on the periodic table
- the second element becomes an '-ide'
- if there is more than one molecule of an element, then prefixes are used (Table 7.15.1).

▼ TABLE 7.15.1 The prefixes used in covalent compounds

Number	Prefix
1	Mono-
2	Di-
3	Tri-
4	Tetra-
5	Penta-
6	Hexa-
7	Hepta-
8	Octa-
9	Nona-
10	Deca-

Therefore:

- NO = nitrogen monoxide
- NO_2 = nitrogen dioxide
- N_2O = dinitrogen monoxide
- N_2O_4 = dinitrogen tetroxide.

When oxygen is used after a prefix, the last 'a' is often removed. It is easier to see this with an example. A molecule XO_4 would have the ending 'tetraoxide' according to the table, but the 'a' in the middle is removed and it becomes 'tetroxide'. The same occurs with pentoxide.

Common names

A lot of covalent substances have common names, many of which you would be familiar with (Table 7.15.2). Common names do not tell you as much about the atoms in the molecule as the scientific name does.

▼ **TABLE 7.15.2** The common names of covalent substances

Formula	Scientific name	Common name
H_2O	Dihydrogen monoxide	Water
NH_3	Nitrogen trihydride	Ammonia
O_3	Trioxygen	Ozone
CH_4	Carbon tetrahydride	Methane

Writing formulas from the name

If you have the covalent molecule name, you simply need to reverse the process above to write the formula.

Carbon monoxide and carbon dioxide are two molecules containing carbon and oxygen. By the rules above:

- the formula for carbon monoxide is CO
- the formula for carbon dioxide is CO_2.

As you progress through your chemistry studies, you will become familiar with some of the common names of chemicals. Unfortunately, there is no trick for writing the formula of these molecules; you just have to learn them.

7.15 LEARNING CHECK

1 Write the formula and name for each of the structures shown in Figure 7.15.1. Use the key provided to **identify** the elements involved.

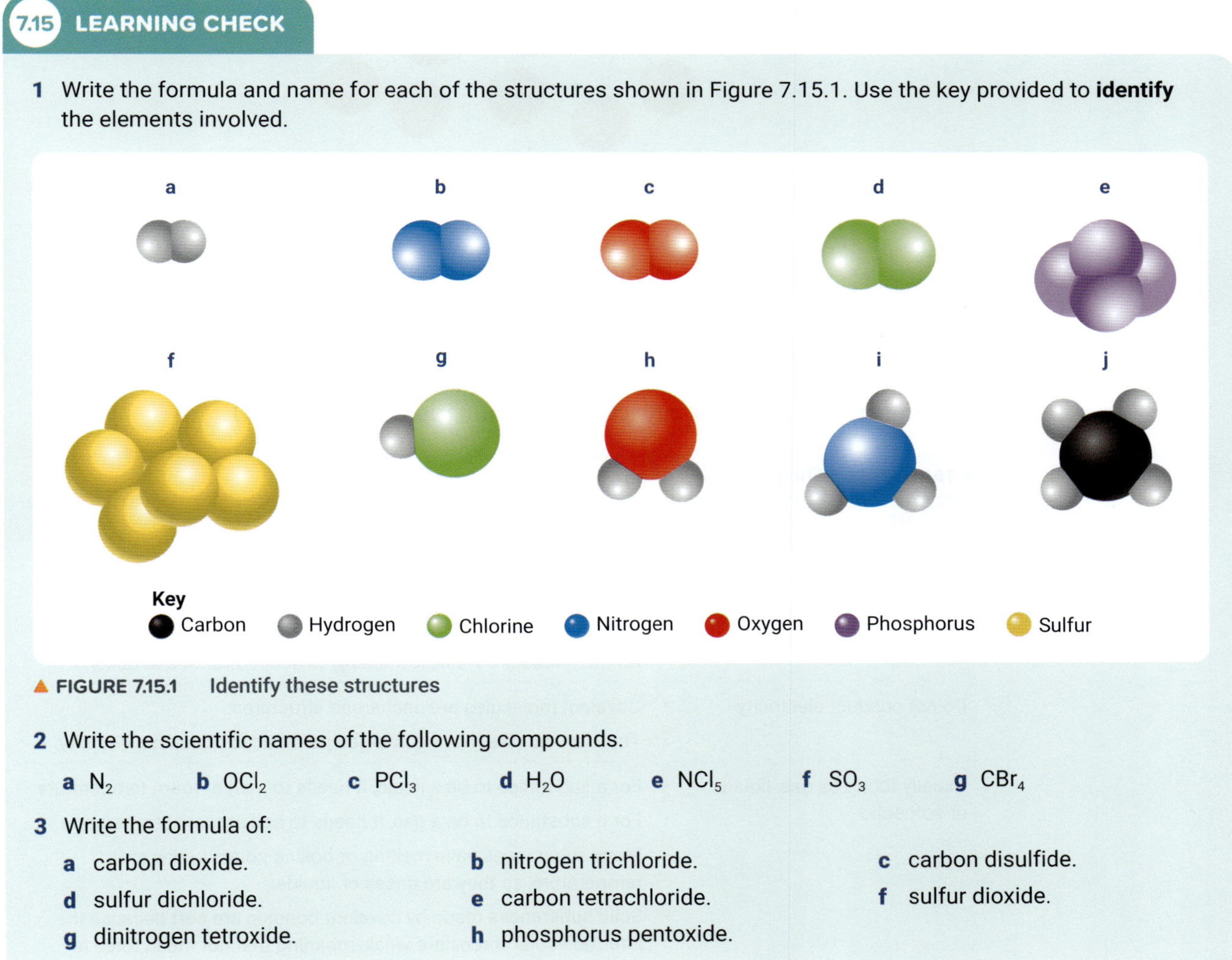

▲ **FIGURE 7.15.1** Identify these structures

2 Write the scientific names of the following compounds.

a N_2 **b** OCl_2 **c** PCl_3 **d** H_2O **e** NCl_5 **f** SO_3 **g** CBr_4

3 Write the formula of:

a carbon dioxide.
b nitrogen trichloride.
c carbon disulfide.
d sulfur dichloride.
e carbon tetrachloride.
f sulfur dioxide.
g dinitrogen tetroxide.
h phosphorus pentoxide.

7.16 Properties of covalent substances

BY THE END OF THIS MODULE, YOU WILL BE ABLE TO:

✓ use the structure of covalent molecules to explain their properties

✓ use the structure of covalent network substances to explain their properties.

GET THINKING

Can you recall the properties of ionic and metallic substances? Are there any properties that are similar? Or different? As you complete this module, compare the properties of covalent substances to ionic and metallic substances.

Properties of covalent molecules

intermolecular forces forces of attraction between covalent molecules

Covalent molecular substances have a common set of physical properties related to their structure. For example, the molecules in water are held together by **intermolecular forces**. 'Inter' means between, so intermolecular forces are between molecules. You can see this in Figure 7.16.1.

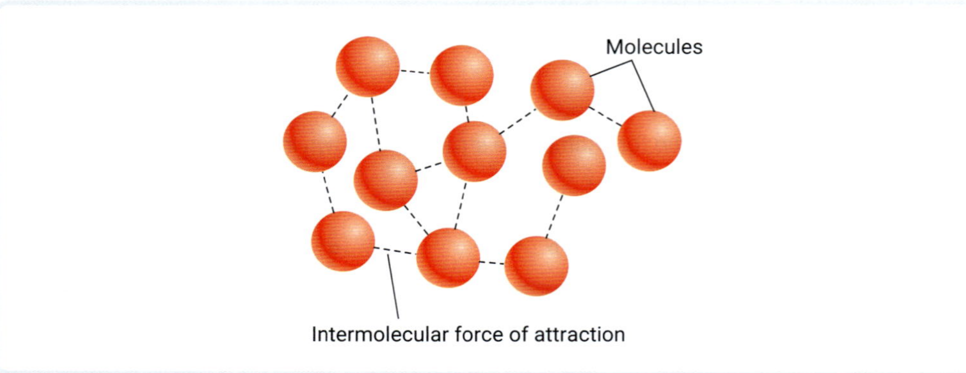

FIGURE 7.16.1 Intermolecular forces hold molecules together and are responsible for most physical properties.

The properties of covalent molecules are summarised in Table 7.16.1.

TABLE 7.16.1 The properties of covalent molecules

Property	Explanation
Low melting and boiling points	• Intermolecular forces are very weak. • To melt or boil a substance, you need to weaken or break these intermolecular forces. • As the forces are weak, low energy is needed for this to occur.
Do not conduct electricity	• Covalent molecules are uncharged structures. • There are no free-moving charged particles to conduct electricity.
Usually found as gas, liquid or soft solid	• For a substance to be a liquid, it needs to melt at room temperature. • For a substance to be a gas, it needs to boil at room temperature. • Many substances have melting or boiling points below room temperature, so they are gases or liquids. • Solid substances made by covalent bonding are soft because the intermolecular forces are weak, meaning that not much force is required to distort them.

Properties of covalent networks

Covalent networks do not have intermolecular forces because they do not have molecules! The properties of covalent network substances arise from the very strong covalent bonds between atoms. As these bonds are strong, they lead to the following properties.

- Covalent networks have high melting and boiling points because the covalent bonds are hard to weaken or break.
- They are hard and tough substances because the strong covalent bonds can withstand a lot of force.
- They do not conduct electricity because all the valence electrons are involved in covalent bonds and there are no charged particles free to conduct.

Diamond (carbon) is the hardest natural substance and cannot be scratched except by another diamond. Saws and drills in industry are diamond-tipped to enable them to cut through very hard and tough materials (Figure 7.16.2).

There are a few exceptions to these general properties. One example is graphite, which is composed of carbon atoms in a lattice structure, but with a different arrangement of the atoms than diamond. It has free electrons and can conduct electricity.

botazsolti/Shutterstock.com

▲ **FIGURE 7.16.2** Due to their hardness, diamonds are sometimes used in drills

7.16 LEARNING CHECK

1. Why do most covalent substances not conduct electricity?
2. **Explain** why covalent molecular substances are usually gases or liquids, while covalent network substances are always solids at room temperature.
3. **Contrast** the properties of covalent molecules and metals.
4. **Compare** the properties of ionic and covalent lattices. Try to find at least one similarity and one difference.
5. **Compare** the strength of the forces in ionic, metallic, covalent molecular and covalent network substances.

WORKING SCIENTIFICALLY

7.17 Report writing

SCIENCE SKILLS IN FOCUS

IN THIS MODULE, YOU WILL FOCUS ON LEARNING AND IMPROVING THESE SKILLS:

- describing observations about the physical properties of covalent substances
- communicating information about an experiment in the correct format for a scientific report.

Scientific reports are used to communicate information about a scientific investigation. The structure of a science report can vary depending on the experiment, but will usually include the following sections.

1 Aim – state the purpose of the investigation.
2 Hypothesis – state the hypothesis. A hypothesis is not always required, especially for simple observation or measurement experiments.
3 Materials and equipment – a detailed list of all equipment and chemicals required, including amounts of substances.
4 Risk assessment – a list of any hazards associated with the investigation and how you will minimise the risk.
5 Method – a detailed series of steps that can be followed to replicate the investigation. This is written in the past tense. It often includes a labelled diagram or photograph of the set-up.
6 Results – tables, graphs, photographs, calculations and written observations of what occurred in the investigation.
7 Discussion – an analysis and evaluation of the experiment. This section will vary depending on the investigation but can include:
   - the trend found, the scientific reason for it, and whether it supports the hypothesis
   - the accuracy, reliability and repeatability of the data
   - the validity of the method
   - ways that the investigation could be improved or extended.
8 Conclusion – state the trend and whether the results support the hypothesis. If you conducted a simple experiment, you can summarise your observations and any conclusions you form.

PROPERTIES OF COVALENT MOLECULES AND COVALENT NETWORKS

AIM

To investigate the physical properties of covalent molecules and covalent networks, including hardness, electrical conductivity and melting point

Safety

Take care when handling an electrical circuit. Always turn the power off at the main switch and unplug the power source when not in use.

Only use the circuit switch to connect the circuit for as long as you need to record the data.

Ensure your hands are dry.

Do not handle circuit components straight after use; they can get hot and may need time to cool down.

9780170491785

MATERIALS AND EQUIPMENT

- ☑ samples of substances with covalent bonds: quartz (covalent network), ethanol, candle wax and a sugar cube
- ☑ hotplate and stirrer (for the melting point test)
- ☑ electrical conductivity set-up (Figure 7.13.1)
- ☑ iron nail

PROCEDURE

HARDNESS TEST

Conduct this test as outlined in Module 7.13, using the iron nail.

ELECTRICAL CONDUCTIVITY TEST

1. For the solids sugar, quartz and candle wax, conduct this test as outlined in Module 7.13.
2. Your teacher will demonstrate the electrical conductivity of ethanol by carefully placing the alligator clips from the circuit into the ethanol liquid without letting the clips touch.

MELTING POINT

1. Place a small amount of the candle wax in a beaker and place it on a hotplate. Heat it until it starts to melt. Record the time taken for this to occur.
2. Repeat step 1 for the sugar cube and quartz sample. If no melting is seen after 3 minutes, end the experiment.

RESULTS

Construct a suitable table(s) to record your observations. You can use a table similar to the one suggested in Module 7.13 (Table 7.13.1).

ANALYSIS

Using the dot points in the Discussion section of the blue 'Science skills in focus' box of this module, write a discussion that **analyses** and **evaluates** this experiment.

CONCLUSION

Based on this experiment, make a justified conclusion about the properties of covalent molecules and networks.

SCIENTIFIC REPORT

Create a scientific report that includes all of the testing of properties that you conducted in Modules 7.11, 7.13 and 7.17. Use the sections outlined in the blue 'Science skills in focus' box in this module.

ABORIGINAL & TORRES STRAIT ISLANDER SCIENCE CONTEXTS

7.18 Developing pigments and dyes

IN THIS MODULE, YOU WILL:

✓ investigate the ways Aboriginal and Torres Strait Islander Peoples developed pigments and dyes through understanding of chemical reactions.

What is in paint?

Aboriginal and Torres Strait Islander Peoples use paint to record and communicate knowledge. We can see this in rock and bark paintings, body decorations and features added to implements. Paints are a liquid mixture containing pigments or dyes that leave a solid film on drying. Binders can be added to thicken paints and fixatives can be added to provide durability once paint is applied to a surface. Many traditional paints used by Aboriginal and Torres Strait Islander Peoples are prepared using chemical processes.

Calcination and pyrolysis

Calcination is the controlled thermal treatment of a solid compound. Calcination involves removing water, carbon dioxide and sulfur dioxide, and oxidising the substance through controlled heating. By calcination, some Aboriginal Peoples produce a paint pigment of yellow ochre (limonite). The yellow limonite is carefully heat treated to chemically convert the pigment into haematite, a red-brown pigment.

Sheralee Stoll/Alamy Stock Photo

▲ **FIGURE 7.18.1** Rock paintings in the Cathedral Caves in Carnarvon Gorge, Queensland, on the lands of the Bidjara and Karingbal Peoples

Pyrolysis is the decomposition of organic matter at high temperatures and with limited oxygen that changes the chemical composition of a material. Many Aboriginal and Torres Strait Islander Peoples produce charcoal by pyrolysis to make a pigment for paint. Particular woods are selected and exposed to high heat with limited oxygen to produce charcoals. Heating causes the water in the wood to evaporate before the wood is converted into carbon. The processes release compounds with a high vapour pressure, including combustible gases such as carbon monoxide, hydrogen and methane. This leaves a carbon-rich solid residue, charcoal. The charcoal is collected, ground into a powder and combined with other products to make paint.

Some examples of paints

Mineral pigments

Table 7.18.1 gives examples of some of the recorded types of mineral pigments used by Aboriginal Peoples.

Joe Sambono

▲ FIGURE 7.18.2 Making red ochre pigment

▼ TABLE 7.18.1 Mineral pigments used by some Aboriginal Peoples

Name and chemical component	Details
White huntite, $Mg_3Ca(CO_3)_4$	• Used by the Ngarinyin Peoples of the north-west Kimberley region (Western Australia) in rock painting • The pigment is powdery and can flake from surfaces • Sites were regularly revisited to repaint illustrations
Yellow ochre (limonite), $FeO(OH)$ Red ochre (haematite), Fe_2O_3	• Used by Tasmanian Aboriginal Peoples and mixed with animal fat, blood, saliva or water • Is more resistant to degradation than some other mineral pigments
Pyrolusite, MnO_2	• Used in rock paintings of the Cathedral Caves in the Carnarvon Gorge by the Bidjara and Karingbal Peoples

Plant pigments

Some Aboriginal and Torres Strait Islander Peoples use the pigments from plants to produce paints that are used for decorating implements or as body paint, as shown in Table 7.18.2.

alybaba/Shutterstock.com

▲ FIGURE 7.18.3 The bright red fruits of the saltbush have long been used to make face paints.

▼ **TABLE 7.18.2** Plant pigments used by some Aboriginal Peoples

Name	Details
Red fruits of saltbush	• Used to make skin paint, e.g. by Arrernte Peoples (Northern Territory) and the Wurundjeri Peoples (Victoria)
Bark of mountain ash tree	• Used to produce a red-brown paint for decorating implements such as spears, e.g. by the Anguthimri People (northern Queensland)

Binders

Binders are chemicals used to stick materials or objects together. Aboriginal and Torres Strait Islander Peoples use a variety of substances as binders, as shown in Table 7.18.3.

▼ **TABLE 7.18.3** Binders used by some Aboriginal Peoples

Name	Details
Sap of the native orchid	• A sticky and viscous substance used as a binder, e.g. by the Anindilyakwa Peoples of Groote Eylandt (Northern Territory)
Wax or honey of native bees Turtle egg yolk	• Used as binders to reduce paint flaking from surfaces, e.g. by Tiwi Peoples (Northern Territory)
Animal fats	• Used by the Barngarla Peoples (South Australia) mixed with charcoal as body paint, which makes the paint less soluble in water

Avalepsap/Adobe Stock Photos

▲ **FIGURE 7.18.4** Nuts of the candlenut tree are used to make oil, which is used as a fixative.

Fixatives

Fixatives are substances used to fix other chemicals, such as paints, so they last longer. Some examples of fixatives used by Aboriginal and Torres Strait Islander Peoples are shown in Table 7.18.4.

▼ **TABLE 7.18.4** Fixatives used by some Aboriginal Peoples

Name	Details
Emu fat	• Used by the Ngaatjatjarra People (Western Australia) as a hydrophobic fixative for rock painting
Resinous material from yellow tea tree	• Used by the Anguthimri People (northern Queensland), who mixed the resin with pigment, warmed the paint and applied it to permanently paint implements
Candlenut oil	• Used by the Walmbaria Peoples (northern Queensland) to fix paints to implements

Making paints

ACTIVITY

Materials and equipment

- ochre or ochre powder
- charcoal
- water
- egg yolk
- lard or oil
- water in a spray bottle
- mortar and pestle
- 3 stones
- paintbrushes
- 3 plastic containers

Procedure

1. Choose either ochre or charcoal as your pigment.
2. Grind the ochre or charcoal in the mortar and pestle until it is a fine powder.
3. Label three containers 'Egg yolk', 'Water' and 'Fat/oil'.
4. Add a teaspoon of the powder to each of three containers.
5. Add 5 mL of water to each container and mix.
6. Add an egg yolk to the container labelled 'Egg yolk'. Mix.
7. Add 5 mL of oil or a tablespoon of lard to the container labelled 'Fat/oil'. Mix.
8. Add 5 mL of water to the container labelled 'Water'. Mix.
9. Paint the entire surface of each of the stones with one of the mixtures. Make sure you label the stones.
10. Allow the paint to dry.
11. Spray the stones with a fine water mist. Record your immediate observations. What happened to each of the mixtures? What does the mist of water represent?
12. Place the stones outside.
13. Make a table for observations and record what you notice every day for a fortnight.
14. Look for changes to the paint, such as flaking, fading, cracking or powdering.
15. Record observations about the weather.

Analysis

At the end of the fortnight, consider your results.

1. Which paint stayed the brightest?
2. Which paint lasted the longest?
3. How did the weather affect each of the paints?
4. What other non-toxic natural paint components could you use from your local environment?
5. What does this experiment help you understand about Aboriginal and Torres Strait Islander Peoples' achievements in producing long-lasting paint?

SCIENCE IN CONTEXT

7.19 Extracting and refining minerals and resources in Australia

BY THE END OF THIS MODULE, YOU WILL BE ABLE TO:

- ✓ identify the different types of resources extracted in Australia
- ✓ describe the environmental issues involved with mining resources
- ✓ understand factors involved in the production of useful materials from minerals and resources

Types of resources

Resources can be categorised into three main types: energy resources, metals and other resources.

The three groups of resources are all finite. This means that once they run out, there won't be any more. Therefore they are also referred to as non-renewable resources. As you will learn elsewhere in your Stage 5 studies, there is a lot of work being done on developing renewable energy sources.

Energy resources

Energy resources include coal, natural gas, crude oil and uranium. Coal is used for energy production in the majority of Australian electricity generation stations. Natural gas is used in heating, cooking and as a vehicle fuel. Uranium is mostly exported overseas for use in nuclear power reactors for electricity generation.

Bjoern Wylezich/Shutterstock.com

small smiles/Shutterstock.com

▲ FIGURE 7.19.1 (a) Uranium is used in nuclear reactors and (b) coal in coal-fired power plants for electricity generation.

Metals

Australia extracts a wide variety of extremely useful metals for use here and for export. Metals you would be most familiar with include aluminium (cans and plane panels), copper (electrical wiring), zinc (fencing and parts for cars), nickel and lithium (batteries) and iron (fencing, building materials). We also produce many other useful metals you may not have heard of, including cobalt, titanium, zircon, molybdenum, tantalum and vanadium.

Other resources

Other resources include such things as the components of fertilisers (potassium and nitrogen), and gemstones such as diamond and opal.

9780170491785

Extraction of resources

The majority of resources and minerals produced in Australia are extracted by mining. There are two main types of mining used in Australia: underground mining and surface mining.

Underground mining

Underground mining involves digging deep tunnels and shafts into the earth (Figure 7.19.2). This is a costly and often dangerous method of mining.

Alexandre.ROSA/Shutterstock.com

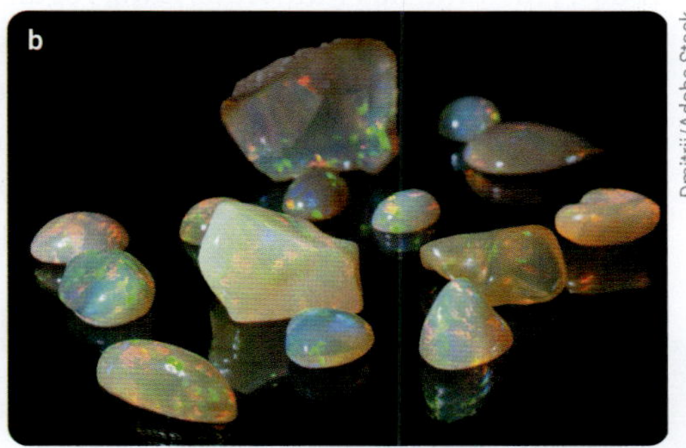

Dmitrii/Adobe Stock

▲ **FIGURE 7.19.2** (a) Opal mining in Coober Pedy, South Australia, is done in underground tunnels. (b) Opal gemstones are primarily used in jewellery.

Surface mining

Surface mining involves extracting resources close to the surface (Figure 7.19.3). This is usually cheaper and safer than underground mining. There are two main types of surface mining. Strip mining removes the top layer of soil to reveal the minerals below. Open-cut mining involves digging a large, open pit to remove the resources.

Jason Bennee/Adobe Stock

▲ **FIGURE 7.19.3** Open cut mining involves digging large holes and removing resources to be processed.

Mining risks

Mining in any form comes with potentially serious environmental and safety risks:

- Mine collapses and accidents can cause serious injury and death to mine workers.
- Leaking of toxic chemicals or materials from the mine site into surrounding waterways and ecosystems can poison soils, wildlife, plant life and/or humans.
- Mine sites often destroy ecosystems such as forests and grasslands when roads and rail lines are built to move in equipment and through the removal of land and resources.

Today, Australia has numerous laws, regulations and guidelines that ensure that mining is conducted in a safe manner. There are also laws that govern potential environmental impacts. For example, when mines are closed down, their operators are required to refill tunnels and pits, replant vegetation and return the area as close as possible to the original ecosystem. There are also strict laws around use and storage of mining products, chemicals and potentially harmful materials that could be accidentally released.

crbellette/Shutterstock.com

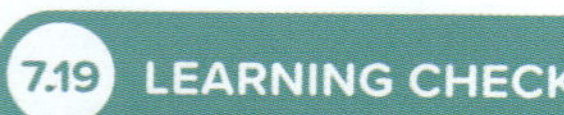

▲ **FIGURE 7.19.4** In Tasmania, hydroelectricity is used to produce power for refining aluminium metal, to reduce the environmental impact of this process.

Production of useful materials

Most materials mined from the earth are not in the pure form required for their final use. As outlined in Module 7.1, aluminium is mined as bauxite ore, then refined to the mineral aluminium oxide, then further refined to aluminium metal. Almost all metals have similar processes that need to occur.

Most of the processes involved in the extraction of final product from ores involve either use of chemical reactions to separate the components, or energy to convert the metal that is in the mineral into a pure metal. The energy required to refine a metal is usually very high, so many producers try to use renewable energy sources such as hydroelectricity where this is available.

7.19 LEARNING CHECK

1 **Outline** the different types of resources Australia produces.
2 **Describe** the possible environmental impacts of mining on the surrounding environment.
3 **Assess** the advantages and disadvantages of the use of mining in Australia to produce everyday materials for our use.

9780170491785

7 REVIEW

REMEMBERING

1 How many electrons fit into the third electron shell?

2 **Describe**, using an example, the difference between an ore and a mineral.

3 **State** the formulas of:
- a sodium chloride.
- b magnesium fluoride.
- c aluminium oxide.
- d copper(II) nitrate.
- e sulfur trioxide.
- f carbon tetrachloride.
- g calcium hydroxide.

4 **State** the names of the following.
- a $LiCl$
- b MgO
- c NCl_3
- d CF_4

5 **Describe** how a covalent bond forms.

UNDERSTANDING

6 **Determine** the electron configuration of:
- a hydrogen.
- b calcium.
- c phosphorus.

7 **Explain** why most elements do not occur as single atoms.

8 **Explain** why ions of the group 2 elements have a 2+ charge.

9 The melting point for most metals varies from around 300°C to over 3000°C. **Explain** why most metals have medium to high melting points and suggest a reason why there is such a range in melting points between different metals.

10 Use diagrams and examples to **summarise** the difference between covalent molecules and covalent network structures.

11 **Explain** the difference between electronegativity and ionisation energy, using sodium and chlorine as examples.

APPLYING

12 Using electron configurations for beryllium, calcium and magnesium, **explain** how many valence electrons are found in group 2 elements.

13 Helium is an unreactive gas. **Identify** one reason why it does fit into group 18, and one reason it doesn't fit into group 18.

14 **Describe** a use of metals where the property of malleability would be required.

15 Carbon dioxide has a low boiling point and is a gas at room temperature. Diamond (carbon) is the hardest natural substance. **Explain** why substances involving the same type of atom (carbon) can have such different properties.

16 By considering the advantages and disadvantages of mining metals such as aluminium and resources such as coal, **explain** why mining is important to Australia.

17 **Draw** an electron transfer diagram to illustrate how electrons move in the formation of sodium oxide.

ANALYSING

18 A student claims the electron configuration of an atom is 2,8,10. The teacher points out there is a mistake in the claim.
- a **Identify** the element they are using and write the correct electron configuration.
- b **Explain** the mistake the student made in writing the configuration.

19 Ionic substances have high melting points, while covalent molecules have low melting points. **Explain** the relationship between bond strength and melting point, using the examples given.

20 Chlorine atoms can either gain electrons or share electrons to achieve a stable electron configuration. **Identify** whether potassium must gain or share electrons to achieve a stable electron configuration and **explain** why this is so.

21 **Differentiate** why an ionic solid cannot conduct electricity, but an ionic solution can.

EVALUATING

22 The table below shows four metals and some of their properties.

Metal	Melting point	Electrical conductivity	Density
A	Medium	Low	Low
B	High	High	Low
C	Low	High	Low
D	Low	Low	High

- **a** **Justify** why metal A is used to build aeroplanes.
- **b** **Predict** and **explain** whether metal B or C would be better for electrical wiring.
- **c** Metal D is used for construction in high-rise buildings. **Describe** what other properties it would be important for this metal to have.

23 A student says that they can **determine** the charge on the ions of any element from the periodic table. **Assess** their claim by commenting on the strengths and weaknesses of this statement.

24 **Choose** one of the items listed below and **judge** whether the materials used to build it are likely to be ionic, metallic, covalent molecular or a covalent network. You must refer to the bonding and physical properties in your response.

surfboard, tennis racquet (frame and strings), chopping knife (blade and handle), plastic bag

CREATING

25 **Create** a mind map that links the following keywords.

atom, ion, cation, anion, periodic table, group, proton, electron, ionic bonding, covalent bonding, metallic bonding, covalent molecule, covalent network

Join related words and include annotations to show why you are joining them.

26 **Construct** an infographic to show the production of a metal in Australia. You should include information on potential environmental issues involved in the production, the uses of the metal and where it is mined and refined.

9780170491785

SCIENCE IN DEPTH STUDY

1 Connect what you've learned

In this chapter you've learned about different resources that are mined and processed in Australia. Pick one of the resources you have learned about and think about how its extraction and/or production might affect the environment.

2 Check your thinking

Do some research into the specific environmental issues surrounding the production of your chosen resource. Research how it is mined and how it is processed.

3 Get into action

Assess the overall impact to the environment of extracting and producing this resource. Are these factors outweighed by its usefulness in society? What criteria will you use to make an overall conclusion?

4 Communicate

Document your findings in a written scientific report. You should include some or all of the following sections:

- information about the resource, where it is mined and how it is used in its final form
- an assessment of the advantages and disadvantages of the mining process
- an explanation of the production and/or refining process, including environmental issues
- an overall judgement about the environmental impact of producing this resource.

8 Organic compounds

8.1 Organic and inorganic substances (p. 298)

Organic substances contain carbon, while inorganic substances are not primarily carbon-based. Inorganic substances include metals, ionic salts and non-metal elements.

8.2 Alkanes (p. 300)

Alkanes are hydrocarbon molecules with only carbon and hydrogen atoms. In an alkane, the carbon atoms are connected by single bonds. Alkanes are named according to a fixed set of rules.

8.3 Types of organic compounds (p. 303)

Alkenes are hydrocarbons with at least one double bond between the carbon atoms. Other types of organic compounds include alcohols and carboxylic acids.

8.4 Uses of hydrocarbons: fossil fuels (p. 306)

Fossil fuels provide a large number of products, including fuel, materials and plastics. Our use of fossil fuels has changed significantly as technology has progressed.

8.5 Production of hydrocarbons from crude oil (p. 308)

Fractional distillation is used to separate the components of crude oil. These components have a variety of uses, including fuel, lubricants and road surfaces.

8.6 Combustion (p. 312)

Combustion is a chemical reaction where fuels combine with oxygen to release energy in the form of heat. Combustion can be either complete or incomplete.

8.7 WORKING SCIENTIFICALLY: **Problem-solving** (p. 315)

Do all fuels produce the same amount of energy?

8.8 Polymers (p. 317)

Polymers are made from building blocks called monomers. Different types of polymers have different uses.

8.9 Properties of polymers (p. 320)

Polymers are specifically manufactured to have a range of physical and chemical properties that make them useful.

8.10 WORKING SCIENTIFICALLY: **Recording observations of physical properties** (p. 323)

Observing and measuring properties of polymers

8.11 Environmental impacts of polymers (p. 325)

Polymer waste can build up in marine animals. Biodegradable polymers are being introduced as alternatives to traditional plastic.

8.12 WORKING SCIENTIFICALLY: **Writing for your audience** (p. 329)

Biodegradability of polymers

8.13 SCIENCE IN CONTEXT: **How can we solve the plastic problem?** (p. 331)

Polymer waste is a serious issue around the world. Scientists are investigating using microbes to break down polymer waste.

9780170491785

SCIENCE IN DEPTH

#8

▲ FIGURE 8.0.1 Polymer pollution is a major environmental concern, with millions of tonnes of plastic thrown away each year.

Polymers are in objects we use in almost every aspect of our lives, such as packaging, building materials and the clothes we wear. We use a wide variety of polymers including plastics, resins and materials such as polyester. The vast majority of polymers are disposed of in landfills or discarded as rubbish once we finish using them.

There are major environmental concerns about the levels of discarded plastic, especially in the ocean. Plastics can take hundreds or even thousands of years to break down, so the levels of waste are constantly increasing.

- ▶ **How does plastic get into the oceans?**
- ▶ **What other environments are significantly impacted by polymer waste?**
- ▶ **Is there plastic pollution in your local area?**
- ▶ **What can be done to minimise the amount of plastic waste?**

#8 DIVE INTO SCIENCE!

At the end of this chapter, you can complete Science in Depth Study #8. You can use the information you learn in this chapter to complete the project.

Assessments
- Prior knowledge quiz
- Chapter review questions
- End-of-chapter test
- Depth study: Field survey

Videos
- Science skills in a minute: Problem-solving in science **(8.7)**; Science communication **(8.12)**
- Video activities: Oxygen and combustion **(8.6)**; The diverse world of polymers **(8.8)**

Science skills resources
- Science skills in practice: Problem-solving **(8.7)**; Science communication **(8.12)**

Interactive and other resources
- Simulation: Modelling hydrocarbons **(8.2)**
- Crossword: Polymers **(8.9)**
- Quiz: Fossil fuels **(8.4)**
- Activity sheets: Making alkanes **(8.2)**

Nelson MindTap

To access resources above, visit **cengage.com.au/nelsonmindtap**

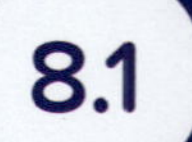

8.1 Organic and inorganic substances

BY THE END OF THIS MODULE, YOU WILL BE ABLE TO:

✓ distinguish between organic and inorganic chemical substances.

GET THINKING

The term 'organic' is a popular word used to describe categories of food. What is the difference between organic vegetables and those that are not labelled as organic? In chemistry, the word 'organic' has a very different meaning compared to when it is used to describe food. As you move through this module, ask yourself whether the term 'organic' is being used correctly when applied to food.

Differences between organic and inorganic substances

organic
covalent molecules that contain carbon

inorganic
substances that are not primarily carbon based, such as metals and ionic salts

In chemical terms, all substances are divided into one of two categories. **Organic** substances contain carbon, often bonded to hydrogen (or other atoms) through covalent bonds. **Inorganic** substances include a broader range of materials, such as ionic salts, metals, non-metal elements and covalent network substances.

It is important to note that these 'rules' are general rules only. For example, carbon dioxide contains carbon and is a covalent molecule but is not considered an organic substance.

Organic substances

You will already be familiar with many organic substances, including DNA, crude oil, glucose, ethanol and methane (Figure 8.1.1). One of the key characteristics of organic substances is that they are all covalent molecules that contain carbon. DNA is a very large covalent molecule containing carbon, hydrogen, oxygen, phosphorus and nitrogen atoms. Glucose and ethanol are both molecules with carbon, hydrogen and oxygen, while methane contains only carbon and hydrogen.

Most fuels we use (including petrol and diesel) come from crude oil, which is a fossil fuel made from the remains of ancient organisms that lived millions of years ago. The process of formation of fossil fuels is explored in Module 8.4. In Module 8.5, we will discuss the process of turning crude oil into fuels.

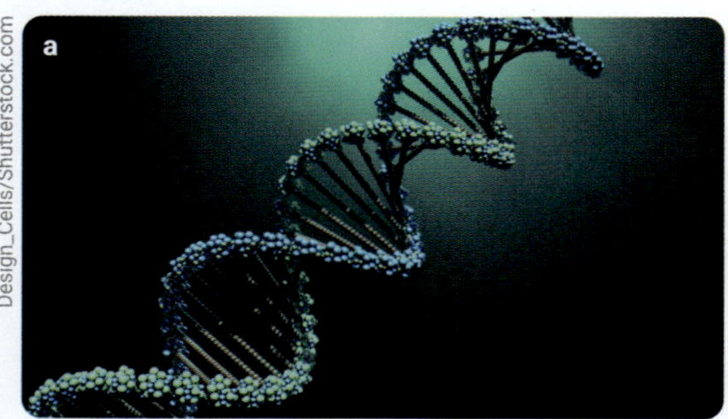

Design_Cells/Shutterstock.com

Anan Kaewkhammul/Shutterstock.com

▲ FIGURE 8.1.1 Some examples of organic substances include (a) DNA and (b) crude oil.

Inorganic substances

Inorganic substances include all metals, ionic salts (such as sodium chloride or calcium carbonate) and non-metal elements (such as chlorine or sulfur, Figure 8.1.2). Some substances are classed as inorganic even though they follow some of the 'rules' of organic substances. For example, diamonds are pure carbon arranged in a covalent network structure. While they contain carbon, they are still classed as an inorganic substance.

RHJPhtotos/Shutterstock.com

▲ **FIGURE 8.1.2** Sulfur, a non-metal element, is an inorganic substance.

8.1 LEARNING CHECK

1. Do an internet search and **list** five organic and five inorganic substances you might use every day.
2. **Define** the 'rules' that **describe** most organic substances.
3. **Describe** one example of an inorganic substance that fits some of the 'rules' for organic substances.

8.2 Alkanes

BY THE END OF THIS MODULE, YOU WILL BE ABLE TO:

- ✓ draw and name a simple alkane that has up to eight carbons
- ✓ represent alkanes using different types of diagrams
- ✓ understand why there are so many different organic molecules.

Interactive resource
Simulation: Modelling hydrocarbons

Other resource
Activity sheet: Making alkanes

GET THINKING

The names of some common alkanes include pentane and octane. The start of each of these names ('pent' and 'oct') should be familiar to you. Where have you heard these terms before? Give at least two examples of where you know these words from.

Hydrocarbons

hydrocarbon
organic molecules that contain only hydrogen and carbon

Hydrocarbons are molecules that contain only hydrogen and carbon. They are used extensively as fuels, including methane (cooking gas), propane (barbecue gas) and octane (in petrol, Figure 8.2.1).

Kim Britten/Shutterstock.com

▲ **FIGURE 8.2.1** Petrol contains octane.

```
        H
        |
    H - C - H
  H     |    H   H   H   H
  |     |    |   |   |   |
H-C  -  C  - C - C - C - C - H
  |     |    |   |   |   |
  H     H    |   H   H   H
         H - C - H
             |
         H - C - H
             |
             H
```

▲ **FIGURE 8.2.2** Carbon can bond to form long chains and branched structures. Each carbon bonds to up to four other atoms.

In a hydrocarbon, carbon atoms bond to other carbon or hydrogen atoms by sharing electrons, often to form long chains (Figure 8.2.2). Each carbon forms up to four bonds with other atoms. This means the carbon could bond with one carbon and three hydrogen atoms, or bond with two carbons and two hydrogen atoms. It could also bond with four hydrogen atoms, or even with four other carbon atoms. The only rule it must follow is that every carbon must have four bonds.

The carbon chains can be anywhere from one to millions of carbons long. They can have no branches (called straight chains), have some branching, or have many branches in different places. The first and last carbons in a chain can even connect, forming cyclic structures. All these combinations mean there is a very large number of different types of organic molecules.

9780170491785

What are alkanes?

Alkanes are a group of organic chemicals that contain only carbon and hydrogen, connected by single bonds. (Carbon atoms can also form double or triple bonds by sharing multiple pairs of electrons to form hydrocarbons known as alkenes and alkynes.)

The simplest alkanes are called **straight-chain alkanes**. Straight-chain alkanes arise when the carbon chain does not have any branches. Figure 8.2.3 shows the first four straight-chain alkanes. The first alkane is methane. The prefix 'meth-' in chemistry means one carbon. As methane has only one carbon, it is bonded to four hydrogens to complete its four bonds, giving rise to the formula CH_4.

Figure 8.2.3 shows different models used to present organic molecules in diagrams. **Space-filling molecules** (Figure 8.2.3c) are the closest representation to what the molecule looks like. A space-filling molecule has atoms represented by spheres, with the size of the atom related to the size of the sphere. **Ball-and-stick models** (Figure 8.2.3a) are used to show how the atoms bond to each other in 3D, with the atoms represented as balls and the bonds between atoms represented as sticks. However, neither of these models clearly shows the type of atoms. Figure 8.2.3b shows **structural formulas**. This is the most commonly used model to represent organic molecules. A structural formula shows the types of atom, represented by their chemical symbols, joined by lines to represent the bonds present.

alkane
a hydrocarbon containing only carbon and hydrogen, connected by single bonds

straight-chain alkane
a single, continuous alkane carbon chain with no branches

space-filling molecule
a representation of molecules that approximates their physical appearance, where atoms are represented as spheres of different sizes

ball-and-stick model
a three-dimensional representation of a molecule that uses balls to represent atoms (often coloured differently for each element) and sticks to represent the bonds between them

structural formula
a two-dimensional representation of a molecule that shows bonds between atoms drawn as lines

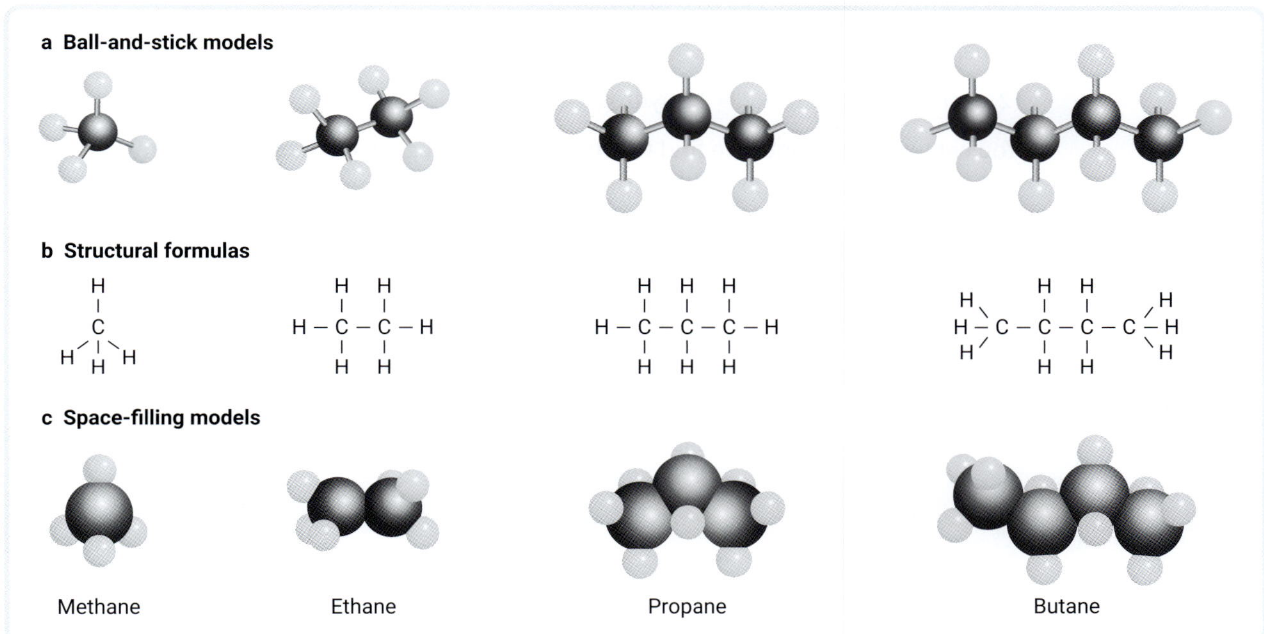

▲ **FIGURE 8.2.3** The first four alkanes represented using (a) ball-and-stick models, (b) structural formulas and (c) space-filling models.

All organic molecules are named according to a set of rules created by the International Union of Pure and Applied Chemistry (IUPAC). These rules are used by scientists all over the world, so that no matter what language you speak, the names of organic substances are consistent. The system of naming chemicals, including organic substances, is known as chemical **nomenclature**.

nomenclature
a system of names; for example, the naming system used for chemicals

Weblink
IUPAC

The first eight alkanes and their chemical formulas are shown Table 8.2.1. The third column shows the prefix that is used to identify the number of carbons. For example, when the alkane has five carbons, it has the prefix 'pent-', while an alkane with eight carbons has the prefix 'oct-'. You will need to learn these prefixes. All alkanes have '-ane' at the end of their name to show they are alkanes.

▼ **TABLE 8.2.1** The first eight alkanes

Alkane	Number of carbons	Name prefix	Chemical formula
Methane	1	Meth-	CH_4
Ethane	2	Eth-	C_2H_6
Propane	3	Prop-	C_3H_8
Butane	4	But-	C_4H_{10}
Pentane	5	Pent-	C_5H_{12}
Hexane	6	Hex-	C_6H_{14}
Heptane	7	Hept-	C_7H_{16}
Octane	8	Oct-	C_8H_{18}

Can you find a pattern in the chemical formula for alkanes? All the alkanes have a chemical formula that fits the rule: C_nH_{2n+2}.

For example, pentane has five carbons.

Thus, using $n = 5$, you can work out that it has:

$(2 \times n) + 2 = (2 \times 5) + 2 = 12$ hydrogens

You can use this formula to find the number of hydrogens for alkanes with any number of carbons. If there were 20 carbons, the number of hydrogens would be $(2 \times 20) + 2 = 42$.

8.2 LEARNING CHECK

1 **Identify** the features all hydrocarbons have in common.
2 Using methane as an example, **describe** the different ways that organic substances can be represented in drawings.
3 **Draw** the structural formula of the following alkanes.
   a Pentane
   b Ethane
   c Heptane
4 **Explain** why alkanes are hydrocarbons, but not all hydrocarbons are alkanes.
5 **Explain** why carbon can form a large number of different molecules.

9780170491785

8.3 Types of organic compounds

BY THE END OF THIS MODULE, YOU WILL BE ABLE TO:

✓ describe different groups of organic molecules
✓ identify the names of some simple organic molecules.

GET THINKING

Organic substances can be classified into different groups based on their structures. Many areas of society, like sports and transport, are divided into groups based on their physical features and actions. Can you think of any other areas of society and/or science where groups are used? Share your ideas with another student.

There are a lot of different groups of organic molecules. These groups include molecules that are essential to human life because they are used to create everyday products, form a part of our food and drinks, or occur naturally in the environment.

Alkenes

In Module 8.2, you learned about alkanes, one of the hydrocarbon groups. There is another major group of hydrocarbons called **alkenes**. Alkenes are similar to alkanes, except instead of containing only single bonds, alkenes have at least one double bond between two of the carbon atoms.

For example, Figure 8.3.1 shows the **ethene** molecule. The carbon atoms are joined by two bonds. Each carbon atom in ethene also bonds to two hydrogens, resulting in four bonds per carbon. Ethene is an important chemical in society, because it is used to create several plastics, such as polyethylene, polystyrene and polyvinyl chloride (PVC), that we use for a wide range of applications. These plastics are all examples of polymers, which will be explored in Module 8.8.

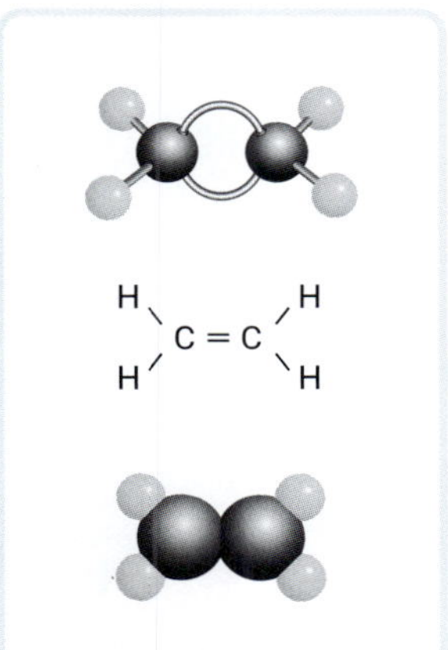

▲ **FIGURE 8.3.1** Alkenes such as ethene have a double bond between carbon atoms.

alkene
a hydrocarbon that contains one or more double bonds between carbon atoms

ethene
an alkane with two carbon atoms

Alcohols

Another group of organic substances is **alcohols**, such as ethanol. Alcohols are not classified as hydrocarbons because they also contain oxygen atoms.

alcohol
a group of organic molecules containing an -OH group, such as ethanol

Alcohols have a fixed structure that classifies them into their unique group: they all contain an -OH group of atoms, as seen in the ethanol molecule in Figure 8.3.2. Ethanol is an alcohol with two carbons. You can tell this from the 'eth-' at the start of the name – the same prefixes are used for naming alcohols as for the alkanes (see Module 8.2). The name of the molecule ends in '-ol', which is the same for all molecules in the alcohol group.

▲ **FIGURE 8.3.2** All alcohol molecules have an -OH group as part of the molecule.

Ethanol is used for cleaning, in vehicle fuel (Figure 8.3.3), as a disinfectant, and is in alcoholic drinks. There are also more complex alcohols that you may be familiar with, including menthol (which is used as a flavouring similar to peppermint) and cholesterol (which can be found in human arteries and can be dangerous to a person's health if allowed to build up).

▲ **FIGURE 8.3.3** Many petrol stations in Australia sell E10 fuel, which contains ethanol.

Carboxylic acids

carboxylic acid
a group of organic molecules containing a -COOH group, such as ethanoic acid

Another common group of organic molecules is the **carboxylic acids**. Every carboxylic acid has a -COOH group of atoms, as shown, for example, in ethanoic acid (Figure 8.3.4).

H
|
H – C – C(=O)–OH
|
H

Carboxylic acid group

▲ **FIGURE 8.3.4** Ethanoic acid. All carboxylic acid molecules contain a -COOH group.

Some common carboxylic acids include ethanoic acid, citric acid and lactic acid. Ethanoic acid is found in vinegar that is used in cooking and for pickling vegetables (Figure 8.3.5). Citric acid is found in citrus fruits such as oranges, lemons and limes (Figure 8.3.6). When you go running or play sport, your muscles create lactic acid, which is responsible for the burning you feel in your muscles during and after strenuous exercise.

9780170491785

zi3000/Shutterstock.com

▲ FIGURE 8.3.5 Vinegar contains ethanoic acid and is used to pickle vegetables.

New Africa/Shutterstock.com

▲ FIGURE 8.3.6 Citrus fruits contain citric acid, a type of carboxylic acid.

8.3 LEARNING CHECK

1 **Contrast** the structure of alkanes and alkenes.
2 **Draw** the structure of ethene and **explain** why it is an important molecule for society.
3 **Explain** why alcohols and carboxylic acids are not hydrocarbons.
4 **Draw** the structural formulas of ethanol and ethanoic acid and **describe** where they are used.
5 **Compare** the structural formulas of ethane, ethene, ethanol and ethanoic acid.

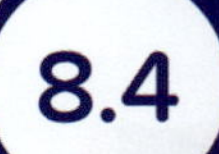

8.4 Uses of hydrocarbons: fossil fuels

BY THE END OF THIS MODULE, YOU WILL BE ABLE TO:

- ✓ define combustion, fossil fuels, natural gas, coal, crude oil and hydrocarbon
- ✓ state the steps involved in the formation of fossil fuels and explain why they are non-renewable.

Quiz
Fossil fuels

GET THINKING

Fossil fuels are frequently in the news because of environmental issues. What environmental problems do you associate with fossil fuels? What fossil fuels are used most in Australia? For what purpose?

What are fossil fuels?

fossil fuel
a substance containing hydrocarbons that is used as an energy source; takes millions of years to form from the remains of dead plants and animals

non-renewable
describes resources that are either non-replaceable or replaceable at a slower rate than they are used

crude oil
an unrefined liquid form of a fossil fuel

natural gas
a gaseous form of a fossil fuel; usually methane, propane or butane

coal
a solid form of a fossil fuel

Fossil fuels are organic substances containing hydrocarbons that are burned to produce energy. Fossil fuels form over millions of years from the remains of ancient living organisms. This process only occurs if:

- the organic matter, such as plants and algae, is rapidly buried after dying
- there is no oxygen, which stops the matter from decomposing
- the organic matter is covered with layers of sediment, usually from oceans, seas and swamps
- the matter is pushed far below the surface of Earth many millions of years later
- the pressure and heat at great depths transform the organic matter into fossil fuels.

Because they take so long to form, fossil fuels are **non-renewable**. This means they cannot be replenished as fast as they are used.

Fossil fuels can be classified into three main categories: **crude oil**, **natural gas** and **coal** (Figure 8.4.1). Crude oil is sometimes called petroleum.

▲ **FIGURE 8.4.1** (a) Crude oil, (b) natural gas and (c) coal are fossil fuels.

How has our use of hydrocarbons changed?

Fossil fuels have been used for thousands of years in various forms. As technology has improved, humans have developed methods to refine and process fossil fuels to make them suited to particular uses.

The first fossil fuel to be widely used was coal. It was used in homes for cooking and heating, as well as in engines and machinery, where it was burned to help generate steam to turn turbines. Coal was used in trains, and even in some early cars, as a vehicle fuel. Over time, coal became less desirable to use because it was messy to handle and produced a lot of ash and soot that caused health problems.

Crude oil is another type of fossil fuel. It began to be used widely in the 1800s. Crude oil can be split into its components by distillation, a process that will be discussed in Module 8.5. One of the first components to be extracted from crude oil and widely used was kerosene, a liquid that can be burned to produce heat and light. Kerosene lamps allowed people to have lighting in their homes and also resulted in the development of the first streetlights (Figure 8.4.2). Over time, lighting, cooking and heating became powered by electricity obtained from coal-fired power stations. This is still the main form of electricity generation in Australia today.

Animaflora PicsStock/Shutterstock.com

▲ **FIGURE 8.4.2** Kerosene lamps were popular in homes until electricity took over as the main form of power for lighting.

8.4

As more components were isolated from crude oil, vehicle fuels changed from coal and kerosene to petrol. The petrol we use today is mostly octane, with some other additives to enhance engine life and performance. However, jet and rocket engines still use kerosene as their main fuel (Figure 8.4.3).

Other components of crude oil are also vital to society. Oils and lubricants, for example, prevent wear on machinery due to friction. Tar, bitumen and asphalt are used to make roads. Aviation fuel and diesel used in heavy vehicles are also derived from crude oil. A by-product of crude oil production is ethene, which is used to make a wide variety of polymers. The use of polymers in plastics and fabrics will be explored in Module 8.8.

Evan El-Amin/Shutterstock.com

▲ **FIGURE 8.4.3** Highly refined kerosene is often used in rocket fuel.

8.4 LEARNING CHECK

1 **Identify** the three categories of fossil fuels.
2 **Describe** the process of formation of fossil fuels.
3 Why are fossil fuels described as non-renewable?
4 **Analyse** how we have changed our methods of lighting and heating our homes over time.
5 **Explain** the importance of crude oil to current society.

8.5 Production of hydrocarbons from crude oil

BY THE END OF THIS MODULE, YOU WILL BE ABLE TO:

✓ describe how hydrocarbons are separated from crude oil
✓ identify uses of the components of crude oil.

GET THINKING

Hydrocarbons are covalent molecules. In Chapter 7, you learned that one of the properties of covalent molecules is a relatively low boiling point. Can you remember the reason for this property? Review your work from Chapter 7 before starting this module.

In Chapter 7, you learned about covalent molecules. All covalent molecules have relatively weak forces between molecules, called intermolecular forces. These forces can be broken when heat is applied. Because intermolecular forces are so weak, it does not require very high temperatures to break them. For this reason, covalent molecules have very low melting and boiling points, meaning many hydrocarbons are liquids and gases at room temperature.

The melting and boiling points of a hydrocarbon depend on its size. Small hydrocarbons, such as methane, have very weak intermolecular forces and are often gases at room temperature. Larger hydrocarbons have stronger forces between molecules, leading to higher melting and boiling points. Larger molecules are more likely to be liquids, such as octane, or even soft solids, like wax (Figure 8.5.1).

Photo Sesaon/Adobe Stock Photos

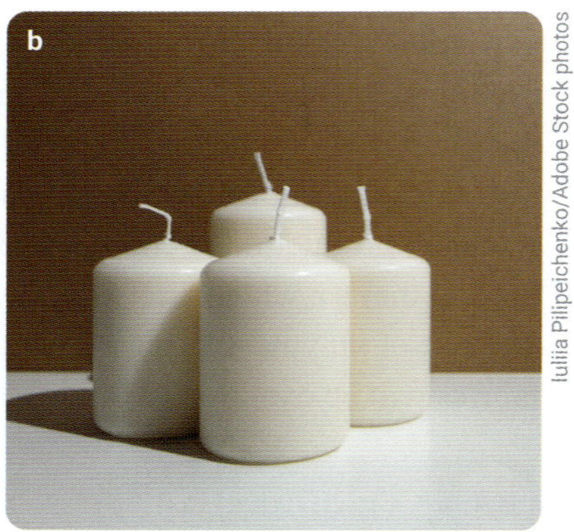

Iuliia Pilipeichenko/Adobe Stock photos

▲ **FIGURE 8.5.1** (a) Some hydrocarbons are liquids, such as engine oil. (b) Larger hydrocarbons are wax-like solids, due to the stronger forces between molecules.

viscosity
a liquid's resistance to flowing

distillation
a process used to separate solutions that collects both the solute and the solvent

fractional distillation
the process of separating a mixture into multiple components based on their boiling points

distillation column
the tower used in crude oil distillation, which has a temperature gradient that separates components as they condense

Fractional distillation

Crude oil is a mixture with high **viscosity**, made up of hydrocarbons of different sizes, each with different boiling points. Hydrocarbons of different sizes can be isolated from crude oil using a type of **distillation** called **fractional distillation** (Figure 8.5.2). This technique uses a **distillation column** – a large tower that is heated to different temperatures at different heights, where the temperature decreases as you move further up the column.

The bottom of the tower is the hottest, around 600°C, while the top of the tower is around 20°C. During fractional distillation, the crude oil is boiled at the bottom of the distillation column and hydrocarbons of different sizes are collected at different points in the column, depending on their boiling point.

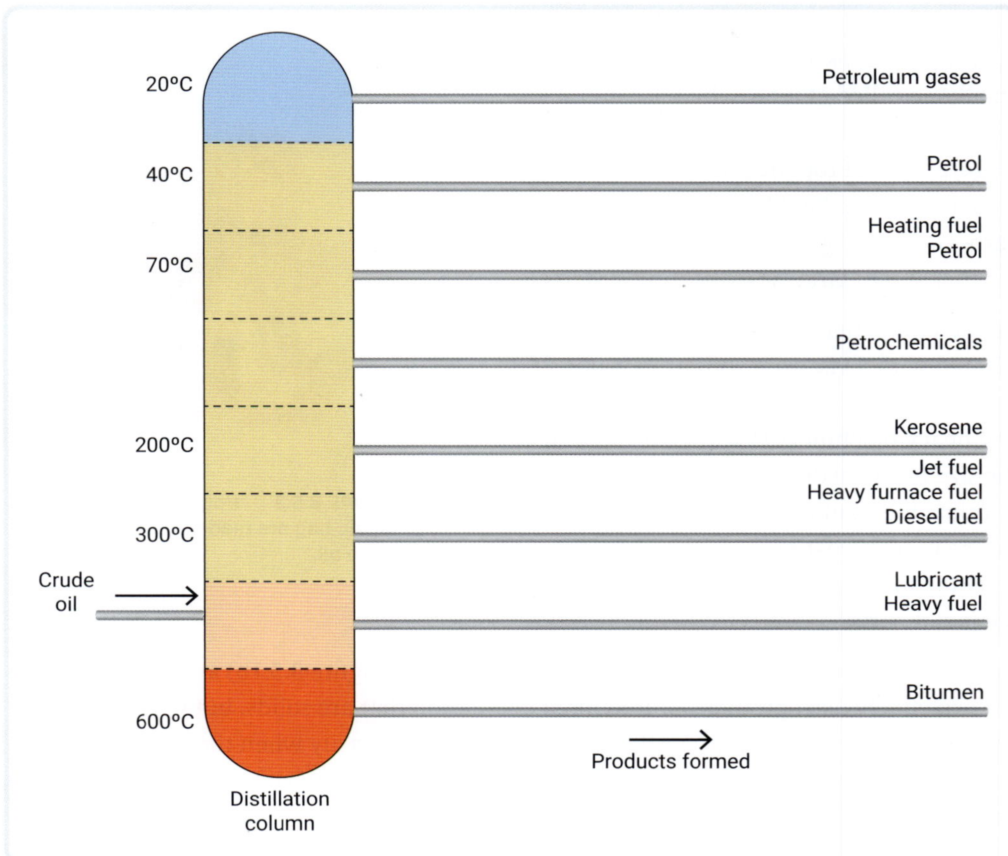

▲ **FIGURE 8.5.2** Crude oil is separated into its components using fractional distillation. Each component is separated according to its boiling point.

When crude oil is boiled, the majority of the hydrocarbon components form a gas and rise up the distillation tower. However, there are some components that do not form a gas at this temperature. These components are removed and collected from the bottom of the tower. As you can see in Figure 8.5.2, this component is asphalt or **bitumen**.

bitumen
a solid used in road surfaces

As the evaporated hydrocarbon components rise up the tower, they start to condense (turn back into a liquid) at their boiling point. For example, diesel fuel and jet fuel condense at approximately 300°C. They are collected at that level in the tower, separated from the other components of the crude oil. This process occurs at all levels of the tower, resulting in each component being collected at different heights and temperatures. Many of the different components that are collected can be seen in Figure 8.5.2.

Uses of crude oil components

Out of all the components of crude oil, the first four alkanes are the smallest molecules and they typically exist as gases (Module 8.2). Methane is commonly used in gas stoves and heaters for heating and cooking. Ethane is used to create an alkene called ethene, which is a building block for making a large range of polymers (Figure 8.5.3). Propane and butane are used in a mixture called **liquid petroleum gas (LPG)** as barbecue gas (Figure 8.5.4), cigarette lighters, car fuel and in large-scale heating systems.

liquid petroleum gas (LPG)
a mixture of propane and butane used as barbecue gas and as vehicle fuel

Liquid components of crude oil vary from petrol, a liquid with a viscosity similar to water, up to heavy oils that are used as lubricants for engines and heavy machinery (Figure 8.5.1a). The larger the size of the hydrocarbon, the more viscous, or thick, the liquid becomes. Heavier liquids can be used to coat moving parts in an engine or machine to prevent friction from wearing away the metal parts.

Janis Smits/Shutterstock.com

▲ **FIGURE 8.5.3** Polymers such as this plastic bag are made from products of crude oil.

Some of the components of crude oil are solids. Waxes (Figure 8.5.1b) and bitumen are two widely used products of fractional distillation that are solids, but with very low melting points. Candle wax, for example, melts very easily when the candle is lit. Bitumen (Figure 8.5.5) is used to make road surfaces. You may have noticed on very hot summer days that the bitumen starts to melt a little. This is because it has reached its relatively low melting point.

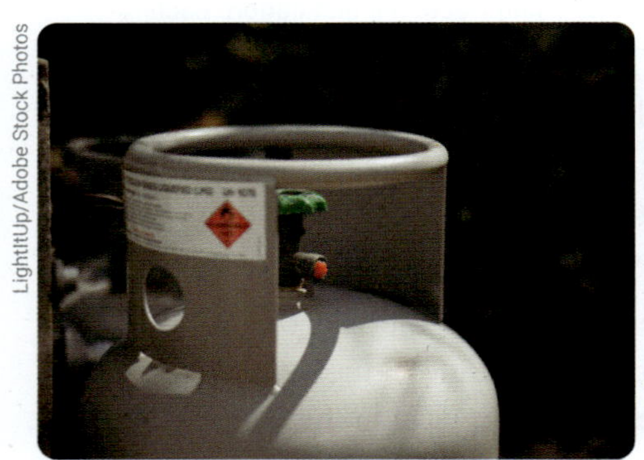
LightItUp/Adobe Stock Photos

▲ **FIGURE 8.5.4** A mixture of propane and butane called liquid petroleum gas (LPG) is used in barbecue gas bottles.

schankz/Shutterstock.com

▲ **FIGURE 8.5.5** Bitumen is a soft solid used in road surfaces.

☆ ACTIVITY

8.5

Distillation

Fractional distillation is a process used to separate crude oil into different, useful substances such as a petrol. In the school laboratory, we can perform a simplified version of fractional distillation, known as simple distillation. This is a separation technique. This can be a teacher demonstration or student activity.

Aim

To separate ethanol from an ethanol–water mixture by simple distillation

Safety

Be extremely careful when lighting ethanol since the flame is blue and very hard to see. Light from the side and move your hand away quickly once you have ignited the sample.

Materials and equipment

- 100 mL of ethanol–water mixture (in a 1:4 ratio of ethanol to water, so 20 mL/80 mL)
- distillation apparatus (round-bottom flask, Bunsen burner or heating mantle, condenser and thermometer)
- thermometer
- Bunsen burner
- tripod stand and clamp to hold distillation flask
- small beaker to capture ethanol
- watch glass

Procedure

1 Put 100 mL of ethanol–water mixture into the distillation (round-bottom) flask.
2 Set up the heating mantle or Bunsen burner and distillation apparatus, as shown in Figure 8.5.6.
3 Heat the distillation flask to maintain 78°C, until the ethanol has evaporated and then condensed and collected in the small beaker.
4 To confirm the substance you have collected, pour 5 drops of ethanol onto a watch glass then ignite it. Ethanol is flammable. Water is not flammable.

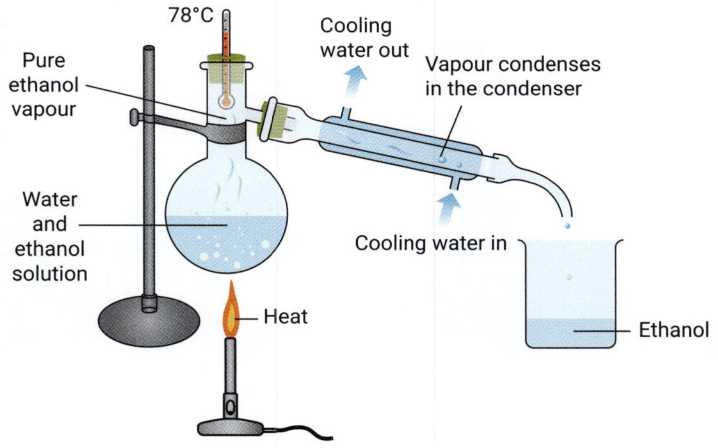

▲ **FIGURE 8.5.6** The experimental set-up

Analysis

Describe the purpose of the condenser. **Explain** why the property known as flammability was used to confirm the separated substance.

8.5 LEARNING CHECK

1 **Identify** the property of covalent molecules that allows them to be separated.
2 **Describe** the process of fractional distillation.
3 **Explain** why some crude oil components are gases, some are liquid and some are solids.
4 **Outline** the uses of at least two components of crude oil.
5 **Explain** why it is important that the distillation tower changes temperature at different levels.

8.6 Combustion

BY THE END OF THIS MODULE, YOU WILL BE ABLE TO:

- ✓ describe the differences between complete and incomplete combustion reactions
- ✓ compare examples of combustion and its products, and the energy produced.

Video activity
Oxygen and combustion

GET THINKING

The combustion of fossil fuels releases energy. Write down a list of activities or events that have happened to you today that required energy that came from fossil fuels. Is there anything you have done today that does not involve energy from fossil fuels?

Combustion of hydrocarbons

combustion
a chemical reaction where a fuel combines with oxygen to produce energy

Many human activities involve the **combustion** of fossil fuels, such as burning coal, natural gas (Figure 8.6.1) and crude oil. In Australia, most of our power stations, manufacturing, construction industries and transport use combustion.

Oil and Gas Photographer/Shutterstock.com

▲ **FIGURE 8.6.1** Combustion of natural gas

When a substance such as a fossil fuel is burned in oxygen, it releases energy, carbon dioxide and water. All fossil fuels release similar products, because fossil fuels are mostly made of hydrocarbons, compounds that are composed only of hydrogen and carbon. The carbon in fossil fuels was taken out of the atmosphere millions of years ago by producers via photosynthesis. When we burn fossil fuels, we are returning huge amounts of carbon to Earth's atmosphere in the form of carbon dioxide.

exothermic
describes a chemical reaction that releases energy

complete combustion
combustion that occurs with plentiful oxygen; it produces carbon dioxide and high amounts of energy

incomplete combustion
combustion that occurs in limited oxygen conditions, producing carbon monoxide, carbon and less energy than complete combustion

Combustion is an **exothermic** chemical reaction, which means it releases energy. Combustion can be complete or incomplete. **Complete combustion** occurs when oxygen is plentiful, while **incomplete combustion** occurs when the oxygen supply is limited.

Complete combustion

The products of complete combustion of a carbon-based fuel are carbon dioxide (CO_2) and water. This is represented by the word equation:

Fuel + oxygen → carbon dioxide + water

Incomplete combustion

When there is incomplete combustion, limited oxygen is available and there is not enough oxygen to completely combust the fuel.

The products of incomplete combustion have less oxygen than the products of complete combustion. The usual products are either carbon monoxide (CO) or carbon (C), or a mixture of both. Water is also produced.

Incomplete combustion can be represented as:

Fuel + oxygen → carbon monoxide + water
Fuel + oxygen → carbon monoxide + carbon + water

Complete and incomplete combustion can be observed with a Bunsen burner (you will investigate this in the activity below).

Observations of combustion

You can see incomplete and complete combustion by looking at the colour of a flame and the emissions produced. You can see in Figure 8.6.2a that complete combustion results in a blue flame, while in Figure 8.6.2b, incomplete combustion leads to a yellow–orange flame.

Another sign of incomplete combustion is black particles being emitted from the flame. These particles are commonly called 'soot' and are particles of carbon (C) (Figure 8.6.3).

Krasula/Shutterstock.com

FotoCuisinette/Shutterstock.com

▲ **FIGURE 8.6.2** (a) Complete and (b) incomplete combustion can be identified by the colour of the flame.

ID1974/Shutterstock.com

▲ **FIGURE 8.6.3** Black soot deposits are evidence of incomplete combustion occurring.

Complete and incomplete combustion in a Bunsen burner

✩ ACTIVITY

Carefully light a Bunsen burner and observe the flame with the airhole fully opened (maximum oxygen). Then observe the flame with the airhole closed (reduced oxygen).

1 **Compare** the two flames. Record the differences you observe between the two flames.
2 **Identify** which flame is showing complete combustion. Use your observations to **justify** your selection.
3 Bunsen burner gas is rich in methane (CH_4). **Write** a word equation for the complete combustion of methane.

Energy and products of combustion

Complete and incomplete combustion not only release different chemical products, but different amounts of energy. Sometimes, you need high levels of heat when cooking or in the laboratory; for example, if you want to boil water. When you have the flame set to high heat, the gas burns with a blue colour and produces significantly more heat than if it was set to low heat (yellow-orange flame). The more oxygen that is present for combustion, the more heat is released per mass of fuel.

When you burn fuels such as wood or coal in a firepit or on a barbecue you can see evidence of incomplete combustion. When wood burns in air, the oxygen supply is limited. While air seems unlimited, it only contains 20 per cent oxygen. As a result, wood fires release a lot of soot (carbon), which you can see deposited on fireplaces, barbecues and in fire pits.

8.6 LEARNING CHECK

1. **Identify** the products of complete and incomplete combustion of hydrocarbons.
2. **Explain** why your teacher might tell you to open the air hole on a Bunsen burner if you want a hotter flame.
3. **Describe** the observations you might make that would indicate incomplete combustion had occurred.
4. **State** the word equations that represent the complete and incomplete combustion of hydrocarbon fuels.

WORKING SCIENTIFICALLY

8.7 Problem-solving

SCIENCE SKILLS IN FOCUS

IN THIS MODULE, YOU WILL FOCUS ON LEARNING AND IMPROVING THESE SKILLS:

- identify problem-solving strategies
- propose a method to determine the energy produced by different fuels.

Solving problems in science

Many scientific investigations pose a problem to be solved. A problem in science could be:

- observing and measuring the properties of an object to determine what they are
- comparing two objects or chemicals for a specific feature; for example, which type of washing detergent cleans better
- testing a theory; for example, whether keeping food in the fridge or freezer slows its decomposition rate.

Identifying strategies

When you are given a problem to solve in science, you have a range of tools you can use to propose a solution. Here is a series of questions you can ask so you can identify which strategies best suit the investigation you are conducting.

1 Will you be making observations?
2 What features will you observe?
3 Will you be taking measurements?
4 What equipment will you need?
5 Will you need to set up a reliable and valid experiment?
6 How will you do this?
7 Do you need a control?

Problem-solving examples

Once you have asked yourself these questions and decided what strategies you are going to use, you need to think about how you will finally solve the problem. Some examples are given below.

- Determining properties of an object: what properties will you include? How will you decide if the object has the property you are testing for?
- Comparing two objects, for example, which detergent is better: what features or properties of the detergents will you compare to make your decision? How many features or properties will you test? How will you make your final decision?
- Testing a theory; for example, does keeping food in the fridge slow decomposition? Do you have a control (something to compare to)? How will you know whether your theory was correct or not? What measurements or observations will help you decide this?

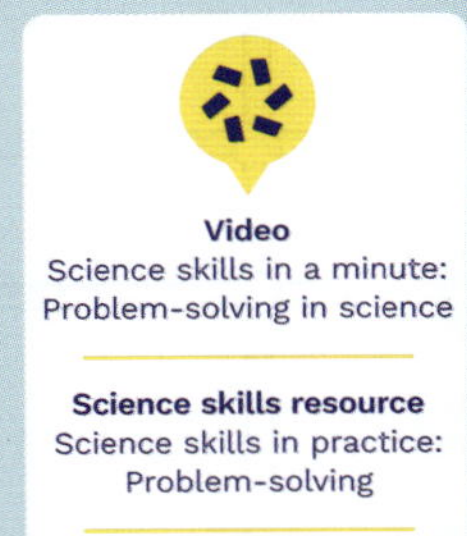

Video
Science skills in a minute: Problem-solving in science

Science skills resource
Science skills in practice: Problem-solving

DO ALL FUELS PRODUCE THE SAME AMOUNT OF ENERGY?

AIM

To compare how much energy different fuels produce

 **Safety**

Using spirit burners will protect you from touching the fuels, which are highly flammable. Use gloves to handle the spirit burners and do not touch the wicks or you will come into contact with the fuel.

Be careful with the open flame; keep all paper and clothes at a safe distance. Extinguish flames with the spirit burner lid, not by blowing them out.

MATERIALS AND EQUIPMENT

- ☑ spirit burners with three different fuels (Your teacher will let you know what fuels you have in your experiment. Try to use ethanol as one of your fuels. Other possible fuels are methanol and kerosene.)
- ☑ 250 mL conical flask
- ☑ 100 mL measuring cylinder
- ☑ thermometer or data logger temperature probe
- ☑ clamps and retort stand
- ☑ matches
- ☑ heatproof mat
- ☑ stopwatch (or phone with timer)

PROCEDURE

Set up equipment as shown in Figure 8.7.1. Keep the conical flask at a fixed height above the spirit burner.

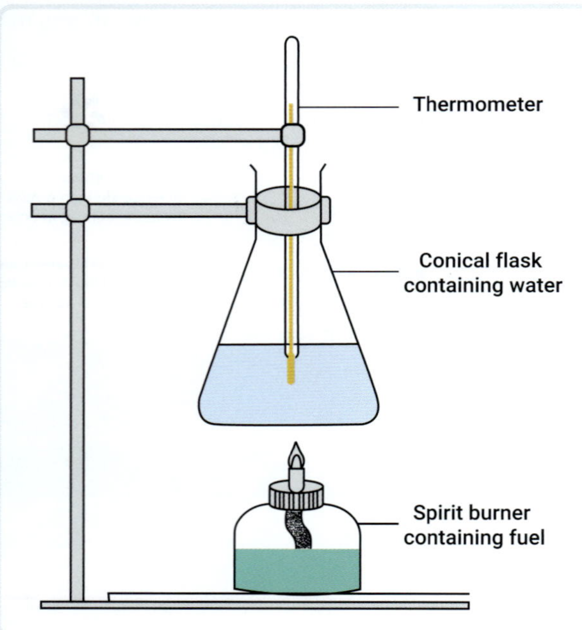

▲ **FIGURE 8.7.1** The equipment set-up

Using the information in the blue Science skills in focus section, you are to design an investigation to compare the energy produced by at least two or three different fuels.

Things to consider in your problem-solving strategies include:

- what volume of water will you use?
- how long will you heat the water for?
- what measurements will you take to compare the energy produced? (Note that you cannot measure energy directly from this experiment: what can you measure that will let you compare energy?)
- will you do trials? If so, how many?
- how many fuels will you use?

RESULTS

Construct a separate table for each fuel that records the measurements you took during your investigation. Ensure you have appropriate column headings, units of any quantities measured and enough space to record the data for all trials you conduct.

ANALYSIS

1 **Describe** the steps taken in the method to ensure that:
   a a valid test was conducted – this means the experiment accurately tested its aim
   b reliable results were produced – this means that each time you did a trial for the same fuel, you should have measured almost the same result.

2 **Explain** why a control is not suitable for this investigation.

3 **Propose** why there was variation between the:
   a different fuels. (Why was the average temperature change different for each fuel?)
   b trials of the same fuel. (Why was the data not identical in your separate trials?)

4 **Explain** what measurement you took to **compare** the energy produced by the fuels. Why did you select this measurement?

CONCLUSION

Review the aim and write a conclusion for this investigation. Refer to specific result data in your conclusion.

8.8 Polymers

BY THE END OF THIS MODULE, YOU WILL BE ABLE TO:

- ✓ identify the raw materials used to make polymers
- ✓ describe examples of polymers commonly used in society.

GET THINKING

Plastic is one of the most well-known polymers. But did you know that plastic is not one type of material, but many different types of materials with different structures? Look around you and identify as many things as you can that are made of plastic. Discuss their properties with a classmate and suggest reasons why different plastics have different properties.

Video activity
The diverse world of polymers

What is a polymer?

Polymers are all around you (Figure 8.8.1). You may be able to easily identify many **plastic** items around you right now. Water bottles, pen casings, folders, storage containers, bags and many other things are made from plastic. Plastics are one class of polymer, but there are lots of other items that you might not realise are also polymers.

For example, you are probably wearing clothes that contain synthetic fibres like **nylon** or **polyester**. Even clothes with natural fibres, like wool or silk, often contain some synthetic fibres. These fibres are also polymers. Most waterproof clothing is made from polyester fibres that are able to repel water.

The transparent walls of aquariums (and other situations where the ability to withstand high pressure is needed) are made from a glass-like polymer called **acrylic**. You might know this as Perspex from some of your laboratory experiments or from cookware in your kitchen. Some saucepans and frypans are coated with a specially invented non-stick polymer called **Teflon** (Figure 8.8.2).

polymer
a large molecule made up of many repeating units called monomers, joined together to form a long chain

plastic
a common name for a group of polymers made from crude oil components

nylon
a synthetic fibre used in clothing

polyester
a synthetic fibre used in clothing

acrylic
an opaque transparent polymer used in place of glass

Teflon
a non-stick polymer coating for uses including cooking equipment, waterproofing fabric and coating medical devices

Mazur Travel/Shutterstock.com

▲ **FIGURE 8.8.1** Polymers are used to make a variety of commonly used items in your home.

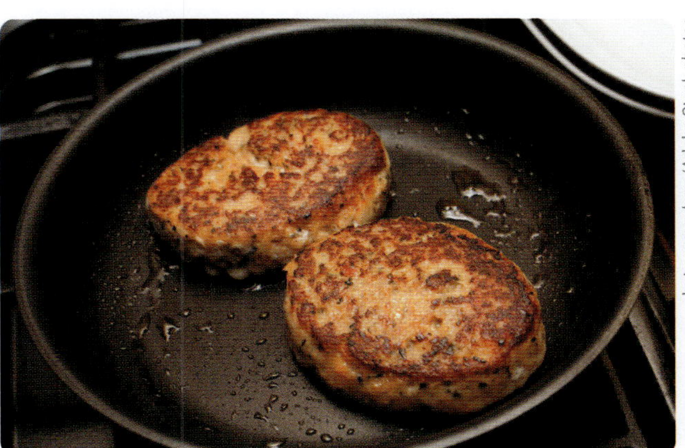

photoeverywhere/Adobe Stock photos

▲ **FIGURE 8.8.2** Teflon-coated frypans ensure food does not stick when it's cooking.

Polymer structure

The word 'polymer' originates from the Greek words for 'many parts', which explains exactly what a polymer is. Small molecules called **monomers** (meaning 'one part') join together and repeat over and over, forming long polymer chains (Figure 8.8.3). A polymer can have hundreds, thousands or millions of monomers in the chain.

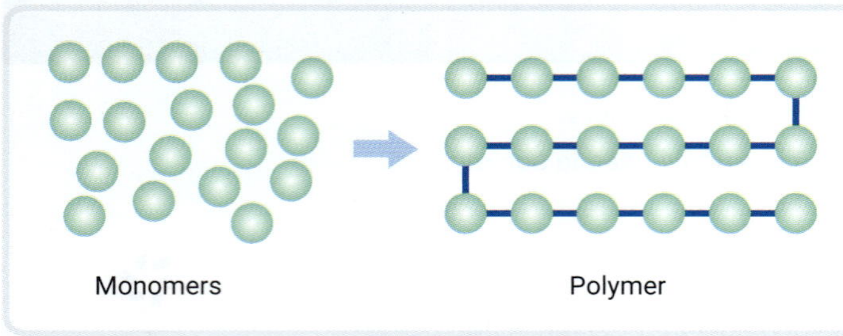

▲ **FIGURE 8.8.3** Small molecules called monomers join to form long chains called polymers.

An example of a polymer: polyethylene

You learned in Module 8.5 that fractional distillation is used to separate components of crude oil using their boiling points. One by-product of this fractional distillation process is the molecule ethene. Ethene is an alkene that has two carbons, joined with a double bond (Figure 8.8.4).

H H
C = C
H H

▲ **FIGURE 8.8.4** A molecule of ethene. Ethene is used to create many polymers you use every day.

monomer
a small molecule building block used to make polymers

polyethylene
a polymer formed from ethene monomers

When multiple ethene monomers join together, they form a long chain polymer that is called **polyethylene** (Figure 8.8.5). This is a common polymer that comes in two forms, high-density polyethylene (HDPE) and low-density polyethylene (LDPE). In HDPE, the chains are closely packed and there is a high number of chains in a given area. In LDPE, the chains are spread out and branched so there is a low number of chains in a given area, as seen in Figure 8.8.6b.

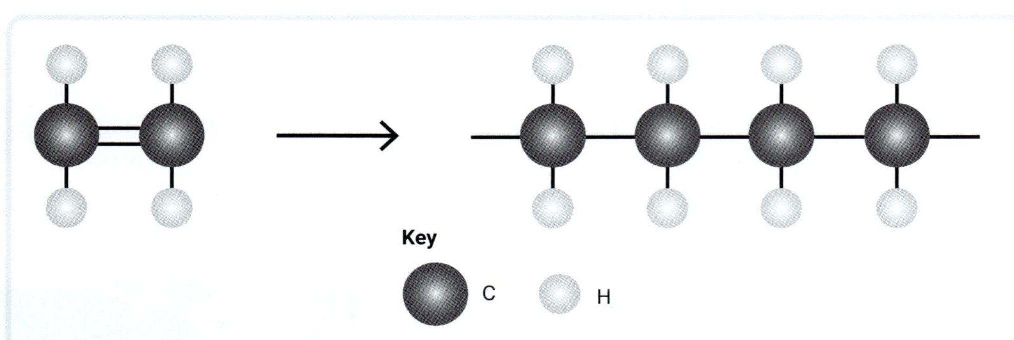

▲ **FIGURE 8.8.5** Ethene monomers can combine to form a polymer called polyethylene.

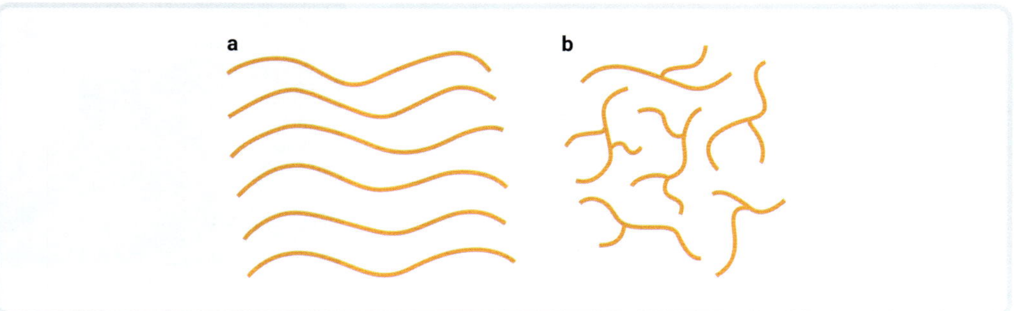

▲ **FIGURE 8.8.6** **(a)** HDPE has closely packed chains. **(b)** LDPE has branched chains that are randomly arranged and more spread out.

Common polymers

Six types of polymers make up about three-quarters of polymers used in Australia. Many of these polymers you would use nearly every day. Table 8.8.1 shows the names of the polymers, their abbreviations and common uses.

▼ **TABLE 8.8.1** Commonly used polymers

Polymer name	Abbreviation	Common uses
Polyethylene terephthalate	PET	Clear bottles, storage containers, carpets
High density polyethylene	HDPE	Opaque bottles, buckets, crates
Low density polyethylene	LDPE	Bags, food wrap, bubble wrap, toys
Polyvinyl chloride (hard and soft forms exist)	PVC	Rigid form – pipes and credit cards Soft form – clothes and hoses
Polypropylene	PP	Bottle caps, yoghurt containers
Polystyrene	PS	Styrofoam packing material, food containers

Many polymers can be recycled. Most plastic objects have a recycling symbol on them, including a number that indicates the type of polymer they are made from, as seen in Figure 8.8.7. The number indicates how easily and where they can be recycled. For example, polystyrene (number 6) is not easy to recycle and would need to be taken to a specialist facility. On the other hand, HDPE (number 2) is easily recycled and converted into new products.

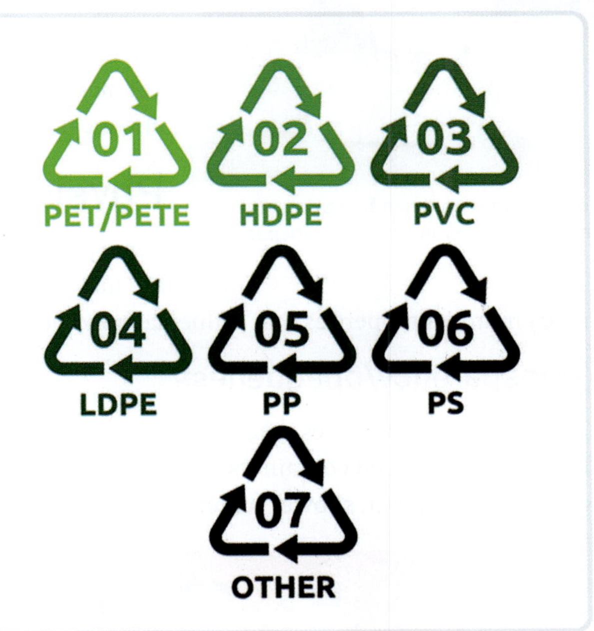

▲ **FIGURE 8.8.7** Many plastic objects are labelled with a number to indicate the type of polymer they contain and to help with sorting before recycling.

8.8 LEARNING CHECK

1 **Identify** three polymers that are commonly used in society.
2 **State** the relationship between monomers and polymers.
3 **Compare** the structures of LDPE and HDPE.
4 **Explain** what the numbers in the triangles on plastic objects mean.
5 A student claims that polymers are made from crude oil. **Assess** this claim.

DATA SCIENCE 

Learn more about supporting scientific claims in **Module 2.2**.

8.9 Properties of polymers

BY THE END OF THIS MODULE, YOU WILL BE ABLE TO:

✓ describe the properties of a range of polymers.

GET THINKING

Polymers have both physical and chemical properties. Recall the difference between the two types of properties. How might you determine either of these types of properties if you were given some polymer samples? What tests might you conduct?

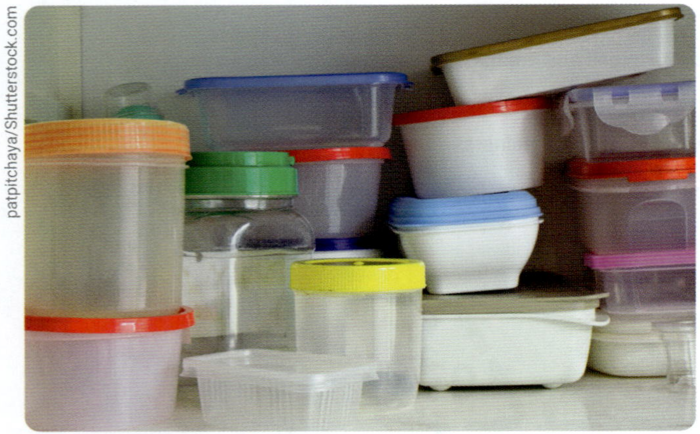

patpitchaya/Shutterstock.com

▲ **FIGURE 8.9.1** Plastic storage containers can be strong and hard or flexible, have a range of colours and can be transparent or opaque.

Polymers are so widely used because they can be manufactured to the exact requirements of the end user (Figure 8.9.1). Consider the wide variety of phone cases available. They can be made **transparent** or **opaque**, flexible or hard, and can be any colour or texture you wish. And this is just one example of a plastic product.

To manufacture polymers for a specific purpose, you need to know what type of physical and chemical properties they can have.

Physical properties

Some physical properties, like colour and texture, are mostly to make plastics look nice. We will focus more on properties that influence how the product functions.

transparent
see-through

opaque
cannot be seen through

Transparency/opaqueness

This property determines whether you can see through the material or not (Figure 8.9.2). Many takeaway food containers are transparent so that you can see what you have ordered; in contrast, shower screens may be opaque for privacy.

Interactive resource
Crossword: Polymers

iStock.com/yuliash

▲ **FIGURE 8.9.2** Polymer containers can be opaque or transparent, depending on their use.

Density

The density of a polymer can be important depending on its use. For example, a pool cleaner designed to clean the bottom of a swimming pool wouldn't be much use if it floated. On the other hand, children's bath toys are designed to float and are usually made from low density polymer materials. Another example is styrofoam for packing (Figure 8.9.3).

giedre vaitekune/Shutterstock.com

▲ **FIGURE 8.9.3** Packing materials are often made from low density, flexible styrofoam pieces that will not add weight to the box but will protect the contents.

Flexibility/hardness

Most people have dropped their phone enough times to understand the value of a solid phone case. Polymers that are hard can be used as protective cases for phones, as water bottles and as storage containers for heavy objects.

However, there is also high demand for more flexible polymers. Plastic bags, food cling wrap, plastic folders and many other objects need to be flexible (Figure 8.9.4).

New Africa/Shutterstock.com

▲ **FIGURE 8.9.4** Flexible food wrap is used to protect food.

Heat resistance

Many polymer containers are deemed 'oven safe' or 'microwave safe,' which is a desirable property when food needs to be cooked or reheated. Heat resistance is also important for drink containers that hold hot liquids, like soup or coffee, or for objects that are constantly out in the sun. An example of this type of polymer is PEI or polyethyleneimine, which is used in circuit boards, food sterilisation equipment and in aircraft parts. It can withstand temperatures of up to 170°C.

Some polymers are specifically designed for colder temperatures, such as those found in freezers or in snowy environments.

Chemical properties

Chemical properties can be important for certain uses. For example, if containers are being used for food storage, they must not react with acids and bases, because many foods are acidic or basic. Plastic containers used to store chemicals must be chemically resistant to avoid reacting with the products put in them, which could make the containers dissolve or disintegrate.

Plastics are often used for outdoor activities. This includes items that must be weather-resistant and water-resistant (Figure 8.9.5), like outdoor furniture, camping and sporting equipment (such as skis and footballs). Specific plastics and coatings are manufactured that repel water.

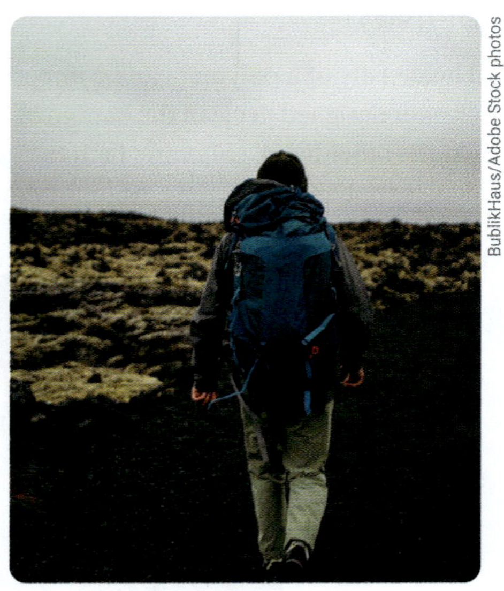
BublikHaus/Adobe Stock photos

▲ **FIGURE 8.9.5** Hiking and camping equipment needs to be water- and weather-resistant, a chemical property of many polymers.

8.9 LEARNING CHECK

1 **List** three physical properties a polymer could have.
2 **Describe** two examples where low-density polymers are required.
3 **Choose** the most appropriate polymer for the following uses from the four provided in Table 8.9.1. **Justify** your choices.
   a A lightweight polymer travel mug that holds both hot and cold liquids
   b Polymer outdoor furniture
   c A polymer container to carry heavy items and keep them out of the light. The container will travel by truck, so will be exposed to the weather.

▼ **TABLE 8.9.1** Properties of common polymers

Property	Polymer			
	LDPE	PET	Polypropylene	Polystyrene
Strength (pressure until polymer breaks) (megapascals)	10–12	55–79	15–27	24–60
Water absorbance by a given area of polymer (millilitres per day)	0.005–0.015	0.1–0.2	0.01–0.1	0.01–4
Transparency	High	Variable – can be opaque or transparent	Poor	Excellent
Chemical resistance	Good	Good	Good	Poor
Density (g/cm³)	0.89–0.93	1.38–1.41	0.85–0.92	0.01–1.06
Melting point (°C)	105–115	235–260	150–160	240–250

WORKING SCIENTIFICALLY 

8.10 Recording observations of physical properties

SCIENCE SKILLS IN FOCUS

IN THIS MODULE, YOU WILL FOCUS ON LEARNING AND IMPROVING THIS SKILL:

- recording observations of physical properties of polymers.

OBSERVING AND MEASURING PROPERTIES OF POLYMERS

AIM

To observe and measure the physical properties of different polymers

MATERIALS AND EQUIPMENT

- ☑ a range of polymers, including objects made from:
  - PET (e.g. soft-drink bottle)
  - HDPE (e.g. plastic bucket)
  - LDPE (e.g. plastic food wrap or plastic bag)
  - PVC (e.g. flexible hose)
  - polypropylene (e.g. bottle cap)
  - polystyrene (e.g. packing pellets or packing foam)
- ☑ 100 mL 1:1 ratio solution ethanol/water, density = 0.94 g/mL
- ☑ 100 mL deionised water, density = 1.0 g/mL
- ☑ 100 mL 10% sodium chloride solution, density = 1.08 g/mL
- ☑ 3 × large test tubes
- ☑ test-tube rack
- ☑ stirring rod

PROCEDURE

TEST 1: PHYSICAL PROPERTIES

1 Copy Table 8.10.1 into your book to record your results. Create a row for each polymer sample you plan to test.

2 For each polymer you have been provided, identify the type of polymer and record it in your table. You should be able to find the recycling numbers on most items.

3 Record whether each polymer is transparent or opaque.

4 Record whether each polymer is flexible or rigid.

5 Try to break the polymer. Does it snap or just bend?

TEST 2: DENSITY

1 Copy Table 8.10.2 into your book to record your results.

2 Your teacher will supply you with three small squares of each polymer for testing.

3 Half-fill one test tube with the ethanol/water mixture, another test tube with deionised water and the third with 10% sodium chloride solution. Ensure each test tube has the same height of liquid.

4 Place one square of polymer in the first test tube, one square in the second test tube and the last square in the third test tube.

5 Push each piece under the surface of the liquid to break the surface tension. Record whether the polymer sample sinks quickly or slowly, or floats.

6 Repeat for each of the types of polymers, recording all results into your table.

RESULTS

Create tables like these to record your results.

▼ **TABLE 8.10.1** Physical properties of various polymers

Polymer type	Transparent or opaque?	Flexible or rigid?	Snap or bend?
Polymer 1			
Polymer 2			

▼ **TABLE 8.10.2** Density of various polymers

Polymer type	1:1 ethanol/water (density 0.94 g/mL)	Deionised water (density 1.0 g/mL)	10% sodium chloride (density 1.08 g/mL)
Polymer 1			
Polymer 2			

ANALYSIS

If a sample floats, it has a density lower than the solution it was placed into. If the sample sinks, it has a density greater than the solution it was placed into.

1 Use your results table for the physical properties test to **identify** any similarities between the polymers. Can you relate these similarities to their uses?

2 Use your results table for the physical properties test to **identify** any differences between the polymers. Can you relate these differences to their uses?

3 Use the results for your density test to rank the polymers in order of their density. **Justify** your ranking using the density results.

4 An unknown polymer floats in the 10% sodium chloride solution but sinks in water. What possible densities could the polymer have? **Justify** your answer.

5 Often, polymers are combined to make a new polymer that will have a mixture of the properties of the original ones. **Predict** how the density of a combined polymer might be affected if a low-density polymer and a high-density polymer were combined.

CONCLUSION

Write a conclusion summarising the properties of the different polymers.

9780170491785

8.11 Environmental impacts of polymers

BY THE END OF THIS MODULE, YOU WILL BE ABLE TO:

✓ describe the extent of the pollution problem related to polymer use

✓ outline current and future initiatives aimed at reducing polymer use.

GET THINKING

In most states in Australia, governments have implemented bans on some plastic items. What items are no longer available in plastic form? What have they been replaced with? What other initiatives have you seen to reduce the amount of plastic used in everyday life? What ideas do you have to reduce your use of plastics?

Polymer waste and biomagnification

In all oceans of the world, including areas far away from land masses, you can find areas of polymer pollution. Any rubbish that washes into rivers and the ocean system will eventually make its way out to sea on ocean currents. There is one area of polymer pollution in the Pacific Ocean that is approximately 1.6 million square kilometres (Figure 8.11.1). To understand how massive this is, for comparison the entire state of New South Wales is only 800 000 square kilometres.

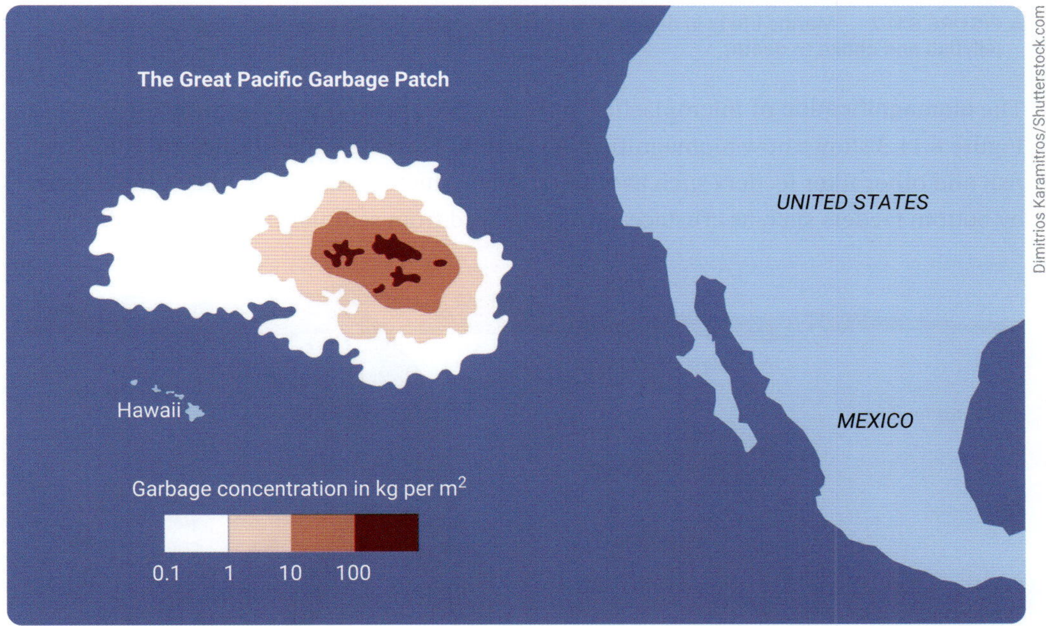

▲ **FIGURE 8.11.1** The area of plastic waste sometimes called the Great Pacific Garbage patch covers approximately 1.6 million square kilometres and contains large amounts of polymer waste.

While waste in the ocean includes a lot of floating debris like bottles, cans, fishing nets, clothes, shoes and containers, another major problem is the small plastics known as **microplastics**. Most polymers do not break down fully, but instead break into smaller and smaller pieces. Many of these pieces of microplastic are small enough to remain suspended in the water rather than fully sinking or floating.

microplastics
very small pieces of plastic debris found in the environment

Larger debris can be very dangerous to marine life. Turtles often mistake floating plastic bags for jellyfish, one of their favourite foods (Figure 8.11.2). Seals, whales, turtles and dolphins get caught in nets discarded from fishing vessels. Birds mistake pieces of plastic for fish eggs and feed the plastic to their chicks. Scientists are increasingly finding dead marine life with large amounts of plastic debris in their stomachs.

▲ **FIGURE 8.11.2** Marine life is in danger from plastic waste. Turtles can mistake plastic bags for jellyfish and choke to death.

biomagnification the increase in the concentration of a substance in organisms as you move up a food chain

The **biomagnification** of microplastics in the marine environment is a growing problem. Figure 8.11.3 shows how biomagnification occurs. Floating microplastics are eaten by fish and other filter feeders such as whales. When microplastics sink, they can be eaten by bottom feeders such as crustaceans as they feed from sediment on the ocean floor.

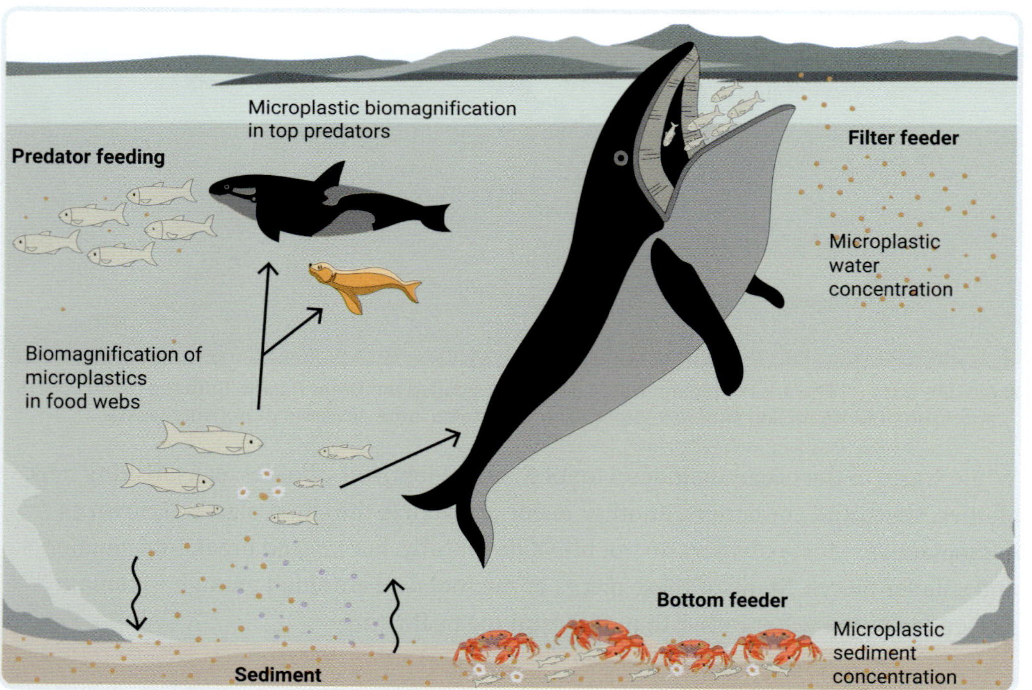

▲ **FIGURE 8.11.3** Biomagnification of microplastics is an increasing problem.

9780170491785

Smaller fish and bottom feeders accumulate microplastics in their system as they feed. They are then fed on by larger fish, which are in turn consumed by larger fish again, and eventually predators like seals, whales and dolphins. As you go up each level of the food chain, the amount of microplastics increases. Large fish eat many smaller fish, so they gain increased amounts of microplastics in their stomachs. By the time a predator is hunting and eating a whole school of larger fish, it is ingesting toxic levels of microplastics. As microplastics continue to break down, they release chemicals, which disrupt bodily functions and can lead to death if the levels ingested are high enough.

Scientists are also investigating the biomagnification of microplastics in humans. As we consume food from the sea and other sources, we are also ingesting substantial amounts of microplastics. How this affects the human body is not yet fully known, but many researchers are working to discover more.

Biodegradable polymers

Many companies and governments have recognised the problem of polymer waste and have brought in new products and laws to combat it. In New South Wales, plastic bags are no longer provided in supermarkets and plastic straws are no longer available with takeaway drinks. Shoppers are expected to bring their own bags when shopping and paper straws are now standard (Figure 8.11.4). Numerous other states and countries have similar laws, with many going further.

Jorge/Adobe Stock Photos

▲ **FIGURE 8.11.4** Paper straws have replaced plastic straws for takeaway drinks in many places in Australia.

The use of more sustainable takeaway food and drink containers and cutlery is one example of a widespread action to reduce the amount of plastic used. In the past, the containers and cutlery were usually made of non-biodegradable plastic, which was then usually discarded as waste. Many companies and food outlets have now moved to using **biodegradable** paper and **biopolymer** materials (Figure 8.11.5).

biodegradable
a substance that can be broken down by bacteria or other living organisms

biopolymer
a polymer made from plant or animal materials

Smart Calendar/Shutterstock.com

▲ **FIGURE 8.11.5** Food containers are increasingly made from plant materials like bamboo, which break down quickly in the environment.

Biopolymers are made mostly from plant and animal sources. Table 8.11.1 shows some examples of biopolymers already in use. Biopolymers only take 3 to 6 months to break down. The products of the breakdown of biopolymers are relatively harmless, unlike microplastics. This means they have much less of an environmental impact than the more traditional plastics, which can take hundreds or thousands of years to completely decompose.

▼ **TABLE 8.11.1** Biodegradable polymers and their uses

Polymer description	Uses
Cellulose – found in the cell walls of plant cells	Used to make paper, textiles, biodegradable plastic bottles and food containers
Starch – found in corn and potatoes	Turned into bioplastics and used in bags, packaging and disposable plates and cutlery
Polylactic acid – made from corn starch or sugar cane	Turned into bioplastics and used in 3D printing, disposable cutlery and plastic cups
Chitin – found in the shells of crustaceans and insects	Used in food packaging and in cosmetics

8.11 LEARNING CHECK

1 **Describe** how microplastics spread throughout the marine environment.
2 **Outline** why large plastic rubbish is dangerous for marine life.
3 **Explain** how biomagnification of microplastics can cause the death of large marine predators.
4 **Outline** an action that governments have taken to reduce the amount of plastic waste produced.
5 **Describe** two biopolymers and their uses.
6 **Compare** the environmental impact of traditional plastics with biopolymers.

9780170491785

WORKING SCIENTIFICALLY

8.12 Writing for your audience

SCIENCE SKILLS IN FOCUS

IN THIS MODULE, YOU WILL FOCUS ON LEARNING AND IMPROVING THESE SKILLS:

- investigating the biodegradability of different plastics
- communicating scientific information clearly and accurately using a format that is appropriate to the audience.

Know your audience

Communication of scientific information is very important so that the audience is clear on the findings and any possible implications of an investigation. Consider scientists conducting investigations into food or medicine quality, or predicting the effect of pollution on our environment. If the audience doesn't clearly understand the information from the investigation, any of these examples could have serious consequences for people, animals or the environment. Scientists' reports must be clear, accurate and use appropriate language for the audience.

Figure 8.12.1 summarises key factors you need to think about to ensure you communicate effectively so your science research can be easily understood and is useful for your intended audience.

Format
- Engaging and interesting
- Relevant to the audience
- Uses more than one method; e.g. words and diagrams

Words
- Language; e.g. English
- Terminology
- Correct use of spelling and grammar

Message
- Clear explanations
- Supported with evidence, examples and/or diagrams

▲ **FIGURE 8.12.1** Factors to consider for effective communication

BIODEGRADABILITY OF POLYMERS

AIM

To compare the biodegradability of different polymers

MATERIALS AND EQUIPMENT

- ☑ samples of different polymers. This can include soft-drink bottles, plastic bags, or clean food containers. Try to use samples of at least one polymer that claims to be biodegradable.
- ☑ large container, similar to a planter box or plant pot
- ☑ soil (enough to fill the container)
- ☑ bamboo sticks to use as markers
- ☑ gloves

Safety

Handle soil in a well-ventilated area.

Do not handle soil with bare hands. Always use gloves and wash hands thoroughly after finishing the investigation.

PROCEDURE

1 Fill the container with soil.

2 Cut your first polymer so you have three squares of each polymer that are 5 cm × 5 cm. It is important to keep all samples the same size. If you can, take photographs of your samples before burying them so you can compare them with the final result.

3 Bury the three samples of the first polymer in a row in your container. Make the depth around 5 cm and keep this consistent for all samples. Mark the locations with bamboo sticks so you can find them again later.

4 Repeat with all the polymers you have available.

5 Leave your polymers buried for at least a week, or longer if you can. During this time, make sure the soil does not dry out by adding water every few days. If possible, leave the container in the sun to simulate the outdoor environment. If not, indoors on a windowsill will work just as well.

6 At the end of the time, dig up each polymer, take photographs for comparison to the original. Record observations of any breakdown or decay you can see.

ANALYSIS

When writing your responses to each of the following questions, consider the following:

- Who is your audience?
- Have you used appropriate scientific terms and language in your response?
- Have you used data, diagrams or images in your answer to make your response clear?

1 Did any of your polymers show signs of breaking down? What observations did you make that show this? Can you rank your polymers in order of level of breakdown?

2 If you chose a biodegradable polymer, did it show more, less or similar signs of decay compared to the others? What does this tell you about biodegradable polymers?

3 It is possible that none of your samples will show signs of decay. Suggest a reason for this.

4 Why was it important that all samples were the same size?

CONCLUSION

Write a conclusion for this investigation that is related to the aim.

8.13 How can we solve the plastic problem?

BY THE END OF THIS MODULE, YOU WILL BE ABLE TO:

- ✓ explain why we need to develop methods for disposal and breakdown of plastics
- ✓ explain how microbes may be able to be used to control plastic waste.

How big is the plastic problem?

More than 200 million tonnes of plastic waste is generated around the world each year. Despite efforts to recycle plastics, less than 20% of waste plastic is recycled. Some of this goes into the oceans (Figure 8.13.1), but most ends up in landfills, where it can take hundreds or thousands of years to break down.

Benny Marty/Shutterstock.com

▲ **FIGURE 8.13.1** There is an area more than twice the size of New South Wales in the Pacific Ocean where plastic waste, including microplastics, has built up.

What can microbes do to help?

In 2016, Japanese scientists visiting a plastics recycling facility observed that a bacteria (*Ideonella sakaiensis*) was breaking down PET plastics (Figure 8.13.2). PET, or polyethylene terephthalate, is a long-chain polymer.

Scientists discovered that the bacteria produces an enzyme called PETase that is able to break the long PET chains into the monomers it was produced from. Enzymes are proteins that cause specific reactions to occur. The monomers were then further broken down by the enzymes to release energy and help the bacteria grow.

Studies are currently underway to improve the efficiency of the bacteria, so they can be used as a method of helping to break down PET plastics. Other scientists are working on using other bacteria to break down different types of plastics.

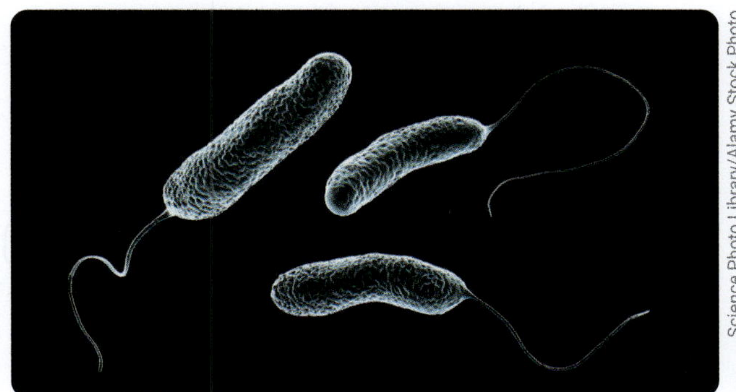

Science Photo Library/Alamy Stock Photo

▲ **FIGURE 8.13.2** *Ideonella sakaiensis* bacteria produce enzymes that break down PET into its monomers.

8.13 LEARNING CHECK

1. What does PET stand for?
2. Conduct research to **describe** impacts of plastics in our oceans.
3. **Describe** the process used by the bacteria *Ideonella sakaiensis* to break down PET.
4. **Explain** how the use of bacteria could help solve our waste plastic problem.

8 REVIEW

REMEMBERING

1 **Outline** the difference between organic and inorganic substances.

2 **Describe** the importance of crude oil to our society, including examples of the different components of crude oil and what they can be used for.

3 **Outline** the differences between an alkane, an alkene, an alcohol and a carboxylic acid.

4 **Identify** the fractions collected at the top and bottom of a distillation tower during fractional distillation of crude oil.

5 Write a word equation for the combustion of a hydrocarbon.

UNDERSTANDING

6 **Explain** how a polymer is formed.

7 What is the name and chemical formula of an alkane with seven carbons?

8 **Explain**, using examples, why we have both flexible and hard polymers.

9 **Explain** the process of fractional distillation of crude oil.

10 What do the numbers in the triangle symbols found on many objects made from different polymers mean?

APPLYING

11 **Identify** why ethane is a hydrocarbon but ethanol is not a hydrocarbon.

12 A flame is yellow–orange and producing black particles. **Determine** whether complete or incomplete combustion is occurring.

13 A plastic bag is lightweight and flexible. **Predict** whether it is more likely to be made from LDPE or HDPE.

14 **Identify** the properties you would expect in a shock-proof polymer phone case.

15 **Describe** which components of crude oil you would use for:

a lubricating moving machinery parts.

b cooking on your barbecue.

c paving roads.

ANALYSING

16 **Explain** why some components of crude oil are gases, some are liquids and some are solids.

17 What conditions need to change for complete combustion to become incomplete combustion?

18 In the Australian Capital Territory, surveys were done in various bushland sites to **determine** the number of plastic shopping bags that had been dumped as rubbish. Two types of bags were identified, lightweight bags and heavier, more durable bags. The results of the surveys are seen in the graph below. **Suggest** what change might have occurred between 2012 and 2013, based on the graph.

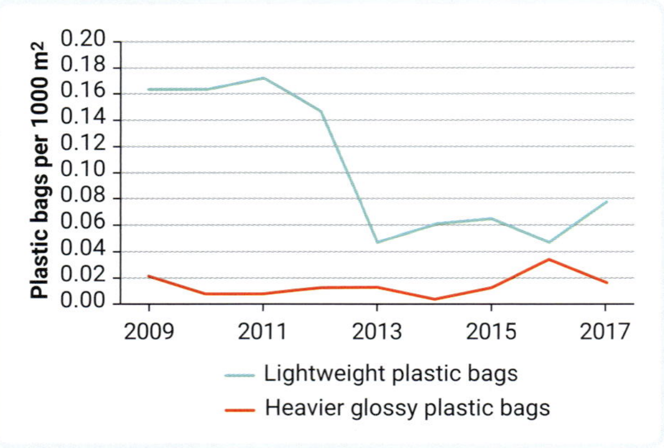

19 Scientists are finding increasing numbers of large marine mammals that have died from toxins associated with plastics. **Explain** why this is happening.

EVALUATING

20 Before the 1800s, coal was widely used for a range of uses such as cooking and heating. Once crude oil became more available, coal was used less. By considering the advantages and disadvantages of coal and crude oil, **analyse** why our use of fuels has changed over time.

21 **Research** a government action designed to reduce plastic waste in your state. **Assess** the pros and cons of the action and **judge** whether you think it is a good idea.

CREATING

22 **Create** a poster to convince people to choose biopolymers over traditional plastics. **Propose** how you will convince them to make this choice.

SCIENCE IN DEPTH STUDY

1 Connect what you've learned

In this chapter, you've learned about polymers, how they are made and used, and the environmental issues associated with polymer waste. Consider and record the types of plastics you use or encounter each day. You can document this as a list, with photographs or with images from the internet. Do some research to identify the type of plastics in each item.

2 Check your thinking

Choose two or three items made of plastic that you use on a regular basis. Explain what properties make them useful. Do you reuse any of these items, or are they thrown away after a single use? Do some research to see if you can determine how long it will take these items to break down.

3 Get into action

Find a local area and conduct a survey to determine the level of polymer pollution. This could be your school yard, a nearby park, your backyard, a nearby beach, river, lakeside or bushland area. To conduct the survey, you can write notes or take photographs to record the waste you find. Be environmentally friendly and clean up the plastic waste as you conduct your survey.

4 Communicate

Create a photo collage or report to document the findings of your survey to your local government to get action taken. Suggest sources of the pollution that you found and the potential consequences of this pollution being in the environment.

9780170491785

ENVIRONMENTAL SUSTAINABILITY

SYLLABUS OUTCOMES

A STUDENT:

- analyses the impact of human activity on the natural world SC5-ENV-01
- analyses data from investigations to identify trends, patterns and relationships, and draws conclusions SC5-WS-06
- selects suitable problem-solving strategies and evaluates proposed solutions to identified problems SC5-WS-07

© 2023 NSW Education Standards Authority

THE CHAPTERS RELATED TO THIS FOCUS AREA ARE:

- CHAPTER 9 – CLIMATE CHANGE
- CHAPTER 10 – SUSTAINABILITY

frozygraphie/Shutterstock.com

Climate change

9780170491785

SCIENCE IN DEPTH 

▲ **FIGURE 9.0.1** A healthy section of the Great Barrier Reef

We all contribute to climate change through our lifestyles, behaviours and choices. An effect of climate change we hardly notice is ocean acidification: the increasing acidity of sea water.

- What causes ocean acidification?
- What effects does a lower pH have on life in the ocean?
- How will ocean acidification affect human societies?
- How will it affect ecosystems, such as coral reefs?
- How can you highlight the effects of ocean acidification and its causes so that your community change their behaviour to reduce the rate of ocean acidification?

 **DIVE INTO SCIENCE!**

At the end of this chapter, you can complete Science in Depth Study #9. You can use the information you learn in this chapter to complete the project.

Assessments
- Prior knowledge quiz
- Chapter review questions
- End-of-chapter test
- Depth study: Research project and podcast

Videos
- Science skills in a minute: Trends and patterns in data **(9.8)**; Cause and effect **(9.10)**
- Video activities: The carbon cycle **(9.2)**; Cape Grim monitoring station **(9.7)**; Carbon farming **(9.11)**

Science skills resources
- Science skills in practice: Trends and patterns in data **(9.8)**; Cause and effect in science **(9.10)**
- Extra science investigations: Greenhouse effect in a bottle **(9.3)**; How the greenhouse effect impacts air and soil temperature **(9.3)**

Interactive and other resources
- Simulation: The greenhouse effect **(9.3)**
- Crossword: Earth's spheres **(9.1)**
- Drag and drop: Evidence of climate change **(9.6)**
- Label: Interactions between Earth's spheres **(9.2)**; Climate change and human health **(9.9)**
- Quizzes: Main climate zones **(9.4)**
- Activity sheets: Campaign for change **(9.11)**; What do the media say? **(9.11)**
- Worksheets: Mapping the global spheres **(9.1)**; Indicators of climate change **(9.6)**; Effects of global warming **(9.9)**

Nelson MindTap

To access resources above, visit **cengage.com.au/nelsonmindtap**

9.1 The Earth system

BY THE END OF THIS MODULE, YOU WILL BE ABLE TO:

✓ describe Earth's atmosphere, biosphere, hydrosphere and geosphere.

GET THINKING

Earth is a complex planet. Scientists model Earth as a system having four distinct spheres. As you read through this module, think of examples of different things around you that are part of each of the four spheres. Can you identify anything that is difficult to place into a single sphere?

climate
the average weather conditions in a particular region over an extended period

Earth system
Earth as a complex whole, made up of the four interacting spheres: atmosphere, biosphere, hydrosphere and geosphere

atmosphere
the gaseous layer surrounding Earth that is retained by Earth's gravity

biosphere
all aspects of Earth and the atmosphere that support life, including all living things

hydrosphere
the parts of Earth containing all forms of water

geosphere
the rocks, minerals and landforms of the surface and interior of Earth

greenhouse gases
gases in the atmosphere that can trap heat and affect global surface temperatures and other aspects of the climate

photosynthetic
able to photosynthesise, converting carbon dioxide and water into sugar and oxygen

cyanobacteria
colonies of blue-green algae capable of photosynthesis

continental ice sheet
an extensive sheet of permanent ice covering a large area of the land surface

Introducing the Earth system

To understand our planet's **climate**, we first need to understand the nature of the **Earth system**. We describe Earth as being made up of four main 'spheres': **atmosphere**, **biosphere**, **hydrosphere** and **geosphere**. The four spheres interact continuously and no single sphere acts in isolation without having an effect on the others.

Atmosphere

The atmosphere is composed of the gases that make up the air surrounding the planet. The force of Earth's gravity keeps the atmosphere in place. Chemical analysis shows that the atmosphere is made up of 78 per cent nitrogen and 21 per cent oxygen, with the remaining 1 per cent made up of trace gases such as argon and a group collectively known as the **greenhouse gases**.

The atmosphere of 3 billion years ago was very different from the one we breathe today. Evidence from the fossil record and ancient rocks indicates that oxygen started to accumulate in the atmosphere about 2.4 billion years ago. **Photosynthetic** microbes called **cyanobacteria**, some types of which still exist today, were responsible for this increase in oxygen. They used the Sun's energy to convert carbon dioxide and water into sugars in a process called photosynthesis.

We know that the composition of the atmosphere changed significantly in its early history, but for the past 2 million years it has been relatively stable. Evidence for the composition changing over time is seen in the chemistry of rock and mineral deposits, such as the iron ore deposits in northern Western Australia, and ancient bubbles of atmospheric gases that have been trapped in ice. Ice cores from Antarctica's **continental ice sheet** provide a record of the atmospheric composition going back as far as 800 000 years. Gases from the past are trapped as bubbles of air trapped within the ice. The ice cores allow scientists to examine the composition of the atmosphere before human activities and compare this with the atmosphere today. Ice cores can also identify major events such as volcanic eruptions (Figure 9.1.1).

Heidi Roop, National Science Foundation (NSF)

NASA's Goddard Space Flight Center/Ludovic Brucker

▲ **FIGURE 9.1.1** (a) An Antarctic ice core containing trapped air and a dark line indicating a major volcanic eruption that deposited a layer of material on the ice. (b) A close-up of an ice core with trapped bubbles of atmospheric gas.

Increases in carbon dioxide concentration coincided with the Industrial Revolution. Societies became more urbanised and industrial, moving away from agriculturally based economies. Inventions such as the steam engine, the development of rail networks and electricity generation led to an increase in the use of coal. Since the 1850s, human activities around the world have continued to increase the amount of carbon dioxide and other greenhouse gases in the atmosphere.

Interactive resource
Crossword: Earth's spheres

Other resource
Worksheet: Mapping the global spheres

Biosphere

The biosphere contains all of Earth's living organisms – an amazing array of species. The different zones of the biosphere contain suitable habitats and ecosystems for the survival of different animals, plants and other species. Some species can live almost anywhere, but others are very restricted to specific locations. For example, Galápagos land iguanas are found on only a few small islands in the Pacific Ocean 1000 km from South America and lay their eggs in burrows in moist sand.

climate change
a change in global or regional climate patterns

While new species are still being discovered regularly around the world, many others are facing extinction. We know from the fossil record that mass extinctions have occurred in the past. Several of these, including the extinction at the end of the Ordovician period that occurred 445 million years ago, are thought to be climate related. In the Ordovician extinction, 60–70 per cent of all species died out due to a short but intense ice age.

In 2019, scientists declared the Bramble Cay melomys to be the world's first mammal to become extinct due to **climate change** (Figure 9.1.2). It lived on a single island in the Great Barrier Reef that was subjected to rising sea levels.

naturepl.com/Bruce Thomson

▲ **FIGURE 9.1.2** The Bramble Cay melomys is the world's first mammal to become extinct due to climate change.

Today, scientists have identified more than 700 species of animals and plants that are under threat from climate change. There are many endangered species in Australia, some of which visit or live in regions like Ningaloo Reef in Western Australia or the Kosciuszko National Park in New South Wales (Figure 9.1.3).

iStock Essentials/Istock.com

Chiara & Sigismondo/Alamy Stock Photo

▲ **FIGURE 9.1.3** Two different parts of the biosphere in Australia: (a) Ningaloo Reef in Western Australia; (b) Kosciuszko National Park in New South Wales, home of the endangered mountain pygmy-possum

Hydrosphere

permafrost permanently frozen soil, sediment or rock

The hydrosphere includes all forms of water on the planet, whether in solid, liquid or gaseous form. The frozen water in glaciers and ice caps is sometimes considered a separate sphere, known as the cryosphere. In this chapter, we have included the cryosphere as part of the hydrosphere. Water occurs as fresh water and salt water. Only 2.5 per cent of the world's water is fresh, and most of this is inaccessible to people because it is locked up in the form of ice and **permafrost**. Humans use fresh water for drinking, bathing, recreation, agriculture and industry, but we can only access about 0.3 per cent of all fresh water on Earth, as shown in Figure 9.1.4.

The ocean plays a major role in storing and transferring heat. The ocean stores approximately 90% of the excess heat trapped by greenhouse gases in the atmosphere. It also is a major store of carbon dioxide. The ocean currently absorbs about 22 million tons of CO_2 per day.

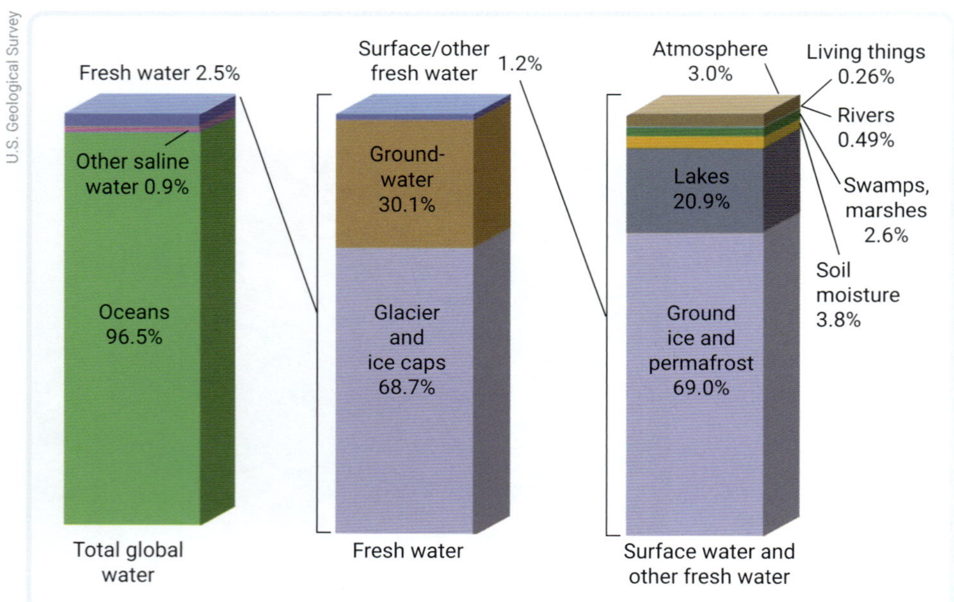

U.S. Geological Survey

▲ **FIGURE 9.1.4** The distribution of Earth's water

Geosphere

The geosphere includes the rocks, minerals and landforms that make up Earth's surface and all the material that makes up the interior of Earth (Figure 9.1.5). Land makes up about 29 per cent of the total surface area of Earth, with the rest (71 per cent) being water, mainly the oceans. The geosphere interacts with the other spheres; for example, it overlaps with the biosphere because many species live on or in this sphere. Although it appears static, the geosphere interacts with the flows of matter and energy from one sphere to another.

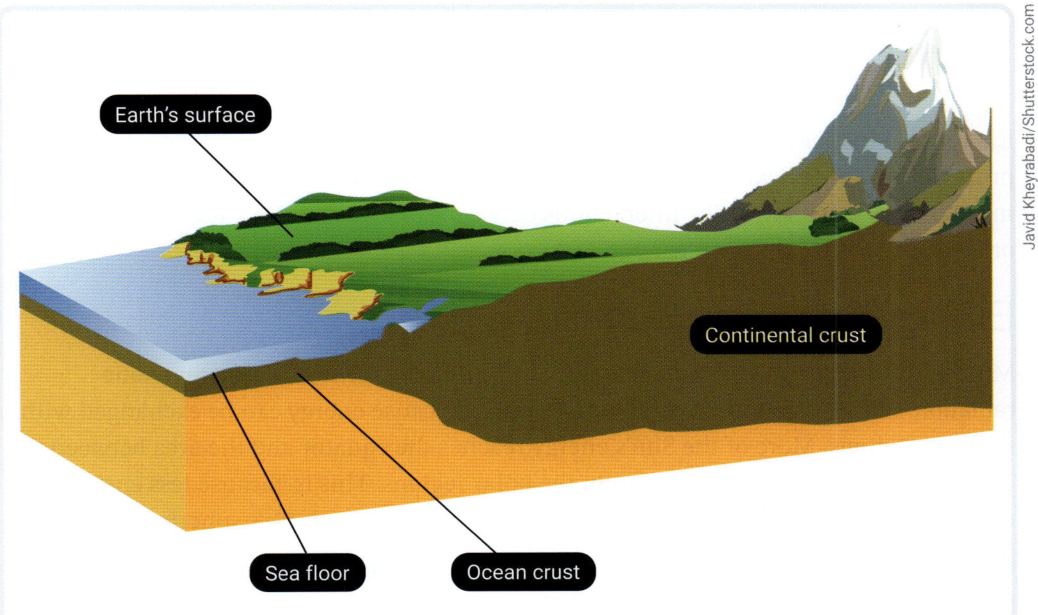

Javid Kheyrabadi/Shutterstock.com

▲ **FIGURE 9.1.5** The geosphere

9.1 LEARNING CHECK

1 **List** the four spheres that make up the Earth system.
2 **Describe** the composition of gases within the atmosphere.
3 **Describe** how the concentration of oxygen in the atmosphere increased billions of years ago.
4 **Explain** the importance of ice cores to scientists' understanding of the atmosphere.
5 **Explain** why it is so important to take care of naturally occurring fresh water.

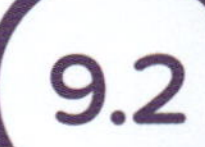

9.2 Interactions between Earth's spheres

BY THE END OF THIS MODULE, YOU WILL BE ABLE TO:

- ✓ describe how energy flows through the Earth system
- ✓ explain the cause of deep ocean currents and how they affect climate and marine life.

Video activity
The carbon cycle

Interactive resource
Label: Interactions between Earth's spheres

GET THINKING

We experience weather every day and can recognise patterns in the seasons. But have you ever looked at a satellite photo or zoomed around in Google Earth? The swirling clouds over the oceans and continents give you a clue about Earth's complexity, with many things interacting with one another. As you read this module, focus on the interactions taking place between the four Earth spheres and how these link to our changing climate.

Energy and matter flow through Earth's four spheres in complex ways through natural processes and chemical cycles. This means that the four spheres are connected and interact with each other. The interactions between spheres are also affected by the direct or indirect activities of people.

Energy from the Sun

electromagnetic radiation
energy that travels in electromagnetic waves, such as infrared and visible light

open system
a system that allows the transfer of matter and energy in and out of the system

Most of the energy on Earth comes from the Sun in the form of **electromagnetic radiation**. Earth's energy flow is an **open system** because energy can be exchanged with its surroundings. More of the Sun's energy reaches the equator and the area between the Tropics of Cancer and Capricorn than the polar regions. This is because less energy per unit area is received at the surface at higher latitudes due to Earth's spherical shape.

Of the energy coming from the Sun, 30 per cent is reflected back out to space without being absorbed by Earth (Figure 9.2.1). The remaining 70 per cent of the energy flows through Earth's four spheres.

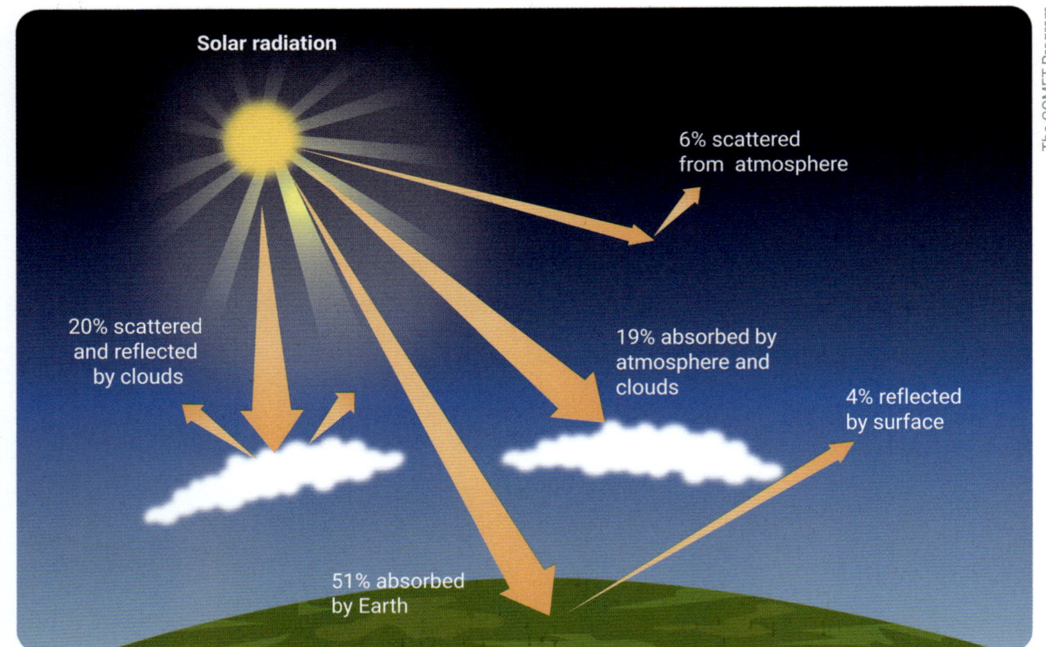

▲ **FIGURE 9.2.1** Earth's energy budget

Energy and matter flow between the spheres

A large amount of energy is absorbed by Earth's surface and atmosphere. The ocean absorbs two-thirds of the energy and, together with land surfaces, transfers some heat back into the atmosphere.

Processes in the spheres redistribute energy through naturally occurring cycles, such as the water cycles, **deep ocean currents**, and wind and **weather** patterns. These interact across multiple spheres and are all driven by the energy of the Sun (Figure 9.2.2). The cycling of energy is complex and often includes the cycling of matter.

deep ocean currents
the water movement occurring below a depth of 400 m caused by changes in water temperature and salinity

weather
the conditions in the lower part of the atmosphere at a given time or place

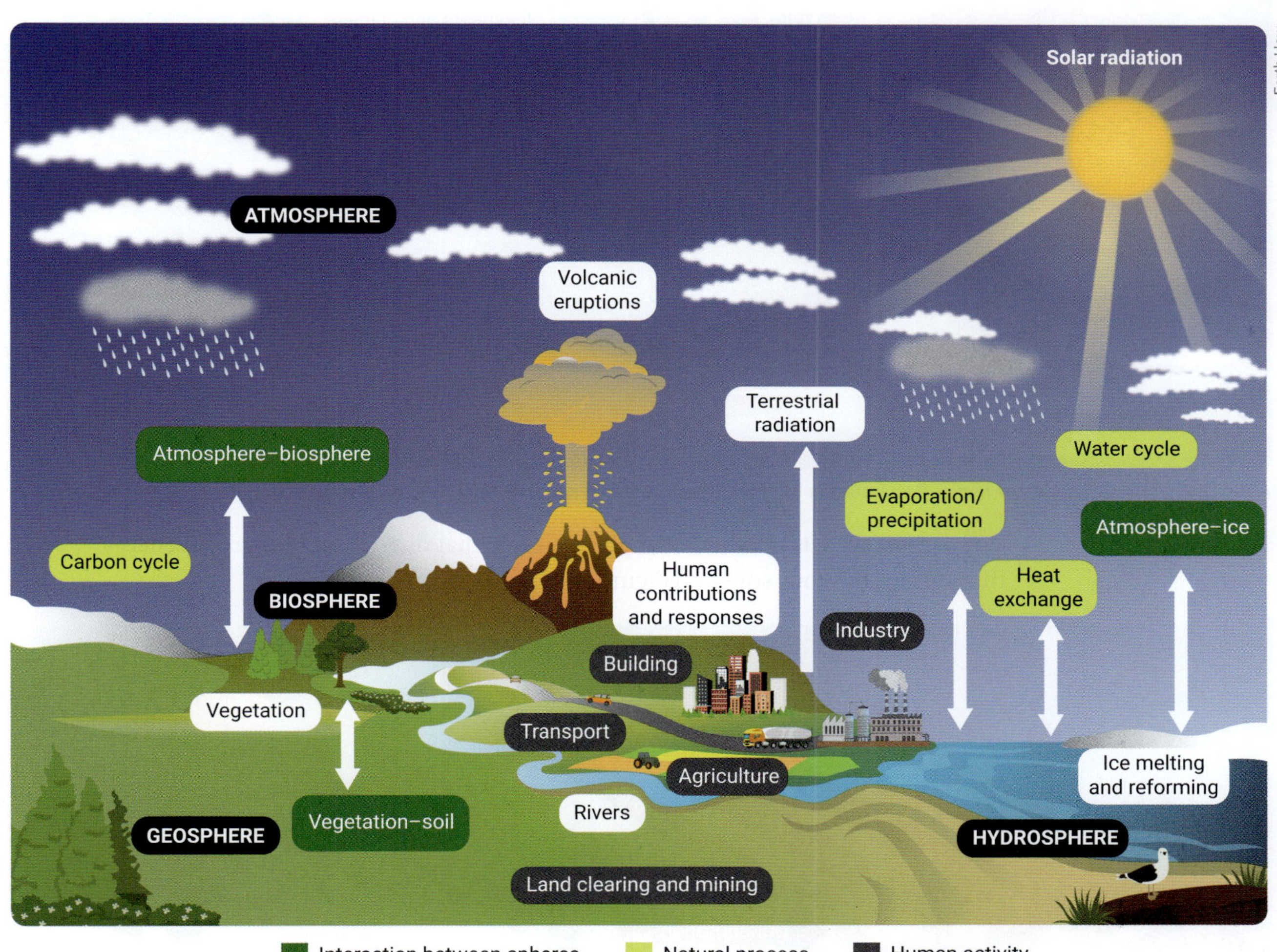

▲ **FIGURE 9.2.2** Some of the interactions between Earth's four spheres

One example of a cycle that interacts with Earth's four spheres is the water cycle (Figure 9.2.3). Energy from the Sun causes water to circulate through the atmosphere, biosphere, hydrosphere and geosphere.

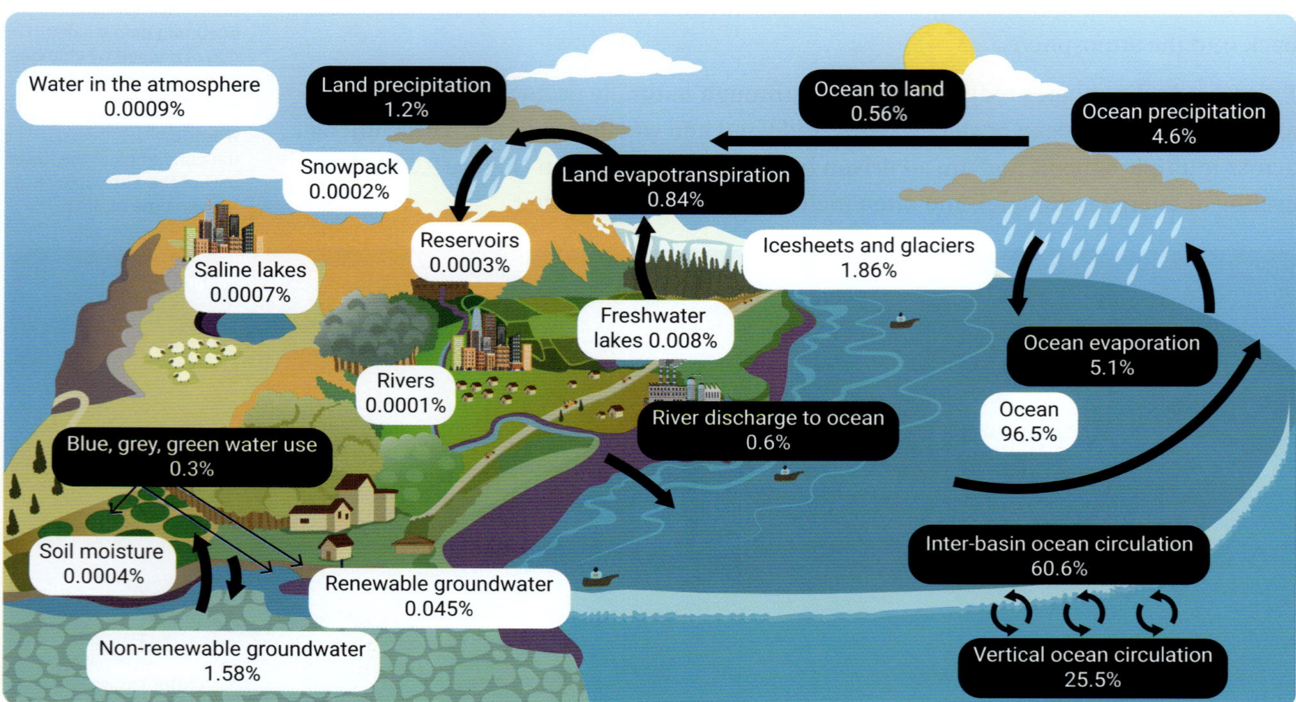

▲ **FIGURE 9.2.3** The water cycle. Percentages indicate the amount of water in each phase.

The carbon cycle is another example of a cycle that distributes matter and energy through Earth's four spheres (Figure 9.2.4).

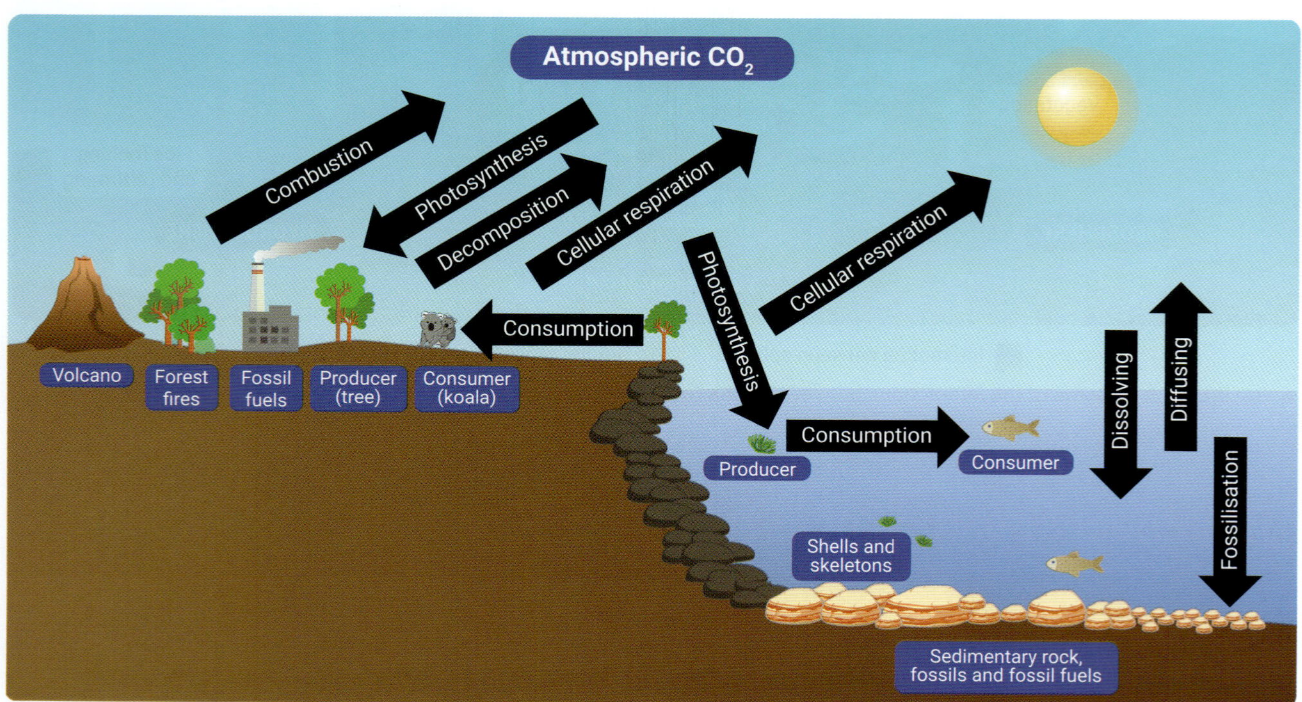

▲ **FIGURE 9.2.4** The carbon cycle

9780170491785

The impact of human activities on the Earth system

Human activities affect the natural resources and cycles within each sphere and, in some cases, disrupt the normal flow of energy and matter. Agriculture, aquaculture, mining and industrialisation have all had significant effects (Figure 9.2.5). For example, Australians have cleared almost 50 per cent of native forests in 200 years. **Land clearing** in Queensland and New South Wales is a serious threat to koalas and many other species that rely on forests for habitat. Land clearing increases the surface runoff of sediment into rivers and coastal regions, damaging coral reefs and marine ecosystems such as seagrass beds.

land clearing the removal of natural vegetation and habitats, such as forests

iStock Signature/Istock.com, Symbiosis Australia/Shutterstock.com, rickyd/Shutterstock.com, Shutterstock.com

▲ **FIGURE 9.2.5** (a) Land clearing, and (b) mining, in Australia destroys the habitat of different species and can disrupt Earth's natural cycles. (c) Land clearing has greatly reduced the habitat of koalas in New South Wales and Queensland. (d) Sediment runoff damages ecosystems such as seagrass beds.

Deep ocean currents – an energy conveyor belt

surface currents
currents in the top 100 m of the ocean

thermohaline
the circulation of dense, cold-water currents deep in Earth's oceans

salinity
the concentration of dissolved salt

Surface currents and deep ocean currents distribute heat and influence weather and climates. The deep circulation currents are called the **thermohaline** circulation and are also known as the 'global conveyor belt' (Figure 9.2.6). The currents are controlled by changes in **salinity** and temperature, which affect water density. The thermohaline circulation begins in polar regions. As sea ice forms, it leaves behind salt in the water below the ice. This makes the cold water denser than the water below it, so it sinks and flows along close to the sea floor.

As the cold water flows in the deep ocean, it absorbs heat from surrounding water. This makes it less dense, causing it to rise. A warmer flow of water then carries heat from the Pacific to the north Atlantic where the circulation begins again. The current moves slowly – it can take up to 1000 years to complete a cycle.

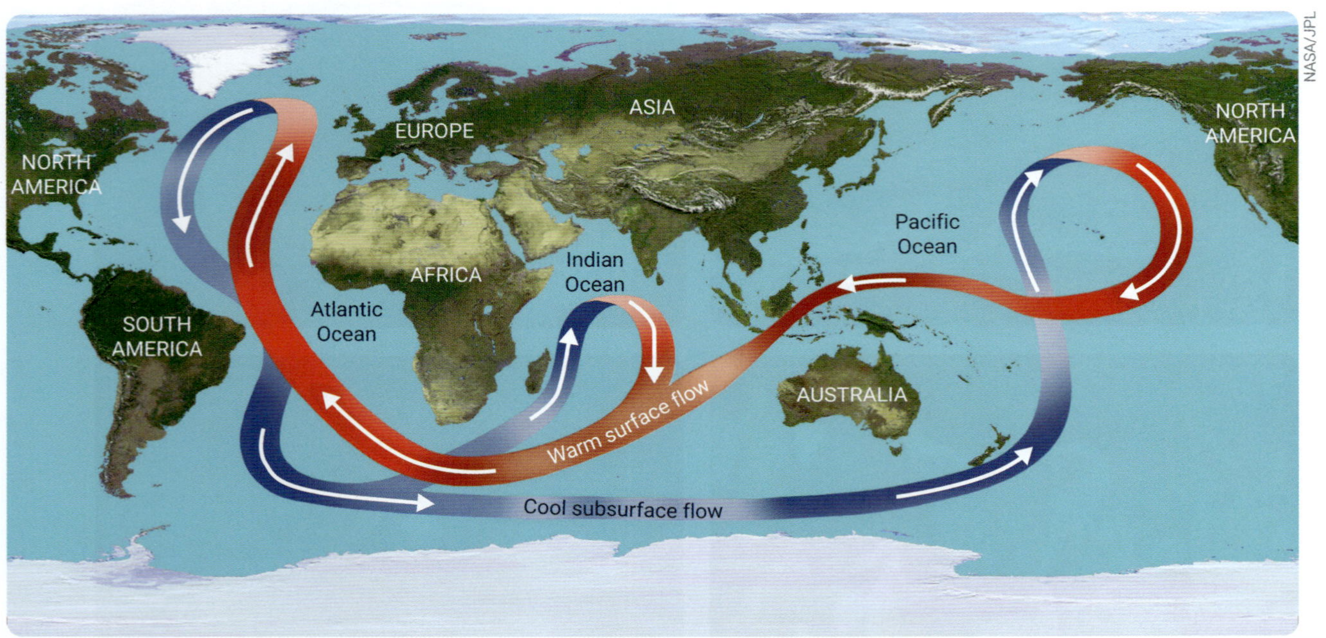

▲ **FIGURE 9.2.6** Global ocean currents

The effect of deep ocean currents on climate and marine life

surface temperature
the measure of the relative hotness or coldness of Earth's surface

Deep ocean currents have major effects on the climate. As the currents move, they gradually transfer heat to the surrounding water and the atmosphere, helping to regulate water and **surface temperatures**. As ocean water evaporates, it moderates the temperature and increases the humidity of the surrounding air. This in turn directly affects the climate near these currents, moderating the climate along the coasts.

The deep ocean currents are also important for marine life. The water in these deep currents is rich in nutrients and carbon dioxide. As the currents rise to the surface, they bring nutrients and carbon dioxide to the warmer surface layers through a process called **upwelling**. This helps the growth of algae and seaweed, supporting food chains in the upper layers of the oceans where most marine organisms live (Figure 9.2.7).

upwelling
the movement of water from the depths of the ocean to the surface, bringing nutrients and carbon dioxide with it

9.2

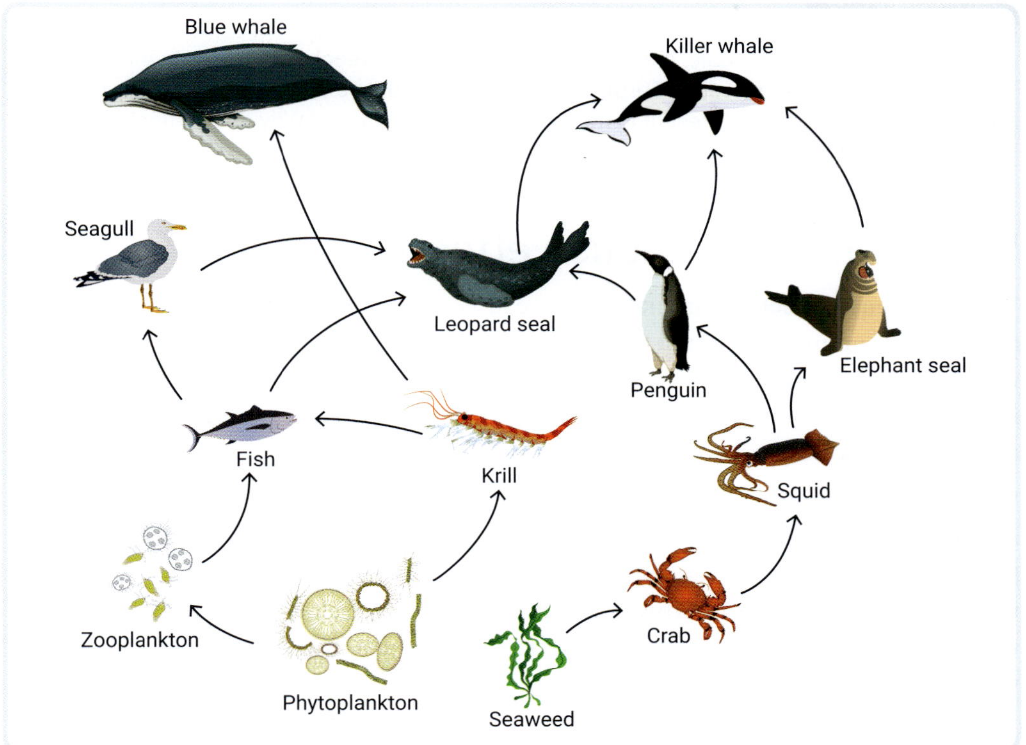

▲ **FIGURE 9.2.7** Food chains in the upper layers of the ocean rely on the nutrients and carbon dioxide moved by deep ocean currents.

9.2 LEARNING CHECK

1. **Describe** how the energy from the Sun drives the water cycle.
2. **List** the types of human activity that interrupt the natural flow of matter and energy in Earth's cycles.
3. **Describe** the path that deep ocean currents take in circling the world.
4. **Describe** an example of how Earth's spheres interact with each other.
5. **Explain** the cause of deep ocean currents.
6. **Explain** how deep ocean currents influence climate.
7. **Compare** and **contrast** the water and carbon cycles. Consider how water and carbon cycle through different spheres.
8. **Create** a mind map that links all the ways Earth's spheres interact with each other, and the human activities that disrupt or change these interactions and cycles.

9.3 The greenhouse effect

BY THE END OF THIS MODULE, YOU WILL BE ABLE TO:

✓ use a labelled diagram to show the greenhouse effect

✓ explain how the greenhouse effect warms Earth.

Interactive resource
Simulation: The greenhouse effect

Extra science investigations
Greenhouse effect in a bottle
How the greenhouse effect impacts air and soil temperature

greenhouse effect
a natural process that traps energy within the atmosphere, raising Earth's surface temperature

electromagnetic spectrum
the range of electromagnetic waves arranged in sequence from high-energy gamma rays to low-energy radio waves

GET THINKING

As you complete this module, think about the presence of greenhouse gases in the atmosphere and their effect on Earth's climate. What would be the situation if these gases were not present at all, or present in much lower concentrations? Why do you think carbon dioxide and methane are so important to Earth's climate?

What is the greenhouse effect?

The **greenhouse effect** is a natural process caused by a group of gases in the atmosphere commonly referred to as 'greenhouse gases'. These gases include water vapour, carbon dioxide, methane, nitrous oxide and ozone. Humans have caused an increase in all of these (other than water vapour) and added new, synthetic ones such as chlorofluorocarbons.

Visible light is one part of the **electromagnetic spectrum**. The spectrum is made up of electromagnetic waves that have different wavelengths but all travel at the speed of light (Figure 9.3.1). All types of radiation can enter the atmosphere. Some of this radiation is reflected back into space. Shortwave radiation includes wavelengths within the infrared, visible and ultraviolet parts of the electromagnetic spectrum. Longwave radiation falls within the infrared region of the electromagnetic spectrum.

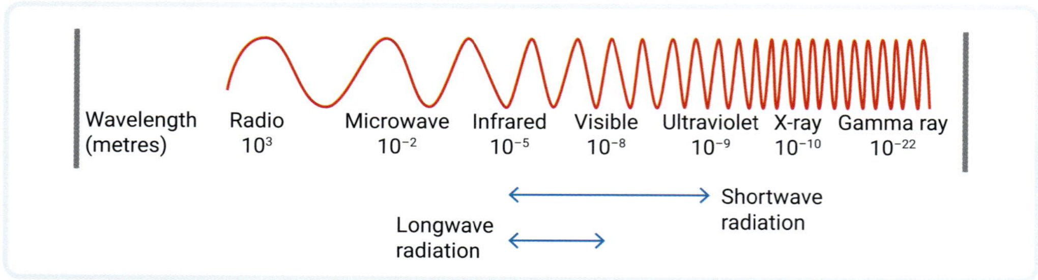

▲ **FIGURE 9.3.1** Shortwave and longwave radiation of the electromagnetic spectrum

Most of the radiant energy from the Sun, in the form of high-energy shortwave radiation, passes through the atmosphere to Earth's surface. As it passes through the atmosphere, some of this radiation is absorbed by clouds and gases. Earth's oceans, rivers, land and plants also absorb energy. Earth's atmosphere and surface re-radiate this back into the atmosphere as longwave radiation or heat. The greenhouse gases absorb and trap the re-radiated energy, warming the atmosphere (Figure 9.3.2).

9780170491785

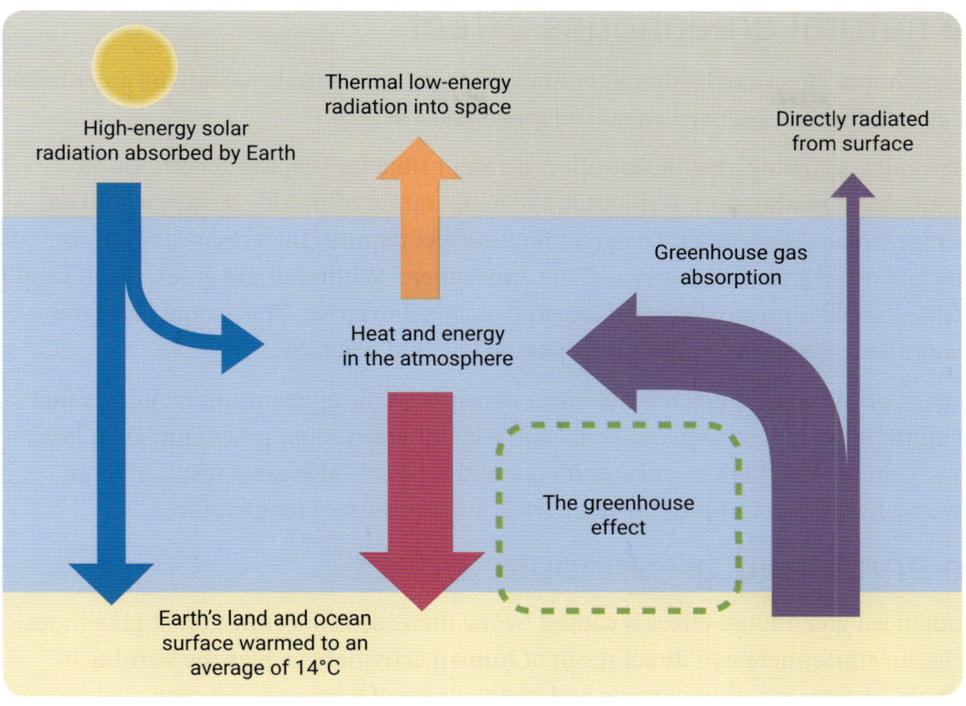

▲ FIGURE 9.3.2 How the greenhouse effect is generated

Modelling the greenhouse effect

☆ ACTIVITY

Materials and equipment

- 2 × 2 L soft-drink bottles (or similar)
- 2 thermometers (or temperature probes, if available)
- plastic wrap
- elastic band
- masking tape
- 100 W light globe and lamp
- scissors or craft knife

 **Safety**

To complete this activity, you will need to cut the top off two soft-drink bottles using scissors or a craft knife. Take care, as the cutting tool and the top of the cut soft-drink bottles will be very sharp and could easily cut you if handled incorrectly.

Procedure

1 Cut the top of the soft-drink bottles off at the 'shoulder', ensuring that they are the same height.

2 Tape the thermometers to the inside of the soft-drink bottles, with the thermometer bulb or probe about 5 cm from the base. Ensure that it is easy to read the temperature from the outside of the bottles.

3 Place one of the two soft-drink bottles 15 cm either side of the light globe.

4 Place the plastic wrap on top of one container and use the elastic band to hold it in place. Immediately take and record a temperature reading from both bottles.

5 Turn on the light, and then take and record measurements every 2 minutes for 30 minutes.

6 Compare the temperatures in the two soft-drink bottles.

Analysis

Was there a difference between the two bottles? If there was a difference, **explain** why this happened.

The natural greenhouse effect

enhanced greenhouse effect
strengthening of the natural greenhouse effect due to an increase in greenhouse gases caused by human activities

The greenhouse effect can be thought of in two ways: the naturally occurring greenhouse effect and the human-induced **enhanced greenhouse effect**.

The naturally occurring greenhouse effect is responsible for keeping conditions on Earth suitable for habitation. Without it, the temperature of Earth's surface would be, on average, about 33°C cooler. The greenhouse effect requires the presence of carbon dioxide and other greenhouse gases in the atmosphere. Without these gases, Earth would be unable to hold onto any heat and it would escape into space. This would leave the planet as an icy wasteland and life as we know it would not exist.

The extent of the natural greenhouse effect depends on the concentration of gases that make up the atmosphere. Too low a concentration of greenhouse gases and Earth loses too much energy. Too high a concentration and the greenhouse gases result in more energy and heat being trapped, raising Earth's average surface temperature.

The enhanced greenhouse effect

The enhanced greenhouse effect is caused by the increased concentration of greenhouse gases in the atmosphere as a direct result of human activities, such as the burning of fossil fuels. The gases carbon dioxide and methane are of particular concern.

The measured concentration of carbon dioxide in the atmosphere has increased by approximately 50 per cent since the beginning of the Industrial Revolution, from around 280 parts per million (ppm) to more than 420 ppm. What is alarming to scientists is the rate of change in concentration. Before the Industrial Revolution, changes in carbon dioxide concentrations occurred over thousands of years. Figure 9.3.3 shows the concentration levels of atmospheric carbon dioxide over the last 800 000 years.

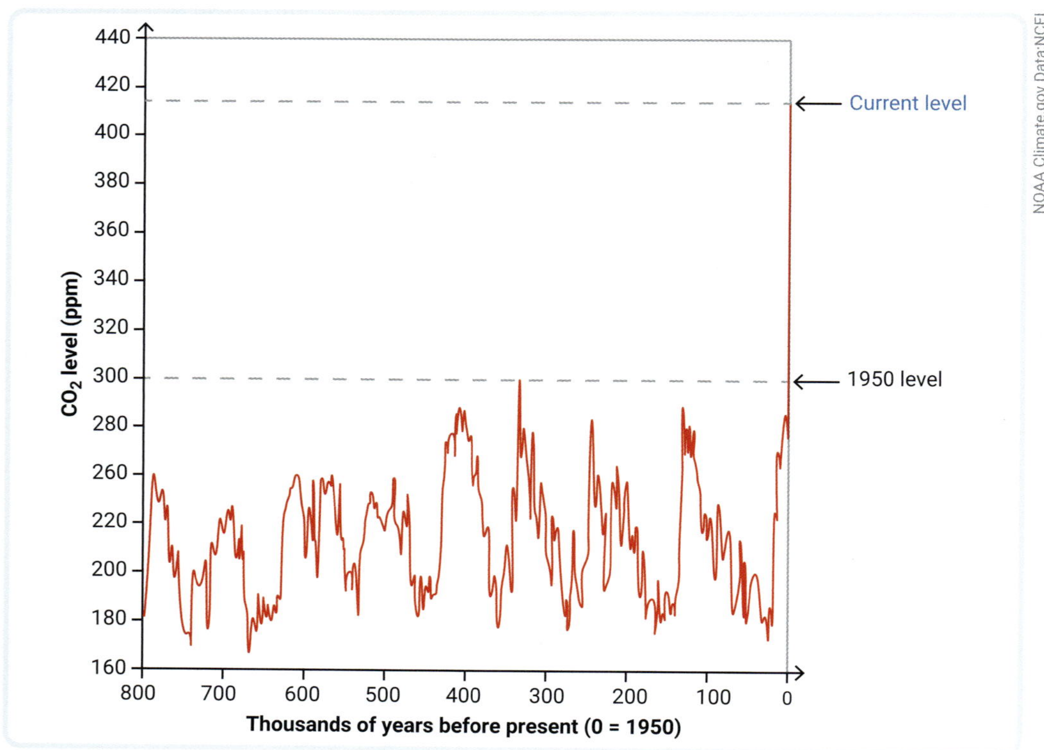

▶ **FIGURE 9.3.3** Atmospheric carbon dioxide concentration (in ppm) over the last 800 000 years

9780170491785

Measurements reported by the Intergovernmental Panel on Climate Change (IPCC) in 1986 showed it had taken 200 years for carbon dioxide levels to increase by 25 per cent. More worrying are the measurements reported for 2021, which show that carbon dioxide levels have increased 21 per cent in the last 35 years.

Methane concentrations have increased by nearly 160 per cent over the same 235-year period (Figure 9.3.4). This is mostly due to the increased farming of livestock, coal and gas production, and waste disposal in landfills.

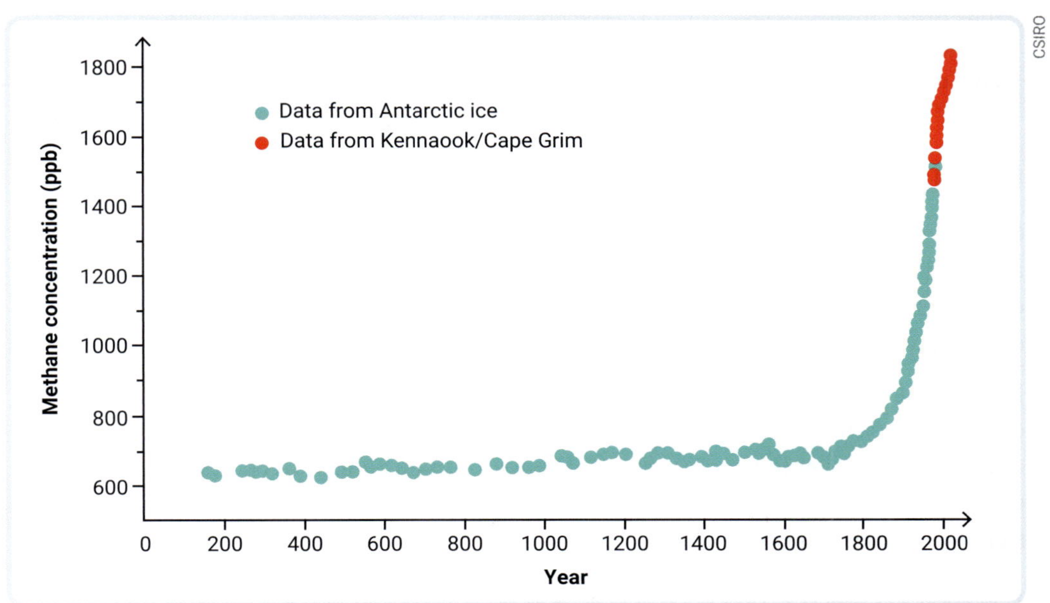

▲ **FIGURE 9.3.4** Atmospheric methane concentrations (in parts per billion) over the last 2000 years

As we emit more greenhouse gases into the atmosphere, the enhanced greenhouse effect is becoming stronger, and the average global temperature is increasing.

9.3 LEARNING CHECK

1 **List** five greenhouse gases.
2 **Describe** characteristics and features of shortwave and longwave radiation.
3 **Explain** why the natural greenhouse effect is so important to life on Earth.
4 **Discuss** how the enhanced greenhouse effect contributes to warming Earth.
5 **Compare** and **contrast** the natural greenhouse effect with the enhanced greenhouse effect.
6 Imagine you have to **explain** the greenhouse effect to Year 7 students. **Create** an infographic or animation to show how the greenhouse effect is generated. Keep your explanations simple and concise, using language your intended audience will understand.

9.4 Global climates

BY THE END OF THIS MODULE, YOU WILL BE ABLE TO:

✓ label a map with the locations of Earth's different climate zones
✓ explain why different locations experience different climates.

GET THINKING

As you work your way through this module, consider the climate where you live. What are the characteristics of your climate? How does the weather vary over the course of the year? Is the weather similar from year to year, or have there been significant changes over the past decade? How would your parents or grandparents respond to the same question?

Climate zones

precipitation
liquid or solid water in the form of rain, snow, sleet or hail

Köppen climate classification
a system used since 1900 that classifies climate zones based on temperature, amount and type of precipitation, and vegetation

Climate refers to long-term weather conditions in a particular region. It includes the average temperature, humidity, type and amount of **precipitation**, and seasons that occur over extended periods of time (usually 30 years or more). Weather is a description of the day-to-day variation that may occur in a location.

It is important to remember that climate is described by averages. Weather may have extremes in rainfall or temperature from day to day, but climate describe the averages of these factors over tens or hundreds of years. Climate zones can be described and categorised in many ways. In this module, the five climate zones are based on the **Köppen climate classification**.

Factors that influence and create climate zones

Coriolis effect
the apparent deflection of large masses of air and water due to Earth's rotation on its axis

Three main factors influence the different types of climate zones: differences in solar radiation, the moisture in the air and the **Coriolis effect**.

Differences in solar radiation

The general distribution of climate zones (Figure 9.4.1) roughly follows a pattern that seems related to latitude. This is due primarily to these locations receiving different amounts of the Sun's energy per unit of area because of Earth's spherical shape, with more energy received at the equator and less energy received at the poles.

Differences in Earth's surface temperature affect the air temperature and how the air moves around the planet. Hot air rises and cold air sinks, which produces areas of low and high air pressure, respectively. Air moves from areas of high pressure to low pressure, creating wind.

NOAA

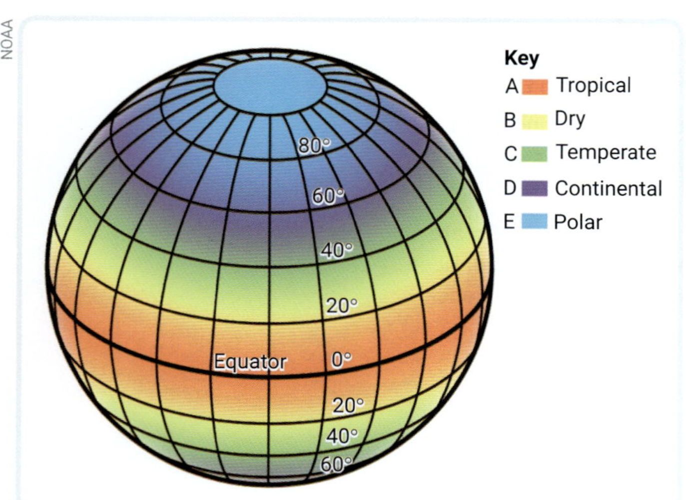

▲ **FIGURE 9.4.1** The distribution of the five zones from the Köppen climate classification

9780170491785

Moisture in the air

The second factor that influences climate is the moisture in the air and the resulting precipitation. The amount of water vapour the air can contain depends on the temperature, which is why latitude also indirectly affects precipitation in climate zones. The amount of water vapour in the air in comparison to the maximum amount the air could hold at that temperature is described as the **relative humidity**. The rate and type of precipitation in a climate zone depend on both temperature and relative humidity. The rate and type of precipitation are key distinguishing factors between different climate zones.

Quiz
Main climate zones

relative humidity
the amount of moisture that air holds compared with the amount it could hold if saturated at a given temperature

The Coriolis effect

The third influence is the Coriolis effect. This effect appears as the deflection of moving objects, such as large air masses, due to the rotation of Earth on its axis. The deflection appears to move towards the left (clockwise) in the southern hemisphere and to the right in the northern hemisphere. The Coriolis effect causes air to flow in different directions at different points on Earth.

Global convective cells

The combination of these three factors creates three types of **global convective cells** in each of the hemispheres: the Hadley cell, the Ferrel cell and the polar cell. These cells make the air circulate around Earth in directions that depend on the location (Figure 9.4.2). The movement of air within these cells is due to **convection currents**.

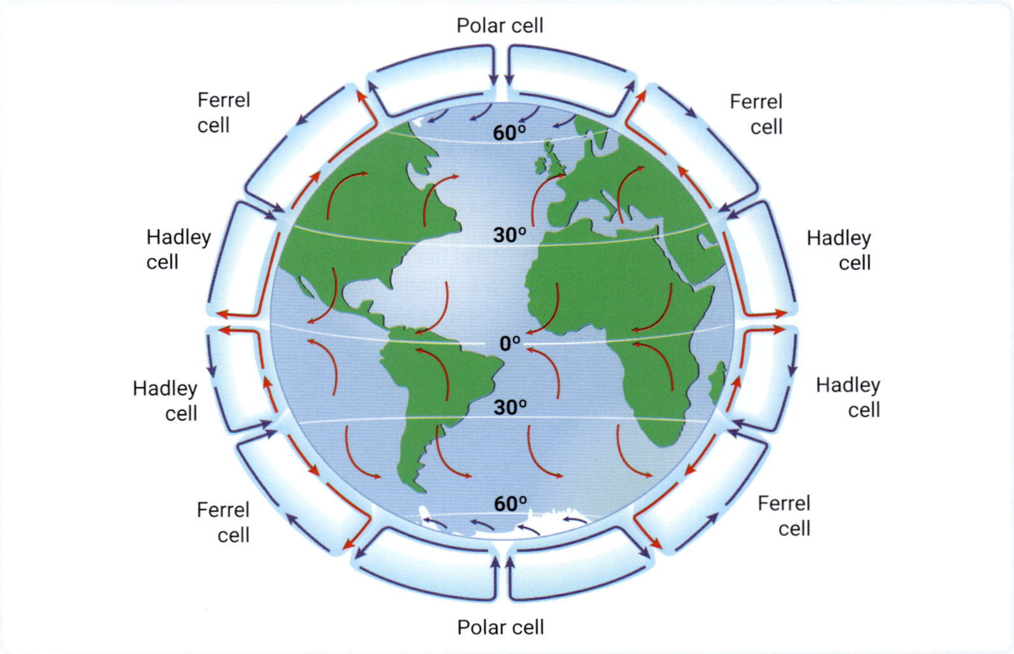

global convective cells
three large atmospheric pressure cells that occur in both the northern and southern hemispheres

convection current
a flow of materials (such as air, water or molten rock) and energy caused by differences in densities due to temperature differences

Alexey Ju. Retejum/Open Journal of Geology (CC BY 4.0)

▲ **FIGURE 9.4.2** Global atmospheric circulation

The global convective cells influence average rainfall and average temperatures. They also circulate energy through the atmosphere in a similar way to the deep ocean currents, bringing warm or cool air to different parts of Earth. The global convective cells also interact with the ocean currents, move moisture and influence the precipitation in different climate zones.

Characteristics of the climate zones

The five climate zones are categorised by the features of temperature, frequency, amount and type of precipitation, and the variety of vegetation (Table 9.4.1 and Figure 9.4.3).

tropical climate
a climate zone where it is hot and humid all year round, often with high precipitation

dry climate
a climate zone where seasonal evaporation exceeds seasonal precipitation

temperate climate
a climate zone with moderate temperature and precipitation that exists between the extremes of tropical climates and polar climates

continental climate
a climate zone with large variation in temperature between seasons and irregular patterns of precipitation

polar climate
a climate zone with ice, snow and temperatures too low to support most vegetation

▼ **TABLE 9.4.1** The five main climate zones

Climate zone	Temperature	Precipitation	Vegetation type
A: Tropical climate	• Average above 18°C all year round	• More than 1500 mm of rain per year	• Tropical rainforest
B: Dry climate	• Large daily temperature range due to reduced cloud cover • Maximum temperatures can be above 40°C • Low night temperatures can get below 0°C	• Unpredictable • Less than 350 mm of rain per year • High rates of evaporation	• Mostly grasslands • Xerophytic (adapted to survive on limited water)
C: Temperate climate	• Variable; experience four seasons • Mean temperature above −3°C and below 18°C in its coldest month	• Moderate: 750–1500 mm of rain per year • Averages 800 mm of rain per year	• Mixed forests • Low shrub • Eucalypt forest, heath and mallee in Australia
D: Continental climate	• Large temperature variation • Summer average above 10°C • Coldest month average below −3°C	• Irregular • Averages 600–1200 mm of precipitation per year, mostly as snow in winter	• Coniferous and deciduous forests
E: Polar climate	• Summer temperature rarely above 10°C • Winter temperatures often reach −30°C	• Less than 250 mm of precipitation per year, mostly as snow	• Very low-lying types • Lichens dominant • Some moss and liverworts

Distribution of climate types in Australia

The Australian Bureau of Meteorology uses a variation of the Köppen climate classification. It divides some Köppen climates and combines others to better reflect the climate conditions and the ecological zones humans experience in these diverse parts of Australia (Figures 9.4.3 and 9.4.4).

Australia has many diverse zones of climate. As shown in Figure 9.4.4, Australia has a tropical climate in the north, with temperate climate zones in the south. Within these broad climate zones there is significant ecological variation due to the natural landscape and the Australian Bureau of Meteorology's modified climate classification takes this into account.

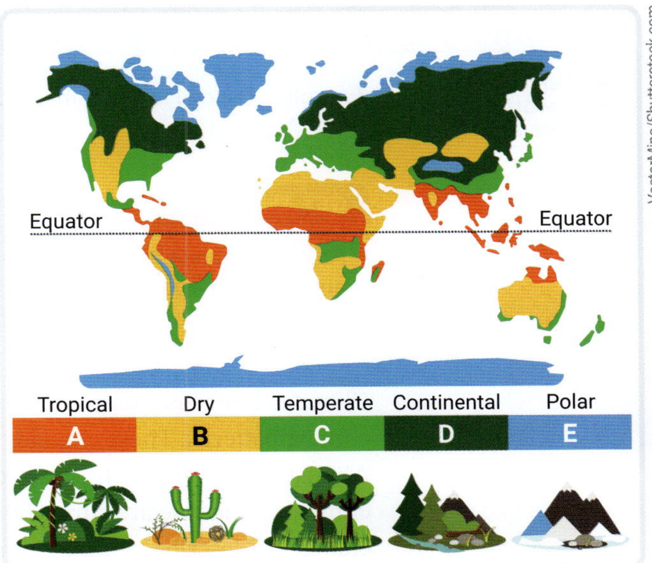

▲ **FIGURE 9.4.3** The distribution of Köppen climate types

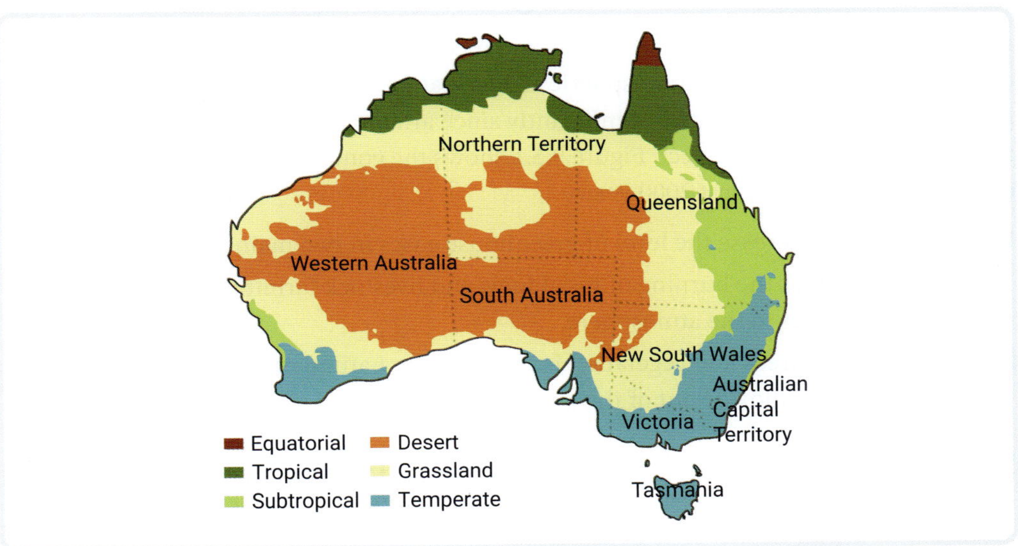

▲ **FIGURE 9.4.4** The distribution of modified climate zones in Australia

9.4 LEARNING CHECK

1. **List** the five main climate zones according to the Köppen classification system.
2. **Explain** the difference between weather and climate.
3. **Describe** where the different climate zones are located around Earth.
4. **Explain** the role that atmospheric convection currents and ocean currents have on the distribution of climate zones.
5. **Explain** how the energy from the Sun has an effect on climate.
6. **Draw** a simple diagram of Earth to show the general location of climate zones. Mark the equator and poles on your diagram and Earth's five climate zones.

9.5 Changes in global climate

BY THE END OF THIS MODULE, YOU WILL BE ABLE TO:

✓ explain the causes of climate change.

GET THINKING

As you complete this module, think about what is causing changes to the climate. Think about the way we live and how that is causing the rate of climate change to accelerate. Start thinking about the simple things we can do to slow the progress of climate change.

Climate change is a term used to describe long-term changes to regional or global climate patterns. These changes can occur naturally or as the result of human activities. Almost all climate scientists agree that human activities are the main cause of the climate change we are currently experiencing. We generally use the term 'climate change' to describe climate change caused by human activities.

Causes of climate change

Climate data collected from observations and records such as tree rings and ice cores clearly shows that the atmospheric concentration of carbon dioxide and other greenhouse gases has increased significantly since around 1850. These are causing an enhanced greenhouse effect. Figure 9.5.1 shows the combined greenhouse gas concentrations over the last 2000 years.

Human activities increase the level of greenhouse gases in two main ways.

1 Activities such as the burning of fossil fuels and livestock farming are adding extra greenhouse gases to the atmosphere.
2 The removal of trees and vegetation is reducing the natural **carbon sinks** or **reservoirs** that take carbon dioxide out of the atmosphere.

carbon sink
something that naturally stores large quantities of carbon dioxide

reservoir
a place where something is collected or stored

Sources of greenhouse gas emissions

A major cause of climate change is the increase of carbon dioxide from burning fossil fuels for energy and transport.

Approximately 90 per cent of the world's carbon emissions are due to the burning of fossil fuels: coal, oil and natural gas. Australian emissions result from the production of electricity and the use of fossil fuels in transport, agriculture and industrial processes. Australia ranks 14th in the world in carbon emissions, contributing 1 per cent of global emissions.

Methane concentrations in the atmosphere have also increased, mainly due to farming, waste disposal in landfills, and gas and oil production. For example, the methane produced by food waste in landfills across the world contributes 11 per cent of all greenhouse gases.

Another greenhouse gas, nitrous oxide, is increasing because of agriculture, fuel combustion, wastewater treatment and industrial processes.

9780170491785

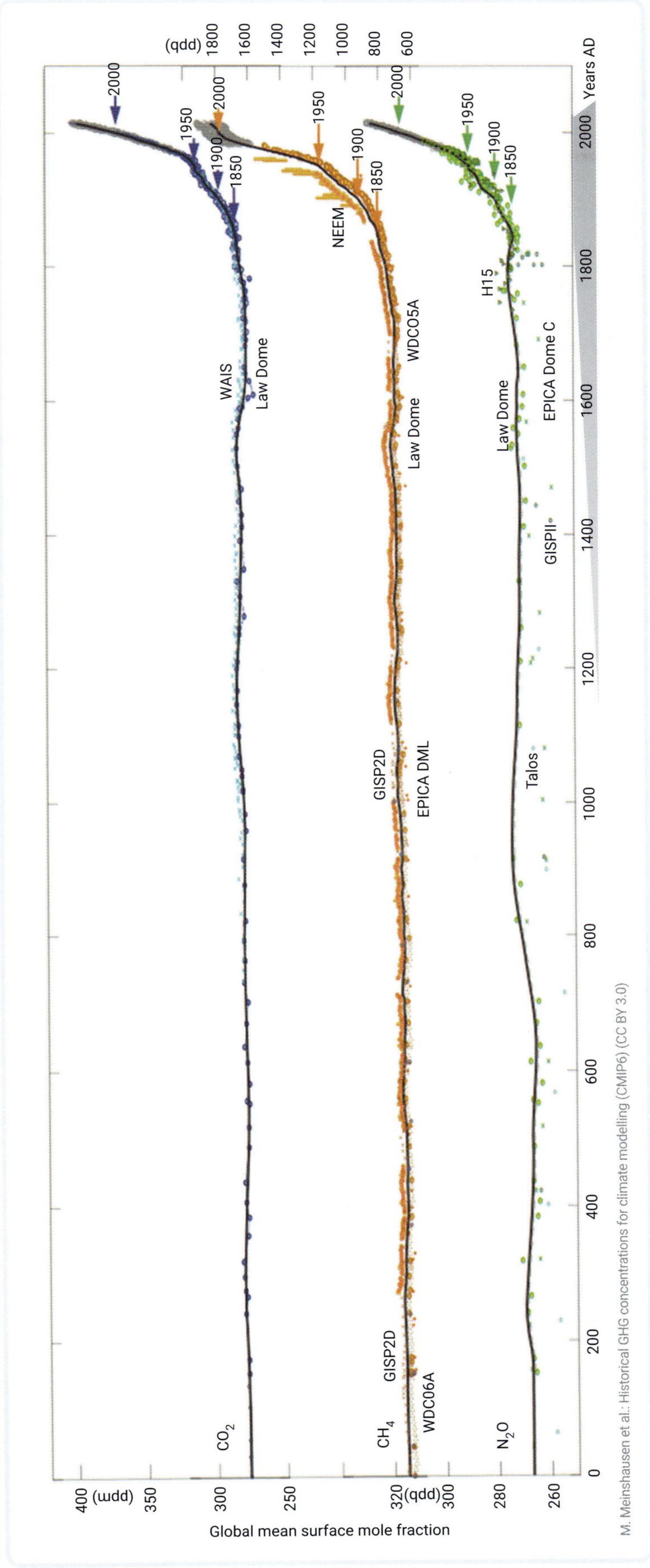

M. Meinshausen et al.: Historical GHG concentrations for climate modelling (CMIP6) (CC BY 3.0)

▲ **FIGURE 9.5.1** Greenhouse gas concentrations over the last 2000 years

Deforestation

Large areas of forest continue to be removed to meet worldwide demand for timber and forest products and to increase the amount of land available for agriculture (Figure 9.5.2).

Some of the main causes of deforestation are:

- agriculture – clearing land for agriculture, including stock grazing and the planting of crops
- urbanisation – clearing land for housing developments and infrastructure such as roads and dams
- logging – the harvesting of trees for wood and paper products
- forest fires.

deforestation
the removal of large areas of forest to enable the land to be used for other purposes

Trees and vegetation store and use carbon dioxide, taking it out of the atmosphere. Fewer trees and plants mean Earth has fewer carbon sinks to help regulate the levels of carbon dioxide in the atmosphere. A further problem with **deforestation** is that when trees are removed or destroyed (e.g. by burning), the carbon stored in them is released, adding even more carbon dioxide to the atmosphere. Deforestation also leads to further land degradation and other environmental problems when deforested areas are subjected to heavy rainfall and flooding. The Amazon rainforest and forests in Asia have the greatest amounts of land clearing.

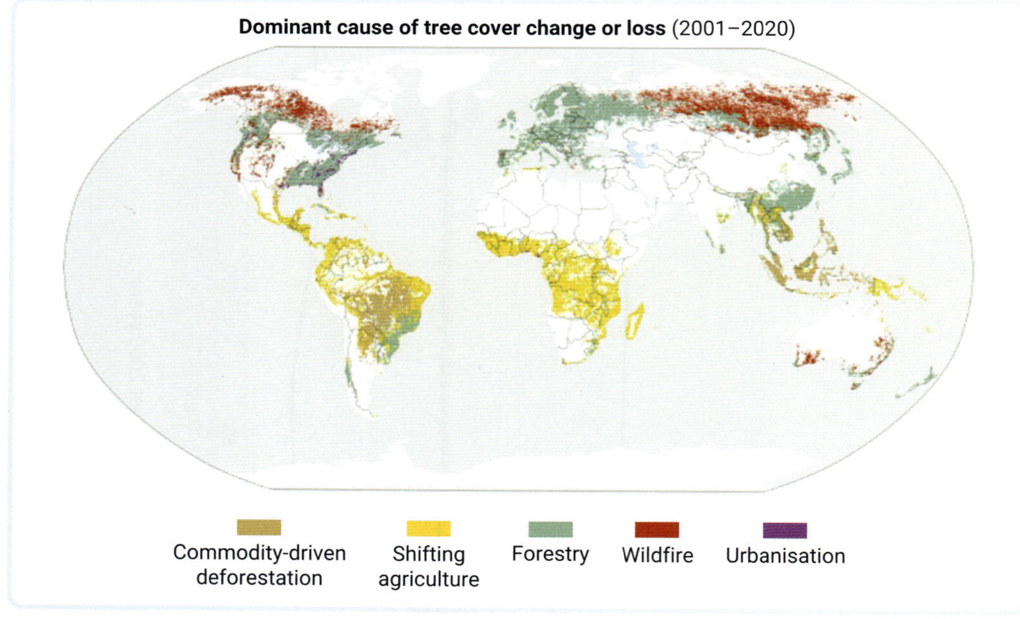

▲ **FIGURE 9.5.2** The reasons for deforestation

9.5 LEARNING CHECK

1. **Identify** two greenhouse gases that contribute to the enhanced greenhouse effect.
2. **Describe** the difference between the natural greenhouse effect and the enhanced greenhouse effect.
3. **Describe** how food waste disposal contributes to greenhouse gases.
4. **Explain** why deforestation is significant for the concentration of atmospheric carbon dioxide.

9780170491785

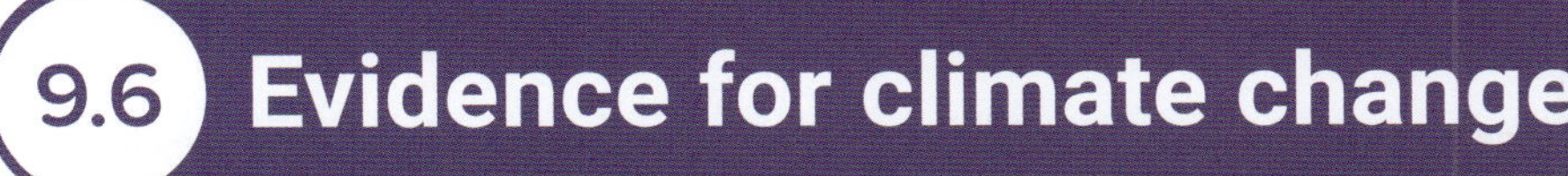

9.6 Evidence for climate change

9.6

BY THE END OF THIS MODULE, YOU WILL BE ABLE TO:

- ✓ describe trends in Earth's climate and biodiversity caused by climate change
- ✓ describe the impact of climate change on climate and ecosystems.

GET THINKING

In recent years, there have been regular reports in the media on sea ice melting in the Arctic Ocean, droughts and floods in South Africa, and coral bleaching on the Great Barrier Reef. Are these natural events or indicators that the climate is changing? Think about other things that you may have seen in the media that might be either natural variability or indicators of climate change.

Interactive resource
Drag and drop: Evidence of climate change

Other resource
Worksheet: Indicators of climate change

Globally, there is an accumulation of different types of scientific evidence that all show the same clear picture: our climate is changing, and Earth is getting warmer.

Who finds the evidence for climate change?

In Australia, the Bureau of Meteorology and CSIRO play key roles in monitoring, analysing and communicating observed and future changes in Australia's climate. Around the world, there are many research scientists working in meteorological and government organisations to monitor changes in climate. The IPCC is responsible for assessing the latest knowledge on human-induced climate change, its impacts and how to respond.

Scientists use many types of evidence to measure and understand climate change, including:

- increases in the concentrations of greenhouse gases
- increases in atmospheric and ocean temperature
- increases in ocean acidity
- increases in global sea levels
- reductions in sea ice and permafrost
- reductions in **biodiversity**.

biodiversity
the variety of living species on Earth, including plants, animals, bacteria and fungi

Atmospheric and ocean temperature

Over the past million years, global average surface temperatures have varied by up to 5°C, as Earth experienced cycles of ice ages and warmer periods approximately every 100 000 years. What is concerning to climate scientists is the current increased rate of change. The enhanced greenhouse effect has caused Earth's average global temperature to increase by around 1.1°C compared with levels in 1850–1900. The year 2023 was the warmest year in the 174 years that records have been kept. The global mean surface temperature was 1.45°C above the 1850–1900 average.

In Australia, climate change means that extreme heat is becoming a more regular occurrence, with temperature records being broken in many places. On 13 January 2022, Onslow, Western Australia, equalled Australia's highest recorded temperature by reaching 50.7°C, previously recorded in Oodnadatta in South Australia on 2 January 1960.

Climate change also means parts of Australia are subjected to unseasonal increases in rainfall and record flooding, as was the situation in south-east Queensland and coastal New South Wales in February and March 2022, in Victoria and New South Wales in October and November 2022 and in South Australia and Western Australia in January 2023.

Ocean temperatures are also important. NASA has determined that the oceans hold 90 per cent of the accumulated heat in the Earth system. Recent measurements of heat in the ocean showed that 2024 was the ocean's warmest year on record and that rates of ocean warming have shown strong increases in the past 20 years. Figure 9.6.1 shows how the amount of heat held in the ocean has increased dramatically. Increasing ocean temperatures cause effects such as coral bleaching and changing distributions of fish.

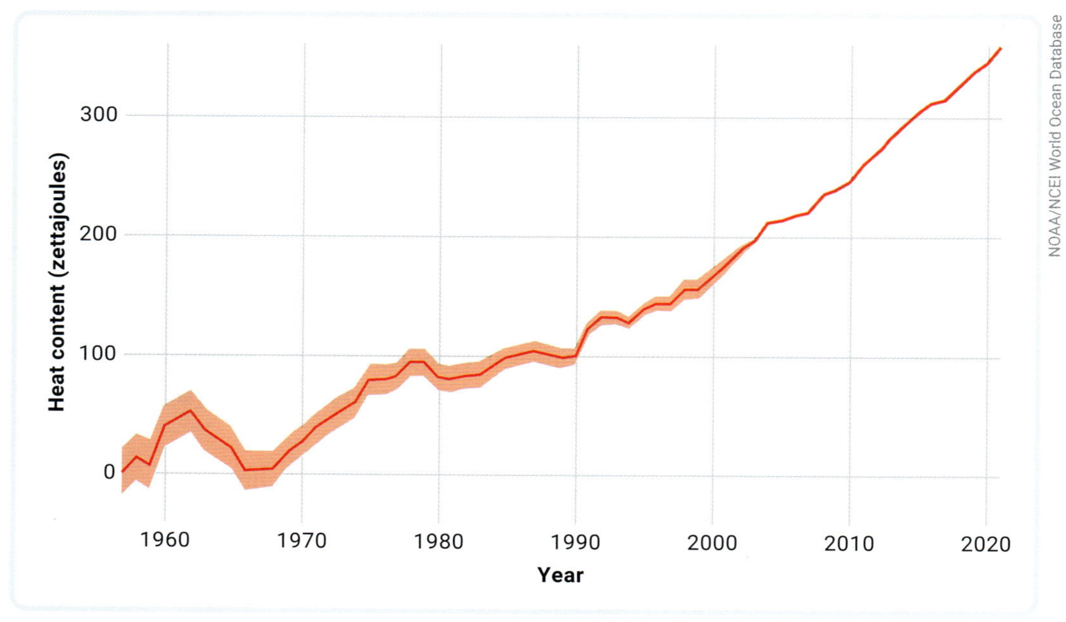

▲ **FIGURE 9.6.1** Estimated change in the global ocean heat content since 1955 (1 zettajoule = 10^{21} joules). The red shading provides an indication of the confidence range of the estimate.

Ocean acidity

Like trees, the ocean is an important carbon reservoir that takes carbon dioxide out of Earth's atmosphere. The ocean absorbs about 23 per cent of annual human-produced carbon dioxide emissions. Increasing ocean temperatures are causing oceans to become more acidic. Measurements of pH show the ocean is the most acidic it has been for at least the past 26 000 years. As the ocean becomes more acidic, it absorbs less carbon dioxide from the atmosphere. Increased ocean acidity makes it difficult for marine calcifying organisms, such as coral and some plankton, to form shells and skeletons. Acidity also can lead to existing shells dissolving and becoming weaker.

Sea levels and sea ice

Higher atmosphere and ocean temperatures are causing the continental ice sheets in Antarctica and Greenland to melt. Higher temperatures are also causing glaciers to melt, adding more water to the oceans. The melting of land-based ice, combined with the **thermal expansion** of water in the ocean, is making sea levels rise. Since 1880, sea levels have risen 21–24 cm, with almost a third of that rise occurring in the past 25 years.

thermal expansion an increase in the volume of materials as they get hotter

9780170491785

Satellites have been used since the early 1990s to determine changes in sea level remotely. Satellite data shows that the rate of sea-level rise is accelerating. Between 1993 and 2002, there was an average global increase in sea level of 2.1 mm per year. Between 2013 and 2021, sea levels increased at more than twice that rate, by an average of 4.4 mm per year (Figure 9.6.2).

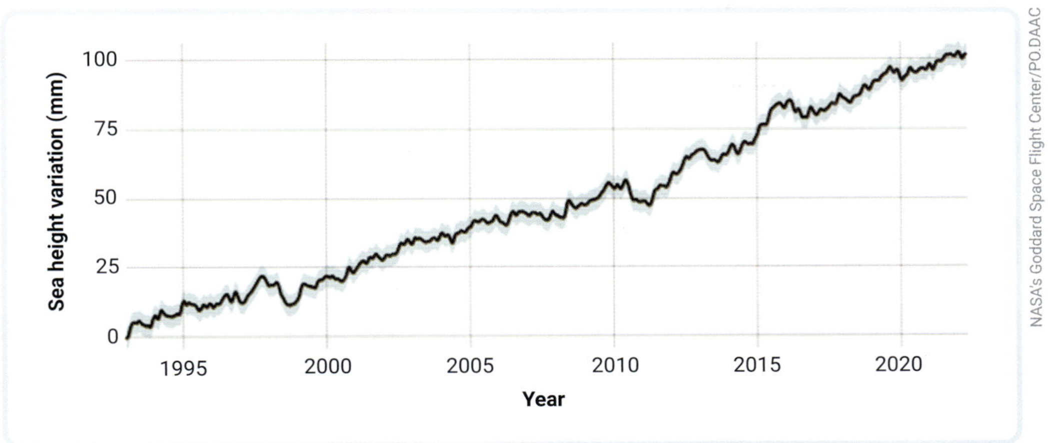

▲ **FIGURE 9.6.2** Global sea-level rise, 1993–2021

A further source of evidence of climate change is the reduction in sea ice in the Arctic and Antarctic regions. Satellite monitoring shows that the amount of Arctic sea ice is now declining at a rate of about 13 per cent every decade.

Permafrost

Permafrost is ground that has been at a temperature of 0°C or below for at least two consecutive years. Permafrost accounts for about 17 per cent of Earth's surface and is very sensitive to rising temperatures (Figure 9.6.3). Data from boreholes drilled between 2007 and 2016 showed that in permafrost regions, the temperatures 10 m below the ground have increased by about 0.3°C.

It is predicted that the continued warming of the planet will ultimately melt permafrost areas. When this happens, the carbon released from these soils will add large amounts of carbon dioxide and methane to the atmosphere.

wikimedia/Boris Radosavljevic (CC BY 2.0)

▲ **FIGURE 9.6.3** Permafrost and ice on Herschel Island, Canada, 2012

Biodiversity

It is estimated that one-eighth of the world's species are threatened with extinction. Climate change poses a significant risk to many of these species. The IPCC has confirmed that climate change and biodiversity are linked. Climate change and the loss of biodiversity are both driven by human activity, and each negatively affects the other.

Changes in climate have altered environmental conditions and habitats. If animals and plants are unable to adapt to these changes, their survival is at risk. Species that live within a restricted range are particularly vulnerable, such as the endangered Carnaby's black cockatoo that is endemic to south-west Australia (Figure 9.6.4). It is very susceptible to heat stress, so more frequent heatwaves are making its future even more precarious. Green turtles on the northern Great Barrier Reef are in danger because rising temperatures are causing 99 per cent of hatchlings to be female.

iStock Essentials/istock.com

▲ **FIGURE 9.6.4** Carnaby's black cockatoo is endangered and susceptible to climate change.

Research has found that the top predators in ecosystems are usually affected the most by climate change. For example, polar bears in the Arctic are declining in large numbers as both their food sources and available habitat are restricted by the reduction in sea ice.

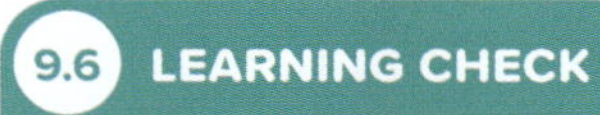

1 **List** five sources of evidence that confirm climate change is occurring.
2 **List** three examples that offer evidence of climate change in Australia.
3 **Describe** the purpose of the IPCC.
4 **Describe** how the Great Barrier Reef is being affected by climate change.
5 **Explain** what causes the acidification of the oceans and suggest the likely outcome if it were to continue.

SCIENCE IN CONTEXT

9.7 Why CSIRO monitors greenhouse gases at Kennaook/Cape Grim

BY THE END OF THIS MODULE, YOU WILL BE ABLE TO:

- ✓ interpret and analyse data to make decisions based on the needs of society
- ✓ explain how the values and needs of society influence the focus of scientific research
- ✓ describe how the need to minimise greenhouse gas emissions has led to scientific and technological advances.

Kennaook/Cape Grim air monitoring site

Video activity
Cape Grim monitoring station

Greenhouse gases are on the rise because of human activities. Scientists have shown that there is a clear link to this fast rise in the atmospheric concentration and the enhanced greenhouse effect. This is leading to global warming and other changes to the climate.

In the early 1970s, Australia's CSIRO made a commitment to the United Nations Environment Programme to monitor greenhouse gases in the atmosphere for climate change research. The monitoring location is Kennaook/Cape Grim, a site on the north-west tip of Tasmania (Figure 9.7.1). This location was chosen because Kennaook/Cape Grim has some of the cleanest air in the world because the air comes from thousands of kilometres of ocean to the west and south-west. Data from Kennaook/Cape Grim Baseline Air Pollution Station is considered 'baseline', which allows scientists to compare current data to historic datasets and get a clear picture of global changes to the atmosphere. CSIRO has been continuously measuring the air at Kennaook/Cape Grim since 1976.

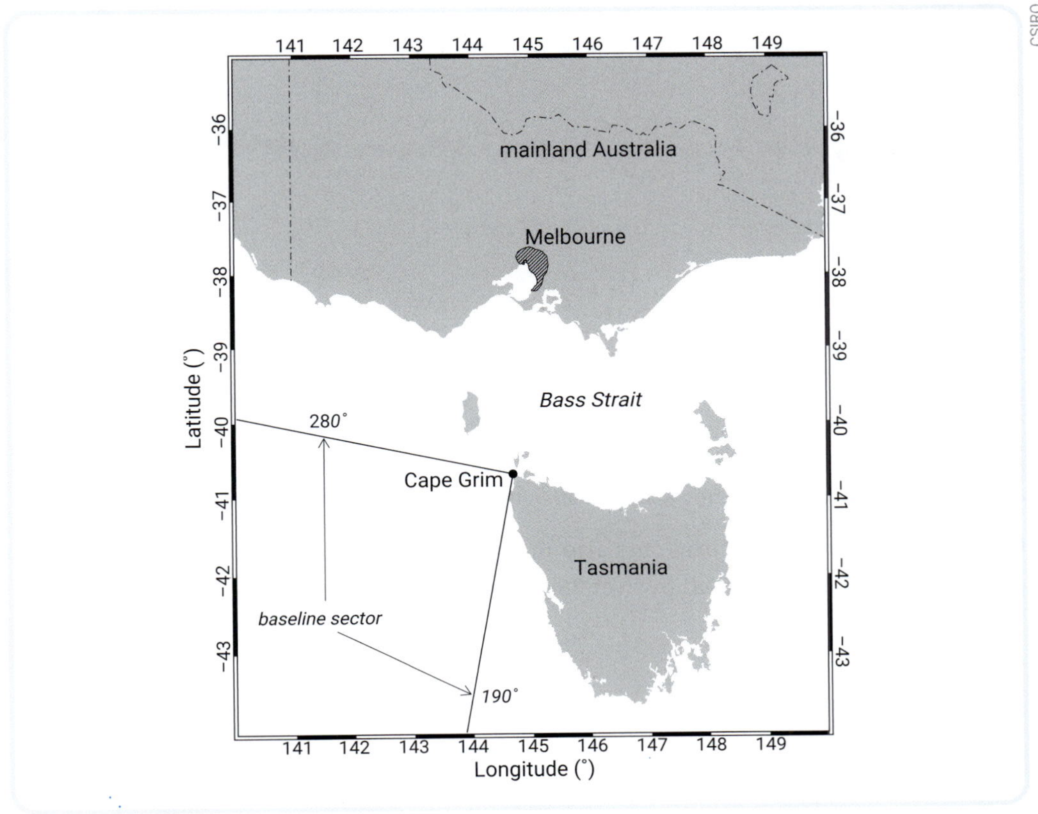

CSIRO

FIGURE 9.7.1 Kennaook/Cape Grim, on the north-west tip of Tasmania, is the site of CSIRO's monitoring station for air pollution.

What is being measured?

DATA SCIENCE

Learn more about analysing datasets for trends and causation in **Modules 2.6** and **2.7**.

CSIRO uses very precise instruments to monitor more than 80 greenhouse gases at Kennaook/Cape Grim, including carbon dioxide (CO_2) and methane (CH_4). Scientists analyse and use the data to improve climate models, which helps them make more accurate predictions about climate change and weather patterns.

Since 1978, the atmospheric methane concentration has increased by about 24 per cent. The increase is attributed to human activities such as coal mining, use of landfills and agricultural practices.

Before the Industrial Revolution (mid-1700s), the concentration of methane in the atmosphere was 700 parts per billion (ppb). This means a dry sample of air containing 1 billion molecules would include 700 methane molecules. Today the value exceeds 1800 ppb.

Figure 9.7.2 shows atmospheric methane concentrations (in ppb) over the last 2000 years. This is based on measurements of air trapped in Antarctic ice and data from Kennaook/Cape Grim.

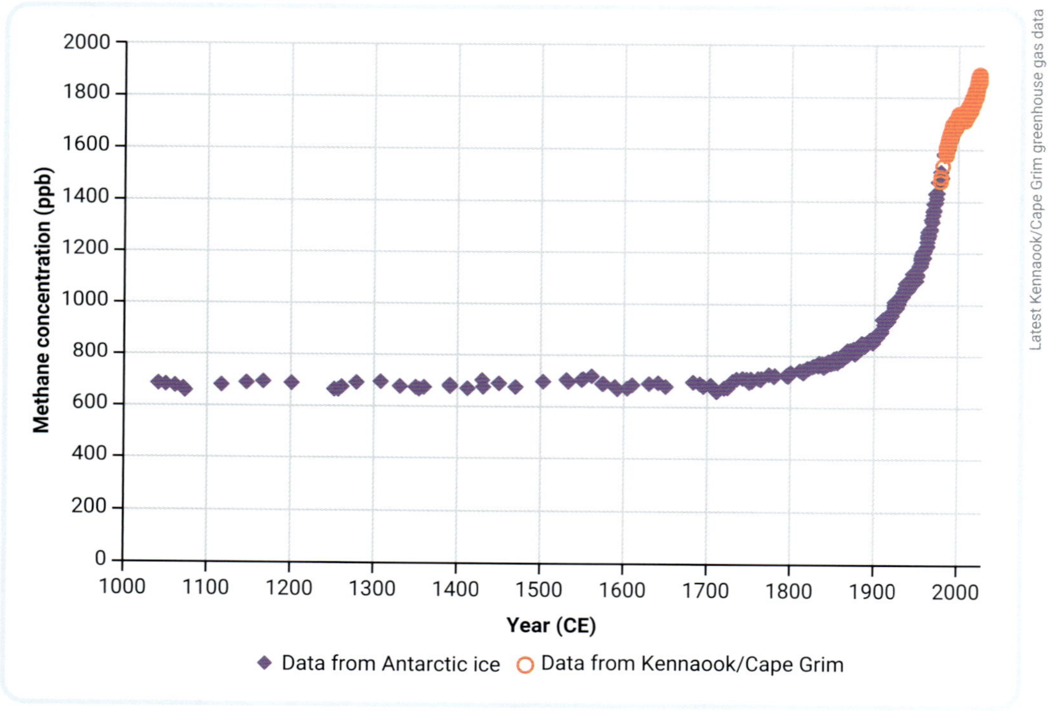

▲ **FIGURE 9.7.2** Atmospheric methane concentrations (in ppb) over the last 2000 years

CSIRO's Kennaook/Cape Grim data also shows that atmospheric concentrations of carbon dioxide have increased by over 25 per cent since 1976 (Figure 9.7.3).

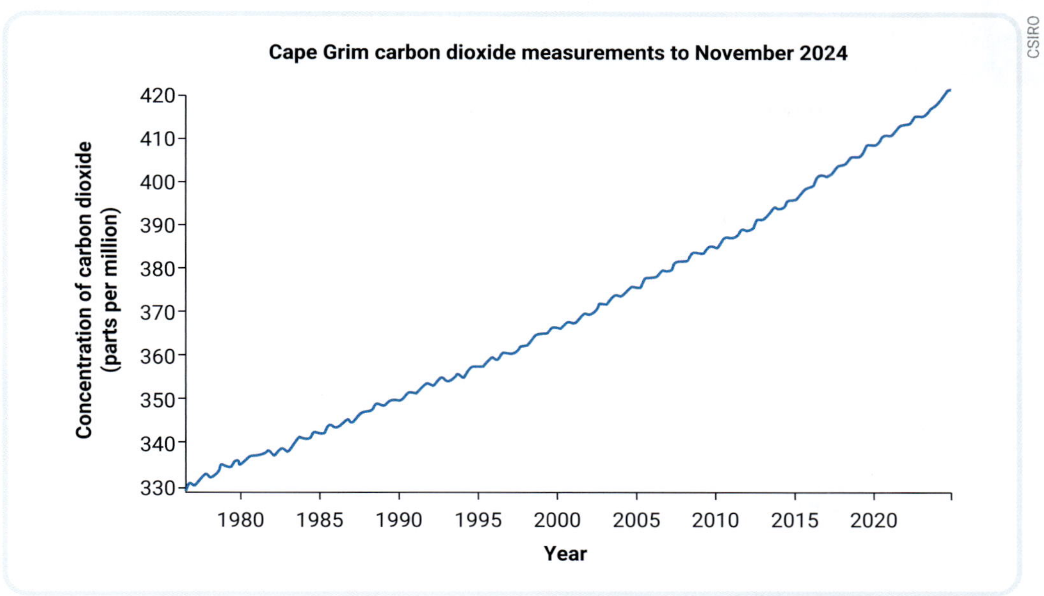

CSIRO

▲ **FIGURE 9.7.3** The atmospheric concentrations of carbon dioxide measured at Cape Grim, to November 2024

How do scientists use the data?

The data from Kennaook/Cape Grim is available on CSIRO's website and from major global data archives. CSIRO's data is used as part of international assessments of climate change, such as those conducted by the IPCC and in Australia's *State of the Climate* reports.

The Australian Government also uses data from Kennaook/Cape Grim to meet international obligations and report on pollution levels. Scientists also use Kennaook/Cape Grim data in hundreds of research papers on climate change and atmospheric pollution.

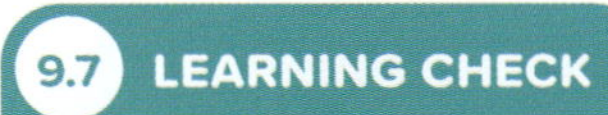

1 **State** three reasons why CSIRO monitors greenhouse gases.
2 **Explain** why Kennaook/Cape Grim was chosen as an important site to conduct air monitoring.
3 Look at Figure 9.7.2. **Describe** the trend in atmospheric methane concentrations and **explain** how this relates to human activities.
4 Look at Figure 9.7.3. What do you notice about how atmospheric concentrations of carbon dioxide are increasing? Does it look as though carbon dioxide levels are increasing at a constant rate?

WORKING SCIENTIFICALLY

9.8 Analysing data to identify patterns and trends

SCIENCE SKILLS IN FOCUS

IN THIS MODULE, YOU WILL FOCUS ON LEARNING AND IMPROVING THESE SKILLS:

- posing questions and making predictions
- analysing representations of data to identify patterns and trends
- analysing models of climate change.

How to analyse data to identify patterns or trends

The purpose of any investigation is to test the validity of a hypothesis. Once the investigation is conducted and data is collected, you need to analyse the data.

1. Check the data. Is there sufficient data to come to a conclusion? How many trials were used? Was that sufficient? If not, collect more data.
2. If repeated trials were conducted, the data would need to be averaged to reduce the effect of variations between trials and make it easier to analyse the data.
3. The averaged data should be presented in a suitable table with column titles and units. This data is what must be analysed to determine whether there is a pattern or trend that will confirm or reject the hypothesis.
4. Plot the data on a graph. A line graph is often used in science, where the independent variable is plotted on the *x*-axis and the dependent variable data on the *y*-axis.
5. To determine if there is any relationship between the dependent and the independent variables, the data may need to be manipulated to create a straight line with the formula $y = mx + c$, where m is the gradient and c is the *y*-intercept (if there is one).

For example, the plotted data may give a curved graph similar to Figure 9.8.1. To check if there is a relationship between A and t in this graph, we can transform the curve into a straight line, by plotting A versus $\frac{1}{t}$, as shown in Figure 9.8.2.

However, the data may not always appear linear. For example, when the data exhibits an exponential relationship, a smooth curve may be a better way of showing the trend.

Software such as Excel will help. You can use the 'chart menu' to display your data in several ways to help analyse for trends.

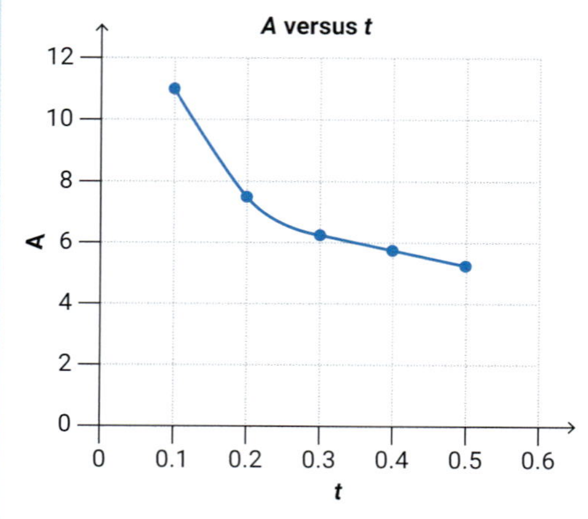

▲ FIGURE 9.8.1 A curved line graph

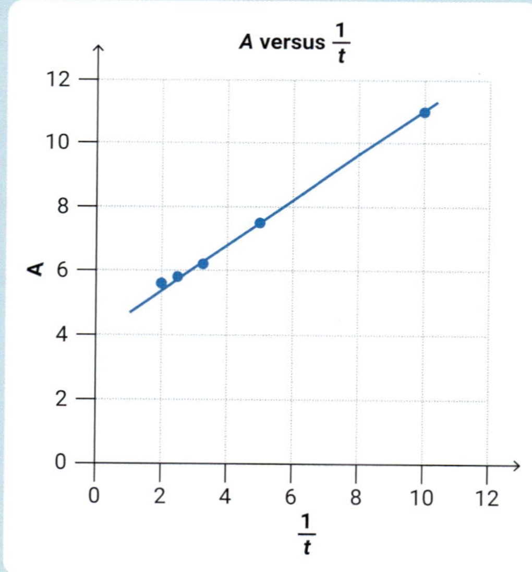

▲ FIGURE 9.8.2 A straight line graph

9780170491785

ANALYSING ICE CORE DATA TO DETERMINE THE RATE OF CARBON DIOXIDE INCREASE IN THE ATMOSPHERE

AIM

To analyse data to identify whether the rate of change in carbon dioxide concentration has increased

MATERIALS AND EQUIPMENT

- ☑ graph paper
- ☑ ruler
- ☑ pencil
- ☑ scientific calculator

PROCEDURE

In this investigation, you will use the ice core data in Table 9.8.1. These have been collected from Antarctica by the Scripps Carbon Dioxide Program. The dataset is an averaged subset of data that begins in 1959. It has been modified to 5-year intervals for graphing.

1. Look at the data in Table 9.8.1 and determine the scale you need to use on your graph.
2. To make the graph easier to analyse, start your concentration axis at 300 ppm.
3. Plot the data and draw a line of best fit.
4. Determine the gradient of the line of best fit. This should be the average rate of increase in carbon dioxide concentration.

Video
Science skills in a minute: Trends and patterns in data

Science skills resource
Science skills in practice: Trends and patterns in data

▼ **TABLE 9.8.1** The average atmospheric concentrations of carbon dioxide in ppm

Year	Carbon dioxide concentration (ppm)
2020	411.99
2015	399.19
2010	388.00
2005	378.16
2000	368.25
1995	359.52
1990	352.84
1985	344.77
1980	337.85
1975	330.34
1970	325.00
1965	319.72
1960	316.68

ANALYSIS

1. Does a line of best fit suit the data or should a smooth curve be drawn instead?
2. If the data fitted a smooth curve rather than a line, what would be the best way of determining the rate of increase?
3. Does using 5-year data points affect the rate of increase of carbon dioxide concentration?

CONCLUSION

Write a statement based on what you were able to determine from this data.

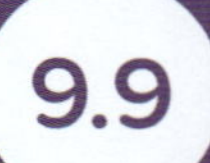

9.9 The effects of climate change

BY THE END OF THIS MODULE, YOU WILL BE ABLE TO:

- ✓ summarise the impacts of climate change using a flow chart
- ✓ explain that extreme weather events are occurring with greater frequency as a consequence of climate change.

Interactive resource
Label: Climate change and human health

Other resource
Worksheet: Effects of global warming

GET THINKING

As you complete this module, think about recent examples of extreme weather events you have experienced or heard about; for example, increases in temperature, severe droughts or heavy rainfall, which lead to disasters such as fires and floods. Is there a link between these events and human activities? What is the role of climate change in these events?

As we saw in Module 9.6, there is a lot of scientific evidence that clearly shows our climate is changing. But what does this mean in terms of impacts? How could climate change affect us in Australia?

Extreme weather

In Module 9.2 and Module 9.4, we looked at how the world's climate and weather patterns are closely linked to the circulation of energy via deep ocean and atmospheric currents. Climate change is affecting and modifying these currents, interrupting the flow of energy around the planet and causing further changes to the climate.

extreme weather event unexpected, unusual, severe or unseasonal weather

Unfortunately, we are already seeing the impacts of these climate changes in the severity and frequency of **extreme weather events** around the world. Extreme weather events include cyclones, flooding, heatwaves and drought. In Australia, bushfires are becoming more intense and are causing more damage. They are also burning in places that do not normally experience bushfires, such as in the ancient rainforests of south-east Queensland and north-east New South Wales (Figure 9.9.1). Climate change is drying out these forests, making them more vulnerable to fires. This, in turn, destroys the habitat of many animals and plants, further threatening Australia's biodiversity.

Nick Fox/Shutterstock.com

Cavan Images/Alamy Stock Photo

▲ **FIGURE 9.9.1** The Gondwana Rainforests in Queensland (a) before and (b) during the severe bushfires of early 2020

9780170491785

Human health

Extreme weather can have immediate impacts on human health, such as heat stroke leading to death. Each year, extreme heat kills more people in Australia than all other natural disasters combined.

Extreme weather can also have far-reaching negative effects on humans, by affecting our food supplies through crop failures or damage from weather events such as storms, floods and droughts. Other examples of the indirect effects of climate change on human health include an increase in air pollution from larger, more intense bushfires and droughts severely reducing our reserves of fresh water. Further links between climate change and human health are shown in Figure 9.9.2.

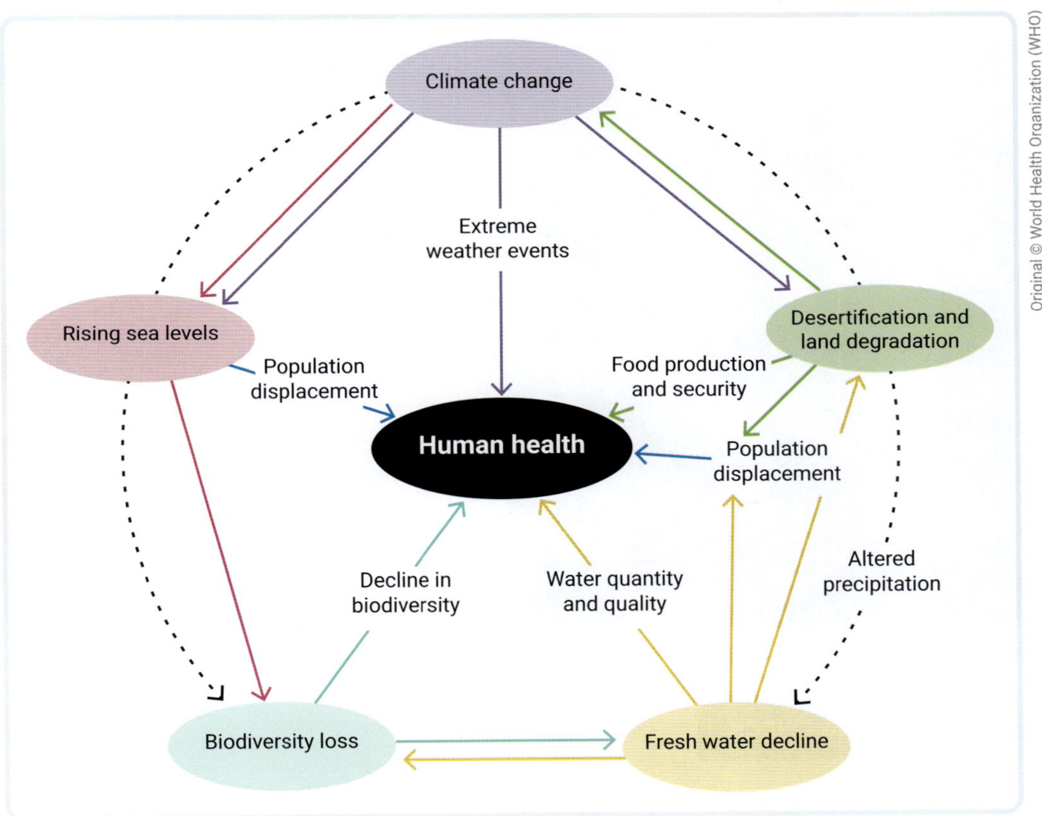

Original © World Health Organization (WHO)

▲ **FIGURE 9.9.2** The links between climate change and health

Disappearing islands and coastlines

Rising sea levels due to climate change are posing a serious threat to millions of people living in low-lying islands and coastlines around the world. Islands in countries such as the Maldives, Solomon Islands, Tuvalu and Papua New Guinea are already dealing with the consequences of rising sea levels, with many people forced to leave their homes and relocate to higher ground or different islands.

storm surge
a brief increase in sea levels due to the combined effect of storms, low air pressure, strong winds and tides

On coastlines, very high tides during extreme weather events such as cyclones can result in **storm surges**. These cause destructive flooding, which can damage buildings, roads, beaches and vegetation (Figure 9.9.3).

Brook Mitchell/Getty Images

▲ **FIGURE 9.9.3** Storm surges can cause huge amounts of damage during extreme weather events.

9.9 LEARNING CHECK

1 **List** five extreme weather events affecting Australia.
2 **Describe** how extreme weather events have an impact on biodiversity.
3 **Explain** how storm surges occur.
4 **Describe** how human health can be affected by climate change.
5 **Explain** why the effects of rising sea levels are a major concern.
6 **Create** a flow chart to show the links between human activities, climate change and the impacts of climate change.

9780170491785

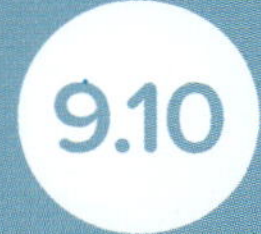

9.10 Modelling cause and effect

SCIENCE SKILLS IN FOCUS

IN THIS MODULE, YOU WILL FOCUS ON LEARNING AND IMPROVING THESE SKILLS:

- analysing data from investigations to identify relationships and draws conclusions
- using models to predict cause and effect.

In Stage 4, you learned that models provide a way to test scientific ideas. You also learned about testing for cause-and-effect relationships.

All models are simplifications of the thing being modelled. This lets us focus on and study a small number of variables, which makes it easier to identify, test for and predict cause-and-effect relationships. Here are some important points to remember when using models to study cause-and-effect relationships between variables.

- You can test for cause-and-effect relationships using graphs and models.
- Correlation means there is a relationship between two or more variables.
- Causation is a relationship in which one variable causes a change in another variable.
- Correlation does not always indicate causation.
- It's valuable to make predictions of what you think the relationship between variables will be before you model.
- You can use physical models to observe the effect of one variable on another, such as the models you will use in Investigations 1 and 2.
- Always clearly record your observations in a systematic way, such as by using a table.
- The easiest way to observe cause-and-effects is to plot the independent and dependent variables on a graph and draw a line of best fit.
- Once you have identified causation, you can make further predictions about the relationship between the variables.

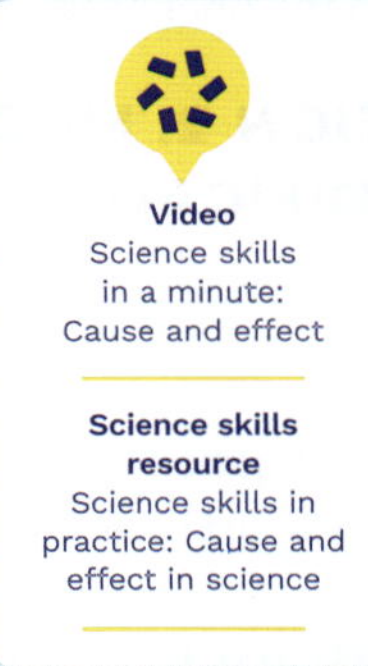

Video
Science skills in a minute: Cause and effect

Science skills resource
Science skills in practice: Cause and effect in science

INVESTIGATION 1: MODELLING SEA-LEVEL RISE

AIM

To model sea-level rise

MATERIALS AND EQUIPMENT

- ☑ 2 × 300 mL cups or beakers
- ☑ 80 mL beaker
- ☑ Blu Tack
- ☑ marker pen
- ☑ 2 ice cubes

PROCEDURE

1. Invert the 80 mL beaker and place it in the centre of one of the larger beakers (or cups). Use the Blu Tack to fix it firmly in place. Label the large beaker as '1: Continental ice'.
2. Fill the 'continental ice' beaker with enough tap water to just reach the height of the smaller inverted beaker.
3. Label the second large beaker as '2: Sea ice'. Place the same volume of water in this beaker.

4 Place an ice cube on top of the 80 mL beaker in Beaker 1 so it is above the water, and the same sized ice cube into the water in Beaker 2 so that it floats in the water.

5 Immediately mark the level of the water in both beakers.

6 Allow the ice in each beaker to completely melt and observe the water level.

ANALYSIS

What did you **observe**? Try to **explain** why this occurred.

CONCLUSION

Write a statement based on what you observed in the model, linking it to the phenomenon of rising sea-levels.

INVESTIGATION 2: MODELLING OCEAN ACIDIFICATION

AIM

To conduct an investigation to show the effect of carbon dioxide on the pH of sea water

BACKGROUND INFORMATION

Carbon dioxide reacts with sea water to produce carbonic acid.

You can determine the pH of a substance by checking the coloured chart that comes with universal indicator.

MATERIALS AND EQUIPMENT

- ☑ small-to-medium conical flask
- ☑ 50 mL measuring cylinder
- ☑ 40 mL of dilute NaOH solution
- ☑ 40 mL of universal indicator
- ☑ reuseable straw (metal straws that can be put into dishwasher) or paper straws
- ☑ safety glasses, lab coat and gloves

Safety

Be careful – acids can splash and burn eyes and skin.

Only blow through straw – do not suck because this may draw acid into your mouth.

PROCEDURE

1 Determine your hypothesis by using an 'If …, then …' statement; for example, 'If I blow carbon dioxide into the solution continuously for 1 minute, then the colour will change to yellow, indicating a pH of 6.'

2 Put on safety glasses, lab coat and gloves.

3 Measure 40 mL of NaOH solution with a measuring cylinder and transfer it to a conical flask.

4 Rinse the measuring cylinder with water. Add 3–4 drops of universal indicator to the conical flask.

5 Create a table, similar to Table 9.10.1, to record your observations.

6 Submerge part of the straw and gently blow air into the solution in the conical flask for 1 minute. Repeat the experiment two or three times to find out if the result is reproduced consistently. Remember to keep all other variables the same.

7 Replicate the experiment by asking other students to do the same experiment.

9780170491785

RESULTS

Copy and complete Table 9.10.1 in your workbook.

▼ **TABLE 9.10.1** Results table

Experiment	Colour before blowing in the straw	Colour change(s) after 30 s of blowing in the straw	Colour after 1 min
1			
2			
3			
4			

ANALYSIS

1 **Identify** the part of your experiment that represented sea water.

2 **Identify** the component of your experiment that represented carbon dioxide emissions.

3 After you complete your table of results, write a paragraph to **summarise** your findings about the colour of universal indicator under acidic and basic conditions.

4 Did you have to alter your hypothesis or method? Why?

CONCLUSION

Write a conclusion for the investigation. Include a statement about whether or not the results supported your hypothesis.

9.11 Solutions to climate change

BY THE END OF THIS MODULE, YOU WILL BE ABLE TO:

- ✓ explain how different strategies may limit climate change or reduce its extent
- ✓ discuss the barriers that may stop climate change strategies from being used.

Video activity
Carbon farming

Other resources
Activity sheets: Campaign for change

What do the media say?

GET THINKING

As you read through this module and reflect on what you know of climate change, think about whether we can slow down or stop it from happening. Think about strategies you and your family can use to help.

In 2021, **COP26** set the aim of achieving **net zero greenhouse emissions** by 2050. It is hoped that achieving this aim will keep global temperature change to 1.5°C above pre-industrial levels. How can this target be achieved?

COP26
the 26th Conference of the Parties to the United Nations Framework Convention on Climate Change

net zero greenhouse emissions
when emissions of greenhouse gas do not exceed the amount of gases absorbed or stored

Reducing carbon dioxide emissions

The most important thing we all need to do is to greatly reduce the amount of carbon dioxide and other greenhouse gases emitted into the atmosphere. There are many different ways we can do this.

Stop using fossil fuels to generate electricity

Phasing out and replacing coal-fired power stations with nuclear power and renewable alternatives such as wind power, geothermal energy, hydroelectricity, hydrogen power and solar energy reduces carbon dioxide emissions (Figure 9.11.1).

FiledIMAGE/Shutterstock.com

myphotobank.com.au/Shutterstock.com

▲ **FIGURE 9.11.1** Due to our sunny, windy climate, solar and wind energy are the most viable and common forms of renewable energy in Australia.

Find alternative ways to power vehicles

Vehicle emissions are a major source of greenhouse gases. Replacing traditional internal combustion engines with electric engines or alternatives that use hydrogen gas is an important step to reducing climate change.

Global shipping and aviation contribute 8 per cent of carbon dioxide emissions. Reducing international movement and finding less polluting fuels or vehicles can reduce emissions from these sources.

Reducing methane emissions

Methane is a potent greenhouse gas, capable of holding up to 25 times more heat than carbon dioxide. Methane is released as a by-product of producing oil and gas and from the decay of material in landfill. However, most comes from farming livestock such as cows and sheep. Livestock are responsible for about 15 per cent of all greenhouse emissions, half of which come from cattle (Figure 9.11.2). There is a number of other ways to reduce methane emissions. Scientists are working on changing diets and the bacteria in livestock stomachs to reduce emissions. Changing our diets to reduce the amount of meat we consume can reduce our **carbon footprint** by almost half.

▲ **FIGURE 9.11.2** Feeding livestock seaweed is one way to reduce methane emissions.

carbon footprint
a measure of the amount of greenhouse gases released into the atmosphere by an individual or group

Carbon sequestration

As well as reducing emissions, we can increase the extraction of carbon from the air and store it away from the atmosphere. Plants naturally remove carbon dioxide from the air through photosynthesis and store it in their leaves, stems, branches, bark and roots. This is a form of **carbon sequestration**.

carbon sequestration
the process of capturing and storing atmospheric carbon dioxide

Some examples of carbon sequestration methods are shown in Figure 9.11.3.

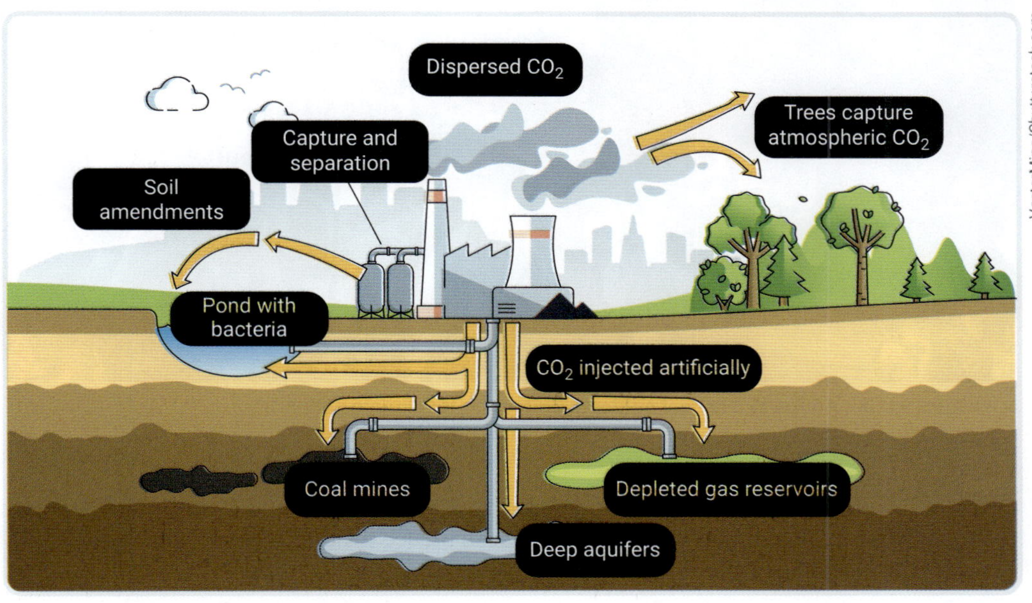

▲ **FIGURE 9.11.3** Carbon sequestration methods

Recycling

When **finite resources** are effectively recycled instead of being manufactured from scratch, it helps reduce both the amount of energy used and the greenhouse emissions produced. You will learn more about recycling in Chapter 10.

finite resource
a limited resource

Barriers to achieving change

There is serious concern among the United Nations and governments around the world that we won't be able to reach our goal of having net zero emissions by 2050. Unless we can make large reductions to our greenhouse gas emissions in the coming decades, global temperature rises of up to 4.5°C could occur by 2100.

Two main barriers could stop us from reaching the goal of net zero emissions by 2050:

1 the economic cost of investing in new technologies and implementing strategies
2 the resistance of some groups in society to adopting **mitigation** policies.

mitigation
reducing the severity of an impact or event

There are substantial costs to finding and implementing solutions to address climate change. An additional complication is that because climate change is a problem affecting the whole planet, many nations and governments need to work together to find solutions. Some industrialised countries are able to deal with the costs and pressures caused by climate change. However, many countries will be slower to implement change because their economies are still developing and they rely on cheap energy, or they don't have the political will or the resources to manage the impacts of climate change.

In Australia, we need to do our bit to slow down or stop climate change (Figure 9.11.4). What do you think governments in Australia could do to address and find solutions to climate change?

▲ **FIGURE 9.11.4** Adults and school students protesting for climate action

9.11 LEARNING CHECK

1 **List** three strategies that individuals can use to reduce their carbon footprint and help reduce greenhouse gas emissions.
2 **Describe** the goal set at COP26.
3 **Describe** three examples of carbon sequestration.
4 **Explain**, using an example, how research is being used to devise means of reducing greenhouse emissions.
5 **Explain** why strategies to reduce climate change might not be implemented.

9.12 Fire management of Country

ABORIGINAL & TORRES STRAIT ISLANDER SCIENCE CONTEXTS

IN THIS MODULE, YOU WILL:

✓ investigate the impact of environmental fire management of Country on reducing carbon emissions.

Managing the environment with fire

Uncontrolled fires in Australia in the last 200 years have cost more than $1.6 billion. Uncontrolled fires have devastating impacts on properties, crops and livestock. They endanger Australia's biodiversity and increase carbon emissions.

Aboriginal and Torres Strait Islander Peoples have used fire to manage the environment for thousands of years before European colonisation.

Early Europeans misunderstood fire management by Aboriginal and Torres Strait Islander Peoples, thinking it was dangerous and destructive, and they prevented the traditional-style burning of the environment. This change in management altered the environment. Introduced weeds and native grasses have been left to grow uncontrollably, while soil erosion and soil salinity have increased. These changes have culminated in an increase in uncontrolled wildfires.

Now, contemporary science is working with traditional management practices of Aboriginal and Torres Strait Islander Peoples to manage the environment. Carefully placed and controlled burns can positively affect the environment, including reducing the risks of wildfires and lowering greenhouse gas emissions (Figure 9.12.1).

Joe Sambono

Joe Sambono

▲ **FIGURE 9.12.1** (a) A low-intensity burn on Lardil Country (Mornington Island, North Queensland); (b) grass and trees after a low-intensity burn

A case study

The North Australian Indigenous Land and Sea Management Alliance has been working to demonstrate the impact of traditional-style burning. Teams of Aboriginal and Torres Strait Islander land managers and science partners select regions of the land to study. They record the types of plants on the ground and in the canopy, as well as the grass cover and the height and width of trees. They measure a number of 1 m^2 plots, collecting and weighing the grass and ground litter. This is the 'before burning' measurement.

Carefully controlled fires are then set to remove the grasses and other fuels on the ground. The sampling is repeated after burning to measure the amount of fuel that has burned. Measurements of the smoke produced provide information about the gases released by burning. The same process is carried out in areas affected by wildfire.

☆ ACTIVITY

Investigating the impact of fire management

Projects such as the ones mentioned in this module demonstrate the potential to reduce harmful greenhouse gases.

1 Table 9.12.1 shows calculations for Matuwa (Lorna Glen National Park in Western Australia). This region, the lands of the Martu Peoples in the desert area, is dominated by dry spinifex grass. The Martu Peoples have Native Title rights to about 20 million hectares of north-western Western Australia. The calculations assume about 7 per cent of the land is managed by fire each year. Martu Peoples set fires in small patches to protect the habitats of native animals. This is known as mosaic burning because of the patch-like pattern seen on the land.

▼ **TABLE 9.12.1** Calculations of fuel and emissions for traditional-style fire management versus those for unmanaged fires in Matuwa

	Fuel burned (t ha^{-1} year)	CH_4 emissions (t ha^{-1} year)	NO_2 emissions (t ha^{-1} year)	CO_2 emissions (t ha^{-1} year)
First Nations (Martu Peoples) traditional-style fire management	5.02	0.0095	0.0003	0.281
Unmanaged fires	6.64	0.0126	0.0003	0.371

Data from https://nailsma.org.au/uploads/resources/Desert-Fire-and-Carbon-Report-230714.pdf

a **Calculate** the difference in potential emissions of each of the gases and the amount of fuel burned.

b What is the difference in emissions between fire-managed land and unmanaged land?

c Martu Country comprises 20 million hectares of land. How does this affect the gas emissions according to the information in Table 9.12.1?

2 Figure 9.12.2 shows areas of Martu Country affected by fire in 1954, 1973 and 2000. Martu Peoples had been absent from their Country from the 1960s and through Native Title determination have returned since 2000.

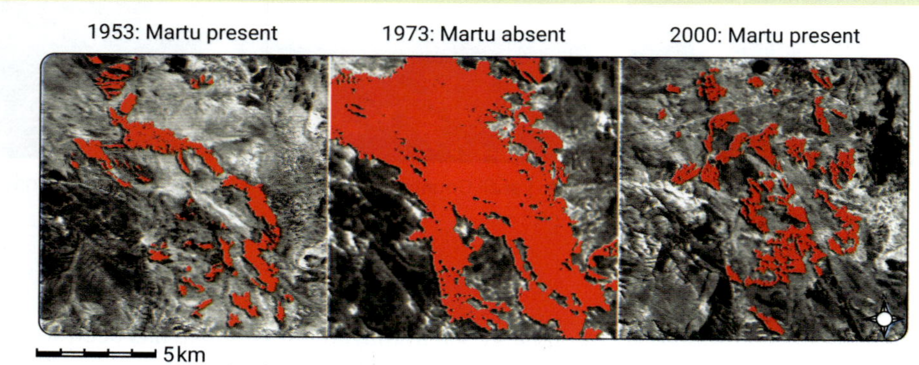

▲ **FIGURE 9.12.2** Areas affected by fire in Martu Country. The middle image was taken when no Martu Peoples were on Country.

Explain why you think there are differences in the area affected by fire on Martu Country. Use information from Table 9.12.1 to help you **construct** an explanation.

SCIENCE IN CONTEXT

9.13 Climate change models

BY THE END OF THIS MODULE, YOU WILL BE ABLE TO:

✓ explain how scientists model climate change.

To understand and model climate change, climate scientists use datasets that contain measurements collected over long periods of time. Scientists use the long-term measurements to construct computer-generated climate models. These datasets provide baseline information that scientists use to compare with the results from climate modelling.

Climate models use similar physics, mathematics and processes as those used by weather forecasting models, but make calculations over a much longer time period. Climate models produce projections of how average conditions will change decades into the future.

Climate scientists use technology to monitor climate change and to help develop models to simulate what will happen to Earth's climate in the future. Scientists monitor the climate using observations from satellites hundreds of kilometres above Earth's surface (Figures 9.13.1 and 9.13.2), as well as measurements on the ground and in the ocean.

DATA SCIENCE 

Learn more about large datasets in **Module 2.5**.

▲ **FIGURE 9.13.1** The Suomi National Polar-orbiting Partnership satellite is an example of a climate satellite.

▲ **FIGURE 9.13.2** An image taken from the climate satellite Suomi National Polar-orbiting Partnership showing smoke and storms due to massive bushfires across eastern Australia in December 2019.

Other measurement activities include ice core drilling in Antarctica and Greenland, monitoring permafrost ground temperatures in deep holes in North America and measuring the extent of coral bleaching on the Great Barrier Reef.

▲ **FIGURE 9.13.3** Data about coral bleaching extent and frequency is used in climate change models.

Climate models can simulate the transfer of water and energy in a climate system. Measurement data is entered into computer models to represent current climate conditions. Starting with this initial data, the model then performs thousands of mathematical calculations based on the physics of the Earth system to produce climate projections. Additional information can be added to the model to prepare different scenarios. For example, reducing methane emissions by 5 per cent in the model will show how that will affect things such as atmospheric and ocean temperatures and sea level.

The three most common types of climate model are:

- energy balance models – monitor changes in Earth's energy to determine the effects of heat accumulation in the oceans and atmosphere in a region or across a continent
- intermediate complexity models – similar to energy balance models but also include impacts from land, oceans and ice sheets. They detect changes in ocean currents, glaciation and atmospheric composition over longer periods
- general circulation models – the most complex and accurate models, used to model climate change by simulating geochemical cycles, atmospheric chemistry, glaciers, ocean circulation and other aspects of the Earth system.

There are many possible outcomes based on each of these models, depending on the information entered (such as future changes to greenhouse gas emissions) and how the models simulate aspects of the climate. Scientists use the range of model results accompanied by a measure of statistical probability.

9.13 LEARNING CHECK

1 **Describe** why climate scientists work with computer simulations to make predictions about the effects of climate change.

2 **Describe** why climate scientists use remote sensing techniques and satellites to collect data.

3 **Explain** why the datasets used in simulations are from as long a timeframe as possible.

4 **Analyse** why the scenarios each model describes would be accompanied by a statistical probability of it occurring.

9 REVIEW

REMEMBERING

1 **Define** the greenhouse effect.

2 **List** the four spheres that make up the Earth system.

3 **Describe** the difference between the greenhouse effect and the enhanced greenhouse effect.

4 **Describe** the difference between climate and weather.

5 **Describe** what is significant about the Bramble Cay melomys.

UNDERSTANDING

6 **Explain** why the greenhouse effect is important to the existence of life on Earth.

7 **Outline** and **describe** the characteristics used by Köppen to **classify** climate types.

8 **Describe** the parts of the electromagnetic spectrum that are:

a shortwave radiation. **b** longwave radiation.

9 **Construct** a table as shown below. **Outline** the sources responsible for the increased concentration of each gas in the atmosphere.

	Carbon dioxide	Methane	Nitrous oxide
Sources of increased atmospheric concentration			

10 **Explain** why the deep ocean current might be referred to as Earth's energy conveyor belt.

11 Use the diagram below to **summarise** how the carbon and water cycles are affected by a warming climate.

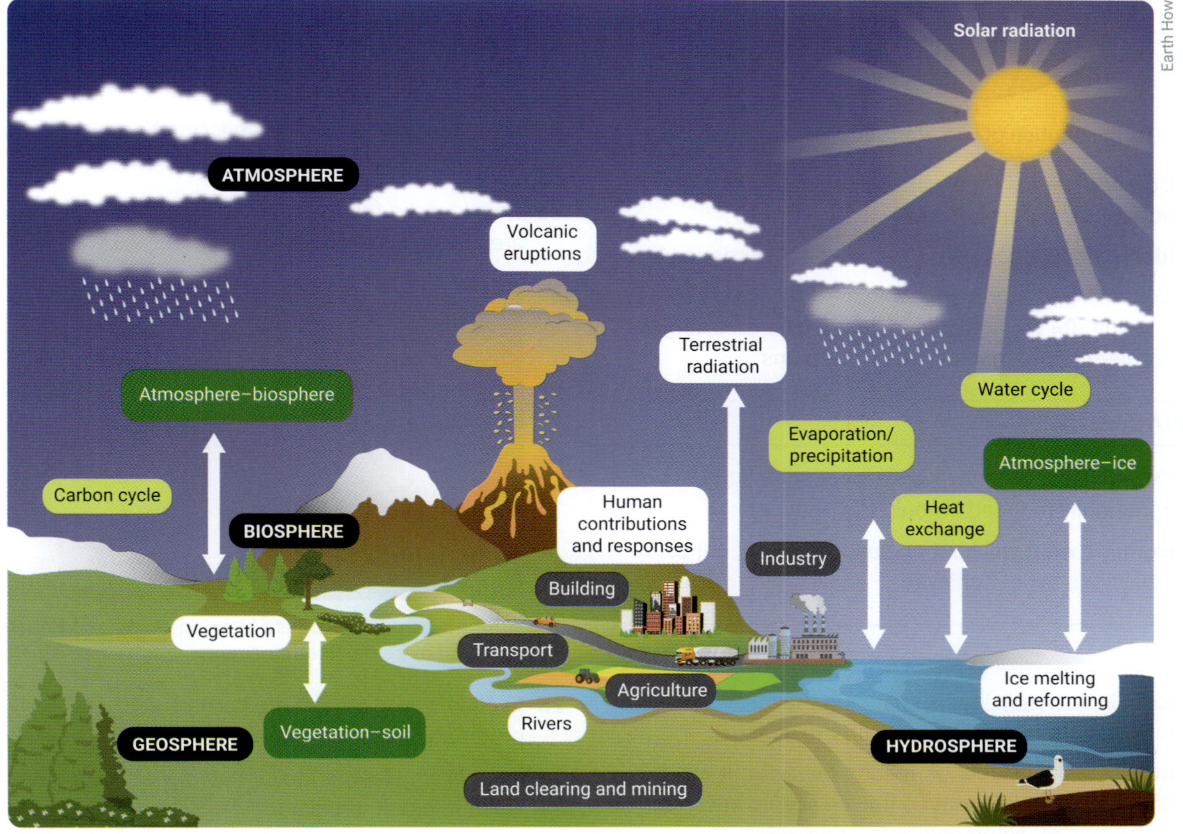

▲ Some of the interactions between Earth's four spheres

APPLYING

12 **Identify** why the acidification of the oceans is occurring.

13 **Compare** the effects of reducing the atmospheric concentrations of carbon dioxide and methane.

14 Using the following diagram, **demonstrate** why there is a relatively small continental climate zone in the southern hemisphere.

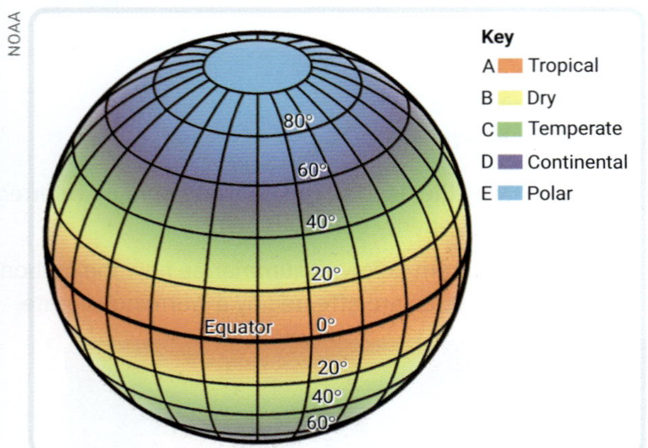

▲ The distribution of climate zones

15 **Compare** deep ocean current circulation to atmospheric convection currents.

16 **Identify** how deep ocean currents affect marine life.

17 **Construct** an argument explaining why biodiversity and climate change should be considered a single problem rather than separate problems.

ANALYSING

18 **Examine** and **list** the likely impact on polar bears of continually reduced amounts of sea ice.

19 **Examine** what will happen beyond 2050 if net zero emissions is not achieved.

EVALUATING

20 **Explain** how individuals changing their behaviour can have a significant effect on climate change.

21 **Compare** the effects of melting permafrost with the effects of melting continental ice sheets such as those on Greenland and Antarctica.

22 **Explain** why forests are a form of carbon sequestration.

23 The following diagram shows the concentration of atmospheric carbon dioxide over 800 000 years.

a **Explain** why there are peaks and troughs in the earlier part of the graph.

b **Determine** why the orange line has risen so sharply.

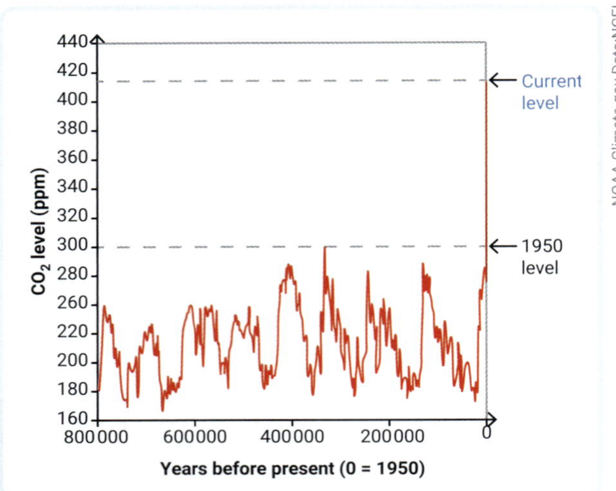

▲ Atmospheric carbon dioxide concentration over the last 800 000 years

24 Consider the following photo. **Justify** that what you are seeing is related to climate change.

▲ Storm surges can cause huge amounts of damage during extreme weather events.

25 **Explain** what some of the barriers are to achieving progress on climate change mitigation.

26 **Explain** why temperate climate zones are located close to coastal regions.

27 **Explain** why a dry climate zone occurs in central Asia.

9780170491785

SCIENCE IN DEPTH STUDY

1 Connect what you've learned

By completing the modules in this chapter, you've learned about the causes and some effects of climate change and what can be done to slow down or reduce its severity. The global plan is to achieve net zero emissions by 2050, but will this be enough? This challenge gets you to reflect on how your actions and behaviours contribute to climate change by examining in more detail the causes and effects of ocean acidification.

2 Check your thinking

First, consider what you know of ocean acidification. List what you know about its cause and effects. What questions does your list suggest you need to answer to understand the issue more fully?

Next, reflect on your daily activities that contribute to greenhouse gas emissions. Start by listing as many activities as possible that you are engaged in that you know are linked to greenhouse gas emissions.

- Which activities generate the greatest amount of greenhouse gases?
- Is it possible to modify any of these activities to reduce your impact?
- Realistically, what changes to lifestyle are people able to make to reduce their contribution to greenhouse gas emissions and ocean acidification?

3 Get into action

Make a list of the things you now know about ocean acidification. Summarise your understanding into four or five key ideas. Next, list the major ways that you and your community can reduce greenhouse gas emissions. Finally, identify the angle you will use to promote reducing greenhouse emissions and reducing the effects of ocean acidification.

4 Communicate

Create a podcast to communicate your ideas about reducing greenhouse emissions and reducing ocean acidification.

A podcast is a prerecorded audio file that can be downloaded and listened to by an audience. You may create the podcast yourself or work with someone else to create a discussion.

A good podcast makes use of storytelling ideas with a beginning, a middle and an ending, which may include a call to action.

- **Beginning:** In the first few seconds, make a statement or ask a question to grab the listener's attention. This is called the hook. Make it concise and memorable.
- **Middle:** Structure the content you have so that you describe your subjects and link them together.
- **End:** A good conclusion will summarise the issues you have discussed for your listeners. Provide the key ideas to be remembered.
- **Call to action:** This is an invitation to your audience to act on what they have just learned. What specific things do your audience need to do to reduce emissions? Do you want your audience to share what they have learned? Can you create a sense of urgency about your call to action?

You can record your podcast on any device that can record your voice. Make it long enough to demonstrate your understanding of the issues you are presenting (but less than 15 minutes).

10 Sustainability

9780170491785

SCIENCE IN DEPTH

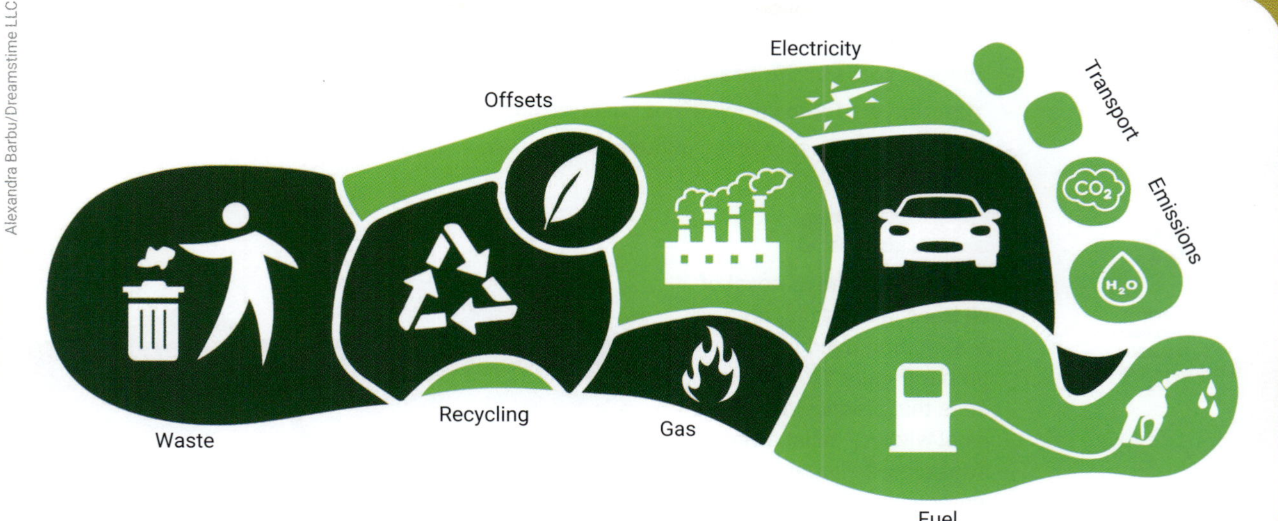

▲ **FIGURE 10.0.1** What is your carbon footprint?

We all contribute to climate change in many ways, whether through our lifestyle, behaviour or choices. We all need to do our bit to reduce the effects of climate change because the alternative is not good for our planet – or us. We can do this by choosing more sustainable options. Even if everyone does just a little bit, the collective change can be significant.

▶ **What types of things could you and your family do every day to live more sustainably and help reduce the effects of climate change?**

#10 DIVE INTO SCIENCE!

At the end of this chapter, you can complete Science in Depth Study #10. You can apply the knowledge and skills you learn in this chapter to complete the project.

Assessments

- Prior knowledge quiz
- Chapter review questions
- End-of-chapter test
- Depth study: Project and scientific text

Videos

- Science skills in a minute: Reproducible methods and valid results **(10.4)**; Evaluating scientific claims **(10.7)**
- Video activities: The United Nations Sustainable Development Goals **(10.1)**; How to reduce your environmental footprint **(10.3)**; Biofuels **(10.9)**

Science skills resources

- Science skills in practice: Reproducible methods and valid results **(10.4)**; Evaluating scientific claims **(10.7)**

Interactive and other resources

- Drag and drop: Renewable or non-renewable energy **(10.5)**
- Quizzes: Causes and consequences of pollution **(10.2)**; Renewable or not? **(10.5)**
- Activity sheets: Design a sustainable house **(10.3)**
- Worksheets: Considering sustainability **(10.3)**; Types of energy **(10.5)**; Identifying types of energy resources **(10.5)**; Renewable and non-renewable resources **(10.5)**; Renewable resources revision **(10.R)**

To access resources above, visit **cengage.com.au/nelsonmindtap**

10.1 Principles of sustainability

BY THE END OF THIS MODULE, YOU WILL BE ABLE TO:

- ✓ define sustainability
- ✓ identify some of the goals of sustainable development
- ✓ describe the three principles of sustainability
- ✓ explain how the principles of sustainability are interconnected.

Video activity
The United Nations Sustainable Development Goals

GET THINKING

Clear-felling, the removal of all trees in a logging area, is a method used to obtain timber (Figure 10.1.1). How does this practice affect the local wildlife? What effect does clear-felling have on waterways? Are the benefits to the community worth the costs that result from timber harvesting? Make a list of the costs and benefits and any questions you have about clear-felling.

Fairfax Media/Fairfax Media Archives/Getty Images

▲ **FIGURE 10.1.1** Native forest after clear-felling (logging)

What is sustainability?

We are part of Earth's complex system. Life, oceans, the lithosphere and atmosphere all provide resources we need to live. A resource is something in our environment that can be used to meet a need. Some resources are physical materials, such as ores, timber, plants, animals, water or air. The knowledge and skills of people around us are also resources.

sustainability
consuming resources in a way that meets our current needs but also allows for the needs of future generations

sustainable development
development that meets our current needs without reducing the ability of future generations to meet their needs

Sustainability means consuming resources in a way that meets our needs but still allows for the needs of future generations. If we consume natural resources faster than they can be regenerated or new sources are discovered, we limit how we can meet our needs and reduce the quality of our lives. **Sustainable development** is using natural resources in a way that allows future generations to meet their needs as well.

The United Nations promotes sustainability through 17 global sustainable development goals (Figure 10.1.2).

iStockPhoto.com/Bigmouse108

▲ **FIGURE 10.1.2** The United Nations sustainable development goals

Each of these goals is important and they are interconnected: addressing one goal affects the others. For example, increasing the availability of fresh water and sanitation (Goal 6) improves health and well-being (Goal 3). Reducing pollution due to poor sanitation improves living conditions and makes communities more sustainable (Goal 11). Increasing the availability of fresh water also increases food production (Goal 2) and assists with employment and economic growth (Goal 8). You can read more about the UN Sustainable Development goals via the weblink.

Weblink
UN Sustainable Development goals

The three principles of sustainability

There are three principles of sustainability: environmental, social and economic (Figure 10.1.3). The three principles of sustainability are interconnected. Addressing, or failing to address, one principle of sustainability affects the other principles.

Environmental sustainability

Environmental sustainability is the preservation and protection of the natural environment so that we can meet our needs and have a healthy environment both now and in the future.

Environmental sustainability is reduced by such things as pollution, biodiversity loss, unsustainable consumption of natural resources and climate change.

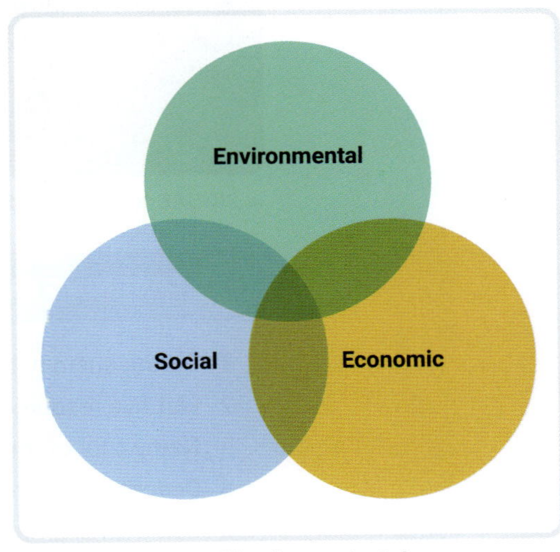

▲ **FIGURE 10.1.3** The three principles of sustainability

Social sustainability

Social sustainability focuses on the long-term well-being of people and their communities. Poverty, lack of access to resources, health care or education and lack of security are all factors that reduce social sustainability.

Economic sustainability

Economic sustainability involves achieving a balance between:

- promoting economic development
- using resources efficiently
- taking care of the social and environmental parts of the economy.

A sustainability case study

Golden perch is a freshwater fish native to the river systems of eastern and South Australia (Figure 10.1.4). A commercial fishery for golden perch operated in New South Wales from the 1880s until 2001. The fishery closed due to several factors: overfishing, disease, habitat degradation and the effects of introduced species. By failing to address the environmental sustainability of the fish population, the fishery reduced its economic sustainability. This in turn affected the social sustainability of small towns where people relied on the fishery for work.

Auscape International Pty Ltd/AUSCAPE/Alamy Stock Photo

▲ **FIGURE 10.1.4** Golden perch, an Australian native fish

10.1 LEARNING CHECK

1 **Define** sustainability.
2 **List** five resources used to make your lunch for today.
3 **Name** four of the UN developmental goals.
4 **Describe** the three principles of sustainability.
5 Using an example, **explain** how a change affecting environmental sustainability can affect both social and economic sustainability.

10.2 Pollution

BY THE END OF THIS MODULE, YOU WILL BE ABLE TO:

- ✓ describe the causes of environmental pollution
- ✓ explain how pollution negatively affects sustainability.

GET THINKING

Scan the tables and images in this module and use what you observe to write a definition of pollution. Use your definition to make a list of the types of pollution you are exposed to on an average day. How is pollution related to the concept of sustainability?

Pollution and pollutants

Pollution is the introduction of harmful substances or effects into the environment. A harmful substance or effect that negatively affects living things is referred to as a **pollutant**. Most pollutants are chemicals, but heat, light and sound can also be pollutants. Ash and some gases from volcanic eruptions are examples of **natural pollutants**, but most pollutants are due to human activities.

Some pollutants are absorbed or broken down, but others can accumulate to dangerous levels. Methane, a greenhouse gas, is produced naturally in ecosystems and will naturally break down in about a decade (Figure 10.2.1). Human activities have introduced methane into the environment faster than it can be broken down though, causing negative effects. Methane's contribution to climate change is described in Chapter 9.

William Edge/Shutterstock.com

▲ **FIGURE 10.2.1** A single cow can produce around 100 kg of methane gas a year, mostly through burping, so increasing cattle farming has a huge effect on atmospheric methane.

Pollution from human activities

There are many forms and sources of pollutants. Natural sources of pollutants include volcanic ash, sea salt, dust, bushfires and pollen. Figure 10.2.2 summarises the main sources of pollutants caused by human activities.

pollution
the introduction of harmful substances into the environment

pollutant
a substance introduced into an environment that can be harmful

natural pollutants
pollutants from natural processes

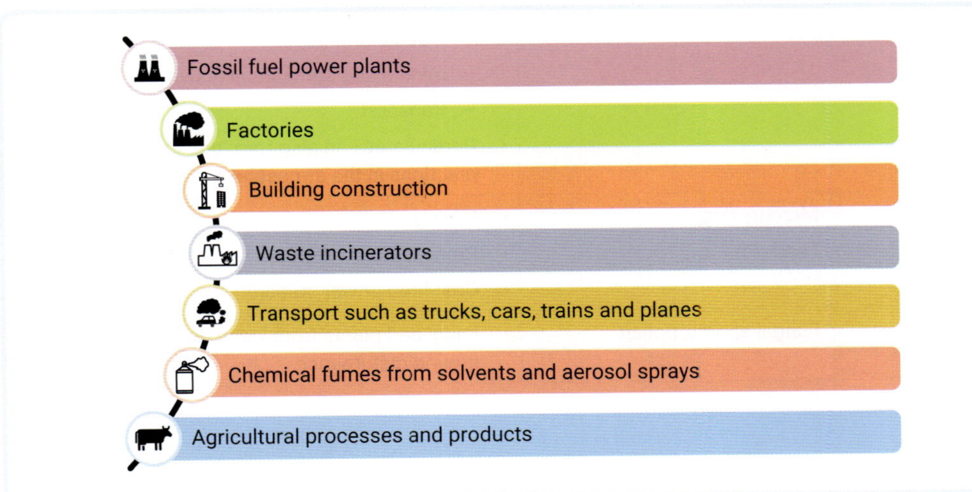

▲ **FIGURE 10.2.2** Major sources of pollutants from human activities

persist
stay or remain

Some pollutants **persist** in the environment for a very long time. Mercury is a heavy metal that was used in gold mining during the second half of the 1800s. From this process, there is still mercury cycling between the atmosphere, the soil and vegetation. Persistent organic pollutants (POPs) such as DDT stay around for a long time, too.

Pollutants can be classified in many ways. Table 10.2.1 presents some of the most common pollutants by type and lists their sources.

TABLE 10.2.1 Sources of pollutants produced by human activities

Pollutant type	Pollutant	Sources
Gases	Carbon monoxide (CO)	• Domestic combustion • Prescribed burns and bushfires • Motor vehicles • Metal manufacturing
	Nitrogen oxides	• Motor vehicles • Coal-fired power stations • Manufacturing industries
	Sulfur dioxide (SO_2)	• Coal-fired power stations • Production of metals such as lead, copper and zinc
	Ozone (O_3)	• Chemical reactions of substances released into the atmosphere
	Volatile organic compounds (VOCs)	• Evaporation from paints, solvents, refrigerants and fuels • Cleaning products
	Persistent organic pollutants (POPs) (can also be attached to particulates)	• Pesticides • PFAS, a large group of different manufactured chemicals • Industrial chemicals • Industrial processes
Particulates	Coarse particulate matter (PM_{10}) Fine particulate matter ($PM_{2.5}$)	• Mostly caused by combustion such as in: – bushfires – wood heaters – mining and agricultural activity – transport • Particulates can also form due to chemical reactions in the atmosphere
Heavy metals	Mercury (Hg) Lead (Pb)	• Mining • Processing of metals • Coal-fired power stations
Allergens	Fungal spores, pollen	• Fungi and plants from weed dispersal and ground disturbance

particulates
very small particles of a substance, especially those produced by burning fuels

Air pollution

Air pollution affects both the natural environment and human health. It can speed up the weathering of rocks, affect plant growth and contribute to water pollution. When carried by the wind, poisonous substances such as lead or pesticides can spread throughout the environment. Sulfur oxides, which dissolve in water to create acids, can also cause **acid rain**, damaging or killing trees (Figure 10.2.3).

Air pollutants create more health problems for people than other types of pollution. The premature death of about seven million people each year is caused by air pollution. Breathing pollutants into our lungs damages our health and can lead to serious conditions such as lung cancer, heart disease and respiratory disease.

During the 2019–20 summer bushfires, 11 million people in south-eastern Australia were exposed to particulates in smoke from the fires (Figure 10.2.4). The health cost of the fires was estimated to have been 1.95 billion dollars. Many people are exposed to air pollution inside their homes. An estimated 2.4 billion people are exposed to air pollution from open fires or stoves using fuels such as kerosene, wood, animal dung or coal.

A.Freund/Adobe Stock photos

▲ **FIGURE 10.2.3** Trees damaged by acid rain in Europe

diyben/Shutterstock.com

▲ **FIGURE 10.2.4** The 2019 bushfires produced air pollution over Sydney and many other parts of New South Wales.

Water pollution

Water pollutants are wastes deposited in water. Sometimes water pollution is intentional, for example, when a factory discharges waste into a river. However, other times, the pollution is unintentional, such as when stormwater carries chemicals or litter into a waterway (Figure 10.2.5).

acid rain
rain that is acidic due to chemicals dissolved in water in the atmosphere

Marmolejos/Shutterstock.com

▲ **FIGURE 10.2.5** If we throw litter into the street, it can become water pollution.

toxic
able to cause harmful health effects

pesticide
a poisonous substance used to control pests

heavy metal
metallic elements with relatively high density and toxic properties such as lead, mercury, arsenic and cadmium

bioaccumulation
the gradual accumulation of substances in an organism

biomagnification
the increase in the concentration of a substance in organisms as you move up a food chain

acid mine water
water from mines that has become acidic through a chemical reaction and contains dissolved minerals

eutrophication
the accumulation of nutrients in a water body leading to the rapid increase in algae and bacteria

Accumulation of pollutants in aquatic ecosystems

Toxic pollutants are poisons such as **pesticides** or **heavy metals**. Some toxic pollutants build up in living cells through a process called **bioaccumulation**. If toxic pollutants build up to high levels, they can produce mutations in the organism or even kill it. When levels are low, organisms can survive, and if they are consumed by another animal, these pollutants pass further up the food chain. The bioaccumulated toxins from the prey become part of the predator (Figure 10.2.6). As you go up the food chain, the amount of toxins in each organism increases, a process called **biomagnification.**

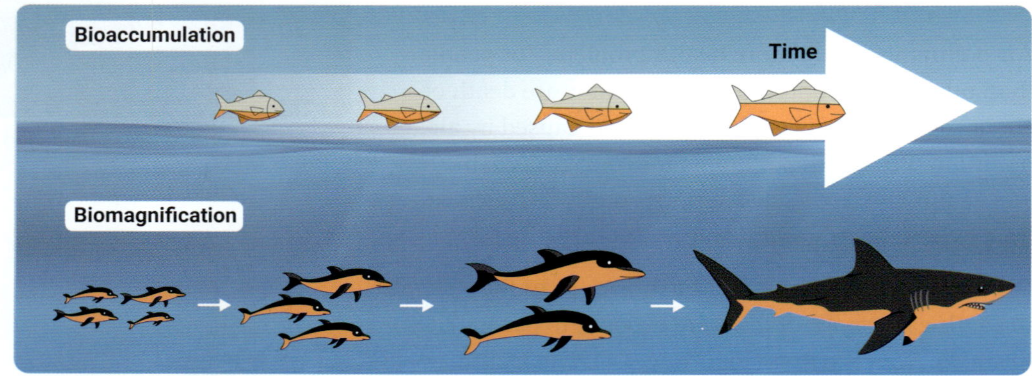

▲ **FIGURE 10.2.6** Bioaccumulation occurs in a single organism. Biomagnification occurs along a food chain.

Jose Arcos Aguilar/Shutterstock.com

▲ **FIGURE 10.2.7** After more than a thousand years of mining nearby, the Rio Tinto in Spain carries acid waters.

Sergey Muhlynin/Shutterstock.com

▲ **FIGURE 10.2.8** Algal blooms can poison a waterway.

Acid mine water

Mine rocks are rich in minerals containing sulfur. When sulfur compounds react with oxygen and rainwater, they produce an acid. This acid can dissolve heavy metals from the surrounding rock and carry them to creeks and rivers. In this way, **acid mine water** can kill a river (Figure 10.2.7), and there are strict rules against untreated water being released from mine sites.

Organic matter and plant nutrients

Farms and some factories generate large amounts of nutrient-rich wastes. In water, these wastes can be broken down by bacteria. But the feeding bacteria use up oxygen from the water, and this low-oxygen environment can kill other aquatic organisms such as fish.

Eutrophication is a similar process. Eutrophication occurs when a water body is polluted with nutrients, such as compounds of nitrogen or phosphorus, causing a rapid growth in bacterial algae. Some algal blooms turn the water bright green (Figure 10.2.8). Some blooms are toxic, killing fish and making the water unfit to drink. As the algae die and decay, bacteria use up oxygen in the water, making it harder for other organisms in the pond to survive.

9780170491785

Hot water from power plants and cold water from dams are also forms of pollution that can have harmful effects on the environment. This form of waste is called **thermal waste**.

thermal waste hot or cold water or air, released into the environment by industry or energy production

soil degradation physical, chemical and biological decline in soil quality

Soil pollution

Soil degradation is the physical, chemical and biological decline in the quality of soil. Pollutants are a major cause of soil degradation. Industrial, mining and agricultural activity, together with the improper disposal of waste, can pollute soils (Figure 10.2.9). The most common soil pollutants are chemicals derived from crude oil, solvents, pesticides and heavy metals, particularly lead.

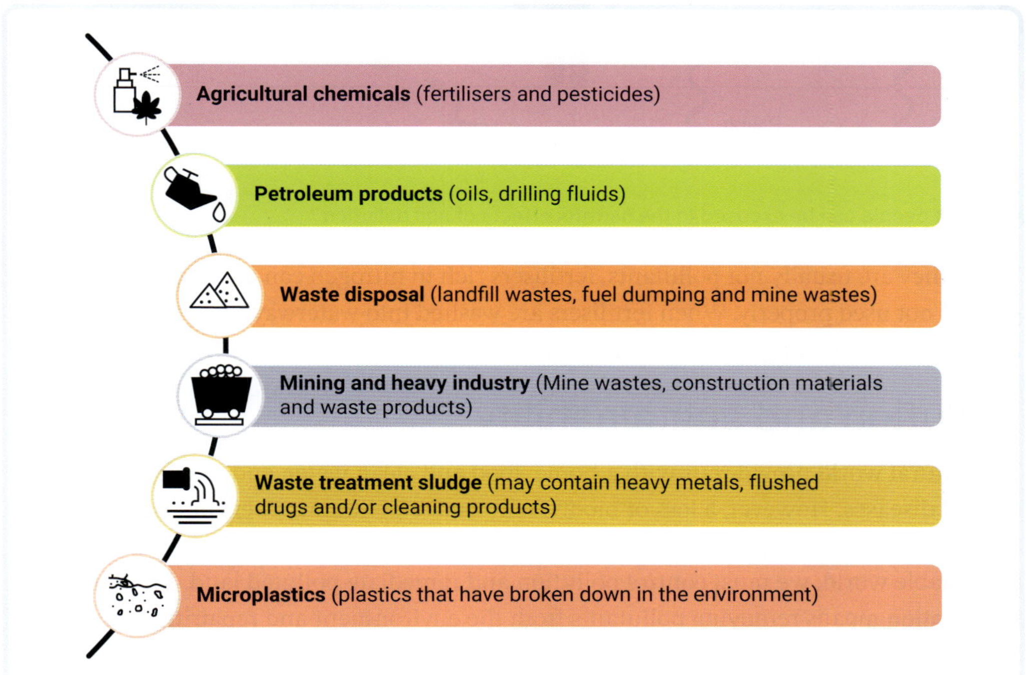

▲ **FIGURE 10.2.9** Some sources of soil pollution

Soil pollution and soil degradation may have serious consequences, including:

- poor human health
- poor harvests, leading to food shortages and famine
- climate change, as degraded soils release stored carbon
- extinction of species.

Farming is a source of pollutants due to the pesticides and fertilisers that are spread across paddocks. Pesticides can contribute to species extinction by reducing biodiversity. For example, the loss of insect populations negatively affects the species that rely on them for food and pollination. Toxic chemicals in soil can also damage people's health in many different ways (Figure 10.2.10).

Quiz
Causes and consequences of pollution

Over time, pesticides and heavy metals can reduce the fertility of soils and crop yields, and this affects economic sustainability. Polluted soil and the famine it can cause affect social sustainability, because people leave the area to seek better places to live.

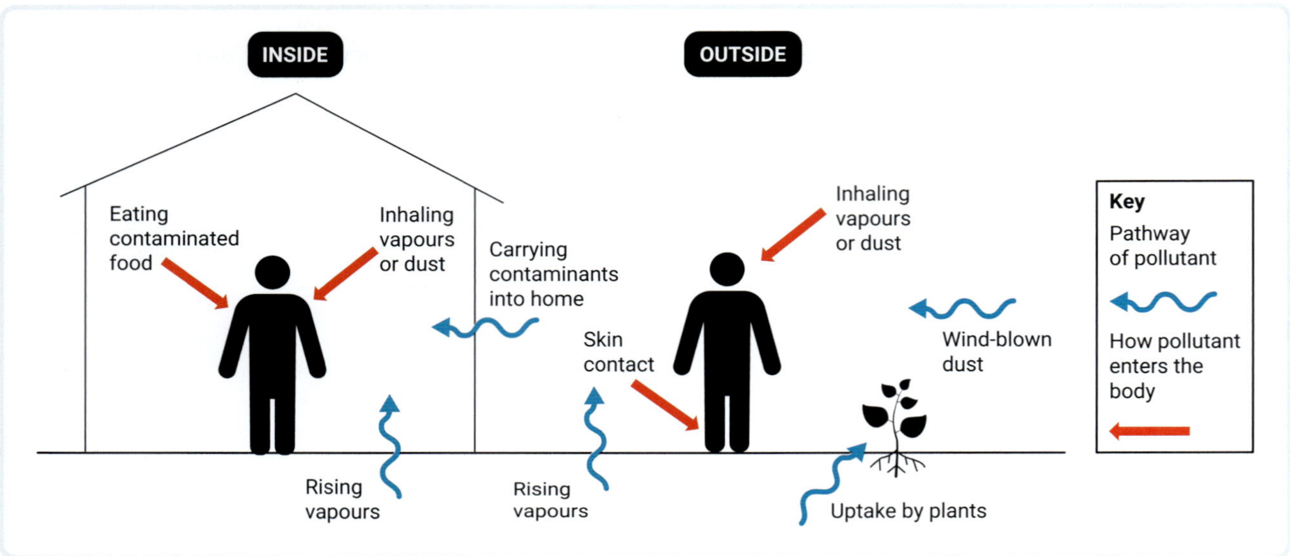

▲ **FIGURE 10.2.10** Some ways that people can be exposed to the harmful effects of soil pollution

While they are usually not pollutants, fertilisers rich in nitrogen can create greenhouse gases if not used properly. When fertilisers are washed into waterways, it can also increase nitrogen levels, leading to imbalanced nutrients, increasing the chance of algal blooms.

Pollution and sustainability

The natural world provides us with a home, food and clean water, and it helps to control diseases. However, a loss of biodiversity due to pollutants alters our world and reduces the services it provides and the lands where people can live safely. To achieve a sustainable world, we must control pollution and remediate polluted land. Environmental **remediation** means removing pollutants from the environment and repairing their effects. If we fail to reduce pollution and remediate polluted land, water and air effectively, we will see a decrease in the standard of living for current and future generations.

remediation
the process of cleaning up pollution or contaminants from the environment

10.2 LEARNING CHECK

1 **Define** pollution.
2 **Describe** three sources of pollutants and the specific pollutants they **produce**.
3 **Create** a table like the one below to **identify** examples of pollutants and their effects on air, water and soil.

Type of pollution	Examples of pollutants	Three effects on environment
Air		
Water		
Soil		

4 **Explain** how pollution negatively affects sustainability.

10.3 Sustainability in action

BY THE END OF THIS MODULE, YOU WILL BE ABLE TO:

- ✓ describe some ways that individuals can live more sustainably
- ✓ explain why alternatives to current resource use increase sustainability.

GET THINKING

Australians like fashion and are the second-highest consumers of textiles in the world (Figure 10.3.1). On average, 23 kg of clothing per person goes to landfill each year. So, what is needed to make our clothing consumption more sustainable?

Nata Sha/Shutterstock.com

▲ **FIGURE 10.3.1** Is fashion sustainable?

Video activity
How to reduce your environmental footprint

Other resource
Worksheet: Considering sustainability

Sustainability and our ecological footprint

To achieve a sustainable world, we need to change what we consume and how much. A person's **ecological footprint** measures how much of the environment is needed to supply the goods and services that support their lifestyle (Figure 10.3.2). It is expressed as the amount of land required to sustain their needs, including the land required to absorb their greenhouse gas emissions and account for their carbon footprint. As shown in Figure 10.3.2, your carbon footprint has the biggest impact on your overall ecological footprint.

ecological footprint the amount of the environment necessary to supply the goods and services that support a lifestyle

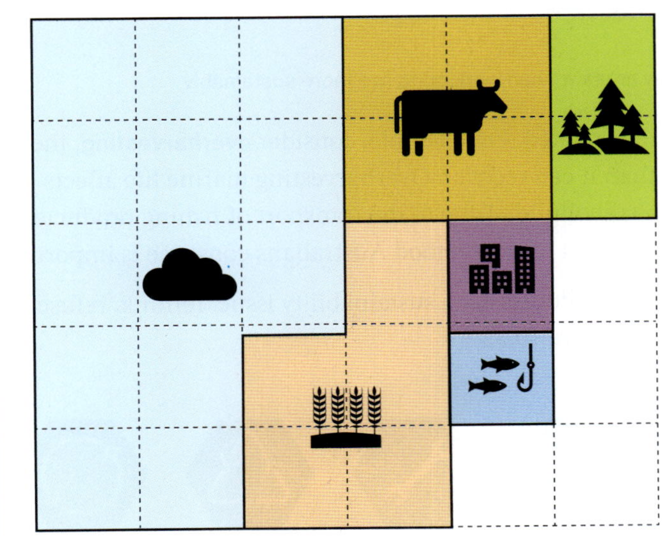

Key

 Carbon footprint

 Crops

 Forestry products

 Grazing

 Built-up land (houses, roads, industry, etc.)

 Fishing

▲ **FIGURE 10.3.2** An ecological footprint is a measure of how much of the environment is needed to support a person's lifestyle. The average ecological footprint of Australians is currently about 2.6 hectares per person.

There are three key parts of our ecological footprint we can change to live more sustainably:

- the food we eat (how much, where it comes from and how it is produced)
- the greenhouse gases we produce (e.g. the energy sources we use, how much we travel)
- the building materials and other products we use (e.g. plastic used, waste produced).

How you can contribute to sustainability

Sustainability is something everyone can work towards (Figure 10.3.3). Living more sustainably will improve the quality of our lives now and in the future.

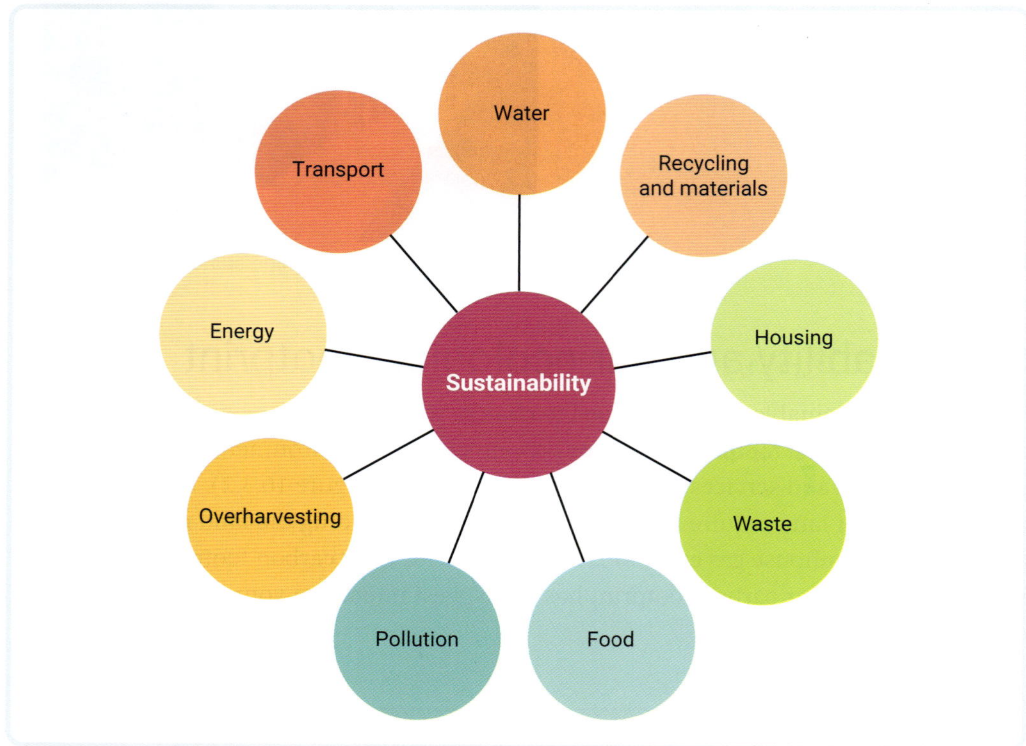

▲ **FIGURE 10.3.3** Nine of the many areas we can work on to live more sustainably

overharvesting harvesting a resource faster than the resource can recover

Many sustainability issues are interrelated. For example, consider **overharvesting**, the harvesting of a resource faster than it can recover. Overharvesting marine life affects both food sustainability and global pollution (due to the transport of fishing products). It is estimated that more than 60 per cent of the seafood Australians consume is imported.

The six Rs are a reminder of ways to address each sustainability issue: rethink, refuse, reduce, repair, recycle and reuse (Figure 10.3.4).

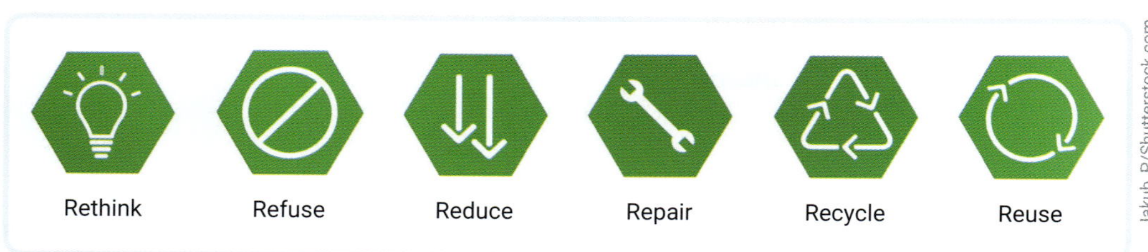

▲ **FIGURE 10.3.4** Following the six Rs can help us to live more sustainably.

Rethink

Consider how what we buy affects the environment. How much energy is needed for the things we do or buy? Can rethinking excess packaging or how waste is disposed of improve sustainability?

Refuse

When we say 'no' to buying something, we send a message to the product's producer. By choosing what we consume, we can influence what producers sell us. Declining heavily packaged items, and explaining why, can change how the product is packaged.

Reduce

We buy more than we need. New South Wales generates about 1.7 million tonnes of food waste each year and 40 per cent of this comes from our households. As a community we typically buy more food, clothing and other household items than we need. By reducing our consumption, we reduce the amount of resources we use.

Recycle

Material does not go to landfill if we recycle it. Recycling also reduces our demand for raw resources. More on recycling is presented in Module 10.6.

Repair

A lot of things we buy are not designed to last. Choosing well-built items and learning how to repair things reduces our costs and the demand for resources. By repairing things, we extend the life of the item and reduce what we contribute to waste.

Reuse

Many items such as shopping bags and food containers can be reused rather than sent to landfill. Choosing to shop at charity shops rather than buying brand new items can give used items like clothes, shoes and crockery a new lease of life, or even to be upcycled. **Upcycling** refers to creating a product of higher value from a recycled item (Figure 10.3.5).

pressmaster/Adobe Stock Photos

▲ **FIGURE 10.3.5** Upcycling can involve the creative reuse of clothing and other items.

Mitigation and adaptation strategies

Mitigation and adaptation are important ways to help sustainability in areas such as climate change. Mitigation means to reduce the impact of something. Finding a way to prevent rubbish from washing into a creek is an example of a mitigation strategy. **Adaptation** means changing how things are done so that the impact of an issue is reduced. In a warming world, having better insulated, passively cooled homes is a way to adapt to hotter conditions.

Mitigation and adaptation strategies conserve resources. They help to maintain a good **standard of living** for communities while ensuring there will be enough resources to provide for generations in the future.

upcycling
the creation of a product of higher value from a used item

adaptation
a change in how something is done so that the impact of an issue is reduced

standard of living
a person's level of wealth and access to food, shelter and safety

PradeepGaurs/Shutterstock.com

▲ **FIGURE 10.3.6** Helping people achieve a better standard of living requires a better distribution of resources.

Mitigation and adaptation will become more urgent as more of the world's population lifts out of poverty. Currently, there are 713 million people living in poverty, which is defined as living on less than $3.20 Australian dollars a day. However, as we continue to support those in poverty and their incomes increase, so will their daily demand for resources. To support a healthy, sustainable world for more people, we need to change how we consume resources today (Figure 10.3.6).

☆ ACTIVITY

Calculating your ecological footprint

Weblink
Global Footprint Network's Ecological footprint calculator

Our ecological footprint measures how much of the environment is needed to supply the goods and services that support our lifestyle. It is a measure that compares our demand for resources against Earth's capacity to renew itself. Calculating an individual's ecological footprint is a good way of understanding our personal impact on the planet and sustainability.

In this activity, you will calculate and review your ecological footprint and describe some ways you might reduce it.

Procedure

1 Use the link to access the Global Footprint Network's Ecological footprint calculator.
2 Answer the questions. Move from one question to the next using the arrow on the right side of the screen.
3 Record the number of Earths needed for everyone to live like you.
4 Press the 'See details' button and summarise your footprint.

Analysis

1 **Identify** which part of your ecological footprint is the largest. **Describe** how this might be reduced.
2 **Identify** the two largest consumption categories in your footprint. **Explain** how the size of the categories might be reduced.
3 **Compare** your ecological footprint with the Australian average of 6 hectares per person.
4 **Explain** how addressing one aspect of your footprint will contribute to sustainability.

10.3 LEARNING CHECK

1 **Define** ecological footprint.
2 **Outline** the three key areas of an ecological footprint that need to be addressed for sustainability.
3 Plastic pollution negatively affects sustainability. **Apply** the six Rs to **outline** ways you might reduce the number of plastics in your life.
4 **Compare** mitigation and adaptation as strategies for reducing the use of resources.
5 **Describe** two sustainability issues and a method that individuals can use to address them.

9780170491785

WORKING SCIENTIFICALLY 

10.4 Reproducible methods and valid results

SCIENCE SKILLS IN FOCUS

IN THIS MODULE, YOU WILL FOCUS ON LEARNING AND IMPROVING THESE SKILLS:

- design an appropriate investigation method to collect reliable and valid data
- conducting a valid, reproducible investigation.

Reproducible methods and valid results

A scientific investigation needs a good, reproducible method to generate valid and reliable results. But what do 'valid' and 'reproducible' actually mean? And what should you do to ensure your methods are up to scratch?

A reproducible method is one where all the steps of an investigation, such as preparing, mixing, observing and measuring, can be repeated to produce the same or very similar results in the same circumstances. In other words, someone else could replicate your methods to get comparable and reliable results.

Valid means that the actions (method), data (results) and inferences (conclusions) you produce are accurate and measure or show what you intended them to measure or show.

To get valid and reliable results, make sure your method is reproducible and valid by considering the following.

Validity

1. Have a clear aim. What are you testing for? What do you want to find out?
2. Identify which variables you will control (dependent variable) and which variable you will change (independent variable).
3. Use a control group – a group in which the independent variable is not included. This is what you compare your other results with.

Accuracy

4. Write the method clearly, step by step.
5. Be precise about what you will measure or observe, and how you will do it.
6. Always state the units of measurement (e.g. seconds, millilitres, grams).

Reliability

7. Use large sample sizes or multiple trials/tests to get as much data as possible.
8. Calculate averages from repeated trials or tests to get a more accurate result (e.g. the sum of the mass of 30 apples divided by the total number of apples gives you the average mass).

Video
Science skills in a minute: Reproducible methods and valid results

Science skills resource
Science skills in practice: Reproducible methods and valid results

SUSTAINABLE DESIGN CHALLENGE

BACKGROUND INFORMATION

In a warming world, people need homes that remain cool without the need for large amounts of energy to cool them. Passive cooling means cooling without using energy. It involves design choices to reduce the heat coming into and increase the heat lost from buildings (Figure 10.4.1).

MAVRITSINA IRINA/Shutterstock.com

▲ **FIGURE 10.4.1** Windcatcher towers in Yazd, Iran, channel cool breezes into and warm air out of homes. This technology is over 3000 years old.

Ventilation, airflow, insulation and shading are all methods of passive cooling. You can find out more about these methods at the Australian Government Your Home website.

In this investigation, you will investigate a passive cooling strategy to gather data and show how the method works.

Weblink
Australian Government Your Home

AIM

To design a method to model and assess the effect of a passive cooling strategy using valid data

MATERIALS AND EQUIPMENT

The materials you need will depend on the method of cooling you wish to investigate. You will need a source of heat and a model building, such as a shoe box.

You will also need an instrument to measure the air temperature.

PROCEDURE

1 Choose the method of cooling you wish to investigate.
2 Identify your variables.
3 Write a hypothesis to test.
4 Design your procedure so that you have a treatment condition and a control condition. Consider the following points when designing your procedure:
   a Besides measuring temperature, what else will you measure in your treatment?
   b What will be the units of measurement?
   c What equipment will you need?
   d Can you measure the temperature in your model building before and after the cooling method is used?
   e Can you make multiple trials?

RESULTS

Create a table of your results, including temperature without a passive cooling method, temperature with the method and average difference.

ANALYSIS

1 Did the passive cooling method work in your model?
2 Review your method and **outline** the features of the method that make it reproducible.
3 **Explain** why your choices in setting up the investigation helped you collect valid data.
4 If other groups in class explore different methods, **suggest** which method appeared to produce the best level of cooling.
5 **Explain** why your results are valid.

CONCLUSION

Write a scientific argument assessing the effect of your passive cooling strategy.

9780170491785

10.5 Renewable resources

BY THE END OF THIS MODULE, YOU WILL BE ABLE TO:

- ✓ distinguish between renewable and non-renewable resources
- ✓ describe sustainable use
- ✓ discuss alternatives to current resource use, including how to reduce, reuse and recycle.

GET THINKING

Think about the clothes you are wearing. What sorts of resources have been used to make your clothes and make them available to you? Make a list and for each item, try to identify whether it can be easily replaced or not.

Quiz
Renewable or not?

Interactive resource
Drag and drop: Renewable or non-renewable energy

Other resources
Worksheets:
Types of energy
Identifying types of energy resources

Renewable and non-renewable resources

Types of resources

Sustainability depends on how we use and conserve resources. **Sustainable use** describes the use of resources in ways that do not lead to long-term degradation of the environment or make resources unavailable to future generations. All the resources we use can be classified as **renewable** or **non-renewable**. Resources include both materials and sources of energy (Table 10.5.1).

sustainable use
using resources in ways that do not harm the environment and leave the resource available for future generations

renewable
describes resources that can be replaced at a rate greater than they are used

non-renewable
describes resources that are either non-replaceable or replaceable at a slower rate than they are used

▼ **TABLE 10.5.1** Examples of renewable and non-renewable resources

	Renewable resources	Non-renewable resources
Matter	• Food plants • Fresh water • Timber • Fresh air • Fertile soil • Animal species	• Metallic ores (e.g. for iron, copper and aluminium) • Non-metallic minerals (e.g. salt, phosphates, natural diamonds) • Biodiversity • Tropical forests • Oil for manufacturing chemicals
Energy sources	• Wind • Tides • Sunlight • Geothermal heat • Hydroelectricity	• Oil • Natural gas • Coal • Uranium

Non-renewable resources

There are two types of non-renewable resources: those that cannot be replaced and those that are replaced at a slower rate than they are used. Biodiversity is a non-renewable resource because it cannot be replaced in an ecosystem. Non-renewable resources are described in terms of **resource** and **reserve**. In this field, the word 'resource' has a particular meaning: it is an estimate of how much of a non-renewable substance exists. A reserve is the amount of the resource that can be mined or harvested at a profit both now and in the future. Extracting and using non-renewable resources reduces reserves and creates pollution.

resource
an estimate of how much of a useful non-renewable substance exists

reserve
the amount of a resource available for mining or harvesting at a profit now and in the future

▲ **FIGURE 10.5.1** Deep-sea drilling extracts oil and gas, both non-renewable resources.

Lithium and crude oil are examples of non-renewable resources. Lithium is a metal used in batteries. It is estimated that worldwide, lithium's resource is around 105 million tonnes, but its reserve is about 28 million tonnes. Crude oil is used worldwide to produce fuel for vehicles. It takes millions of years to form from the remains of ancient marine organisms and cannot be replaced as fast as we are using it (Figure 10.5.1). In 2024, it was estimated that crude oil production would be roughly 103 million barrels per day (a barrel of oil has a volume of approximately 159 litres). World oil reserves are approximately 47 times the world's current annual consumption.

Renewable resources

Renewable resources are those that can be replaced faster than they are being used. Timber can be a renewable resource if forests are managed carefully. However, forestry can become unsustainable when old trees are harvested faster than they are replaced. Examples of renewable energy are wind, solar and geothermal energy.

Conserving resources

Figure 10.5.2 shows the traditional way that resources are treated. It follows a linear process of extraction, production, use and disposal.

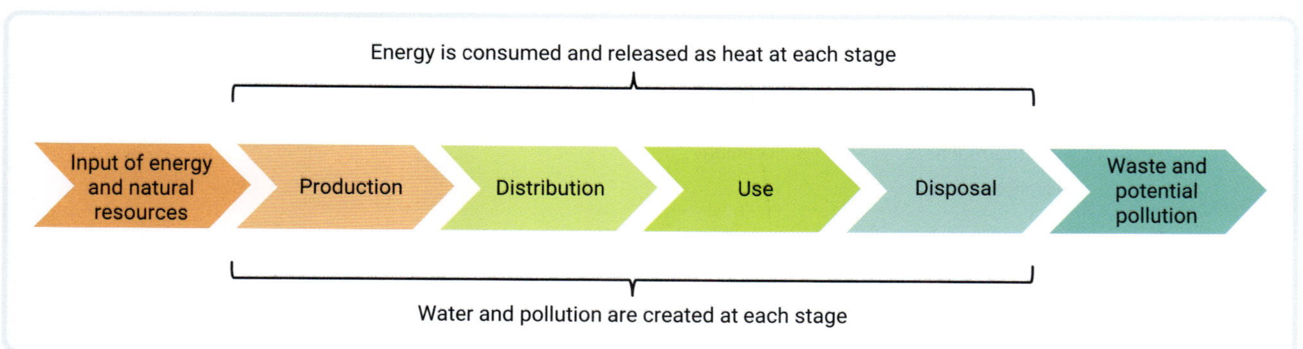

▲ **FIGURE 10.5.2** The traditional path for resources is linear and generates unsustainable levels of waste and pollution.

circular economy an economy using behaviours that reduce the rate of resource use and waste creation

Many renewable and non-renewable resources are important in our lives, but we need to use them in more sustainable ways. As Figure 10.5.2 shows, traditionally the lifespan of a resource was limited and always ended in the resource being disposed of as waste. But if we reuse and recycle resources throughout their lifespan, this process becomes more like a cycle. This is called a **circular economy** (Figure 10.5.3).

9780170491785

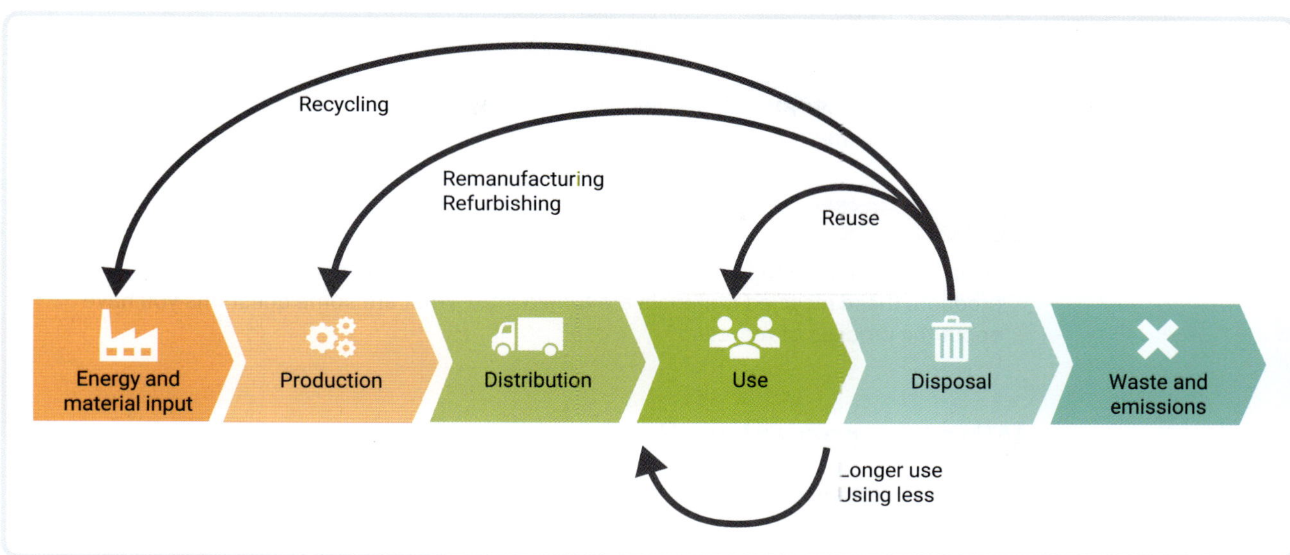

▲ **FIGURE 10.5.3** A circular economy cycles resources and reduces waste production. (Remanufacturing means manufacturing using reused materials.)

A circular economy can reduce the rate at which we consume resources by changing people's behaviour, reducing the rate of production and extending the lifespan of products (Table 10.5.2). A circular economy can also reduce the amount of waste and potential pollution from that waste.

▼ **TABLE 10.5.2** Methods of slowing consumption and using products for as long as possible

Behaviour	Consequence
Mindful choice	Carefully considering such things as packaging, length of use and potential for recycling can reduce what we buy and the resources we consume.
Consuming less	Reducing the quantities of materials used to serve a need.
Sharing	Owning a product jointly with others means fewer products are needed for a community.
Repairing	Repairing a product increases its life and reduces waste in landfills.
Refurbishing	Painting, repairing and cleaning extend the life of a product and increase its value as a second-hand item.
Reusing	Reusing some products, such as shopping bags, reduces the resources needed to manufacture new ones.
Hiring	Paying to use a product for a period before returning it means fewer products are needed in a community.
Recycling	Converting waste into useful products conserves resources and reduces landfill waste.

10.5 LEARNING CHECK

1 **Contrast** renewable and non-renewable resources.
2 **Identify** the following resources as renewable or non-renewable.
  a Natural gas b Copper minerals c Timber
  d Fresh water e Oil
3 **Explain** why sustainable use is important now and for the future.
4 **Describe** three ways of reducing the amount or rate of resources our society consumes.

10.6 Recycling

BY THE END OF THIS MODULE, YOU WILL BE ABLE TO:

- ✓ identify the advantages and disadvantages of recycling
- ✓ outline some methods for recycling materials.

GET THINKING

Does your school recycle materials? What are they? Make a list. What are some of the products these recycled materials are turned into? What are some questions you have about the process of recycling?

What is recycling?

recycling
the process of converting waste into new materials or products

Recycling is the process of converting waste into new materials or products. What we can and cannot recycle varies between council areas, but there are many household wastes that can be recycled (Figure 10.6.1). Often councils will have recycling schemes, targeting metals, glass, hard plastics and paper. Some councils will include food waste as part of the Food Organics and Garden Organics (FOGO) recycling. There are also schemes to recycle building materials and electronic waste in many regions. In New South Wales in 2022–23, 66 per cent of recyclable waste was recovered, but an equivalent of 2.64 tonnes of waste per person was still directed to landfill.

MicroOne/Shutterstock.com

▲ **FIGURE 10.6.1** Things that can be recycled

Advantages and disadvantages of recycling

Recycling conserves valuable materials and often requires less energy than manufacturing things from the unprocessed, raw resources. For example, to make 1 tonne of new glass requires 1.2 tonnes of raw materials, but recycling glass requires only 1 tonne of used glass. The energy required to melt and recycle glass is also far less than the energy needed to create new glass from silica and sand. This means fewer greenhouse gases are produced. Recycling glass also produces less waste, and less air and water pollution.

Metal recycling is effective for metals such as aluminium and steel. Scrap metal is melted in a furnace and recast. Less energy is required to remelt scrap metal compared to **smelting** new metal, and the metals produced have the same properties as those they were made from.

smelting
the process of extracting a metal from the ore that contains it

However, some other metals are hard to extract from the materials that contain them, such as the metals in electronic components. These products are often classified as e-waste, and they need to be shredded into small pieces and then sorted according to their properties. There are some rare metals that are increasingly being used in technology, such as in electric cars (Figure 10.6.2). These metals are only recycled in small amounts because there is currently no easy way to do so.

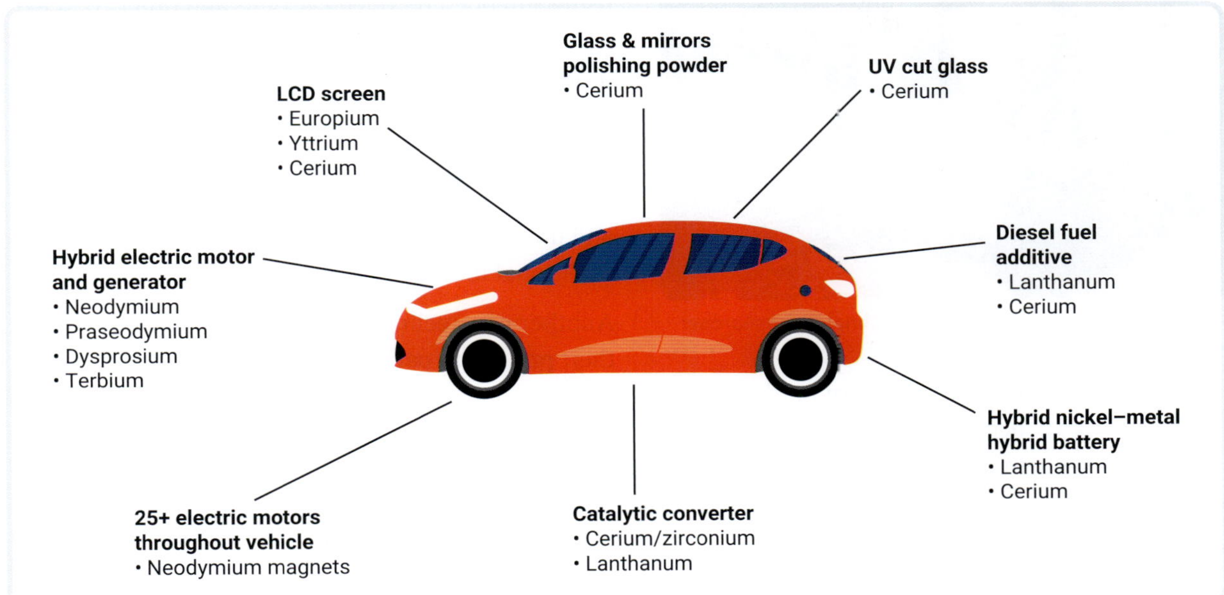

▲ **FIGURE 10.6.2** Electric cars use rare metals that are very difficult to recycle.

While recycling is an important part of a circular economy it does have some disadvantages. The collection and processing of wastes consumes energy and therefore emits greenhouse gases. Workers, particularly those in developing countries, can be exposed to hazardous materials while handling waste (Figure 10.6.3). Poor handling of waste and the fossil fuels used for processing them can create pollution. Poorly-sorted waste may not be recyclable and ends up in landfill, where it may further pollute the environment.

MOHAMED ABDULRAHEEM/Shutterstock.com

▲ **FIGURE 10.6.3** Sorting wastes can expose people to hazardous materials.

Some innovations in recycling

Finding new methods of recycling and reusing waste is important to creating a sustainable world. The SMaRT (Sustainable Materials Research & Technology) Centre at the University of New South Wales develops innovative solutions to waste recycling. The centre was founded by the 2022 New South Wales Australian of the Year, Professor Veena Sahajwalla (Figure 10.6.4).

University of New South Wales

▲ **FIGURE 10.6.4** Waste and recycling technology researcher Professor Veena Sahajwalla

One of the Centre's innovations is the production of 'green steel' through a process called polymer injection technology. This process produces steel using shredded recycled car tyres and other recycled polymers instead of using coking coal. It also produces hydrogen, so less energy is needed and less carbon is emitted.

The steel produced is high quality and the method of production is both cost-effective and more environmentally friendly than the traditional method.

Another new technology from the SMaRT Centre is their project, in partnership with a company called Renew IT, to turn hard plastic waste into the material used by 3D printers. This work began in July 2024 and is valuable because making new plastic uses a lot of energy and materials derived from oil. Reusing old hard plastics saves up to 90 per cent of the energy required to make the same item from oil **feedstock**.

feedstock
raw material that is used in an industrial process

10.6 LEARNING CHECK

1 **Define** recycling.
2 **List** five types of waste your household can recycle.
3 **Describe** two advantages of recycling waste.
4 **Outline** how steel or aluminium is recycled.
5 **Describe** a contribution to recycling made by Professor Veena Sahajwalla and her colleagues at SMaRT.
6 **Discuss** two potential disadvantages of recycling.

10.7 Evaluating scientific claims

SCIENCE SKILLS IN FOCUS

IN THIS MODULE, YOU WILL FOCUS ON LEARNING AND IMPROVING THESE SKILLS:

- evaluating a claim using scientific knowledge and findings from an investigation
- using cause-and-effect relationships and models to explain ideas and make predictions
- creating a written text to present a scientific argument supporting a claim.

Scientific arguments

A good scientific argument has three key parts: claim, evidence supporting the claim, and reasons why the evidence supports the claim. In writing a good discussion of results you will sometimes need to make more than a single claim about what you have discovered. Each claim requires its own supporting evidence and reasoning.

Claim

A claim is a statement about what you have learned from the data in the investigation. It may be about the differences between a treatment group and a control group; a correlation you have identified; clusters in your data; a univariate analysis of a variable; or another aspect of the data and what it means.

Evidence supporting the claim

Evidence supporting your claim may be summary evidence from your data, a reference to calculations or a graph you produced as part of your analysis. The amount of evidence you provide must be sufficient to support the claim.

Reasons why the evidence supports the claim

It is not enough to present evidence and leave it to a reader to make the connection between the claim and data. This part of the argument must explain how or why the evidence supports the claim.

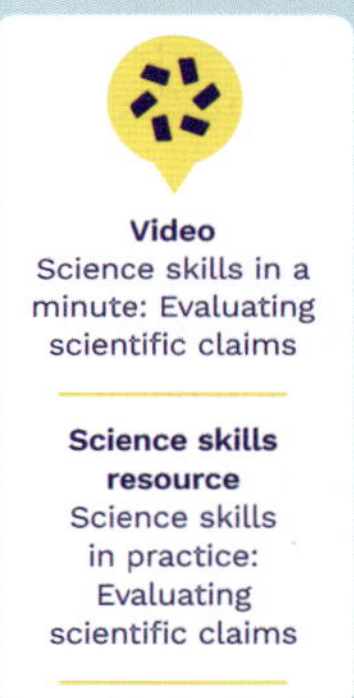

Video
Science skills in a minute: Evaluating scientific claims

Science skills resource
Science skills in practice: Evaluating scientific claims

EVALUATING RECYCLING CONTAMINATION

BACKGROUND INFORMATION

Recycling contamination happens when unwanted waste materials are included with genuine recyclables. For example, oil or fat on a pizza box can make paper waste unrecyclable because the oils cannot be removed from the paper fibres and this affects the quality of recycled paper. Similarly, when plastics are put in with green waste, composted material can be contaminated with chemicals or microplastics.

In this investigation, you will evaluate how food waste has contaminated recyclable plastic and paper in your school or home waste. Using the data you collect, you will propose methods you can use to reduce recycling contamination and write scientific arguments to support your proposed methods.

AIM

To identify types and amounts of recycling contamination and propose strategies to reduce rates of contamination

MATERIALS AND EQUIPMENT

- ☑ four or five waste bins selected randomly around the school and labelled with their origin
- ☑ large sheet of plastic
- ☑ face mask
- ☑ safety glasses, lab coat and gloves
- ☑ tongs

 **Safety**

In this investigation you are working with potentially hazardous materials. Use personal protection equipment during the procedure and wash your hands after packing away the waste.

PROCEDURE

1. Review the procedure and draw up a suitable table for your results. Remember to record qualitative observations as you carry out the investigation.
2. Lay out the sheet of plastic in an area where the wind will not disturb it.
3. Carefully tip the bin of waste onto the plastic sheet.
4. Randomly select up to 20 paper items from the waste. Count the number of pieces that show evidence of food contamination. Record your results.
5. Randomly select up to 20 plastic items from the waste. Count the number of these pieces that show evidence of food contamination. Record your results.
6. Carefully replace the waste in the bin.
7. Repeat steps 2–6 for the other waste bins.

RESULTS

1. Calculate the percentage contamination for paper and plastics from each bin (100 × contaminated items/number of items). For example, if 100 items are counted and 32 of them are contaminated, the percentage contamination is

$$100 \times \frac{18}{50} = 36\%$$

2. Summarise your observations from the procedure.

ANALYSIS

1. **Calculate** the average percentage of contamination for paper and plastic from all the bins.
2. **Calculate** the range of contamination (Range = largest contamination percentage – smallest contamination percentage).
3. Do some bin locations have a higher rate of contamination than others? Suggest possible reasons for this.
4. Working with the other students in your group, **develop** three strategies that might be used to reduce the amount of waste contamination you have measured.
5. Write an argument for each strategy using the claim–evidence–reasoning model described in the blue Science skills in focus box above.

CONCLUSION

Summarise the contamination you identified in the investigation.

Outline the strategy you identified as being potentially successful in reducing waste contamination.

DATA SCIENCE 

To learn more about the claim–evidence–reasoning model, go to **Modules 2.2** and **2.4**.

ABORIGINAL & TORRES STRAIT ISLANDER SCIENCE CONTEXTS

10.8 Sustainable use of plants

IN THIS MODULE, YOU WILL:

- ✓ recognise that Aboriginal and Torres Strait Islander Peoples have successfully and sustainably harvested plant materials for thousands of years
- ✓ examine examples of how Aboriginal and Torres Strait Islander Peoples applied their traditional knowledge to support sustainability of plant resources.

Harvesting plant materials sustainably

Aboriginal and Torres Strait Islander Peoples accumulated a wealth of information and experience concerning the biology, ecology and physical properties of plants over many thousands of years. Although plants form a crucial part of the traditional diet, they were also collected and harvested for many other purposes. As well as being used as a source of medicines and water, plants are important in the construction of tools, weapons, ornaments, musical instruments and toys, and even as an aid to catching animals.

Because plants were essential, Aboriginal and Torres Strait Islander Peoples understood that it was important to harvest plant resources sustainably, so that future generations would still have them.

Harvesting bark

The outer bark of specific types of trees has been carefully harvested by Aboriginal and Torres Strait Islander Peoples for millennia to be used in the construction of canoes, shields, tools, implements and as a canvas for painting or to expose heartwood for carving. To ensure the ongoing survival of the tree, great care was taken not to cause too much damage to the plant's system when the bark was removed. Across southern and eastern Australia, you can still see living, scarred trees. This shows Aboriginal Peoples' in-depth knowledge of plant biology.

Joe Sambono

▲ **FIGURE 10.8.1** A canoe scar-tree in Yugambeh Country, south-east Queensland. Careful removal of the outer bark to construct a canoe ensures the tree's survival.

The Alyawarre Peoples of the Central Desert Region carefully cut pieces of outer bark from opposite sides of the ghost gum (*Corymbia aparrerinja*) to make trays that are used to

Joe Sambono

▲ **FIGURE 10.8.2** A shield tree that was harvested to make a Kaurna shield, South Australia

Joe Sambono

▲ **FIGURE 10.8.3** The bark from trees such as the paperbark can be used to make blankets.

process resin or gum. In the Hunter Valley region of New South Wales, there is a very old tree that bears scars believed to be more than 100 years old, and is consistent with the removal of bark to manufacture canoes. The Dyirbal Peoples of the northern Queensland rainforest region have long collected bark up to 12 metres above the ground from the banana fig (*Ficus pleurocarpa*) to prevent damage to the root system and ensure survival of the tree. The bark was used to manufacture blankets.

ACTIVITY 1

1 Use what you have learned about plant systems to **explain**:
   a which plant system(s) is being affected by the removal of bark.
   b why the careful removal of bark does not kill the tree.
   c why a scar forms.

Sustainable cultivation

Aboriginal and Torres Strait Islander Peoples have long used controlled fires to promote the growth, propagation and germination of certain plants and seeds. This agricultural technique is known as fire-stick farming.

On the Australian mainland, cycads have historically been an important source of carbohydrates for many Peoples. Carefully controlled fires are used to promote the distribution and growth of cycads. In certain areas, discrete groves of cycads are grown and fire is used to improve the quality and yield of crops as well as trigger fruit production when required.

Like the cycads, many other Australian native plants have evolved to become fire tolerant. Some species of acacia and banksia require the heat and/or smoke of a fire to start the germination of seeds. Other species such as eucalypts have developed thick bark that covers the lower sections of their trunks.

Joe Sambono

FIGURE 10.8.4 A grove of cycads, Belyuen, Northern Territory

ACTIVITY 2

1 **Describe** how fire-stick farming is similar to traditional Western farming techniques.
2 a **Explain** what adaptations plants need to thrive by fire-stick farming.
   b **Describe** what would happen to plants that don't have these adaptations.
3 Suggest why controlled low-temperature fires are used by Aboriginal and Torres Strait Islander Peoples when fire-stick farming.

10.9 Bioethanol as a more environmentally friendly fuel

BY THE END OF THIS MODULE, YOU WILL BE ABLE TO:

- ✓ explain the need for the development of bioethanol
- ✓ explain why bioethanol is considered to be more environmentally friendly than fossil fuels.

Why do we need environmentally friendly fuels?

Video activity
Biofuels

In Chapter 9, you looked at the issue of climate change due to the enhanced greenhouse effect. This is a result of the release of heat-trapping gases into the air, such as carbon dioxide from fuel combustion. Scientists are racing to develop environmentally friendlier fuels to try to combat climate change.

Why is bioethanol considered more environmentally friendly than fossil fuels?

Bioethanol is produced from crops and added to petrol. In Australia, this is sold in many places as E10 fuel. This means it contains about 90 per cent regular petrol and 10 per cent ethanol. Ethanol burns 'cleaner' than regular petrol. This means fewer pollutants such as tiny particles, soot and carbon monoxide form.

When fossil fuels are combusted, they produce carbon dioxide. Carbon dioxide was originally taken out of the atmosphere by plants millions of years ago when fossil fuels formed from plant and animal remains. This means we add *extra* carbon dioxide to the atmosphere when we burn fossil fuels. As you learned in Chapter 9, extra carbon dioxide in the atmosphere is a major contributor to climate change.

At first glance, it seems bioethanol isn't any different. When manufactured and combusted, it still produces carbon dioxide, but ethanol is a renewable fuel. The crops used in bioethanol production only recently absorbed carbon dioxide from the atmosphere by photosynthesis, and new crops are continuously grown to replace them. This means the carbon dioxide produced during combustion of bioethanol is just replacing and not adding to the carbon dioxide in the atmosphere, as shown in Figure 10.9.1. When combusted, biofuels also often produce slightly less carbon dioxide than regular petrol.

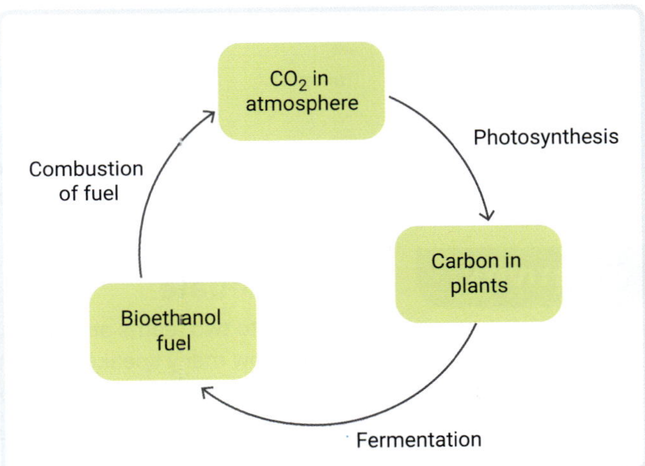

▲ **FIGURE 10.9.1** The cycle of carbon in bioethanol. Carbon dioxide from the atmosphere is taken up by plants during photosynthesis. The carbon moves from glucose to ethanol in fermentation, then back to carbon dioxide during combustion.

Ethanol can be produced from agricultural feedstocks such as sugarcane, and also from forestry wood wastes and agricultural residues.

Bioethanol in vehicle fuel is now used in many countries and is available at many Australian service stations.

What other factors affect the production of bioethanol?

When deciding whether bioethanol is a more sustainable and environmentally friendly fuel than other options like fossil fuels, several factors need to be considered. Some of these factors are listed in Table 10.9.1

▼ **TABLE 10.9.1** Factors to consider when assessing the production of bioethanol

Factor	What does it have to do with bioethanol?	Why is this an issue?
Land use	A large amount of land is required to grow the crops used to make bioethanol.	Land clearing for agriculture means cutting down trees that take in carbon dioxide. Fewer trees taking in carbon dioxide contributes to climate change. It also means that less land is available for growing food crops or grazing animals. Overall, more land is used for agriculture (food) and bioethanol than for regular agriculture alone.
Use of crops as fuel	The crops used for bioethanol cannot be used as food.	A large part of the world population currently does not have access to enough food. Many countries cannot grow enough crops to provide food for their people because of droughts, floods, fire, war and other recent world events. Some people think it is unethical to use food as fuel when so many people cannot get enough food to eat.
Industrial production	The production of bioethanol requires large amounts of energy for the different processing stages.	The energy for the conversion of crops to bioethanol requires several processes that need either fuel or electricity. Most fuels are fossil fuels such as gas or coal. Most electricity comes from fossil-fuel-powered power stations. This releases additional carbon dioxide into the atmosphere, compared to what is normally produced during combustion.
Transport	The crops need to be transported to the processing facility. The bioethanol is then transported around the world and across countries where it is sold.	Most transport within Australia is done by trucks or trains, while overseas transport relies on ships. These vehicles usually run on fossil fuels, which release carbon dioxide to the atmosphere.

ACTIVITY

Class debate

Split up into two teams and choose speakers to represent your team. Discuss with your teacher how many speakers will make up each team. Depending on the number of speakers, speeches could be 2–5 minutes long.

The debate question is: 'Should Australia produce and use more bioethanol?'

Your team needs to come up with arguments for or against (depending on your side). You should use the ideas in Table 10.9.1 and do some extra research to find facts and examples that support your arguments.

9780170491785

10.9 LEARNING CHECK

1 A scientist produced the graph in Figure 10.9.2 after testing petrol and E10 in a series of experiments. The aim was to give drivers more information about both fuels so they could make a more informed choice. The graph shows the fuel consumption (litres of fuel used to travel 100 km) of petrol and E10 (ethanol + petrol).

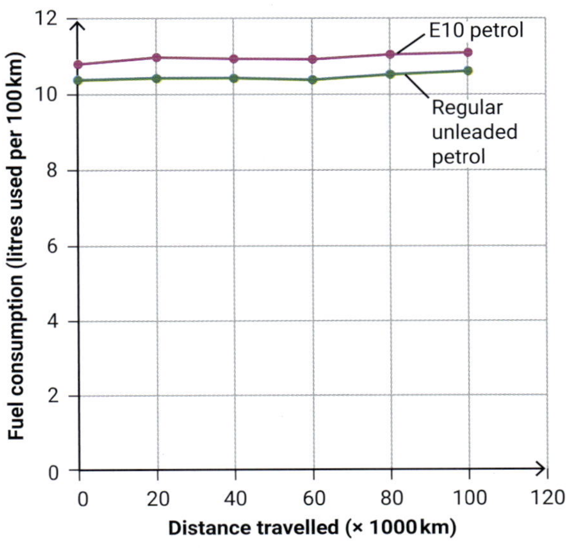

▲ FIGURE 10.9.2 A comparison of regular unleaded petrol and E10 fuel consumption

a **Describe** what the 10 means in E10 fuel.

b **Write** a conclusion discussing the difference in fuel consumption of the two fuels, based on the graph.

2 Petrol and bioethanol both release carbon dioxide into the atmosphere. **Explain** why bioethanol is considered to be better environmentally than fossil fuels.

Istimages/Shutterstock.com

▲ FIGURE 10.9.3 Many petrol stations in Australia sell E10 fuel, which contains bioethanol.

3 In pairs, groups or individually, **create** a poster or advertisement that explains why E10 is a more environmentally friendly fuel than petrol.

4 Conduct **research** to find out why 100 per cent ethanol cannot be sold as car fuel. Write a few sentences to **summarise** what you found.

10 REVIEW

REMEMBERING

1 **Define** sustainability.

2 **List** the three principles of sustainability.

3 **Describe** how nutrient-rich wastes can damage rivers and lakes.

4 **Outline** four ways of addressing sustainability in our homes.

5 **Describe** what makes a resource renewable.

6 **List** five things that can be recycled.

7 **Describe** an advantage and a disadvantage of using bioethanol as a fuel.

UNDERSTANDING

8 The figure above right shows the three aspects of sustainability as pillars holding up a roof. Use the figure to **explain** why all three aspects need to be addressed for sustainability to be achieved.

Dmitry Kovalchuk/Shutterstock.com

9 **Explain** how pollution negatively affects sustainability.

10 The graph below shows the contribution of different types of food transport to global greenhouse gas emissions in 2015. **Explain** why choosing food grown close to where you live and reducing the amount of food you consume will contribute to sustainability.

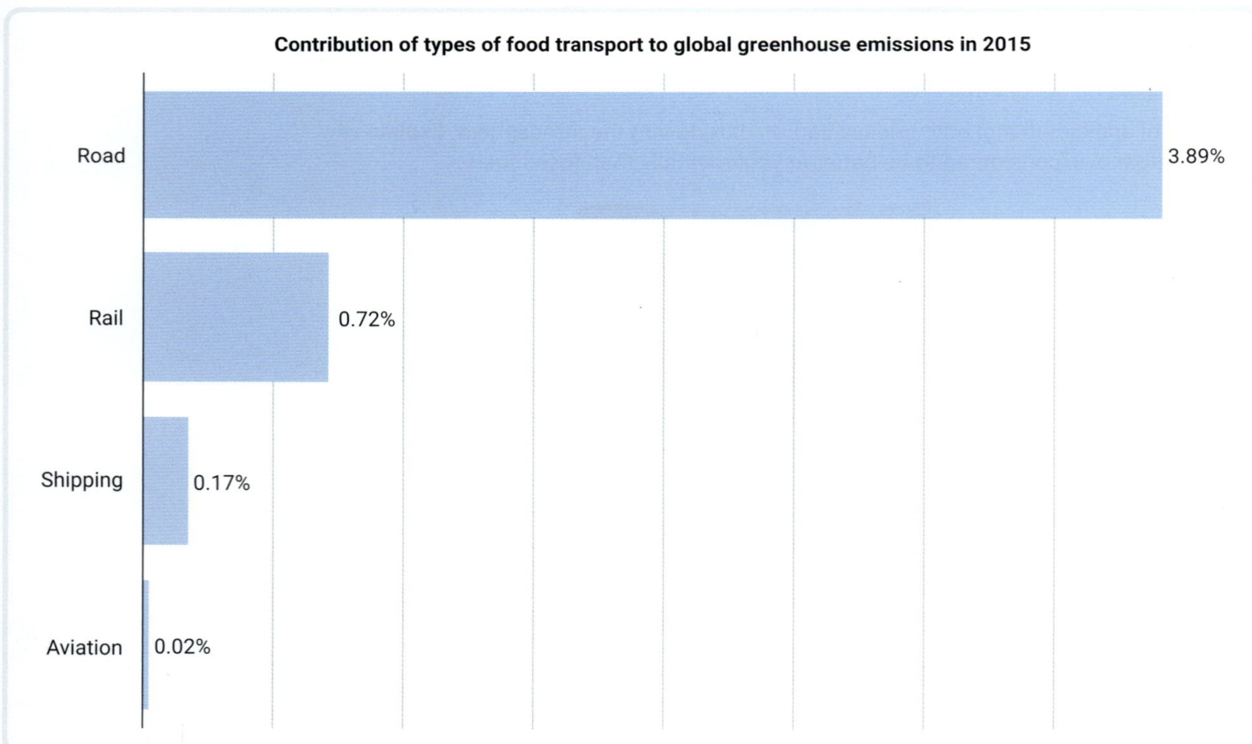

Other resou
Workshee
Renewable res
revision

11 **Summarise** how pollution from fertilisers can reduce the biodiversity of the river.

12 a **Classify** the following resources as renewable or non-renewable.

i Wind

ii Coal

iii Sunlight

iv Uranium

v Hydroelectricity

b **Explain** why using renewable sources of energy is more sustainable than using fossil fuels.

APPLYING

13 **Identify** why it is important to limit particulate pollution in the environment.

14 **Determine** how effective reducing metal use and recycling is in making metal consumption more sustainable.

15 **Explain**, using the graph below, why increasing the use of renewable sources of energy is necessary for sustainable energy use.

16 **Explain** why fire-stick farming is considered a sustainable land-management practice.

ANALYSING

17 **Examine** and **describe** the potential impacts of a circular economy on sustainability.

18 **Explain** how the work of researchers such as Dr Veena Sahajwalla is important for making New South Wales a more sustainable state.

EVALUATING

19 **Compare** the effects of pollution and recycling on environmental sustainability.

20 **Justify** the energy consumption of recycling by outlining its advantages.

CREATING

21 **Design** and record a 30-second podcast explaining a strategy to reduce household waste.

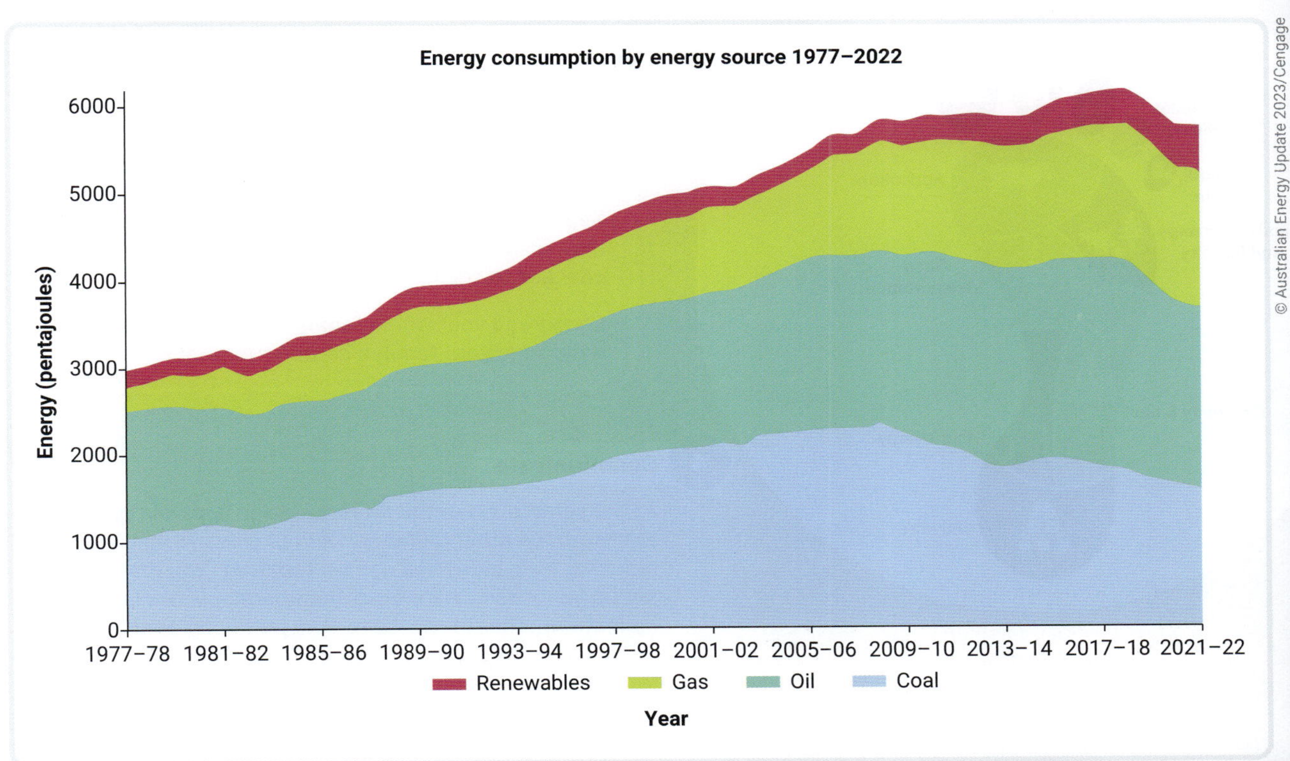

© Australian Energy Update 2023/Cengage

SCIENCE IN DEPTH STUDY

1 Connect what you've learned

By completing the modules in this chapter, you've learned about the principles of sustainability, the impacts of pollution and the positive difference sustainability can make. Being more sustainable is everyone's responsibility. This challenge gets you to reflect on how your actions and behaviours contribute to climate change by determining your individual carbon footprint and understanding how this fits into your overall ecological footprint. You can find out more by asking your family and friends to also undertake a carbon footprint analysis. But let's start with you.

2 Check your thinking

The first thing you need to do is reflect on your daily activities that contribute to greenhouse gas emissions. Start by listing as many activities as possible that you participate in and know contribute to greenhouse gas emissions.

- Is it possible to modify any of these activities to reduce your impact?
- What changes to your lifestyle are you willing to make that are realistic and will reduce your individual carbon footprint?
- To help, there are many free online carbon footprint calculators that you can use to calculate your carbon footprint. Do an online search to locate a suitable Australian carbon footprint calculator.

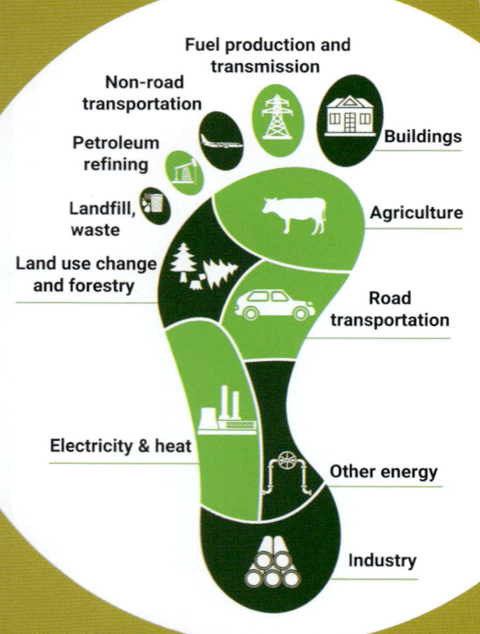

▲ What is your carbon footprint?

3 Get into action

Use an online carbon footprint calculator to determine your current carbon footprint. Make a note of the result.

Now identify the changes you are willing to make. Repeat your calculation and compare the difference. You may wish to do this several times with a different mix of changes to see the difference it makes to your footprint.

Encourage your whole family to be involved and determine your family's carbon footprint.

4 Communicate

At the completion of the project, design a simple pamphlet to encourage other members of your class to think about reducing their individual carbon footprint. You may like to set up a process where other members of your class can indicate whether they wish to take part in an ongoing project to measure the collective change to your class's carbon footprint. There is nothing stopping you encouraging others beyond your class to get involved. The more people who participate, the better off the planet will be!

9780170491785

GENETICS AND EVOLUTIONARY CHANGE

SYLLABUS OUTCOMES

A STUDENT:

- describes the relationship between the diversity of living things and the theory of evolution SC5-GEV-01
- explains how DNA is responsible for the transmission of heritable characteristics and can be manipulated through genetic technologies SC5-GEV-02
- selects and uses a range of tools to process and represent data SC5-WS-05
- communicates scientific arguments with evidence, using scientific language and terminology in a range of communication forms SC5-WS-08

© 2023 NSW Education Standards Authority

CHAPTERS RELATED TO THIS FOCUS AREA ARE:

- CHAPTER 11 – GENETICS
- CHAPTER 12 – EVOLUTION

11 Genetics

SCIENCE IN DEPTH 

Lopolo/Shutterstock.com

▲ FIGURE 11.0.1 Identical twins

Have you ever wondered why you look more like one parent than the other?

Perhaps you have already heard of genetic screening and genetic counselling that prospective parents may go to before they start having children.

You may already know the terms 'cloning' and 'genetically modified'.

All these questions and issues link back to the field of genetics: a branch of biological science concerned with inheritance and variety in living things.

- **What do you think are the advantages of understanding more about your genes?**
- **What is your opinion on changing genes to produce a favourable characteristic? Or to make a product that can help treat a disease?**
- **If you could 'cure' cancer through gene editing, do you think this would be a medical breakthrough?**

 **DIVE INTO SCIENCE!**

At the end of this chapter, you can complete Science in Depth Study #11. You can use the information you learn in this chapter to complete the project.

Assessments
- Prior knowledge quiz
- Chapter review questions
- End-of-chapter test
- Depth study: Research project and podcast

Videos
- Science skills in a minute: Modelling DNA **(11.2)**; Using a microscope **(11.8)**
- Video activities: Rosalind Franklin **(11.1)**; Sex determination **(11.10)**; Cancer and genetics **(11.14)**; Gel electrophoresis **(11.15)**

Science skills resources
- Science skills in practice: Modelling DNA **(11.2)**; Using a microscope to examine cells **(11.8)**
- Extra science investigations: Karyograms **(11.3)**; Looking at chromosomes **(11.3)**; Inheritance and chance **(11.9)**

Interactive and other resources
- Drag and drop: Punnett squares **(11.11)**
- Label: DNA molecule **(11.1)**; Chromosomes **(11.3)**; Phases of mitosis **(11.6)**; Meiosis I and II **(11.7)**; Pedigree charts **(11.12)**
- Quizzes: Types of asexual reproduction **(11.5)**
- Activity sheets: Modelling mitosis **(11.6)**; Sex and 'chance' **(11.10)**
- Worksheets: Mendel's results **(11.9)**; The Mendelian ratio **(11.9)**; Punnett squares **(11.11)**; Pedigrees **(11.12)**

Nelson MindTap

To access resources above, visit **cengage.com.au/nelsonmindtap**

11.1 The structure of DNA

BY THE END OF THIS MODULE, YOU WILL BE ABLE TO:

- ✓ describe the structure of DNA
- ✓ label the key components of DNA: the nucleotides (each containing a sugar, phosphate and nitrogenous base) and the hydrogen bonds
- ✓ define 'nucleotide', 'nitrogenous base' and 'hydrogen bond'
- ✓ list the nitrogenous bases in DNA and their complementary base pairs (Chargaff's rule).

Video activity
Rosalind Franklin

Interactive resource
Label: DNA molecule

GET THINKING

Did you know you have approximately 2 metres of DNA in each of your cells? Could you calculate the total length of DNA found in your entire body?

What is DNA?

DNA stands for **deoxyribonucleic acid**. It is a complex **molecule** made up of two chains coiled together, like a twisted ladder (**double helix**). It was successfully described in 1953 by two scientists, James Watson and Francis Crick, who determined its shape using an X-ray diffraction image of DNA taken by Raymond Gosling, working under the supervision of Rosalind Franklin (Figure 11.1.1a). The image, named *Photo 51*, showed a clear cross in the centre (Figure 11.1.1b). This indicated to Watson and Crick that a DNA molecule has a double helix shape.

deoxyribonucleic acid
the molecule that makes up the genetic material inside the nucleus of cells and in some viruses

molecule
a group of atoms bonded together

double helix
the shape of DNA, similar to a twisted ladder

Despite the great diversity shown in living things, the basic structure of DNA is the same in the cells of all living things and can be shown as a simple model (Figure 11.1.2).

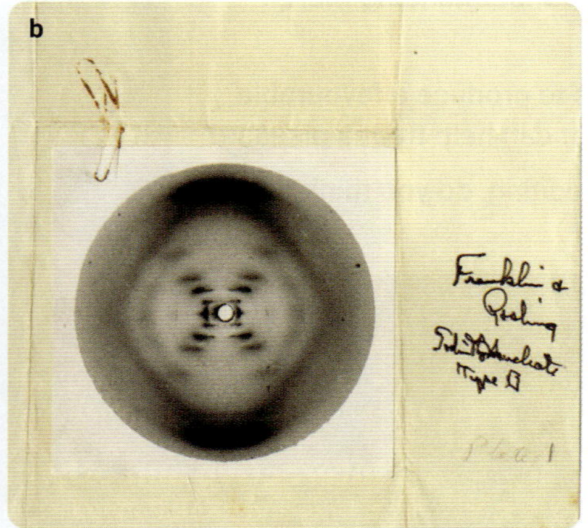

▲ **FIGURE 11.1.1** (a) Rosalind Franklin; (b) *Photo 51* displaying the centre cross, indicating a helix shape

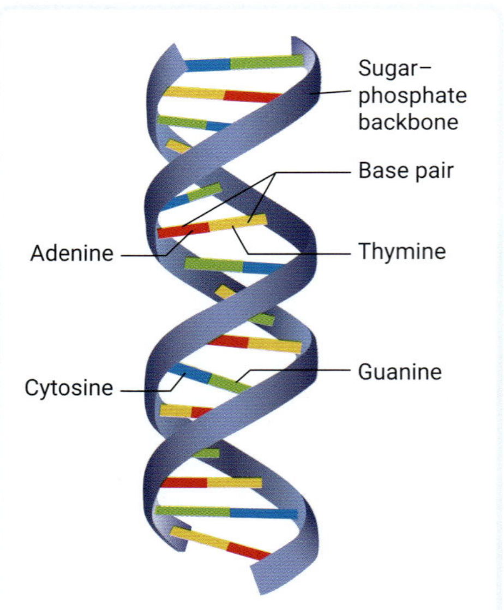

▲ **FIGURE 11.1.2** The double helix shape of DNA

9780170491785

The structure of DNA

DNA's double helix consists of two strands that twist around each other. Each strand of DNA is composed of repeating **nucleotide** units, each consisting of a **phosphate group** and a **deoxyribose sugar** attached to a **nitrogenous base** (Figure 11.1.3).

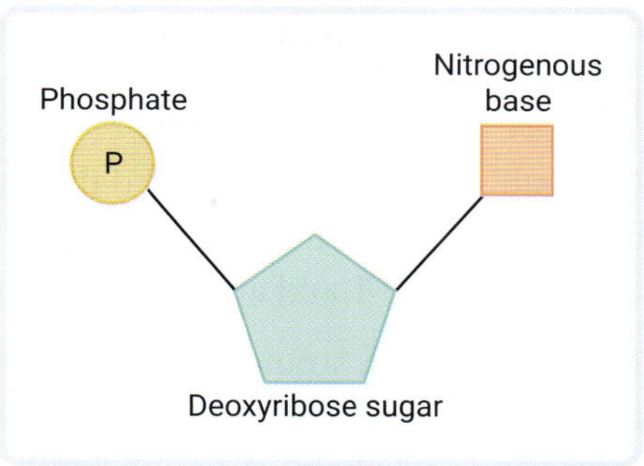

▲ **FIGURE 11.1.3** The three components of nucleotides

The sugars and phosphate groups form the backbone of the strand, while the nitrogenous bases (represented with the letters A, T, C and G) are attached to the sugar molecules. Hydrogen bonds are formed between the nitrogenous bases on opposite strands and these bonds hold the strands together. This explains why DNA is referred to as two complementary strands.

nucleotide
the building block of DNA

phosphate group
one of the components of a nucleotide in DNA

deoxyribose sugar
one of the components of a nucleotide in DNA

nitrogenous base
a base that contains nitrogen: adenine (A); thymine (T); cytosine (C) and guanine (G)

You can think of a DNA molecule as a ladder; the alternating sugar–phosphate molecules form the side rails, and the nitrogenous bases form the rungs. Each nucleotide is attached to the one before and the one after it, creating a long chain that is the DNA molecule. This is shown in Figure 11.1.4.

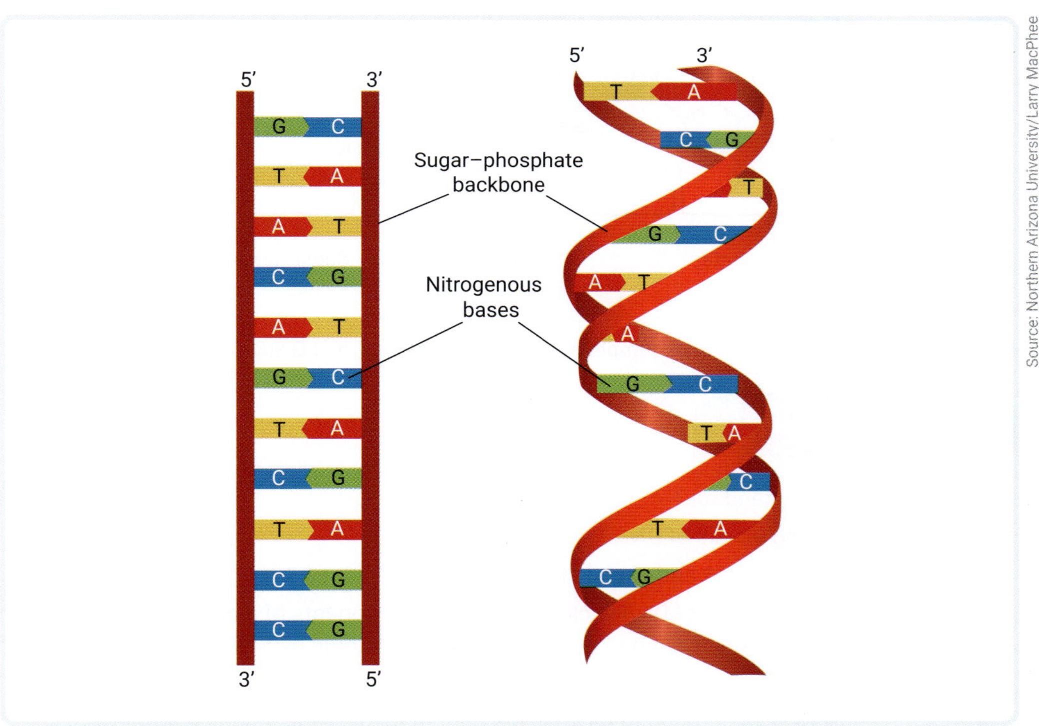

▲ **FIGURE 11.1.4** DNA as a ladder and twisted into the double helix shape

The four bases of DNA

There are four nitrogenous bases, adenine (A), thymine (T), guanine (G) and cytosine (C). The sequence of these bases forms the **genetic code** (e.g. GTACAGTCTCC), as shown in Figure 11.1.4.

genetic code
the sequence of nitrogen-rich bases in an organism's DNA

In the late 1940s, the scientist Erwin Chargaff observed that, in a given DNA sample, the amount of adenine always equalled the amount of thymine, and the amount of cytosine always equalled the amount of guanine.

Extension: Chargaff's rule

Chargaff's observation provided the clue behind how the nitrogenous base pairs form and led to 'Chargaff's rule'. This states that in DNA, there is always an equal quantity of the bases A and T and an equal quantity of the bases G and C.

hydrogen bond
a type of attraction between atoms of different molecules

For the two DNA strands to join, **hydrogen bonds** form between complementary nitrogenous bases (Figure 11.1.5). These bonds effectively 'stick' the strands together to form the double helix. Adenine bonds with thymine and uses two hydrogen bonds to join. Cytosine bonds with guanine and uses three hydrogen bonds to join.

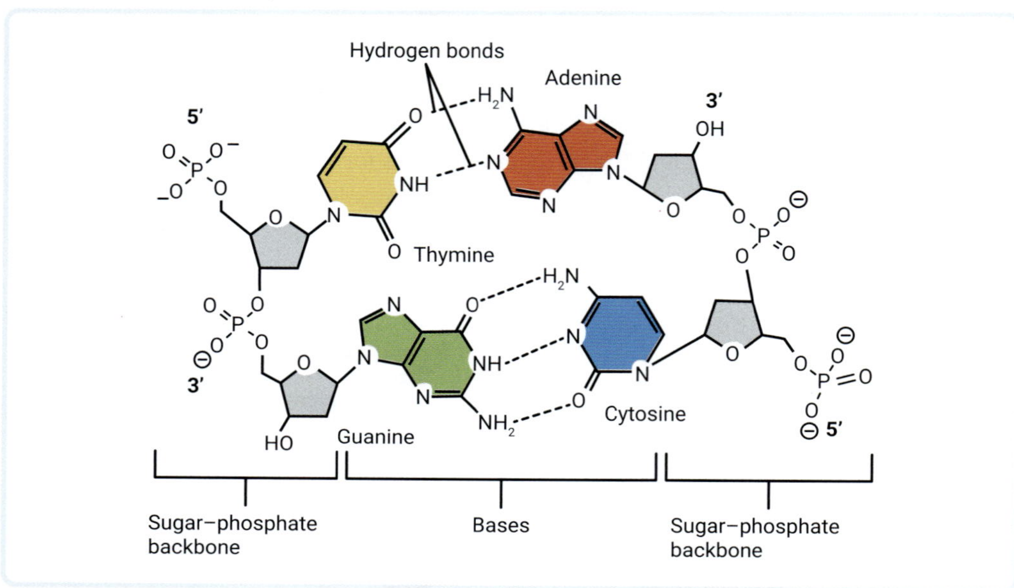

▲ **FIGURE 11.1.5** DNA is made up of complementary bases: A and T, and G and C.

11.1 LEARNING CHECK

1 What does DNA stand for?
2 **State** the building blocks of DNA.
3 **Name** the shape of DNA and **explain** how DNA is like a ladder.
4 **Draw** a labelled diagram of DNA, including the key components of a deoxyribose sugar, phosphates, the four nitrogenous bases, a nucleotide and the hydrogen bonds.

Extension

5 If a molecule of DNA is found to have 35 per cent of its bases as guanine, what would the percentage of thymine be? **Explain** your answer using Chargaff's rule.

WORKING SCIENTIFICALLY

11.2 Modelling DNA

SCIENCE SKILLS IN FOCUS

IN THIS MODULE, YOU WILL FOCUS ON LEARNING AND IMPROVING THIS SKILL:

- building a model to represent DNA.

Models are used in science to explain or predict a scientific concept or phenomenon. Models are especially helpful when we want to show or explain something we can't see with the naked eye. For example, we use models to represent very small structures, such as a molecule of DNA.

A good model is clear and interactive and helps communicate ideas and information clearly. Look around your science classroom. There will be models on display that your teacher refers to or passes around for students to use. Can you pick your favourite model? What is it about that model that attracts you? Is it:

- colourful?
- clearly visible in the room?
- simple?
- able to communicate an idea to a range of students with different ages and abilities?

When you make your DNA model, consider the audience and purpose.

- Who are you making your model for?
- What do you want them to understand or learn?

When you are creating a model of DNA, there are some simples rules you should follow.

- Use simple materials that represents clearly the parts of the model.
- Differentiate each part of the model by using different colours and textures.
- Once your model is ready, take a picture or draw it and include labels to indicate what parts of the molecule are represented in the model.

Once you have built your model, you need to assess it for advantages and disadvantages by answering questions like:

- Did the model represent what was intended? In what ways?
- What are the limitations of the model?
- What are the advantages of making this model?

Video
Science skills in a minute: Modelling DNA

Science skills resource
Science skills in practice: Modelling DNA

BUILDING A DNA MOLECULE

AIM

To build a model DNA molecule

MATERIALS AND EQUIPMENT

- ☑ 4 × 30 cm long pipe cleaners of two different colours
- ☑ 60 assorted beads of six different colours. Plastic beads with holes work best for this activity
- ☑ 30 cm ruler

PROCEDURE

1. Cut one pair of pipe cleaners into strips that are 5 cm in length. You will have 12 short strips in total. You will use 10 of these in the activity (the other two are spares). Leave the other pair of pipe cleaners at their full length (Figure 11.2.1).
2. Using two different colours of beads, thread the beads in alternating colours down each full-length pipe cleaner. Ensure the two strands match. Leave about 1.5 cm between successive beads (Figure 11.2.2).

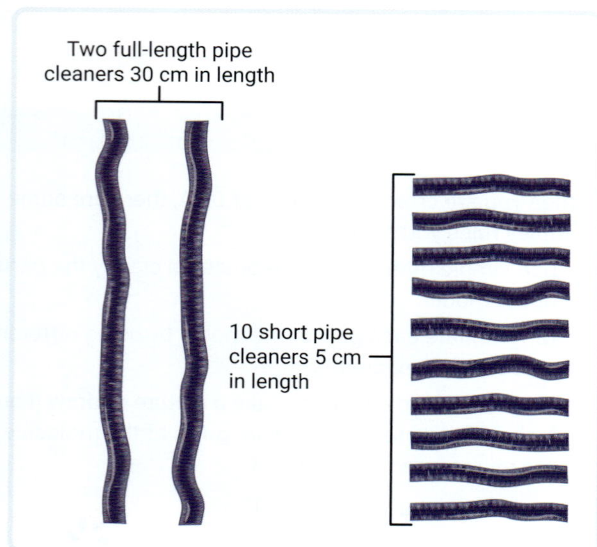

▲ **FIGURE 11.2.1** Step 1

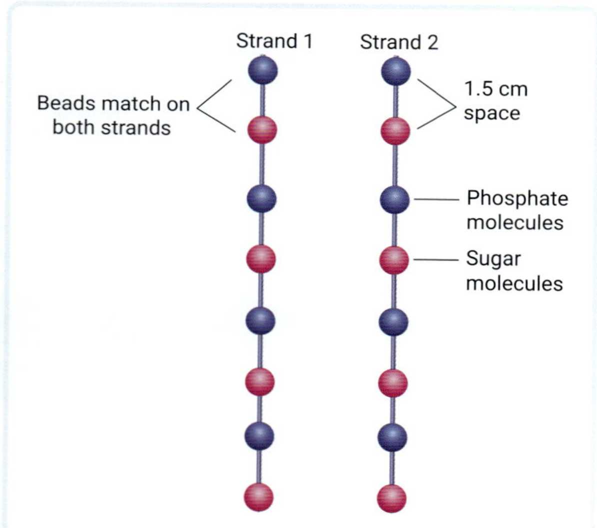

▲ **FIGURE 11.2.2** Step 2

3 Pair up the other four colours of beads so that the same two colours always match. Thread the pairs onto the 5 cm lengths of pipe cleaner (Figure 11.2.3).

4 Attach the 5 cm lengths to the long pipe cleaner strands. Do this by hooking each end of the 5cm pipe cleaner behind the 'sugar molecule' bead on the long pipe cleaner strand (Figure 11.2.4). Ensure you are attaching them to the same colour bead on each long strand.

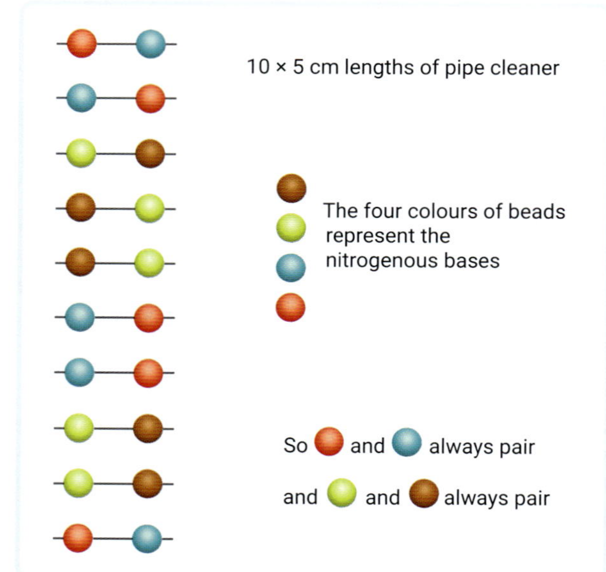

▲ **FIGURE 11.2.3** Step 3

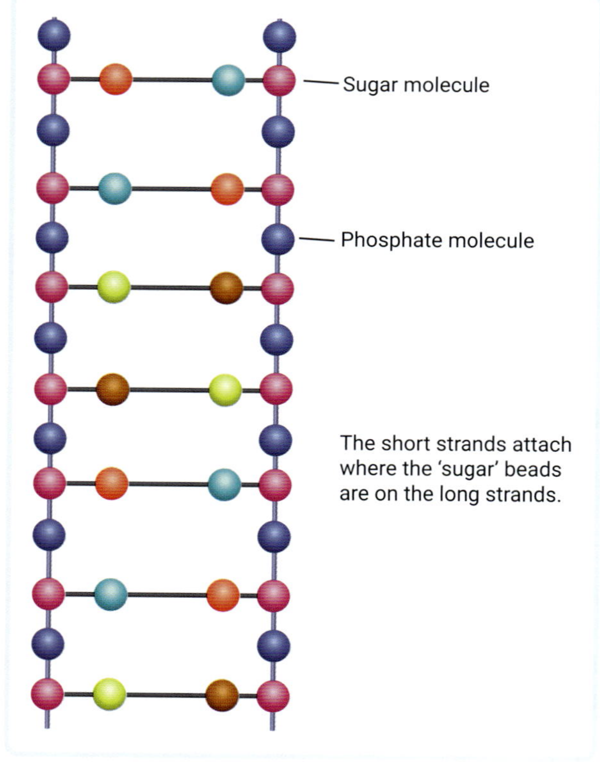

▲ **FIGURE 11.2.4** Step 4

9780170491785

5 Once all the small pieces have been attached, make the long strands into a double helix shape by twisting them in an anticlockwise direction (Figure 11.2.5).

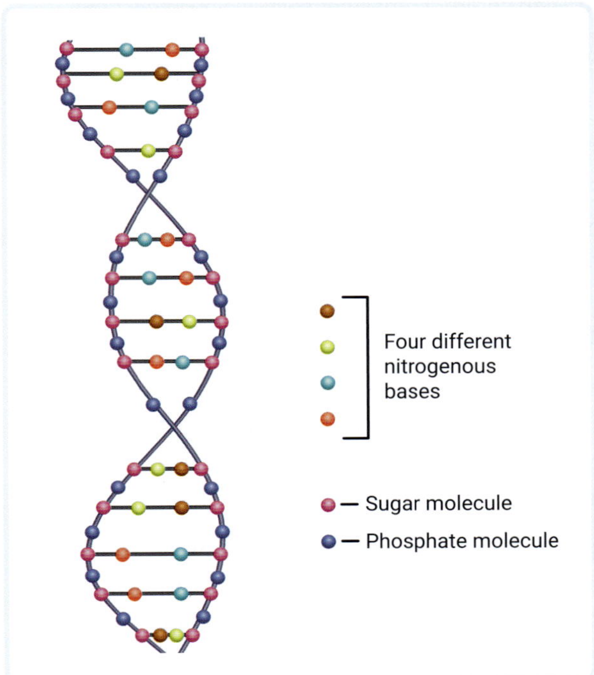

▲ **FIGURE 11.2.5** Step 5

ANALYSIS

1 What part of your model represents the sugar and phosphate backbone?

2 How did you represent the nitrogenous base pairs?

3 Which colour bead in your model represents the deoxyribose sugar? Why?

4 **Draw** a labelled diagram of your model in your workbook.

5 **Examine** the other students' models. Are any strands identical? How does this represent DNA in your cells?

6 **Assess** the limitations of your DNA model.

11.3 Genetic material

BY THE END OF THIS MODULE, YOU WILL BE ABLE TO:

✓ define chromosome, gene, genome and karyotype
✓ describe the relationship between chromosomes, genes and DNA.

Interactive resource
Label: Chromosomes

Extra science investigation
Karyograms

Other resource
Activity sheet: Looking at chromosomes

GET THINKING

The organism with the most chromosomes is the Adder's tongue fern (*Ophioglossum reticulatum*). This plant has an amazing 1440 chromosomes per cell! Do you know how many chromosomes humans have per cell?

The role of DNA

DNA is a molecule found in all living things. Often called the blueprint of life, DNA holds the instructions to build proteins, provides the instructions for the cell's activities and contributes to the characteristics of an organism.

How does DNA fit into a cell?

DNA is found in the nuclei of most cells. The exception is red blood cells, which do not contain DNA, and many single-celled organisms such as bacteria, where DNA exists simply within the cell walls in a central area called a nucleoid. In this module, we'll be looking at DNA in the nuclei of cells in multi-celled organisms.

chromatin
unpackaged DNA found within the nucleus of a non-dividing cell

chromosome
a thread-like structure found in the cell, composed of DNA

chromatid
one half of a duplicated chromosome

When a cell is not dividing and multiplying, DNA can be found as **chromatin** and looks like spaghetti in a bowl. During cell division, DNA condenses and takes the form of **chromosomes**. Chromosomes are thin, thread-like structures of tightly coiled chromatin found in the nucleus of a cell. They are made of two **chromatids**, as shown in Figure 11.3.1.

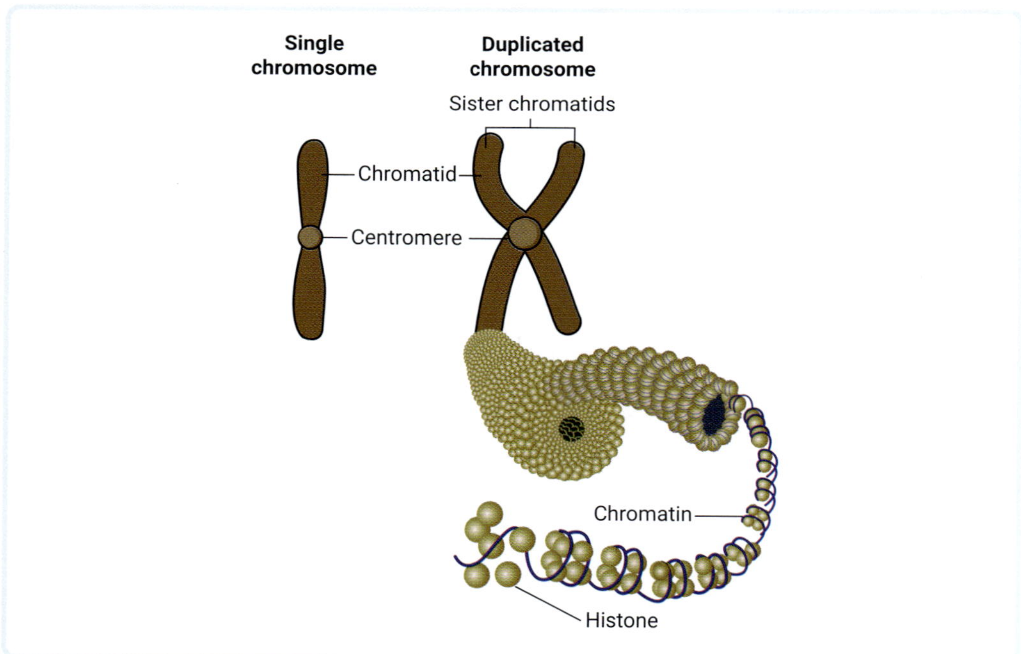

▲ **FIGURE 11.3.1** DNA coils to form chromosomes made from two chromatids attached at the centromere.

9780170491785

Each chromosome is one long DNA molecule coiled around proteins called histones. One way to think about chromosomes is like cotton wrapped around a spool, as shown in Figure 11.3.2. The cotton is the DNA, and the spool is the histone proteins. This allows the long molecule to take up much less space in the nucleus and the cell.

Olga Danylenko/Shutterstock.com

▲ **FIGURE 11.3.2** Chromosomes consist of DNA wrapped around histones, a bit like thread wrapped around a spool.

The number of chromosomes in a cell varies greatly between different types of organisms. A higher number of chromosomes doesn't relate to an organism's size or complexity. For example, elephants have 56 chromosomes in their cells, while the Atlas blue butterfly has between 448 and 452 chromosomes.

How many chromosomes do humans have?

Your cells have 46 chromosomes; half came from your mother and the other half from your father. This combining of chromosomes happens at fertilisation.

Chromosomes form X-shapes when they duplicate, and are visible under a light microscope just before cell division. They can be arranged and numbered from longest to shortest and then photographed, forming a **karyotype**. The karyotype allows doctors to check for serious abnormalities in the chromosomes (sizes, numbers and shapes).

There are 46 chromosomes, or 23 pairs; 22 of those pairs are called **autosomes**. The final pair are the **sex chromosomes**. These chromosomes determine the sex of the body you are born with and whether you are biologically female or male. Figure 11.3.3 shows a female karyotype with two XX chromosomes. Figure 11.3.4 shows a male karyotype, arranged with one X and one Y chromosome in the same location.

In the karyotype, the chromosomes are arranged in **homologous pairs**. The chromosomes in each homologous pair have the same **genes** at the same location. They are the same length and have their **centromere** located at the same position. Consider Figure 11.3.4, showing a male karyotype. The chromosomes are in numbered, homologous pairs. Note the difference in size between the X and Y chromosomes. The complete genetic material in an organism is called the **genome**.

karyotype
a picture of an organism's complete set of chromosomes arranged using size, genetic composition, and centromere positions

autosomes
all of the chromosomes in a cell, except for the sex chromosomes

sex chromosomes
a pair of chromosomes that determine the sex of an individual

homologous pairs
maternal and paternal chromosomes with genes found at the same location

gene
a section of DNA that codes for a protein or a certain trait

centromere
the point on a chromosome where the two chromatids are joined

genome
the complete set of genetic material present in an organism

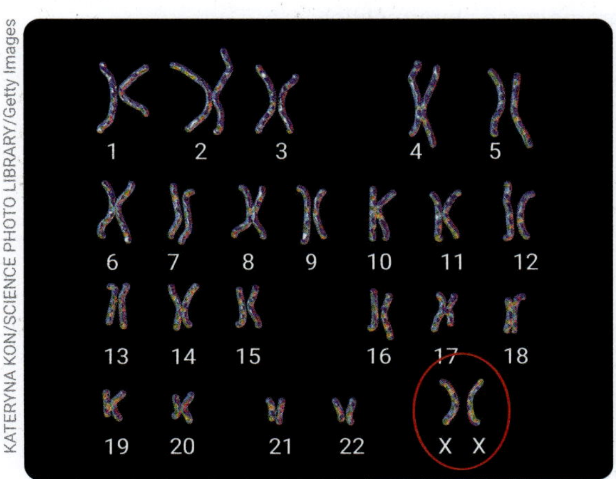

KATERYNA KON/SCIENCE PHOTO LIBRARY/Getty Images

▲ **FIGURE 11.3.3** A female karyotype with the XX chromosomes circled

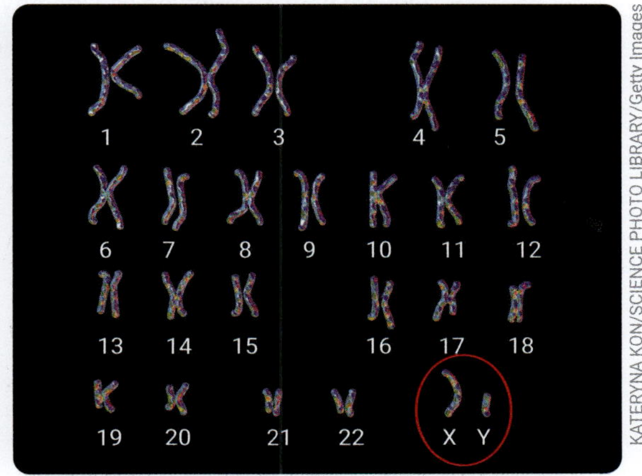

KATERYNA KON/SCIENCE PHOTO LIBRARY/Getty Images

▲ **FIGURE 11.3.4** A male karyotype with the XY chromosomes circled

Genes and chromosomes

Genes are the instructions to build a protein. They are sequences of bases within the DNA and found along the length of the chromosomes, as shown in Figure 11.3.5. Each chromosome can hold hundreds to thousands of genes.

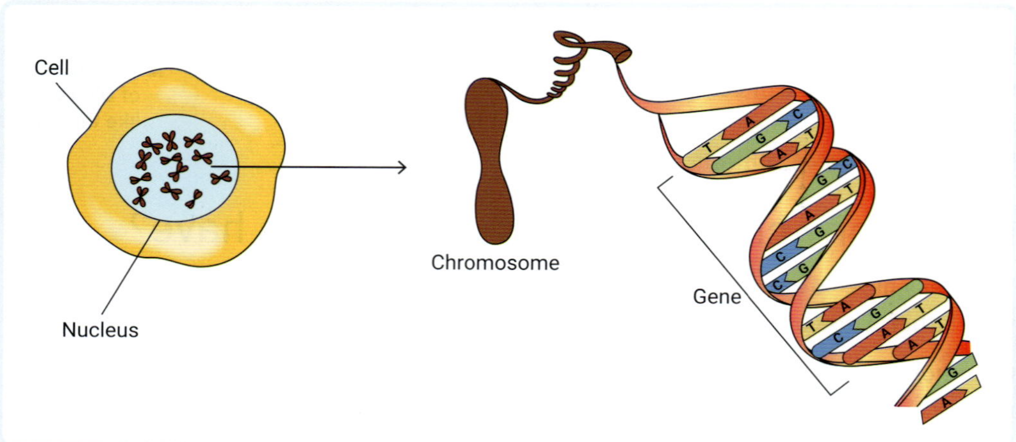

▲ **FIGURE 11.3.5** Each chromosome can have many thousands of genes, which consist of sequences of bases within DNA.

Genes differ in the sequence and number of nitrogenous bases they contain. The sequence of bases is the genetic code. The proteins created from the instructions in the genes are needed to complete cell activities and produce the cell's and organism's characteristics.

11.3 LEARNING CHECK

1. **Define** genetic code.
2. **Define** chromosome and chromatin.
3. **Identify** the term used to describe all of the genetic material in an organism.
4. **Describe** how DNA is packaged into a cell's nucleus.
5. **Explain** why DNA is sometimes called the 'blueprint of life'.
6. **Explain** how a karyotype can identify chromosome abnormalities.
7. **Compare** the structure and function of genes and chromosomes.

11.4 Observing DNA from cells

SCIENCE SKILLS IN FOCUS

IN THIS MODULE, YOU WILL FOCUS ON LEARNING AND IMPROVING THIS SKILL:

- extracting DNA from an organism.

Extracted DNA can be purified and used to study the genetic make-up of an individual, identify possible genetic mutations and diagnose genetic disorders.

DNA is found in the cell nucleus and you will need to break cell membranes to reach it. The buffer solution helps to protect the DNA from breakage.

DNA precipitates (separates from solution) in the presence of ethanol, and because ethanol is less dense than water, the layer containing the DNA will float at the top.

In Module 11.15 you will learn about the next steps scientists take to identify the genes in extracted DNA.

EXTRACTING DNA

AIM

To extract DNA from an organism

MATERIALS AND EQUIPMENT

- ☑ biological tissue containing DNA, such as strawberry, kiwi fruit, banana, wheat germ or onion
- ☑ buffer solution containing water, salt, dishwashing detergent and meat tenderiser
- ☑ zip-lock bag
- ☑ filter funnel with filter paper
- ☑ ice-cold ethanol
- ☑ test tube
- ☑ 250 mL beaker
- ☑ hooked Pasteur pipette or paperclip

Safety

If you have allergies to any of the fruits or vegetables that are used in this investigation, inform your teacher and avoid using them.

PROCEDURE

1 Place the biological tissue into the zip-lock bag. Note the size of the biological tissue that you start with. Break up the material by massaging it through the plastic bag for 1 minute.

2 Place the buffer solution into the zip-lock bag. Continue to massage the bag for another 2 minutes.

3 Using the filter funnel and filter paper, filter the contents of the zip-lock bag into the beaker.

4 Pour the filtrate into the test tube until it is about a third full.

5 Holding the test tube at an angle, slowly add the ice-cold ethanol to form a layer on top of the filtrate. Add roughly the same volume of ethanol as filtrate, so the test tube is two-thirds full. DNA is insoluble in alcohol, so it will form visible strands in the top layer.

6 Lift the DNA out of the test tube using the hooked Pasteur pipette or paperclip.

ANALYSIS

1 What material did you extract the DNA from?

2 What is the purpose of using a buffer solution?

4 Why does the ethanol remain on top of the filtrate in the test tube?

5 **Compare** the volume of the DNA extracted with the volume of the original tissue. **Estimate** the percentage of DNA present in the tissue.

11.5 Asexual and sexual reproduction

BY THE END OF THIS MODULE, YOU WILL BE ABLE TO:

- ✓ describe several different methods of asexual reproduction and give examples of each type
- ✓ describe sexual reproduction and explain how it results in genetic variations in offspring
- ✓ compare sexual and asexual reproduction.

Quiz
Types of asexual reproduction

GET THINKING

During the school holidays, you visit the Sydney Aquarium. You notice that a starfish is missing an arm, but a small starfish is growing close by, and a staff member pointed out that it is the missing arm of the first starfish. Also, you see a pair of penguins with chicks that have just hatched. What is the difference between the reproduction of the starfish and the penguins? What is the benefit for the penguin species of having two parents, compared to a starfish coming from a single parent?

▲ **FIGURE 11.5.1** Methods of reproduction differ in (a) star fish and (b) penguins.

Ian Waldie/Getty Images

▲ **FIGURE 11.5.2** A koala parent with offspring joey, which was produced by sexual reproduction.

Methods of reproduction

Reproduction is the process that allows a species to continue to exist. When a living thing reproduces, it is known as the parent. The new organism produced through reproduction by one or two parents is the **offspring**. When the offspring is a cell, it is usually called a daughter cell.

There are two main types of reproduction: sexual and asexual. For example, two koalas produce offspring, a koala joey (Figure 11.5.2), by **sexual reproduction**, whereas bacteria, honey bees and some lizards produce offspring by **asexual reproduction**.

offspring
a new organism produced by asexual or sexual reproduction

asexual reproduction
a method of reproduction that involves one individual producing an identical copy of itself

sexual reproduction
a method of reproduction that involves two parents producing offspring that are not identical to the parent or each other (except for identical twins or triplets)

Asexual reproduction

What do bacteria and sea cucumbers have in common? They produce offspring by asexual reproduction. This is a method of reproduction that involves a single parent making one or more copies of itself to form offspring identical to the parent and to each other. Usually, numerous offspring are produced. The type of cellular division that occurs in asexual reproduction is mitosis. You will learn about mitosis in detail in Module 11.6.

Asexual reproduction has these features:

- there is only one parent, whose genetic material is passed to its offspring
- there is no genetic variation: the offspring are identical to the parents
- an organism can reproduce without needing to find a mate or use energy to transfer sperm to an egg.

The different types of asexual reproduction are described in Table 11.5.1.

TABLE 11.5.1 Different types of asexual reproduction

Asexual reproduction method	Description	Examples	Diagram
Binary fission a method of asexual reproduction in bacteria in which a parent cell splits into two identical daughter cells	The parent cell splits into two identical daughter cells.	Bacteria and paramecia (single-celled protists)	Parent cell → Two identical daughter cells
Budding a method of asexual reproduction in which a bud (a growth on the parent body) forms, grows and then separates or spreads	A bud forms, grows and then separates from the parent to form new offspring.	Hydra	Parent body; Bud (Science Photo Library/Alamy Stock Photo)
Fragmentation a method of asexual reproduction in which a body part breaks off from a parent body and then develops into an offspring	A body part or a fragment from a parent's body breaks off to form a new independent offspring.	Sea star	**1** A fragment of the parent sea star breaks off. **2** Both the parent and fragment grow to form two genetically identical individuals.
Parthenogenesis a method of asexual reproduction in which an unfertilised egg matures and develops into an offspring without fertilisation by sperm	The spontaneous development of an embryo from an unfertilised egg cell.	Bees, ants, wasps, aphids, some lizards, birds and fish	Queen bee → Cell division → Egg cells → No fertilisation / Parthenogenesis → Drone (Aldona/Adobe Stock Photos; Courtesy of the USGS Bee Inventory and Monitoring Lab)

Sexual reproduction versus asexual reproduction

Asexual reproduction requires one parent and results in identical offspring. In comparison, in sexual reproduction there are two parents, and genetic information from each is contributed to the offspring, as shown in Figure 11.5.3. This mixing of genetic information explains why offspring are not identical to the parents or to each other and is what gives us genetic variation within a species. Sexual reproduction requires more energy than asexual reproduction because an organism may need to travel to find a partner and then mate. In general, sexual reproduction produces fewer offspring than asexual reproduction.

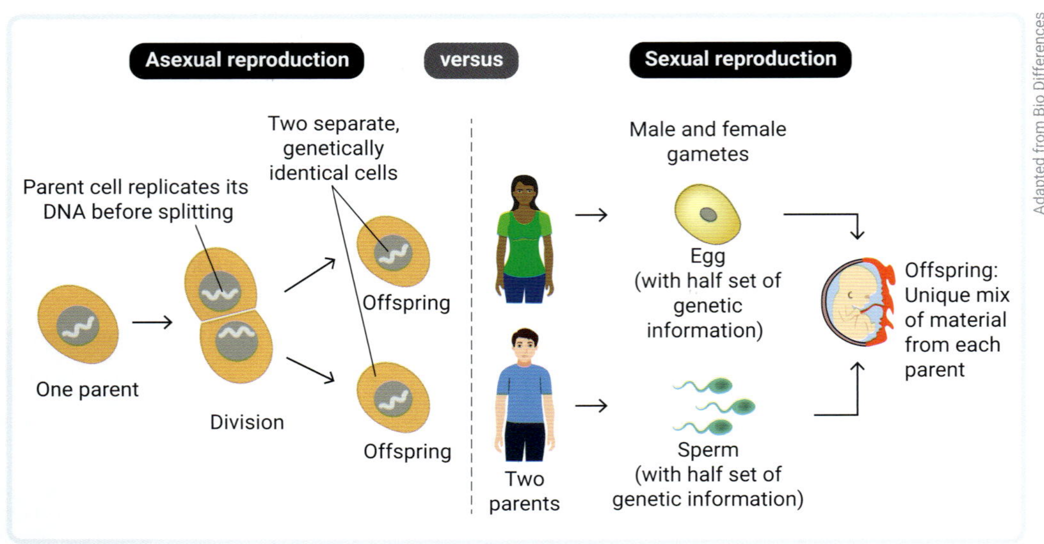

▲ **FIGURE 11.5.3** Asexual reproduction and sexual reproduction have differences in the number of parents involved and the type of offspring produced.

▲ **FIGURE 11.5.4** The puppies in a litter may look different because of genetic variation.

Sexual reproduction and variation

The most important benefit of sexual reproduction is genetic variation within the species. Just as puppies in a dog litter may look different (Figure 11.5.4), we look different from our parents and siblings because of genetic variation.

Genetic variation is due to the mixing of genetic material from each parent when gametes, the male and female reproductive cells, join. Genetic variation is also produced when gametes form in the process called meiosis. You will learn more about meiosis and gamete formation in Module 11.7.

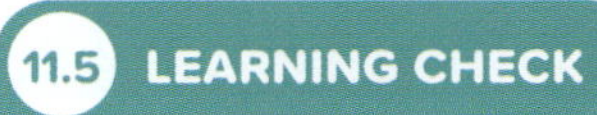

11.5 LEARNING CHECK

1 **Define** asexual reproduction and sexual reproduction.

2 Copy and complete Table 11.5.2 to **compare** asexual and sexual reproduction in animals.

▼ **TABLE 11.5.2** Summary of differences between asexual reproduction and sexual reproduction

Factor	Asexual reproduction	Sexual reproduction
Number of parents		
Are offspring identical to parent?		
Number of offspring		
Type of cell division		
Are gametes needed?		
Three examples of organisms that use this method		

3 **State** the main advantage of sexual reproduction.

4 **Choose** two organisms from the following list. For each, **research** and **explain** its method of reproduction, including whether it is sexual or asexual.

- Komodo dragon
- Yeast
- Coral
- Squid
- Aphid
- Echidna

11.6 Mitosis

BY THE END OF THIS MODULE, YOU WILL BE ABLE TO:

- ✓ describe the cell cycle
- ✓ list the stages of mitosis
- ✓ recognise the stages of mitosis
- ✓ recall the importance of mitosis for growth and repair.

Interactive resource
Label: Phases of mitosis

Other resource
Activity sheet: Modelling mitosis

GET THINKING

The cells in your stomach lining have an average life span of 3 days. However, the cells in your brain have an average life span of more than 80 years! Do you know the name of the process organisms undergo to repair and replace cells?

Cell division for growth and repair

Imagine if every time you cut your finger or broke a bone, your body could not repair itself. **Mitosis** is an important process that takes place in the cell's nucleus and drives the growth, repair and replacement of cells. Mitosis occurs in all **somatic** cells of organisms.

mitosis
cell division for growth, replacement and repair of somatic cells

somatic
relating to the cells that make up the body other than the reproductive cells

cell cycle
the series of events that takes place in a cell as it grows and divides

interphase
the resting phase of the cell cycle

The cell cycle

Each cell in your body is at some point along its **cell cycle** (Figure 11.6.1). This is a series of events that allows your cells to perform normal cellular functions and prepare for cell division.

The cell cycle can be broken down into two major phases: **interphase** and mitosis. During interphase, the cell grows and functions normally (G1 phase). Cells spend most of their time in interphase. In this stage, the DNA can be found as chromatin, unwound in the nucleus. Throughout the S phase, the DNA is duplicated and then condensed or folded into chromosomes containing two identical chromatids.

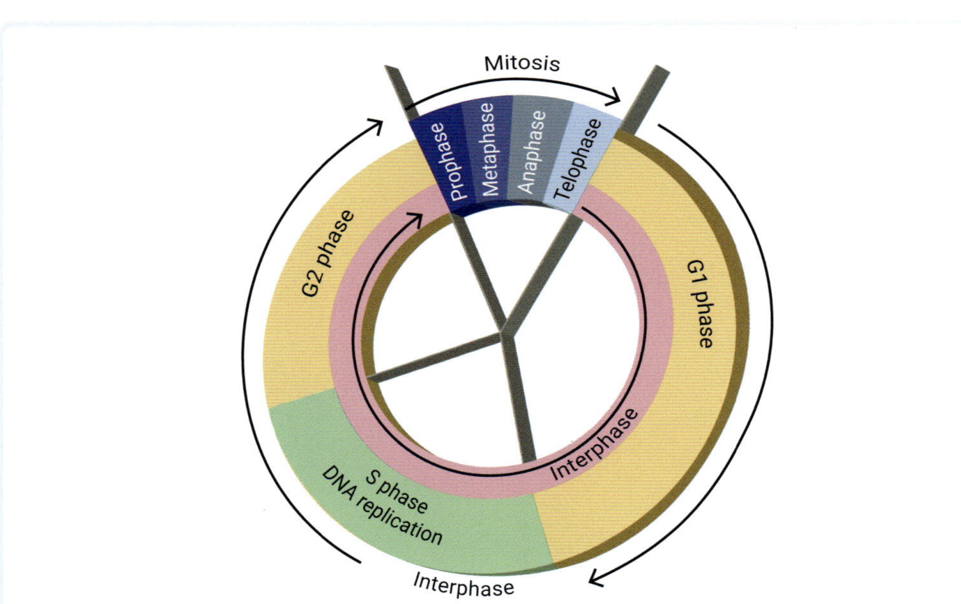

▲ **FIGURE 11.6.1** The cell cycle

9780170491785

The phases of mitosis

Mitosis is the process of nuclear division. This process is further broken down into four successive phases, as shown in Figure 11.6.2.

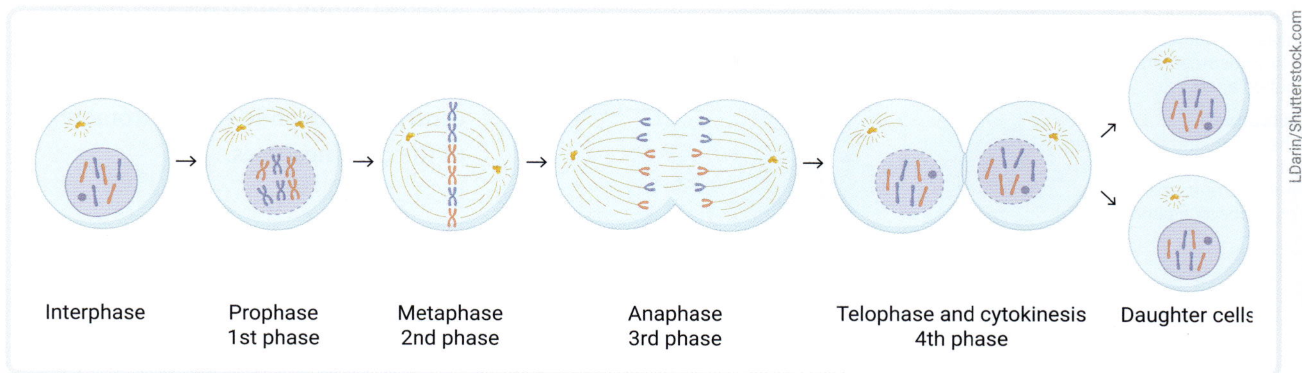

▲ **FIGURE 11.6.2** The process of mitosis consists of four phases and produces two genetically identical daughter cells.

Prophase is the first phase of mitosis. During this phase, the chromosomes duplicate and condense, forming visible 'X' shapes made from identical chromatids, as seen in Figure 11.6.3. Other important organelles in the cell, the **centrioles**, also duplicate and start to move to the opposite poles of the cell. The role of the centrioles is to produce **spindle fibres**. The nuclear membrane breaks down during prophase.

prophase
the first phase of cell division, when chromosomes duplicate and condense

centrioles
organelles in the cell that produce spindle fibres

spindle fibres
protein structures that separate the chromosomes during cell division

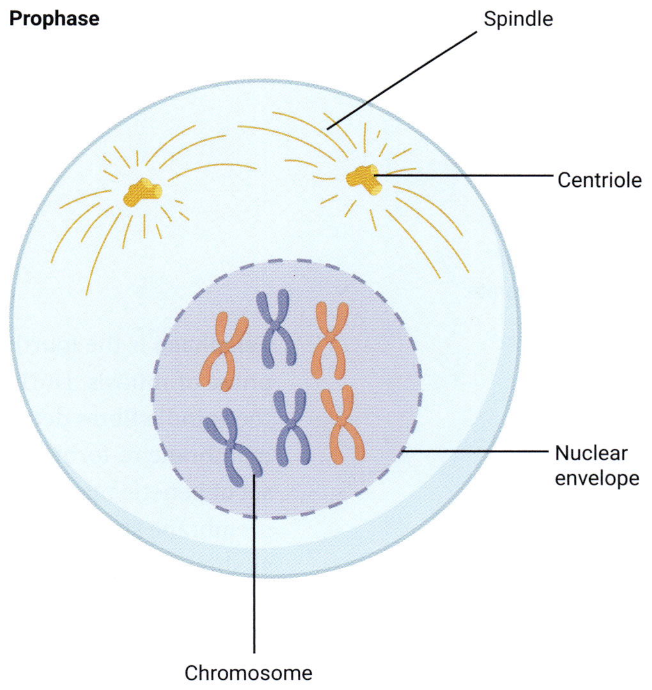

▲ **FIGURE 11.6.3** During prophase, the chromosomes duplicate and condense.

metaphase
the second phase of cell division, when chromosomes line up in the centre of the cell

anaphase
the third phase of cell division, when the chromosomes are pulled apart

telophase
the fourth and final phase of cell division, when the nucleus re-forms and the chromosomes unravel

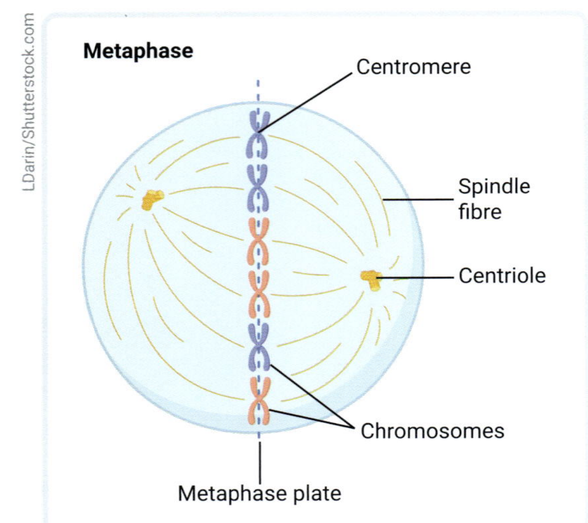

▲ **FIGURE 11.6.4** During metaphase, chromosomes line up on the metaphase plate.

Metaphase is the second phase of mitosis. During metaphase, the chromosomes line up on the equator of the cell along an imaginary line called the metaphase plate, as seen in Figure 11.6.4. The spindle fibres, extending from the centrioles, attach to the centromeres of the chromosomes.

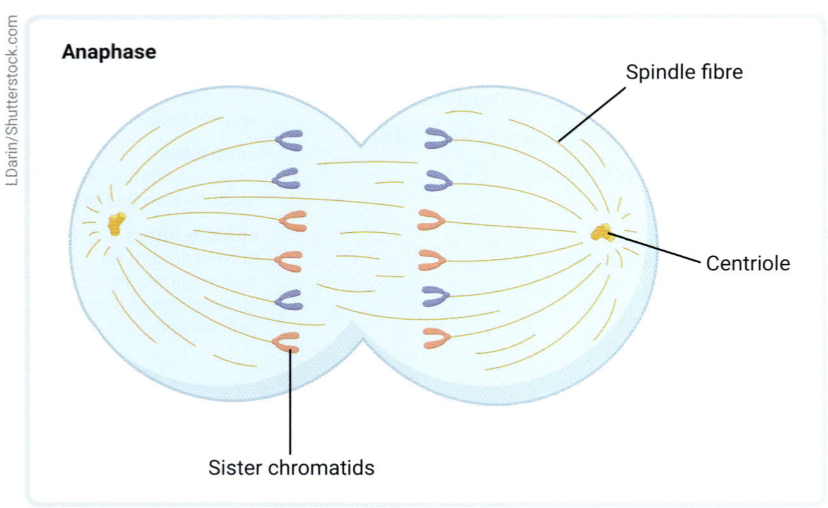

▲ **FIGURE 11.6.5** During anaphase, the chromosomes separate.

Anaphase is the third phase of mitosis. During anaphase, the spindle fibres retract, pulling the chromosomes apart at the centromeres. This drags the chromosomes to opposite poles of the cell, as seen in Figure 11.6.5.

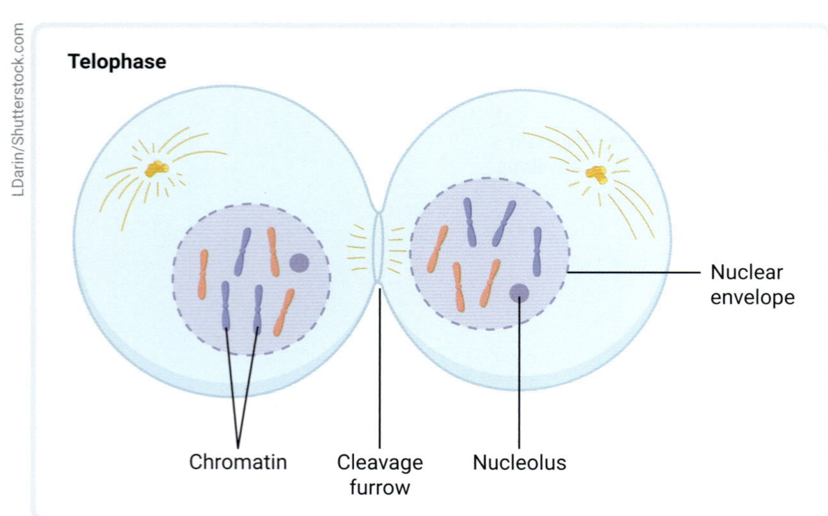

▲ **FIGURE 11.6.6** Telophase is the last phase of mitosis. The division of the cytoplasm, known as cytokinesis, is the final step of mitosis.

Telophase is the fourth and final phase of mitosis. During telophase, the spindle fibres detach, the nuclear membrane re-forms around each set of genetic material, and the cell membrane starts to pinch inwards, as shown in Figure 11.6.6. The chromosomes unravel, forming chromatin that will be encased in the nucleus of each daughter cell.

9780170491785

At the completion of telophase, **cytokinesis** begins. Cytokinesis is the division of the cytoplasm, resulting in two **daughter cells** (Figure 11.6.7). Each daughter cell has an identical copy of the parent DNA and a full complement of paired chromosomes with only one chromatid each. This condition is called **diploid** ($2n$). The cells are genetically identical because the DNA was exactly replicated during interphase (Figure 11.6.7).

cytokinesis
the division of the cytoplasm after mitosis

daughter cells
the cells that are produced as a result of cell division

diploid
the full complement of DNA, represented as $2n$

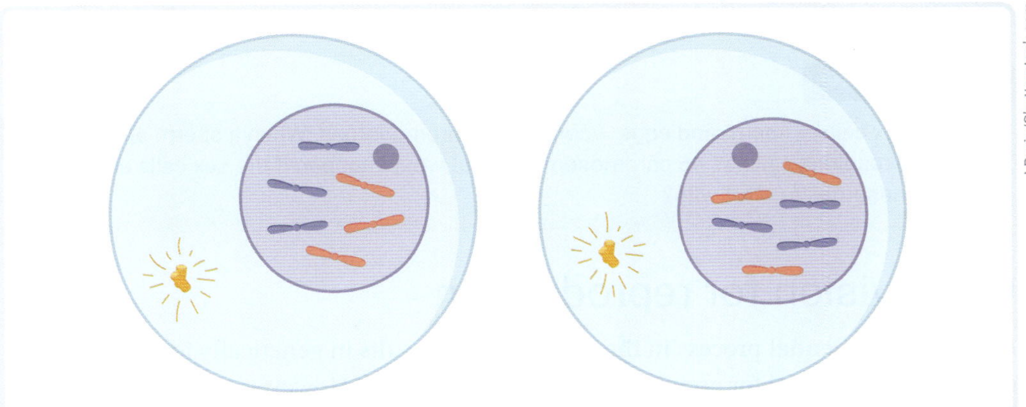

LDarin/Shutterstock.com

▲ **FIGURE 11.6.7** Two identical daughter cells are produced by the process of mitosis.

Mitosis is essential in the growth of organisms, the repair of damaged tissue and the replacement of dead cells in the body.

11.6 LEARNING CHECK

1 **Describe** interphase and the four phases of mitosis and **draw** a simple diagram of each.
2 **Explain** the role of the centrioles in mitosis.
3 **Define** diploid.
4 **Explain** why it is important that all multicellular organisms undergo the process of mitosis.
5 **Design** and make a flip book showing the phases of the cell cycle. Each page should have the name of the phase and a labelled diagram representing the condition of the cell during that phase.

11.7 Meiosis

BY THE END OF THIS MODULE, YOU WILL BE ABLE TO:

- ✓ list the stages of meiosis
- ✓ describe the stages of meiosis
- ✓ recall the importance of meiosis to produce cells needed for sexual reproduction
- ✓ compare and contrast the processes of mitosis and meiosis.

Interactive resource
Label: Meiosis I and II

GET THINKING

Human sex cells – sperm and eggs – contain 23 chromosomes. When a sperm and an egg unite, the resulting cell has 46 chromosomes. What would happen if the sex cells each had 46 chromosomes to start with?

Cell division for reproduction

Mitosis is an essential process in the cell cycle that results in genetically identical daughter cells needed for growth, replacement and repair. However, many organisms also need to make cells that can be used in sexual reproduction. As you learned in Module 11.3, humans have 46 chromosomes in each cell, 23 from the mother and 23 from the father. **Meiosis** explains how this is possible and how chromosome numbers are maintained in a species from one **generation** to the next.

meiosis
cell division producing cells that will specialise into gametes with half the number of chromosomes of the parent cell

generation
a group of organisms that start life and reproduce at around the same time; also, the time between one group and their offspring reproducing

gonads
the sex organs of an organism; where meiosis occurs

germline cells
the cells that form the ovum and the sperm

gametes
the sex cells of a sexually reproducing organism

zygote
a diploid (2*n*) cell resulting from the joining of two haploid gametes

Where does meiosis occur?

Meiosis occurs in the **gonads** of sexually reproducing organisms. The gonads contain **germline cells** that are responsible for the creation of **gametes**. In humans, the gonads are the ovaries (female, which produce ova) and testes (male, which produce sperm). Gametes are also called sex cells. When the sex cells of a male and a female combine at fertilisation, the cell they form has the potential to develop into a new individual.

Meiosis is a form of cell division that produces daughter cells with half the chromosome number of the parent cell, so that when two gametes combine to form a **zygote**, the correct number of chromosomes are present. Meiosis also results in variation between individuals because it produces sex cells that are genetically different from each other. The stages of meiosis are shown in Figure 11.7.1.

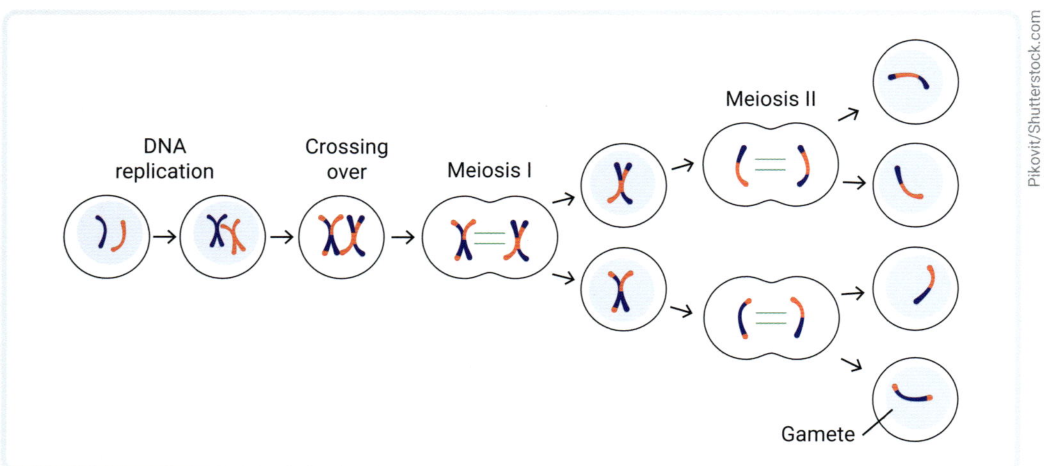

▶ **FIGURE 11.7.1** Meiosis

The phases of meiosis

Meiosis has more phases than mitosis does. In meiosis, the cell replicates its DNA once but divides twice. Meiosis is divided into two stages: meiosis I and meiosis II.

The parent cell grows during interphase, replicates its DNA and prepares the chromosomes for cell division. This is similar to mitosis. After interphase, the chromosomes are in their X-shape, consisting of two sister chromatids. At this stage, the chromatids are genetically identical.

Meiosis I

The cell progresses into meiosis I, which has four phases – prophase I, metaphase I, anaphase I and telophase I – as shown in Figure 11.7.2. Cytokinesis occurs at the completion of telophase I. There is a short interphase before the cell begins meiosis II.

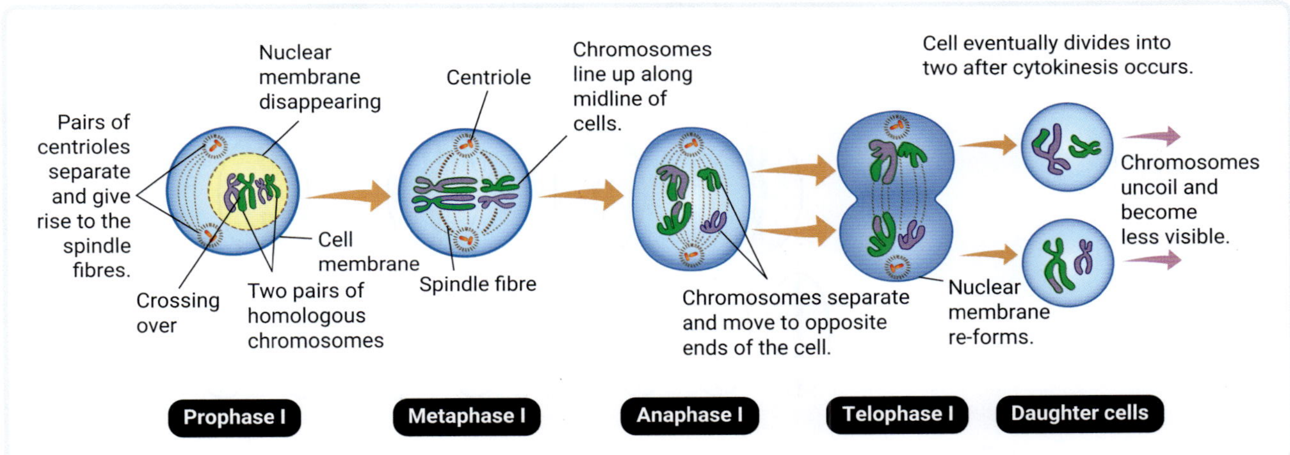

▲ **FIGURE 11.7.2** The phases of meiosis I

During prophase I, the nuclear membrane disappears, the centrioles duplicate and migrate to the opposite sides of the cell, and the chromosomes are visible in their X-shape. The chromosomes move into their homologous pairs and **crossing over** occurs, as shown in Figure 11.7.3.

crossing over
the exchange of genetic material between non-sister chromatids on homologous chromosomes

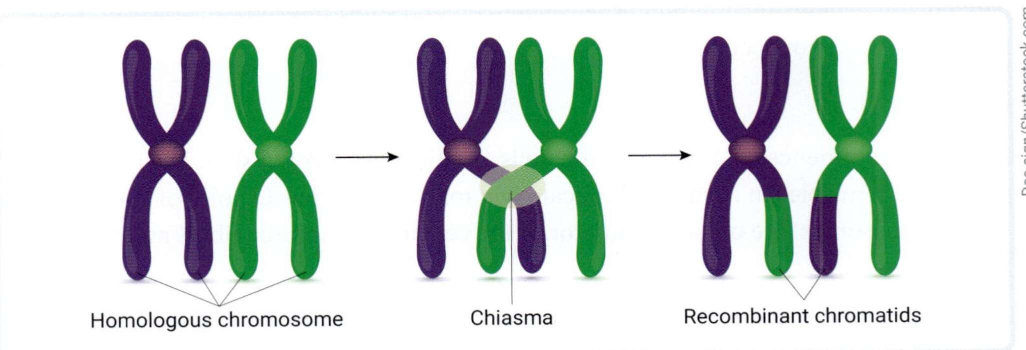

▲ **FIGURE 11.7.3** Crossing over occurs between non-sister chromatids.

chiasma
a point on the chromatids where crossing over occurs

random assortment
the way chromosomes line up during metaphase I of meiosis, resulting in random combinations of genes

In crossing over, the non-sister chromatids exchange genetic material with each other at a point known as the **chiasma** (plural: chiasmata). This results in a recombination of genetic material and is a major source of genetic variation.

During metaphase I, the chromosomes line up at the equator of the cell, and the spindle fibres attach to the centromeres. The chromosomes line up randomly in a process called **random assortment**, as shown in Figure 11.7.4. This is another source of genetic variation, as the resulting cells receive different combinations of chromosomes.

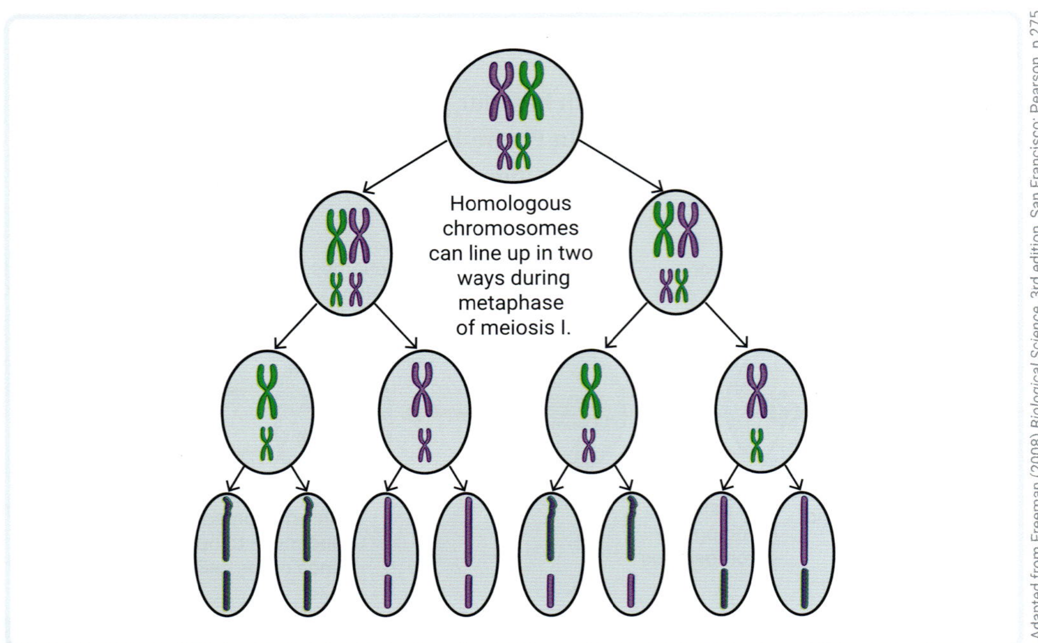

Adapted from Freeman (2008) *Biological Science*, 3rd edition, San Francisco: Pearson. p 275

▲ **FIGURE 11.7.4** Chromosomes line up randomly in metaphase I, resulting in their random assortment in the daughter cells.

homologous
carrying the same genes for characteristics at the same locations on a chromosome

haploid
having one copy of each chromosome, represented as *n*

During anaphase I, the **homologous** chromosomes are pulled to the opposite poles of the cells. The chromatids are not separated at this point, and the chromosomes are not in homologous pairs. Each chromosome consists of two chromatids.

Telophase I and cytokinesis result in the production of two daughter cells that are **haploid** and genetically different. The cell will enter a short interphase; however, DNA replication does not occur.

Meiosis II

Meiosis II commences in both daughter cells, starting with prophase II. During this phase, the centrioles in each cell duplicate and migrate to opposite poles of the cell. The chromosomes line up at the equator of the cell and the spindle fibres attach to the centromeres.

Anaphase II occurs when the spindle fibres contract, and the chromatids are separated and pulled to the opposite poles of the cells. Telophase II and cytokinesis follow, resulting in four haploid cells that are genetically different. Each chromosome now consists of one chromatid. The phases of meiosis II are depicted in Figure 11.7.5.

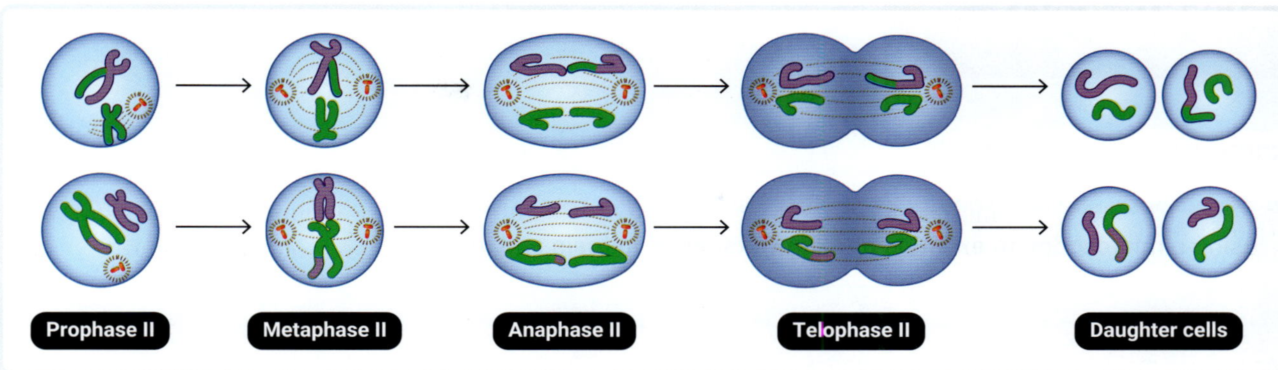

▲ **FIGURE 11.7.5** The phases of meiosis II

Mitosis versus meiosis

Mitosis and meiosis are similar, but key differences in some of the steps and the number of phases result in very different outcomes. Table 11.7.1 provides a comparison of mitosis and meiosis.

▼ **TABLE 11.7.1** A comparison of mitosis and meiosis

Characteristic	Mitosis	Meiosis
Type of cells that use this process	Body cell	Germline cell
Number of nuclear divisions	1	2
Number of daughter cells produced	2	4
Chromosome number of the parent cell	Diploid (2*n*)	Diploid (2*n*)
Chromosome number of the daughter cells	Diploid (2*n*)	Haploid (*n*)
Involves the separation of homologous chromosomes?	No	Yes, during anaphase I
Involves the separation of sister chromatids?	Yes, during anaphase	Yes, during anaphase II
Purpose of the process	Growth, replacement and repair of cells	Production of haploid gametes for reproduction

11.7 LEARNING CHECK

1. **Recall** the location where meiosis occurs.
2. **Define** haploid.
3. **Identify** two sources of genetic variation due to meiosis and the phases they occur in.
4. **Draw** a labelled diagram showing the process of crossing over.
5. **Distinguish** between the terms 'chromosome' and 'chromatid'.
6. **Draw** a Venn diagram to **compare** and **contrast** mitosis and meiosis.

WORKING SCIENTIFICALLY

11.8 Observing and drawing cells under the microscope

SCIENCE SKILLS IN FOCUS

IN THIS MODULE, YOU WILL FOCUS ON LEARNING AND IMPROVING THIS SKILL:

- using a microscope to examine and draw stages of mitosis and meiosis.

Mitosis and meiosis consist of several stages (prophase, metaphase, anaphase, telophase in mitosis; and similar stages with additional steps, such as crossing-over, in meiosis). Microscopes can be used to identify and study each stage, providing insights into the dynamics of cell division.

When observing mitosis and meiosis specimens under the microscope, you need to apply these skills:

- Focus the specimen under the microscope, first with the coarse focus and then the fine focus to see details.
- Always start with the lower magnification, 4X, and then increase to the higher magnification, 100X.
- Adjust the light intensity by using the diaphragm, the specimens showing mitosis and meiosis specimens can be dark due to the stains.
- Draw a clear and simple diagram of what you are observing under the microscope, label the chromosomes, and indicate the mitosis and meiosis phases that the specimen is showing.
- Indicate the magnification in the drawing.

SCIENCE SKILLS

Need a reminder of how to safely use a microscope? Check out the science skills resources in **Module 5.6**.

MITOSIS AND MEIOSIS UNDER THE MICROSCOPE

AIM

To identify key stages of mitosis and meiosis in plant cells using a compound microscope

MATERIALS AND EQUIPMENT

- ☑ compound microscope
- ☑ microscope slides showing meiosis (e.g. lily anthers)
- ☑ microscope slides showing mitosis (e.g. onion root tip)

Safety

Microscope slides are fragile and can break easily. If they break, ask your teacher to dispose of the glass safely.

PROCEDURE

1 Turn on your microscope and position it so that you can easily look through the lens. Ensure it is on the lowest magnification setting.

2 Carefully place the mitosis slide into position and centre it under the microscopic field. Rotate to the high-power objective and refocus to observe the cells in greater detail.

3 Once you have located the cells, draw a scientific diagram of what you can see in your workbook. Remember to include the magnification.

4 Identify the stage of mitosis each cell is undergoing and label the diagram accordingly.

5 Once your diagram is complete, move your microscope back to the lowest setting and remove the mitosis slide. Replace it with the meiosis slide.

9780170491785

6 Increase the magnification as before and refocus to see the cells in clearer detail.

7 Complete a scientific diagram of what you can see on the slide in your workbook. Identify the stage of meiosis each cell is undergoing and label the diagram accordingly. Remember to include the magnification.

ANALYSIS

1 Which phases of mitosis were you able to **identify**? What are the key processes occurring during each phase?

2 Which phases of meiosis were you able to **identify**? What are the key processes occurring during each phase?

3 From your observations, how are meiosis and mitosis different?

4 From your observations, how are meiosis and mitosis similar?

5 Why is it important to **record** the magnification of the microscope?

6 **Explain** the limitations of observing mitosis and meiosis under the light microscope.

11.9 Mendelian inheritance

BY THE END OF THIS MODULE, YOU WILL BE ABLE TO:

- ✓ define the common terms used in genetics
- ✓ recognise dominant and recessive traits
- ✓ classify genotypes as homozygous dominant, heterozygous or homozygous recessive.

Extra science investigation
Inheritance and chance

Other resources
Worksheets:
Mendel's results

The Mendelian ratio

GET THINKING

Why do you look more like one parent than the other? Why are some characteristics seen in all your family members, whereas others occur randomly? The answer is genetics!

Mendel: the father of genetics

wikimedia/Professor William Bateson

▲ **FIGURE 11.9.1** Gregor Mendel

Gregor Mendel was an Austrian monk who was interested in meteorology, mathematics and biology (Figure 11.9.1). His breeding experiments with pea plants (*Pisum sativum*) in his monastery in the 1860s resulted in the development of the principles of inheritance. Mendel cross-pollinated his pea plants and meticulously recorded the statistics of the resulting offspring. This data and his observations were published but largely ignored until the 1900s when other scientists discovered his work and linked it to their own studies in genetics.

Discoveries that explain the mechanisms of inheritance

The later discovery of chromosomes by Walther Flemming in 1879, and the link between chromosomes and heredity described by Theodor Boveri in 1902, helped scientists further understand the mechanisms behind inheritance. In addition, Walter Sutton's observations in 1903 of chromosome behaviour during cell division and gamete formation was consistent with Mendel's work.

Mendel's laws of inheritance

Mendel proposed three principles of inheritance, which are now known as Mendel's laws of inheritance.

- The law of segregation: each inherited trait is defined by a single gene pair. The parent genes are randomly separated into the gametes. Offspring inherit one **allele** from each parent when the gametes join at fertilisation.
- The law of independent assortment: traits are inherited separately from one another. The inheritance of one trait is not dependent on the inheritance of another.
- The law of dominance: where an organism has two alternative forms of a gene, the **dominant allele** will be expressed.

allele
an alternative form of a gene

dominant allele
the allele that will be expressed in the phenotype

Mendel's pea plant experiments

Mendel studied seven characteristics of pea plants, including plant height, pea colour, flower colour, pod shape and seed shape (Figure 11.9.2). He worked with plants that self-pollinated and consistently produced the same characteristics from one generation to the next. Mendel used these observations to conclude that the parental lineage of these pea plants was **purebred**.

purebred
having the same alleles for a given gene; see homozygous dominant/recessive

Characteristics of pea plants Gregor Mendel used in his inheritance experiments						
Seeds		Flower	Pod		Stem	
Form	Cotyledons	Colour	Form	Colour	Position of inflorences	Size
round roundish	yellow	white	full	yellow	axial	long
wrinkled	green	violet-red	constricted between the seeds	green	terminal	short

LadyofHats, reworked by Sciencia58

▲ **FIGURE 11.9.2** Mendel studied seven of the characteristics of pea plants.

Mendel cross-pollinated one variety of purebred plant with another and discovered that the offspring looked like either one of the parent plants, not a blend of the two. For example, crossing a long stem plant with a short stem plant resulted in offspring having long stems, not stems somewhere between long and short. In general, if the offspring (or **progeny**) of crosses between purebred plants looked like only one of the parents, Mendel called the expressed trait the **dominant** trait. The trait that was not seen in the progeny, and was masked by the dominant trait, was called the **recessive** trait.

Mendel designated the two pure-breeding parental generations as P_1 and P_2 and identified the progeny as the **filial** or F_1 generation. Although the F_1 generation looked uniformly like one parent of the P generation, they had inherited a different allele from each parent plant. This type of breeding is known as a **monohybrid cross**, as only one gene is being investigated (Figure 11.9.3).

Mendel then crossed the F_1 generation by allowing the F_1 generation to self-pollinate, creating the F_2 generation. The F_2 generation had offspring that were either long or short in stem length, showing that the recessive trait had been carried down in the F_1 generation. The ratio of long to short stems was roughly three long stems to one short stem.

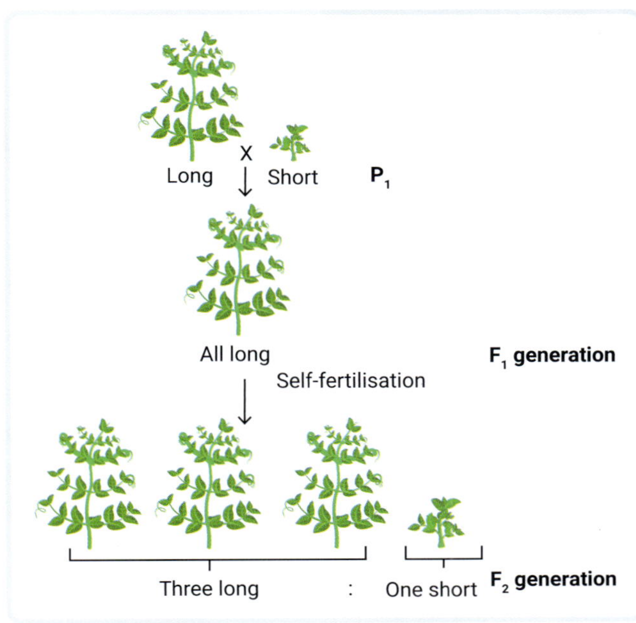

▲ **FIGURE 11.9.3** A monohybrid cross over two generations

progeny
offspring

dominant
an inheritance that identifies the dominant allele in a genotype

recessive
an inheritance that identifies the recessive allele in a genotype

filial
first set of offspring from a cross between parents

monohybrid cross
a cross between two organisms with two alleles at one gene location

DATA SCIENCE

Learn more about how evidence is used to support scientific claims in **Module 2.2**.

DNA, genes and alleles

Mendel's data suggested that each parent contributes some particulate matter to their offspring, and he called this hereditary material 'elementen'. Today we know this as DNA and genes. We also know that these forms of genes are called alleles. For example, when considering stem height, there is an allele for long and an allele for short stem height.

Mendel's genetic notation

recessive allele
the allele that is masked by a dominant allele and is only expressed in the homozygote

homozygous dominant
having two of the same dominant alleles on homologous chromosomes

homozygous recessive
having two of the same recessive alleles on homologous chromosomes

heterozygous
having two different alleles on homologous chromosomes

genotype
the combination of alleles for a specific gene

phenotype
the observable characteristics of the genotype

Mendel used a specific notation to represent his data. For a dominant allele he used a capital letter, and he used a lowercase letter to represent the **recessive allele**. Purebred long stem plants, which are **homozygous dominant**, have a notation of LL. Pure-breeding short stem plants, which are **homozygous recessive**, have a notation of ll. The hybrids in the F_1 generation, which are **heterozygous**, have a notation of Ll. Heterozygous plants carry the recessive allele, despite physically showing the dominant allele.

This notation is known as the **genotype** of the individual, as it is a representation of the combination of alleles found for a specific gene. The observable expression of the genotype is called the **phenotype**. For example, Figure 11.9.4 shows a pea plant with the phenotype of a long stem, which could have the genotype LL or Ll.

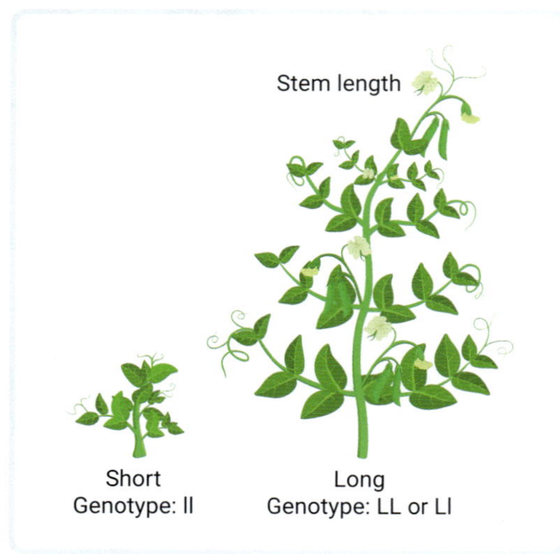

▲ **FIGURE 11.9.4** A pea plant with the phenotype short stem must have the genotype ll. A pea plant with phenotype long stem can have the genotype LL or Ll.

11.9 LEARNING CHECK

1 **Define** gene and allele.
2 **Describe** the difference between a dominant allele and a recessive allele, using an example from Mendel's investigations.
3 Figure 11.9.5 shows the characteristics of pea plants and the classification of dominant and recessive alleles.
   a Assign a key to each characteristic of a pea; for example, seed shape R = round and r = wrinkled.
   b **Draw** this as a table in your workbook.
4 Write the genotypes for the following pure-breeding plant characteristics:
   a wrinkled seeds.
   b inflated pod shape.
   c white flower.
   d tall stem height.

9780170491785

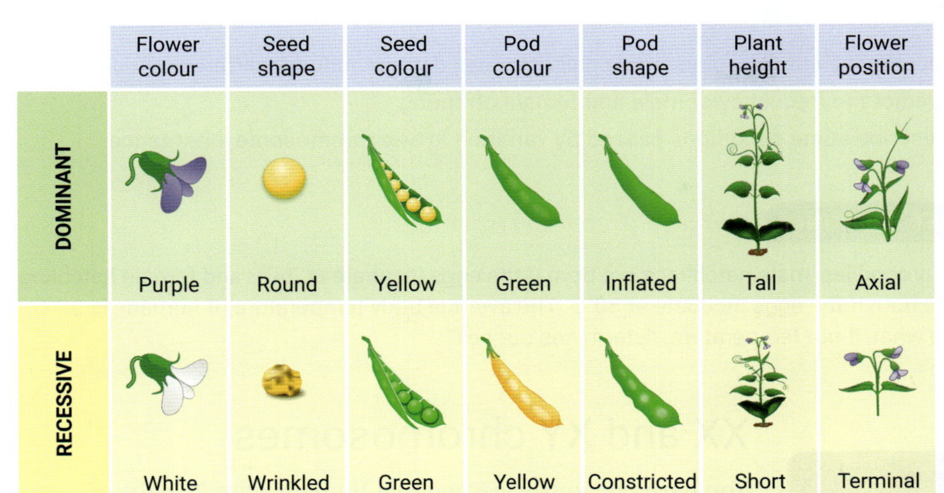

▲ **FIGURE 11.9.5** The dominant and recessive traits of pea plants

5 **Predict** the phenotypes of the offspring in the following crosses. Use Figure 11.9.5 to help you.

Example: A purebred inflated seed pod crossed with a purebred constricted seed pod will produce offspring that all have inflated seed pods because inflated seed pod is the dominant trait.

a A purebred purple flower crossed with a purebred white flower

b A purebred yellow seed crossed with a purebred green seed

c A purebred tall stem crossed with a purebred short stem

6 **List** the genotypes for the following phenotypes into your workbook:

a homozygous green pod.

b homozygous terminal flower position.

c homozygous wrinkled seed.

d heterozygous purple flower.

e heterozygous inflated pod shape.

7 **Describe** Mendel's three laws.

8 **Define**:

a purebred.

b monohybrid cross.

c genotype.

d phenotype.

9 Conduct some research on the internet to **determine** if the following human characteristics are dominant or recessive.

a The ability to roll your tongue into a U shape

b Free earlobes

c Interlocking fingers with the left thumb on top

d A widow's peak hairline on the forehead

11.10 Sex determination

BY THE END OF THIS MODULE, YOU WILL BE ABLE TO:

- ✓ identify the chromosomes responsible for sex determination in different species
- ✓ predict the frequency of male and female offspring
- ✓ describe some conditions caused by variation in sex chromosome inheritance.

Video activity
Sex determination

Other resource
Activity sheet: Sex and 'chance'

GET THINKING

In crocodiles, male hatchlings are born if the eggs incubate at 34°C and female hatchlings are born if the eggs incubate at 30°C. The average body temperature of humans is 37°C. So what, if not temperature, determines our sex?

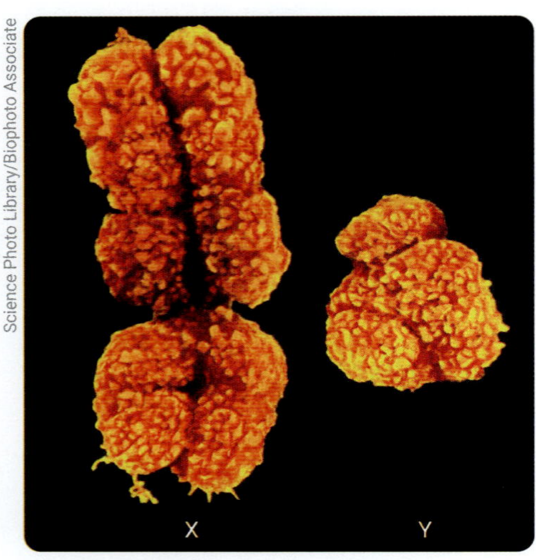

Science Photo Library/Biophoto Associate

▲ **FIGURE 11.10.1** A scanning electron micrograph showing the X chromosome (left) and Y chromosome (right)

XX and XY chromosomes

You have previously learned that humans have 23 pairs of chromosomes, of which 22 are the autosomes and the 23rd pair are the sex chromosomes. In humans, and most mammals, these sex chromosomes are labelled X and Y (Figure 11.10.1). The combination of the sex chromosomes determines the sex of the offspring: XX for female and XY for male.

How is sex determined?

The sex of offspring is determined by what happens during the separation of the chromosomes during meiosis (Figure 11.10.2). In meiosis, the chromosome number halves, with each daughter cell receiving a haploid number of chromosomes from the parent cell. In humans, all the eggs cells (ova) produced by the female carry one X chromosome. Half the sperm cells produced by the male carry one X chromosome and the remaining half carry the Y chromosome.

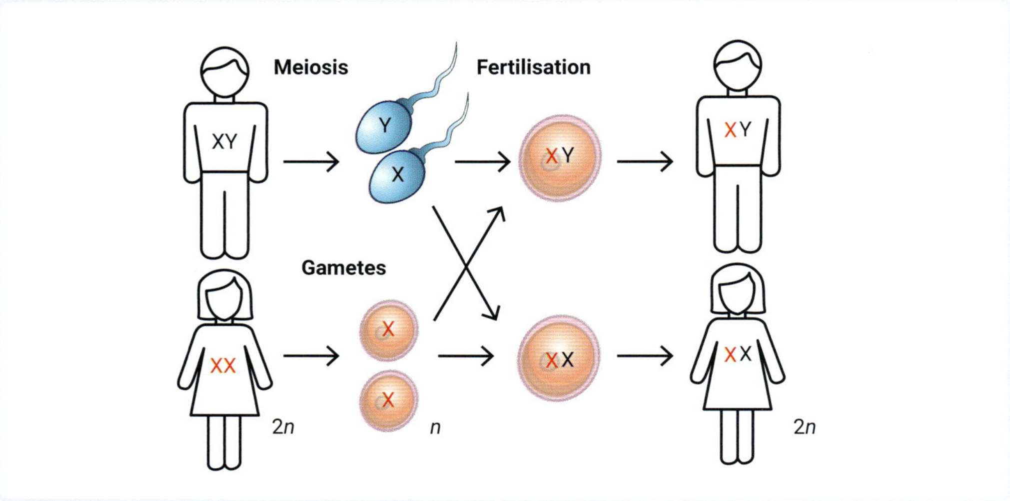

▲ **FIGURE 11.10.2** How sex is determined in humans

At fertilisation, there is a 50 per cent probability that the female egg cell will be fertilised by an X-carrying sperm and a 50 per cent probability that the female egg cell will be fertilised by a Y-carrying sperm. The resulting zygote will be diploid ($2n$) and will have a full complement of sex chromosomes.

Sex chromosomes in other animals

Other animals have different sex-determining systems from most mammals. For example, birds have a ZW system, where males are ZZ and females are ZW (Figure 11.10.3). Reptiles and amphibians have a mixture of XX/XY and ZZ/ZW depending on the species. These animals, in particular turtles and crocodiles, are influenced by the environment, usually temperature, which can influence the sex of the offspring. For example, turtle eggs that are incubated below 27.7°C will result in male hatchlings and eggs incubated at temperatures above 31°C will result in female hatchlings.

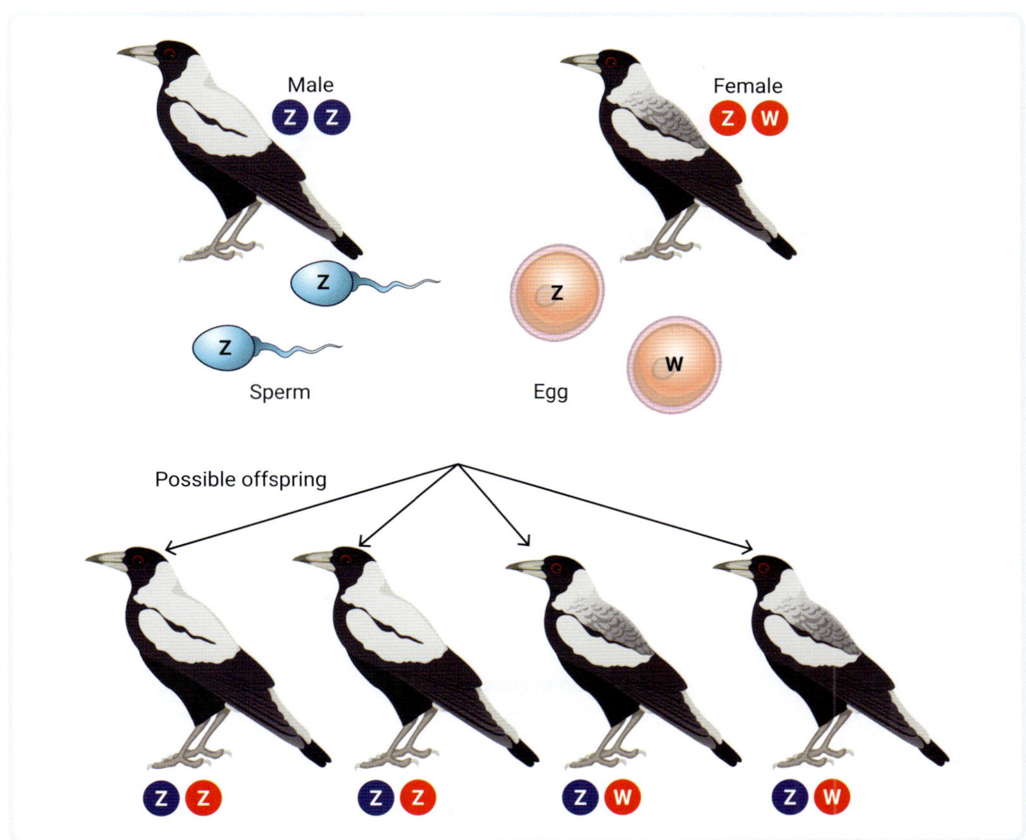

▲ **FIGURE 11.10.3** The inheritance of sex chromosomes in birds

Insects also inherit sex chromosomes, although they do so differently from vertebrate animals. Honey bees are an interesting example of a different system, where all drones (males) are formed from unfertilised eggs and are, therefore, haploid. Female worker bees develop from fertilised eggs and are diploid. The queen bee is a female (diploid) that was fed royal jelly in the hive. She lays all the eggs in the hive. Sex determination in honey bees is shown in Figure 11.10.4.

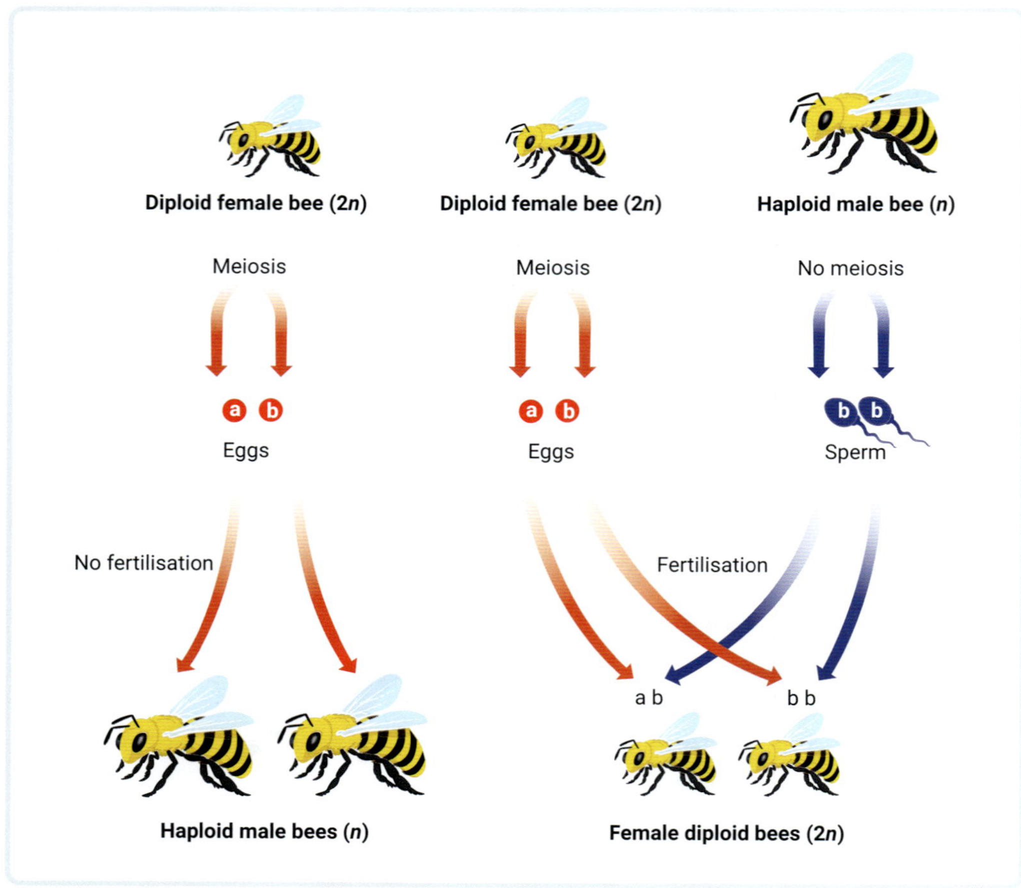

▲ **FIGURE 11.10.4** In honey bees, sex is determined by being haploid or diploid.

Extra or missing sex chromosomes

non-disjunction the incorrect separation of chromosomes or chromatids at the centromere, resulting in gametes with an unusual chromosome number

Sometimes the chromosomes do not separate normally during meiosis, resulting in gametes that have one extra or one fewer sex chromosome. This is called **non-disjunction** and can be seen in Figure 11.10.5.

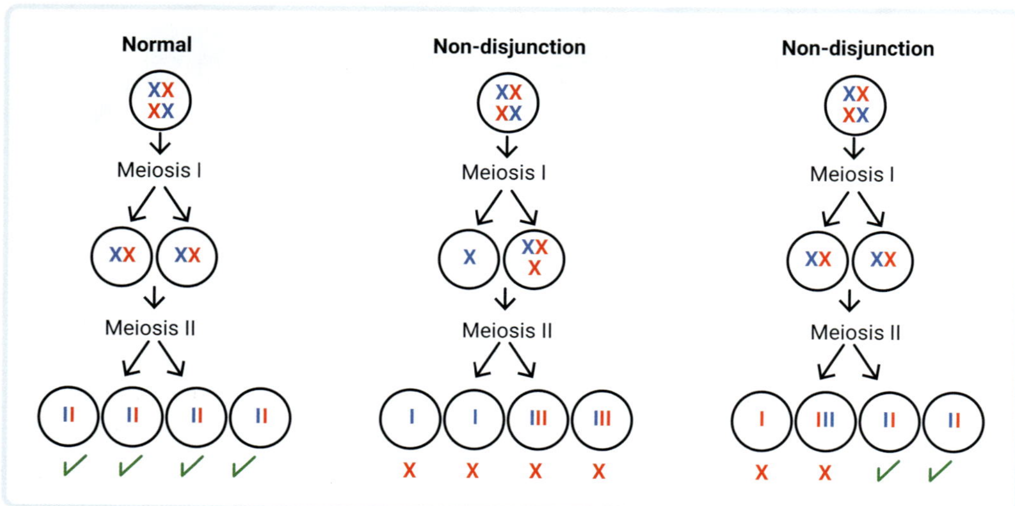

▲ **FIGURE 11.10.5** If chromosomes do not separate normally, gametes can end up with more or fewer chromosomes than normal.

9780170491785

Klinefelter syndrome occurs in individuals who have XXY as their sex chromosomes. They are male, taller than average with less facial and body hair. Their testes are smaller and produce less testosterone and sperm. A karyotype of a person with Klinefelter syndrome is shown in Figure 11.10.6.

Turner syndrome occurs in females who inherit only one X chromosome. Their genotype is XO. Characteristics of the condition can include infertility, a webbed neck, short stature and irregularities in the development of their reproductive organs. A karyotype of a person with Turner syndrome is shown in Figure 11.10.7.

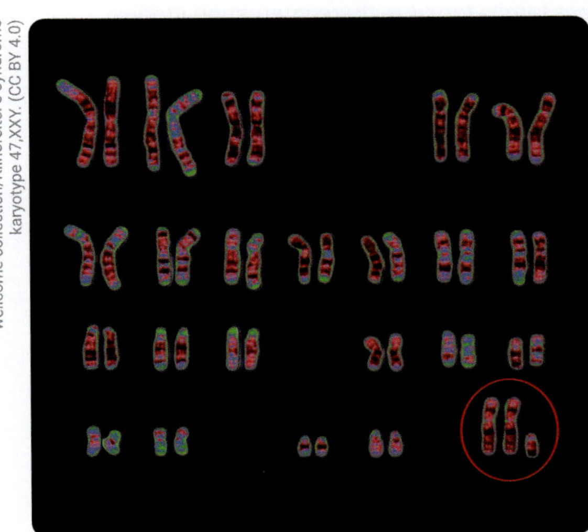
wellcome collection/Klinefelter's syndrome karyotype 47,XXY. (CC BY 4.0)

▲ **FIGURE 11.10.6** A karyotype of a person with Klinefelter syndrome (XXY).

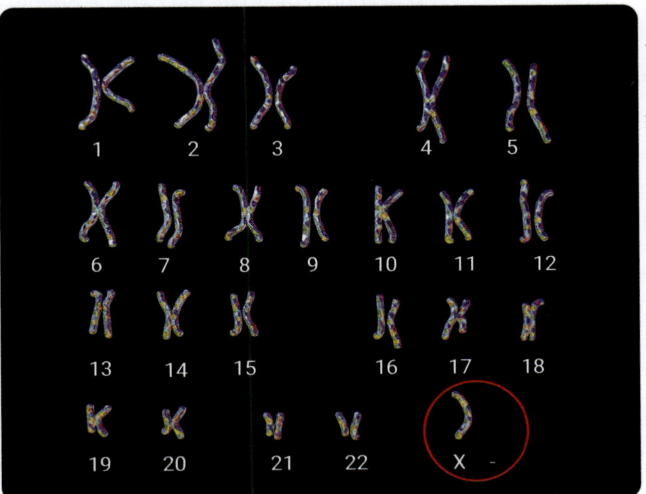

Kateryna Kon/Shutterstock.com

▲ **FIGURE 11.10.7** A karyotype of a person with Turner syndrome (XO)

11.10 LEARNING CHECK

1 **Recall** the sex chromosomes for human males and females.
2 Use a diagram with correct notation to **predict** the probability of male offspring in:
   a mammals.
   b birds.
3 Using the internet, **research** the SRY gene, located on the Y chromosome in male mammals. **Identify** what role the SRY gene plays in determining the maleness of a mammal.
4 **Research** and **describe** the full characteristics of Klinefelter syndrome and Turner syndrome. In your description include the frequency of occurrence of both syndromes in Australia and the phase(s) of meiosis where incorrect separation of chromosomes can occur.

11.11 Punnett squares

BY THE END OF THIS MODULE, YOU WILL BE ABLE TO:

✓ define Punnett squares and explain how to use them

✓ use Punnett squares to predict the genotypes and phenotypes for monogenic traits in offspring from crosses.

Interactive resource
Drag and drop: Punnett squares

Other resource
Worksheet: Punnett squares

GET THINKING

Studying your ancestry can reveal some amazing stories about your family history. What if we could examine the genetic history of your family too? Are there interesting stories hidden in your genes?

What are Punnett squares?

monogenic trait a trait that is determined by a single gene

Punnett squares are used to show the possible genotypes and phenotypes for **monogenic traits** in offspring produced from crosses. As you learned in Module 11.9, Mendelian inheritance shows that alleles (versions of genes) occur in pairs, and these allele pairs separate independently during meiosis so that each allele for the trait appears in a gamete. At fertilisation, gametes unite randomly to produce potential offspring.

Punnett squares are a simple way to show this process. Capital letters are used for dominant traits, and lowercase letters are used for recessive traits. It is conventional to keep the letters the same for the trait being examined; for example, AA, Aa, or aa.

Punnett squares in action

first filial generation the first set of offspring from a parent cross

As Mendel observed in peas, round seed shape (R) is dominant to wrinkled seed shape (r). Let's cross a homozygous round seed (RR) with a homozygous recessive wrinkled seed shape (rr) to determine the offspring produced in the **first filial generation** (F_1) generation.

1 Show the parent genotype cross:

$$P_1 \times P_2 \text{ cross} = RR \times rr$$

2 Complete the Punnett square by placing each parent allele into the top and side boxes. You can then fill in the boxes, showing the probability of the offspring. A completed Punnett square showing P_1 and P_2 genotypes and the F_1 genotypes is shown in Figure 11.11.1.

3 State the genotypes of the first filial generation (F_1) generation.

One hundred per cent of the genotypes for the F_1 are Rr, or heterozygous.

4 State the phenotypes of the F_1 generation.

One hundred per cent of the progeny will display the dominant phenotype trait of round, as the round dominant allele (R) will mask the wrinkled recessive allele (r).

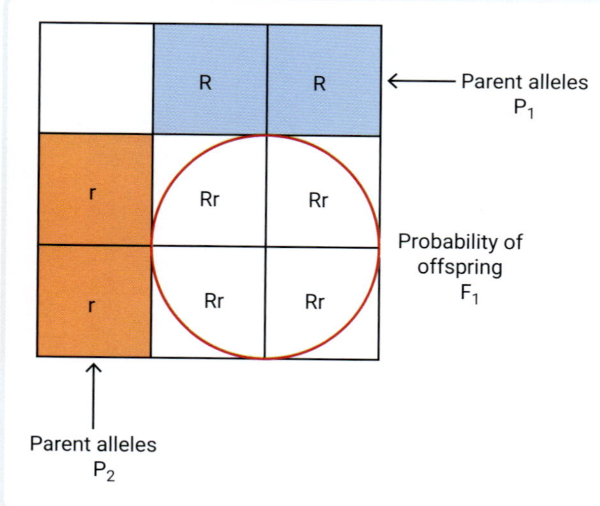

▲ **FIGURE 11.11.1** A completed Punnett square showing P_1 and P_2 genotypes and the F_1 genotypes

9780170491785

By separating the parent genotypes into the boxes at the top and side of the Punnett square, you are representing the independent assortment of chromosomes and production of gametes from meiosis.

Now, if you cross the F_1 generation you will see the Mendelian ratio appear in the **second filial generation** (F_2 generation). Follow the same four steps from the first cross. Table 11.11.1 shows a completed Punnett square demonstrating a cross between two F_1 genotypes and the resulting F_2 genotypes.

second filial generation the set of offspring from the first filial parent cross

$$F_1 \times F_1 = Rr \times Rr$$

▼ **TABLE 11.11.1** A completed Punnett square showing two F_1 genotypes and the resulting F_2 genotypes

	R	**r**
R	RR	Rr
r	Rr	rr

▲ **FIGURE 11.11.2** A wrinkled pea seed and a round pea seed

Twenty-five per cent of the genotypes are RR (1 out of 4), homozygous dominant, 50 per cent are Rr (2 out of 4), heterozygous and 25 per cent are rr, homozygous recessive (1 out of 4). Phenotypically, 75 per cent of the offspring (3 out of 4) will display round seeds and 25 per cent will show wrinkled seeds (1 out of 4). As a ratio, the genotype can be expressed as 1:2:1 and the phenotype as 3:1. Figure 11.11.3 demonstrates the potential seed shapes produced in the F_2 generation.

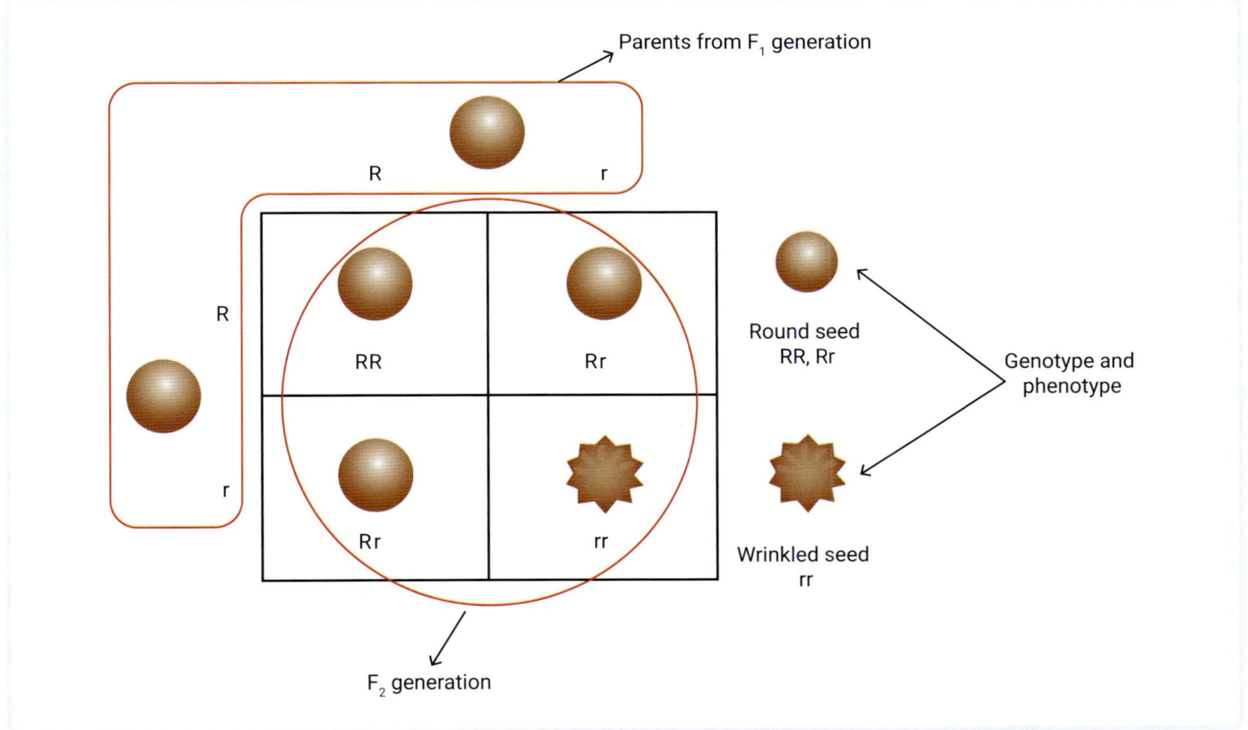

▲ **FIGURE 11.11.3** A Punnett square showing the potential seed shapes produced in the F_2 generation

11.11 LEARNING CHECK

1 In your workbook, copy and match the terms to the correct definitions.

Term	Definition
Homozygous	• Another term for offspring
Heterozygous	• The trait masked in the heterozygote
Dominant	• Having two different alleles for a gene
Recessive	• Having two of the same alleles for a gene
Allele	• An alternative form of gene
Progeny	• The trait expressed in the heterozygote

2 For each genotype listed below, **determine** whether it is homozygous dominant, homozygous recessive or heterozygous.

a AA b Tt c tt d Gg e Mm f ss

3 For each of the genotypes listed below, **determine** the phenotype given the following information.

a Purple flowers are dominant to white flowers.

A PP B Pp C pp

b Brown eyes are dominant to blue eyes.

A Bb B bb C BB

c Grey fur is dominant to white fur.

A GG B gg C Gg

4 Complete the following crosses for these monogenic traits. Remember to set a key, show the parent cross, complete the Punnett square and **calculate** the probabilities of both genotype and phenotype of the offspring.

a In seals, the gene for the length of whiskers has two alleles. Long whiskers are dominant to short whiskers.

i **Determine** the genotypes and phenotypes of the offspring from a cross between a homozygous dominant seal and a heterozygous seal.

ii If one parent seal is heterozygous and the other is short whiskered, **determine** the probability that their offspring will have short whiskers.

b Curly hair is dominant in humans and straight hair is recessive.

i A woman with curly hair has children with a man who is homozygous for straight hair. **Predict** the genotypes and phenotypes of their children.

ii A man with straight hair, whose mother was curly haired, has children with a woman with curly hair. **Predict** the genotypes and phenotypes of their children.

Djavan Rodriguez/Shutterstock.com

c In humans, right-handedness is dominant and left-handedness is recessive.

i Two parents, both heterozygotes, have children. What are the predicted genotypes and phenotypes of their children?

ii If those parents have already had one left-handed child, what is the probability of a second left-handed child?

11.12 Pedigree charts

BY THE END OF THIS MODULE, YOU WILL BE ABLE TO:

- ✓ define, interpret and draw pedigree charts
- ✓ predict genotypes and phenotypes of individuals based on a pedigree chart
- ✓ justify the possible genotypes of individuals in a pedigree chart
- ✓ model the occurrence of a certain trait over several family generations.

Interactive resource
Label: Pedigree charts

Other resource
Worksheet: Pedigrees

GET THINKING

When you hear the word 'pedigree', what do you think of? For many, it means dogs or cats that have papers showing the parents of the offspring to prove a 'pure line'. Pedigrees can also be applied to humans, as they display the lineage of inheritance over many generations. How many generations do you know of in your family tree?

Inheritance in families

pedigree chart
a diagram showing patterns of inheritance over generations; also called a family tree

In Module 11.11 you learned how Punnett squares can be used to predict the probability of genotypes and phenotypes from parent crosses. A **pedigree chart**, also known as a family tree, is a visual representation of the individuals in a family. It can be used to determine the frequency of a trait and the likelihood of inheriting that trait in subsequent generations.

How to draw a pedigree chart

There are some conventions you need to follow when drawing and interpreting pedigree charts. Figure 11.12.1 shows the symbols commonly used.

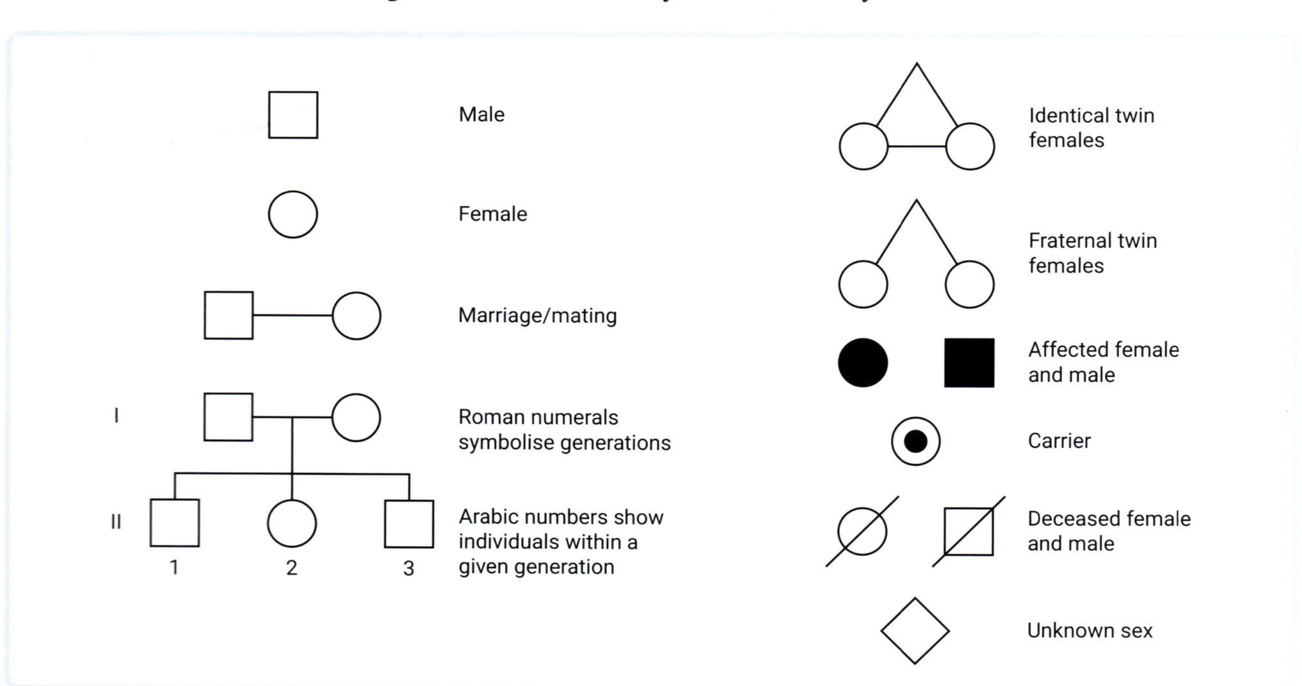

▲ **FIGURE 11.12.1** Common symbols used in drawing and interpreting pedigree charts

How to interpret pedigree charts

Following these conventions allows **geneticists** to interpret pedigrees and predict genotypes and phenotypes of individuals. This information can be particularly important if there is a history of a genetic condition in your family and you are considering having your own children. Analysing your pedigree can help you to better understand the chances of passing on a genetic condition.

geneticists scientists who study genetics and inheritance

KIKE CALVO/Alamy Stock Photo

▲ **FIGURE 11.12.2** A family with one child affected with albinism

Albinism is a genetic condition that makes a person unable to produce the protein that makes melanin. Melanin is a pigment that gives skin, eyes and hair their colour. Most people with albinism have pale skin and eye conditions and are sensitive to sunlight (Figure 11.12.2).

Consider the pedigree shown in Figure 11.12.3, which demonstrates the inheritance of albinism in a family. Each row (labelled I, II etc.) represents a different generation.

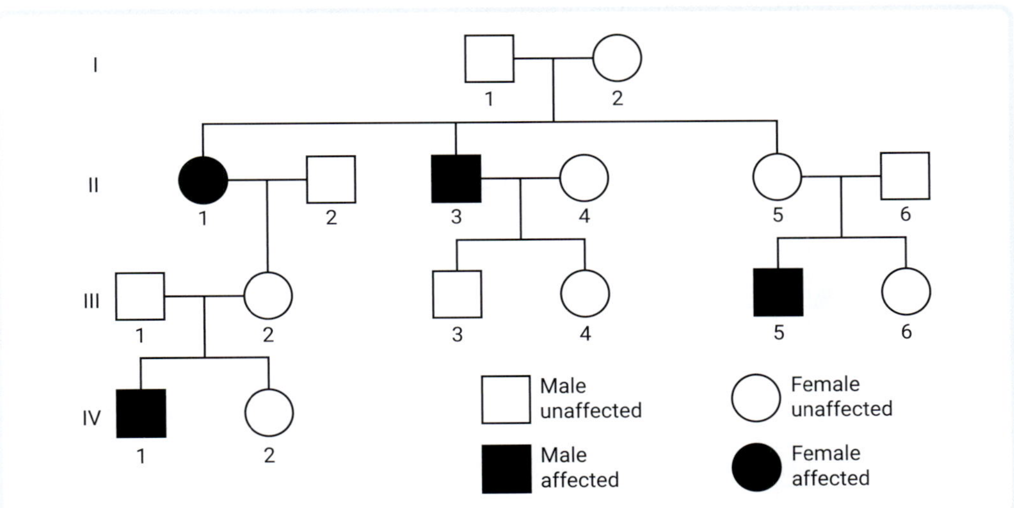

▲ **FIGURE 11.12.3** A pedigree showing the occurrence of albinism in four generations of a family

Here are some facts we can pull from this pedigree by simple observation.

- Generation II individuals 1 and 3, generation III individual 5 and generation IV individual 1 are all affected with albinism.
- Generation I individuals 1 and 2 had three children, two females and a male.
- Generation III individuals 1 and 2 had two children. The first-born son has albinism and the second-born daughter does not.

Determining mode of inheritance

A quick look at the pedigree in Figure 11.12.3 reveals some clues about the **mode of inheritance** of albinism. Albinism is not present in every generation (it is absent in generation I). This is because the trait is masked in some individuals. This shows us that albinism is a recessive trait.

mode of inheritance the manner in which a genetic trait or disorder is passed from one generation to the next

Albinism has affected males and females in this pedigree. This indicates that the gene for albinism is carried on an autosome, rather than a sex chromosome. We call this **autosomal** inheritance.

autosomal
the inheritance of genes on autosomes

Determining genotype

The next step is to assign genotypes to the individuals of the pedigree.

1 Set a key. For example, A = normal skin pigmentation and a = albinism. The individuals that are shaded on the pedigree have albinism. As albinism is a recessive condition, their genotype must be homozygous recessive, or aa.

2 Work out the parents of each affected individual. As they do not have albinism, they must possess one dominant allele (A). However, because they have a child with albinism both parents must have the recessive allele in their genotype (a). This makes them heterozygotes, or Aa.

3 Finally, there are some individuals in the pedigree chart whose genotype we cannot be sure about; for example, because they have married into the family or have not produced offspring. These individuals are generation II 4, generation III 6 and generation IV 2. They do not have albinism, so must possess at least one dominant allele (A), but there is not enough information to determine if they are heterozygotes (Aa) or homozygote dominant (AA). As you are unsure, you need to annotate these examples with both possibilities, as shown in Figure 11.12.4.

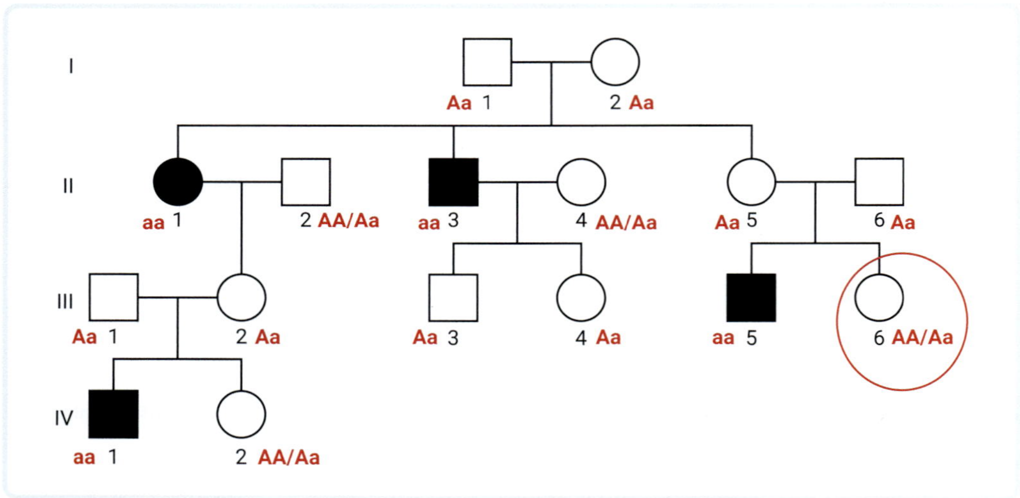

▲ **FIGURE 11.12.4** An albinism pedigree chart with completed genotypes. Individuals with two possible genotypes are annotated (for example, see Individual III 6).

Pedigree charts and autosomal dominant traits

genetic disorder
disease symptoms produced when a DNA sequence is different from normal

Pedigree charts can also be used to show autosomal dominant traits. Consider the pedigree showing Huntington's disease in Figure 11.12.5. Huntington's disease is a rare **genetic disorder** that results in the progressive breakdown of neurons in the brain, affecting a person's cognitive, social and physical abilities.

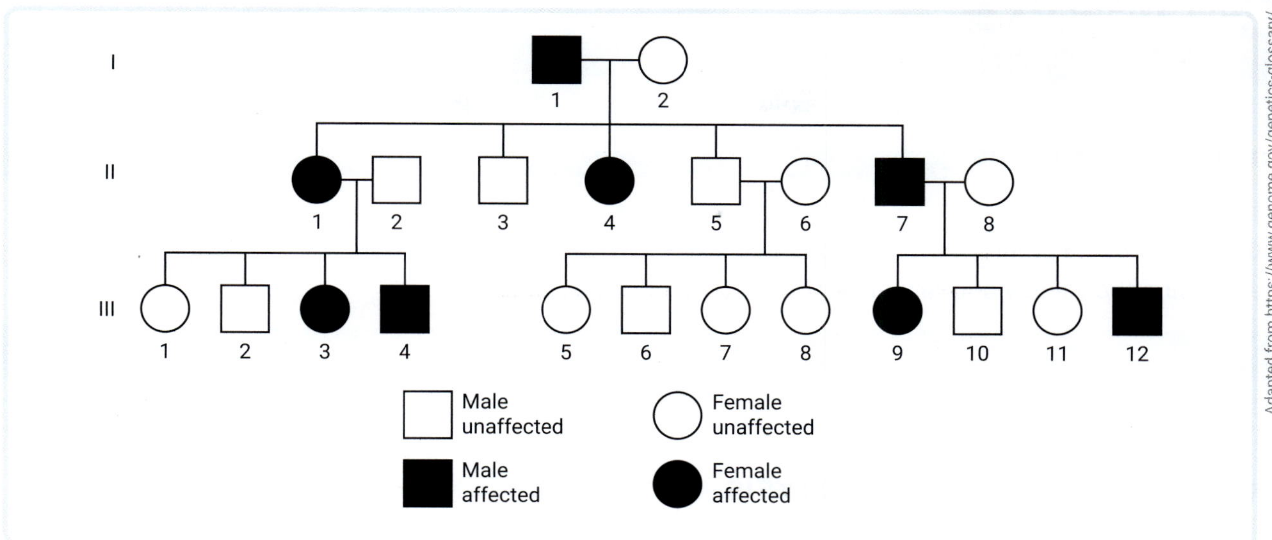

▲ **FIGURE 11.12.5** A pedigree showing the inheritance of Huntington's disease

In this pedigree, you can observe that Huntington's disease appears in every generation, and that every affected offspring has at least one affected parent. This is due to the dominant allele that is expressed in both the homozygous dominant genotype and the heterozygous genotype. Equal numbers of males and females are affected. The mode of inheritance for Huntington's disease is autosomal dominant.

11.12 LEARNING CHECK

1. **Draw** the correct symbols for:
   a. a male without the trait.
   b. a female with the trait.
   c. a deceased male with the trait.
   d. identical twins.
2. Consider the pedigree of a family with near-sightedness shown in Figure 11.12.6. Near-sightedness is a recessive condition. Use the letters N for normal sight and n for near-sightedness to **annotate** each individual's genotype.
   a. How many generations are shown in this pedigree?
   b. How many children did individuals 1 and 2 have?
   c. How many sets of partners/marriages are shown in this pedigree?
   d. What is the sex and genotype of individual 2?
   e. What are Jane's possible genotypes?
   f. What is the probability that individuals 5 and 6 will have a child that is near-sighted?
   g. What is Jane's relationship to individual 1?

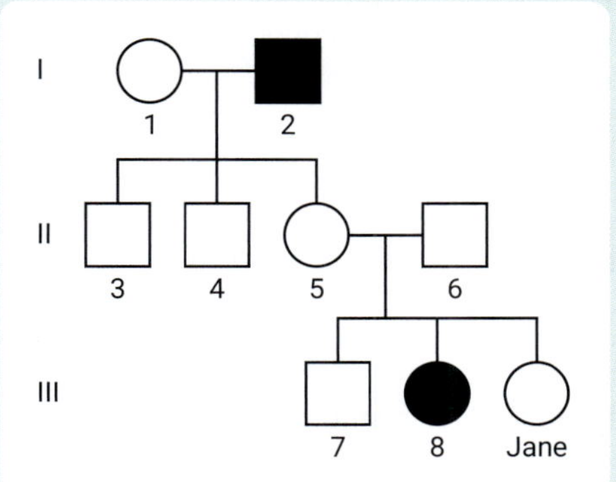

▲ **FIGURE 11.12.6** A pedigree chart showing near-sightedness

WORKING SCIENTIFICALLY

11.13 Analysis of gene frequency

SCIENCE SKILLS IN FOCUS

IN THIS MODULE, YOU WILL FOCUS ON LEARNING AND IMPROVING THESE SKILLS:

- using a model to predict genotypes and phenotypes
- assessing the validity and reliability of the collected data.

PHENYLTHIOCARBAMIDE ANALYSIS

Phenylthiocarbamide, or PTC, is a natural chemical found in many vegetables, such as kale and broccoli. PTC tastes bitter to some people but is tasteless to others. The ability to taste PTC is a dominant trait.

AIM

To measure the frequency of the PTC gene in a population

Safety

Wear appropriate personal protective equipment. Follow the laboratory guidelines for the disposal of PTC papers. Ensure you are aware of any food allergies. PTC is safe when eaten in small amounts; do not exceed one strip. Wash your hands thoroughly before and after this activity.

MATERIALS AND EQUIPMENT

- ☑ tap water
- ☑ PTC papers
- ☑ lab coat, gloves

PROCEDURE

1 Drink some plain tap water, then place a strip of PTC paper on your tongue and let it sit for a few minutes.

2 Note any taste or sensation that you feel.

▼ **TABLE 11.13.1** PTC sensitivity

Bitterness score	PTC sensitivity	Genotype
2 or less	Low	pp
3 or 4	Mid-level	Pp
5 or more	High	PP

3 On a scale of 1–9, record how bitter the PTC tastes. Use Table 11.13.1 to determine the PTC sensitivity category you belong to.

ANALYSIS

The PTC gene has two common alleles: a tasting allele and a non-tasting allele.

1 Using the letters P/p, assign genotypes to your group members based on their results in the taste-test.

2 Collate the class data of genotypes in a table like Table 11.13.2. Remember to also **identify** the sex of the individual if appropriate.

▼ **TABLE 11.13.2** Sample genotype frequency table

	Male	Female	Prefer not to say
PP			
Pp			
pp			

3 Use class data to **calculate** each genotype's frequency.

4 Does the class data reflect Mendel's ratios for single gene inheritance? If so, **explain** what this implies about the mode of inheritance.

5 Conduct a further study by asking your classmates if they like eating broccoli, pepper and cabbage (which contain PTC). Use the data collected to **determine** if your class has food preferences that correlate with having the PTC tasting allele.

CONCLUSION

Write a conclusion that refers to your results and the PTC gene's inheritance pattern.

9780170491785

11.14 Cancer and genetic disorders

BY THE END OF THIS MODULE, YOU WILL BE ABLE TO:

- ✓ define and provide examples of cancer and genetic disorders
- ✓ predict whether a genetic disorder is dominant or recessive based on a pedigree chart
- ✓ investigate the incidences of genetic disorders in families by using pedigree charts
- ✓ explain the role of DNA in cancer and genetic disorders.

GET THINKING

Nearly half of all Australian men and women will be diagnosed with cancer by the age of 85. Cancer is a leading cause of death in Australia, with almost 50000 deaths caused by cancer in 2021. These statistics are taken from the Cancer Council website. What are some positive steps you can take to reduce your risk of cancer?

Video activity
Cancer and genetics

Cancer is uncontrolled cell division

apoptosis
programmed cell death

cancer
uncontrolled cell division resulting in a growth or tumour

DNA is an incredible molecule! It stores all the information needed to build the proteins in a cell, produces the enzymes needed for normal cellular function, and controls your height and hair colour and even your blood type. But what happens when the code in DNA produces a protein that can be damaging to the health of an individual?

In Module 11.6 you learned about mitosis, the cell division that is essential for the growth, replacement and repair of somatic cells. This important process is controlled by genes that help regulate the cell cycle. This ensures that the cell's DNA is copied accurately, that any errors in the DNA are repaired and that each daughter cell receives a full set of chromosomes. The cell cycle has checkpoints during interphase that allow certain genes to check for errors throughout the process. If a cell has a DNA error, it means it can't be repaired. This can lead to programmed cell death, known as **apoptosis** (Figure 11.14.1). This removes damaged cells from the body.

However, if the body is unable to regulate the cell cycle and the cell division continues unchecked, it can lead to diseases such as **cancer**. Cancer can be defined as uncontrolled cell division. There are many different types of cancer.

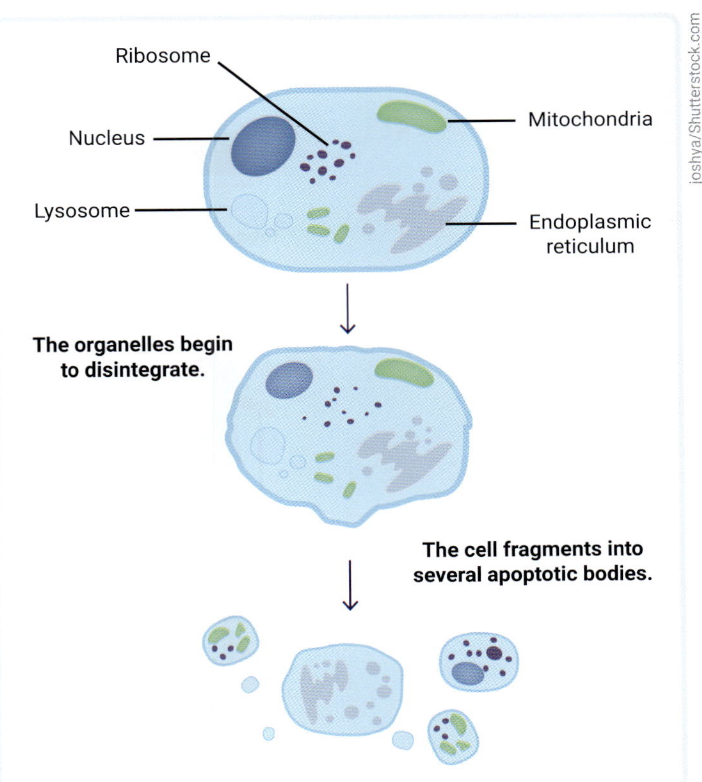

▲ **FIGURE 11.14.1** A cell undergoing apoptosis

Genetic link to cancer

Sometimes faulty genes can increase the risk of getting a specific type of cancer. For example, 5–10 per cent of breast cancers are due to faulty *BRCA1* and *BRCA2* genes. The normal versions of these genes are responsible for repairing cell damage and for keeping breast and ovarian cells growing normally. However, if there is a **mutation** present in these genes, the risk of breast and ovarian cancers increases. These faulty genes can be inherited.

mutation
a spontaneous and permanent change to a DNA sequence

Other cancers that can have a strong family link include melanoma, colon, pancreatic, uterine and prostate cancers. If you have a family history of cancers, your doctor might recommend you undergo genetic testing to see if you have a mutated gene.

Other causes of cancer

carcinogen
an agent that increases the likelihood of developing cancer

Cancers can also be caused by exposure to **carcinogens**. These are cancer-causing agents like UV radiation, chemicals in cigarettes and X-rays. Carcinogens can damage DNA or cause the cell cycle to continue unregulated, leading to cancers. Many studies have shown that exposure to carcinogens can increase the likelihood of developing cancers by up to 70 per cent. Things like smoking, drinking alcohol and leading a sedentary lifestyle can increase the chance of developing cancer, as can be seen in Figure 11.14.2.

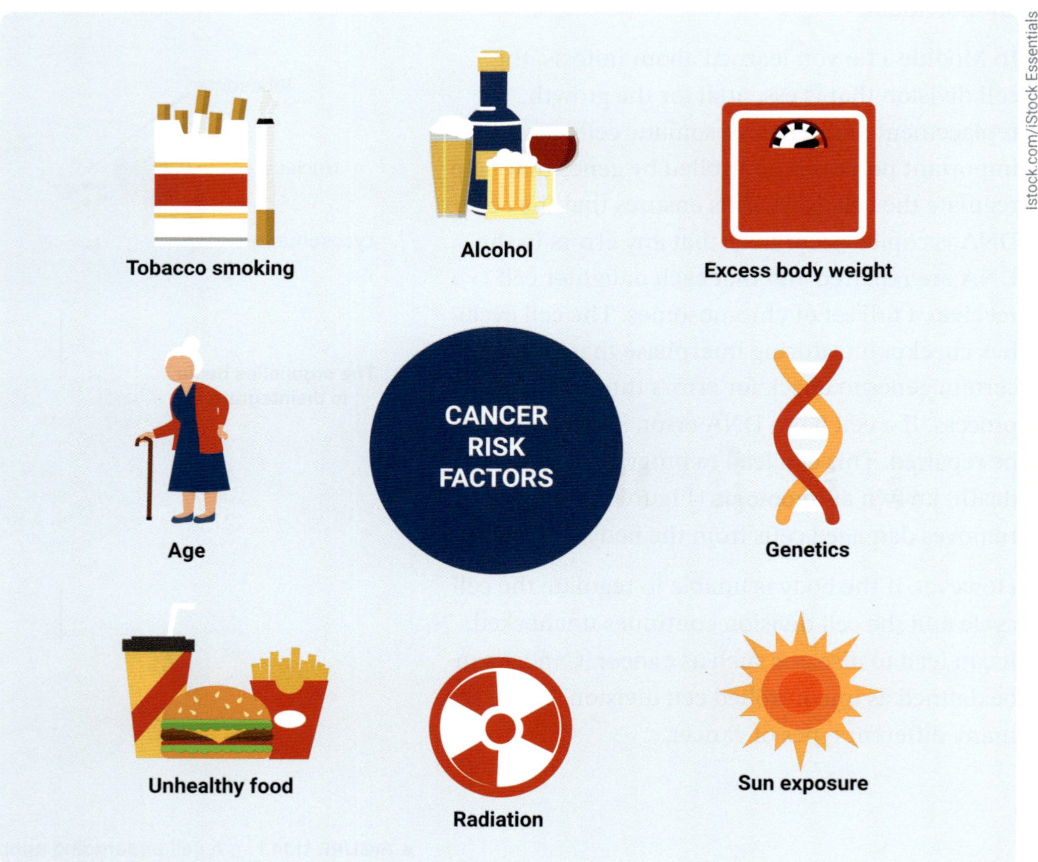

▲ **FIGURE 11.14.2** Factors that can increase the risk of cancer

Mistakes in the duplication and division of chromosomes lead to mutations. A mutation can either be inherited or spontaneously appear in an individual. Mutations can be passed on to new cells produced from the abnormal cell. While some mutations can be beneficial and important in the process of evolution, many mutations result in conditions that are not favourable to the individual or the population. Mutations will be discussed in further detail in Chapter 12.

11.14

Genetic disorders

Genetic disorders arise when the DNA sequences are changed, resulting in the sequence producing malfunctioning proteins, or no proteins at all. Genetic disorders can be caused by:

- a mutation in one gene or many genes
- damage to chromosomes
- abnormal chromosome number.

Mutations can also be influenced by environmental factors, including exposure to harmful chemicals or radiation.

Examples of genetic disorders caused by a single gene include cystic fibrosis, phenylketonuria (PKU), albinism and sickle-cell anaemia. Examples of genetic disorders at the chromosome level include Klinefelter syndrome, Turner syndrome and Down syndrome (trisomy 21) (Figure 11.14.3).

Zuzanae/Shutterstock.com

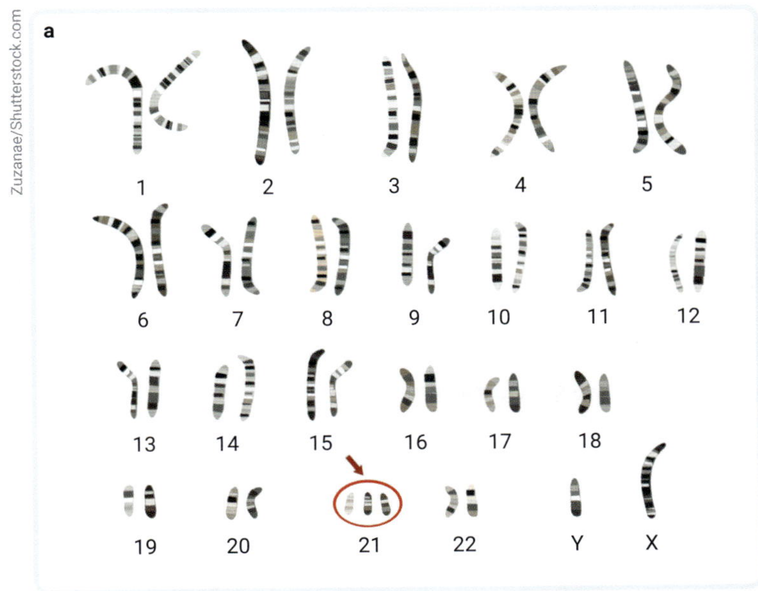

Monkey Business Images/Shutterstock.com

▲ **FIGURE 11.14.3** **(a)** A karyotype of person with Down syndrome, with three copies of chromosome 21 (trisomy 21); **(b)** A boy with Down syndrome and his family

11.14 LEARNING CHECK

1 In your workbook, copy and match each term to the correct definition.

Term	Definition
Cancer	• An allele that is masked in the heterozygote condition
Mutation	• Programmed cell death
Recessive	• A spontaneous and permanent change to a DNA sequence
Apoptosis	• Uncontrolled cell division

2 Consider the pedigree in Figure 11.14.4, showing the incidence of PKU in a family. PKU is a rare genetic disorder that causes an amino acid called phenylalanine to build up in the body. If left untreated, this can lead to brain damage, intellectual disabilities, behavioural symptoms or seizures.

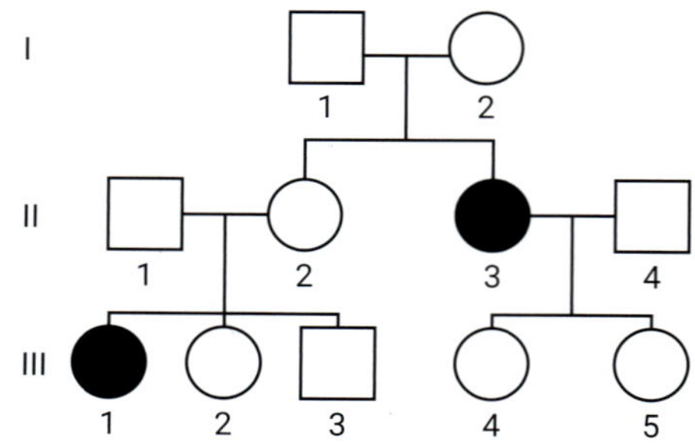

▲ **FIGURE 11.14.4** A pedigree showing the incidence of PKU in a family

a Using evidence from the pedigree, **determine** the mode of inheritance for PKU.

b Complete the genotypes for the following individuals:

i I-1 **ii** II-3 **iii** II-2 **iv** III-5

c **Calculate** the probability of II-3 and 4 having a child with PKU. To calculate the probability of the gene you will need to **create** a Punnett square and see the ratios of the genotype.

d What is the probability of III-2 being a heterozygote for this condition?

3 Use the internet to **research** the following cancers. Include the causes (genetic or environmental), the likelihood of developing the cancer, ways to prevent the cancer occurring and any treatments available. Present your findings in a table.

- Breast cancer
- Melanoma
- Prostate cancer
- Cervical cancer
- Bowel cancer

9780170491785

11.15 Biotechnology

BY THE END OF THIS MODULE, YOU WILL BE ABLE TO:

✓ define and explain the processes of polymerase chain reactions and gel electrophoresis
✓ explain how polymerase chain reactions and gel electrophoresis produce DNA profiles.

GET THINKING

Crime shows often show an image like the one seen in Figure 11.15.1 to help convict a criminal.

Have you ever wondered how this process works or what the bands represent? DNA profiling is an important component of biotechnology.

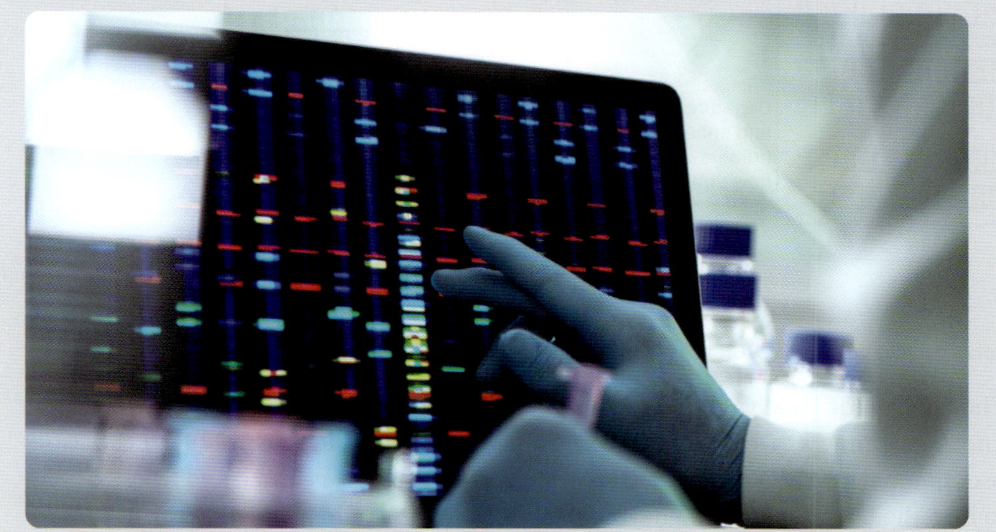

TEK IMAGE/SCIENCE PHOTO LIBRARY/Getty Images

▲ **FIGURE 11.15.1** A DNA profiling image showing matches (red lines) across different individuals

Video activity
Gel electrophoresis

What is biotechnology?

Biotechnology is a branch of science that uses living organisms, biological systems or DNA to create, repair or modify products. Although it might sound like modern technology, humans have used biotechnology for centuries, such as using yeast to make bread, cheese and alcohol. Since the mid-1900s, advances in DNA knowledge have resulted in manipulating DNA and genes to treat and even cure some diseases.

biotechnology
a branch of science that manipulates living organisms, systems or DNA to produce products that are useful to humans

One of the most important breakthroughs in biotechnology was due to the Human Genome Project, an international research project from 1990 to 2003 that mapped the human genome. The project's goal was to identify the structure of the entire human genome, including its 3 billion base pairs and approximately 22 000 genes. Scientists' knowledge of the human genome will help them prevent and treat many of the illnesses resulting from genetic errors.

DATA SCIENCE

Learn more about large datasets in **Module 2.5**.

Polymerase chain reaction

Polymerase chain reaction, also known as PCR, is a technique used to multiply fragments of DNA. This is known as amplification and can generate millions of copies of DNA from a small original amount, as seen in Figure 11.15.2.

polymerase chain reaction (PCR)
a technology used to make many copies of DNA

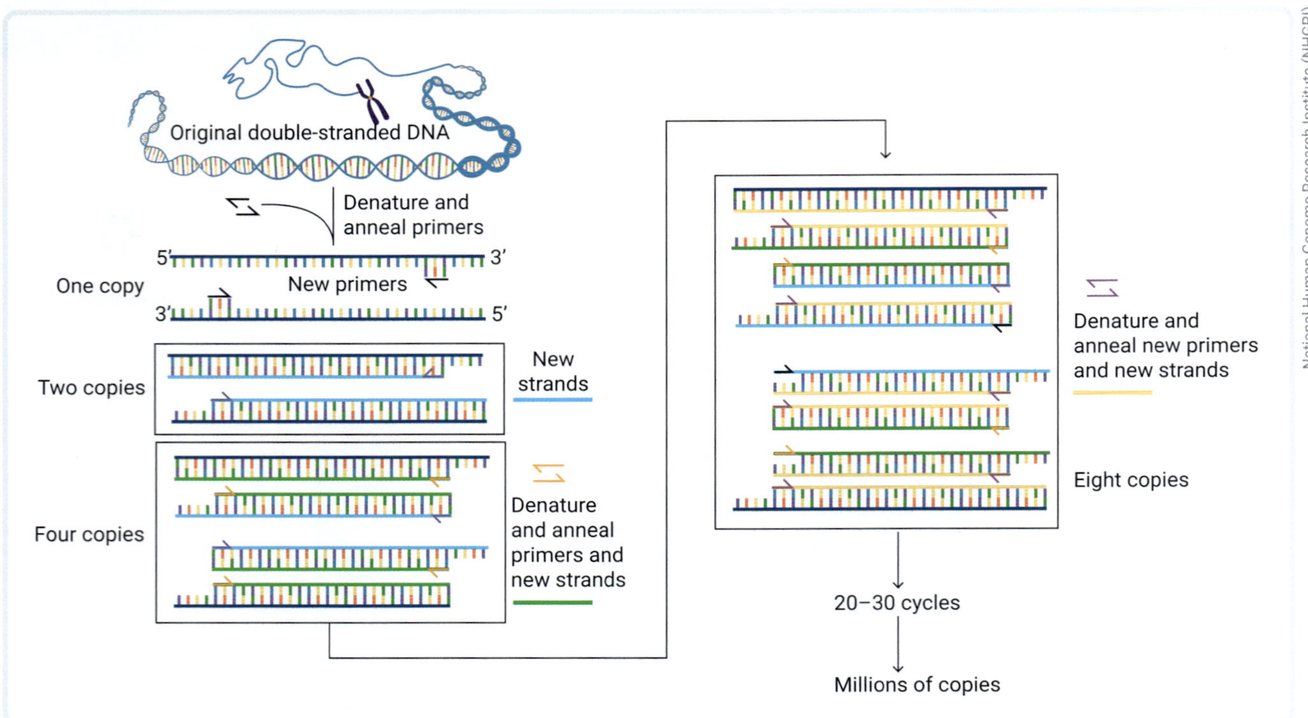

▲ **FIGURE 11.15.2** The repeated cycling process of PCR to amplify a fragment of DNA

thermal cycling the repeated process of heating and cooling required in PCR

master mix a premixed solution used in PCR techniques to make copies of DNA

The method, originally developed in 1983, involves using short DNA sequences known as primers to select the area of the genome needed to be amplified. PCR also uses **thermal cycling**, where the temperature of the process is raised and lowered to help the DNA replication enzyme (DNA polymerase) add nucleotides in a complementary sequence.

PCR can be simplified into three main steps. It requires a set of ingredients called a **master mix** that is used each time. The master mix must contain DNA primers, DNA nucleotide bases, DNA polymerase (specifically *Taq* polymerase), a buffer and the original sequence of DNA needed to be copied.

The three main steps are:

1 Denaturing: The original DNA strand, extracted from a cell, is heated to 96°C to separate the double strands into two single strands.

2 Annealing: The temperature is lowered to about 65°C, and the DNA primers are attached to either side of the DNA to be copied.

3 Extension: The temperature is raised to 72°C for *Taq* polymerase to synthesise two new strands of DNA based on the exposed templates.

These steps result in two strands being produced from the original. The new fragments are then used as templates for the next cycle to produce four strands. Each cycle doubles the number of DNA fragments produced (Figure 11.15.3). After 32 cycles of PCR, more than 4 million copies of the original strand can be produced. This can be calculated using 2^n, where n is the number of cycles of PCR.

Heating the strands to 96°C breaks the hydrogen bonds between the nitrogen-rich bases and results in two strands with exposed bases to act as templates for synthesis.

DNA primers are short sequences of DNA (20–30 bases) designed to be complementary to short sequences of DNA on either side of the sequence to be copied. They serve as a point for *Taq* polymerase to commence synthesis.

Taq polymerase is a specific DNA polymerase extracted from a bacterium called *Thermus aquaticus*. This is a heat-stable polymerase, able to withstand the high temperatures of denaturing, which is essential in thermal cycling.

It takes approximately 1 minute to copy 1000 DNA bases.

Cycle 1 Cycle 2 Cycle 3

Genome Research Limited

▲ **FIGURE 11.15.3** Three complete cycles of PCR, resulting in eight copies of DNA from the original fragment

Gel electrophoresis

Electrophoresis, or **gel electrophoresis**, is a technique used in the laboratory to separate DNA or RNA strands of different lengths. This is useful because a **DNA profile** can be produced to visualise the DNA fragments. DNA profiles are used in things such as forensics cases and paternity testing, to check the success of a PCR reaction, to examine evolutionary relationships between organisms or to test for genes linked to a particular disease.

Electrophoresis uses a porous gel (usually made of agarose), a buffer solution and an electric current to separate fragments of genetic material according to size. The buffer solution contains salts and conducts an electrical current. An example of electrophoresis equipment can be seen in Figure 11.15.4.

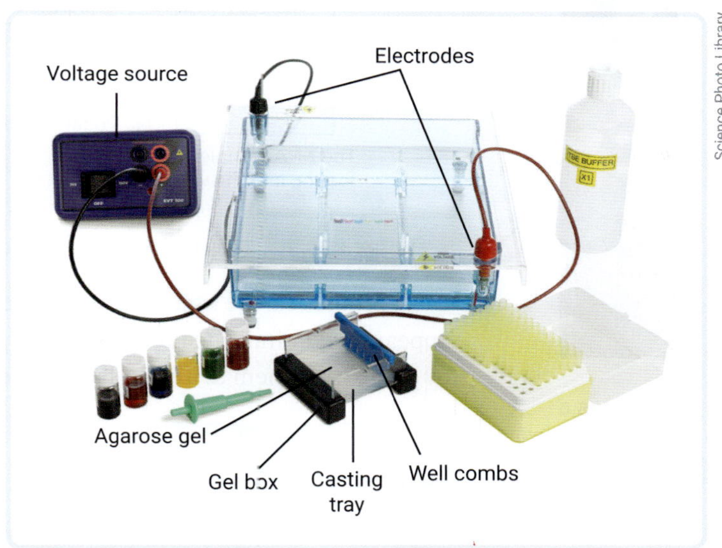

Science Photo Library

▲ **FIGURE 11.15.4** Labelled electrophoresis equipment

gel electrophoresis
a technology used to separate DNA by size to produce a DNA profile

DNA profile
an image used to determine an individual's DNA characteristics

How does it work?

DNA samples are collected, stained (or have radioactive tags attached) and pipetted into the agarose gel wells. The buffer solution is added, and electrodes are attached to a power pack that, when turned on, generates an electrical field. DNA has a negative charge, so the fragments are attracted to the positive end of the gel. The agarose gel provides resistance to the DNA, and because smaller fragments encounter less resistance, they move further through the gel, with larger fragments moving a shorter distance. The result is a series of bands containing DNA molecules of a particular size. These can be compared against each other, or a ladder can be created by known lengths of DNA pieces. This process can be seen in Figure 11.15.5.

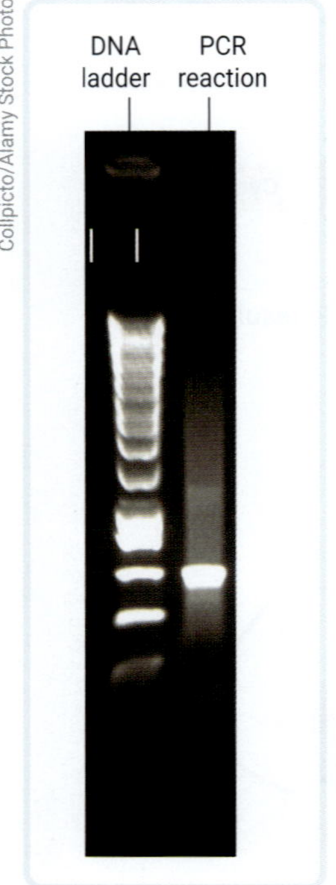

▲ **FIGURE 11.15.6** A DNA profile showing DNA fragments of 700 base pairs in length

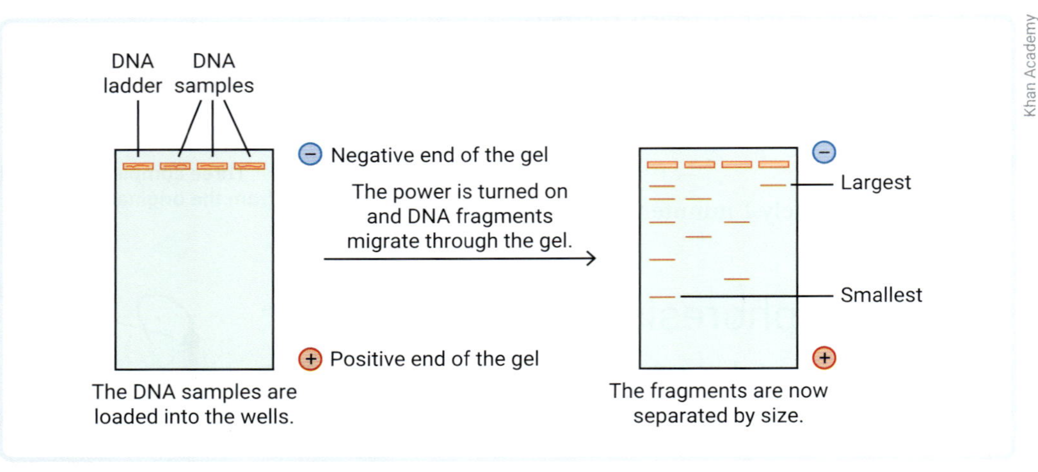

▲ **FIGURE 11.15.5** The steps of electrophoresis and the resulting DNA profile

DNA profiles can be used to check if the PCR reaction has amplified the DNA. Bands that appear thicker indicate more fragments of DNA of that size. The size can then be estimated by comparing them to a DNA ladder added as a reference guide (Figure 11.15.6).

11.15 LEARNING CHECK

1 **Define** PCR.
2 **Calculate** the number of copies of DNA made after 30 cycles of PCR. Remember 2^n, where n equals the number of cycles of PCR.
3 Why is *Taq* polymerase and not a different DNA polymerase used in PCR?
4 **Explain** how, in electrophoresis, the electric current and agarose gel separate DNA of different sizes.
5 DNA has an overall negative charge. **Explain** the significance of this in electrophoresis.

11.16 Manipulating DNA

BY THE END OF THIS MODULE, YOU WILL BE ABLE TO:

- ✓ define and explain the processes of gene therapy and recombinant DNA
- ✓ explain how biotechnology can be used to treat diseases.

GET THINKING

People with type 1 diabetes rely on daily insulin injections to help them regulate their blood sugar levels. Prior to the development of recombinant DNA in 1978, the insulin they were given was extracted from the pancreases of cattle and pigs. Can you identify any problems with this source of insulin? Another important hormone, growth hormone, used to be extracted from cadavers (dead bodies) and given to children with stunted growth. Thankfully, due to the technology of recombinant DNA, we can now produce these hormones in a laboratory.

gene therapy
the introduction of functional genes into cells to replace defective or missing genes to treat genetic disorders

Gene therapy

Gene therapy has been used to varying levels of success since 1990. It can be defined as the introduction of functional genes into cells to replace defective or missing genes to treat genetic conditions. It is most effective when used as a treatment for people with single-gene conditions such as cystic fibrosis. A vector, usually a virus, can be used to transfer a healthy gene into a patient's cells. Once the gene has been transferred, the person can make the product of that gene. This process is demonstrated in Figure 11.16.1.

'Laura Olivares Boldú / Wellcome Connecting Science'

▲ **FIGURE 11.16.1** The introduction of a functional gene into the nucleus of a cell using a viral vector

Recombinant DNA

Recombinant DNA technology was first developed in the 1970s when scientists used special enzymes to cut DNA from different organisms at specific sites and used different enzymes to glue them back together. Recombinant DNA has now been developed to join DNA molecules from two different species, which is then inserted into a host to create a **transgenic organism**. The new combination of DNA allows the transgenic organism to create products that are either useful to the organism or can be harvested and used in medicine. Insulin was first produced using this technology in 1978, and it is now commercially produced through recombinant DNA and bacteria. This has significantly improved the quantity and quality of insulin available to people with type 1 diabetes. Bacteria are used because they contain **plasmids** and can reproduce quickly, given the right conditions in a laboratory.

recombinant DNA
DNA that has been manipulated by combining DNA from other species

transgenic organism
an organism that contains DNA sequences from an unrelated organism that have been artificially introduced

plasmid
a small circular fragment of DNA in the cytoplasm of bacteria

The process of producing recombinant DNA involves a number of steps (Figure 11.16.2).

1. The gene of interest is located and isolated using special enzymes called **restriction enzymes**.
2. A bacterial plasmid is isolated and cut with the same restriction enzyme.
3. The restriction enzymes cut the DNA with a **staggered cut** and produce sticky ends.
4. The desired gene is introduced into the plasmid, where DNA ligase will 'glue' them together.
5. The transgenic plasmid is introduced into another bacterium, which allows the bacterium to create the modified gene. This can then be collected, purified and distributed.

restriction enzymes special enzymes that cut DNA at specific recognition sites; isolated from bacteria

staggered cut a cut from a restriction enzyme that results in 'sticky ends', overhanging unpaired nucleotides

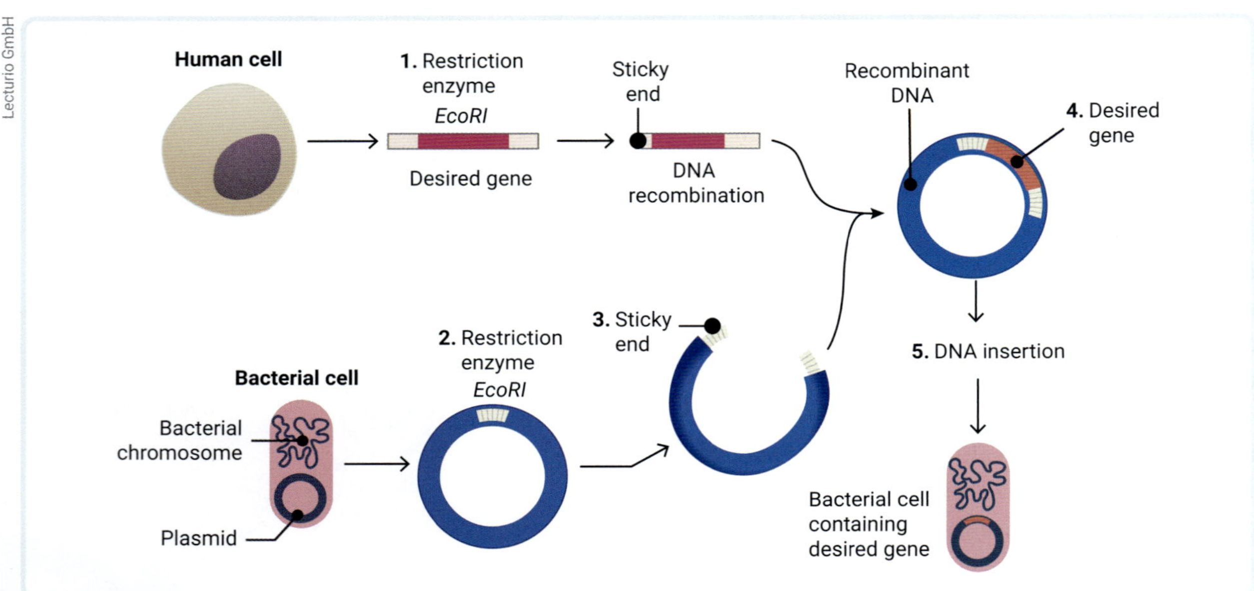

▲ **FIGURE 11.16.2** The steps of recombining DNA

11.16 LEARNING CHECK

1. **Define**:
   a. gene therapy.
   b. recombinant DNA.
   c. a transgenic organism.
2. Prior to recombinant DNA technology, human growth hormone was collected from the pituitary glands of human cadavers.
   a. **Determine** what problems might arise from this process of extraction.
   b. **Explain** why bacteria are useful in recombinant DNA technology.
3. **Research** cystic fibrosis and **identify** how gene therapy can be used as a possible treatment for this disease.

9780170491785

11.17 Kinship structures

IN THIS MODULE, YOU WILL:

✓ explore the significance of Aboriginal and Torres Strait Islander Peoples' kinship and family structures.

Kinship is fundamental to Aboriginal and Torres Strait Islander Peoples' societal structure. Aboriginal and Torres Strait Islander Peoples' cultures have complex societal structures. These determine how people relate to each other socially, ceremonially and spiritually. An important aspect of these structures is they inform who is allowed to marry whom. These structures have existed long before British colonisation. In part, they function to prevent relationships between people who are too closely related. Offspring born to closely related people may inherit detrimental traits. Aboriginal and Torres Strait Islander Peoples' kinship systems ensure that such relationships do not occur.

Kinship structures include:

- moiety: two halves of society divided along matrilineal (mother) and patrilineal (father) lines of descent
- totem: the natural resources a person is accountable for, ensuring their protection for passing to the next generation
- skin name: a person's bloodline and connections across generations.

Skin names are the element of kinship systems that indicates a person's bloodline. It also communicates information about how generations are linked and how they should interact. Husbands and wives do not share a skin name, and children do not have the skin name of their parents. The system is sequential in that the name is given based on a person's position in the cycle. Males and females of the same skin name are also distinguished by variations to the skin name.

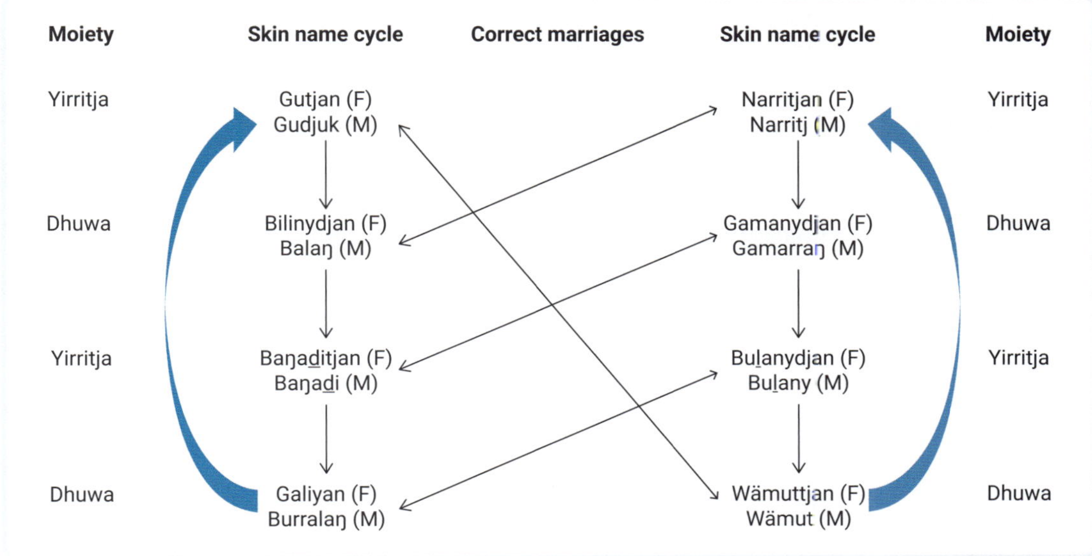

▲ **FIGURE 11.17.1** The skin name cycles of the Gurrutu system. The skin name cycles are only one component of Gurrutu system.

The Gurrutu system is a complex societal structure of kinship used by the Yolngu People of north-east Arnhem Land. In the Gurrutu system, the two moieties are *Yirritja* and *Dhuwa*. A person's moiety is always opposite to their mother. Each moiety has four skin groups, and each skin group has a male (M) and female (F) version of the name, as seen in Figure 11.17.1.

The arrows show the sequential cycles of the skin names. For example, children of a person with the skin name *Gamanydjan* would have the skin name *Bulanydjan/Bulany* and their children would take the skin name next in the cycle. The moiety *(Yirritja* [Y]/ *Dhuwa* [D]) changes with each generation.

Correct marriages connect people of opposite moiety. These systems are underpinned by an understanding of the biological principles of heredity and the transmission of heritable characteristics. The diagonal arrows in Figure 11.17.1 show the correct marriages between the two cycles of skin names.

ACTIVITY

Mapping the Gurrutu system of kinship

Procedure

1 Take 16 icy pole sticks.
   a Write a Gurrutu system skin name on each icy pole stick.
   b Write a Y or a D to indicate the moiety.
   c Colour the icy pole sticks with male skin names one colour and the icy pole sticks with female skin names another colour.
2 Use these icy pole sticks to model the generations of the Gurrutu system.
   a Begin by selecting a skin name from the cycle. Identify the correct marriage of this person.
   b Then identify the skin name of their child. Choose either the male or the female version of the skin name.
3 Continue this process to model the Gurrutu skin name system. Note that in the Gurrutu system, the children take the skin name next in the cycle following matrilineal lines (i.e. the child takes the name next in the cycle from the mother's skin name).

Analysis

1 What does this show you about the purpose of skin name structure as a societal organisation system?
2 How do such systems prevent the transmission of inheritable harmful traits?
3 How does this system relate to Mendelian genetics?

11.18 Genetically modified food

BY THE END OF THIS MODULE, YOU WILL BE ABLE TO:

- ✓ investigate key factors that contribute to scientific knowledge and practices being adopted more broadly by society
- ✓ understand why agricultural practices have changed to include the widespread use of genetically engineered crops.

Humans have been altering the genetics of organisms for over 30 000 years. Although our ancestors had no concept of genetics, they were able to influence the DNA of organisms through **artificial selection**. You will learn about this in Chapter 12. While artificial selection is not what we typically consider **genetic engineering** today, it is the precursor to modern processes.

Genetically modified organisms, or GMOs, refer to a modern process of altering the genetics of organisms.

artificial selection the process whereby humans breed organisms for desired traits

genetic engineering the deliberate modification of an organism's DNA

genetically modified organism (GMO) an organism whose genes have been altered in the laboratory to produce a desired trait

The history of GMOs

In 1973, scientists Herbert Boyer and Stanley Cohen together engineered the first successful genetically engineered organism (GE organism). They developed a method to cut out a gene for antibiotic resistance from one bacterium and paste it into another, making the GE organism resistant to antibiotics. One year later, another pair of scientists, Rudolf Jaenisch and Beatrice Mintz, followed a similar procedure and introduced foreign DNA into mouse embryos.

Despite the incredible possibilities of these new techniques, there was immediate concern about the possible consequences on human health and Earth's ecosystems. By the middle of 1974, a universal prohibition of GE projects was observed, allowing experts time to meet and consider the next steps for the safety of GE experiments. The meeting, known as the Asilomar Conference of 1975, resulted in a set of guidelines with defined safety and containment regulations.

The cooperation seen at the Asilomar Conference gave government bodies around the world confidence to support GE research and launched a new era of modern genetic modification.

GMOs have the green light

In 1980, the US Supreme Court allowed ownership rights over GMOs, giving large companies the incentive to rapidly develop GMO tools that were both useful and profitable. In 1982, the US Food and Drug Administration approved the first human medication produced by a GMO. Bacteria had been genetically engineered to produce human insulin, allowing the hormone to be purified, packaged and prescribed to diabetes patients as the drug Humulin.

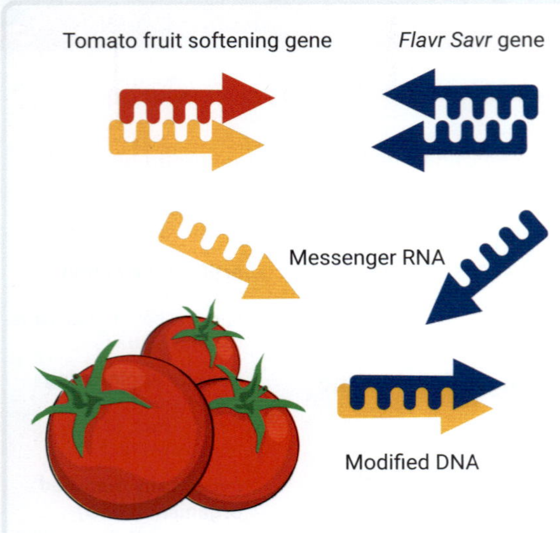

▲ **FIGURE 11.18.1** The Flavr Savr tomato has been genetically engineered to delay rot.

International Rice Research Institute (IRRI) (CC BY 2.0)

▲ **FIGURE 11.18.2** A comparison of normal rice and Golden Rice

The first experiments with food crops that had been genetically modified began in 1987. After 5 years of extensive health and environmental testing, the Flavr Savr tomato was the first food crop to be approved for commercial production. These tomatoes were modified to include a gene that inhibits a natural tomato protein, thereby increasing the firmness and shelf-life of the Flavr Savr variety (Figure 11.18.1).

In 1995, the first pesticide-producing crop was approved, and in rapid succession, herbicide-resistant crops were also engineered. This made it easier for farmers to control unwanted plants in their fields.

Scientists have also genetically engineered crops to increase their nutritional value. In 2000, Golden Rice was developed to combat vitamin A deficiency, which is estimated to kill over 500 000 people each year (Figure 11.18.2).

The future of GMOs

A recent review of over 150 studies concluded that genetically modified technology has increased crop yields and farmers' profits over the past 20 years. Today GMOs of soybeans, maize and cotton have been associated with a 22 per cent increase in yield, a 37 per cent decrease in pesticide use and a 68 per cent increase in farmer profits, despite the increase in the cost of GM seeds.

The United Nations has predicted that by 2050 humans will need to produce 60 per cent more food than we currently do to meet the needs of the global population. Innovative approaches, including GMOs, will be required to solve this problem.

11.18 LEARNING CHECK

1 **Define** genetic engineering.
2 **Develop** a timeline to show the development of genetically modified organisms.
3 Using the internet, **research** GMOs that have been developed in the areas of:
   a drought-resistant plants.
   b enhanced growth in animals.
   c disease-resistant crops such as the Rainbow papaya.
4 What do you think the future of GMO technology is going to be? Write down some of your ideas and share your thoughts with your classmates.

11 REVIEW

REMEMBERING

1 **Recall** what DNA stands for.

2 **Explain** why DNA is known as the 'blueprint of life'.

3 **Define**:

a chromosome.

b karyotype.

c autosome.

d gene.

4 The shape of the DNA molecule is known as a double helix.

a **Draw** a diagram to represent this shape.

b **Explain** how *Photo 51* confirmed the shape of DNA.

5 **Name** the four bases found in a DNA molecule and **explain** how and why they pair up.

UNDERSTANDING

6 **Explain** why it is important that DNA replicates during interphase, before the cell divides.

7 **Examine** the figure below, which shows the phases of mitosis.

Then copy and complete the table that follows, explaining what is occurring during each phase.

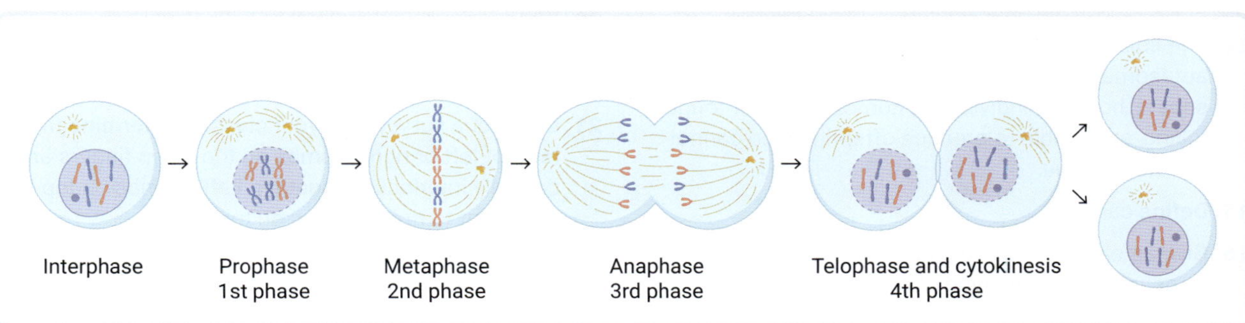

LDarin/Shutterstock.com

Phase	Description
Interphase	
	The chromosomes form, nuclear membrane breaks down and the centrioles replicate.
Metaphase	
	The chromatids are pulled to opposite poles of the cell.
Telophase	
	The cytoplasm divides and two daughter cells are produced.

8 **Describe** the result of meiosis.

9 **Demonstrate**, using a colour-coded and labelled diagram, how crossing over results in genetically different chromatids.

10 **State** the benefits of Aboriginal and Torres Strait Islander Peoples' kinship structures.

11 **Identify** the following statements as true or false. Modify any false statements in your workbook so they read true.

a Homozygous individuals have two different alleles for one gene.

b A dominant allele will mask a recessive allele in the heterozygote genotype.

c The phenotype is the representation of the alleles in an individual.

d Lowercase letters are used to denote the dominant trait in the genotype.

12 **Examine** the karyotype shown in the figure below.

a What sex does this karyotype show?

b There is an abnormal number of chromosomes in this karyotype. **Identify** which chromosome has an abnormal number.

c What is the number of chromosomes found in a normal human body cell?

d What is the number of chromosomes found in a normal human sex cell?

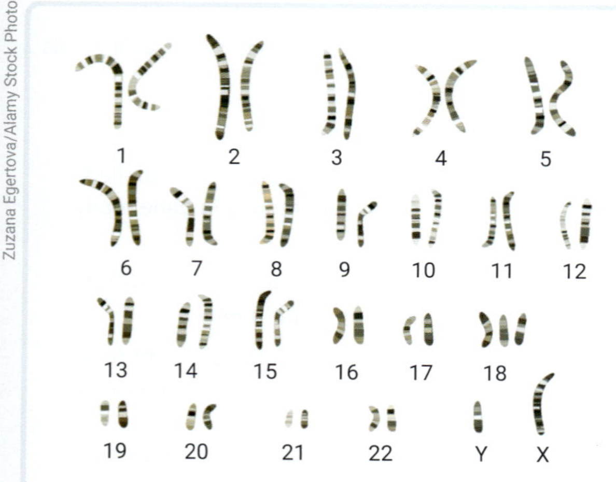

Zuzana Egertova/Alamy Stock Photo

APPLYING

13 Random assortment is the chance alignment of chromosomes at the metaphase plate during metaphase I. If a cell has a diploid (2*n*) number of 58, **calculate** the total number of combinations of chromosome pairs that can occur.

14 The ability to roll your tongue into a U shape is an autosomal dominant trait. **Justify**, using a Punnett square, how two individuals, homozygous for tongue rolling, are unable to produce offspring who cannot roll their tongue.

15 **Define** cancer.

16 Give an example of one type of cancer caused by genetics and one type of cancer caused by environmental factors.

ANALYSING

17 Belle does not have albinism; however, her maternal grandfather did, and her maternal aunt does too. On her father's side, there is no family history of albinism.

a What are the genotypes for Belle's:

i grandfather?

ii mother?

iii father?

b What is the probability of Belle being a heterozygote for albinism? **Justify** your response by showing your working out.

c **Draw** a pedigree of Belle's family, showing the occurrence of albinism. Include all labels and genotypes of every individual mentioned in this question.

d **Classify** the mode of inheritance of albinism.

18 Melanoma is a skin cancer that can develop from exposure to UV rays. **Identify** five steps that you could take to limit your chance of developing skin cancer.

19 During anaphase II, chromosome pair 21 failed to separate into chromatids in one of the cells.

a **Draw** the resulting daughter cells from this meiotic division.

b **Explain** what will occur if these cells are fertilised with another gamete and **name** this genetic condition.

EVALUATING

20 Write a short paragraph that links the terms nucleotide, nitrogenous base, DNA, genes and chromosomes.

21 Conduct a survey with your class to **determine** the ratio of left or right thumbs on top when fingers are casually interlocked together. Left thumbs on top is dominant to right thumbs on top.

a Does the ratio in your classroom reflect one of Mendel's laws?

b Is there a correlation between each person's dominant writing hand and which thumb naturally rests on top?

9780170491785

SCIENCE IN DEPTH STUDY

1 Connect what you've learned

In this unit you've learned about DNA, its role in cellular activities, how it is inherited and how the instructions it holds can influence the characteristics of future generations. Using the key terms from each module, create a mind map to show how the information that you have learned is connected.

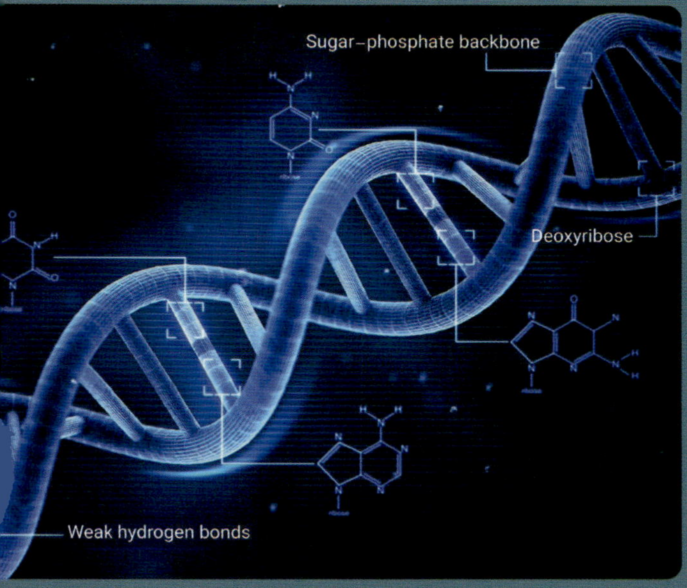

2 Check your thinking

Explain why recessive genetic conditions such as albinism or cystic fibrosis can be unknowingly passed down from generation to generation.

How do these types of conditions compare to autosomal dominant disorders that may cause premature death, such as Huntington's disease?

3 Get into action

Conduct research into a genetic condition of your choice. In your research you will need to determine the:

- mode of inheritance
- frequency of the condition in Australian populations
- signs and symptoms of the disorder
- treatment or prevention of the disorder.

Some suggestions to consider include cystic fibrosis, Huntington's disease, sickle-cell anaemia, melanoma, Duchenne muscular dystrophy, fragile X syndrome, Klinefelter syndrome, neurofibromatosis and von Willebrand disease.

4 Communicate

Create a podcast presenting your research on a genetic condition for a medical centre's website. It should be clear and concise to effectively inform patients about the disease.

12 Evolution

9780170491785

SCIENCE IN DEPTH

TOON WELL/Shutterstock.com

▲ **FIGURE 12.0.1** Hominins in order of evolution, ending in *Homo sapiens*.

It has taken millions of years for humans to evolve into the species *Homo sapiens*. During that time, our ancestors were influenced by many evolutionary mechanisms that have resulted in us. But are we still capable of evolving? *Homo sapiens'* biological and cultural evolution has resulted in our species not feeling the influence of the environment as strongly as we once did; we can treat diseases that would previously have killed us, and we can even manipulate DNA to aid in our survival.

- **What advances in medical technology have been made to extend the life span of humans?**
- **How do you think these technological advances have affected the evolution of *Homo sapiens*?**
- **Will there be a species after *Homo sapiens*? Or are we the 'ultimate species'?**

#12 DIVE INTO SCIENCE!

At the end of this chapter, you can complete Science in Depth Study #12. You can use the information you learn in this chapter to complete the project.

Assessments

- Prior knowledge quiz
- Chapter review questions
- End-of-chapter test
- Depth study: Research project and article

Videos

- Science skills in a minute: Controlling variables **(12.8)**
- Video activities: Charles Darwin **(12.3)**; Natural selection **(12.4)**; Speciation **(12.6)**; Fossil evidence **(12.9)**; Evolution: The evidence **(12.13)**

Science skills resources

- Science skills in practice: Controlling variables **(12.8)**
- Extra science investigations: Modelling natural selection **(12.4)**; Modelling selection pressures **(12.4)**; Modelling fossilisation processes **(12.9)**

Interactive and other resources

- Simulation: Natural selection: **(12.4)**
- Crossword: Evolution **(12.7)**
- Drag and drop: Types of mutation **(12.2)**; Structural evidence of evolution **(12.11)**
- Quizzes: Why is there variation between individuals? **(12.1)**
- Worksheets: Evolution case study **(12.11)**; Comparing homologous proteins **(12.13)**

Nelson MindTap

To access resources above, visit **cengage.com.au/nelsonmindtap**

12.1 Variation between individuals

BY THE END OF THIS MODULE, YOU WILL BE ABLE TO:

- ✓ define variation, mutation, crossing over and random assortment
- ✓ explain how mutations and meiosis contribute to variation.

Quiz
Why is there variation between individuals?

GET THINKING

Imagine walking through a rainforest or snorkelling above a coral reef. How many different organisms can you see? If you take a closer look and only focus on one type of plant or one type of fish, do you notice small differences between them? Perhaps not every leaf is the same shape, or the fish have slightly different markings. These differences are known as variation. Without variation, the process of natural selection cannot occur, and there is a greater risk of extinction. In your workbook, write down your explanation of why variation is important. After reading this module, come back and see if you need to adjust your understanding.

What is variation?

variation
a difference in characteristics due to different genes

species
a group of organisms capable of reproducing under natural conditions to produce fertile offspring

evolution
the gradual change in characteristics of a species over many generations resulting in a new and different species

population
a group of individuals of the same species living in the same place at the same time

Have you ever noticed that even within a group of people of similar age or ethnicity, no one looks the same? Even within a family, siblings with the same parents can look similar or quite different. The reason behind this is **variation**. Variation can be described as the differences seen in the phenotypes that are the expression of genotypes of individuals of the same **species**.

Why is variation so important? The key here lies in understanding the processes of change that result in **evolution**. Imagine if every individual within a species was genetically identical. What would happen to the species if the environment was no longer favourable to individuals with this specific genetic make-up? The species would become extinct.

An example is the small **population** of cheetahs in Africa. The cheetahs show characteristics of inbreeding and are highly vulnerable to infectious diseases carried by domestic cats. Cheetahs struggle to cope with these pressures partly because they lack the genetic variation to respond to them effectively.

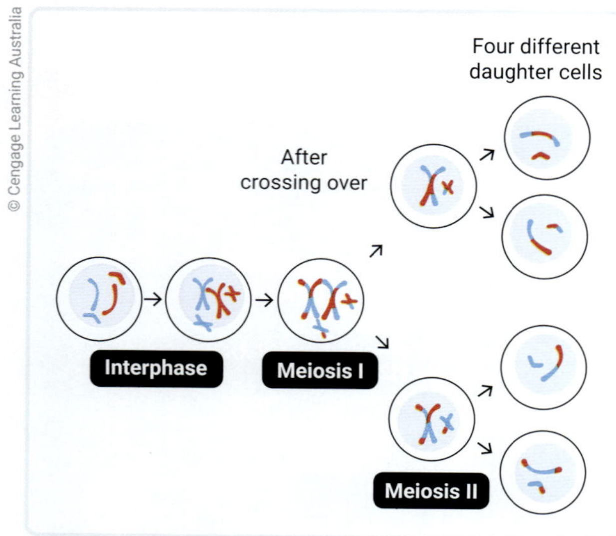

© Cengage Learning Australia

▲ **FIGURE 12.1.1** The result of meiosis is four genetically different haploid daughter cells.

Variation in a species has several sources. Meiosis, fertilisation, mutations and random mating are the main sources of variation within species.

Variation from meiosis

Meiosis is a primary source of variation. In meiosis, cell division of germline cells produces daughter cells that are genetically different from each other and from the parent cell, as shown in Figure 12.1.1. Crossing over during prophase I results in the recombination of alleles between non-sister chromatids of homologous pairs of chromosomes. The non-sister chromatid arms entangle and, at the chiasma, can detach and re-attach, exchanging genetic material. The amount of genetic material exchanged during crossing over varies, so the resulting recombination of alleles can be small or extensive.

9780170491785

Random assortment occurs during metaphase I. This involves the random lining up of the chromosome pairs at the cell's equator (Figure 12.1.2). The paternal or maternal chromosomes line up randomly on either side of the equator. So, when the chromosomes split during anaphase I, there is a random chance of paternal and maternal chromosomes occurring in the resulting daughter cells.

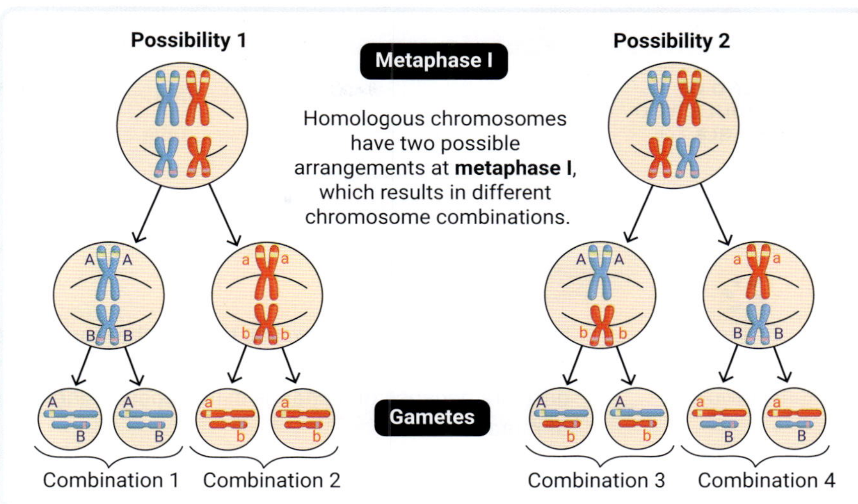

▲ **FIGURE 12.1.2** The possible combinations of chromosomes lining up during metaphase I

The number of combinations in which chromosomes can line up at the equator is 2^n, where n is the haploid number of the organism. For example, humans have a haploid number (n) of 23, making the total number of possible combinations of paternal/maternal chromosomes $2^{23} = 8\,388\,608$.

Other sources of genetic variation

Random fertilisation and random mating are other sources of variation in species that reproduce sexually. The different types of gametes produced through meiosis and the fact that most animal species mate with multiple individuals result in a wide variety of allele combinations in the offspring.

Mutations are spontaneous and random changes in the DNA sequence. If a mutation is inherited, it can change how frequently an allele appears in a population, resulting in many individuals with affected phenotypes. We will look at mutations in more detail in Module 12.2.

12.1 LEARNING CHECK

1 **List** four sources of variation in a species.

2 **Calculate** the number of combinations in which the chromosomes can line up during metaphase I if the diploid number of an organism is 14.

3 **Examine** the paired homologous chromosomes in Figure 12.1.3. Write out all the possible genotypes found in the gametes if crossing over occurred at the marked chiasma.

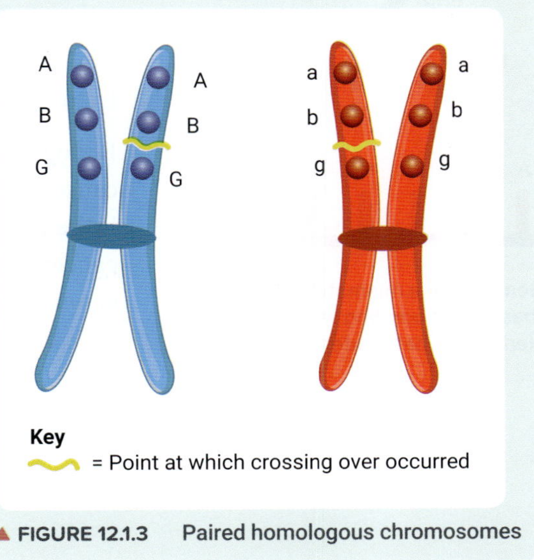

▲ **FIGURE 12.1.3** Paired homologous chromosomes

12.2 Mutation

BY THE END OF THIS MODULE, YOU WILL BE ABLE TO:

- ✓ define and provide examples of mutation, mutagens, genes and chromosomal mutations
- ✓ describe the differences between gene and chromosomal mutations
- ✓ recognise the changes to the DNA sequence resulting from the types of mutations within these categories.

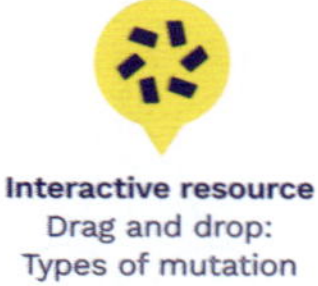

Interactive resource
Drag and drop: Types of mutation

GET THINKING

What comes to mind when you think of the word 'mutant'? Perhaps the X-Men are the basis of your understanding of a mutant. In biology, a mutant is an organism that is physically different because of a change in its genes. Using this definition, write down some examples of mutants.

Mutations and variation

Mutations are a significant contributor to variation in a population and are important mechanisms of change in evolution. Even though mutations are spontaneous and random, exposure to certain chemicals or agents can increase the likelihood of a mutation in the genome. These agents can be physical or environmental and are known as **mutagens**. Mutagens can be further classified by their origin and effects on the DNA.

mutagen
an agent that increases the likelihood of mutation

proteins
a group of large, complex molecules vital to cell replication, growth and repair

Physical, chemical and biological mutagens

Physical mutagens include ultraviolet and gamma radiation. These create free radicals that cause unusual bonds between bases in a DNA molecule. This prevents specific **proteins** from being produced. Figure 12.2.1 demonstrates how UV radiation can cause a mismatch in a sequence of DNA.

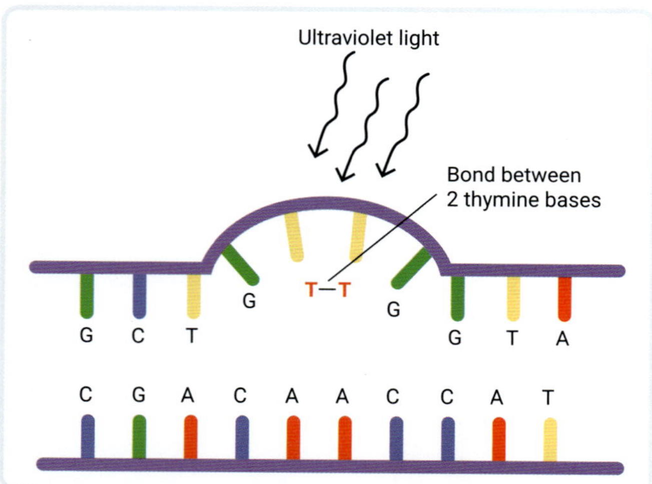

▲ **FIGURE 12.2.1** UV radiation causes a mismatch in a sequence of DNA, with two thymine bases (Ts) bonding with each other instead of with their complementary base, adenine (A).

Chemical mutagens are found in toxic compounds; for example, mustard gas. Exposure to these chemicals cause the DNA double strands to break, affecting the cells' ability to divide and multiply. Chemical mutagens are commonly used in chemotherapy to prevent cancerous cells from reproducing.

Biological mutagens are bacteria or viruses that can cause changes in the DNA sequence. An example is the human papillomavirus, which can result in cancerous cells in infected tissues, commonly the cervix or the throat.

9780170491785

Gene mutations

Mutations can be categorised as gene or chromosomal.

Gene mutations occur in a single gene because of changes to the nucleotide sequence. Occasionally they occur during DNA replication. **Insertion**, **deletion** and **substitution** are types of gene mutation. They can change a single nucleotide, causing a point mutation that can result in changes to the protein.

gene mutation
a change in the DNA sequence of a gene

insertion
the addition of one or more nucleotides into a DNA sequence

deletion
the removal of one or more nucleotides from a DNA sequence

substitution
the swapping of a nucleotide within a DNA sequence

In Chapter 11, we learned that the DNA consists of four bases (A, T, G and C). Each group of three bases (called a triplet) instructs or codes for an amino acid to be added when a protein is being built (for example, ATG codes for the amino acid methionine). When the DNA sequence is correctly read, the correct sequence of amino acids will form a functioning protein. Keeping this in mind, consider the sequence of words: THE CAT ATE THE RAT. Think of each word as representing a triplet of nucleotides in DNA and the sentence representing a sequence of DNA. Changing just one letter, such as the fifth letter, A, to an O, makes the sequence THE COT ATE THE RAT. You can still read the new sentence, but it no longer makes sense. This is what happens in a substitution mutation. Although the gene can still produce a protein, it may not be the protein required for normal cell function.

An insertion mutation occurs when an additional nucleotide is inserted into a gene sequence. For example, adding a second C in the fifth position in the original sentence gives us THE CCA TAT ETH ERA T.

A deletion mutation occurs when a nucleotide is removed from a gene sequence. For example, removing the sixth letter in the sentence gives us the sequence of THE CAA TET HER AT.

Both an insertion and a deletion result in a frameshift mutation. This is when the sequence of the nucleotides is read differently, which interferes with protein production.

Figure 12.2.2 shows the three different types of gene mutations.

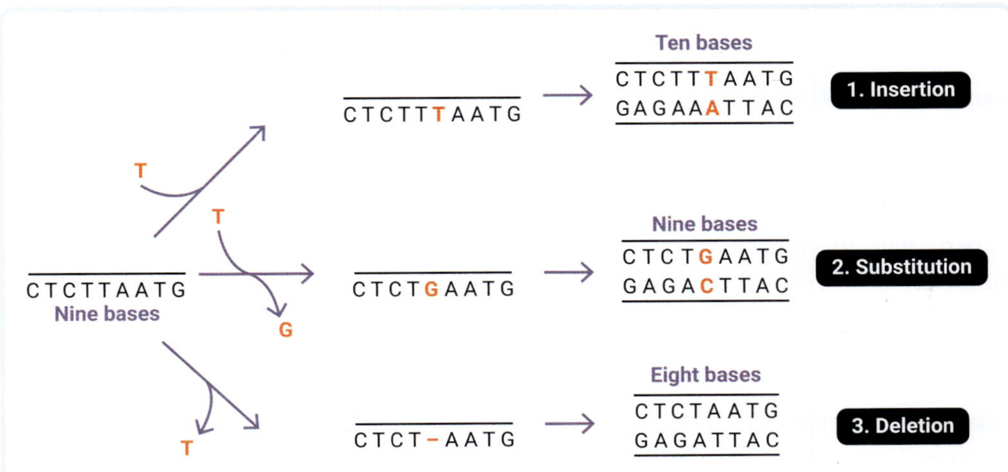

▲ **FIGURE 12.2.2** The three different types of gene mutations

Gene mutations may result in a dysfunctional protein or no protein being produced, which can lead to a genetic condition or a cancerous growth.

Chromosomal mutations

chromosomal mutation a change in the number of chromosomes or parts of a chromosome

Chromosomal mutations occur at the chromosome level and involve more than one gene. They occur because of an error in the cell division during mitosis or meiosis.

Deletion mutations occur when part of a chromosome is removed. As a result, the genes that part of the chromosome should carry cannot be read, and no proteins are produced. An example in humans is cri-du-chat syndrome, where a portion of chromosome 5 is deleted (Figure 12.2.3).

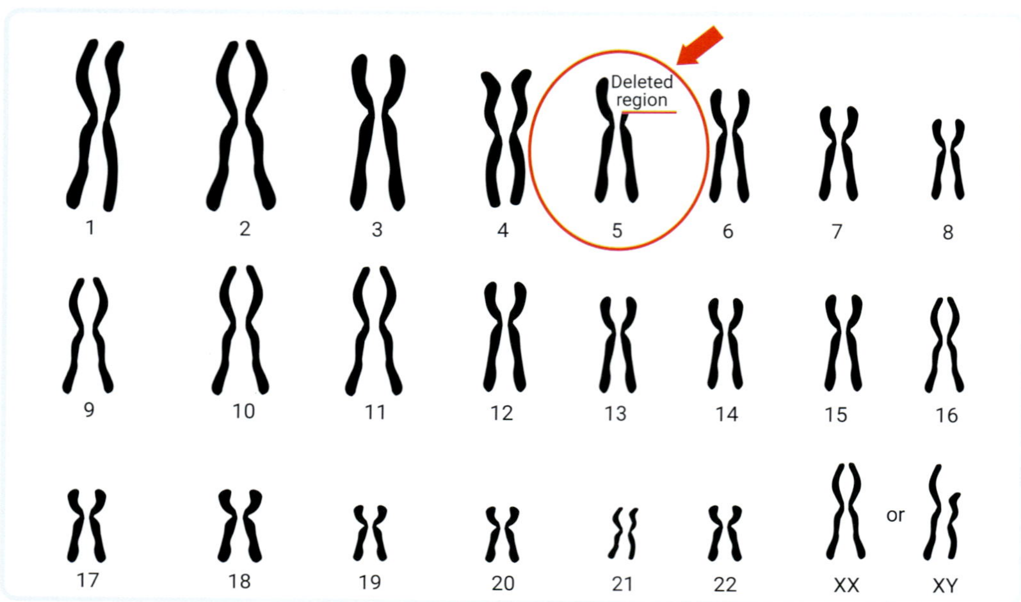

▲ **FIGURE 12.2.3** Cri-du-chat syndrome (or cat-cry syndrome) occurs when part of chromosome 5 is deleted.

inversion mutation a chromosomal mutation in which part of a chromosome is reversed end-to-end

Inversion mutations occur when the arm of a chromosome breaks off and reattaches upside down. While no genes are lost, the sequence of the nucleotides is different, and functional proteins may not be produced. For example, haemophilia A results when a portion of the blood coagulation factor VIII is dysfunctional, and blood cannot clot.

translocation the result when part of a chromosome detaches and reattaches to a different chromosome

Translocation results when part of a chromosome detaches and reattaches to a different chromosome pair. This results in part of a chromosome going missing or an extra part of a chromosome in the nucleus. In males, the translocation of the *SRY* gene from a Y chromosome onto an X chromosome can occur when sperm is produced during meiosis. If sperm with this X chromosome fertilises an egg, the individual will develop male characteristics, despite not having a Y chromosome.

Figure 12.2.4 shows the types of chromosomal mutations.

aneuploidy having additional or missing chromosomes

Another type of chromosomal mutation is non-disjunction. This results from the incorrect separation of the chromosome or chromatid pairs during anaphase I or anaphase II in meiosis. The resulting daughter cells (gametes) display **aneuploidy**, an unusual number of chromosomes. An example of aneuploidy is trisomy 21, or Down syndrome, a genetic disorder that results when the chromosome 21 pair do not separate as normal, and the resulting zygote has three copies of chromosome 21 (Figure 12.2.5).

9780170491785

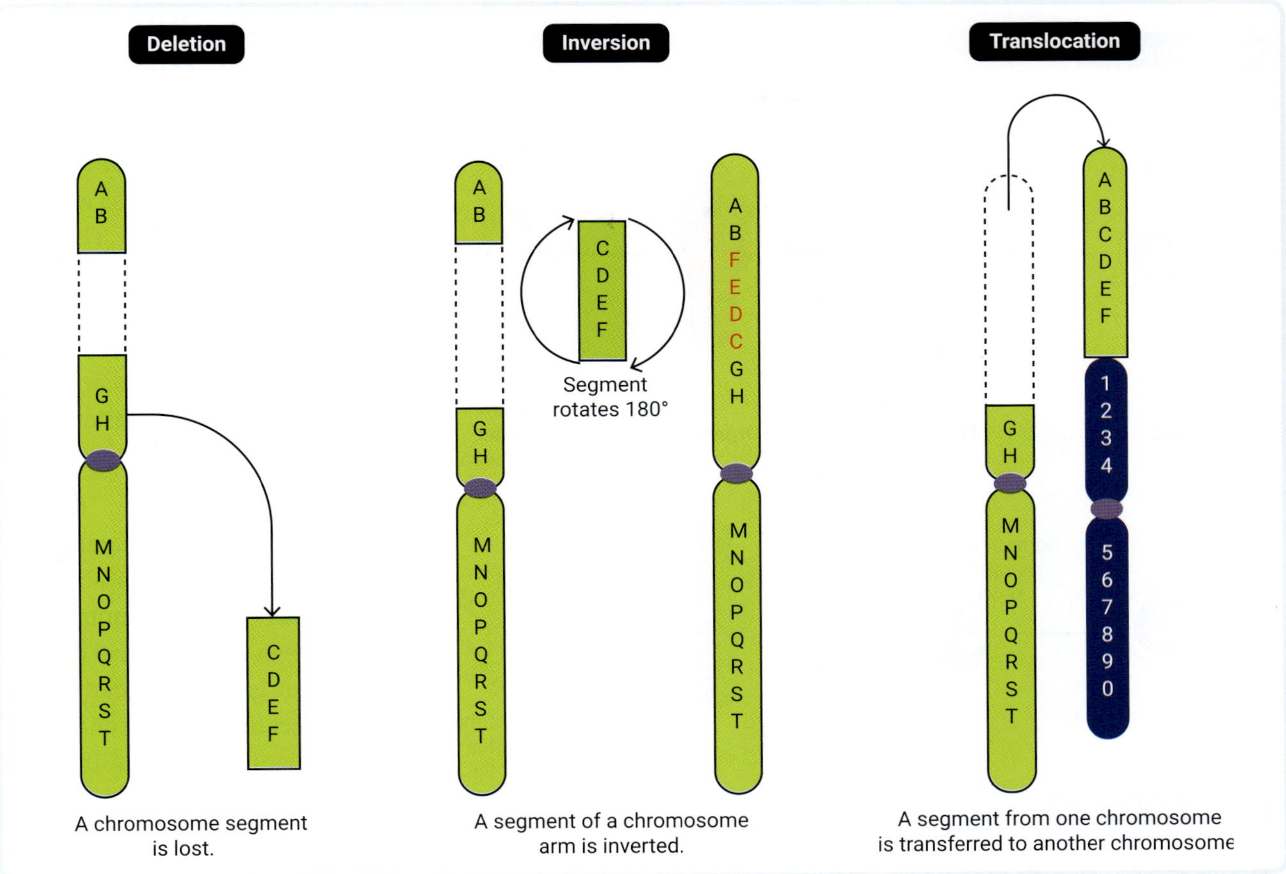

Deletion

A chromosome segment is lost.

Inversion

Segment rotates 180°

A segment of a chromosome arm is inverted.

Translocation

A segment from one chromosome is transferred to another chromosome

▲ **FIGURE 12.2.4** The types of chromosomal mutations

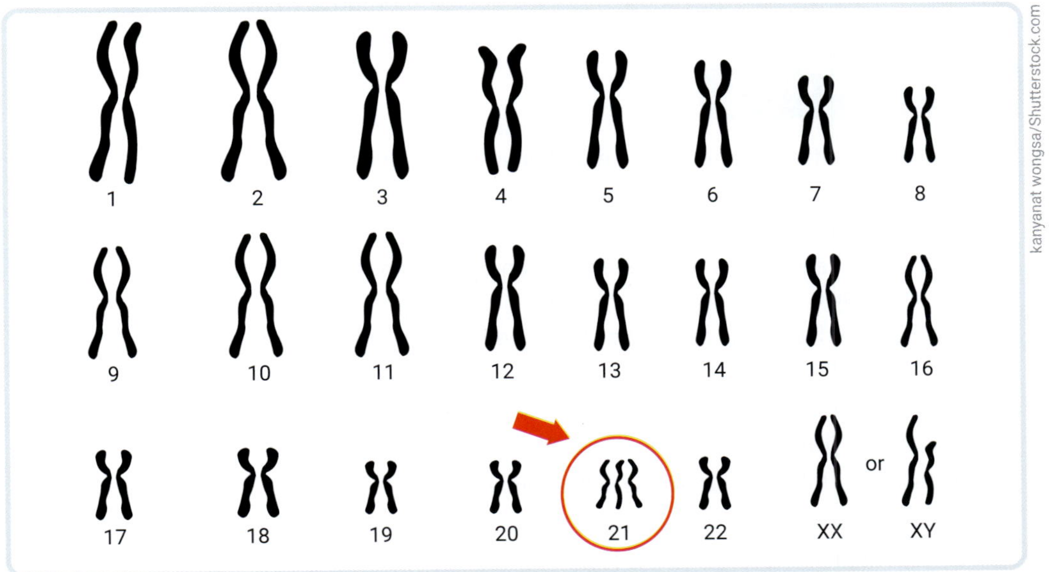

kanyanat wongsa/Shutterstock.com

▲ **FIGURE 12.2.5** Down syndrome occurs when there are three copies of chromosome 21 (trisomy 21).

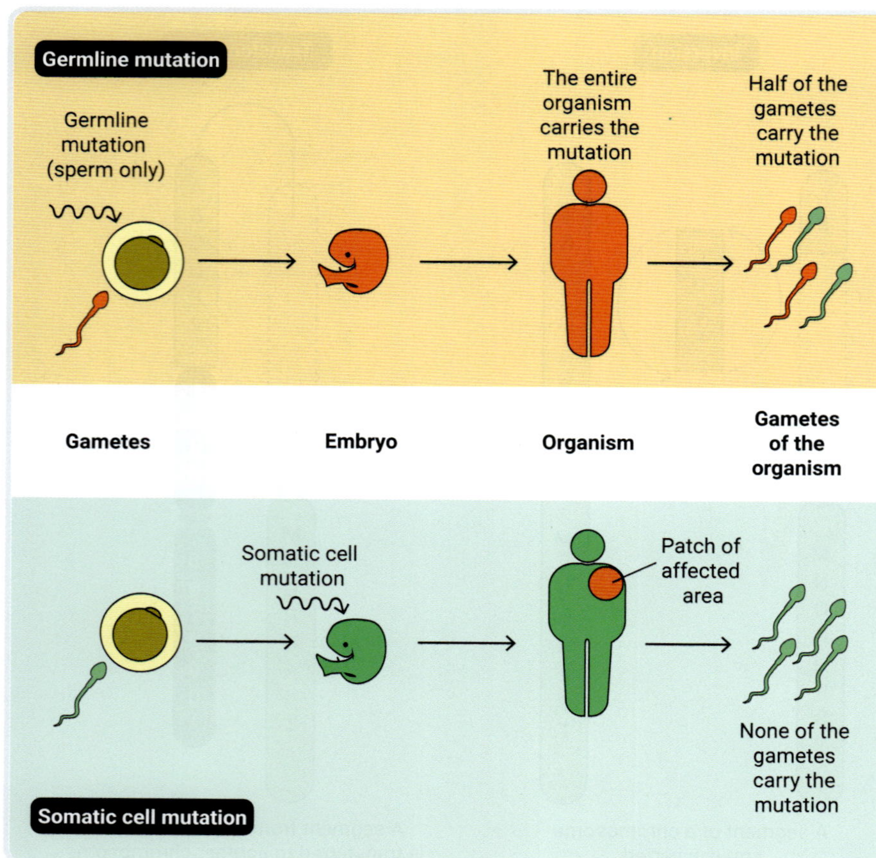

▲ **FIGURE 12.2.6** Somatic and germline mutations

Somatic and germline mutations

Many mutations occur in somatic cells rather than genes or chromosomes and so are not passed on to offspring. These are called somatic mutations. However, if the mutation occurs in the gametes (sex cells), these can be passed on and are deemed **germline mutations** (Figure 12.2.6).

From an evolutionary perspective, somatic mutations do not impact a species' ability to evolve because the mutation ends when the individual dies. However, germline mutations can affect the offspring and subsequent generations. If the mutation is beneficial to the survival of the individual, it can be maintained in the population. If it is not beneficial, it may reduce in frequency or be removed over time.

germline mutations
mutations that occur in sperm or ova

12.2 LEARNING CHECK

1 **Explain** the difference between a mutation and a mutagen.
2 A mutation resulted in a person being unable to make a protein to assist in blood clotting. The condition this person has is called haemophilia.
   a **Classify** haemophilia as a gene mutation or a chromosomal mutation and **justify** your choice.
   b The offspring of this parent also had blood that did not clot properly. Using this information, **classify** haemophilia as either a somatic or a germline mutation.
3 Consider the DNA sequence TAC–GCA–AAA–CGA–GTC–ATT.
   Rewrite the DNA sequence after the following mutations occurred.
   a Deletion of the 5th nucleotide in the sequence
   b Insertion of adenine (A) after the 7th nucleotide in the sequence
   c Substitution of every thymine (T) with a guanine (G)
4 Sometimes, a nonsense mutation results from the deletion or insertion of a nucleotide. Use the internet to **research** what 'nonsense mutation' means and how it might affect the protein produced by that gene.

9780170491785

12.3 History of evolutionary theory

BY THE END OF THIS MODULE, YOU WILL BE ABLE TO:

✓ describe Darwin's observations and inferences on natural selection.

GET THINKING

Look at Figure 12.3.1.

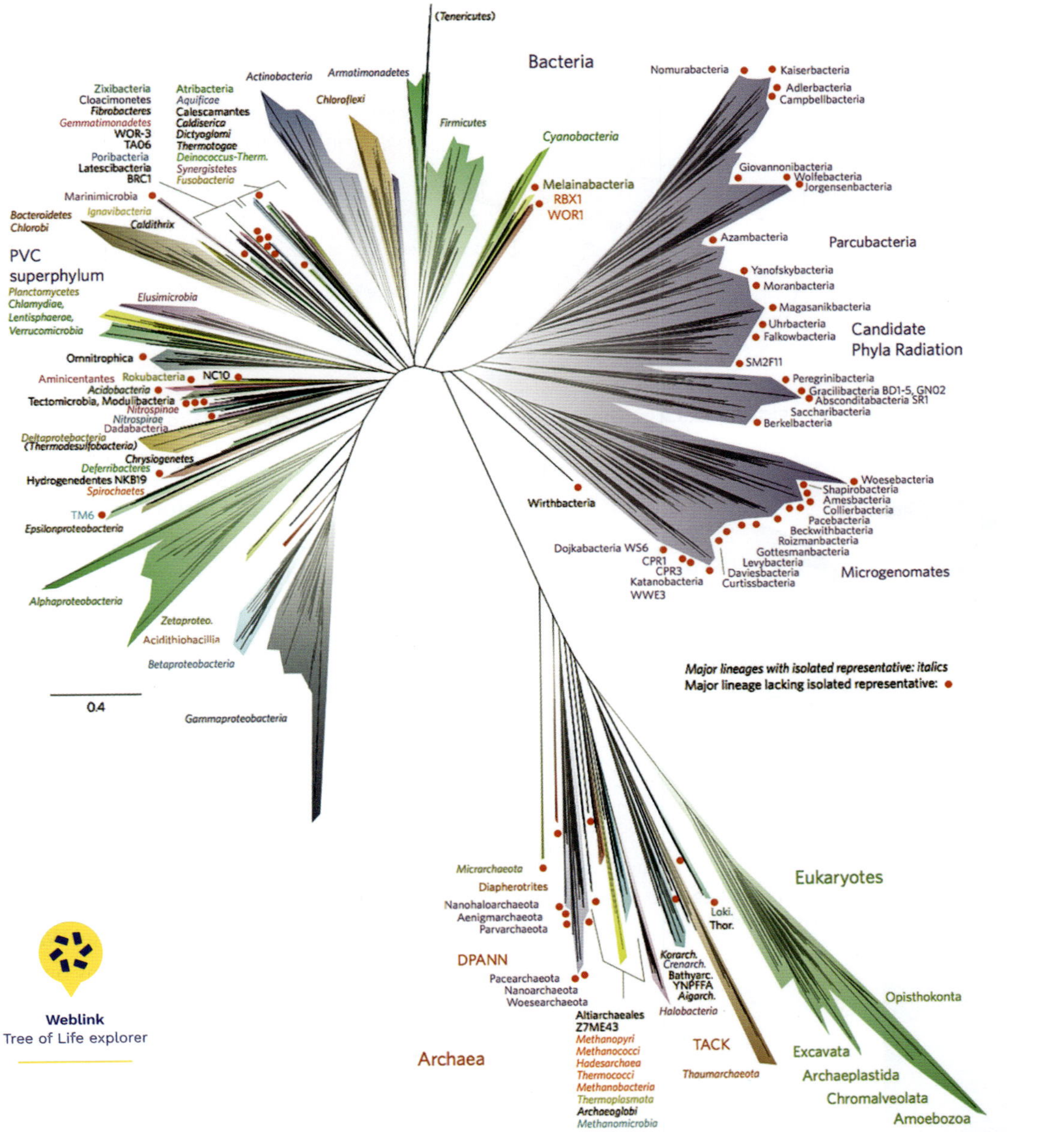

Weblink
Tree of Life explorer

▲ **FIGURE 12.3.1** The universal tree of life

Upon first glance, it resembles a bird in flight. However, on closer inspection, it is a detailed diagram representing the hypothesis that all life descended from the last universal common ancestor. It is known as the 'tree of life' and is used both as a metaphor and as a research tool to help understand the evolution of life and the relationships between organisms, living and extinct. In your workbook, write two or three sentences in your own words to describe how the tree of life could represent evolution.

Video activity
Charles Darwin

survival of the fittest
the idea that individuals with the best suited characteristics will survive, reproduce and pass their traits on to the next generation

natural selection
the process in which an environmental factor acts on a population, resulting in some individuals being more likely to survive and reproduce

DATA SCIENCE

Learn more about investigating scientific questions and supporting scientific claims in **Modules 2.1** and **2.2**.

You might have heard the expression '**survival of the fittest**' used to explain evolution, but what does it actually mean?

Charles Darwin – the father of evolution

In 1831, a young naturalist by the name of Charles Darwin left England for South America on the HMS *Beagle* as part of a survey of the South American coastline (Figure 12.3.2).

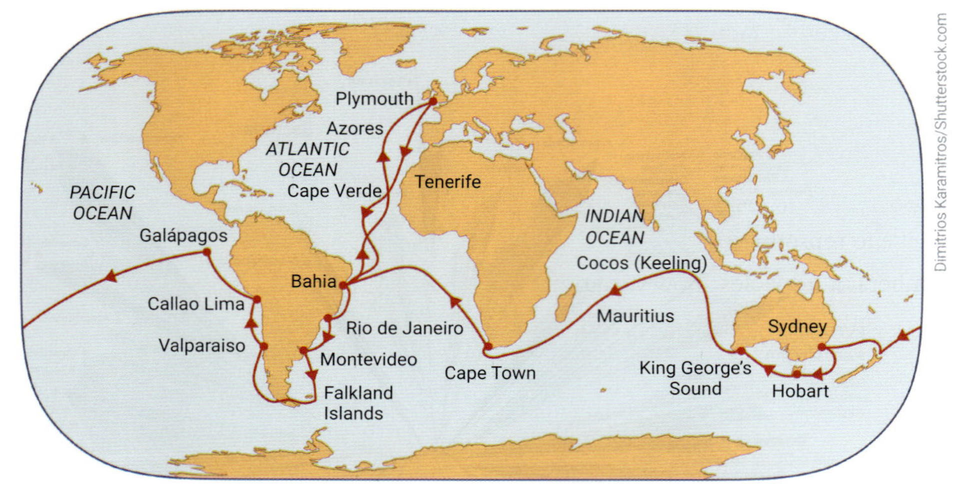

▲ **FIGURE 12.3.2** The route of the HMS *Beagle* voyage from 1831 to 1836

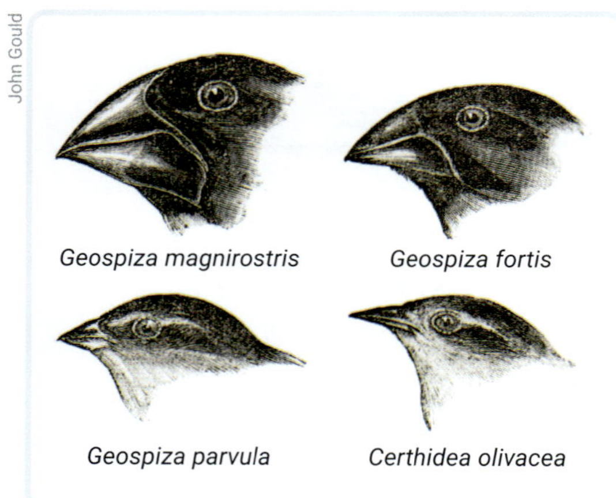

▲ **FIGURE 12.3.3** Some of Darwin's finches and the variety of beak shapes observed in the Galápagos Islands

During the nearly 5-year voyage, Darwin studied many countries' geology and natural history and gathered an immense private collection of specimens. He took careful notes in his journals and sent many specimens to Cambridge University for further study. On his return to England in 1836, Darwin worked as a self-funded 'gentleman scientist', free to explore his own collections and publish his research. Darwin's notes on the tortoises and the finches of the Galápagos Islands (Figure 12.3.3), coupled with the information given to him by a renowned ornithologist, allowed Darwin to speculate on the possibility that 'one species does change into another'.

Darwin felt no urgency to publish his ideas until another British naturalist and explorer, Alfred Russel Wallace (Figure 12.3.4), wrote about his observations on species distribution in the Malay Archipelago. Incredibly, Wallace independently conceived a very similar theory of evolution through **natural selection**. Wallace wrote a paper, 'On the Law which has Regulated the Introduction of New Species', which was read by Darwin's colleague. This led to Wallace's paper being jointly published with some of Darwin's writings in 1858. The positive reception from the scientific community encouraged Darwin to publish *On the Origin of Species by Means of Natural Selection or the Preservation of Favoured Races in the Struggle for Life* in 1859.

▲ **FIGURE 12.3.4** (a) Alfred Russel Wallace and (b) Charles Darwin

9780170491785

Darwin's key observations

On the Origin of Species is considered the foundation of evolutionary biology. The book presents a body of evidence that explains the diversity of life from a **common ancestor**. Darwin explains this descent with modification through a branching pattern of evolution, as seen in Figure 12.3.5.

In his book, Darwin identified his key observations:

- traits are inherited
- all species produce more offspring than will survive to reproduce
- members of a species show variation.

These observations gave rise to Darwin's two main inferences.

- There is a **struggle for existence** due to high reproduction rates and limited resources in the environment.
- The organisms with the traits to help them survive best will produce more offspring, passing on the genes that helped them survive. Over generations, the population will have more individuals with the genes best suited to their environment. This is coined survival of the fittest.

This is the mechanism of evolution known as natural selection.

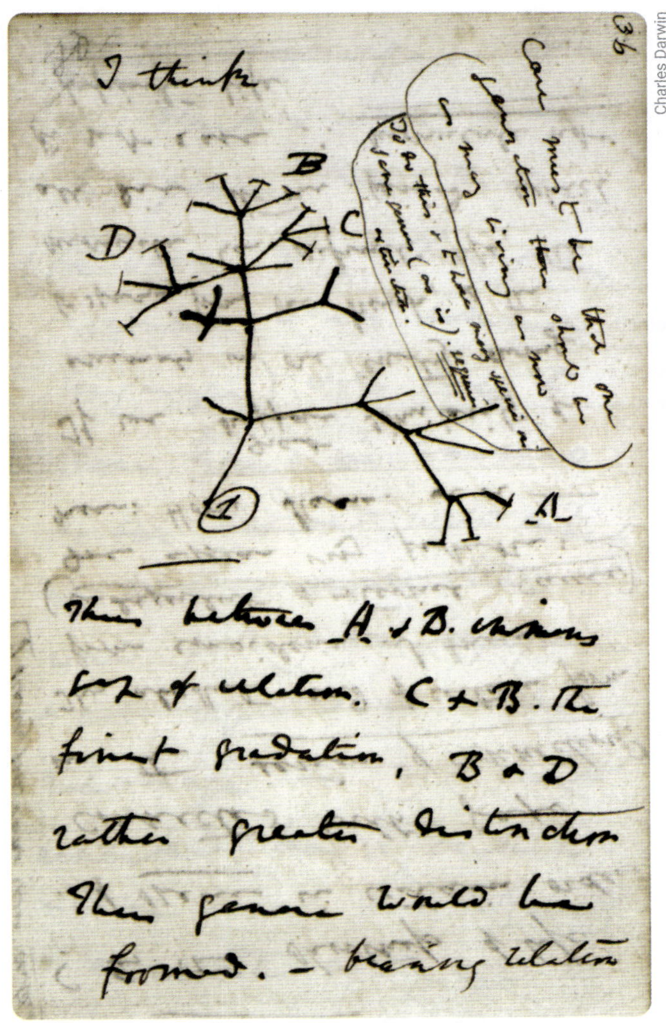

Charles Darwin

▲ **FIGURE 12.3.5** Darwin's first evolutionary tree sketch, complete with 'I think'

common ancestor
the ancestor that two or more descendants have in common

struggle for existence
the competition between individuals for required resources such as food, water or space

12.3 LEARNING CHECK

1 **Define** natural selection.

2 a Who was Charles Darwin?

  b What were his key observations made on the HMS *Beagle* voyage?

  c **List** the two inferences he made from these observations.

3 Use the internet to find information about Charles Darwin's scientific findings and **create** a timeline from the beginning of his journey on the HMS *Beagle* to the publication of *On the Origin of Species*.

12.4 Natural selection

BY THE END OF THIS MODULE, YOU WILL BE ABLE TO:

- ✓ define and describe the processes of natural selection and artificial selection
- ✓ define the term 'selective agent' and describe the effect of a selective agent on a population
- ✓ compare and contrast natural selection and artificial selection.

selective agent
the environmental factor acting on the population

gene pool
the total range of genetic material available in a population

GET THINKING

Imagine there is a population of rabbits living in a forest. These rabbits vary in fur color – some have white fur and others have brown fur. This variation is due to genetic differences. Think about which environmental conditions will favour the brown-furred rabbits. What happens to the population if, due to a sudden change in climate, there is snow for a long time? Discuss how this could be an example of natural selection.

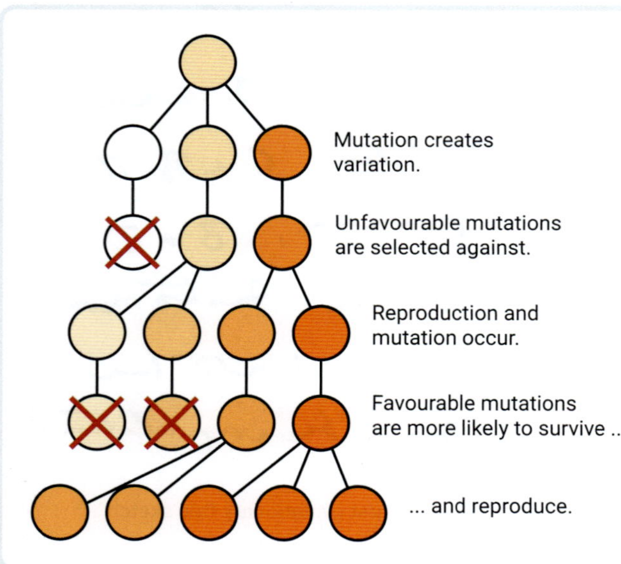

▲ **FIGURE 12.4.1** A representation of natural selection

▲ **FIGURE 12.4.2** The two variants of the peppered moth on a tree. The light-body variant is well camouflaged on this pollutant-free tree.

Natural selection as a mechanism of evolution is now widely regarded as the clearest explanation for the diversity of organisms on Earth. A visual representation of natural selection is provided in Figure 12.4.1.

Peppered moths: a natural selection case study

Natural selection can be seen in peppered moths. Peppered moths exhibit variation in the colour of their body and wings, ranging from a light-coloured body with black speckles on the wings to a variant form in which both the body and wings are mostly black, as seen in Figure 12.4.2. Both these populations lived in rural England and in the cities prior to the Industrial Revolution.

During the Industrial Revolution, the cities became much dirtier due to the pollution produced by engines. The walls of buildings became covered in dark soot. The light-body variant of the peppered moth was not camouflaged against the darkened walls. The **selective agent**, predatory birds, could easily find and eat the light-coloured moths, removing them and their alleles from the **gene pool**. The black moths were hidden against the black walls, allowing them to survive, reproduce and pass the black allele on to their offspring. Over time, the population of peppered moths in industrial areas became, predominantly, the black form variant.

9780170491785

Natural selection in humans

Natural selection can also be seen in human populations. Sickle-cell anaemia is a genetic disorder caused by a mutation in the gene coding for the beta-haemoglobin chain. Individuals with two recessive alleles have sickle-cell anaemia, a condition that results in red blood cells having a sickle shape and being unable to carry sufficient oxygen.

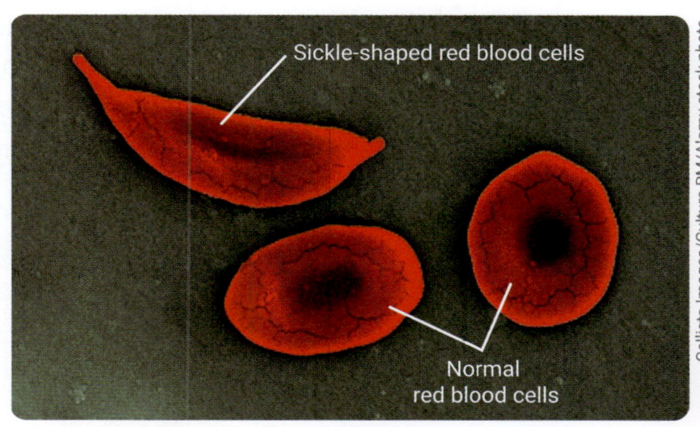

Callista Images/Cultura RM/Alamy stock photo

▲ **FIGURE 12.4.3** Normal red blood cells and sickle-shaped red blood cells

Individuals with one allele for the disease have the sickle-cell trait but suffer no major ill effects, provided they live at sea level. Living above sea level, where there is less oxygen in the atmosphere, causes some of the red blood cells to become sickled, which can become stuck in the blood vessels, creating a clot. The difference between normal red blood cells and sickle-shaped red blood cells is shown in Figure 12.4.3. The advantage to humans is seen in malaria zones, where the heterozygotes have a resistance to malaria, survive and reproduce, passing the sickle cell allele on to the next generation.

Artificial selection

Darwin also observed farmers selectively breeding stock to produce desired characteristics in their offspring. He saw this as an analogy for natural selection, whereby humans were the selective agent choosing the best traits to be passed on to subsequent generations. This is called artificial selection.

Artificial selection is a faster mechanism of evolution than natural selection, which takes many generations. Artificial selection is driven by human choice, while natural selection is environmentally driven.

Humans have selected desired characteristics in plants and animals since the Neolithic age, 10 000 years ago. For example, humans have selected animals such as sheep, goats and cows for breeding based on their particular traits, such as good milk production, thicker wool or more meat, so these desirable traits were passed on to the next generations. Belgian Blue cattle, for instance, are bred to produce twice the amount of meat (Figure 12.4.4).

Leitenberger Photography/Shutterstock.com

▲ **FIGURE 12.4.4** Belgian Blue cattle are bred to produce twice the amount of meat.

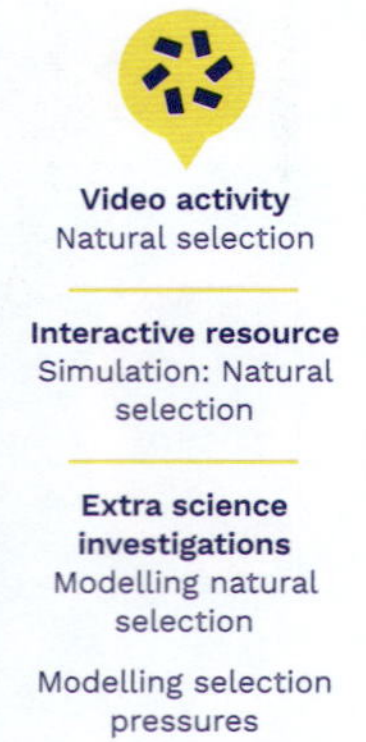

Video activity
Natural selection

Interactive resource
Simulation: Natural selection

Extra science investigations
Modelling natural selection

Modelling selection pressures

Today, the results of artificial selection are clearly seen in domestic cats and dogs, with the careful selection and breeding of traits in individuals used to produce a wide variety of breeds. For example, cattle dogs are selected for their hard work, loyalty and innate ability to herd, and Siamese cats are selected for their light grey fur and blue eyes (Figure 12.4.5).

Sari ONeal/Shutterstock.com

▲ **FIGURE 12.4.5** Over thousands of years, humans have selected traits in domesticated animals to suit their needs.

Humans also use artificial selection in plants. Many of the fruits and vegetables you purchase from the supermarket look nothing like their ancestral wild forms. *Brassica oleracea* (Figure 12.4.6) is the wild form of cabbage and has been carefully cultivated over many years to produce vegetables such as cabbage, cauliflower, broccoli, kale, kohlrabi and brussels sprouts (Figure 12.4.7).

Paul Martin/Alamy Stock Photo

▲ **FIGURE 12.4.6** *Brassica oleracea*, commonly known as wild cabbage, grows in many parts of Europe.

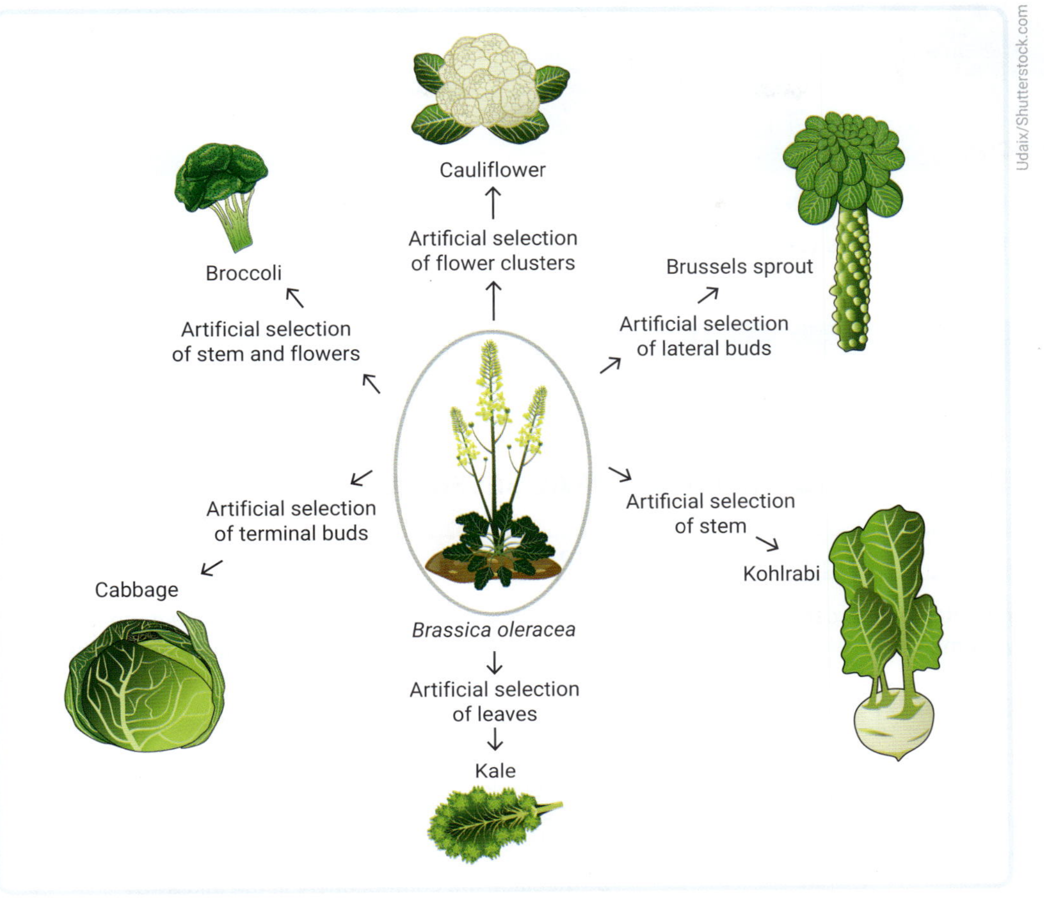

▲ **FIGURE 12.4.7** Many vegetables we eat today originate from *Brassica oleracea*.

12.4 LEARNING CHECK

1. **Define**:
   a. natural selection.
   b. artificial selection.
2. Considering the case study of the peppered moth:
   a. **identify** the two variants.
   b. **name** the selective agent acting on the moths.
   c. **describe** how the different variants of moth in the city regions were affected by the selective agents and **explain** what impact this had on their gene pool.
3. Antibiotic-resistant bacteria are an increasingly common problem. Many bacterial infections (e.g. 'golden staph') no longer respond to antibiotics. Using your understanding of natural selection, **explain** how antibiotic-resistant bacteria have formed.
4. **Draw** a Venn diagram to **compare** and **contrast** the processes of natural and artificial selection.

WORKING SCIENTIFICALLY

12.5 Modelling natural selection

SCIENCE SKILLS IN FOCUS

IN THIS MODULE, YOU WILL FOCUS ON LEARNING AND IMPROVING THIS SKILL:

- investigating environmental factors (selection pressures) influencing survival of different individuals in a species.

BIRD BEAKS – AN INVESTIGATION INTO NATURAL SELECTION

AIM

To investigate how environmental factors can influence the shape of bird beaks over time

MATERIALS AND EQUIPMENT

- ☑ 6 varieties of confectionery
- ☑ tools to represent bird beaks: spoon, 2 chopsticks, toothpick, icy pole stick
- ☑ takeaway food container
- ☑ stopwatch
- ☑ sheet of A3 paper

Safety

This activity uses edible confectionery. If that is not an option, replace the confectionery with coins, marbles, toothpicks, small pieces of gravel, pieces of string or table tennis balls.

PROCEDURE

1. Predict which of the tools listed in the materials section (representing a bird's beak) would be best for picking up each type of confectionery.
2. Place the confectionery into the takeaway food container. In 30 seconds, use one tool to collect as much food as possible. Copy Table 12.5.1 into your workbook and record the number of pieces you were able to pick up.
3. Replace the confectionery in the takeaway food container.
4. Repeat Step 2 with each tool and record the totals in Table 12.5.1.
5. Apply a selective agent to your food collection; for example, one of your foods is now toxic, and if you touch it with your 'beak', that beak can no longer collect food. Repeat steps 2–4 and record your totals.
6. Apply a second selective agent to your food collection; for example, another of your foods is now toxic.
7. Extension: Using only the tools that can still collect food, cover your left eye and try to collect as much food as possible in 30 seconds.

9780170491785

RESULTS

▼ **TABLE 12.5.1** 'Bird beak' results

Tool	Trial 1	First selective agent applied	Second selective agent applied	Closed left eye
Spoon				
Chopsticks				
Toothpick				
Icy pole stick				

ANALYSIS

1 **Define** natural selection.

2 **Compare** columns 2 and 3 in your table.

  a Which tool(s) were still able to collect food?

  b What impact did the selective agent have on collecting food?

3 **Compare** columns 3 and 4 in your table.

  a Which tool(s) were still able to collect food?

  b What impact did the second selective agent have on collecting food?

4 **Compare** the results in column 5 to the results in columns 2–4.

  a What effect did closing your left eye have on the effectiveness of the tools?

  b Which of the three selective agents had the greatest impact on collecting food? Use the results to **explain** which type of beak would be the most likely to be passed on to the next generations.

5 Each tool represents a different bird beak. Using the internet, **identify** which tool best matches a bird's beak and **determine** what that bird's diet is like.

CONCLUSION

Write a conclusion for this investigation based on your observations and the concept of natural selection.

12.6 Speciation

BY THE END OF THIS MODULE, YOU WILL BE ABLE TO:

✓ describe the process of speciation to form a new species.

Video activity
Speciation

GET THINKING

In 2021, 23 species of animals and plants were declared extinct worldwide. Largely, these extinctions resulted from human impacts on habitat, including climate change. However, Professor Andrew Pask, a researcher from the University of Melbourne, is working towards making the Tasmanian tiger de-extinct. The technology he will be using is at the heart of the project to bring the woolly mammoth back from extinction by 2027.

What animals or plants would you bring back from extinction if you could? What do you think will need to change in our global environment to support the de-extinction of these animals and plants?

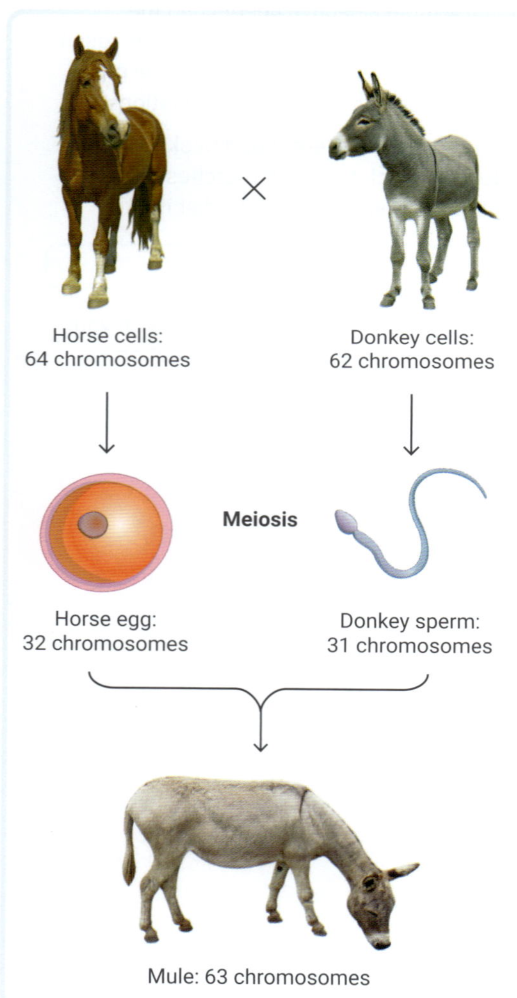

▲ **FIGURE 12.6.1** A cross between a horse and a donkey results in an infertile hybrid mule.

What is a species?

As humans, we have the biological name *Homo sapiens*. This means we belong to the genus of *Homo* and are part of the species *sapiens*. There are many other species in the *Homo* genus that existed before us and that are now extinct. What determines a species, and why is it important in the process of evolution?

Biologically, a species is defined as a group of individuals that breed under natural conditions to produce fertile offspring. For example, the domestic cat belongs to the species *Felis catus*. While there are many different breeds of cats, they can all produce fertile offspring if they mate. If members of different species mate, any successful offspring will likely be infertile.

Consider the hybrid mule, as shown in Figure 12.6.1. A mule is produced when a horse (*Equus caballus*) reproduces with a donkey (*Equus asinus*). Even though the horse and donkey are from the same genus, their different species names indicate that they would not be able to produce fertile offspring if they mate. This is because each species has a different and incompatible number of chromosomes, resulting in offspring with an odd number of chromosomes.

9780170491785

How speciation occurs

The process of **speciation** is when two populations become so genetically different that they can no longer breed with each other to produce fertile offspring. Speciation relies on variation and **selection pressures** on a population. For speciation to occur, variation must exist within a population. That's because different phenotypes allow for selective agents to act on the population, providing an advantage to the individuals that are the 'most fit' for survival in that specific environment.

For speciation to occur, there must also be a degree of **isolation** between the original population and the break-away group(s). Isolation prevents the movement of alleles between population groups, often described as 'gene flow'. Isolation can be classified as geographical or sociocultural in human populations (Table 12.6.1). When isolation occurs over an extended period, it can lead to the groups becoming reproductively isolated, meaning the different populations cannot reproduce with each other.

speciation
a process in which two groups become so genetically different they can no longer breed with each other under natural conditions to produce fertile offspring

selection pressure
the effect the selective agent has on the population

isolation
a mechanism or barrier that separates breeding populations

▼ TABLE 12.6.1 Examples of the different ways a population can become isolated

Geographical isolation	Sociocultural isolation in human populations
Large bodies of water, such as the ocean, lakes or rivers	Language
Mountain ridges	Cultural beliefs
Canyons	Economic status
Deserts	Education
Icebergs	Sexual selection

As groups become increasingly isolated due to limited gene flow, different selective agents begin to act on the separated populations. These selective agents provide an advantage to the individuals with the phenotype best suited to a specific environment. Over many generations, the separated gene pools will become significantly different from each other. Once individuals in the different gene pools can no longer produce fertile offspring, they are considered different species. This process is demonstrated in Figure 12.6.2.

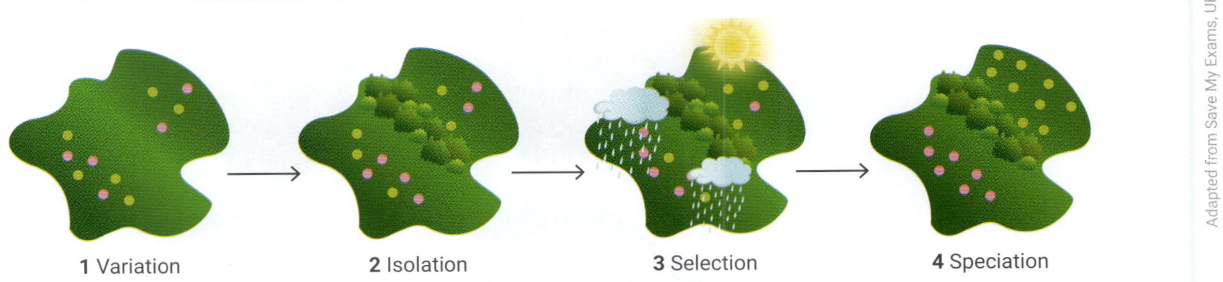

1 Variation → 2 Isolation → 3 Selection → 4 Speciation

Adapted from Save My Exams, UK

▲ FIGURE 12.6.2 The process of speciation

Types of speciation

allopatric speciation speciation due to a barrier

Allopatric speciation is defined as speciation from a geographical barrier, resulting in reproductive isolation. When the barrier is removed, the two groups can no longer breed because they are now genetically distinct species. Charles Darwin saw this on the Galápagos Islands while studying the islands' finches, as shown in Figure 12.6.3.

sympatric speciation speciation due to reproductive isolation

Sympatric speciation is defined as speciation occurring without a geographical barrier. It is more commonly seen in plants as they self-fertilise more easily. An example is the cultivated wheat plant species of emmer, durum and common wheat.

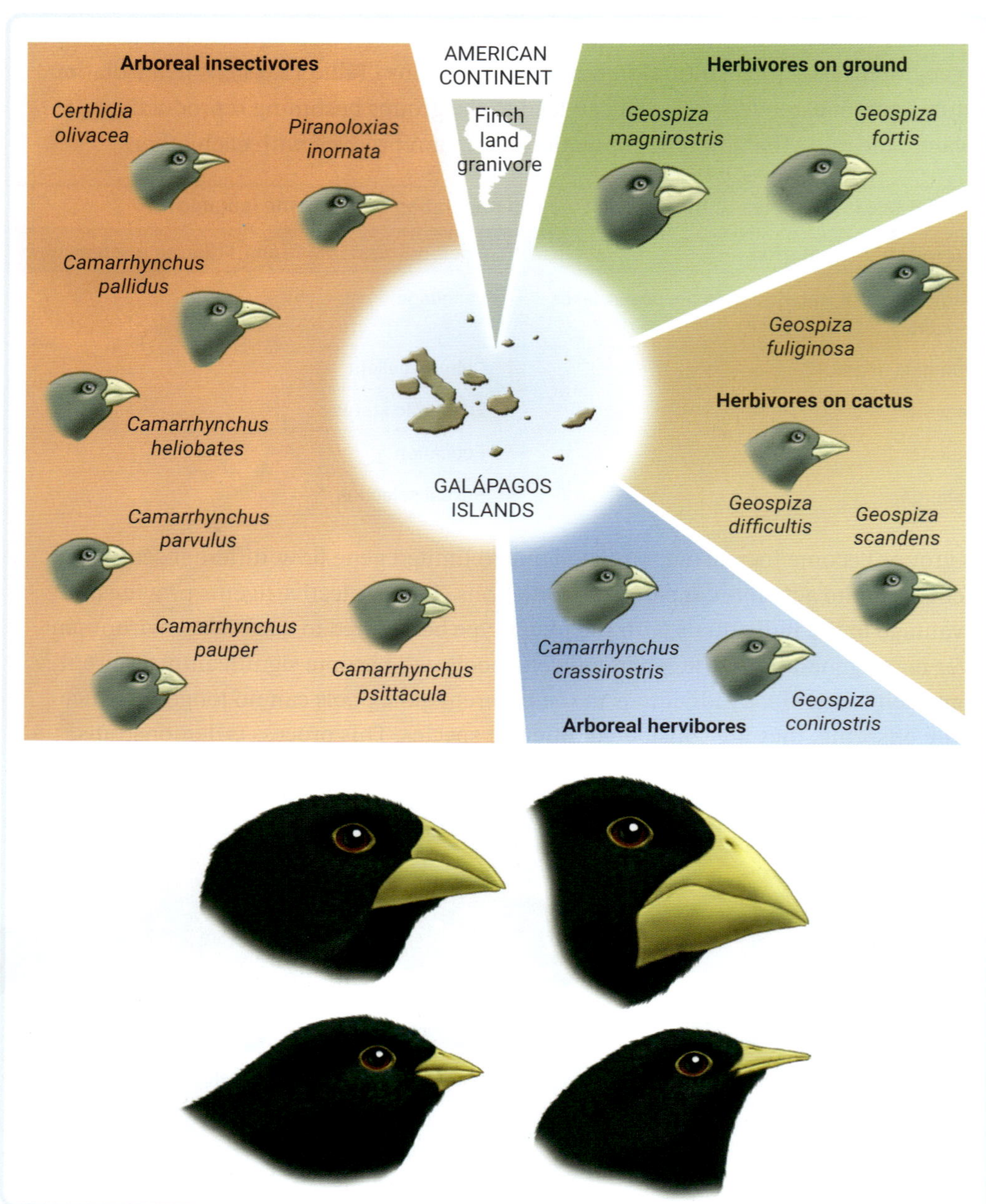

▲ **FIGURE 12.6.3** Darwin's finches are an example of allopatric speciation. An ancestral species spread to different islands, where natural selection led each population to evolve differences, including different beak shapes suited to the best food source in each environment.

12.6 LEARNING CHECK

1 **Explain** why mules are not considered to be a species.
2 **Outline** the process of speciation.
3 **Compare** allopatric speciation with sympatric speciation.
4 Ligers and tigons are hybrid great cats (Figures 12.6.4 and 12.6.5). They are the offspring from breeding tigers and lions in captivity. Despite being hybrids, there have been successful offspring when tigons or ligers are bred.
   a Tigons and ligers have a chromosome number of 38. **Explain** why these hybrids are not sterile, unlike the mule.
   b Despite tigons and ligers being capable of breeding, it is a discouraged practice. Why do you think conservationists do not advocate for this breeding?

videohouse/Shutterstock.com

▲ **FIGURE 12.6.4** A liger

One World Images/Alamy stock photo

▲ **FIGURE 12.6.5** A tigon

5 **Create** a table to **summarise** the main speciation processes for Darwin's finches.

12.7 Evolution

BY THE END OF THIS MODULE, YOU WILL BE ABLE TO:

- ✓ define evolution and describe the mechanisms that result in evolution
- ✓ define gene drift and allele frequency
- ✓ justify the theory of evolution using specific examples and evidence.

Interactive resource
Crossword: Evolution

GET THINKING

Pingelap is a small island in the South Pacific, sometimes described as the 'colour blindness' island. Between 4 and 10 per cent of its inhabitants carry the gene for total colour blindness. In comparison, the incidence of total colour blindness is 0.003 per cent globally. Why do you think there is such a high frequency of the colour blindness gene in this population?

FOTOGRIN/Shutterstock.com

▲ FIGURE 12.7.1 A population of flamingos

Mechanisms of evolution

Evolution can be defined as the gradual change in the characteristics of a species from earlier forms over many generations. An individual does not evolve: the population evolves.

A population is a group of individuals capable of breeding who are living in the same place at the same time, such as the group of flamingos shown in Figure 12.7.1. The genetic information the population possesses can be collectively called the gene pool.

Evolutionary mechanisms such as natural selection and mutation affect individual organisms in a gene pool. However, it is the collective change in **allele frequency** that results in the population evolving.

allele frequency
the measure of how common an allele is in a population

Allele frequency

Allele frequency is a measure of how common an allele is in a population. It can be calculated by determining how many times the allele appears in the population and dividing by the total number of copies of the gene.

$$\text{Frequency of allele A} = \frac{\text{number of copies of allele A in population}}{\text{total number of copies of gene in population}}$$

If a gene pool is small, the frequency of some alleles might be disproportionately high when compared with a larger population. Smaller gene pools tend to undergo greater change when the mechanisms of evolution are applied to them.

Genetic drift

genetic drift
the change in allele frequency seen in small populations due to chance events from one generation to the next

When small populations of a species are isolated, the small number of individuals with rare alleles may fail to transmit them, leading to the disappearance of the allele and the rise of a new species. This effect is known as **genetic drift,** or the Sewall Wright effect, after the US geneticist who proposed the concept.

9780170491785

Genetic drift can be modelled using coloured balls (Figure 12.7.2). Consider a population of 100 individuals represented by 50 red and 50 blue balls. The gene is represented by the colour, and the alleles are red and blue. If you were to place the 100 balls in a bag and randomly select 10 individuals to form the gene pool, the selection of red to blue is unlikely to represent the 50 : 50 ratio seen in the original group. If those individuals are then allowed to breed to bring the new population back up to 100, the resulting allele frequency after one generation will not match the frequency that existed in the original population.

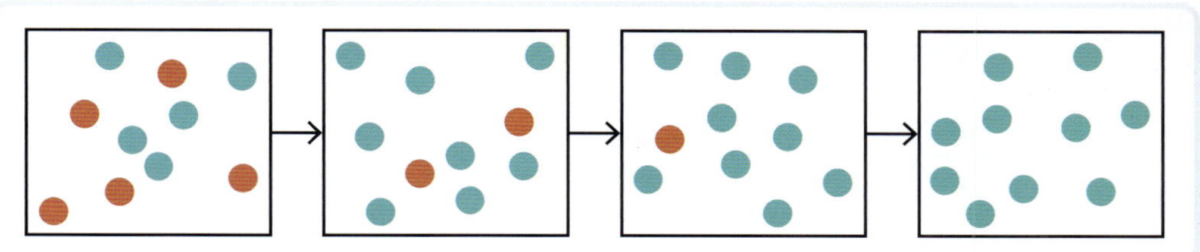

▲ **FIGURE 12.7.2** Genetic drift happens when random events cause changes in allele frequencies over generations, sometimes even leading to the loss of an allele. The effect is strongest in small populations.

DATA SCIENCE 

Learn more about presenting and analysing data in **Module 2.6**.

Genetic drift can occur in populations that are populating a new area or repopulating after an event that resulted in the death of many individuals. The bushfires on Kangaroo Island in 2020 reduced the koala population by up to 90 per cent. Twelve koala orphans were moved to a wildlife park in the Adelaide Hills, where they will form a new population, free of diseases including chlamydia and koala retrovirus. This new population will be used to help build koala populations in the rest of Australia. However, the Kangaroo Island sub-sample was bred from another small sample in the 1920s. Due to genetic drift, these koalas have a condition that is otherwise rare – testicular aplasia, meaning they have only one testicle.

12.7 LEARNING CHECK

1 Read the following statements and **determine** if they are true or false. Rewrite any false statements so that they read correctly.
   a Evolution is a quick change in characteristics resulting in a new species.
   b A gene pool is all available genes in a population at any one time.
   c Allele frequency can change randomly in large populations.

2 A scientist called Lamarck incorrectly hypothesised that giraffes developed long necks by stretching towards branches on trees. By continually doing this, they stretched their necks and their offspring were born with long necks as well. Using your understanding of natural selection, speciation and evolution, **explain** why this hypothesis is incorrect.

WORKING SCIENTIFICALLY

12.8 Controlling variables

SCIENCE SKILLS IN FOCUS

IN THIS MODULE, YOU WILL FOCUS ON LEARNING AND IMPROVING THESE SKILLS:

- investigating environmental factors influencing species' success
- controlling variables for a valid investigation.

Types of variables

There are three categories of variables in science investigations (Figure 12.8.1).

Controlling variables

In experiments where you have an independent and a dependent variable, you need to make sure that a change in the independent variable is the only cause of change in the dependent variable. To do this, you must control all variables other than the independent variable. This ensures your experiment is valid.

For example, if you are testing the effect of different fertilisers on plant growth, the independent variable you are testing is the type of fertiliser. So, this must be the only variable you change. The dependent variable is the growth of the plants, which you might choose to measure as plant height or the number of new leaves on each plant.

To control the other variables in the fertiliser experiment, you need to use the same type of plant, apply the same amount of fertiliser, ensure the plants receive the same amount of sunlight and water, and are being kept in the same location.

Keeping all these variables controlled ensures you are actually comparing the different types of fertiliser. If you changed the amount of fertiliser and the type of fertiliser, then you could never be sure if it was the type or the amount that caused the change.

Some variables can be more difficult to control than others. In the fertiliser example, there might be differences in the plants that are difficult to observe or quantify that could affect the results. For example, a plant could be affected by a disease in its roots or there could be differences in the plants' individual genotypes that meant some plants could grow more than others. It is important to note these possible challenges when you evaluate your investigation.

As you conduct the investigation on factors influencing the viability of brine shrimp. make sure you note which factors are the controlled variables and which factor is the independent variable.

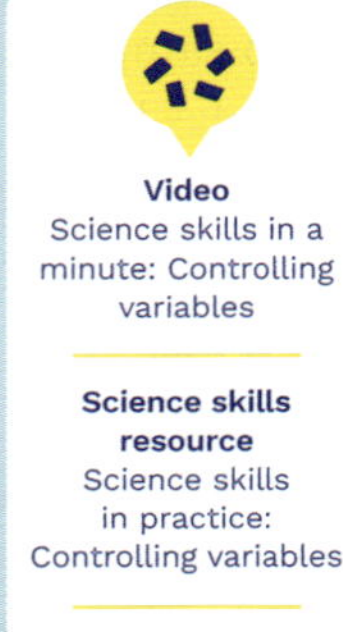

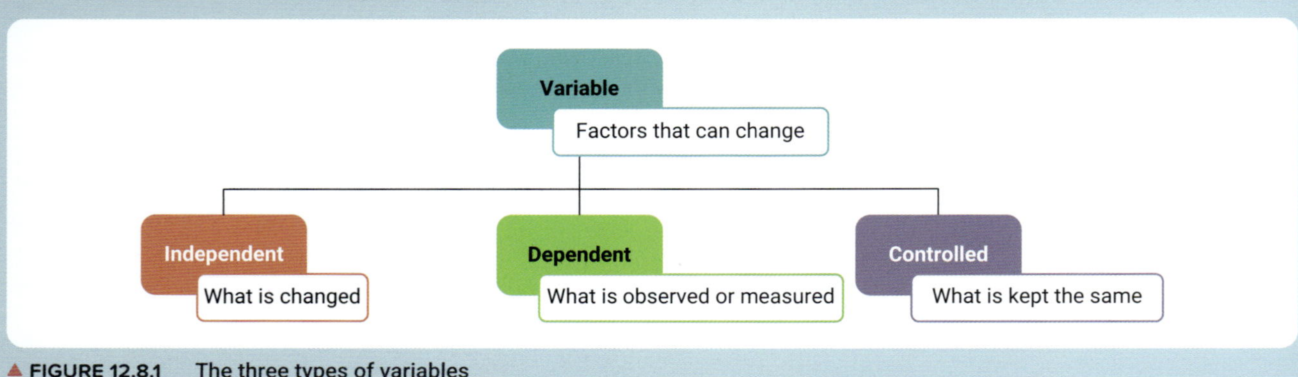

▲ FIGURE 12.8.1 The three types of variables

ENVIRONMENTAL FACTORS INFLUENCING THE HATCHING VIABILITY OF BRINE SHRIMP

BACKGROUND INFORMATION

Brine shrimp (*Artemia salina*) is a species of small crustaceans found in saline environments, specifically salt lakes, worldwide. Brine shrimp do not inhabit oceans due to the high presence of predators. Brine shrimp can avoid predators by living in very high saline environments that other aquatic life cannot. This makes them an excellent model for the study of natural selection and adaptations.

Salinity levels within the environment greatly affect the population growth of brine shrimp. In this investigation, you explore how different saline level environments affect hatching viability. To do this, you will attempt to hatch cysts in four different salt concentrations and measure the hatching viability by counting how many nauplii (brine shrimp larva) emerge from the cysts.

AIM

To understand how environmental factors can affect the hatching viability of brine shrimp and consider how this relates to the theory of natural selection

Safety

Wear appropriate personal protective equipment.

Wash hands thoroughly before and after working with any organic materials.

Know and follow all regulatory guidelines for the disposal of laboratory wastes.

Alternatively, you may like to keep the brine shrimp for further observation.

MATERIALS AND EQUIPMENT

- ☑ brine shrimp eggs (cysts)
- ☑ 4 Petri dishes
- ☑ 3 saltwater solutions (0.5%, 1.0% and 2.0%)
- ☑ fine brush
- ☑ 4 microscope slides
- ☑ double-sided tape
- ☑ magnifying glass
- ☑ permanent marker
- ☑ graduated cylinder
- ☑ distilled water
- ☑ safety glasses, lab coat and gloves

PROCEDURE

PREPARING THE CYSTS FOR HATCHING (DAY 1)

1. Using a permanent marker, label four Petri dishes: 0%, 0.5%, 1.0%, 2.0%.
2. Identify the independent and dependent variables in this investigation. Then formulate your hypothesis.
3. Using the graduated cylinder, measure 30 mL of each saline solution and pour it into the appropriately labelled Petri dish.
4. Collect four microscope slides. Measure and cut four 1.5 cm strips of double-sided tape and gently adhere one of them to each of the microscope slides.
5. Lightly touch the fine brush to the side of the dish containing the brine shrimp eggs. Collect 20–30 eggs on the brush. Do not collect too many eggs because you will be required to count them.
6. To adhere the eggs to the double-sided tape, lightly press the brush onto the tape on the first microscope slide. Repeat this step for the remaining three microscope slides.
7. Using a magnifying glass, count the number of eggs on the first slide. Copy Tables 12.8.1 and 12.8.2 into your workbook and record this information in Table 12.8.1.
8. Once the eggs have been counted, place this slide into the 0% salt solution Petri dish, ensuring that you place the slide with the tape side facing up.
9. Count the eggs on each slide and place them in the respective salt solutions. Record the egg count information in the corresponding row in Table 12.8.1.
10. Place the Petri dishes under a light bank for 24 hours at room temperature.

DATA COLLECTION (DAY 2 AND DAY 3)

1. After 24 hours, examine the contents of each Petri dish with the magnifying glass. You should see that some brine shrimp have hatched and are swimming in the salt solution. Record the number of eggs, the number of dead or partially hatched eggs and the number of swimming brine shrimp in Table 12.8.1.

2 After 48 hours, examine the contents of the Petri dishes again and record your observations in Table 12.8.1. Calculate the hatching viability of each dish at 48 hours by dividing the number of shrimp swimming by the initial number of eggs in the Petri dish. Round up your calculations to the nearest hundredth and add this information to the class results table (Table 12.8.2).

3 Draw a line graph that shows the sample means from the class results.

RESULTS

Copy and complete Tables 12.8.1 and 12.8.2 to record your results.

ANALYSIS

1 Which Petri dish had the highest hatching viability? Which had the lowest? Suggest possible reasons for these results.

2 What is the selective agent in this study?

3 Did the results support your hypothesis?

4 What variables did you control in this investigation? Were there some variables that were harder to control than others?

5 Imagine a hypothetical scenario wherein the salinity of the water lived in by wild brine shrimp dropped to 0.5% and remained that way for a decade. If we repeated this experiment in a decade, do you think the hatching viability percentages would change? **Explain** your answer.

6 What other conditions may affect the hatching viability of brine shrimp? **Design** an experiment to **investigate** another environmental factor that may affect hatching viability.

CONCLUSION

Based on your data and the class data, write a conclusion that explains whether your hypothesis was supported or refuted.

▼ **TABLE 12.8.1** The hatching viability of brine shrimp in varying levels of salinity

% NaCl	0 hours	24 hours			48 hours			
	Eggs	Eggs	Dead or partially hatched	Swimming	Eggs	Dead or partially hatched	Swimming	Hatching viability percentage
0%								
0.5%								
1.5%								
2%								

▼ **TABLE 12.8.2** The hatching viability of brine shrimp in varying levels of salinity; results of different student groups (add rows as required)

Class group	Hatching viability at salinity:			
	0%	0.5%	1.5%	2%
1				
2				
3				

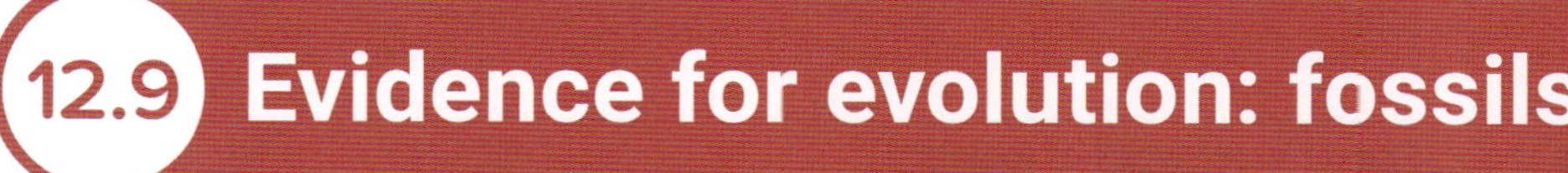

12.9 Evidence for evolution: fossils

BY THE END OF THIS MODULE, YOU WILL BE ABLE TO:

✓ define and provide examples of fossils, absolute dating and relative dating

✓ explain how fossils can be used as evidence for evolution.

GET THINKING

In 1974, American palaeoanthropologist Donald Johanson discovered hundreds of small bone fragments belonging to an individual female hominin of the species *Australopithecus afarensis*. The fossil was dated to 3.2 million years and nicknamed 'Lucy'. She was 40 per cent complete and an important fossil discovery. Examine Figure 12.9.1. In your workbook, identify the characteristics of the Lucy skeleton that led scientists to believe she could stand upright and walk bipedally (on two feet).

Video activity
Fossil evidence

Extra science investigation
Modelling fossilisation processes

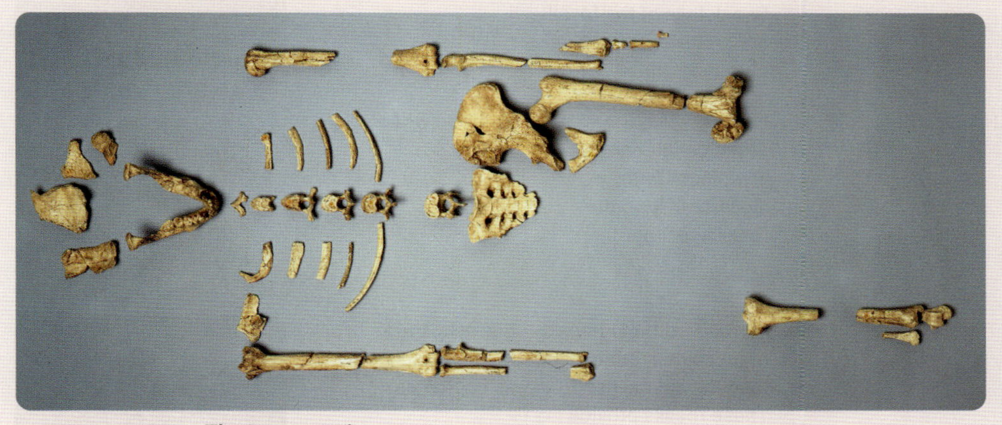

The Natural History Museum/Alamy stock photo

▲ **FIGURE 12.9.1** The Lucy specimen

What is the fossil record?

In Stage 4, you learned that **fossils** are the remains or traces of living things preserved in rocks. The collection of all known fossils and the information that they provided about Earth's past life and environments is called the **fossil record**.

fossil
the remains or traces of living organisms, commonly preserved in sedimentary rocks

fossil record
the collection of all known classified fossils and the information they provide about past life

Not all organisms fossilise well. The remains or traces of many organisms have decayed, been destroyed or are inaccessible. This has created large gaps in the fossil record, which scientists can fill using only assumptions and inferences based on the discoveries on either side of the gap.

Table 12.9.1 describes the different categories of fossils.

▼ **TABLE 12.9.1** The different types of fossils

Type of fossil	Description	Example	Image of fossil
Original	Formed when the chemical composition remains, similar to when the organism was alive. The fossils are recognisable.	Small insects trapped in amber or whole-body fossils	
Carbon film	Formed when a single layer of leaves or seeds is pressed between layers of sediment. The fine details of the fossil are preserved as a black carbon image.	Commonly plant fossils	
Replacement	Formed when the organic material is replaced with silica, petrifying the fossil specimen.	Petrified wood	
Trace	Formed from the trace of an animal rather than the animal itself. These fossils provide information about the organism's environment, diet and life.	Burrows, tracks or faeces (also known as coprolites)	The Laetoli footprints in Olduvai Gorge, Tanzania

absolute dating
determining the age of a fossil in years

relative dating
determining if a fossil or rock is older or younger than another

radiometric dating
a dating method that measures the decay of radioactive isotopes to determine the age of fossils

Absolute dating methods

Once a fossil has been discovered, it needs to be classified and dated to place it on the fossil record. Dating fossils can be done in two ways: **absolute dating** or **relative dating**.

Absolute dating determines the actual age of the fossil in years. This is achieved by using **radiometric dating**, where age can be calculated based on the decay of radioactive isotopes found in the fossil. Carbon-14 dating and potassium–argon dating are examples of radiometric dating.

Carbon-14 dating is useful for determining the actual age of an organic fossil because all living things contain carbon-14. This type of dating is used when the age of the fossil is less than 70 000 years.

Potassium–argon dating is an example of absolute dating suitable for older fossils at least 200 000 years old.

Relative dating

Relative dating is used to organise fossils on a scale to determine if a fossil is older or younger than another fossil, without determining an exact age. Examples of relative dating are **stratigraphy** and the use of **index fossils**.

stratigraphy
comparing strata or layers of rock to determine the relative age of fossils

index fossil
a fossil that can be used to compare the relative age of rock strata from different locations

Stratigraphy is the study of strata, or rock layers (Figure 12.9.2). It uses the superposition principle, which states that layers that are lower in strata are older than layers above (Figure 12.9.3). Fossils that are found in different layers can then be relatively dated as older or younger than each other.

Istock.com/iStock Essentials

▲ **FIGURE 12.9.2** Rock layers

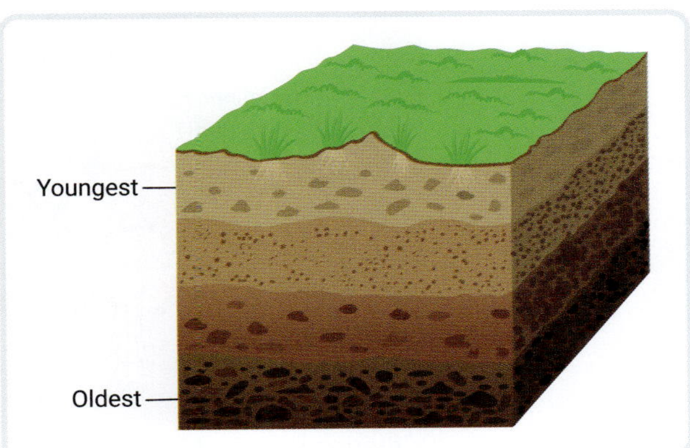

▲ **FIGURE 12.9.3** The principle of superposition in rock strata

Index fossils are fossils that can be classified to a species level and are geographically widespread over a short timeframe. Identical fossils were formed at the same time at the same location, so finding them widespread across Earth helps reinforce the movement of Earth's crust at various times in geological history. Fossils found alongside index fossils can then be relatively dated or correlated to strata from different areas of the world. Index fossils are frequently marine organisms, such as trilobites and ammonites (Figure 12.9.4).

Mardoz/Shutterstock.com

Wlad74/Shutterstock.com

◀ **FIGURE 12.9.4** Common index fossils: (a) trilobite and (b) ammonite

The fossil record as evidence for evolution

The fossil record is an important set of data that provides evidence for evolutionary change because it shows the timeline of life on Earth and how species have changed over millions of years. The fossil record also provides insights into past environments. For example, the fossil record shows how horses evolved over time from a small, multi-toed animal to a large single-toed animal, suited to running in open grasslands (Figure 12.9.5). It also includes marine fossils found in modern-day deserts, which indicate the region was once under water.

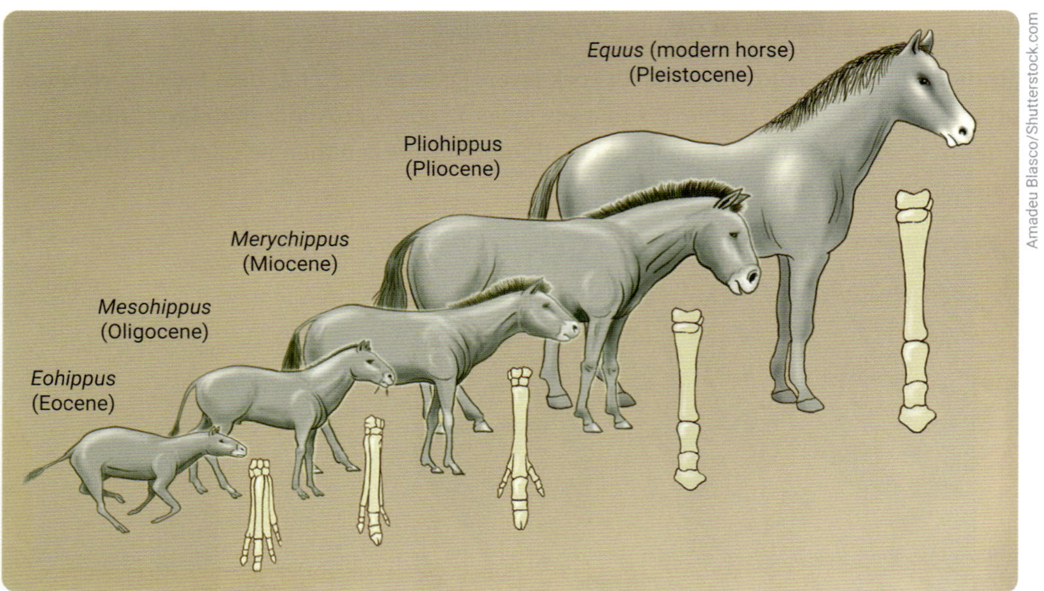

▲ **FIGURE 12.9.5** Evolution of the horse from multi-toed to single-toed animal

▲ **FIGURE 12.9.6** *Archaeopteryx*: a transitional fossil between dinosaurs and birds

Transitional fossils are important fossils that provide further evidence of evolution. *Archaeopteryx* was a 'bird-like' dinosaur that bridges the gap between the non-avian dinosaurs and birds, as it had both feathers and scales. It was first discovered in 1861, and there have been a further 11 body fossil specimens discovered (Figure 12.9.6).

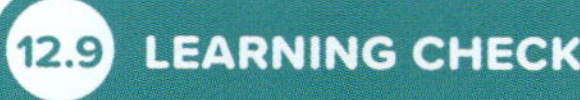

1 **Define:**
   a the fossil record.
   b absolute dating.
   c relative dating.
   d index fossils.

2 Consider the following organisms. **Determine** which one(s) would most likely form a fossil. **Explain** your choices.
   - Earthworm
   - Clam
   - Bird
   - Jellyfish
   - Sea urchin
   - Leech

3 In your workbook, copy and match the type of fossil to its description.

Type of fossil	Description
Trace	• Minerals replace the organic material in the hard parts of an organism
Carbon film	• Unchanged parts of an animal or plant
Original	• A thin black deposit of carbon that shows the fine details of fish scales or plant leaves
Replacement	• The footprint, trail, burrow or faeces of an organism

4 The fossil record is incomplete. **Describe** three reasons why there are gaps in the fossil record.

5 **Explain** why a fossil dated to 100 000 years cannot have been dated using carbon-14.

6 **Compare** and **contrast** absolute and relative dating methods.

7 **Discuss** why the fossil record is evidence for evolutionary change.

WORKING SCIENTIFICALLY

12.10 How to examine fossils

SCIENCE SKILLS IN FOCUS

IN THIS MODULE, YOU WILL FOCUS ON LEARNING AND IMPROVING THIS SKILL:

- observing and recognising different types of fossils to create a small fossil record.

Tips for examining fossils

The fossil record is the collection of known classified fossils and the information they reveal about the history of life on Earth. To observe and create a small fossil record from the ones provided in this investigation you will need to:

- Carefully examine the fossils for features and patterns that will give you insights about the creature, such as shape, ridges, presence of teeth, bone type, body segmentation, leaves or footprints.
- Sort the fossils into small groups that have similar characteristics. This will help you to make an initial classification into the major groups: vertebrates, invertebrates or plants.
- Compare your fossils with modern organisms. This will make it easier to identify them when you are doing your research. It will also help you to recognise evolutionary changes over time or to suggest a common ancestor.

The geological time scale will help you to place your fossils in a timeframe and identify the order they lived. You can link them to the mass extinctions or major evolutionary events in the era of the invertebrates.

As an extension of your investigation, you can create a phylogenetic tree (a diagram that shows evolutionary relationships) to trace the evolution of the fossils presented in this investigation.

Weblink
Britannica: Geological time scale

OBSERVING FOSSILS

AIM

To observe different types of fossils

MATERIALS AND EQUIPMENT

- ☑ a range of fossils consisting of original, carbon film, replacement, casts, moulds and amber-preserved fossils
- ☑ hand lens
- ☑ access to the internet

PROCEDURE

1 Examine a fossil. Describe its appearance and suggest what kind of organism it was and what it may have looked like when it was alive.

2 Identify the way the fossil was formed.

3 Using the internet, research the name of the organism and its age (how long ago it was formed).

4 Repeat steps 1–3 for each fossil.

RESULTS

Record your observations in a table. Remember to give your table a title.

ANALYSIS

1 After determining the age of each fossil, **create** a timeline and place your fossils along it.

- a Are there any gaps in your fossil record? Suggest reasons why this might have occurred.
- b **Research** the geological timescales and overlay these eras and epochs on your timeline.
- c **Research** when *Australopithecus afarensis* first showed in the fossil record and add them to your timeline.

2 **Explain** how fossils are used as evidence for evolution.

CONCLUSION

Write a brief conclusion for this investigation based on your observations.

12.11 Evidence for evolution: comparative anatomy

BY THE END OF THIS MODULE, YOU WILL BE ABLE TO:

- ✓ describe how comparative anatomy can be used as evidence for evolution
- ✓ define and provide examples of homologous structures and vestigial organs
- ✓ explain how embryology is used to provide evidence for evolution
- ✓ explain how comparative anatomy indicates common ancestry.

GET THINKING

The appendix in humans has long been thought of as a remnant organ that does not serve a purpose. New research shows that the role of the appendix is to store beneficial bacteria for good gut health. During early development, the appendix plays a role in maturing white blood cells and producing antibodies for the immune system. Humans have a collection of these remnant organs; for example, the muscles that control the outer ear (pinna). Conduct some research and write a short statement about why these muscles may become 'useful' again in time.

Interactive resource
Drag and drop: Structural evidence of evolution

Other resource
Worksheet: Evolution case study

Fossils provide great evidence for evolution as they show the gradual change in characteristics of organisms. As you learned in Module 12.9, fossils don't often form, and can be hard to find and classify, so looking for other structural pieces of evidence that do not require fossilisation to support evolution is useful.

Structural evidence for evolution

Comparative anatomy is the study of body structures of organisms to show the adaptive changes that have occurred in species that share a common ancestor. Examples of comparative anatomy include:

- **homologous structures**
- **vestigial organs**
- **embryology**.

comparative anatomy the study of the body structures of species to show the adaptive changes made from a common ancestor

homologous structures body parts that can be found in a range of species that have similar structures but different functions

vestigial organs organs that are retained in a species despite no longer being functional as they were in the ancestral species

embryology the study of the early stages of development

Homologous structures

Homologous structures are body parts that can be found in a range of species but have different functions to suit the species' way of life. An example is the vertebrate forelimb. Look at Figure 12.11.1.

In the vertebrate animals shown, the forelimbs show similar structure, called the pentadactyl limb ('penta' means five and 'dactyl' means fingers). However, each limb is adapted for different purposes. For example, a bat's pentadactyl limb is modified for flying, and a human limb is modified for grasping and using tools. The similar structure indicates a common ancestor between all vertebrates. All physical structures are encoded by DNA, which is inherited across generations. The instructions to build the pentadactyl limbs had to exist in a common ancestor, and, over time, different adaptations evolved so that each species developed a form of limb that made it best suited to its environment.

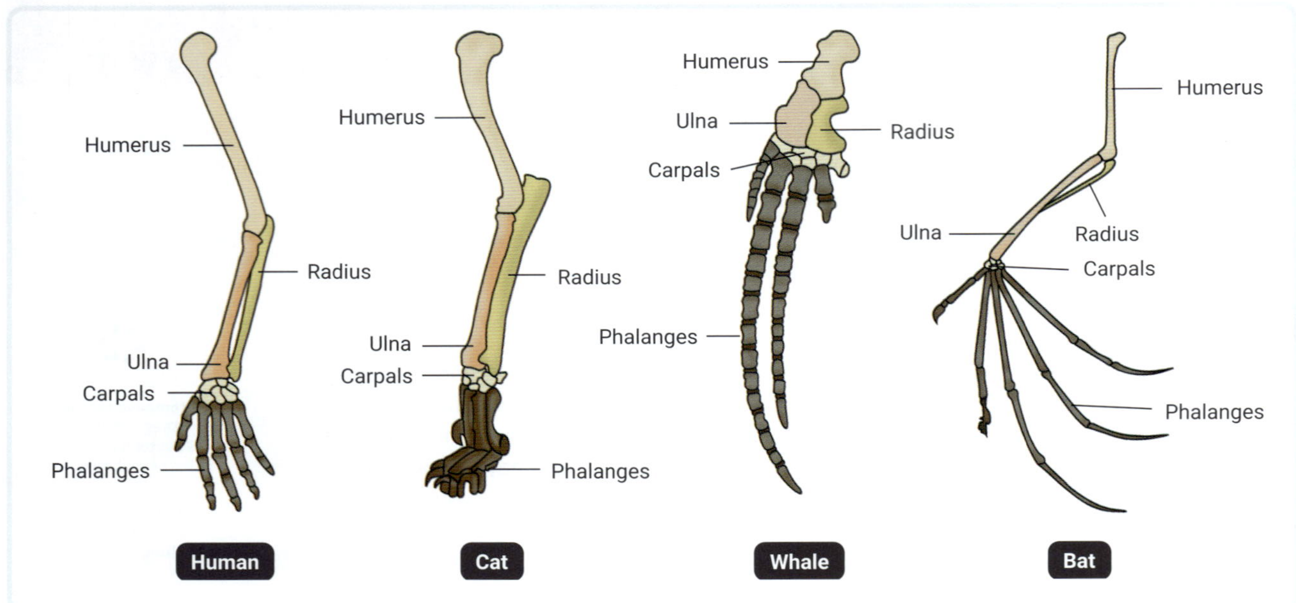

▲ **FIGURE 12.11.1** The pentadactyl forelimb structures in four different vertebrates

analogous structures structures in different species that have similar function but are anatomically different

In contrast to homologous structures, **analogous structures** are features of species that are similar in function but not in structure. An example is the wings of an insect and a bird (Figure 12.11.2). Both animals require wings to survive in their environment, but the underlying structure of the wing is very different, and there is not a recent common ancestor between insects and birds.

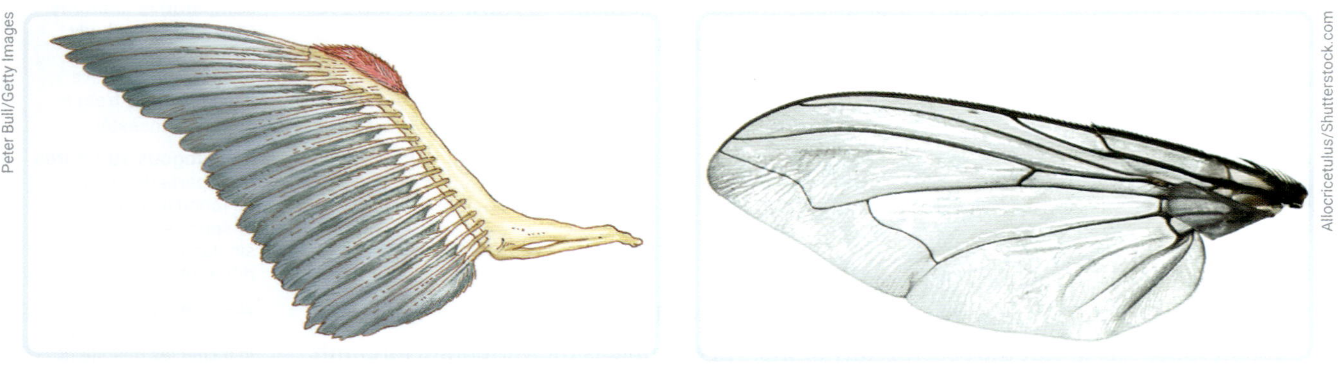

▲ **FIGURE 12.11.2** Bird wings and insect wings have similar functions but different structures.

Vestigial organs

Vestigial organs are organs in living species that no longer serve a purpose. However, these organs were once useful in ancestral species. The DNA that codes for the structures is still present, having been passed down the generations, indicating the modern-day species has evolved from the ancestral one. Vestigial structures tend to be of reduced size, so they do not require as much energy in the living organism. Over time, some organisms are born without these structures.

Humans have a few examples of vestigial organs, which are evolutionary leftovers from our primate ancestors (Figure 12.11.3).

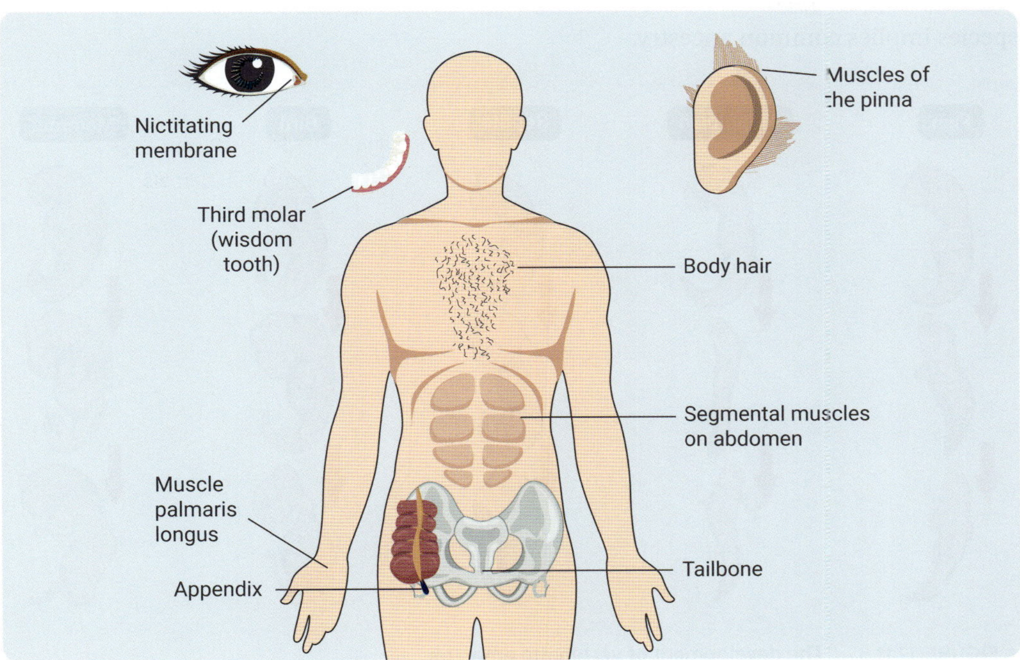

▲ **FIGURE 12.11.3** The vestigial organs present in humans

An easy one to check is the presence of the muscle palmaris longus (Figure 12.11.4). This muscle in the wrist and hand was used to help primates move around the trees. Place your hand flat on a surface and touch your pinkie finger to your thumb. If you see a raised band in your wrist, that is the vestigial muscle. It may not be present in both wrists, and about 14 per cent of humans do not have the muscle anymore.

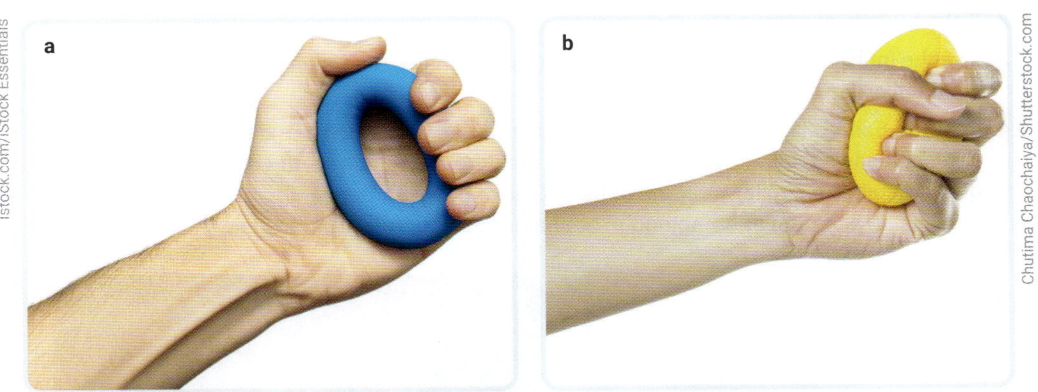

Istock.com/iStock Essentials

Chutima Chaochaiya/Shutterstock.com

▲ **FIGURE 12.11.4** The (a) presence and (b) absence of the palmaris longus muscle

Embryology

Embryology is the study of embryos and can be used to show structural evidence for evolution. An embryo is the name given to an organism early in its development. During this time, the body plan is determined based on the instructions coded in DNA. Many vertebrate embryos show similar developmental patterns and can be indistinguishable from each other in the early stages of development.

All vertebrate embryos have gill slits, although this feature is lost or changed in the adult form (Figure 12.11.5). All human embryos have a tail that becomes the tail bone during later development. The similarity in embryological development between different species implies common ancestry.

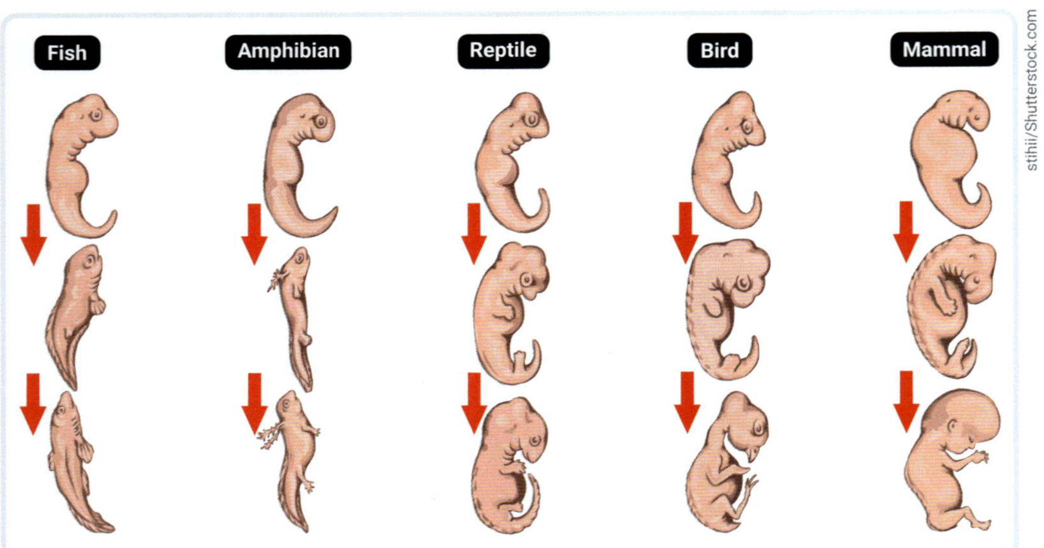

▲ **FIGURE 12.11.5** The development of vertebrate embryos

Comparative anatomy provides further evidence for evolution that can be seen in living organisms, not just fossils.

12.11 LEARNING CHECK

1 **Define** comparative anatomy and provide three examples.
2 **Define** vestigial organs.
3 Use the internet to **identify** the vestigial organ in whales. **Explain** why whales have this vestigial organ.
4 In South-East Asia, there are flying lemurs: primates with a flap of skin between their front and back paws that allows the animals to glide from tree to tree. In Australia, the sugar glider displays a similar webbing to help it move between trees (Figure 12.11.6). **Identify** if this is an example of homologous or analogous structures and **explain** your choice.

▲ **FIGURE 12.11.6** (a) A flying lemur from Madagascar and (b) an Australian sugar glider

12.12 Identifying homologous structures

SCIENCE SKILLS IN FOCUS

IN THIS MODULE, YOU WILL FOCUS ON LEARNING AND IMPROVING THIS SKILL:

- identifying trends and patterns seen in homologous structures.

To analyse homologous structures, focus on identifying patterns and relationships.

To identify patterns, you need to:

- identify common features across species by recognising them as homologous
- compare the size of homologous structures, which may vary depending on their function in the species.

To analyse relationships, you need to:

- understand how specific homologous structures are adapted to the different environmental needs of that species
- relate the structure and its function; for example, wings are structures adapted to flight.

EXAMINING VERTEBRATE SKELETONS

AIM

To compare different vertebrate skeletons

MATERIALS AND EQUIPMENT

☑ skeletons of a human, cat, bird, frog, bony fish and lizard (if not available, use X-ray images from the internet)

RESULTS

▼ **TABLE 12.12.1** A comparison of homologous structures

Structure	Most alike	Least alike
Skull		
Vertebral column		
Ribs		
Front limbs		
Rear limbs		
Pelvic girdle		

PROCEDURE

Carefully observe each skeleton. Copy and complete Table 12.12.1 by recording which animals are most alike and which are least alike for the named structures.

ANALYSIS

1 Which two animals are the most alike, and which two are the least alike?

2 One of these animals does not live on land. **Describe** how its skeleton differs from the others.

3 Most of these animals have similar skeletons, often with the same bones in the same places. **Name** this type of evidence of evolution and **describe** how it occurred.

4 Build a phylogenetic tree based on the similarities recorded in Table 12.12.1. On it, show when the features of scales, feathers, fur/hair and leathery eggs would have appeared. Carry out further research if needed.

CONCLUSION

Write a brief conclusion for this investigation based on your observations.

12.13 Evidence for evolution: DNA and proteins

BY THE END OF THIS MODULE, YOU WILL BE ABLE TO:

- ✓ define and give examples of ubiquitous proteins and bioinformatics
- ✓ predict how recent the common ancestor is based on the similarity of the DNA or proteins
- ✓ explain how phylogenetic trees model evolution.

Video activity
Evolution: The evidence

Other resource
Worksheet: Comparing homologous structures

GET THINKING

Humans and chimpanzees, including the bonobos, share 98.8 per cent of the same DNA. Yet we look very different, have different behaviours and live in very different environments. The 1.2 per cent difference equates to about 35 million differences in the nucleotides, and we understand that the genes in chimpanzees operate at different levels compared to the same genes in humans. Using the internet, research how similar humans are to rats, pigs and bananas. How does comparing our genetic code help scientists understand evolution? Write a sentence to explain the importance of DNA to evolution.

So far, you have learned about structural evidence that supports the theory of evolution. Homologous structures, vestigial organs and embryology imply that these structures are similar because of shared DNA from a common ancestor.

biochemical evidence evidence of evolution based on the fact the same enzymes are found in the cells of most organisms

Biochemical evidence for evolution

Biochemical evidence of evolution uses the premise that proteins, particularly enzymes, are found in the cells of nearly all life on Earth. To understand biochemical evidence for evolution, let us first revisit what DNA is and how it controls the characteristics of an organism.

As you know from Chapter 11, DNA holds the sequence of nucleotides (or base pairs) that code for amino acids, the building blocks of proteins (Figure 12.13.1). Proteins are essential for cell processes and characteristics and, ultimately, the traits of the whole organism. DNA is inherited equally from two parents, and offspring will express the traits of the parents in their phenotype.

Natural selection, as a mechanism for evolution, states that organisms with the traits best suited for their environment will survive and reproduce, passing on the DNA for advantageous or neutral changes to the next generation, resulting in the gradual change of the species.

This means that the DNA and proteins can also indicate relationships between species and be used to show common ancestry.

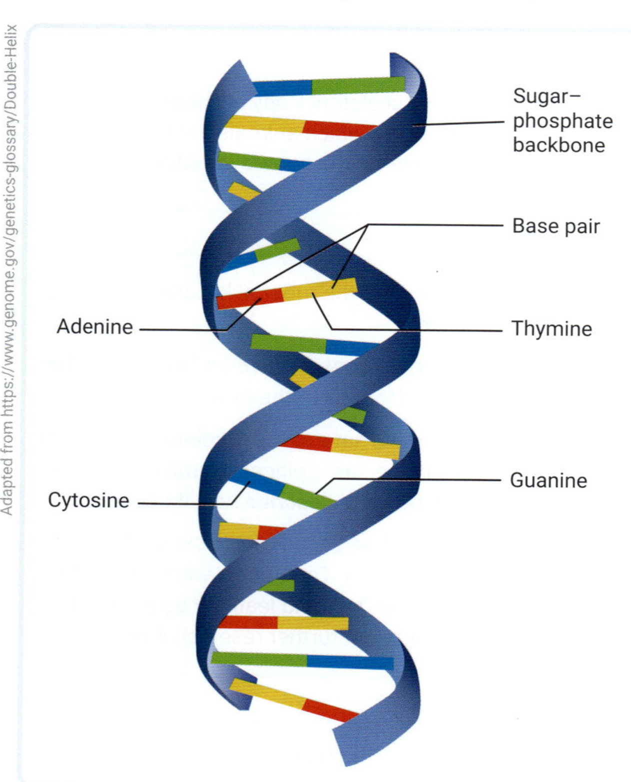

▲ **FIGURE 12.13.1** A DNA molecule showing base pairs: guanine and cytosine, and adenine and thymine

9780170491785

Biointormatics

12.13

Bioinformatics is the use of technologies to collect and analyse biological data, including genetic code. It is necessary because the genome of an organism can be billions of base pairs in length.

Certain proteins are used in bioinformatics because they can be found in the cells of nearly all living things and perform the same function. These are known as **ubiquitous proteins**. An example is cytochrome C, which is a protein necessary in the mitochondria to synthesise adenosine triphosphate (ATP) during cellular metabolism. It is made up of 100–104 amino acids and is largely unchanged between species (Table 12.13.1).

bioinformatics the use of technology to collect and analyse biological data, such as a DNA or amino acid sequence

ubiquitous proteins proteins that are found in nearly all organisms and carry out the same function

Comparing the sequence of amino acids in the protein of two organisms indicates the closeness of the relationship between the organisms. The more differences seen in the sequence, the more time has passed since divergence from the common ancestor. Conversely, the more similar the sequence of amino acids, the more recent the common ancestor.

▼ TABLE 12.13.1 Comparing the amino acid sequence in cytochrome C in humans to other organisms

Organism	Differences in amino acid sequence
Chimpanzee	0
Rhesus monkey	1
Rabbit	9
Pig	10
Dog	10
Penguin	11
Horse	12
Moth	24
Yeast	38

Note: Data taken from Margoliash, E. and Finch, W.M. 1967. Construction of phylogenetic trees. *Science* 155: 279–284.

The alpha and beta chains in haemoglobin are two further examples of ubiquitous proteins. Haemoglobin is found in red blood cells. It binds with oxygen to enable red blood cells to transport oxygen to the body's cells. When comparing the amino acid sequences of both the alpha and beta chains, we can observe there are no differences in the sequences between chimpanzees and humans, and only one difference in each chain comparing humans to gorillas.

GUDKOV ANDREY/Shutterstock.com

Claire E Carter/Shutterstock.com

▲ FIGURE 12.13.2 (a) Chimpanzees and humans have identical amino acid sequences in the alpha and beta chains of their haemoglobin. (b) Humans and gorillas have only one difference in each chain.

phylogenetic tree
a diagram representing lines of evolutionary descent from a common ancestor

Data on the relationship between organisms can be represented pictorially as a **phylogenetic tree**. Phylogenetic trees are diagrams that show the evolutionary descent of different species from a common ancestor. In the tree, two species are drawn closer together with a more recent common ancestor if they are more related. Less related species are drawn further apart with a more distant common ancestor.

Phylogenetic trees are used to show the best hypothesis about how a set of species evolved from a common ancestor. They can be drawn in various ways but follow a basic pattern.

1 The species of interest are found at the tips of the branches.
2 Each node, or branching point, represents a point of divergence from the common ancestor.
3 Each branch represents the series of ancestors leading up to the species at the end.
4 The trunk of the tree is the common ancestor between all species.

An example of a phylogenetic tree based on the data from Table 12.13.1 can be seen in Figure 12.13.3.

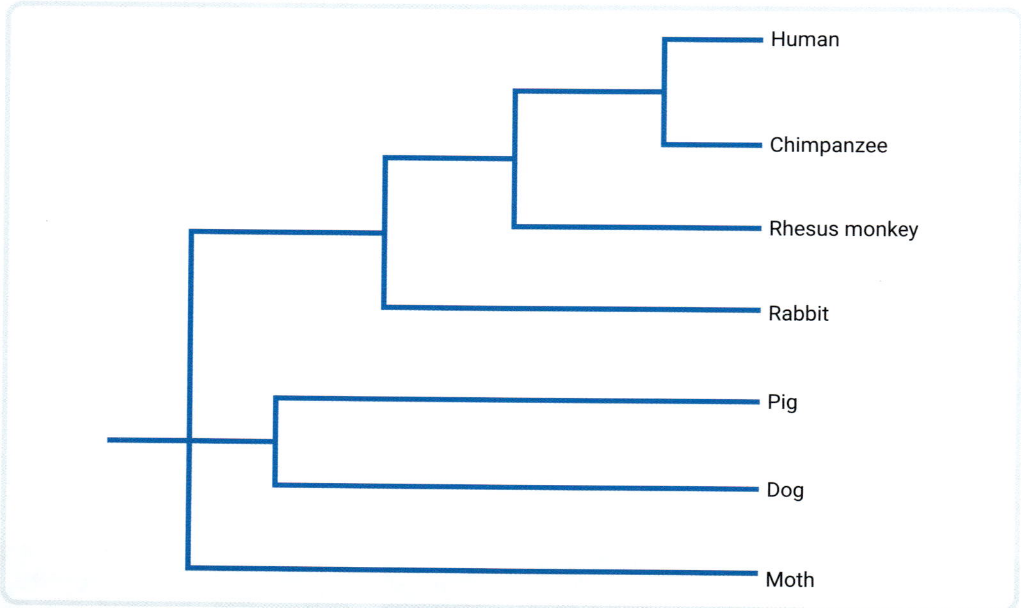

▲ **FIGURE 12.13.3** A phylogenetic tree showing the relatedness of organisms using amino acid sequencing of cytochrome C

9780170491785

DNA sequences can also be used to show relatedness. Three DNA nucleotides code for one amino acid, and there are 64 different combinations to code for the 20 amino acids that make up proteins. More than one triplet of nucleotides may code for the same amino acid. An example is the amino acid valine, which can be coded for in the DNA by CAA, CAG, CAT or CAC.

As such, there could be a difference in the DNA sequence that is not represented by a change in the amino acid sequence. Comparative genomics uses bioinformatics to compare sequences of nucleotides to determine relatedness between species and can result in a more accurate determination of relatedness.

12.13 LEARNING CHECK

1 **Explain** how DNA provides evidence for evolution.
2 **Define** ubiquitous proteins and provide three examples.
3 Use the data in Table 12.13.1 to **explain** how closely related humans are to chimpanzees compared to a penguin and a moth.
4 DNA analysis compares the nucleotide sequence, and protein analysis compares the amino acid sequence. Which is more accurate? **Justify** your choice.

CRISPR: ethics and evolutionary impact

BY THE END OF THIS MODULE, YOU WILL BE ABLE TO:

- ✓ explain what CRISPR is and how it works
- ✓ discuss the implications of CRISPR for evolutionary change
- ✓ assess the ethical implications of CRISPR.

What is CRISPR?

CRISPR stands for **C**lustered **R**egularly **I**nterspaced **S**hort **P**alindromic **R**epeats. It is a groundbreaking technology that allows scientists to edit genes. CRISPR is very precise and is used to target specific genes. The technology is used in many areas of science, including agricultural engineering, biotechnology and medical research.

CRISPR is made up of two parts:

- a protein that can cut DNA in specific locations, called CAS9
- a short sequence of **RNA**, known as guide RNA, that recognises the sequence of DNA that needs to be edited.

RNA
ribonucleic acid (RNA), a single-stranded nucleic acid

Scientists can use CRISPR to insert, delete or change DNA sequences within an organism's genome (Figure 12.14.1).

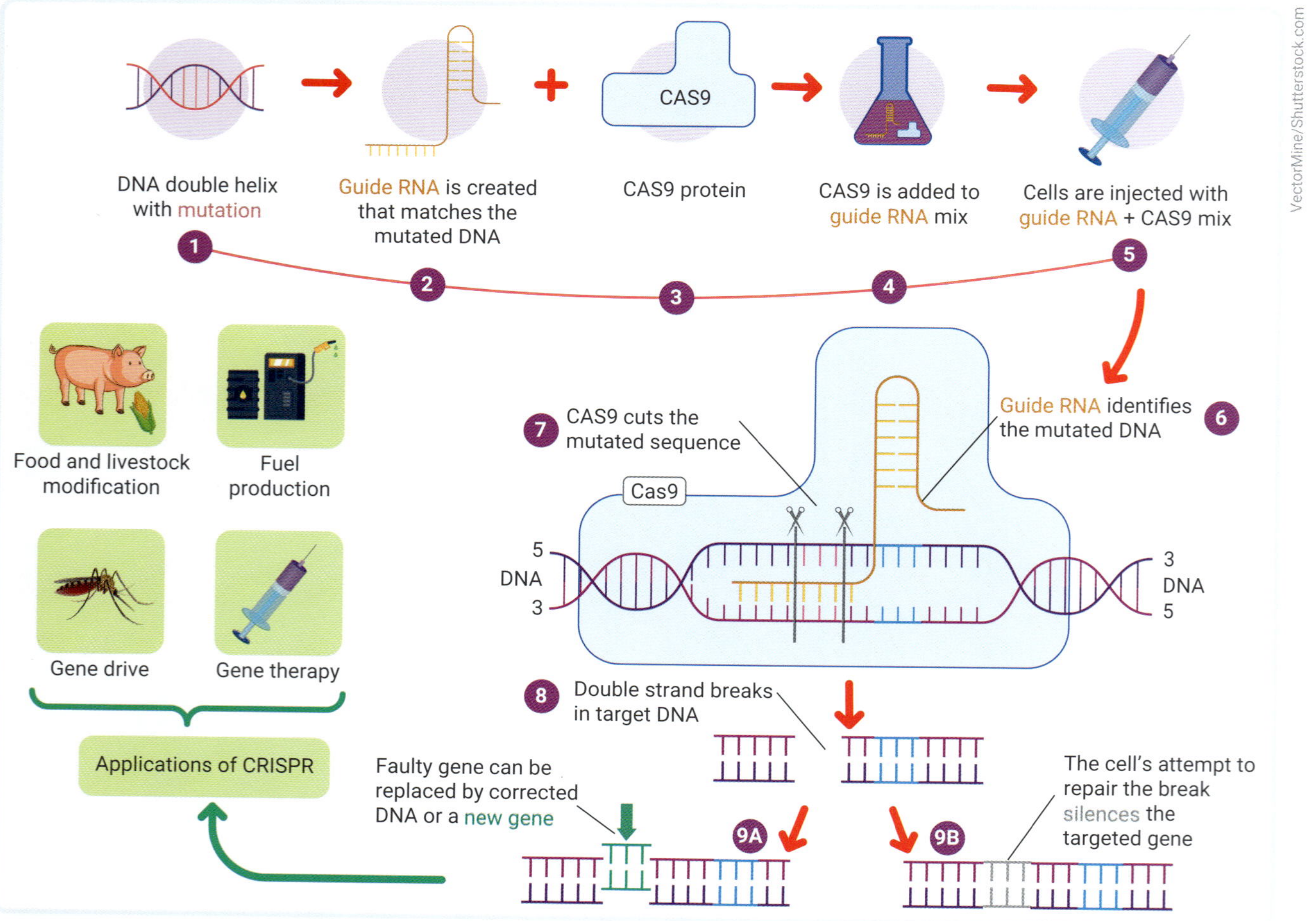

1 DNA double helix with mutation
2 Guide RNA is created that matches the mutated DNA
3 CAS9 protein
4 CAS9 is added to guide RNA mix
5 Cells are injected with guide RNA + CAS9 mix
6 Guide RNA identifies the mutated DNA
7 CAS9 cuts the mutated sequence
Cas9
5 DNA 3
3 DNA 5
8 Double strand breaks in target DNA
9A Faulty gene can be replaced by corrected DNA or a new gene
9B The cell's attempt to repair the break silences the targeted gene
Food and livestock modification
Fuel production
Gene drive
Gene therapy
Applications of CRISPR
VectorMine/Shutterstock.com

▲ **FIGURE 12.14.1** CRISPR is used to edit genes.

9780170491785

CRISPR and evolution

Scientists have been using biotechnology to manipulate the DNA of organisms for decades (Module 11.15). But now with CRISPR, it is easier and faster to edit genes. Scientists can use it to introduce traits that don't occur naturally in the species. For example, in agriculture, scientists can alter the DNA of crops so they are resistant to pests, can better tolerate extreme weather, or grow faster. Similarly, scientists can genetically engineer animals to enhance desired traits, such as disease resistance or increased productivity. CRISPR may help treat people with genetic illnesses and cancer by providing functional genes or therapies that specifically attack cancer cells. These therapies are used on body cells, so any genetic changes are not inherited.

Human germline engineering is currently illegal. The suggestion of using CRISPR to modify human germlines has sparked debates about its impact on human evolution and diversity. Some fear that if it was used, it could lead to a reduction in genetic variation, which could make populations more vulnerable to environmental change. However, the technology offers the possibility to eradicate genetic conditions such as cystic fibrosis and sickle cell anaemia (Figure 12.14.2).

human germline engineering
editing the genome of germ cells so that genetic changes can be passed on to the next generation

In theory, CRISPR could also be used to create entirely new organisms or hybrids. For example, researchers can edit genes related to adaptations, such as those involved in camouflage or disease resistance, in species with short life spans such as bacteria or insects. They can then observe how these traits evolve over generations in controlled environments. This can provide direct insights into how natural selection and other evolutionary mechanisms work at the genetic level.

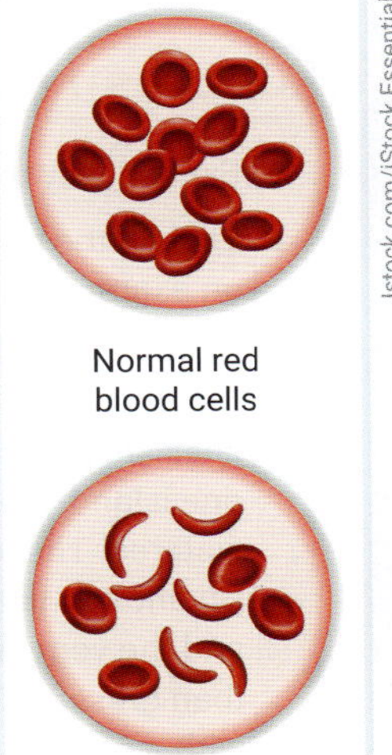

▲ **FIGURE 12.14.2 CRISPR could be used to eradicate genetic conditions such as sickle cell anaemia.**

Ethical and evolutionary considerations

While CRISPR offers incredible power to influence evolution, it also carries risks of unintended consequences. Changes made to one species may have unforeseen effects across entire ecosystems. For example, editing one gene in an organism might change how it interacts with other species, which in turn could affect food chains, habitats and biodiversity. Research organisations must meet strict rules to ensure they don't produce ecologically damaging organisms or release them into the environment. In most countries, studies also must be approved by ethics committees.

CRISPR could lead to genetic inequality in society. People in wealthy, developed countries will have more access to technologies than people in undeveloped countries. These ethical challenges need to be addressed as CRISPR becomes more widely used.

12.14 LEARNING CHECK

1. What does the acronym CRISPR stands for?
2. What are the applications of CRISPR in different fields of science?
3. **Describe** the process of CRISPR in simple steps.
4. **Outline** the advantages of CRISPR.
5. **Discuss** the implications of CRISPR on evolution.
6. **Assess** the ethical considerations of CRISPR.

12 REVIEW

REMEMBERING

1 What is the difference between a gene mutation and a chromosomal mutation?

2 **Identify** an example of a physical mutagen, a chemical mutagen and a biological mutagen.

3 **Draw** a representation of a:

- **a** deletion mutation.
- **b** translocation mutation of chromosome 5.
- **c** substitution mutation.

4 **Define** the fossil record.

5 **Define** evolution by means of natural selection.

UNDERSTANDING

6 **Explain** why there are gaps in the fossil record.

7 **Explain** the process of speciation by using an example.

8 On his trip to the Galápagos Islands, Charles Darwin observed the giant tortoises and how their shell shape differed between different islands. On one island, the vegetation was lush, green and well watered. The tortoises here had rounded shells. On another island, the available water was less, and the vegetation grew higher on bushes rather than grasses. These tortoises had a flat shell with a peak at the front of the shell. **Explain** these observations of tortoise shells using Darwin's two inferences.

9 Is genetic drift more likely to affect a large population or a small population? **Explain** your answer, giving an example.

10 **Examine** the three strata shown in the diagram at the top of the next column.

- **a** **Determine** the youngest layer and outcrop.
- **b** **Determine** the oldest layer and outcrop.
- **c** **Explain** what conditions must be achieved for a fossil to be classified as an index fossil.
- **d** **Propose** how index fossils would be useful in dating strata from different areas in the world.

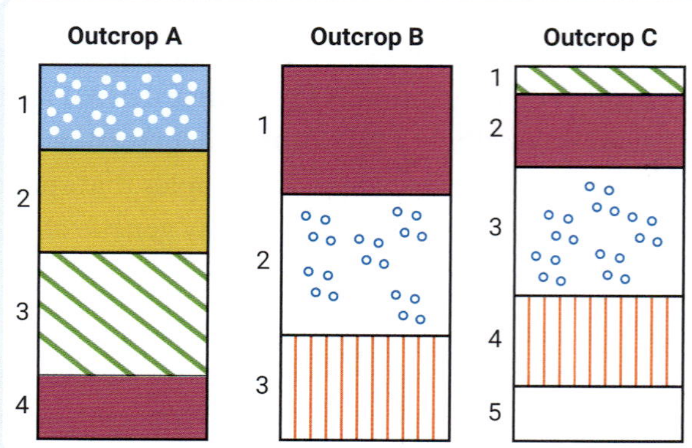

APPLYING

11 **Recall** the calculation to **determine** the combinations that can be formed when chromosomes line up in metaphase I.

- **a** Using this calculation, **determine** the combination of chromosomes that can be formed during meiosis in a human.
- **b** Some organisms, such as aphids, reproduce through asexual reproduction, resulting in genetically identical offspring. **Propose** what would happen to a population of aphids if a new selective agent was introduced into the population.

12 Sharks and dolphins both live in the ocean and require fins or flippers to help them swim. **Research** whether shark fins and dolphin flippers are homologous or analogous structures and **justify** your reasoning.

13 Consider the following information on the domesticated dog, *Canis familiaris*.

- The ancestor of the domesticated dog is an ancient extinct wolf.
- Domestication of the dog began 15 000 years ago, before the development of agriculture.
- All current dog breeds are of the same species.
- Pedigree dogs, such as German shepherds and French bulldogs, are expensive and have many genetic conditions such as hip dysplasia.

- **a** **Identify** what traits the wolves may have displayed for humans to want to domesticate them.

9780170491785

b **Explain** what the humans would have done for the wolf to become tame.

c **Define** species.

d What do you think the term 'pedigree' means?

e **Explain** how artificial selection has increased the frequency of genetic diseases in pedigree dogs.

EVALUATING

14 The amino acid sequence of cytochrome C of five different species of vertebrates was analysed. Table 12.15.1 shows the number of differences in the sequence between each pair of species.

a Using the data in the table below, **construct** a phylogenetic tree to reflect the evolutionary relationships of the organisms.

b **Identify** the organism that is the least related to the others.

c In your phylogenetic tree, **determine** when feathers and fur may have appeared in the ancestral forms.

15 Antibiotics are used to treat bacterial infections and have been prescribed to patients since 1910. However, many strains of bacteria no longer die upon exposure to antibiotics. They have become resistant and are considered 'superbugs'.

a Using the principle of natural selection, **determine** how these antibiotic-resistant bacteria came to be.

b **Suggest** how antibiotic use could change to reduce the occurrence of antibiotic-resistant bacteria.

c Can you think of a way to use DNA technology to develop a new antibiotic?

16 *Archaeopteryx* is a transitional fossil, showing the characteristics of a reptile and a bird. **Explain** why transitional fossils occur and their importance in providing evidence for evolution.

17 **Compare** and **contrast** artificial selection and natural selection.

CREATING

18 **Create** a crossword with the glossary terms in this chapter.

	Wild horse	Black mamba snake	Red jungle fowl	Emperor penguin	African wild donkey
Wild horse	0	21	11	13	1
Black mamba snake		0	18	17	20
Red jungle fowl			0	3	10
Emperor penguin				0	12
African wild donkey					0

SCIENCE IN DEPTH STUDY

1 Connect what you've learned

Consider this statement: 99.9 per cent of everything that has ever lived on Earth has died. Apply your understanding of natural selection and evolution to that statement.

2 Check your thinking

Use the internet to research the following predecessors to modern humans. For each extinct hominin, identify the temporal distribution, their geographical distribution, their cranial capacity and a summary of any cultural advances scientists have evidence of. Display your findings in a table.

- *Australopithecus afarensis*
- *Paranthropus robustus*
- *Homo habilis*
- *Homo erectus*
- *Homo neanderthalensis*

3 Get into action

Imagine you are an anthropologist working on a small island in the Malay Archipelago, and you discover some fossilised bones that look like *Homo habilis*. On closer examination, you can see that they are much younger than the temporal distribution of *Homo habilis* – they are much closer in age to *Homo sapiens*. You also know the accepted geographical distribution of *Homo habilis* has never extended out of Africa.

What dating technique would you use to determine the age of the fossils? What other evidence (structural and biochemical) would you need to collect to determine if this is a new species or a transitional species?

4 Communicate

Write an article that can be distributed on social media and television to announce the discovery of this new hominin. You may have discovered a previously unknown species on the pathway to *Homo sapiens*, so be clear in your descriptions and the discussion of the evidence.

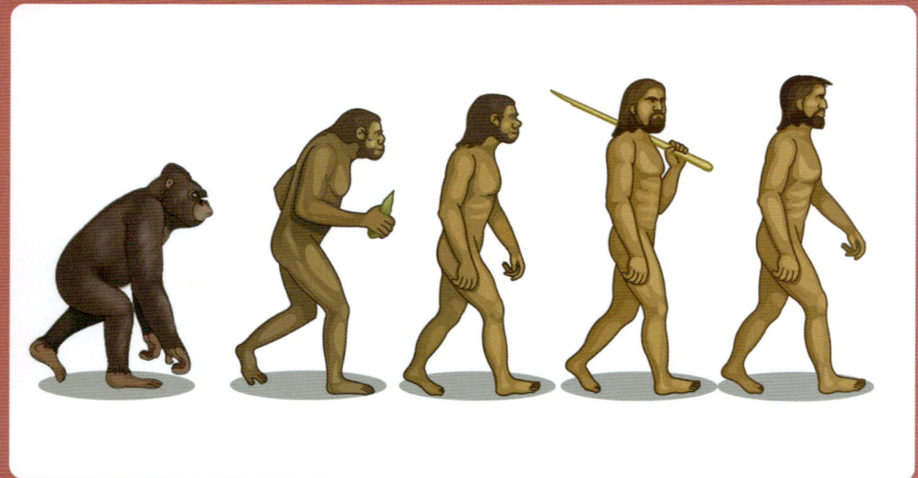

REACTIONS

SYLLABUS OUTCOMES

A STUDENT:

- describes a range of reaction types SC5-RXN-01
- explains the factors that affect the rate of chemical reactions SC5-RXN-02
- selects and uses scientific tools and instruments for accurate observations SC5-WS-01
- develops questions and hypotheses for scientific investigation SC5-WS-02
- designs safe, ethical, valid and reliable investigations SC5-WS-03
- follows a planned procedure to undertake safe, ethical, valid and reliable investigations SC5-WS-04

© 2023 NSW Education Standards Authority

CHAPTERS RELATED TO THIS FOCUS AREA ARE:

- CHAPTER 13 – CHEMICAL REACTIONS
- CHAPTER 14 – NUCLEAR REACTIONS

SariSyno/Shutterstock.com

13 Chemical reactions

SCIENCE IN DEPTH 

▲ **FIGURE 13.0.1** A mining disaster occurred at Pike River in New Zealand in 2010.

Figure 13.0.1 shows an explosion that occurred at Pike River mine in New Zealand in 2010. Coal mines can explode for many reasons. The two most common are a build up of methane gas and a build up of coal dust. Methane is highly flammable, so a spark can set off an explosion. Coal is difficult to ignite in solid form. When the coal is being drilled, coal dust forms and remains suspended in the air throughout the mine. This ignites very easily when exposed to a flame. So why is solid coal hard to ignite when coal dust is explosive with a single spark? The answer lies in a series of chemical reactions and an understanding of rates of reaction.

- **What chemical reactions could be occurring to cause this damage?**
- **How could coal pieces and coal dust be the same chemical, yet so different in their explosive potential?**
- **How could an understanding of chemistry prevent these disasters?**

#13 DIVE INTO SCIENCE!

At the end of this chapter, you can complete Science in Depth Study #13. You can apply the knowledge and skills you learn in this chapter to complete the project.

Assessments

- Prior knowledge quiz
- Chapter review questions
- End-of-chapter test
- Depth study: Research project and poster

Videos

- Science skills in a minute: Qualitative versus quantitative data **(13.7)**; Predictions, questions and hypotheses **(13.9)**; Methods **(13.15)**
- Video activities: Oxidation reactions **(13.5)**; Reactivity series **(13.8)**; What are acids and bases? **(13.12)**

Science skills resources

- Science skills in practice: Qualitative versus quantitative data **(13.7)**; Predictions, questions and hypotheses **(13.9)**; Writing a method **(13.15)**
- Extra science investigations: Comparing mass loss in chemical reactions **(13.1)**; Extraction of iron **(13.4)**; Comparing the corrosion of different metals **(13.5)**; Making sherbert **(13.12)**; Reactions between acids and metals **(13.13)**; Neutralising an acid **(13.14)**; Comparing reaction rates **(13.16)**; Effect of temperature and concentrations on reactions **(13.18)**

Interactive and other resources

- Simulation: Balancing chemical equations **(13.3)**; pH scale **(13.12)**
- Crossword: Types of reactions **(13.14)**
- Drag and drop: Displacement versus precipitation reactions **(13.10)**; Factors that speed up reactions **(13.18)**
- Label: Collision theory **(13.17)**
- Quizzes: Synthesis reactions **(13.4)**; Displacement reactions **(13.8)**; Rate of reaction **(13.16)**
- Activity sheet: Reactions and the metal series **(13.13)**
- Worksheets: Conservation of mass **(13.1)**; Chemical reactions **(13.2)**; Practise balancing equations **(13.3)**; Acids in the home **(13.12)**; Understanding pH **(13.12)**

Nelson MindTap

To access resources above, visit **cengage.com.au/nelsonmindtap**

13.1 Law of conservation of mass

BY THE END OF THIS MODULE, YOU WILL BE ABLE TO:

✓ explain how the law of conservation of mass relates to chemical reactions.

Extra science investigation
Comparing mass loss in chemical reactions

Other resource
Worksheet: Conservation of mass

reactants
chemical substances that, when added together, react to form products in a chemical reaction

products
chemical substances that form in chemical reactions

law of conservation of mass
the total mass of reactants and products in a chemical reaction is equal

GET THINKING

This module is about a law in chemistry. What is your understanding of a law in science? How can scientific laws be used?

What is the law of conservation of mass?

In the 1700s, the French scientist Antoine Lavoisier conducted experiments and proposed a theory about chemical reactions. After many careful observations, he noticed that the mass of the **reactants** in a reaction was always the same as the mass of the **products**.

This became known as the **law of conservation of mass**:

Total mass of reactants = total mass of products

Figure 13.1.1 shows an example you might be familiar with. When you watch a campfire burning in winter you are observing a chemical reaction. This is represented as:

Wood + oxygen → carbon dioxide + carbon + water

Wood contains several substances that burn in oxygen from the air. The products of this reaction are ash (carbon), carbon dioxide gas and water vapour. The combined reactants – wood and oxygen – have a certain mass. If you were to measure the mass of the carbon, carbon dioxide and water formed, it would be the same as the mass of the combined reactants, even though you might not be able to see or capture them easily.

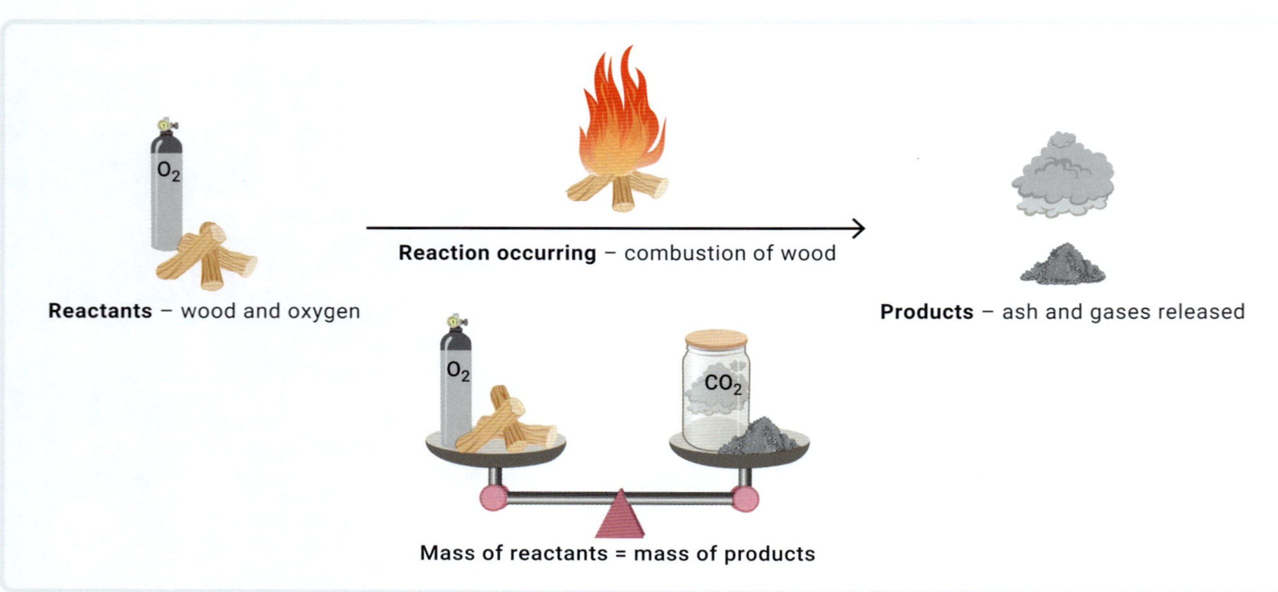

▲ FIGURE 13.1.1 Conservation of mass when burning firewood

9780170491785

How does conservation of mass apply to observations in a chemical reaction?

Figure 13.1.2 shows an experiment in which silver nitrate solution is added to sodium chloride solution. If you were to conduct the experiment as shown, you would record the same mass before and after the reaction. In the first image, the chemicals are separated; the silver nitrate is in the test tube separate from the sodium chloride in the conical flask.

When the test tube is inverted, the silver nitrate pours out and mixes with the sodium chloride. The products formed are a white silver chloride solid, and a sodium nitrate solution.

Silver nitrate (aq) + sodium chloride (aq) → silver chloride (s) + sodium nitrate (aq)

Note that the mass of the reactants is the same as the mass of the products.

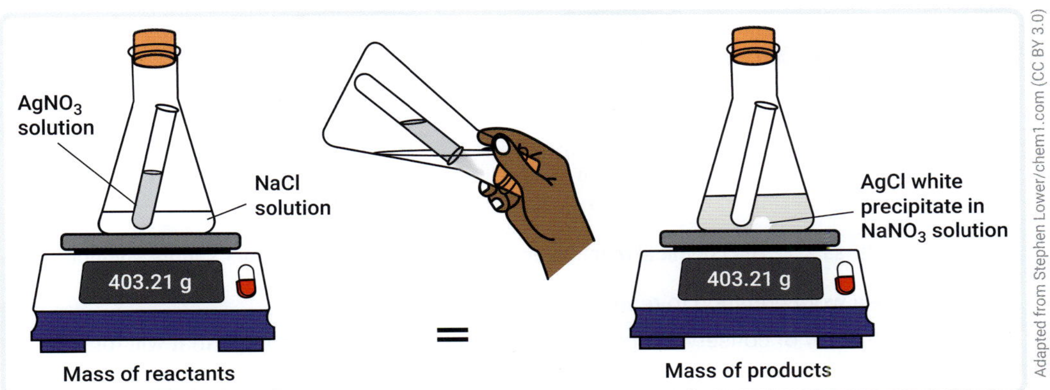

▲ **FIGURE 13.1.2** The chemical reaction between silver nitrate ($AgNO_3$) solution and sodium chloride (NaCl) solution

Conservation of mass

☆ **ACTIVITY**

You will perform a simple conservation of mass experiment. Follow the instructions below, record your results, then answer the questions.

Procedure

1. Measure 20 mL of potassium iodide (KI) solution and pour it into a 50 mL beaker.
2. Measure 20 mL of lead nitrate solution ($Pb(NO_3)_2$) and pour it into a different 50 mL beaker.
3. Place both beakers on an electronic balance at the same time and record the initial mass of the beakers and reactants.
4. Pour one beaker into the other and replace the empty beaker back on the electronic balance.
5. Record the final mass and any observations that show a chemical change.

Analysis

1. Did this experiment **demonstrate** the law of conservation of mass? **Justify** your answer.
2. **Explain** why the empty beaker needed to be placed on the electronic balance. What would have happened to the results if it had been left on the bench?
3. **State** at least two observations that could be used as evidence for a chemical reaction.

Conservation of mass and closed systems

A closed system is one in which mass cannot enter or leave. When conducting experiments involving the law of conservation of mass, it is best to look at closed systems.

If a gas can escape a system, or become a reactant in the experiment, then mass is entering or leaving. In this case, it might appear as if mass is not conserved, which may cause an observer to make an incorrect conclusion about the situation.

13.1 LEARNING CHECK

1 When solid magnesium is added to copper sulfate solution, as shown in Figure 13.1.3, solid copper and magnesium sulfate solution are formed.

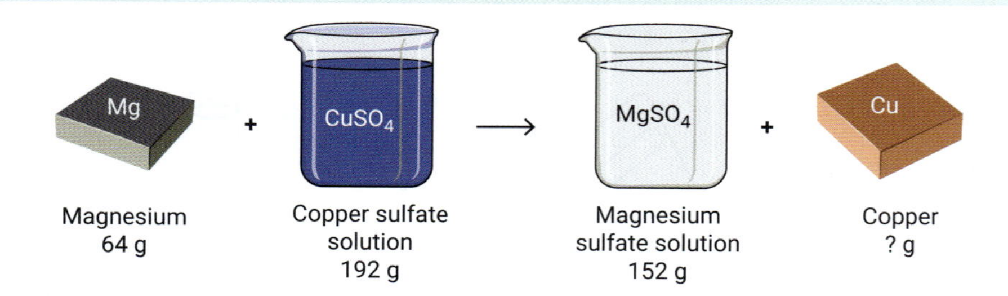

▲ FIGURE 13.1.3 The chemical reaction between magnesium and copper sulfate solution

a Write a word equation to represent this reaction.

b Use the law of conservation of mass to **predict** the mass of copper that will form.

2 Burning a candle is a chemical reaction (Figure 13.1.4). Wax burns in oxygen to form carbon dioxide and water vapour.

a Write a word equation to represent this reaction.

b The mass of the candle at the start does not equal the mass of the candle at the end. **Explain** why this does not make the law of conservation of mass wrong. Consider the nature of the products formed in the reaction in your answer.

Aiisa/Shutterstock.com

▲ FIGURE 13.1.4 Burning a candle is a chemical reaction.

9780170491785

13.2 Review of chemical equations

BY THE END OF THIS MODULE, YOU WILL BE ABLE TO:

- ✓ identify reactants and products in a word or balanced equation
- ✓ write word and balanced equations to represent a chemical reaction.

GET THINKING

Examine the equations in this module. Compare them to the equations you normally see in maths. How are chemical equations similar to maths equations? How are they different?

Other resource
Worksheet: Chemical reactions

What is a chemical equation?

Chemical equations are used to represent the reactants and products in a chemical reaction. Chemical reactions occur if a chemical change occurs when substances are added together. This means a new substance has formed. Chemical changes can be indicated by observations including colour changes, gas formation (bubbles), a solid forming or heat/light energy being released or absorbed.

chemical equations word equations or balanced equations that are used to represent the substances in chemical reactions

Chemical equations can be represented by word equations:

$$\text{Hydrogen} + \text{oxygen} \rightarrow \text{water}$$

They can also be written as balanced equations, using chemical formulas:

$$2H_2(g) + O_2(g) \rightarrow 2H_2O(l)$$

Both types of equation show reactants on the left side. These are the chemicals required to start the reaction. Products that form in the reaction are found on the right. An arrow separates the reactants and products. Chemical equations never include an equals sign (=).

Balanced equations

A balanced equation provides detailed information about the atoms in a chemical reaction. Consider Figure 13.2.1, showing the reaction between methane (CH_4) and oxygen (O_2) to form carbon dioxide (CO_2) and water (H_2O).

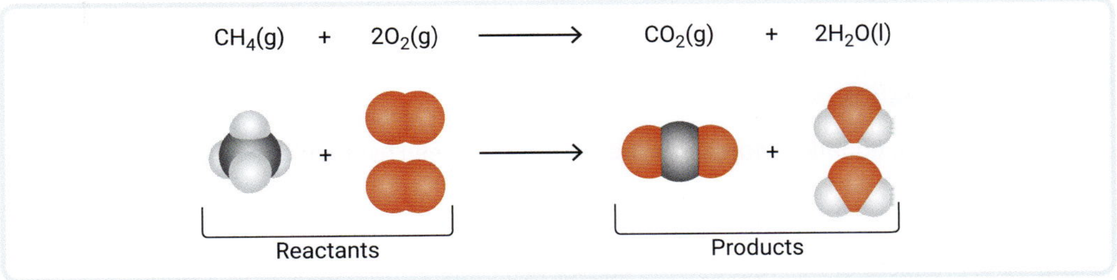

FIGURE 13.2.1 The reaction between methane and oxygen

The first thing you can see from the equation is the type and number of atoms in each reactant and product. For example, CH_4 is methane (the gas used for cooking and heating), and each molecule contains one carbon atom and four hydrogen atoms.

The equation also shows how the law of conservation of mass is being followed in chemical reactions. As we saw in Module 13.1, the law of conservation of mass states that the number and type of atoms are the same in the reactants and products.

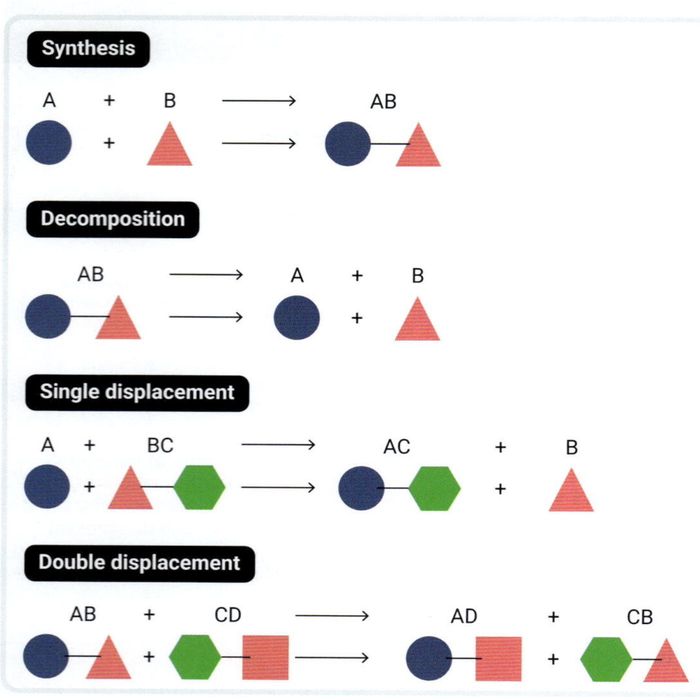

▲ **FIGURE 13.2.2** Types of chemical reactions

Balanced equations use coefficients – numbers in front of the formulas – to ensure that the atoms on each side of the equation are balanced. In Figure 13.2.1, one carbon atom, four hydrogen atoms and four oxygen atoms occur on each side of the arrow. Therefore, this reaction is balanced and the law of conservation of mass is being followed.

Types of reactions

In this chapter, you will be exploring different types of chemical reactions: synthesis, decomposition, single displacement and double displacement reactions.

1 **Synthesis reactions**: two or more elements or compounds combine to form one or more complex products. Synthesis reactions, also called combination or addition reactions, will be explored in Module 13.4 and 13.5.

2 **Decomposition reactions**: a single chemical breaks down into two or more simpler substances. Decomposition reactions will be explored in Module 13.6.

3 Single **displacement reactions**: to displace something means to replace or take the place of another object. In these reactions, one atom replaces an atom in a compound. In Figure 13.2.2, you can see that atom A (blue circle) replaces atom B (pink triangle) in the reactant BC. In Modules 13.8, 13.9 and 13.13, you will explore metal displacement reactions and the reaction between acids and metals.

4 **Double displacement reactions** occur when both reactants have two parts that 'swap'. As both substances replace part of the other substance, two displacements occur. Figure 13.2.2 shows AB splitting, with A replacing C in CD. The two products AD and CB are the 'swaps' from the reactants AB and CD. Precipitation reactions and neutralisation reactions will be explored in Modules 13.10, 13.11 and 13.14.

synthesis reaction
a reaction in which two or more elements and compounds combine to form a more complex substance

decomposition reaction
a reaction involving the breakdown of compounds into simpler elements or compounds

displacement reaction
a reaction involving the replacement of atoms in compounds

double displacement reaction
a reaction in which atoms of the reactants displace each other from their compounds

13.2 LEARNING CHECK

1 The following equation represents the reaction between hydrochloric acid and calcium carbonate:

$$2HCl(aq) + CaCO_3(s) \rightarrow CaCl_2(aq) + CO_2(g) + H_2O(l)$$

a The products of this reaction are calcium chloride, carbon dioxide and water. Write a word equation for this reaction.

b **Demonstrate** how the law of conservation of mass applies to this reaction.

2 **Identify** the four types of chemical reactions.

3 Write a one-sentence summary to **describe** each type of chemical reaction.

9780170491785

13.3 Balanced chemical equations

13.3

BY THE END OF THIS MODULE, YOU WILL BE ABLE TO:

- ✓ write a balanced chemical equation to represent a chemical reaction when given the formula of reactants or products
- ✓ explain why a chemical reaction needs to be balanced by using the law of conservation of mass.

GET THINKING

This module uses a lot of chemical formulas. A common formula you might know is H_2O. Brainstorm and write down as many chemical formulas that you have heard of. Think about what all the letters and numbers mean, then write down what you think a chemical formula represents.

Interactive resource
Simulation: Balancing chemical equations

Other resource
Worksheet: Practise balancing equations

Chemical formulas

So far, we have looked at word equations to represent chemical reactions. Chemists have a system of representing chemicals that describes their structure and composition. Some of these you will already be familiar with.

H_2O is the **chemical formula** for water. This tells you it has two hydrogen atoms and one oxygen atom in its structure. CO_2 is another you might be familiar with. A carbon dioxide molecule has one carbon atom and two oxygen atoms.

chemical formula
symbols and numbers used to represent the composition of a chemical

In this chapter, you will be given any formulas that you need.

What do balanced chemical equations show?

The word equations you reviewed in Module 13.2 have some uses, but don't tell us a lot of information about the chemicals involved in the chemical reaction. Table 13.3.1 shows some simple word equations and what they look like as **balanced chemical equations**.

balanced chemical equation
a chemical equation in which there are the same number of atoms of each element on each side

▼ **TABLE 13.3.1** Word equations and corresponding balanced chemical equations

Word equation	oxygen + hydrogen → water
Chemical equation	$O_2(g) + 2H_2(g) \rightarrow 2H_2O(l)$
Word equation	magnesium (s) + hydrochloric acid (aq) → magnesium chloride (aq) + hydrogen (g)
Chemical equation	$Mg(s) + 2HCl(aq) \rightarrow MgCl_2(aq) + H_2(g)$
Word equation	wax + oxygen → carbon dioxide and water
Chemical equation	$C_{23}H_{48}(s) + 35O_2(g) \rightarrow 23CO_2(g) + 24H_2O(l)$

Balanced chemical equations show:

- The number of atoms in a structure – an O_2 molecule has two oxygen atoms, $MgCl_2$ has one magnesium atom for every two chlorine atoms.
- The number of each type of structure that is required – one O_2 molecule reacts with two H_2 molecules to form two H_2O molecules.

Using the law of conservation of mass

coefficient
a number placed in front of a chemical formula to balance an equation

The numbers in front of the chemical formulas are called **coefficients**. They show the ratio of the chemicals that are needed to react. Recall from Module 13.1 that mass is conserved in chemical reactions. If we look at this statement in terms of the atoms in a chemical reaction, it tells us that you must have the same number and type of each atom in the reactants and in the products.

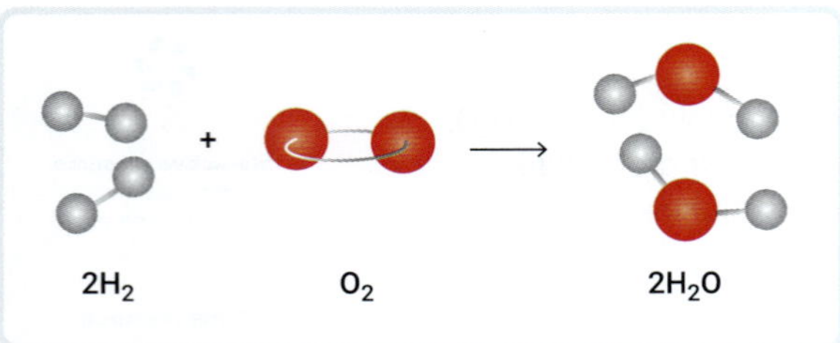

▲ **FIGURE 13.3.1** The chemical reaction between hydrogen and oxygen to form water

Look at the reaction $2H_2(g) + O_2(g) \rightarrow 2H_2O(l)$, shown in Figure 13.3.1.

Table 13.3.2 shows the number of atoms present in the chemical reaction between hydrogen and water. The number of oxygen atoms is the same in the reactants and products. The number of hydrogen atoms is also the same in the reactants and products.

▼ **TABLE 13.3.2** Atoms present in the reaction between hydrogen and oxygen to form water

	Reactants	Products
Formulas	$O_2 + 2H_2$	$2H_2O$
Number of oxygen atoms	2 – there is one O_2 molecule, which has two oxygen atoms – so two in total	2 – each H_2O molecule has one oxygen atom – so two in total
Number of hydrogen atoms	4 – there are two H_2 molecules, each of which has two atoms, so $2 \times 2 = 4$ or four in total	4 – each H_2O molecule has two hydrogen atoms – so four in total

The coefficients are used to make sure the number of each type of atom is equal on both sides of the arrow.

How to balance equations

Figure 13.3.2 shows balanced and unbalanced equations. It uses the example of the addition of hydrogen and oxygen to produce water.

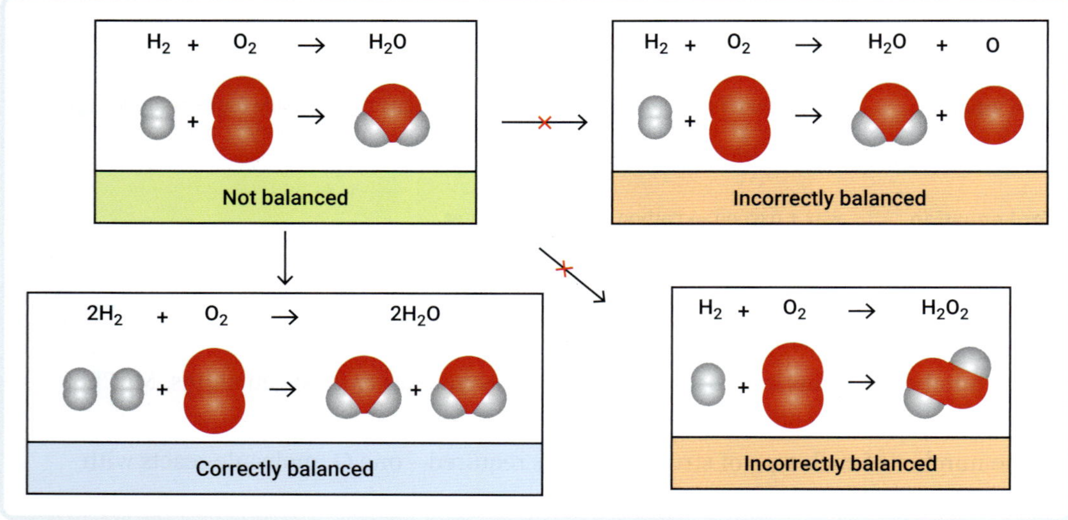

▲ **FIGURE 13.3.2** Equations must be balanced in the correct way; for example, you should not change a chemical formula in order to balance the equation.

Table 13.3.3 shows the steps involved in balancing an equation. In step 3, the instruction says to not change formulas. This is very important. For example, we want to balance the equation $O_2 + H_2 \rightarrow H_2O$. It is tempting to just add an oxygen to the product to balance the equation. This would give us a product of H_2O_2 and the equation would be balanced.

But when you change the formula, you can often end up with a very different chemical. H_2O is water. H_2O_2 is hydrogen peroxide, or bleach – a very different substance!

Another example of this is O_2 and O_3. With only one oxygen atom different, how different can they be? O_2 is the oxygen gas we need to breathe to survive. O_3 is ozone, a toxic pale blue gas.

So, we only balance equations by placing coefficients in front of the chemicals in the equation.

TABLE 13.3.3 Steps to balancing an equation

Step	Example
1 Write the correct chemical formula for each reactant and product.	$CH_4 + O_2 \rightarrow CO_2 + H_2O$
2 Count the number of atoms of each type on both sides of the arrow (both sides of the reaction).	• 1 atom of C on each side • 4 atoms of H on the left, 2 atoms of H on the right • 2 atoms of O on the left, 3 atoms of O on the right
3 If the numbers on each side are not the same, place numbers in front of the formulas until they are balanced. You may need to go back and forth a few times until you get the numbers right. *Do not change any of the formulas!* • The numbers in front apply to all atoms in the formula. *Do not take numbers away!* • You can only balance equations by adding numbers in front of the formula. You can't take numbers away.	• Place a 2 in front of H_2O to get $2H_2O$. • This makes 4 atoms of H on each side (2 molecules with 2 H in each molecule – $2 \times 2 = 4$ H). • Place a 2 in front of the O_2 to get $2O_2$. • This makes 4 O on each side (2 molecules with 2 atoms of O in each – $2 \times 2 = 4$ O). Balanced equation: $CH_4(g) + 2O_2(g) \rightarrow CO_2(g) + 2H_2O(l)$
4 Do a final check to make sure the equation is balanced.	• 1 atom of C on each side • 4 atoms of H on each side • 4 atoms of O on each side

☆ ACTIVITY

Modelling chemical equations

You will need some plasticine of different colours or a chemical modelling kit. You are going to **model** chemical formulas and use the model to help you balance the equation. Follow the instructions listed below to show you how this works.

1 $N_2 + H_2 \rightarrow NH_3$ (nitrogen + hydrogen → ammonia)

- Make a nitrogen molecule (N_2) by making two balls of colour 1 and placing them next to each other or joining them with a toothpick.
- Make a hydrogen molecule (H_2) by making two balls of colour 2 and placing them next to each other or joining them with a toothpick.
- Make an ammonia molecule (NH_3) by making one ball of colour 1 and three balls of colour 2 and joining them so they look exactly like the structure in Figure 13.3.3.
- Decide how many of each molecule you need to make so that the nitrogen and hydrogen atoms are balanced in the equation.
- Make all the molecules you need until the atoms are balanced.
- Write the balanced equation.

2 **Repeat** the modelling process to balance the following equations. The structure of unknown molecules is not important here, just the number of atoms in the molecule.

a $C_2H_4 + O_2 \rightarrow CO_2 + H_2O$

b $N_2 + O_2 \rightarrow NO_2$

Mr.Yankittaphak Phoyalo/Shutterstock.com

▲ **FIGURE 13.3.3** A model of an ammonia molecule

13.3 LEARNING CHECK

1 For each of the following equations, **complete** a table similar to the one shown here. You can use Table 13.3.2 to help guide you.

	Reactants	Products
Formula		
Number of atom type 1		
Number of atom type 2		

a $Mg(s) + 2HCl(aq) \rightarrow MgCl_2(aq) + H_2(g)$

b $C_{23}H_{48}(s) + 35O_2(g) \rightarrow 23CO_2(g) + 24H_2O(l)$

2 Balance the following equations. Use the modelling approach from the activity in this module if necessary or try to balance the equations without making a model.

a $C_3H_8(g) + O_2(g) \rightarrow CO_2(g) + H_2O(l)$

b $Fe(s) + HCl(aq) \rightarrow H_2(g) + FeCl_2(aq)$

c $P(s) + O_2(g) \rightarrow P_4O_{10}(s)$

d $HCl(aq) + CaCO_3(s) \rightarrow CaCl_2(aq) + CO_2(g) + H_2O(l)$

13.4 Synthesis reactions

BY THE END OF THIS MODULE, YOU WILL BE ABLE TO:

- ✓ describe at an atomic level what occurs during a synthesis reaction
- ✓ predict the products and write equations to represent synthesis reactions.

GET THINKING

The word 'synthesise' can mean to create something from lots of pieces of information or ideas. How do you think this definition relates to chemical synthesis?

Quiz
Synthesis reactions

Extra science investigation
Extraction of iron

What is a synthesis reaction?

In a synthesis reaction, two or more chemicals combine to form a more complex product. Atoms from both the reactants can be found in the product. Figure 13.4.1 shows a general representation of a synthesis reaction. A and B are the reactants; AB is the product when they combine.

Predicting the products of simple synthesis reactions

Many simple synthesis reactions involve the combination of elements such as magnesium, hydrogen, aluminium or oxygen.

A + B ⟶ AB

▲ **FIGURE 13.4.1** A synthesis reaction

One type of synthesis reaction where you can predict the products involves the addition of a metal element to a non-metal element (Table 13.4.1). The product is a combination of the two elements called an **ionic salt**. An ionic salt has a metal ion and a non-metal ion and is named by:

- keeping the metal name
- changing the non-metal element ending to '-ide'.

ionic salt
a chemical containing a metal ion and a non-metal ion

corrosion
the breakdown of a metallic substance due to chemical reactions with substances in the environment, such as oxygen or water

For example, magnesium and oxygen combine to form magnesium oxide.

The general equation for this is:

Metal + non-metal → ionic salt

▼ **TABLE 13.4.1** Synthesis reactions involving metal and non-metal elements

Element 1	Element 2	Product	Word and balanced equation
Potassium	Oxygen	Potassium oxide	potassium + oxygen → potassium oxide $4K(s) + O_2(g) \rightarrow 2K_2O(s)$
Iron	Sulfur	Iron sulfide	iron + sulfur → iron sulfide $Fe(s) + S(s) \rightarrow FeS(s)$
Calcium	Chlorine	Calcium chloride	calcium + chlorine → calcium chloride $Ca(s) + Cl_2(g) \rightarrow CaCl_2(s)$

One common example of this type of synthesis reaction is **corrosion** or rusting. Iron metal reacts with oxygen in the air to form iron oxide, which we call rust. This is the red, flaky material you find on iron, as shown in Figure 13.4.2.

PTZ Pictures/Shutterstock.com

▲ **FIGURE 13.4.2** A rusted iron nail. The iron oxide forms from a synthesis reaction.

Examples of more complex synthesis reactions

There are many other types of synthesis reactions. Medicines are generally complex chemicals. They are formed by reacting many simpler chemicals, usually in a series of chemical reactions. The structures of some medicines formed in synthesis reactions shown in Figure 13.4.3.

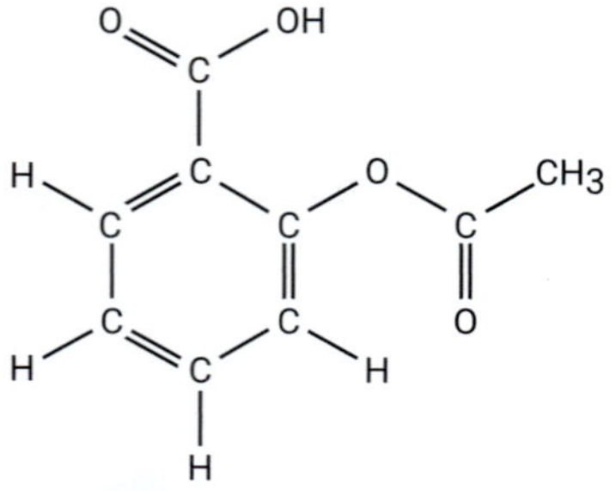

Aspirin 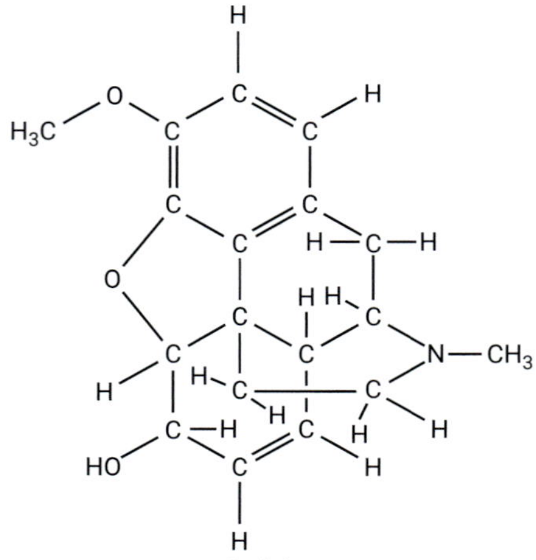 **Codeine** 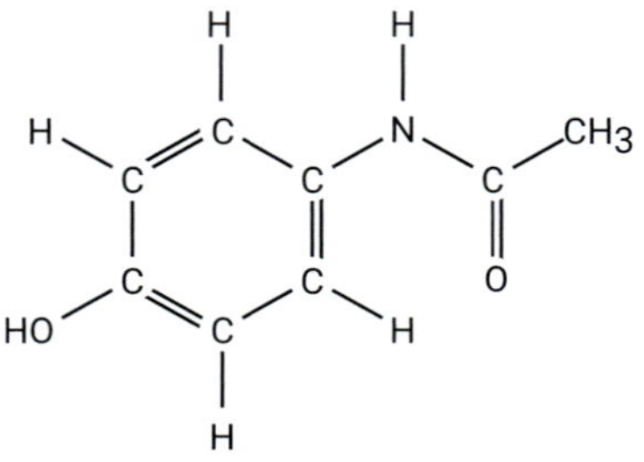 **Paracetamol**

▲ **FIGURE 13.4.3** Examples of medicines formed from synthesis reactions

polymer
a large chemical made in a synthesis reaction from repeating, simpler chemicals called monomers

Polymers, which you learned about in Chapter 8, are large chemicals made by synthesising small chemical structures called monomers (Figure 13.4.4). Polymers include plastics, foam and fibres. Some examples are:

- plastic carry bags – polythene
- non-stick coating on frypans – Teflon
- soft-drink bottles – polyethylene terephthalate (PET)
- clothes – polyester, nylon and acrylic.

Shutterstock.com/molekuul_be

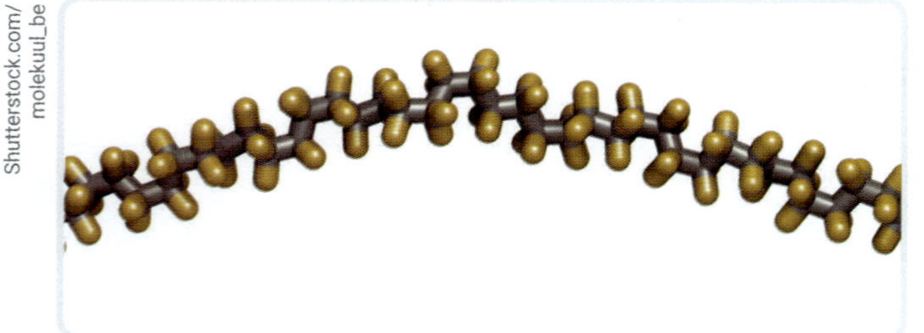

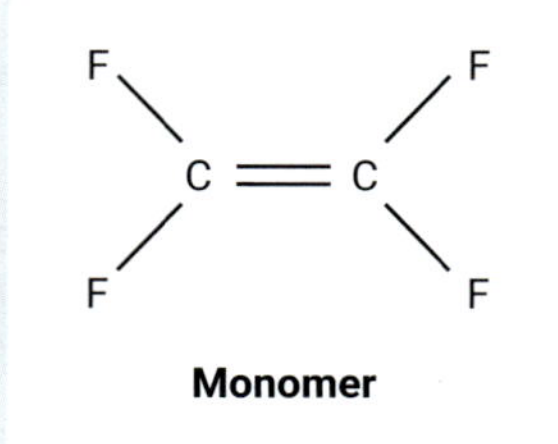

▲ **FIGURE 13.4.4** Polymers, like this polytetrafluoroethylene (PTFE) molecule, are made of many repeating units (monomers) joined together.

Simple synthesis reactions

ACTIVITY

Materials and equipment

- 5 cm strip of magnesium and a small sample of steel wool
- crucible
- heating apparatus
- metal tongs
- safety glasses, lab coat and gloves

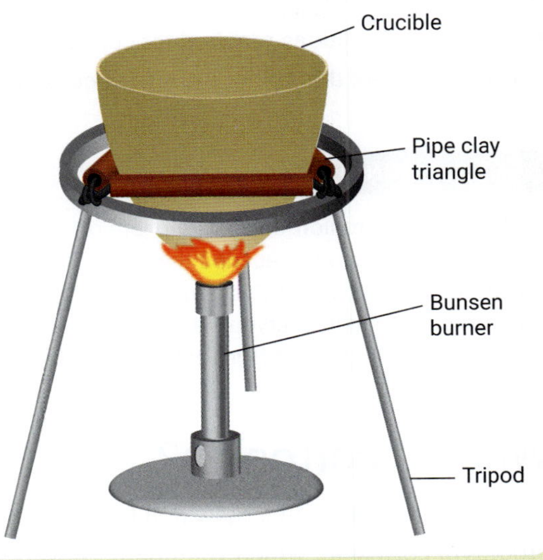

▲ FIGURE 13.4.5 The experimental set-up

Procedure

1 Set up the equipment as shown in Figure 13.4.5.
2 Place the magnesium strip in the crucible and heat it.
3 Using metal tongs, hold the steel wool in the Bunsen burner flame for about 60 seconds. It will glow red while the reaction is occurring.
4 Place the steel wool on a heatproof mat to cool and observe the product.

Analysis

1 **Explain** why these are both synthesis reactions. **Identify** the other element involved in both reactions.
2 Write word equations for both reactions, predicting the name of the ionic salt product.

 **Safety**

When the magnesium ignites, it will release a very intense white light. DO NOT look directly at the light. Observe the products only when the reaction is finished.

To avoid burns, leave equipment to cool before packing it up.

13.4 LEARNING CHECK

1 **Predict** the name of the products when the following are combined.
  a Aluminium + oxygen
  b Calcium + oxygen
  c Lithium + fluorine
2 For each of the examples in Question 1, write a balanced equation to show the full reaction. Your teacher may give you the formulas for the products in Question 1. Alternatively, you can research the formulas of the products you named.
3 **Describe** a common example of a synthesis reaction.
4 **Describe** examples of products that can be formed from more complex synthesis reactions.

13.5 Chemical reactions in our lives: corrosion reactions

BY THE END OF THIS MODULE, YOU WILL BE ABLE TO:

✓ justify the conditions required for corrosion to occur

✓ explain factors that affect the rate of corrosion.

Video activity
Oxidation reactions

Extra science investigation
Comparing the corrosion of different metals

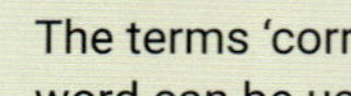

GET THINKING

The terms 'corrosion' and 'rusting' are often used interchangeably. This means that either word can be used to describe the same thing. In science, there are often pairs of words like this where one word is a more scientific term (corrosion), and the other word is more commonly used in everyday language (rusting). Make a list of as many of these pairs of words you can think of from any area of science.

What is corrosion?

You have probably seen a rusty nail like the one in Figure 13.5.1. The rust is the result of the process of corrosion, which you might know better as rusting, the most common form of corrosion. Any object made from iron can rust, such as fencing, roofing and bridges.

fhm/Getty Images

▲ **FIGURE 13.5.1** Rusting occurs when iron reacts with oxygen in the presence of water.

Corrosion is the **oxidation** of a metal. This means that oxygen is added to a metal to form a **metal oxide**. This can be seen in the general word equation:

Metal + oxygen → metal oxide

The most common example you would have encountered is the rusting of iron:

Iron + oxygen → iron oxide

$$4Fe(s) + 3O_2(g) \rightarrow 2Fe_2O_3(s)$$

When iron corrodes, the iron oxide product can weaken items made from iron, such as structures and furniture.

oxidation
a chemical process in which oxygen is added to a substance

metal oxide
a chemical substance made up of a metal and oxygen

Other metals corrode, but not always as obviously as iron. You may have used magnesium ribbon in experiments in your science studies. Figure 13.5.2 shows magnesium ribbon uncleaned (left) and cleaned (right). Magnesium ribbon reacts with oxygen in the air to form magnesium oxide. This is seen as a darker layer on the surface of the magnesium metal.

▲ **FIGURE 13.5.2** Corrosion of magnesium ribbon leaves a dark layer on the surface of the metal.

Factors affecting corrosion

In most cases, metals (e.g. magnesium) react directly with oxygen in the air to form a metal oxide. However, some metals undergo a more complex corrosion process. For example, iron corrodes in a series of steps involving both oxygen and water. Without water being present, iron will not corrode very easily, if at all. This is why you often see rusted iron structures around water. Iron furniture and structures used outdoors are often coated with plastic or another protective material such as paint. This stops the iron from coming in contact with water and oxygen, and prevents corrosion.

Exposure to salt water speeds up the rate of corrosion of iron. Salt water contains sodium chloride as charged particles (ions). These ions speed up the rusting process. Saltwater exposure is a leading cause of the rusting of metals.

Peter Ptschelinzew/Alamy Stock Photo

▲ **FIGURE 13.5.3** Corrosion is a problem in mine sites and gas refineries across Australia. The Karratha gas plant in Western Australia is routinely checked for corrosion problems during shutdown inspections.

☆ ACTIVITY

Factors that affect the rusting of iron

Set up the experiment shown in Figure 13.5.4. This works best if you use ungalvanised nails. Each test tube has a different set of conditions. Oil is used to keep oxygen out of the water. Anhydrous calcium chloride absorbs water from the air, so the environment is dry. Leave the test tubes overnight if possible.

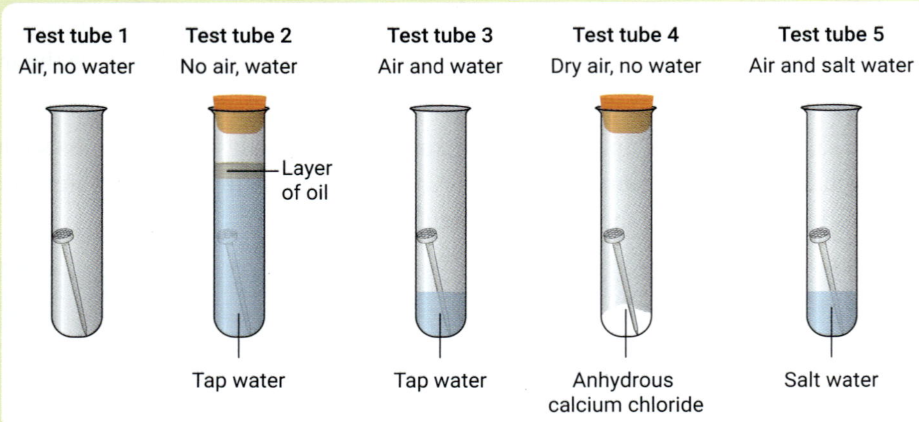

▲ **FIGURE 13.5.4** The experimental set-up

1 Copy and **complete** the following table by circling the conditions (oxygen, water, salt) that were present in each test tube.

2 In the final column of the table, rank the corrosion level of the nails. Because there are five test tubes, you should use 1–5, where:
1 = least corroded
5 = most corroded

3 Use your results to **state** a conclusion about the conditions required for rusting iron.

4 Use your results to **state** a conclusion about the effect of salt water on corrosion.

Test tube	Conditions (circle those present)	Corrosion ranking (1 to 5)
1	oxygen water salt	
2	oxygen water salt	
3	oxygen water salt	
4	oxygen water salt	
5	oxygen water salt	

13.5 LEARNING CHECK

1 Why might you clean metals with sandpaper before using them in experiments?

2 Many outdoor adventurers take their four-wheel drive vehicles and metal boats through salt water. Cars and boats that contact salt water should be washed thoroughly at the end of the day to prevent corrosion. **Explain** why this precaution is necessary.

13.6 Decomposition reactions

BY THE END OF THIS MODULE, YOU WILL BE ABLE TO:

- ✓ describe at an atomic level what occurs during a decomposition reaction
- ✓ predict the products and write equations to represent decomposition reactions.

GET THINKING

Decomposition reactions can be initiated by three different types of energy. This gives three types of decomposition: thermal, photo and electrolytic decomposition. Identify at least one other everyday use for each of the three words 'thermal', 'electric' and 'photo' that are related to energy.

What is decomposition?

Decomposition is the breaking down of chemical substances into two or more smaller products. A simple example of this is the decomposition of water. Water (H_2O) can be broken down into hydrogen (H_2) and oxygen (O_2). This can be seen in the equations:

$$\text{Water} \rightarrow \text{hydrogen} + \text{oxygen}$$

$$2H_2O(l) \rightarrow 2H_2(g) + O_2(g)$$

Figure 13.6.1 shows the breakdown of a chemical containing A and B (AB) into each of A and B individually as products.

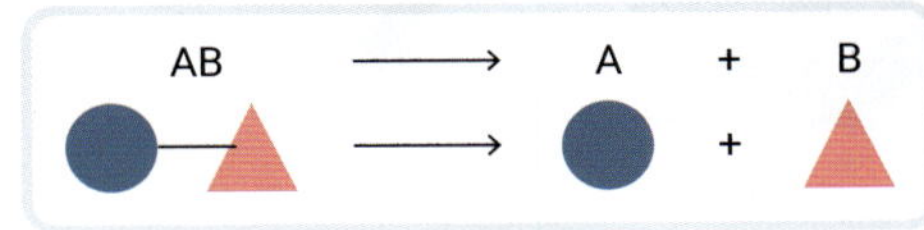

▲ **FIGURE 13.6.1** A decomposition reaction

Types of decomposition

Some decomposition reactions can happen at room temperature, but most require energy input to start. This gives three types of decomposition (Table 13.6.1).

▼ **TABLE 13.6.1** The three types of decomposition

Type of decomposition	Example
Thermal – heat is applied	Heating of copper carbonate to form copper oxide and carbon dioxide $CuCO_3(s) \rightarrow CuO(s) + CO_2(g)$
Photo – light is applied	Decomposition of silver chloride in sunlight into silver and chlorine gas $2AgCl(s) \rightarrow 2Ag(s) + Cl_2(g)$
Electrolytic – electricity is applied	Applying electricity to water to form hydrogen and oxygen $2H_2O(l) \rightarrow 2H_2(g) + O_2(g)$

Predicting the products of simple decomposition reactions

Ionic salts consisting of a metal and non-metal can be decomposed into their elements. You can predict the products by looking at the name of the product. Sodium chloride will decompose into sodium and chlorine. Potassium oxide will decompose into potassium and oxygen.

Metal extraction

ore
a rock that contains one or more minerals containing valuable substances

The process of decomposition can produce useful chemicals. One example of this is the extraction of metals from **ores**. Figure 13.6.2 shows the ores of copper, aluminium and iron and the metal that is extracted. Ores are a combination of metal and other elements such as oxygen, sulfur or hydrogen. Most metals are too reactive to exist in their elemental form. Gold is one of the few metals that can be found in nature in its pure form. Most other metals need to be extracted chemically before they can be used.

▲ **FIGURE 13.6.2** Some metals and their ores

▼ **TABLE 13.6.2** Some metals extracted using high temperatures or electricity

Metal	Ore and formula
Iron	Haematite Fe_2O_3 Magnetite Fe_3O_4
Lead	Galena PbS
Calcium	Dolomite $CaMg(CO_3)_2$
Aluminium	Bauxite $Al_2O_3.2H_2O$
Tin	Cassiterite SnO_2

Most metal extraction is thermal or electrolytic. Very high temperatures or electricity are required to break the bonds and allow the metallic elements to form.

Some examples of metals that are produced in this way are shown in Table 13.6.2.

Other examples of decomposition

Airbags in cars contain a chemical called sodium azide (NaN_3). When a car is involved in an accident of sufficient force, an electric circuit is turned on. The electricity causes the decomposition of the solid sodium azide:

$$2NaN_3(s) \rightarrow 2Na(s) + 3N_2(g)$$

The nitrogen gas formed rapidly inflates the airbag, providing protection for the people in the car, as shown in Figure 13.6.3.

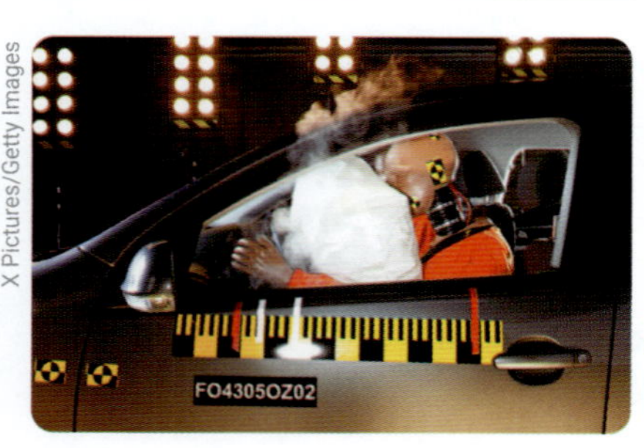

fStop Images - Caspar Benson/Brand X Pictures/Getty Images

◀ **FIGURE 13.6.3** Airbags use a decomposition reaction to inflate.

Electrolytic decomposition of water

☆ ACTIVITY

13.6

The apparatus in Figure 13.6.4 is a Hoffman voltameter. It allows you to pass an electric current through a liquid. In this experiment, water will be decomposed into two different gases. The water has approximately 1 mL sulfuric acid added to help conduct the electricity.

Set up the apparatus as shown in Figure 13.6.4 and set the power pack to 12 volts. Ensure you are using DC electricity and that all wires are connected securely and safely.

When you turn on the electricity, you should see gas forming in the two side cylinders. One of the cylinders will have double the volume of gas of the other one. Each cylinder contains a different gas.

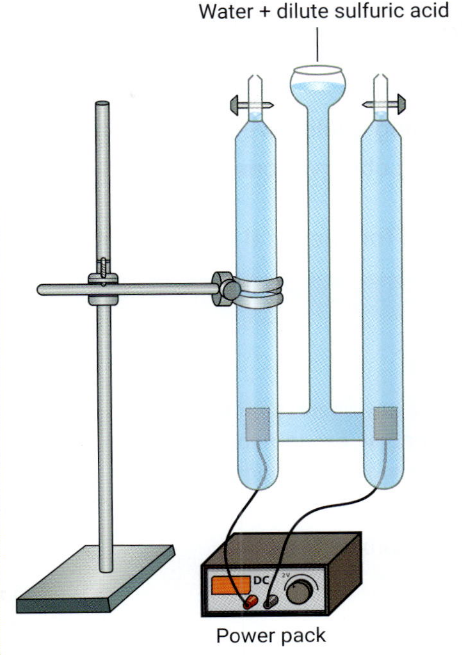

▲ **FIGURE 13.6.4** A Hoffman voltameter

- **a** **Predict** the products of the decomposition of water and write a word equation for this reaction.
- **b** Examine the balanced equation and suggest why one gas has twice the volume of the other. Then, **identify** which gas is in which cylinder.
- **c** **Recall** (or **research**) how you could test to confirm the identity of the two product gases. **Write** a method to show how you could test the gases. (Your teacher may demonstrate this for you.)

13.6 LEARNING CHECK

1. Write word and full balanced equations for the decomposition of:
   - **a** potassium sulfide.
   - **b** calcium oxide.
   - **c** magnesium bromide.
   - **d** aluminium fluoride.
2. Using Table 13.6.2, **predict** the two products from the decomposition of:
   - **a** galena.
   - **b** magnetite.
   - **c** cassiterite.
3. **Discuss** why energy is needed to start most decomposition reactions.
4. Hydrogen peroxide (H_2O_2) does not need much energy to undergo decomposition. It must be stored in a dark, non-transparent bottle. If stored in glass or transparent plastic, it decomposes into water and oxygen. **Explain** why hydrogen peroxide is stored in a dark bottle and write an equation for its decomposition.

13.7 Qualitative observations versus quantitative measurements

SCIENCE SKILLS IN FOCUS

IN THIS MODULE, YOU WILL FOCUS ON LEARNING AND IMPROVING THESE SKILLS:

- gathering and organising qualitative observations about chemical reactions
- describing how making observations can provide information about the products formed in chemical reactions
- writing word equations for chemical reactions observed in experiments.

Qualitative observations

Qualitative observations are any observations you can make with your senses during an experiment. This could be sight, smell, touch or hearing. These might include:

- colours or colour change
- heat being released or absorbed
- bubbles forming
- odours being produced.

Quantitative measurements

Quantitative measurements are any experimental results you can collect with a numerical value (i.e. they can be measured). This might include quantities such as mass or time, which are both very useful.

The rate of reaction is an example where quantitative measurements are important. Collecting data about mass and time allows you to calculate the rate of reaction in an experiment.

There are also many experiments where it is impossible to collect quantitative data. For example, the experiments in this chapter involve qualitative observations alone. Examples of this include precipitation and metal displacement reactions. In these experiments, there is nothing to measure. You are only determining whether a reaction occurred or not. This can be seen through a simple observation such as a solid precipitate forming.

Limitations of qualitative observations and quantitative measurements

There are potential problems with both qualitative and quantitative measurements. Qualitative observations are usually affected by subjectivity. Consider a colour change occurring. If you ask several people to observe a colour change, there will often be different opinions. When colours are 'in between' clear colours, such as a colour between yellow and orange, the judgement is subjective. For example, this colour might be seen as yellow by one person but orange by another person. This may lead to an incorrect positive identification of a reaction, or misidentification of a product if the observations are subjective.

Quantitative measurements can be inaccurate if equipment is not read or used correctly. To avoid problems, it is important that you know how to use the equipment before you start an experiment.

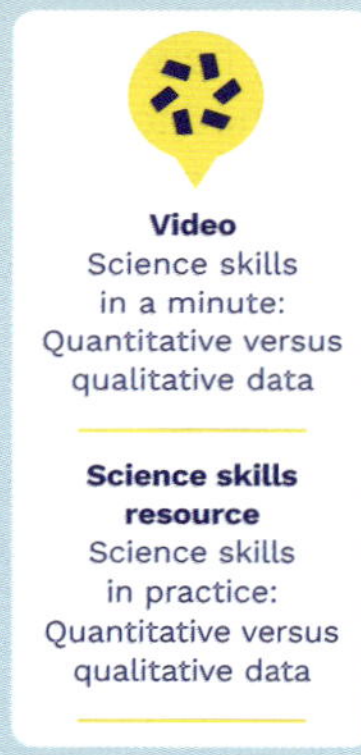

9780170491785

DECOMPOSITION OF COPPER CARBONATE AND TEST FOR CARBON DIOXIDE

AIM

To investigate the decomposition of copper carbonate

MATERIALS AND EQUIPMENT

- ☑ side arm test tube with rubber stopper
- ☑ rubber or plastic hose
- ☑ Bunsen burner heating set-up
- ☑ large test tube half full of limewater (calcium hydroxide)
- ☑ matches
- ☑ copper carbonate powder
- ☑ spatula
- ☑ safety glasses, lab coat and gloves

 **Safety**

Hot substances and objects can cause burns. Allow equipment to cool before packing it up. Wear safety glasses, lab coat and gloves at all times.

PROCEDURE

1. Place two heaped spatulas of copper carbonate powder into a side arm test tube and clamp the test tube to a retort stand.
2. Set up the equipment as shown in Figure 13.7.1, ensuring the hose from the side arm test tube is submerged in the limewater (calcium hydroxide).
3. Light the Bunsen burner and gently heat the copper carbonate until it entirely changes colour.

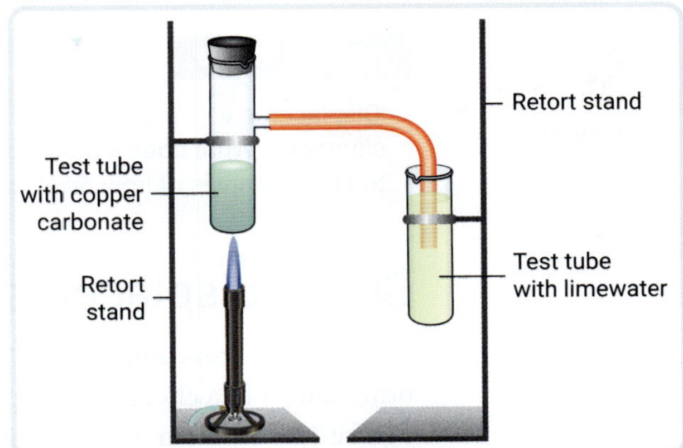

▲ **FIGURE 13.7.1** Experimental set-up

RESULTS

Record your observations by taking photographs or making notes. You should record any colour changes, how long the reaction took to occur, any odours you noticed and whether bubbles of gas formed.

ANALYSIS

1. What did you **observe** that indicated a gas formed in this reaction?
2. Write a word equation and a balanced chemical equation for this reaction.
3. The limewater test detects the presence of carbon dioxide. **Describe** the observations you made and **state** whether they indicate if carbon dioxide was present.
4. **Explain** whether the observations made in this experiment are qualitative or quantitative.
5. **Identify** two observations you made in this experiment that confirmed a chemical reaction took place.

CONCLUSION

Write a conclusion for this experiment.

13.8 Metal displacement reactions

BY THE END OF THIS MODULE, YOU WILL BE ABLE TO:

- ✓ describe at an atomic level what occurs during a metal displacement reaction
- ✓ predict the products and write equations to represent metal displacement reactions.

Video activity
Reactivity series

Quiz
Displacement reactions

GET THINKING

'Displacement' is a commonly used word in science but is also used in many other situations. What does it mean to displace someone or something? Do a quick online search to find examples of how displacement can be used in areas outside science.

Single displacement reactions

Displacement reactions can be 'single' or 'double'. Both types involve atoms replacing other atoms in a chemical compound. A single displacement reaction is illustrated in Figure 13.8.1. When A reacts with BC, it displaces B. This means it replaces B in the reactant so that the end products are AC and B.

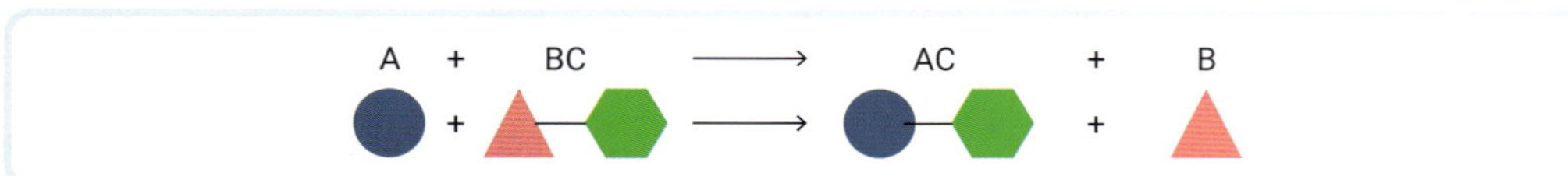

▲ **FIGURE 13.8.1** A single displacement reaction

Figure 13.8.2 shows this in the reaction of aluminium metal with iron oxide. The aluminium displaces the iron from the iron oxide, forming aluminium oxide and iron. The iron has been displaced, so it ends up as an element on its own in the products.

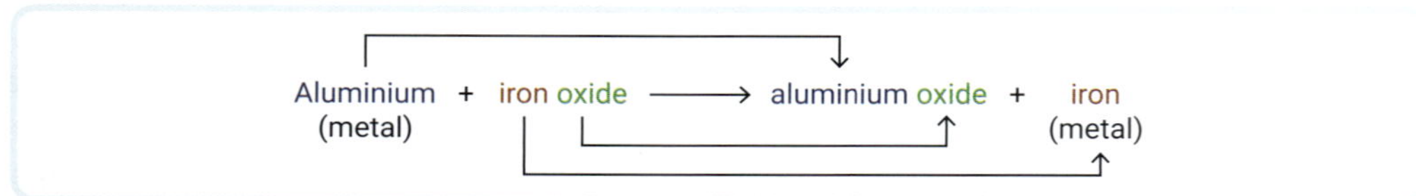

▲ **FIGURE 13.8.2** The metal displacement reaction of aluminium metal with iron oxide

This is an example of a metal displacement reaction. To explore this further, you need to understand the different forms in which metals can exist.

Metals and metal ion solutions

You may recall that almost all metal elements are solids. In chemistry, they are represented by their element symbol: magnesium solid is Mg, aluminium solid is Al and copper solid is Cu.

Metals can also exist in ion form in solution. When an ionic salt is added to water and it is soluble, the ions separate. This means metal ions are present in solution. Metal ions are positively charged ions represented by the chemical symbol and charge. Magnesium ions are Mg^{2+}, aluminium ions are Al^{3+} and copper ions are Cu^{2+}.

Remember, you can predict the charge on the ions of metals in groups 1, 2 and 13 from their position on the periodic table.

9780170491785

Reactivity of metals

Many metals are reactive and so are rarely found as pure metal elements. Chemical reactivity generally refers to how easily substances react with other chemicals. Metals show a wide range of reactivity.

Group 1 and 2 metals are generally the most reactive metals. Figure 13.8.3 shows a lithium battery, which would react explosively with water. Group 1 metals have vigorous reactions with other elements. Several of them are so reactive, they are explosive in air and must be kept in oil or in a vacuum to prevent explosions while they are being stored.

wk1003mike/Shutterstock.com

▲ **FIGURE 13.8.3** Water should never be used to put out a lithium battery fire.

Other metals, such as platinum, copper, silver and gold, are far less reactive. Gold is one of the few metals that occurs in its pure form in nature. This is due to its low reactivity with other elements. Gold, silver and platinum are often used for jewellery because of their lack of reactivity. The skin contains acids, and your jewellery comes into contact with sweat, cleaning materials and other chemicals every day. This is why jewellery needs to be made from non-reactive metals.

Activity series of metals

Scientists rank the reactivity of metals in the **activity series**. Figure 13.8.4 shows a metal activity series indicating that potassium and calcium are two of the most reactive metals. Copper, silver and gold are among the least reactive metals.

activity series
a list of metals ranked by their chemical reactivity

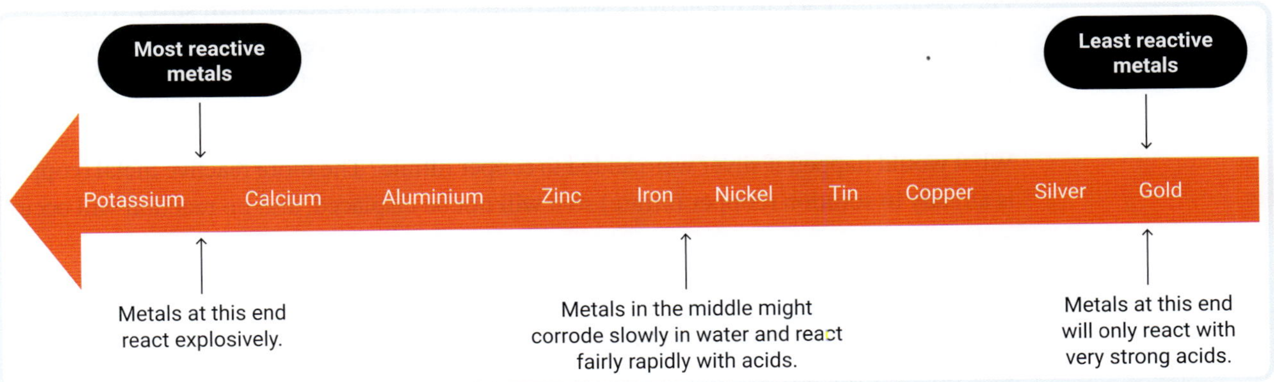

▲ **FIGURE 13.8.4** An activity series of metals

This ranking can be used to predict how metals and metal ion solutions will react when they come into contact. Figure 13.8.5 shows a piece of copper metal that has been placed into a silver nitrate solution containing silver ions. What can you see?

The following observations are made.

- A solid metal (silver) has formed on the piece of copper metal.
- The solution has turned blue. Chemists know this is due to the formation of copper ions in solution.

Charles D. Winters/Science Source

▲ **FIGURE 13.8.5** Copper metal reacts with silver nitrate.

These observations allow conclusions to be made. First, the formation of silver solid means that silver ions have formed silver metal. Second, the formation of copper ions means the copper metal has formed copper ions. This lets us write the word equation:

Copper + silver nitrate → copper nitrate + silver
(silver ions) (copper ions)

The copper has displaced the silver from the silver nitrate solution. This is a metal displacement reaction. As only one substance has been displaced, it is a single displacement reaction.

You can see this clearly from the balanced equation:

$$Cu(s) + 2AgNO_3(aq) \rightarrow Cu(NO_3)_2(aq) + 2Ag(s)$$

The copper (Cu) has displaced the silver (Ag) in the reactant compound. The silver nitrate ($AgNO_3$) reactant becomes copper nitrate ($Cu(NO_3)_2$) in the products.

You may recall the use of state symbols in equations. In balanced equations such as metal displacement, it is important to show what is solid (s) and what is a solution (aq). Other state symbols show gases (g) and pure liquids (l). You should include state symbols in balanced equations if you know the state of the chemicals.

Predicting the products of metal displacement reactions

Let's relate this back to the activity series. Copper is more reactive than silver. There is a rule about metal activity that states that 'the more reactive metal will displace the less reactive metal from a solution'. In this case, the more reactive metal (copper) has displaced the less reactive metal (silver) from the silver nitrate solution.

What would happen if silver were added to copper nitrate? The more reactive metal (copper) is already in solution (copper nitrate) so it will not be displaced, and no reaction will occur.

Example 1

If zinc metal is added to copper nitrate, will a reaction occur?

- Zinc is the more reactive metal.
- Copper is the less reactive metal and is in solution (copper nitrate).
- The more reactive metal will displace the less reactive metal from the solution.
- Zinc will displace copper from the solution.

Yes – a reaction will occur.

The word equation is:

Zinc + copper nitrate → copper + zinc nitrate

The balanced equation is:

$$Zn(s) + Cu(NO_3)_2(aq) \rightarrow Cu(s) + Zn(NO_3)_2(aq)$$

Example 2

If aluminium metal is added to iron nitrate, will a reaction occur?

- Aluminium is the more reactive metal.
- Iron is the less reactive metal and is in solution (iron nitrate).
- The more reactive metal will displace the less reactive metal from the solution.
- Aluminium will displace iron from the solution.

Yes – a reaction will occur.

The word equation is:

Aluminium + iron nitrate → iron + aluminium nitrate

The balanced equation is:

$$2Al(s) + 3Fe(NO_3)_2(aq) \rightarrow 3Fe(s) + 2Al(NO_3)_3(aq)$$

13.8 LEARNING CHECK

1. Ionic salts include sodium chloride, magnesium fluoride and aluminium bromide. Write the formula of the metal ions present in these metal salts.
2. **Draw** a diagram to show what is meant by a single displacement reaction.
3. A student was given a range of metals to conduct experiments. They were given a jar with a silvery metal stored in oil, like the example shown in Figure 13.8.6. **Predict** which group in the periodic table this metal would belong to. **Justify** your answer.
4. Some people get black marks on their skin from wearing silver jewellery. The black substance is silver oxide, formed when silver reacts with chemicals on the skin. Very few people have this reaction when wearing gold jewellery. By referring to the activity series of metals, **explain** this observation.

MARTYN F. CHILLMAID/SCIENCE PHOTO LIBRARY

▲ **FIGURE 13.8.6** Some metals are stored in oil.

5. In which of the following will reactions occur? **Justify** your answer.
   Calcium metal in tin nitrate solution OR tin metal in calcium nitrate solution
6. **Predict** whether the following combination of metals and ionic salts will react, and write word equations for any reactions that occur.
   a. Aluminium metal and silver nitrate solution
   b. Iron metal and calcium nitrate solution
   c. Copper metal and zinc nitrate solution
   d. Tin metal and copper nitrate solution
   e. Zinc metal and silver nitrate solution

13.9 Predictions, questions and hypotheses

SCIENCE SKILLS IN FOCUS

IN THIS MODULE, YOU WILL FOCUS ON LEARNING AND IMPROVING THESE SKILLS:

- developing questions, predictions and hypotheses
- writing word and balanced equations for chemical reactions observed in experiments.

Developing questions and making predictions

When scientists conduct their research, they always start with a question.

The research question describes the purpose of your investigation. Your question should be **testable**, meaning it can be investigated, and should have an answer. An example of a testable question is: 'How does temperature affect how much salt dissolves in water?'

Once you have developed your question, you should use your scientific understanding to make a prediction of the outcome. For example, you might predict that increasing the temperature will allow a greater mass of salt to dissolve in water.

Hypothesis

A hypothesis is a testable prediction for something based on existing knowledge of the scientific theory related to the investigation. In this experiment you will use your knowledge of metal reactivity and metal displacement reactions to predict the outcomes of a series of chemical reactions.

A hypothesis can take two different forms. The most common involves the link between an independent variable and a dependent variable. A hypothesis written this way often takes the form of an 'If…., then….' statement. This gives your prediction of how the dependent variable will change as the independent variable changes.

Sometimes investigations do not have independent and dependent variables. For example:

- conducting an experiment to measure the value of the density of an object – when you are just measuring something, a hypothesis is not required
- conducting an investigation to determine if one thing is 'better' or 'more reactive' than another thing – when you are comparing different things, a hypothesis in the 'If…., then….' form is not appropriate.

In this experiment you will be ranking metals based on their reactivity, which does not have an independent or dependent variable. In investigations like this, your hypothesis could take the form 'Object A is better than Object B for cleaning' or 'The metals ranked in order of reactivity are X, Y, Z from most to least reactive'.

METAL DISPLACEMENT REACTIONS

AIM

To investigate metal displacement reactions involving magnesium, zinc, copper and iron

MATERIALS AND EQUIPMENT

- ☑ test tubes and test-tube rack
- ☑ samples of magnesium, zinc, copper and iron
- ☑ sandpaper to clean the metals
- ☑ solutions of magnesium nitrate, zinc nitrate, copper nitrate and iron nitrate
- ☑ safety glasses, lab coat and gloves

Safety

Wear safety glasses, lab coat and gloves at all times.

PROCEDURE

1. Write a hypothesis for this experiment.
2. Set up four test tubes in a test-tube rack.
3. Place the zinc, magnesium, copper and iron into the four different test tubes, with one metal per test tube.
4. Cover each metal with about 2–3 cm of magnesium nitrate and record observations.
5. Repeat steps 3 and 4 using all the metals and the zinc nitrate solution.
6. Repeat again using copper nitrate, and then again with iron nitrate.

RESULTS

Copy Table 13.9.1 to record your results.

▼ **TABLE 13.9.1** Metal displacement experiment observations

	Zinc	Magnesium	Copper	Iron
Zinc ions				
Magnesium ions				
Copper ions				
Iron ions				

ANALYSIS

1. **Explain**, using a word equation from the results, whether this experiment is demonstrating single or double displacement reactions.
2. Write word and balanced chemical equations for any reactions that occurred in this experiment.
3. Use the results to rank the metals from most to least reactive. **Justify** your rankings.
4. **Describe** any issues you had in deciding whether a reaction had occurred.
5. Use your results to **justify** whether your hypothesis was supported or refuted.

CONCLUSION

Write a conclusion for this experiment. Remember to state if your results support your hypothesis.

13.10 Precipitation reactions

BY THE END OF THIS MODULE, YOU WILL BE ABLE TO:

- ✓ describe at an atomic level what occurs during a precipitation reaction
- ✓ predict the products and write equations to represent precipitation reactions.

Interactive resource
Drag and drop:
Displacement versus precipitation reactions

GET THINKING

In Chapter 7 you learned about the formation of positive and negative ions. Review this work to see why metal ions are always positive and non-metal ions are always negative.

Double displacement reactions

In a double displacement reaction, the atoms of the reactants displace each other from their compounds. Figure 13.10.1 shows reactants AB and CD. When they react, the pairs of atoms 'swap' or replace each other. This forms products AD and CB.

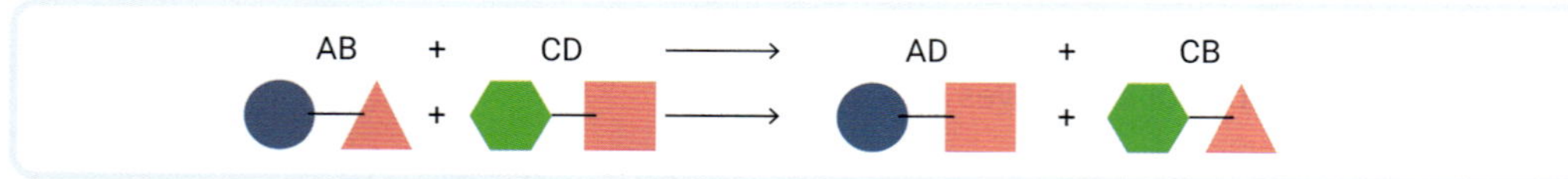

▲ **FIGURE 13.10.1** A double displacement reaction

Figure 13.10.2 shows an example with the reactants lead nitrate and potassium iodide. You can follow the colours to see that the lead and potassium displace each other to form two new products.

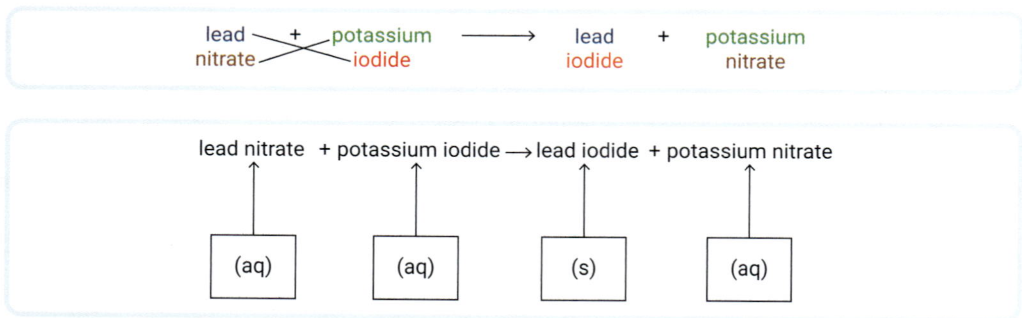

▲ **FIGURE 13.10.2** The reaction between lead nitrate and potassium iodide is an example of a double displacement reaction.

Salts, solutions and precipitates

precipitate
a solid formed from certain combinations of positive and negative ions

A common double displacement reaction forms a **precipitate**. In precipitation reactions, ionic salt solutions react together to form at least one solid ionic compound.

In Chapter 7 we looked at how metals form positive ions and non-metals form negative ions. Sometimes those ions attract to form an ionic lattice or ionic salt.

Figure 13.10.2 shows four ionic salts. Potassium iodide has a metal ion (the potassium ion) and a non-metal ion (the iodide ion). Lead nitrate has a metal ion (the lead ion) and a non-metal ion (containing nitrogen and oxygen).

The nitrate ion (NO_3^-) is a negative ion with more than one type of non-metal atom. This type of ion is called a polyatomic ion. To revise polyatomic ions, go to Module 7.9.

9780170491785

Other polyatomic ions are the hydroxide ion (OH^-), sulfate ion (SO_4^{2-}) and carbonate ion (CO_3^{2-}).

Some ionic salts are soluble. When soluble ionic salts are dissolved in water, the ions separate into positive and negative ions. You can see lead nitrate and potassium iodide depicted in solution with separate ions in Figure 13.10.3.

When the two solutions are added, four ions are present: lead ions, potassium ions, nitrate ions and iodide ions.

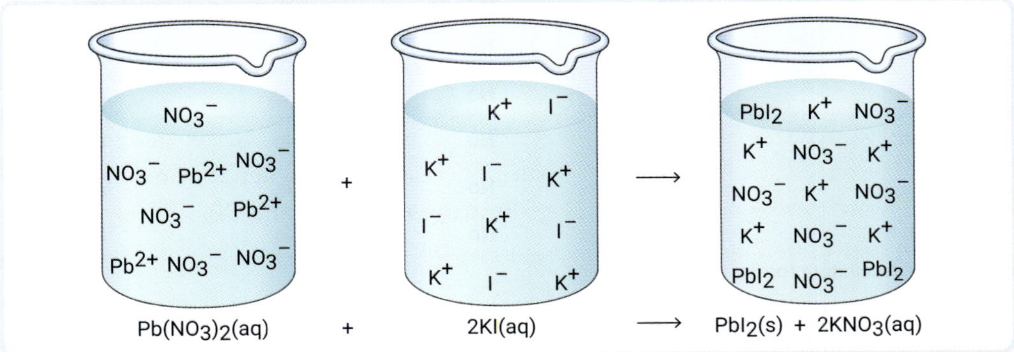

▲ **FIGURE 13.10.3** Lead nitrate and potassium iodide in solution with separate ions

When some pairs of ions are in solution together, they can form an insoluble solid, a precipitate. In the example in Figure 13.10.4, the combination of lead ions and iodide ions forms an insoluble lead iodide precipitate. Lead iodide is a bright yellow solid. The balanced chemical equation for this reaction is:

$$Pb(NO_3)_2(aq) + 2KI(aq) \rightarrow PbI_2(s) + 2KNO_3(aq)$$

Dorling Kindersley ltd/Alamy Stock Photo

▲ **FIGURE 13.10.4** When lead ions and iodide ions combine, they form a bright yellow precipitate, lead iodide.

Predicting the products of precipitation reactions

To predict whether a precipitate will form, we need to know which combinations of ions make an insoluble precipitate. Table 13.10.1 shows the combinations of ions that will form a precipitate.

Using Table 13.10.1, you can see that if silver ions and hydroxide ions are added, then a precipitate will form. To determine whether a precipitate will form:

1. write the word equation
2. check the products against Table 13.10.1
3. identify which product is a precipitate
4. if neither product is a precipitate, then none will form in the reaction.

▼ **TABLE 13.10.1** Ion combinations that form precipitates when added to a solution

Positive ions	Negative ions			
	Hydroxide OH^-	Nitrate NO_3^-	Sulfate SO_4^{2-}	Chloride Cl^-
Magnesium Mg^{2+}	Yes	No	No	No
Copper Cu^{2+}	Yes	No	No	No
Strontium Sr^{2+}	No	No	Yes	No
Zinc Zn^{2+}	Yes	No	No	No
Lead Pb^{2+}	Yes	No	Yes	Yes
Calcium Ca^{2+}	Yes	No	Yes	No
Sodium Na^+	No	No	No	No
Silver Ag^+	Yes	No	Yes	Yes
Potassium K^+	No	No	No	No

Note: Yes = precipitate forms.

For example, if calcium nitrate is added to sodium sulfate, the word and balanced chemical equations are:

Calcium nitrate + sodium sulfate → calcium sulfate + sodium nitrate

$$Ca(NO_3)_2(aq) + Na_2SO_4(aq) \rightarrow CaSO_4(s) + 2NaNO_3(aq)$$

Checking the table, you can see that sodium nitrate does not form a precipitate, but calcium sulfate does. So, this reaction will form a precipitate.

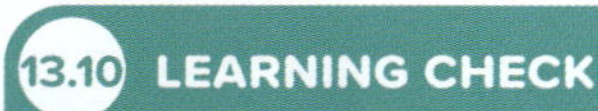

1 For the following combinations of ionic salts, write the word equation and balanced chemical equation for their addition using Table 13.10.1 to **determine** if any insoluble products are present and circling any precipitate that forms.
   - a Magnesium nitrate and sodium hydroxide
   - b Zinc hydroxide and potassium chloride
   - c Sodium sulfate and lead nitrate

2 What combination of ionic salts could you add to form the following precipitates? The other product must be soluble. Write a word equation and a balanced chemical equation for the reaction. The first one is partially completed for you.
   - a Calcium sulfate—reactant 1 could be calcium nitrate, and reactant 2 could be sodium sulfate. The soluble product is sodium nitrate. (Hint: Sodium ions, potassium ions and nitrate ions are always soluble, so they would make a soluble product when combined). Now write the word and balanced chemical equations.
   - b Silver hydroxide
   - c Lead chloride

WORKING SCIENTIFICALLY

13.11 Observing precipitation reactions

SCIENCE SKILLS IN FOCUS

IN THIS MODULE, YOU WILL FOCUS ON LEARNING AND IMPROVING THESE SKILLS:

- observing precipitation chemical reactions
- writing word and balanced equations for chemical reactions observed in experiments.

PRECIPITATION REACTIONS

AIM

To observe the results of precipitation reactions and write equations for the reactions that occurred

Safety

Wear safety glasses, lab coat and gloves at all times.

MATERIALS AND EQUIPMENT

- ☑ dropper bottles of copper nitrate, zinc nitrate, silver nitrate, calcium nitrate and magnesium nitrate (Group A)
- ☑ dropper bottles of potassium iodide, sodium hydroxide, sodium carbonate, sodium chloride and sodium sulfate (Group B)
- ☑ laminated black sheet of cardboard
- ☑ safety glasses, lab coat and gloves

PROCEDURE

1 On the laminated cardboard, create a grid of large squares with five columns and six rows, as shown in Table 13.11.1.

2 Above each vertical column, write one of the chemicals from Group B.

3 To the left of each horizontal row, write one of the chemicals from Group A.

4 Add 3 drops of each chemical to the relevant grid boxes. When you add the second chemical, ensure you add the drops directly on top of the first chemical so they mix.

RESULTS

Take a photograph of your results.

Create a table recording whether a precipitate formed. If a precipitate formed, record its colour.

ANALYSIS

1 **Explain** whether precipitation reactions are single or double displacement reactions.

2 Write word and balanced chemical equations for all reactions that formed a precipitate.

3 Were all the precipitates easy to see? How might this have affected your results?

4 Why do you think black cardboard was used instead of white cardboard?

CONCLUSION

Write a conclusion for this experiment.

▼ **TABLE 13.11.1** Precipitate set-up for laminated cardboard

	Potassium iodide	Sodium hydroxide	Sodium carbonate	Sodium chloride	Sodium sulfate
Copper nitrate					
Zinc nitrate					
Silver nitrate					
Calcium nitrate					
Magnesium nitrate					

13.12 Chemical reactions in our lives: acids and bases

BY THE END OF THIS MODULE, YOU WILL BE ABLE TO:

- ✓ describe properties and everyday uses of acids and bases
- ✓ explain what pH shows about acidic and basic solutions
- ✓ use information about indicators to identify acidic and basic solutions.

Video activity
What are acids and bases?

Interactive resource
Simulation: pH scale

Extra science investigation
Making sherbert

Other resources
Worksheets:
Acids in the home
Understanding pH

GET THINKING

Examine the pictures of indicators in this module. Colours are important in chemistry to identify substances and perform tests. Think about how colours can give you information in everyday life. One example is traffic lights. What others can you think of?

What is an acid?

Acids are commonly occurring substances. The vinegar you use as a salad dressing or for cooking contains an acid called ethanoic acid (Figure 13.12.1). Your body produces hydrochloric acid to help digest food. Citrus fruits such as oranges, limes and lemons contain citric acid. The burning sensation you feel in your muscles after intense exercise is due to lactic acid. In Science classes, you have probably performed experiments with acids such as hydrochloric acid.

acid
a substance that can donate a hydrogen ion when in solution

focal point/Shutterstock.com

▲ **FIGURE 13.12.1** Vinegar contains ethanoic acid.

Acids have common physical properties. All acids:

- taste bitter (Note: You will know the taste of lemon and orange juice, but *never* taste acids or any food in a laboratory.)
- are soluble in water
- conduct electricity when in solution.

Acids and bases involve the formation of ions, including hydrogen or hydroxide ions.

Acids have different chemical structures, but always have one thing in common – acids produce hydrogen ions when dissolved in water. This means that all acids contain hydrogen. For example, when hydrochloric acid (HCl) is in solution, there are hydrogen ions (H^+) and chloride ions (Cl^-).

What is a base?

You can find **bases** in your home. Bases such as sodium bicarbonate are used for cooking, whereas bleach and sodium hydroxide occur in many cleaning materials (Figure 13.12.2). You might be familiar with bases in the school laboratory, including carbonates, hydroxides and oxides.

Bases are bitter and feel slippery or soapy to touch. Like acids, bases conduct electricity when in solution.

iva/Shutterstock.com

▲ **FIGURE 13.12.2** In the home, bases are often used for cooking or cleaning.

Neutralisation

The chemical reaction between an acid and a base is called neutralisation. When a base dissolves in water, it produces hydroxide ions (OH^-), carbonate ions (CO_3^{2-}) or oxide ions (O^{2-}). These ions react with acids in a number of useful reactions around the home. Module 13.14 goes into more depth about neutralisation.

base
a substance that can produce hydroxide, oxide or carbonate ions in solution

When the stomach produces excess acid, people can suffer from heartburn or indigestion. Antacid tablets contain a base to neutralise the excess stomach acid.

Indicators

Indicators are chemical substances that change colour when added to acidic or basic solutions. Figure 13.12.3 shows universal indicator and Figure 13.12.4 shows bromothymol blue. Both are indicators you might use in a school laboratory. The indicators show how acidic or basic a solution is by their colour.

indicator
a chemical that changes colour in acidic and basic solutions

Bjoern Wylezich/Shutterstock.com

Rabbitmindphoto/Shutterstock.com

▲ **FIGURE 13.12.3** Universal indicator is red, pink/orange or yellow in acids. It turns green in a neutral solution and blue or purple in a basic solution.

PHIL DEGGINGER / SCIENCE PHOTO LIBRARY

◄ **FIGURE 13.12.4** Bromothymol blue is a simpler indicator than universal indicator. It turns yellow in acids, blue in bases and green in a neutral solution.

The pH scale

pH
a measure of how acidic or basic a substance is

The **pH** scale shows how acidic or basic a substance is. Most substances have a pH between 0 and 14.

- A substance with a pH less than 7 (pH < 7) is acidic. Substances that are more acidic have a pH closer to 0.
- A substance with a pH of 7 (pH = 7) is neutral. It is neither an acid nor a base.
- A substance with a pH of greater than 7 (pH > 7) is basic. Substances that are more basic have a pH closer to 14.

The pH scale is not a regular scale, where 2 is just 1 more than 1. Instead, it is a logarithmic scale. This means that each pH level is 10 times more than the previous pH level (Figure 13.12.5).

- pH 1 is 10 times more acidic than pH 2.
- pH 3 is 10 times less acidic than pH 2.
- pH 13 is 10 times more basic than pH 12.
- pH 10 is 10 times less basic than pH 11.

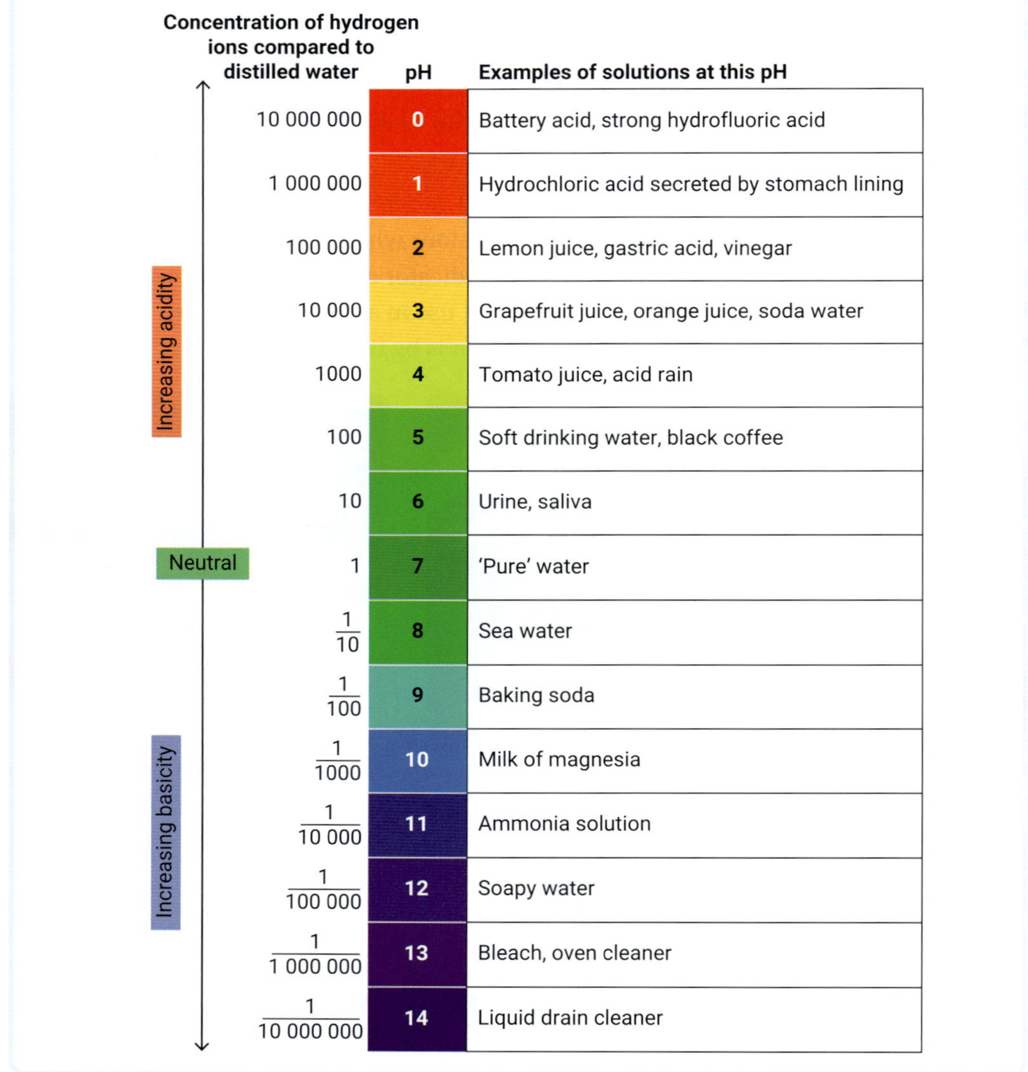

Concentration of hydrogen ions compared to distilled water	pH	Examples of solutions at this pH
10 000 000	0	Battery acid, strong hydrofluoric acid
1 000 000	1	Hydrochloric acid secreted by stomach lining
100 000	2	Lemon juice, gastric acid, vinegar
10 000	3	Grapefruit juice, orange juice, soda water
1000	4	Tomato juice, acid rain
100	5	Soft drinking water, black coffee
10	6	Urine, saliva
1	7	'Pure' water
$\frac{1}{10}$	8	Sea water
$\frac{1}{100}$	9	Baking soda
$\frac{1}{1000}$	10	Milk of magnesia
$\frac{1}{10\,000}$	11	Ammonia solution
$\frac{1}{100\,000}$	12	Soapy water
$\frac{1}{1\,000\,000}$	13	Bleach, oven cleaner
$\frac{1}{10\,000\,000}$	14	Liquid drain cleaner

▲ **FIGURE 13.12.5** The pH scale. Each level is different from the next by a factor of 10.

9780170491785

Identifying acids and bases by using indicators

☆ ACTIVITY

13.12

You will identify the colours of indicators in acidic and basic solutions. You will then use this knowledge to classify common household substances as acids or bases.

Procedure

Part 1: What colour are indicators?

1. Set up three small test tubes with approximately 1 mL of 0.1 mol L^{-1} hydrochloric acid solution (test tube 1), deionised water (test tube 2) and 0.1 mol L^{-1} sodium hydroxide solution (test tube 3).
2. Add 2–3 drops of universal indicator to each test tube and record the colour of the indicator in acidic, neutral and basic solutions.
3. Extension: Your teacher may ask you to prepare your own indicator from red cabbage. Chop the cabbage, add boiling water from a kettle then stir the mixture for 10 minutes until the water is a purple colour. Strain the cabbage and keep the liquid to use as an indicator.

Part 2: Common substances – acid, base or neutral?

1. Gather a selection of household chemicals or drinks, such as juice, soft drink, bleach, shampoo, milk or dishwashing liquid (anything you can find).
2. Place approximately 1 mL of each substance into their own test tube.
3. Add 2–3 drops of universal indicator to each test tube.
4. Repeat with the other indicators and fresh 1 mL samples of the same household chemicals and drinks.
5. Record all colours observed referring to the universal indicator colour chart.

Analysis

1. **Create** a results table showing the colours of each indicator in acid, a base and neutral solutions.
2. Use this table to **identify** whether each household substance is an acid, a base or neutral. **Create** a summary table to show your results.
3. Were any of the household chemical results hard to **determine**?
4. **Describe** two difficulties you encountered while conducting this experiment that may affect the confidence you have in your results.
5. **List** any patterns in your household substances results. Do all the foods have the same acid/base nature? Do all the cleaning materials have the same acid/base nature?

13.12 LEARNING CHECK

1. The following acids are commonly found in nature and/or synthesised for human use.
   - Citric acid
   - Ascorbic acid
   - Nucleic acid
   - Uric acid
   - Ethanoic acid
   - Folic acid
   - Amino acid

   Choose three acids from the **list** and **research** their formulas and uses.
2. **Describe** the properties of acids and bases.
3. **Explain** what information is shown by the pH scale.
4. Phenolphthalein is colourless in an acid and pink in a base. What colour would you expect phenolphthalein to turn when added to:
   a. juice from a lemon?
   b. bleach from your laundry?

13.13 Metal and acid reactions

BY THE END OF THIS MODULE, YOU WILL BE ABLE TO:

✓ describe at an atomic level what occurs during a reaction between acids and metals

✓ predict the products and write equations to represent a reaction between acids and metals.

Extra science investigation
Reactions between acids and metals

Other resource
Activity sheet: Reactions and the metal series

GET THINKING

You had an introduction to acids in Module 13.12. What do all acids have in common? Can you identify some common acids you might find in your house? What are some uses for acids in your home or everyday life?

Acids in reactions

This module will look at one of the chemical reactions that acids undergo.

When acids are in solution, they form hydrogen ions (H^+) and a negative ion, as shown in Table 13.13.1. When acids react in chemical reactions, they are always in solution, so they always form these ions.

▼ **TABLE 13.13.1** Some common acids, their formulas and the ions they form

Acid name	Chemical formula	Ions that form in solution
Hydrochloric acid	HCl	H^+ and Cl^- (chloride ion)
Sulfuric acid	H_2SO_4	H^+ and SO_4^{2-} (sulfate ion)
Nitric acid	HNO_3	H^+ and NO_3^- (nitrate ion)
Carbonic acid	H_2CO_3	H^+ and CO_3^{2-} (carbonate ion)

What happens when acids and metals react?

Acids react with some metals in a single displacement reaction. The products formed when an acid and a metal react are an ionic salt and hydrogen gas (H_2). A general equation for all acid–metal reactions is:

Acid + metal → ionic salt + hydrogen gas

You can use the acid and metal names to predict the name of the ionic salt that will form. As shown in Figure 13.13.1, the ionic salt takes the name of the metal and adds the negative ion from the acid.

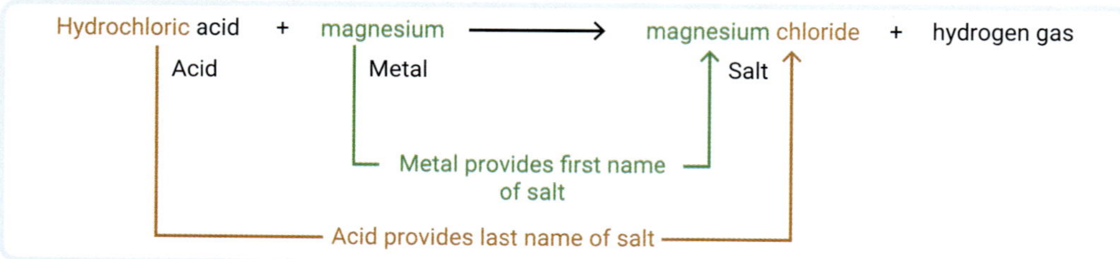

▲ **FIGURE 13.13.1** Using acid and metal names to predict the name of ionic salts that form in acid and metal reactions

9780170491785

If zinc were added to sulfuric acid, the ionic salt would be zinc sulfate. Aluminium and nitric acid would form aluminium nitrate. These reactions can be represented by the following word and balanced chemical equations:

Equation 1: Zinc + sulfuric acid → zinc sulfate + hydrogen

$$Zn(s) + H_2SO_4(aq) \rightarrow ZnSO_4(aq) + H_2(g)$$

Equation 2: Aluminium + nitric acid → aluminium nitrate + hydrogen

$$2Al(s) + 6HNO_3(aq) \rightarrow 2Al(NO_3)_3(aq) + 3H_2(g)$$

When these reactions occur, bubbles of gas form. This is the hydrogen gas formed in the reaction shown in Figure 13.13.1. The amount of bubbling can indicate the reactivity of the metal. As seen in Figure 13.13.2, the bubbling when magnesium is added to hydrochloric acid is vigorous. A lot of bubbles form very quickly. Less reactive metals would form fewer bubbles.

MARTYN F. CHILLMAID/SCIENCE PHOTO LIBRARY

▲ **FIGURE 13.13.2** Magnesium reacting with hydrochloric acid. The bubbles of hydrogen gas appear white.

Acid rain

In many places, **acid rain** falls due to high levels of acidic pollutants in the air. The acid can react with metals to form ionic salts. Metal structures can be weakened in this way (Figure 13.13.3).

acid rain
rain that is acidic due to chemicals dissolved in water in the atmosphere

Istock.com/iStock Essentials

▲ **FIGURE 13.13.3** Acid rain damage to a metal structure

Metals can be protected from acid rain by different methods that prevent the acid reacting with the metal. Most of these methods also protect against corrosion by providing a barrier to oxygen and water. Some common methods are shown in Table 13.13.2.

▼ **TABLE 13.13.2** Protecting metals from acid rain and corrosion

Method	How it works	Examples
Painting	• Covering the metal with paint means acid rain attacks the paint instead of the metal. • The paint needs to be reapplied often, as it wears away. • The paint also helps protect the metal against regular corrosion.	Bridges and fences are often painted regularly as a form of protection.
Galvanising	• The metal is covered in a layer of zinc so that any acid rain is prevented from contacting the metal underneath. • Over time, the zinc can be eaten away, and the metal may need to be recoated.	Nails and roofing materials are commonly galvanised substances.
Alloy metals	• The original metal (e.g. iron) is combined with another substance such as nickel or silicon, giving the metal protection so that it does not react with the acid rain.	Alloys of iron with nickel (e.g. steel) are commonly used to make metal containers and pipes that transport and store acids.

ACTIVITY

Protecting metal from acid

Individually, or in pairs, **design** a scientific method to test the effectiveness of one of the protection methods listed in Table 13.3.2. Consider the following points in your scientific method.

- What material will you use? For example, you could paint a piece of metal. Alternatively, galvanised nails are an easy way to find some galvanised metal, or some nuts and bolts for high-stress jobs are made from stainless steel with nickel.
- How will you know if it provides protection for the original, unprotected metal? How will you compare the two?
- What type and concentration of acid will you use? Consult your teacher regarding what is most appropriate.

If time and resources allow, you might be able to **test** some of your methods. Check with your teacher to organise this.

13.13 LEARNING CHECK

1 What do all acids have in common in their chemical structure?
2 **Identify** the ions present in a solution of:
   a nitric acid.
   b carbonic acid.
3 **Explain** why the reaction of an acid and a metal is a single displacement reaction.
4 Write word and balanced chemical equations for the reaction between the following acids and metals.
   a Calcium and hydrochloric acid
   b Magnesium and sulfuric acid
   c Lithium and carbonic acid
5 **Predict** the metal and acid you would need to form the following ionic salts and, for each salt, write a word and balanced chemical equation for its formation.
   a Zinc nitrate
   b Lead sulfate
   c Aluminium chloride

13.14 Acid and metal hydroxide reactions

BY THE END OF THIS MODULE, YOU WILL BE ABLE TO:

- ✓ describe at an atomic level what occurs during a reaction between acids and metal hydroxides
- ✓ predict the products and write equations to represent a reaction between acids and metal hydroxides.

GET THINKING

Several of the modules in this chapter have referred to 'general equations'. Find the general equation in the previous module and in this one. What do you think the purpose of a general equation is? How can it be used to write word equations to represent specific chemical reactions?

Interactive resource
Crossword: Types of reactions

Extra science investigation
Neutralising an acid

Neutralisation reactions

When an acid and a base react together, it is called a neutralisation reaction.

The general equation for a neutralisation reaction is:

Acid + base → salt + water

Farmers and gardeners use neutralisation reactions. Soils can be acidic or basic, depending on the rocks or minerals present and the source of the soil. Some plants grow best in acidic soils; others grow best in basic soils. Farmers and gardeners often add bases or acids to the soil to change its pH so they can grow crops or flowers successfully (Figure 13.14.1).

Swimming pools that are too acidic can make your eyes burn and itch. Pools that are not acidic enough grow algae and turn green. To properly maintain the pH of a swimming pool, acids or bases are added until the right pH is reached (Figure 13.14.2).

Istock.com/iStock Essentials

▲ **FIGURE 13.14.1** Bases are used to neutralise soils that are too acidic.

Istock.com/iStock Signature

▲ **FIGURE 13.14.2** Acids and bases are used to control the pH of swimming pools.

What is a metal hydroxide?

A **metal hydroxide** contains a metal ion and a hydroxide ion (the polyatomic ion OH^-). Many metal hydroxides are bases. Examples of metal hydroxides are magnesium hydroxide, $Mg(OH)_2$, and sodium hydroxide, $NaOH$.

metal hydroxide
an ionic salt containing a metal ion and a hydroxide ion

Predicting the products of neutralisation reactions

Figure 13.14.3 shows two neutralisation reactions involving metal hydroxides. The ionic salt that forms in such a neutralisation reaction is a combination of the metal from the metal hydroxide and the negative ion from the acid (see the previous module for the list).

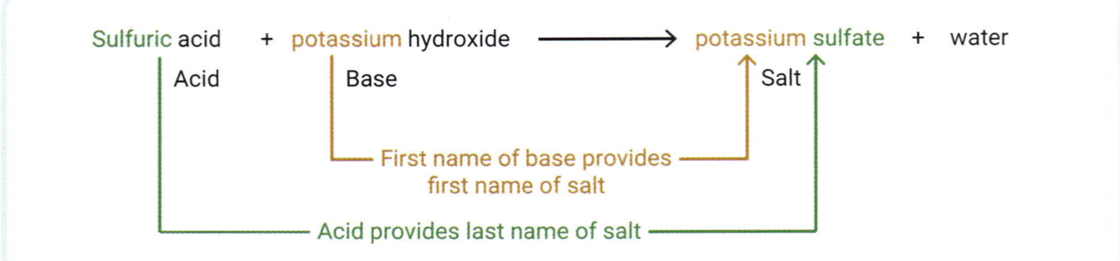

▶ **FIGURE 13.14.3** You can use the names of the acids and bases to predict the products of neutralisation reactions.

☆ **ACTIVITY**

Demonstrating pH change in a neutralisation reaction

▲ **FIGURE 13.14.4** Universal indicator is red, pink or yellow in acids. It turns green in a neutral solution and blue or purple in basic solutions.

- Set up two test tubes.
  - Test tube 1: 5 mL of 0.1 mol L^{-1} hydrochloric acid and 2 drops of universal indicator
  - Test tube 2: 5 mL of 0.1 mol L^{-1} sodium hydroxide and 2 drops of universal indicator
- Carefully pour the contents of Test tube 2 into Test tube 1 a few drops at a time, until it is all transferred.

  a Using Figure 13.14.4 as a guide, **describe** the colour changes of the reactants and products.

  b How do the colours show that neutralisation has occurred?

  c **Write** a word equation to represent this reaction.

13.14 LEARNING CHECK

1 Write word and balanced chemical equations to represent the reactions between the following acids and metal hydroxides:

  a Hydrochloric acid and calcium hydroxide

  b Sulfuric acid and aluminium hydroxide

  c Carbonic acid and zinc hydroxide

  d Nitric acid and sodium hydroxide

2 Two common uses of neutralisation reactions were given in this module. **Research** another common use of neutralisation reactions in everyday life.

13.15 Writing procedures and methods

SCIENCE SKILLS IN FOCUS

IN THIS MODULE, YOU WILL FOCUS ON LEARNING AND IMPROVING THESE SKILLS:

- planning and writing a procedure
- writing word and balanced equations for chemical reactions observed in experiments.

Procedure versus method

A procedure and a method are often mistaken for being the same thing in a science report. While both of them describe the steps taken in conducting an experiment, they are written quite differently.

A procedure outlines the steps you expect to take during the investigation. In most scientific investigations, the steps are altered, so a method is written afterwards with the exact steps that were actually performed.

Similarities

Both a procedure and a method have the following:

- numbered steps
- specific equipment used, including the size, such as '100 mL beaker', not just 'beaker'
- specific details with units of measurement, such as 'measure 5 mL of water using a 10 mL measuring cylinder'
- (where appropriate) the independent variable and how you will measure the dependent variable
- any controlled variables
- the number of repetitions
- any measurements and observations that need to be recorded.

Differences

The main difference between a procedure and a method is the tense they are written in. A procedure is written as a set of instructions, such as 'Add 100 mL water to a 250 mL beaker'. A method is written in past tense and in the third person, such as '100 mL of water was added to a 250 mL beaker'.

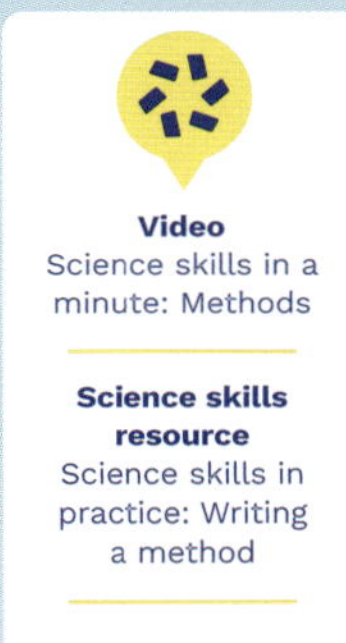

REACTIONS INVOLVING ACIDS

AIM

To investigate the reaction between metals and acids

Safety

Wear your safety glasses, lab coat and gloves at all times.

Take care when using matches and handling chemicals.

MATERIALS AND EQUIPMENT

- ☑ samples of a range of metals. This could include copper, zinc, magnesium and aluminium
- ☑ 2 mol/L hydrochloric acid (HCl)
- ☑ test tubes and test-tube rack
- ☑ 10 mL measuring cylinder
- ☑ matches
- ☑ cottonwool
- ☑ safety glasses, lab coat and gloves

PROCEDURE

1 Place a small piece of each metal in the bottom of a test tube.
2 Add 10 mL of hydrochloric acid to each test tube.
3 Place a small plug of cottonwool loosely in the top of each test tube.
4 Record your observations. Focus on the level of bubbling that is occurring in each test tube.
5 To test for the gas produced, light a match, remove the cottonwool and hold the match to the top of the test tube. A 'pop' sound indicates that hydrogen gas has been produced.

RESULTS

Create a table to record your observations for each metal.

ANALYSIS

1 Write word and balanced chemical equations for each metal's reaction with the hydrochloric acid.
2 Use the equation you wrote in Question 1 to **explain** why the 'pop' test in step 5 of the method was conducted.
3 Did all the metals react with the same intensity?
4 How do the results of this experiment show you the reactivity of metals? Rank the metals you used in order of their reactivity.
5 The observations you made in this experiment are qualitative. **Describe** any difficulties you had in ranking the metals based on your observations alone.
6 **Suggest** an improvement to the method that would allow you to make quantitative measurements.

CONCLUSION

Write a conclusion for this experiment.

9780170491785

13.16 Rate of reaction

BY THE END OF THIS MODULE, YOU WILL BE ABLE TO:

- ✓ describe examples of chemical reactions with high and low rates of reaction
- ✓ describe methods of observing and measuring the rate of a reaction.

GET THINKING

Science experiments can use both qualitative observations and quantitative measurements. What is the definition of 'qualitative' and 'quantitative'? What words do they sound like? Which of these two words applies when measuring rate of reaction?

Quiz
Rate of reaction

Extra science investigation
Comparing reaction rates

Fast and slow reactions

In chemistry, the **rate of reaction** is used to describe how fast or how slow reactions take place. When we apply this to chemical reactions, we examine how much reactant is used up in a period of time. Alternatively, you can examine how much product forms in a period of time.

rate of reaction a measurement of how fast a reaction is proceeding

Fireworks involve chemical reactions that are very fast (Figure 13.16.1). Many chemical reactions are completed after a few seconds or minutes.

Rusting (or corrosion) of iron is an example of a synthesis reaction between iron and oxygen. It is a very slow reaction, with rust sometimes taking years to form on an iron structure (Figure 13.16.2).

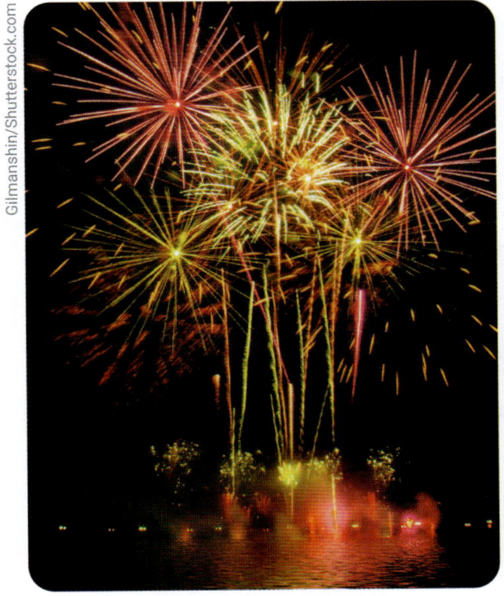

Gilmanshin/Shutterstock.com

▲ **FIGURE 13.16.1** Fireworks are an example of a fast chemical reaction.

bajars/Shutterstock.com

▲ **FIGURE 13.16.2** The synthesis reaction involving iron and oxygen (corrosion or rusting) is a very slow chemical reaction.

Rate of reaction graphs

The progress of a chemical reaction can be measured and plotted onto a graph.

Consider the progress of a reaction involving reactant A forming B (A → B). As A is used up, the amount of A decreases. As B forms, the amount of B increases. Figure 13.16.3 shows two graphs representing the reaction A → B. Graph **a** shows the change in B over time. Graph **b** shows the change in the amount of A over time.

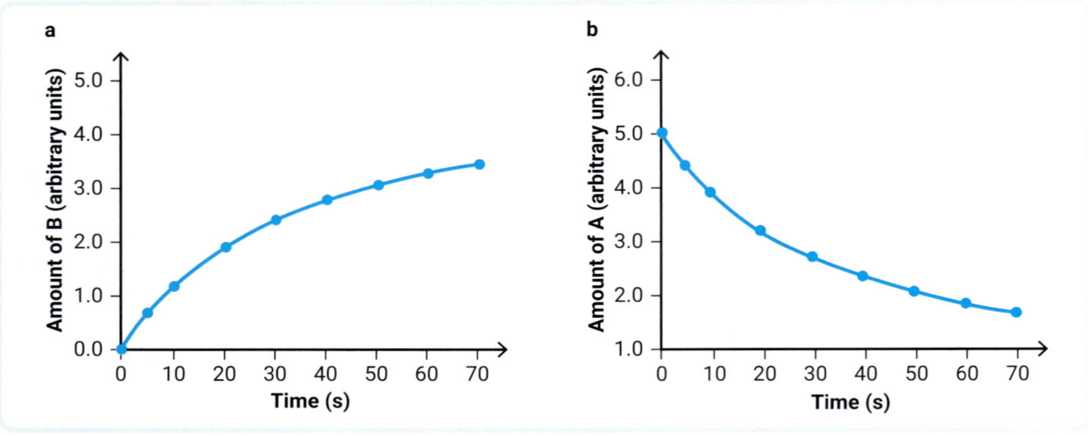

▲ **FIGURE 13.16.3** Rate of reaction graphs showing the change in the amount of (a) B and (b) A

How can the rate of reaction be measured?

The rate of reaction can be calculated quantitatively using the formula:

$$\text{Rate of reaction} = \frac{\text{amount of reactant used}}{\text{time}}$$

$$\text{Rate of reaction} = \frac{\text{amount of product formed}}{\text{time}}$$

For example, magnesium reacts with hydrochloric acid to form magnesium chloride and hydrogen gas:

Magnesium + hydrochloric acid → magnesium chloride + hydrogen

1.3 g of magnesium reacted with hydrochloric acid. The time required for all the magnesium to react was 3 minutes.

$$\text{Rate} = \frac{\text{mass used}}{\text{time}} = \frac{1.3\text{ g}}{3\text{ min}} = 0.043\text{ g/min}$$

The unit is g/min or grams per minute. This means that 0.043 grams of magnesium was used in each minute.

The rate of reaction can also be calculated using the slope or gradient of a graph. You will probably recall from your maths studies that the gradient of a graph can be calculated by using the $\frac{\text{rise}}{\text{run}}$ formula. As the graph has the amount of product on the vertical axis and time on the horizontal axis, the gradient is $\frac{\text{amount of product}}{\text{time}}$, which is the same as the formula for the rate of reaction. Thus, the gradient is the rate of reaction.

Figure 13.16.4 shows you how to use a graph to calculate the rate of formation of a product in a chemical reaction.

13.16

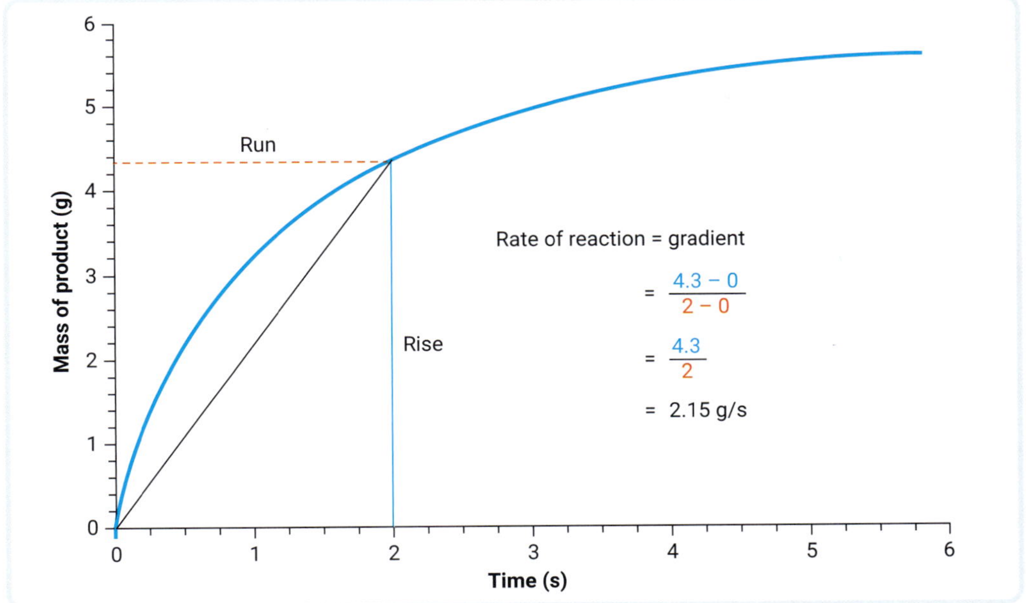

▲ **FIGURE 13.16.4** Using a graph to calculate the rate of reaction

DATA SCIENCE

Learn more about using graphs to interpret data in **Module 2.6**.

☆ **ACTIVITY**

Measuring the rate of reaction

Materials and equipment

- electronic balance
- 20 mL of 2.0 mol/L hydrochloric acid
- small piece (approximately 1 cm strip) of magnesium metal
- 100 mL beaker
- stopwatch

Procedure

1. Measure and record the mass of the magnesium strip.
2. Measure 20 mL of the hydrochloric acid and place it into a 100 mL beaker.
3. Place the beaker on the electronic balance.
4. Add the magnesium piece and at the same time zero the electronic balance. At the same time have another student start a stopwatch.
5. Record the time taken for the magnesium piece to completely react so there is no visible sign of it remaining.
6. Record the reading on the electronic balance at this time.

Analysis

1. Write an equation for the reaction that is occurring in the beaker.
2. **Explain** why the final mass on the electronic balance was negative. What does the negative mass represent?
3. **Calculate** the rate of reaction of the magnesium metal, using the data.
4. **Calculate** the rate of formation of the product, using the data.

13.16 LEARNING CHECK

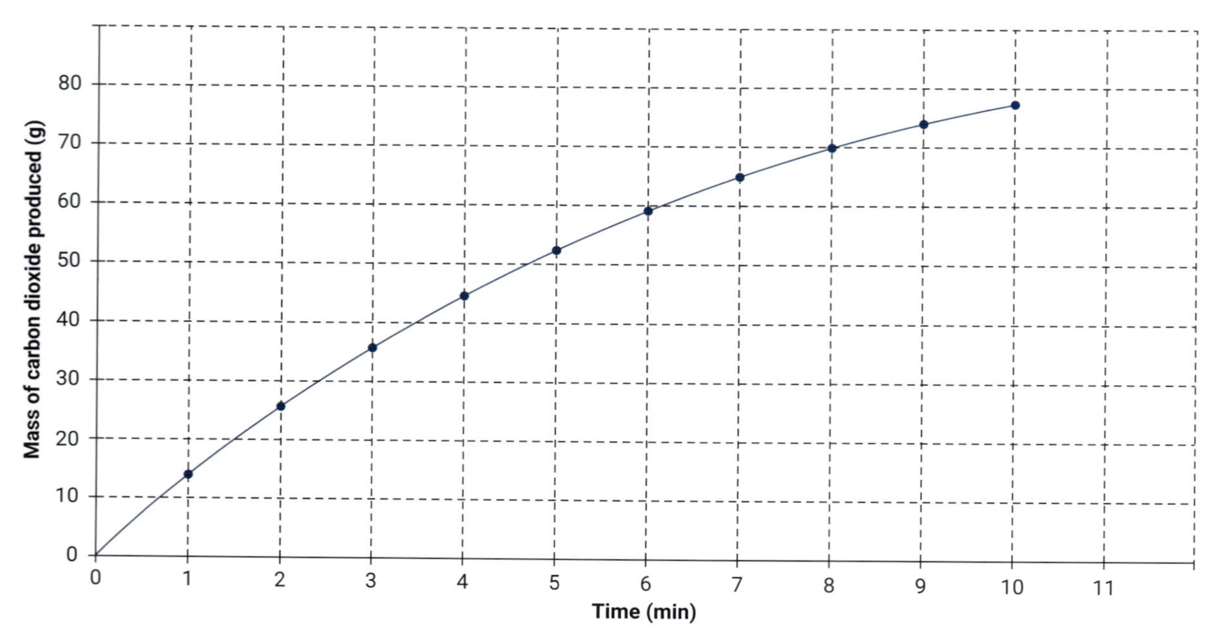

▲ **FIGURE 13.16.5** The formation of carbon dioxide in a chemical reaction

1 Figure 13.16.5 shows the formation of carbon dioxide in a chemical reaction.
   a **Calculate** the rate of formation of the carbon dioxide in the first 8 minutes of the reaction.
   b How does the graph show that carbon dioxide is a product and not a reactant?
2 In a chemical reaction, 2.5 grams of a chemical takes a time of 83 seconds to fully react. **Calculate** the rate of reaction for this chemical in grams per second.
3 If a chemical is produced in a reaction at a rate of 10 grams per minute, **calculate** the mass of chemical that would form if the reaction were conducted for 6 minutes.
4 **Justify** whether the following involve fast or slow rates of reaction:
   a cooking food using a high temperature
   b combusting fuel in a car engine
   c the breaking down of rocks as they react with chemicals in the environment.

9780170491785

13.17 Collision theory

BY THE END OF THIS MODULE, YOU WILL BE ABLE TO:

✓ explain the role of collision theory and activation energy in rates of reaction.

GET THINKING

What do you remember about different types of energy? What is kinetic energy? Particles need energy to react. Why do you think particles need kinetic energy to react?

Interactive resource
Label: Collision theory

What is collision theory?

Collision theory is used to explain why reactions occur. This theory will also be used in the next module to explain how and why we can make reactions faster or slower.

collision theory
the theory that a reaction will only occur if particles come into contact with sufficient energy and at the correct orientation to cause a successful collision

Collision theory states that a reaction will only occur if particles come into contact (collide) with sufficient energy and at the correction orientation to cause the bonds to break and products to form.

Particles moving around or vibrating in solids, liquids and gases have kinetic energy. All particles demonstrate kinetic energy.

If a collision occurs where the particles have more than the required amount of energy and the correction orientation, it is called a **successful collision**. Collisions where the particles do not have enough energy to form products are called unsuccessful collisions. In an unsuccessful collision, the reactants bounce off each other and do not undergo any change. Both types of collisions are shown in Figure 13.17.1.

successful collision
a collision that results in products forming

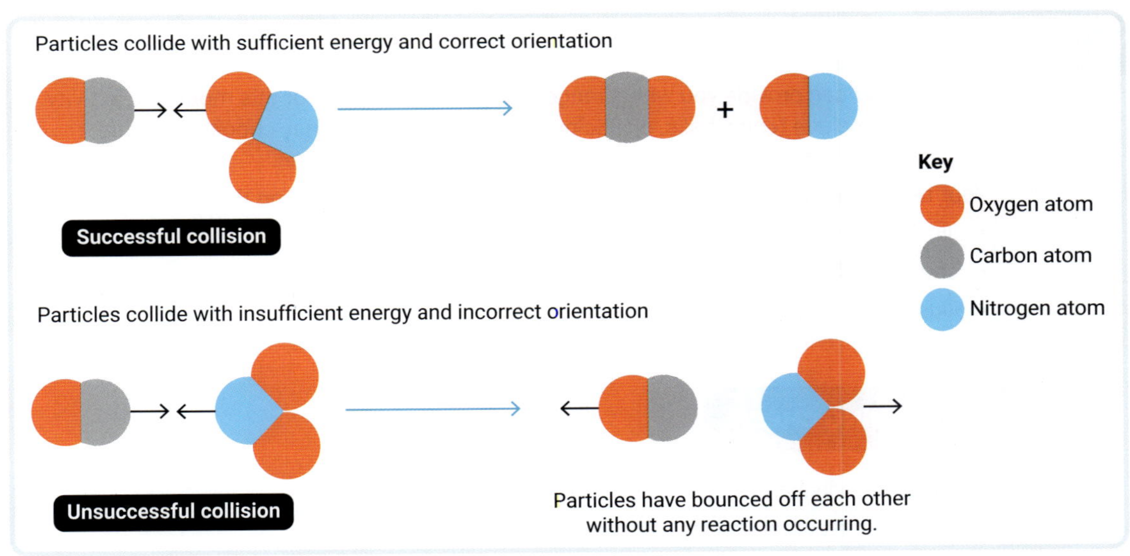

▲ **FIGURE 13.17.1** Collision theory

Activation energy

Activation energy is the term used to describe the minimum amount of energy required for a successful collision to occur and products to form. Each chemical reaction has a particular activation energy that is unique to that reaction.

activation energy
the minimum energy particles need to have for a successful collision to occur

Some reactions have a very low activation energy. Sodium metal and chlorine gas react explosively at room temperature (Figure 13.17.2). Not a lot of energy is required to start the reaction because it has a low activation energy. Even at room temperature, the particles have sufficient kinetic energy for a successful collision.

The reaction between magnesium and oxygen is much harder to start (Figure 13.17.3). In Module 13.4 you heated magnesium with oxygen in a synthesis reaction. The fact you had to heat it with a Bunsen burner indicates the particles needed a great deal of kinetic energy to provide the activation energy for the reaction.

Science Photo Library

▲ **FIGURE 13.17.2** Sodium metal and chlorine gas reacting

sciencephotos/Alamy Stock Photo

▲ **FIGURE 13.17.3** Magnesium ribbon in contact with oxygen gas

☆ **ACTIVITY**

Designing a model to show collision theory

Your teacher can provide you with a range of suitable materials for this modelling activity. This might include tennis balls, plasticine or cardboard, or you may choose to create a stop-motion animation, short video or other electronic presentation.

Your task is to **create** a series of photos or an electronic presentation that:

- illustrates what is required for a successful collision to occur
- shows the before and after of a successful and an unsuccessful collision
- illustrates the idea of activation energy.

13.17 LEARNING CHECK

1 **State** the collision theory.
2 **Describe** the conditions for a successful collision to occur.
3 **Describe** how an unsuccessful collision could become a successful collision.
4 If a reaction occurs at room temperature, does it have a high or low activation energy? **Justify** your answer.

13.18 Factors affecting rate of reaction

BY THE END OF THIS MODULE, YOU WILL BE ABLE TO:

✓ explain how altering the concentration of reactants, surface area, temperature and catalysts affect the rate of a chemical reaction.

GET THINKING

Speeding up chemical reactions is a key part of the chemical industry. Production of medicines, food and materials is more profitable if the chemical reaction is faster. Using the idea of collision theory from the previous module, brainstorm ways that chemical reactions might be made faster.

Interactive resource
Drag and drop: Factors that speed up reactions

Extra science investigation
Effect of temperature and concentration on reactions

How can we speed up reactions?

Speeding up (or slowing down) chemical reactions is based on the two ideas of collision theory.

1 Particles must come into contact to have a collision.

2 Particles must have enough energy to provide enough activation energy.

There are four ways of speeding up reactions:

- increasing the concentration of reactants
- increasing the surface area of reactants
- increasing the temperature of reactants
- adding a **catalyst**.

catalyst
a substance that increases the rate of a chemical reaction by lowering the activation energy of the reaction without itself being changed

Concentration

When you increase the concentration of a solution or a gas, you have more particles in the same space. You can see this effect in Figure 13.18.1. In a crowded room, you are more likely to collide with someone. More collisions occur between reactants when there is a higher concentration, so more product is able to form in a given time.

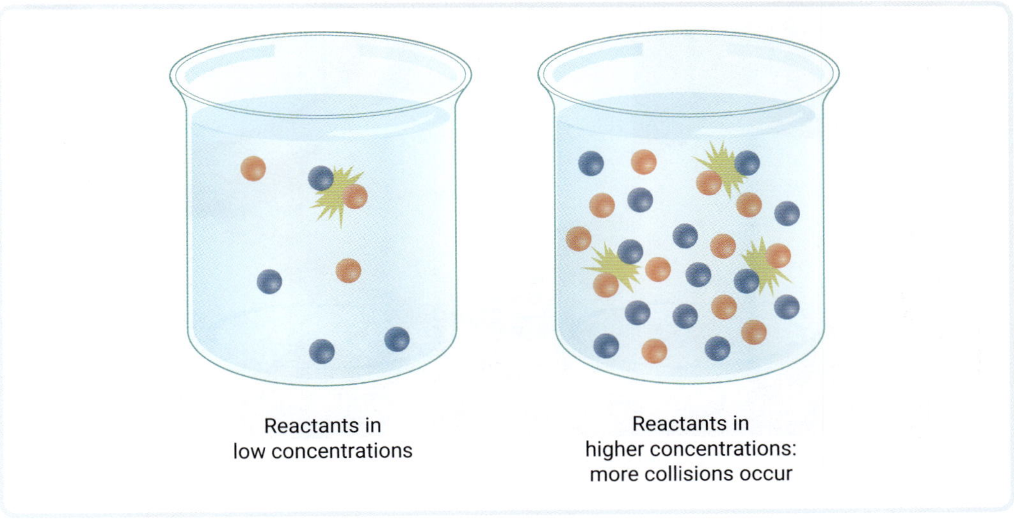

▲ **FIGURE 13.18.1** Higher concentrations result in more particle collisions.

Surface area

Increasing the surface area of a chemical involves chopping or breaking it into smaller pieces. Figure 13.18.2 shows how increasing the surface area makes more of the material accessible to another reactant. This also increases the number of collisions. As a result, the rate of reaction is increased.

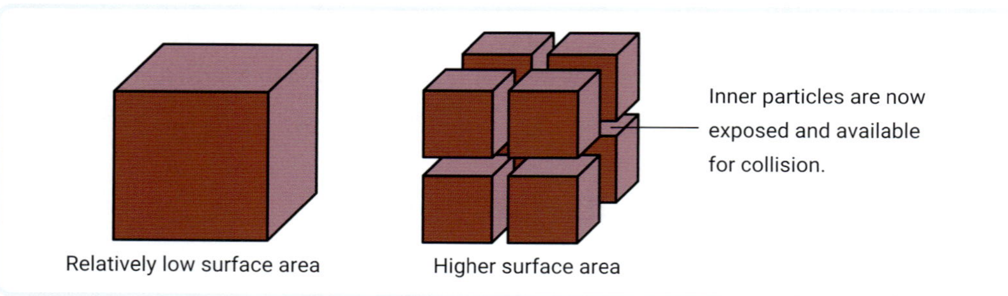

▲ **FIGURE 13.18.2** Increasing the surface area results in more particle collisions.

☆ ACTIVITY 1

The surface area of sugar

This activity does not show a chemical reaction, but it does show the effect of surface area through a model.

Add a solid sugar cube to 50 mL of water and stir it to dissolve. Repeat with the same amount (weight) of sugar grains (larger surface area).

a **Explain** why the grains of sugar dissolve more quickly than the sugar cube.

b **Assess** this model as a method of demonstrating surface area. **Describe** at least one way it is a good model and one way it is a poor model. **Justify** your overall opinion on the usefulness of the model.

Cozine/Shutterstock.com

▲ **FIGURE 13.18.3** Sugar cubes and grains of sugar dissolve at different rates.

9780170491785

Temperature

When a reactant is heated, all the particles gain more kinetic energy. This increases the rate of reaction in two ways.

1 As the particles gain energy, there are now more particles with **kinetic energy** above the activation energy for the reaction, as shown in Figure 13.18.4. This results in more successful collisions.
2 The particles are moving faster and so will collide more often. As a result, more successful collisions occur.

kinetic energy
the energy an object has because of its motion

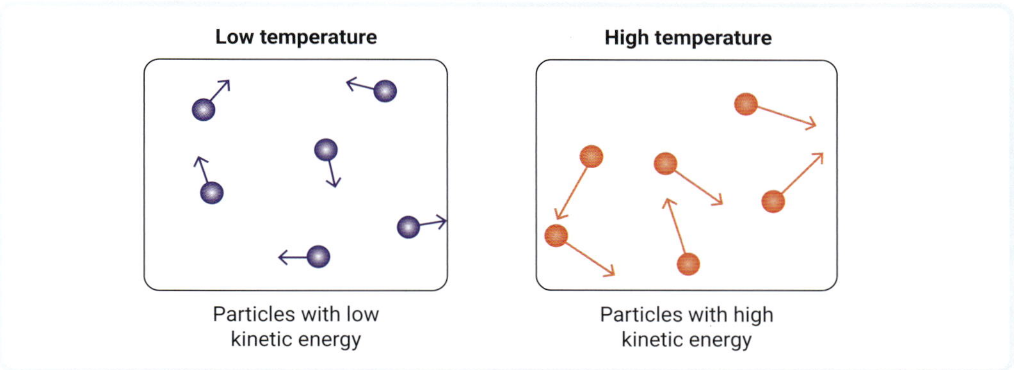

▲ **FIGURE 13.18.4** Increasing temperature gives particles more kinetic energy (represented by the longer and different coloured arrows).

Catalysts

A catalyst is a chemical that speeds up the rate of reaction but is not used up or changed in a chemical reaction. A catalyst reduces the activation energy, meaning more particles will have enough energy to react. As a result, more particles exceed the activation energy and there are more successful collisions. The rate of reaction is increased. The conical flasks shown in Figure 13.18.5 have increasing amounts of catalyst added. The increased rate of reaction can be seen by the formation of a greater volume of bubbles. Using a catalyst does not result in more product being formed; it just makes the reaction happen quicker.

Science Photo Library

▲ **FIGURE 13.18.5** Increasing amounts of catalyst have been added to these conical flasks from left to right. The catalysts were added at the same time.

☆ ACTIVITY 2

Examining the effect of a manganese dioxide catalyst on hydrogen peroxide

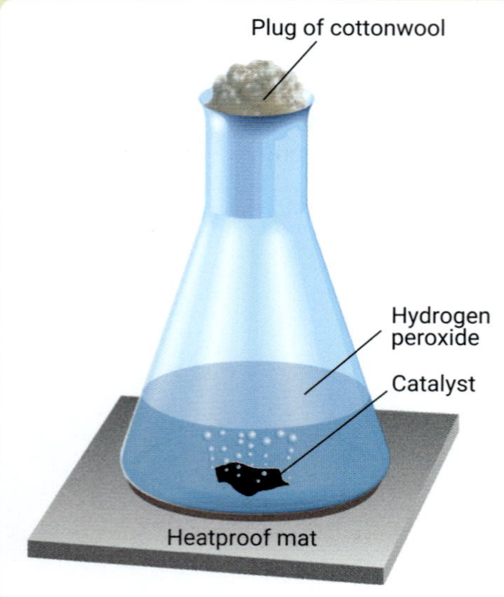

▲ **FIGURE 13.18.6** Equipment set up for the decomposition of hydrogen peroxide

Your teacher will most likely demonstrate this for you. It is a very quick reaction, so you might want to film it so that you can watch it back later.

Add 75 mL of 100 volume hydrogen peroxide solution to a 250 mL conical flask, as shown in Figure 13.18.6. Observe it for a moment. While it may not look like it, there is a reaction going on. The hydrogen peroxide is decomposing into water and oxygen.

a Why does it appear as though no reaction is occurring?

b Add 0.5 g powdered manganese dioxide and record all observations.

c If you were to collect the manganese dioxide at the end, what mass would you expect to collect? **Justify** your answer.

d **Explain** how the manganese dioxide increased the rate of reaction.

e **Compare** how a catalyst and heating can increase the rate of a chemical reaction.

13.18 LEARNING CHECK

1 **Identify** the four ways of increasing the rate of reaction.

2 a **Describe** what happens to the number of particles when the concentration of a solution is increased.

  b **Explain** how this increases the rate of reaction.

3 **Identify** what happens to the kinetic energy of particles when a chemical is heated.

4 **Explain** how heating a chemical reaction increases the rate of reaction.

5 Heating a reaction and adding a catalyst both increase the rate of reaction. **Explain** how these two different methods each work to **produce** the same result.

9780170491785

WORKING SCIENTIFICALLY

13.19 Observing rates of reaction

SCIENCE SKILLS IN FOCUS

IN THIS MODULE, YOU WILL FOCUS ON LEARNING AND IMPROVING THESE SKILLS:

- using an electronic balance to make quantitative measurements
- writing word equations for chemical reactions observed in experiments.

Safety

Wear your safety glasses, lab coat and gloves at all times.

Take care when using matches and handling chemicals.

RATE OF THE REACTION BETWEEN MAGNESIUM AND HYDROCHLORIC ACID

AIM

To investigate how changing the concentration of hydrochloric acid affects the rate of the reaction between hydrochloric acid and magnesium metal

MATERIALS AND EQUIPMENT

- ☑ 9 × 3 cm strips of magnesium (Mg)
- ☑ 2 mol/L hydrochloric acid (HCl)
- ☑ 3 × 100 mL conical flasks
- ☑ electronic balance
- ☑ 20 mL measuring cylinder
- ☑ distilled water
- ☑ safety glasses, lab coat and gloves

PROCEDURE

1 Measure 20 mL of hydrochloric acid using a measuring cylinder and place it into one of the conical flasks.

2 Measure 15 mL of hydrochloric acid and make up to a total of 20 mL with distilled water. Place this in the second conical flask.

3 Measure 10 mL of hydrochloric acid and make up to a total of 20 mL with distilled water. Place this in the third conical flask.

4 Place the first conical flask on the electronic balance. Add a 3 cm strip of magnesium and immediately zero the electronic balance.

5 After 60 seconds, note the mass reading. Since gas is being formed and escaping, the value will be negative. Ignore the negative sign and record the value of the mass.

6 Repeat step 5 with the second and third conical flasks.

7 If time and resources allow, repeat steps 1–6 twice more and average your results. Alternatively, collect data from other groups in your class and calculate the average.

RESULTS

1 Create a suitable table to record your data. For each trial (repetition) of steps 1–6 of the method, include the volume of hydrochloric acid used in the conical flask, the mass lost and the average mass lost.

2 The mass lost is the mass of hydrogen gas formed in 60 seconds. Use this information to calculate the average rate of reaction for each of the three concentrations. Show all workings.

ANALYSIS

1 Write a word equation to represent the reaction that occurred.

2 Use your answer to Question 1 to **explain** why mass is lost in this reaction.

3 **Suggest** a different method of measuring the rate of reaction that could be used instead of the loss of mass.

4 Write a conclusion about the effect of the concentration of hydrochloric acid on the rate of this reaction. Use data from your experiment to **support** your conclusion.

5 **Describe** two errors that could have occurred during this experiment and **explain** how they could have affected the results.

CONCLUSION

Write a conclusion for this experiment.

13.20 Use of chemical reactions to produce a range of products

ABORIGINAL & TORRES STRAIT ISLANDER SCIENCE CONTEXTS

IN THIS MODULE, YOU WILL:

✓ investigate some of the chemical reactions used by Aboriginal and Torres Strait Islander Peoples to produce products for consumption.

Fermentation of plant products

Aboriginal and Torres Strait Islander Peoples have long used chemical reactions to process plant products for consumption. Fermentation is a chemical process that converts carbohydrates to cellular energy using micro-organisms. The products of the reaction are alcohol or organic acids.

The Noongar People of Western Australia are recognised as producers of a fermented liquid called mangaitch, produced from the nectar of a species of banksia tree. The banksia flowers containing nectar were soaked in water and then fermented in a bark vessel taking advantage of native yeasts. In Tasmania, Aboriginal Peoples (Palawa) tapped *Eucalyptus gunnii* trees and allowed the sweet sap to pool in tree hollows. Native yeasts fermented the sap to produce an alcoholic beverage called wayalinah.

Istock.com/iStock Essentials

▲ **FIGURE 13.20.1** *Macrozamia* cone with flesh-covered seeds

It has been suggested that Noongar People also fermented toxic *Macrozamia* seeds, a type of cycad plant. The seeds, including their fleshy outer covering, were soaked and then buried underground, creating an anaerobic environment for the fermentation process. This produces an extremely nutritious, fat-rich food source that is safe for consumption. Uniquely, in this region, some groups were recorded to consume the red outer flesh of the cycad, rather than the seed.

© Joe Sambono

▲ **FIGURE 13.20.2** A Jawun basket made by the Jirrbal Peoples of Tully, North Queensland

Detoxification of cycads

In other parts of Australia, poisonous cycads were detoxified by many different Aboriginal Peoples using chemical processes other than fermentation to produce an edible product.

Soaking the cycad seeds in water hydrolyses the toxin, cycasin, so that the water-soluble chemical bonds are broken. This allows the toxic component to be washed away. Under normal conditions, this reaction proceeds very slowly. However, historical accounts show how the Aboriginal Peoples of the rainforest region in North Queensland increased the rate of the reaction by thinly slicing the seeds to increase the surface area, so more of the seeds were exposed to the action of hydrolysis. This also ensured that all of the cycad seed material was exposed to the action of hydrolysis. The reaction was allowed to proceed in special baskets for sufficient time to ensure complete detoxification. This process resulted in an abundant food resource that was safe to eat. Figure 13.20.2 shows a specialised colander-like basket used by Aboriginal Peoples of North Queensland to detoxify cycads.

ACTIVITY

Investigating how processing food products causes chemical changes

Aboriginal and Torres Strait Islander Peoples have extensive knowledge of the chemical reactions used to process toxic food products and make them edible.

It is not safe to use toxic cycads to investigate chemical reactions in this activity. Instead, you will use pineapples to model how processing foods can cause chemical changes. Pineapples contain an enzyme, bromelain, that digests collagen proteins. This is also a hydrolysis reaction.

Materials and equipment

- fresh pineapple
- frozen pineapple
- canned pineapple
- powdered gelatine
- water
- kettle
- ice water bath
- 4 test tubes or cups
- scales
- measuring cylinder

Procedure

1. Cut equal-sized pieces of fresh, canned and frozen pineapple. Ensure these pieces are small enough to fit into the test tubes (or cups).
2. Weigh each piece and record the weight.
3. Prepare the gelatine according to the instructions on the packet, using the kettle to boil the water.
4. Use the measuring cylinder to pour equal amounts of hot gelatine solution into each test tube.
5. Place each type of pineapple into separate test tubes, leaving one without fruit as a control. Label the test tubes.
6. Mix the test tubes gently and place them into the ice water bath.
7. Check the control tube every few minutes. When the control tube has solidified, remove all the tubes from the ice bath and observe the consistency of the gelatine. Record your observations.

Analysis

1. What did you **observe**? Why do you think this happened?
2. How does processing pineapple through freezing and canning cause chemical changes to the enzyme bromelain?
3. How does this experiment reflect the chemical practices employed by Aboriginal and Torres Strait Islander Peoples in processing foods?

SCIENCE IN CONTEXT

13.21 Production of ammonia

BY THE END OF THIS MODULE, YOU WILL BE ABLE TO:

- ✓ describe the production of ammonia
- ✓ describe the importance of ammonia as a useful chemical.

How do we use ammonia?

wk1003mike/Shutterstock.com

ZikG/Shutterstock.com

▲ **FIGURE 13.21.1** Ammonia is used to make (a) fertilisers and (b) cleaning solutions.

Ammonia, NH_3, is one of the most commonly produced chemicals in the world. It does exist naturally in the environment, but it is produced industrially in large quantities.

The most popular use of ammonia is to produce fertilisers for use in agriculture and gardening (Figure 13.21.1a). Ammonia contains nitrogen, which is vital for good plant growth. Many agricultural crops like wheat, corn and soybeans require high levels of nitrogen to produce healthy crops for harvest and sale.

Ammonia is also very useful as a cleaning agent. It is very effective at breaking down stains from animal fats and vegetable oils like grease (Figure 13.21.1b). It is also used as a coolant gas in some refrigerators and air-conditioners as it absorbs heat easily. Ammonia is also used in the production of other chemicals such as plastics, textiles and dyes.

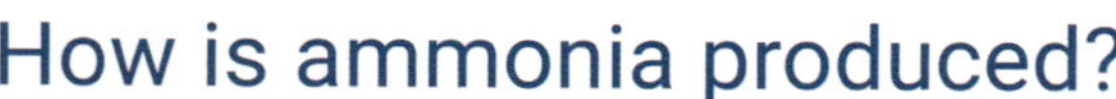

How is ammonia produced?

Ammonia is produced by the Haber process, in which hydrogen and nitrogen gases react at high temperatures, with a catalyst (Figure 13.21.2). The chemical reaction for this is:

$$N_2(g) + 3H_2(g) \rightarrow 2NH_3(g)$$

The catalyst lowers the activation energy of the reaction, allowing more particles to react and speeding up the reaction. The high temperatures give all the particles more energy so there are more successful collisions. Both of these factors increase the rate of reaction. Chemical companies aim to make more product in less time, because this increases their profits.

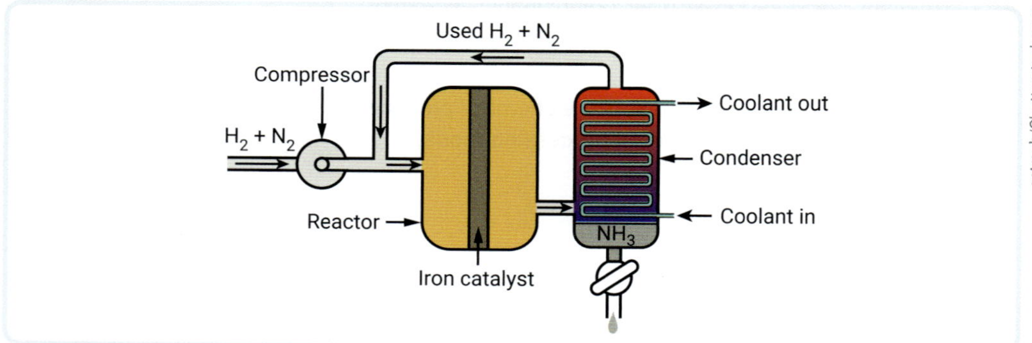

gstraub/Shutterstock.com

▲ **FIGURE 13.21.2** The Haber process is used to produce ammonia.

13.21 LEARNING CHECK

1. **Describe** two uses of ammonia.
2. Write a balanced chemical equation for the formation of ammonia.
3. **Explain** two factors that are used in the production of ammonia to increase the rate of reaction.

13 REVIEW

REMEMBERING

1 Write the general word equation for:
  - **a** an acid and a metal.
  - **b** an acid and a metal hydroxide.

2 A piece of magnesium reacted with sulfuric acid. Bubbles of gas formed. **Identify** the gas that was forming the bubbles.

3 **List** four methods of increasing the rate of reaction.

4 **State** the law of conservation of mass.

5 **Describe** two examples of useful synthesis reactions.

6 **Describe** the three ways that energy can be supplied in a decomposition reaction.

UNDERSTANDING

7 The graph below shows the volume of oxygen gas produced during the decomposition of hydrogen peroxide.
  - **a** **Calculate** the rate of reaction after 4 minutes and after 14 minutes.
  - **b** **Suggest** a reason why there is no change in the volume of oxygen after 16 minutes.

8 **Describe** how a catalyst increases the rate of reaction.

9 **Explain** how an understanding of acids, bases and pH can assist a gardener to grow flowers more successfully.

10 **Describe** how you could confirm through an experiment that the product(s) of an acid and metal hydroxide reaction was neutral.

11 For the following pairs of ionic salts, write word and balanced chemical equations and **predict** whether a precipitate will form.
  - **a** Lead nitrate and sodium hydroxide
  - **b** Magnesium nitrate and potassium chloride
  - **c** Calcium chloride and magnesium sulfate

12 Using an example, **describe** one way Aboriginal and Torres Strait Islander Peoples have used their knowledge of chemical reactions in food preparation.

APPLYING

13 Some hydrochloric acid has been spilled in a laboratory. The teacher states that it would be risky to wipe it up directly and suggests adding some sodium carbonate to the spill to make it safer to wipe up. Use a balanced chemical equation to help **explain** why this suggestion makes it safer to wipe up the acid.

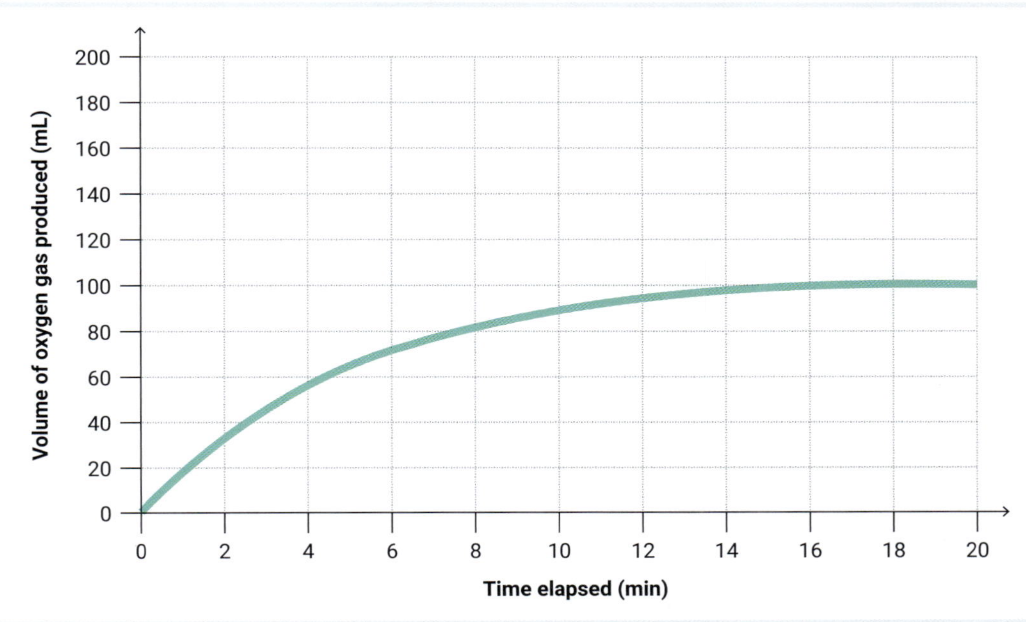

14 **Identify** whether the following reactions are examples of synthesis (S), decomposition (D), precipitation (P), metal/acid reaction (M) or neutralisation (N), and for each reaction (except e), write word and balanced chemical equations to show the products formed.

a Lead nitrate and sodium chloride
b Iron reacting with oxygen
c Heating copper carbonate and forming black copper oxide powder and carbon dioxide gas
d Magnesium reacting with hydrochloric acid
e Adding together monomers to make polystyrene
f Potassium sulfate reacting with silver nitrate
g Zinc metal reacting with sulfuric acid

15 Consider the experiment shown below. Rank the five beakers in order of the expected rate of reaction from fastest to slowest. **Justify** your ranking.

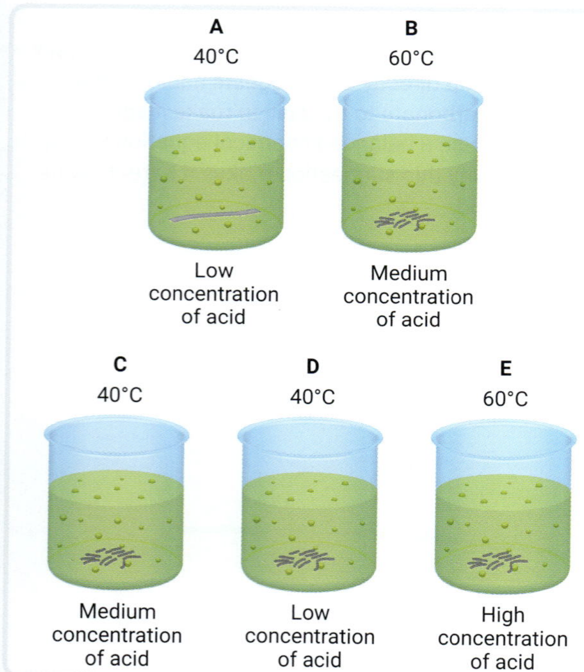

ANALYSING

16 The human stomach contains hydrochloric acid. Sometimes excess acid moves upwards, causing a burning sensation. Antacids containing chemicals such as magnesium hydroxide can remove the excess acid. Write a balanced chemical equation to show how antacids work.

17 When silver nitrate is mixed with potassium chloride, a white solid is formed.

a What type of reaction is this?
b Write the word and balanced chemical equations for this chemical reaction.
c Which of the products is the white solid?

18 A student conducted an experiment where 5 g of calcium metal was added to hydrochloric acid at different temperatures. The time taken for all the calcium to react was recorded in the following table.

a Copy and **complete** the table by calculating the rate of reaction at each temperature.
b **Explain** these results using collision theory.

Student experiment results for calcium metal and hydrochloric acid reaction

Temperature (°C)	Time taken for calcium to react (min)	Rate of reaction (g/min)
40	9.1	
50	7.3	
60	4.6	
70	2.1	

EVALUATING

19 'Chemical reactions have a positive impact on society.' **Discuss** this statement.

CREATING

20 **Create** a summary in poster form, with general equations and specific examples that show the four types of reactions studied in this chapter. Include everyday examples of each type, find photographs to illustrate them or **draw** your own models to represent the reactions.

9780170491785

SCIENCE IN DEPTH STUDY

1 Connect what you've learned

At the start of the chapter, you were asked why solid coal is hard to ignite but coal dust can explode with a spark. Use what you have learned in this chapter to provide a chemical explanation for this phenomenon.

2 Check your thinking

Australia has a number of coal mines. Coal is a major export for Australia and provides fuel for coal-fired power stations.

Research ways that miners in coal mines can be kept safe from a build-up of coal dust.

Research the cause of a coal mine disaster. Suggest ways it could have been prevented. Examples of disasters are those at the Westray mine in Canada in 1992, the Pike River mine in New Zealand and the Upper Big Branch mine in the United States, both in 2010.

3 Get into action

Chemical disasters are the accidental release of chemicals that can potentially be harmful to people or the environment. Three examples of major disasters that happened overseas are the Bhopal disaster in India, the ammonium nitrate explosion in Beirut, and the Deepwater Horizon oil spill disaster in the United States.

Choose one of these disasters (or identify another that interests you) and research the details of the event. This could include the cause of the disaster, the consequences for people or the environment or how it was cleaned up.

4 Communicate

Think about the chemicals that you use around your home. This might include gas for cooking or heating, cleaning chemicals, garden chemicals or other items in your shed or garage. Create a safety poster for your family that shows the safe use of some of these chemicals.

14 Nuclear reactions

SCIENCE IN DEPTH

▲ **FIGURE 14.0.1** These signs warn about the dangers of radioactivity.

When most people hear the word 'radioactivity', images such as the warning sign in the photo above usually spring to mind. However, it is also important to consider how we use radioactivity for our benefit. Radiation is used in medicine, to manufacture materials and even to produce some food.

- **What do you know about radioactivity?**
- **What have media, books, movies or comic books told you about radiation?**
- **How do the warning signs in the image above make you feel?**
- **Where do you think you encounter radiation every day?**
- **Do you know any examples of how radiation is used in medicine or industry?**

DIVE INTO SCIENCE!

At the end of this chapter, you can complete Science in Depth Study #14. You can use the information you learn in this chapter to complete the project.

Assessments
- Prior knowledge quiz
- Chapter review questions
- End-of-chapter test
- Depth study: Presentation

Videos
- Science skills in a minute: Modelling data **(14.5)**; Representing data **(14.8)**;
- Video activities: Radioactive half-life **(14.4)**; Radioactive dating **(14.7)**; Uses of radiation **(14.11)**

Science skills resources
- Science skills in practice: Modelling data **(14.5)**
- Extra science investigations: Modelling isotopes **(14.3)**; Simulating radioactive decay **(14.3)**

Interactive and other resources
- Simulation: Isotopes and atomic weight **(14.3)**
- Drag and drop: Stages of the Big Bang **(14.2)**; Alpha, beta and gamma radiation **(14.6)**
- Quiz: Nuclear reactions and the environment **(14.12)**

Nelson MindTap

To access resources above, visit **cengage.com.au/nelsonmindtap**

14.1 Introduction to nuclear reactions

BY THE END OF THIS MODULE, YOU WILL BE ABLE TO:

- ✓ understand the differences between chemical reactions and nuclear reactions
- ✓ represent nuclear equations using correct atomic format.

GET THINKING

Using numbers and chemical symbols gives important information about the subatomic parts of an atom. When you see scientific notation like C-12 or U-238, what questions do you have?

What is a nuclear reaction?

nuclear reactions
reactions involving a change in composition of the nucleus of an atom

In Chapter 13, you looked at chemical reactions where electrons are shared, donated and accepted to make new products. In all chemical reactions, the types of atoms you start with are the same as those you finish with. For example, when hydrogen and oxygen combine to form water, you start with hydrogen and oxygen atoms, and finish with hydrogen and oxygen atoms. Only the arrangement of the atoms changes, not the *type* of atom.

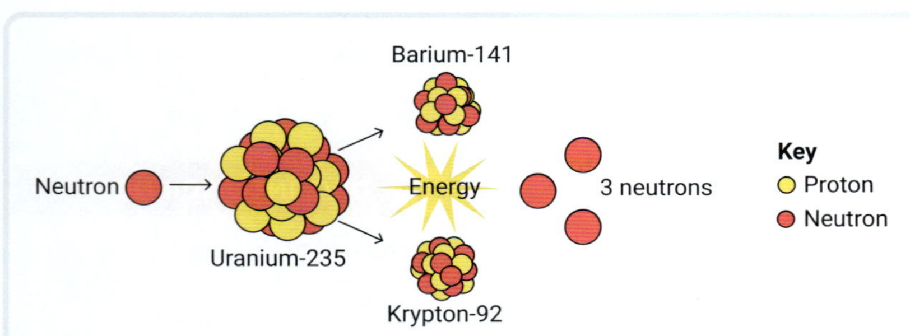

▲ **FIGURE 14.1.1** Nuclear reactions can have different elements as reactants and products. In uranium fission, uranium splits to form barium and krypton.

Nuclear reactions involve changes to the nucleus of the atom, not to the electrons (as occurs in chemical reactions). In a nuclear reaction, the reactant **elements** are often different from the product elements (Figure 14.1.1).

element
a pure substance made up of only one type of atom; it cannot be broken down into a simpler substance

nuclear fission
nuclear reactions where a large atom splits into two or more smaller atoms with a release of energy

nuclear fusion
nuclear reactions where two or more smaller atoms combine to form a larger atom with a release of energy

neutron
an uncharged particle in the nucleus of an atom

robert_s/Shutterstock.com

▲ **FIGURE 14.1.2** Stars undergo nuclear fusion, mostly converting hydrogen into helium. This releases vast amounts of energy.

Two types of nuclear reactions are fission reactions and fusion reactions. **Nuclear fission** is when larger atoms split into smaller ones. Nuclear power reactors and the production of medical isotopes use fission (examined further in Module 14.9). **Nuclear fusion** is when smaller atoms combine to form larger atoms. The fusion of hydrogen into helium is the main source of energy in stars (Figure 14.1.2) (explored in Module 14.10).

When uranium is used in a nuclear reaction, the reactants are uranium and **neutrons**. The products are barium and krypton. This reaction can be represented as:

$$^{235}_{92}\text{U} + ^{1}_{0}\text{n} \rightarrow ^{141}_{56}\text{Ba} + ^{92}_{36}\text{Kr} + 3^{1}_{0}\text{n}$$

You might notice this reaction has some similarities to a chemical equation, but it also has a lot of differences. Before we examine nuclear reactions in detail, we will review how scientists represent atomic information.

Atomic number and mass number

Recall from Module 7.2 that atoms of the same element have the same **atomic number (Z)**. The atomic number is given the symbol Z. Recall also that, in an atom, the number of **protons** is the same as the number of electrons. Remember that the **mass number (A)** is the total number of particles in the nucleus; that is, the number of protons and neutrons added together.

atomic number (Z)
the number of protons in the nucleus of an atom, which is the same for every atom of the same element

proton
a positively charged particle in the nucleus of an atom

mass number (A)
the total number of protons and neutrons in the nucleus of an atom

Mass number = number of protons + number of neutrons

Elements are often referred to by their mass number together with their name or symbol. Carbon-12 or C-12 is a carbon atom with a mass number of 12. Uranium-235 shows that uranium has a mass number of 235.

The atomic number and mass number can be placed next to the element symbol as shown in Figure 14.1.3. This notation can be used to determine the number of protons, neutrons and electrons. See Module 7.2 to review this concept.

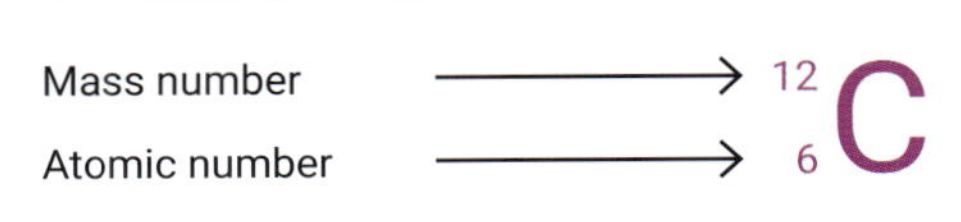

▲ **FIGURE 14.1.3** Notation showing mass number and atomic number for carbon-12

Writing nuclear reactions

Look again at the nuclear reaction for the fission of uranium.

$$^{235}_{92}\text{U} + {}^{1}_{0}\text{n} \rightarrow {}^{141}_{56}\text{Ba} + {}^{92}_{36}\text{Kr} + 3{}^{1}_{0}\text{n}$$

Nuclear reactions have a number of similarities to chemical reactions. Both have reactants and products separated with an arrow. Both use chemical symbols and have to be balanced to be correct.

There are also several significant differences. Nuclear reactions show the atomic and mass numbers next to the symbols, whereas a chemical reaction does not. The balancing of a nuclear reaction works slightly differently to a chemical reaction. In the reaction above, the top line (mass numbers) must add up on both sides of the arrow. You can see the left side adds up to 236 (235 + 1) and the right side also adds up to 236 (141 + 92 + 3 × 1). The same applies for the bottom line (atomic numbers). The left side adds up to 92, as does the right side (56 + 36). This is the process for balancing a nuclear reaction.

14.1 LEARNING CHECK

1. What information is provided by the atomic number and the mass number of an element and how are they different?
2. **Explain** why you cannot get the full information about atomic structure from the term 'oxygen-16' without any other information.
3. **Identify** the number of protons, neutrons and electrons in the following atoms.
   a $^{56}_{26}\text{Fe}$ **b** $^{235}_{92}\text{U}$
4. **Compare** chemical and nuclear reactions. **Describe** at least two similarities and two differences.
5. **Explain** how you would balance a nuclear equation.

14.2 The Big Bang

BY THE END OF THIS MODULE, YOU WILL BE ABLE TO:

- ✓ describe the Big Bang theory and explain how the Big Bang theory is used to describe the origin and evolution of the universe
- ✓ describe the process of element formation after the Big Bang.

Interactive resource
Drag and drop: Stages of the Big Bang

GET THINKING

The term 'Big Bang theory' gives the impression that the universe originated from a massive explosion. As you complete this module, think about whether this is the best way to describe how the universe formed. Why do you think it could be misleading? See if you can think of an alternative name for this theory that might help to correct the misconception that the name implies.

The origin of the universe

The best-supported theory by scientists of the origin of the universe centres on an event known as the Big Bang. It is generally accepted that the universe came into existence approximately 13.8 billion years ago. The Big Bang theory explains that the universe originated from an extremely small point called a **singularity** that underwent a massive expansion in an infinitely small period called **inflation**. The singularity contained everything that now makes up the universe: matter and energy.

singularity
a point where density and matter are infinite

inflation
the initial rapid expansion of space-time just after the Big Bang

The stages of the Big Bang

The **Big Bang theory** suggests that there were two main stages in the universe's evolution, representing key events that shaped the universe. The first stage was dominated by radiation and lasted for about 300 000 years. The second stage was dominated by matter and has lasted most of the 13.8 billion years. These main stages are further broken down into key events, which are outlined in Figure 14.2.1. These key events have shaped the universe from the beginning of time until the present.

Big Bang theory
the generally accepted theory for the formation of the universe

Within the first second of the Big Bang

The universe came into existence from a singularity that contained everything that was formed during the Big Bang. The singularity was immensely dense and hot, and in a 10 million-trillion-trillion-trillionth part of one second (10^{-43} s) it began to expand. At this point, the universe, time and space came into existence. As the universe continued to expand, the temperature decreased.

It took less than 1 millisecond after the Big Bang for the temperature to cool sufficiently to form the fundamental particles known as **quarks**. These combined to form the subatomic particles of protons and neutrons. The conditions present at this time resulted in fusion, which formed the nuclei of the elements hydrogen, deuterium (an isotope of hydrogen), helium and a small amount of the larger nuclei of lithium.

quark
a type of elementary particle that is the fundamental component of matter

Three minutes to 300 000 years after the Big Bang

Three minutes later, fusion ceased, leaving a universe made up of a 'fog' of about 75 per cent hydrogen nuclei, 25 per cent helium nuclei, and a small number of

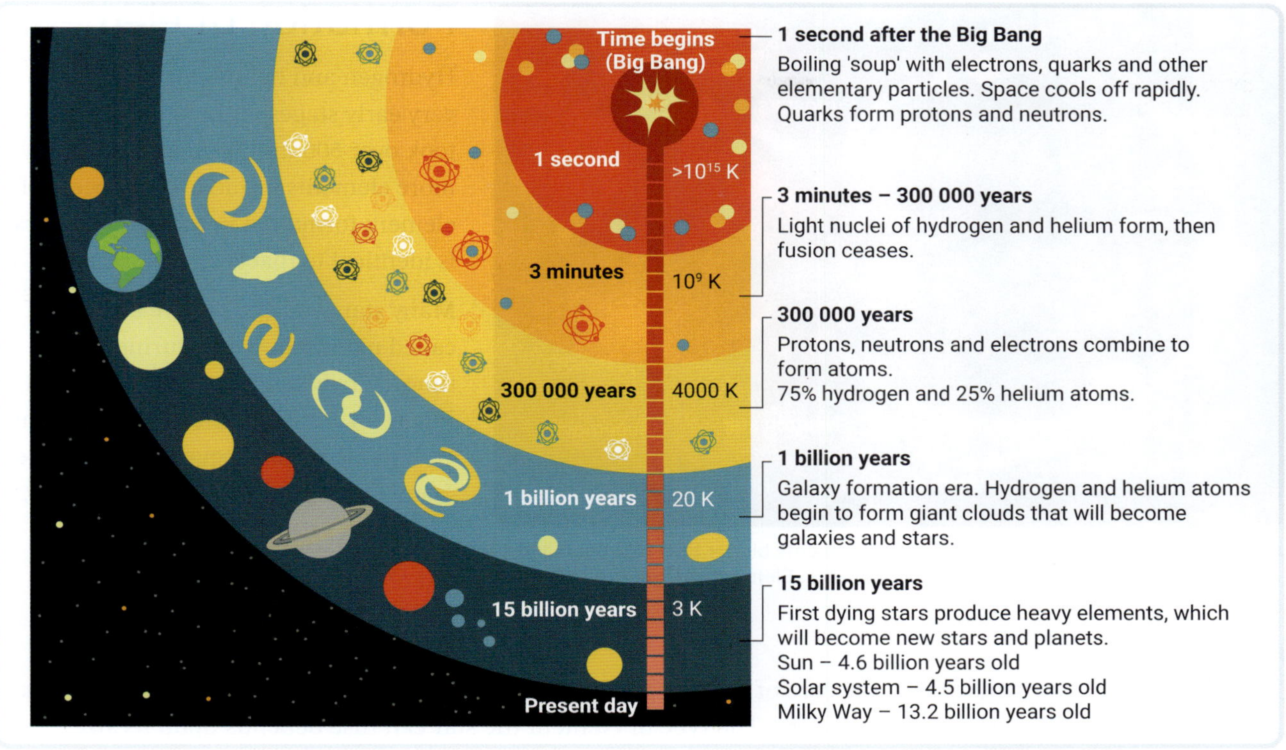

Tashal/Shutterstock.com

▲ **FIGURE 14.2.1** The timeline of the Big Bang

lithium nuclei. For the next 300 000 years, the universe gradually cooled, but it was still very hot compared with the universe today.

300 000 years after the Big Bang

Around 300 000 years after the Big Bang, temperatures dropped to 3000–4000 K (2726–3726°C). This allowed nuclei to capture electrons and form atoms, and **photons** were free to move. When this happened, the universe became transparent to light and entered the stage where matter took over from radiation to dominate the universe.

photon
a form of elementary energy particle

As the universe continued to expand, the force of gravity started to pull these newly formed atoms together to form clouds of hydrogen and helium gas, which became denser and collapsed under gravity, becoming hot enough to start nuclear fusion. We do not know exactly when and how the first stars and galaxies formed, but there is evidence of star and early galaxy formation around 500 million to a billion years after the Big Bang.

A billion years after the Big Bang

Gravity continued to attract stars to form the first primitive galaxies. Galaxies clustered, and swirling clouds of dust and gas formed more stars as the universe continued to cool and expand. Our Sun formed along with our solar system some 4.6 billion years ago.

Today, the universe is still cooling and expanding. The temperature of the universe is currently 2.725 K, just above **absolute zero**. The universe is a sphere that has expanded from a point where the distance from Earth to the edge of the visible universe is calculated to be about 46.5 billion light-years, or about 14.26 billion parsecs in any direction.

absolute zero
the lowest possible temperature (0 K or −273.15°C); the point at which there is no particle movement and no energy

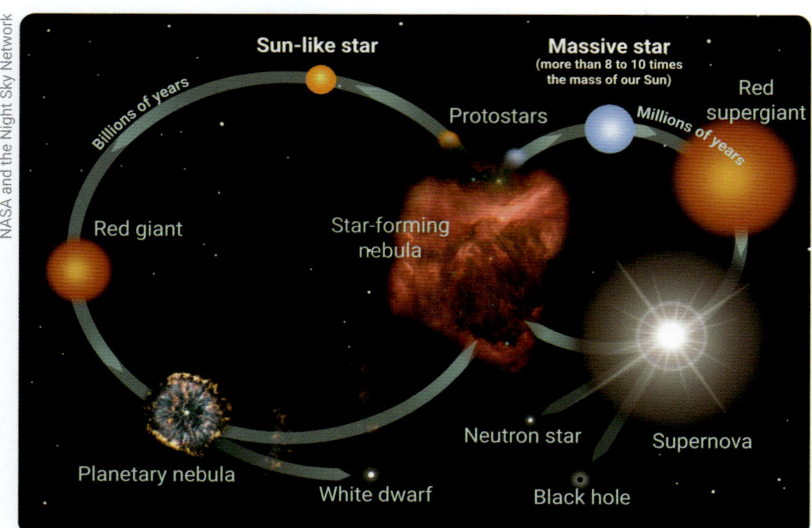

▲ **FIGURE 14.2.2** The life cycle of stars. Elements bigger than hydrogen and helium are made in red giants and supergiants. Elements bigger than iron are made in supernovas.

Formation of elements

Hydrogen and helium formed in the very early stages of the universe. It took over 500 million years before any more elements formed. The process of element formation in the early universe is identical to that seen in stars today.

Many stars you can see are in a phase called main sequence. During this time, stars **fuse** hydrogen into helium. Figure 14.2.2 shows the life cycle of stars. Our Sun is a **main sequence star**, as are many that you can see in the night sky.

When a main sequence star runs out of hydrogen, it transitions into either a **red giant** (smaller stars) or a **red supergiant** (larger stars). During this phase, the core of the star becomes hotter and denser and starts to fuse elements heavier than helium. The type of element the star can fuse depends upon its size. However, no star can fuse any element larger than iron. This process happened in the early universe after stars and galaxies formed (500 million to 1 billion years ago), and continues today.

When a red giant dies, it explodes in a **supernova**. This provides extremely hot conditions that allow elements larger than iron, up to uranium, to form. Uranium is the largest element that is produced naturally anywhere in the universe.

All the elements found naturally on Earth were produced in stars.

fuse
to undergo the process of nuclear fusion

main sequence star
a star that is fusing hydrogen to helium

red giant
a star that is fusing smaller elements larger than helium

red supergiant
a star that is fusing elements up to iron

supernova
an explosion at the end of the life cycle of a star that produces heavy elements larger than iron

14.2 LEARNING CHECK

1 **Describe** how the universe came into existence, according to the Big Bang theory.
2 **Create** a table to **summarise** the stages of the Big Bang listed in Figure 14.2.1. For each stage, provide information on when the stage occurred, what the temperature was and how different components of the universe were forming.
3 **Explain** why the fundamental force of gravity is so important in the formation of the universe.
4 Using what you know from reading the module, **explain** how you would rename the Big Bang theory to better reflect the formation of the universe.
5 **Describe** the differences between the two main stages of the Big Bang.
6 Hydrogen and helium formed early in the evolution of the universe.
   a **Describe** where all the other elements that make up the universe formed.
   b **Explain** how these elements were formed.
7 True or false? The Big Bang theory is an accurate description of how the universe formed. **Justify** your response using the information you have learned in this module.

9780170491785

14.3 Isotopes

BY THE END OF THIS MODULE, YOU WILL BE ABLE TO:

- ✓ compare different isotopes of an element
- ✓ create a model to show the isotopes of a specific element.

GET THINKING

Look at Figure 14.3.1. From what you learned in the previous module, what is the difference between carbon-12, carbon-13 and carbon-14?

Review of isotopes

As you learned in your earlier studies of subatomic particles (Module 7.2), **isotopes** are atoms of an element with the same number of protons but different numbers of neutrons. Three isotopes of carbon are shown in Figure 14.3.1.

isotopes
atoms of an element with the same number of protons but different numbers of neutrons

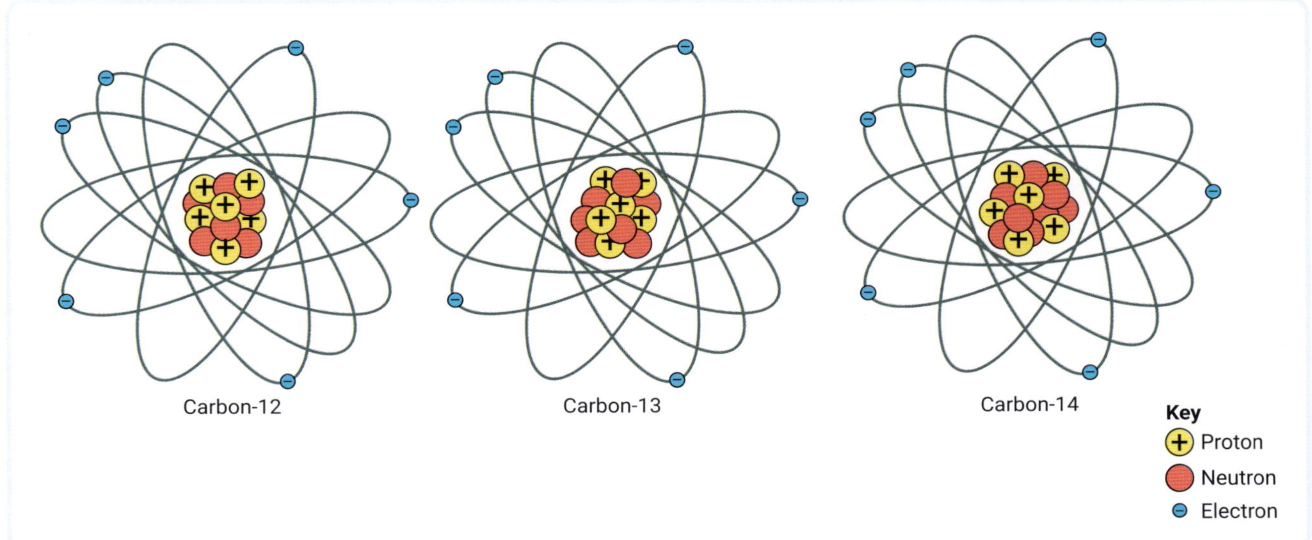

▲ **FIGURE 14.3.1** The three isotopes of carbon

If you look carefully at the diagrams, you can see some similarities and differences between the three carbon isotopes (also summarised in Table 14.3.1).

Similarities include having the:

- same number of protons in the nucleus (same atomic number)
- same number of electrons around the nucleus.

Differences include having a:

- different number of neutrons in the nucleus.

Interactive resource
Simulation: Isotopes and atomic weight

Extra science investigation
Modelling isotopes

▼ **TABLE 14.3.1** Similarities and differences between carbon isotopes

Element	Atomic number (*Z*)	Number of protons	Number of neutrons	Mass number (*A*)
Carbon-12	6	6	6	12
Carbon-13	6	6	7	13
Carbon-14	6	6	8	14

Isotopes have a different mass number because of the different number of neutrons.

Which elements have isotopes?

Many elements have multiple isotopes that can be either natural or artificial (Figure 14.3.2).

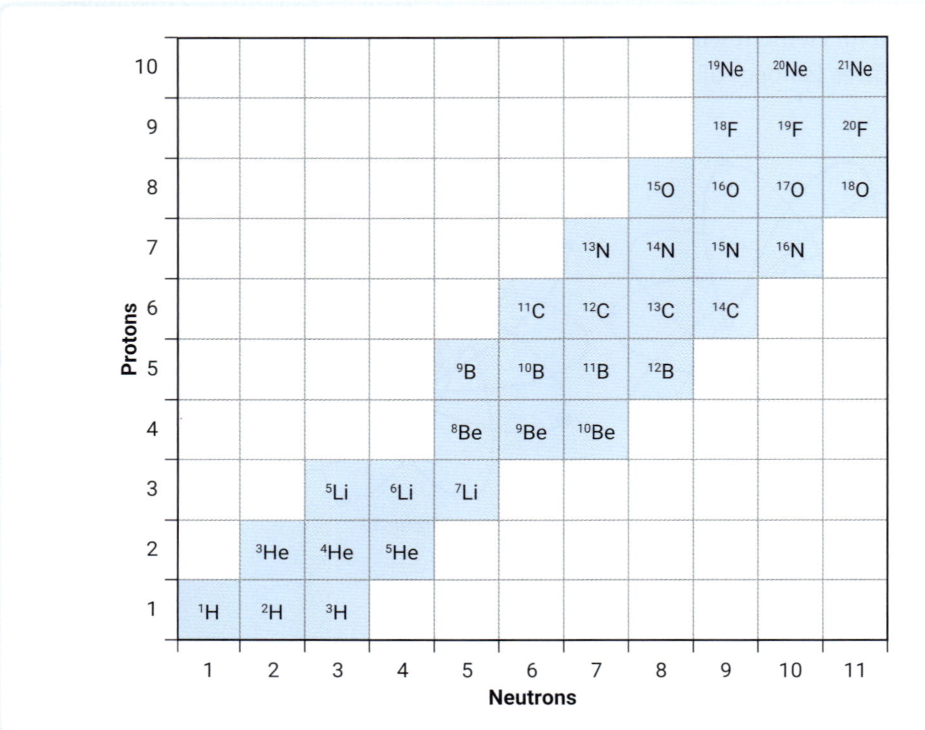

▲ **FIGURE 14.3.2** Some of the isotopes of common elements

natural isotope
an isotope of an element found in nature

Elements with an atomic number of 93 or greater have no **natural isotopes** and are produced in laboratories. These can be seen in pink on the periodic table in Figure 14.3.3. Some elements with atomic numbers less than 93 have isotopes that do not exist in nature but can also be made artificially.

artificial isotope
an isotope of an element produced in a nuclear reactor or particle accelerator

particle accelerator
a machine that enables high-speed, high-energy collisions between atoms to produce artificial isotopes and new elements

One way these **artificial isotopes** can be produced is in a nuclear reactor by adding particles to the nucleus of large elements such as uranium and thorium. This process is known as bombardment (collision) and can occur with protons or neutrons. For example, plutonium-239 can be converted to americium-240 by adding a neutron through neutron bombardment.

$$^{239}_{94}\text{Pu} + ^{1}_{0}\text{n} \rightarrow ^{240}_{95}\text{Am}$$

Artificial elements can also be produced in **particle accelerators** by high-speed, high-energy collisions of particles.

Period	1	2	3	4	5	6	7	8	9	10	11	12	13	14	15	16	17	18
1	1 H																	2 He
2	3 Li	4 Be											5 B	6 C	7 N	8 O	9 F	10 Ne
3	11 Na	12 Mg											13 Al	14 Si	15 P	16 S	17 Cl	18 Ar
4	19 K	20 Ca	21 Sc	22 Ti	23 V	24 Cr	25 Mn	26 Fe	27 Co	28 Ni	29 Cu	30 Zn	31 Ga	32 Ge	33 As	34 Se	35 Br	36 Kr
5	37 Rb	38 Sr	39 Y	40 Zr	41 Nb	42 Mo	43 Tc	44 Ru	45 Rh	46 Pd	47 Ag	48 Cd	49 In	50 Sn	51 Sb	52 Te	53 I	54 Xe
6	55 Cs	56 Ba	*	72 Hf	73 Ta	74 W	75 Re	76 Os	77 Ir	78 Pt	79 Au	80 Hg	81 Tl	82 Pb	83 Bi	84 Po	85 At	86 Rn
7	87 Fr	88 Ra	**	104 Rf	105 Db	106 Sg	107 Bh	108 Hs	109 Mt	110 Ds	111 Rg	112 Cn	113 Nh	114 Fl	115 Mc	116 Lv	117 Ts	118 Og

* Lanthanoids	57 La	58 Ce	59 Pr	60 Nd	61 Pm	62 Sm	63 Eu	64 Gd	65 Tb	66 Dy	67 Ho	68 Er	69 Tm	70 Yb	71 Lu
** Actinoids	89 Ac	90 Th	91 Pa	92 U	93 Np	94 Pu	95 Am	96 Cm	97 Bk	98 Cf	99 Es	100 Fm	101 Md	102 No	103 Lr

▲ **FIGURE 14.3.3** The periodic table with elements with no naturally occurring isotopes shown in pink (current as of November 2024)

14.3 LEARNING CHECK

1 Lithium-5, lithium-6 and lithium-7 are three isotopes of lithium. **Identify** one thing the isotopes have in common, and one way they are different.

2 **Describe** the difference between natural and artificial isotopes.

3 Helium-3, helium-4 and helium-5 all have two protons in the nucleus. **Draw** diagrams or **create** a physical model (using plasticine or coloured balls) to show the arrangement of protons, neutrons and electrons in the three helium isotopes.

4 Use an example to **explain** one way that artificial isotopes are produced.

5 The images A–F show atoms that may or may not be isotopes of each other.

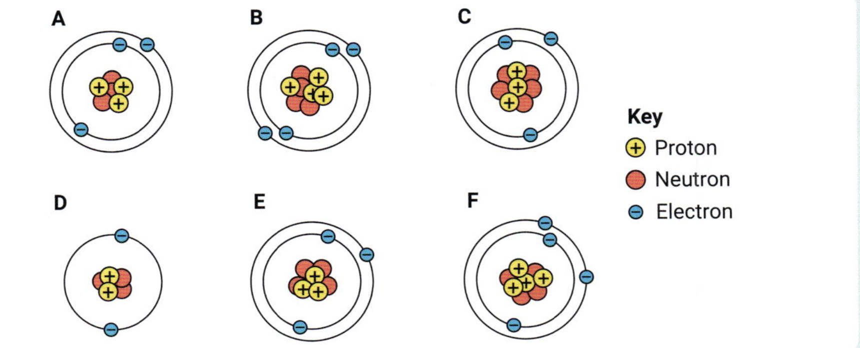

a **Identify** three atoms that are isotopes of each other. **Justify** your selection.

b **Identify** the atom that does not have any isotopes. **Justify** your selection.

c Are there any other isotopes (different from in parts **a** and **b**) in the image? **Explain** your answer.

14.4 Radioactive decay

BY THE END OF THIS MODULE, YOU WILL BE ABLE TO:

- ✓ compare stable and unstable isotopes
- ✓ explain the process of radioactive decay
- ✓ use graphs and data to calculate the half-life of an isotope.

Video activity
Radioactive half-life

Extra science investigation
Simulating radioactive decay

GET THINKING

'Radioactive' is a word that is misunderstood by a lot of people. Write down what you would think if someone told you there were radioactive elements in the same room as you!

Stable and unstable isotopes

As you know, carbon has a number of isotopes. Some of these isotopes, such as carbon-12 and carbon-13, are stable. Other isotopes, such as carbon-14, are unstable. Unstable isotopes are known as **radioactive isotopes** and release **radiation**.

radioactive isotope (radioisotope)
an isotope that is unstable and undergoes radioactive decay to become more stable

radiation
a stream of particles and/or energy from a radioactive source

electrostatic force
a force of attraction between oppositely charged particles

strong nuclear force
a force of attraction between particles in the nucleus of an atom

How do you get stable and unstable isotopes?

Inside all nuclei two opposing forces are acting. You may recall that when you have two identical charges (negative and negative or positive and positive), they will experience a repulsive force. This **electrostatic force** acts between the protons in the nucleus as they push each other away because of their identical, positive charges. There is also a force called the **strong nuclear force** that attracts and holds the protons and neutrons together in the nucleus.

When these two forces are balanced, the isotope is stable. This happens when the attractive and repulsive forces are about the same strength. When one of the forces is much stronger than the other force, the unbalanced forces cause the isotope to be unstable.

Proton/neutron balance

The strong nuclear force is attractive and acts between all particles in the nucleus. It acts between protons, which have a positive charge, and neutrons, which have no charge. This may seem strange because like charges usually repel. However, that is the *electrostatic* force. The *strong nuclear* force is another type of force that works in a different way. It is attractive between all particles in close contact, regardless of charge. It only operates in the nucleus of atoms (Figure 14.4.1).

Therefore, to make a nucleus stable, you need neutrons to provide stronger nuclear force and oppose the repulsive force between protons. This is why stable isotopes have a certain ratio of protons and neutrons that allow the nucleus to be stable.

For smaller elements, such as the first 20 elements, the stable ratio is approximately 1 proton to 1 neutron. For larger elements, more neutrons are required to balance the number of protons, so you see the ratio change.

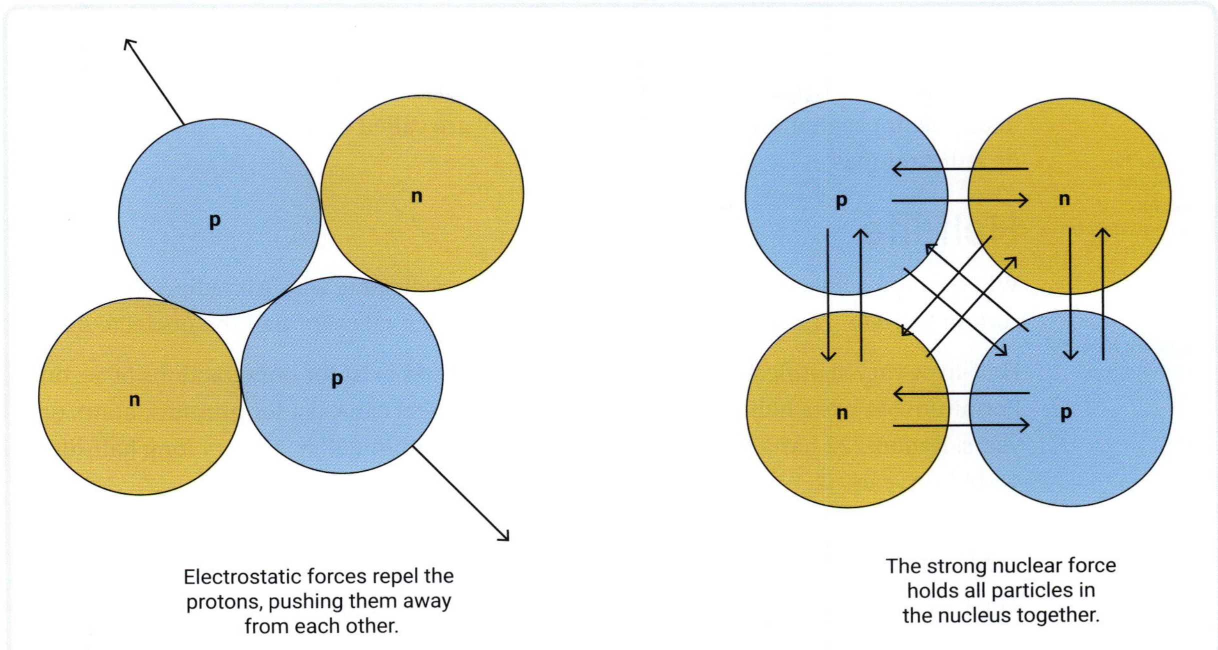

▲ **FIGURE 14.4.1** Isotopes are stable when electrostatic and nuclear forces are balanced.

Lead is the largest stable element with 82 protons (atomic number 82). Atoms larger than lead do not have enough neutrons to balance the forces, so will never be stable. This means elements with an atomic number greater than 82 have no stable isotopes. Table 14.4.1 shows some stable isotope proton-to-neutron ratios.

What is radioactive decay?

Unstable isotopes cannot stay in their unstable form and will release particles or energy from their nucleus to become stable isotopes. Because the reason for instability is an unbalanced proton-to-neutron ratio, there must be changes to this ratio to make the nucleus stable again. In Module 14.6 we will look at three common processes that occur in unstable isotopes.

▼ **TABLE 14.4.1** Some stable isotope proton-to-neutron ratios. As the atom gets larger, the ratio increases

Isotope	Protons	Neutrons	Proton-to-neutron ratio
Helium-4	2	2	1:1
Carbon-12	6	6	1:1
Calcium-40	20	20	1:1
Scandium-44	21	23	1:1.1
Silver-107	47	60	1:1.3
Lead-206	82	124	1:1.5

The process of changing the particles in the nucleus often results in the production of particles and/or the emission of energy from the isotope. This nuclear change is spontaneous (happens without any external factor) and is called **radioactive decay**. Isotopes that undergo radioactive decay are called radioactive isotopes, or radioisotopes.

radioactive decay
the spontaneous disintegration of certain atomic nuclei accompanied by the emission of alpha particles, beta particles or gamma radiation

Half-life

When a radioactive isotope decays, it does so in a predictable way. All radioisotopes have a certain **half-life**. The half-life of an isotope is the time it takes for half the nuclei to decay.

half-life
the time it takes for half of the nuclei in a sample of an isotope to decay

Half-lives vary significantly. Some radioactive isotopes exist for only fractions of seconds. Fermium-244 has a half-life of 3.3 milliseconds. Others can exist for minutes, hours or years. Radon-222 has a half-life of 3.82 days and uranium-238 has a very long half-life of 4.5 billion years.

Calculating half-life

The half-life of cobalt-60 is 5.27 years. This means that if you start with 10 grams of cobalt-60, there would only be 5 grams (half) left after 5.27 years. As seen in Figure 14.4.2, every 5.27 years another half of the sample has decayed. So, after 10.54 years (2 half-lives) only 2.5 grams will remain.

DATA SCIENCE
Learn more about graphing and data analysis in **Module 2.6**.

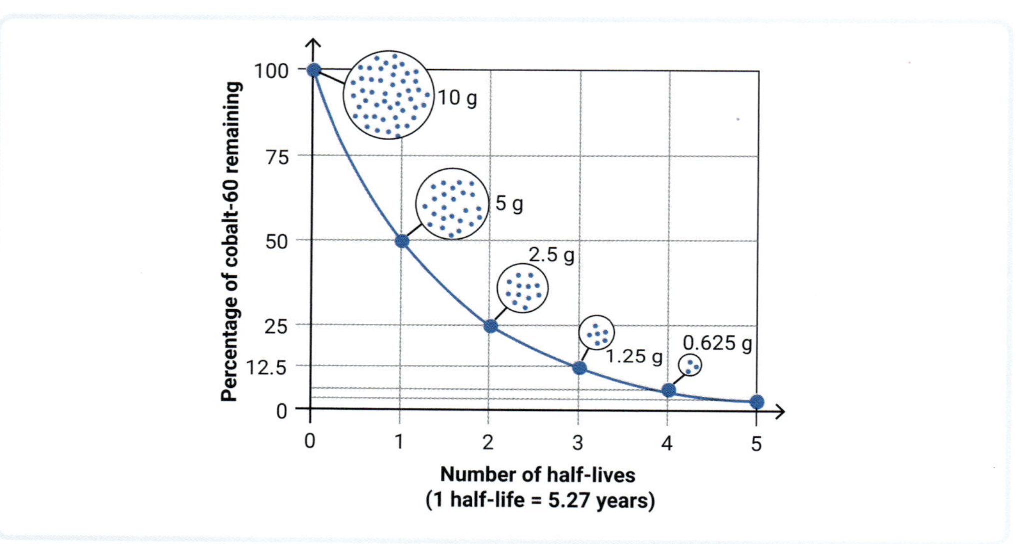

▲ **FIGURE 14.4.2** A decay curve for cobalt-60. A decay curve can be plotted with the number of half-lives or time on the horizontal axis.

If you have a decay curve, you can determine the half-life of a radioactive isotope, or find out how long it takes to decay to a certain level. Look at Figure 14.4.3. You could use this decay curve in a number of ways.

9780170491785

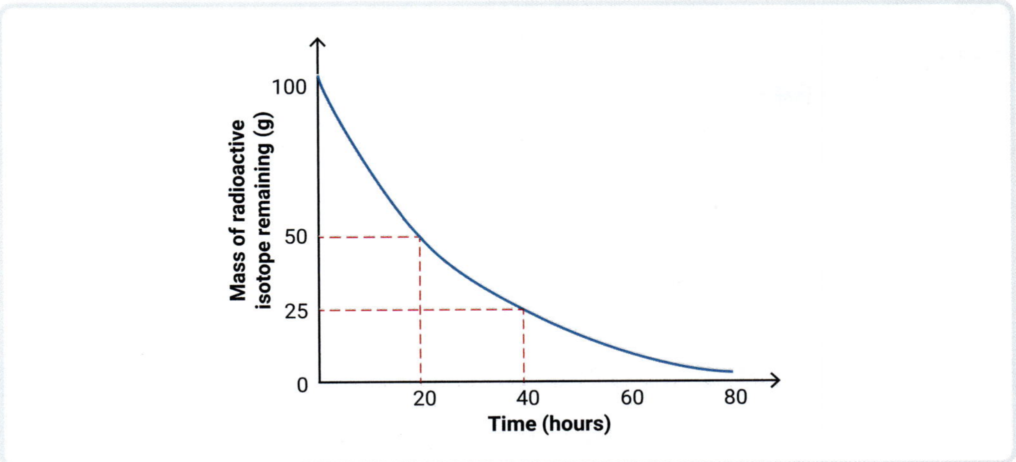

▲ **FIGURE 14.4.3** A decay curve for a radioisotope

1 Calculating the half-life by finding half the original mass (50 grams) and using the graph to find out the time this takes:
   - rule a horizontal line from mass = 50 grams on the vertical axis to the curve
   - rule a vertical line down to the horizontal axis to determine the time. This gives a value of 20 hours.

2 Determining how much mass would be left after a period of time. If you wanted to know how much of the sample was left after two half-lives, you could use the half-life of 20 hours to calculate this. Two half-lives are 40 hours. Rule lines (as described above) but this time from 40 hours on the horizontal axis up to the curve, then across to the vertical axis to find the mass remaining. In this case, 25 grams of the sample remain.

You can also calculate half-life information from simple data. For example, sodium-20 has a half-life of 15 hours. If you start with 100 g, how much will remain after 45 hours?

- $\frac{45 \text{ hours}}{15 \text{ hours}} = 3$ half-lives
- One half-life reduces the sample from 100 g to 50 g.
- The second half-life reduces the sample to 25 g.
- The third half-life reduces the sample to 12.5 g.
- Thus, after 45 hours (or 3 half-lives), 12.5 g of the sample will remain.

14.4 LEARNING CHECK

1 **Define** radioactive isotope.
2 In terms of the strong nuclear and electrostatic forces in the nucleus, **explain** what makes an isotope unstable.
3 If an atom of magnesium had 12 protons and 15 neutrons, **explain** whether it would be stable or unstable.
4 A sample of the radioactive isotope caesium-134 has a half-life of 2.06 years. If you started with a 200 g sample, how long would it take for 6.25 g of the sample to remain?
5 **Create** a diagram to show examples of how the proton-to-neutron ratio increases as the atomic mass increases.

WORKING SCIENTIFICALLY

14.5 Using models to make observations

SCIENCE SKILLS IN FOCUS

IN THIS MODULE, YOU WILL FOCUS ON LEARNING AND IMPROVING THESE SKILLS:

- collecting data from a simulation
- using a spreadsheet to present the data in tables and graphs
- using and analysing a model of radioactive decay.

Why we use models

A scientific model is a simplified representation of something complicated. We can use physical models and simulations to make observations when an experiment cannot be performed, or something cannot be directly observed due to high levels of risk, lack of equipment, or complexity. Examples of physical models and simulations we use in science include representations of radioactivity (too dangerous), atomic structure (too small to be seen) or features of the universe (not able to be experimented on).

Models and data

Models can also be mathematical, where equations and graphs are used to show relationships between variables. Many models are based on large datasets. Models can also be used to generate data to make predictions.

Physical models

When you make a physical model, you usually use everyday objects to represent other objects. An example of this is making a model of the atom. You might use polystyrene balls or plasticine to make the protons, neutrons and electrons. You should try to make the model represent features of the atom. In this example, you might consider the relative sizes of protons, neutrons and electrons, and how they are arranged in the atom.

Choosing when to use models

Analysing and evaluating a physical model or simulation is important. Some phenomena are easy to represent in physical models; others are more difficult. An example of a difficult model is representing the scale of the solar system. The distances in the solar system are so vast, it is difficult, but not impossible, to represent using everyday materials and spaces. When creating or using physical models, you should consider features of the model that are good representations, and features that are not correct representations. Incorrect representations could lead to misconceptions or a wrong idea about a scientific concept.

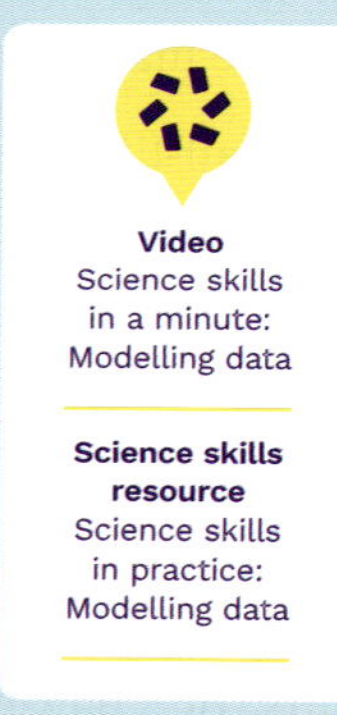

9780170491785

SIMULATING RADIOACTIVE DECAY

AIM

To model the process of radioactive decay and assess how accurate your model is at showing the process of decay

MATERIALS AND EQUIPMENT

- ☑ 50 items to model atoms, such as 50 coins, 50 discs, 50 cards or 50 pieces of paper. One side should be different from the other; for example, by putting an 'X' on one side of the paper/disc
- ☑ container to shake or shuffle the 50 'atoms'

PROCEDURE

1. Place your 'atoms' into your container and shake the container well for about 10 seconds.
2. Pour out the atoms on a large, flat surface so that they randomly fall with one side facing up.
3. Choose one side (you will stick with this throughout the experiment). For example, if you have discs with an X on one side, pick the X. Remove all atoms with the X. Count how many atoms remain and record this number.
4. Repeat steps 1–3 with the remaining atoms until you have no atoms left.

RESULTS

1. Construct an appropriate table to record the number of atoms remaining after each 'shake'. Include the original number of atoms at shake 0.
2. Using Excel, or graph paper, plot a graph of the number of atoms remaining after each 'shake'. Make sure you include the initial number of atoms at 0 shakes. Sketch a curve of best fit to show the trend.

ANALYSIS

1. What is represented by the 'atoms' having two sides?
2. **Describe** how the concept of half-life is represented in this simulation.
3. **Examine** your graph and data. Using evidence such as the shape of the graph or the number of atoms remaining, **describe** the trend you observed in this simulation. Include the concept of half-life in your description.
4. **Explain** two features of this simulation that accurately represent the process of radioactive decay.
5. **Explain** two features of this simulation that do not accurately represent the process of radioactive decay.

CONCLUSION

Write a conclusion about the data collected from the model. Make a judgement about the overall accuracy of the model at representing radioactive decay.

14.6 Alpha, beta and gamma radiation

BY THE END OF THIS MODULE, YOU WILL BE ABLE TO:

✓ describe alpha, beta and gamma radiation

✓ compare the structures and properties of alpha, beta and gamma radiation.

Interactive resource
Drag and drop:
Alpha, beta and gamma radiation

GET THINKING

Look at the diagrams of the types of nuclear decay in this module (Figures 14.6.1–14.6.3). What similarities and differences do you see? List at least two similarities and two differences.

Radioactive decay

In Module 14.4, you learned that the proton-to-neutron ratio of a nucleus determines the stability of the isotope. When the proton-to-neutron ratio is either too high or too low, the isotope is radioactive and will decay.

When a radioactive isotope decays, its proton-to-neutron ratio becomes closer to that of a stable isotope. Not all radioactive decay ends with a fully stable isotope, but the result will always be a more stable isotope than the starting isotope.

Three main types of radiation are emitted during radioactive decay – alpha particles, beta particles and gamma radiation.

Alpha particle production

Alpha particles are primarily produced when very large atoms decay. You may recall that no element of atomic number greater than 82 can be stable. To become more stable, these large radioisotopes lose protons and neutrons to reduce the mass of the nucleus. The most energy-efficient method of doing this is for the nucleus to emit an alpha particle. Alpha particles consist of two protons and two neutrons; they are helium nuclei (helium atoms without any electrons).

An alpha particle is represented in nuclear equations by either ${}^{4}_{2}\text{He}$ or α (the symbol for alpha).

Emitting a particle with multiple protons and neutrons means the isotope (parent nucleus) loses a lot of nuclear mass quickly. The resulting product (daughter nucleus) is more stable than the initial isotope because its nucleus is smaller.

An example of alpha particle production is the decay of radon-222 to polonium-218 (Figure 14.6.1 and Table 14.6.1).

$${}^{222}_{86}\text{Rn} \rightarrow {}^{218}_{84}\text{Po} + {}^{4}_{2}\text{He}$$

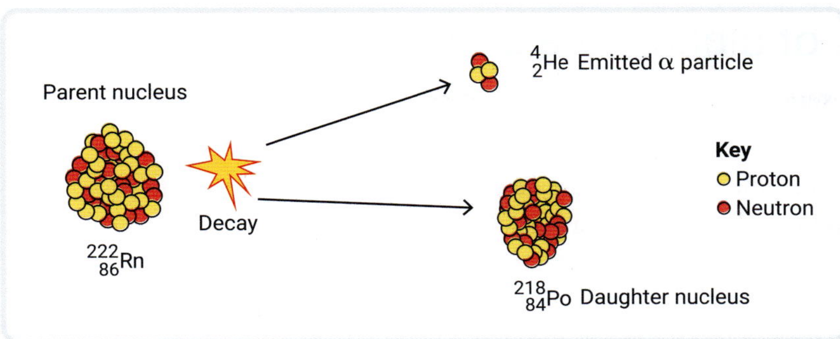

▲ **FIGURE 14.6.1** The alpha decay of radon-222 to polonium-218

▼ **TABLE 14.6.1** The alpha decay of radon-222 to polonium-218

Isotope	Atomic number	Mass number
Radon-222	86	222
Polonium-218	84	218

14.6

Beta particle production

beta particle
a negatively charged particle identical to an electron, emitted when an unstable nucleus decays

Beta particles are negatively-charged particles (electrons) emitted from an unstable atom's nucleus during radioactive decay. The resulting nucleus has a more stable proton-to-neutron ratio than the original atom. Beta radiation can come from unstable nuclei with an excess of neutrons or protons.

An example of beta decay is the conversion of carbon-14 into nitrogen-14 with the emission of a beta particle (Figure 14.6.2).

A beta particle is represented in nuclear equations by either $^{0}_{-1}$e or β (the symbol for beta).

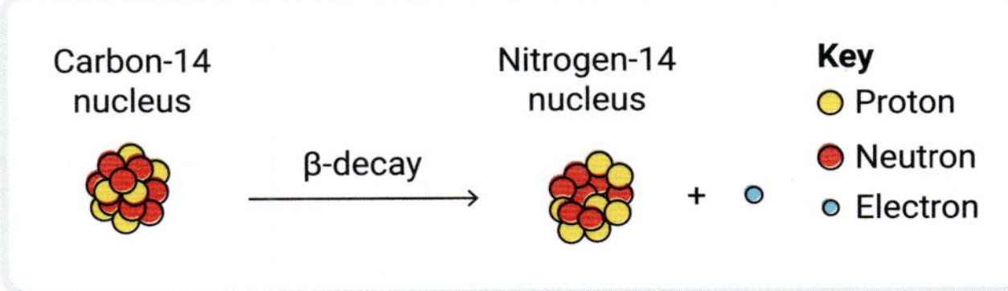

▲ **FIGURE 14.6.2** The beta decay of carbon-14 to nitrogen-14

$$^{14}_{6}\text{C} \rightarrow {}^{14}_{7}\text{N} + {}^{0}_{-1}\text{e}$$

Gamma radiation production

gamma radiation
high-energy electromagnetic energy emitted when an unstable nucleus decays

Gamma radiation is high-energy radiation emitted during radioactive decay. Gamma radiation does not involve particles. It is just energy. An example of gamma radiation production involves cobalt-60. This isotope decays to produce beta particles. The resulting high-energy nickel-60 daughter nuclei emits this extra energy as gamma radiation, forming a low-energy nickel-60 nucleus (Figure 14.6.3).

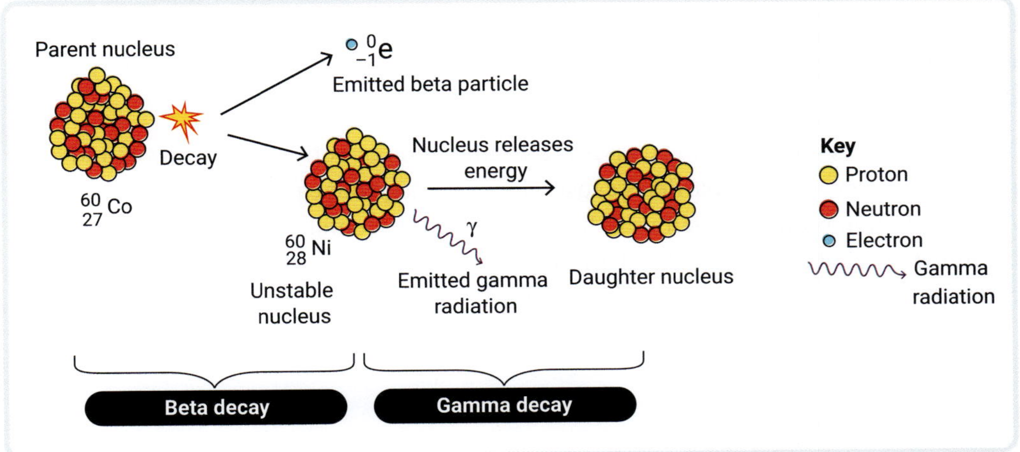

▲ **FIGURE 14.6.3** The beta and gamma decay of cobalt-60

Properties of alpha, beta and gamma radiation

The three types of radiation have different properties, which allow them to be used for different purposes.

penetrating power
the ability of radiation to pass through materials

One property of all radiation is **penetrating power**. This describes its ability to pass through materials (Figure 14.6.4). Alpha particles are large and highly charged so they don't pass through materials easily. Beta particles are smaller in mass and have a smaller charge. They can pass through materials more easily than alpha particles. Gamma radiation has no mass and no charge because it is electromagnetic radiation. It can pass through some dense materials.

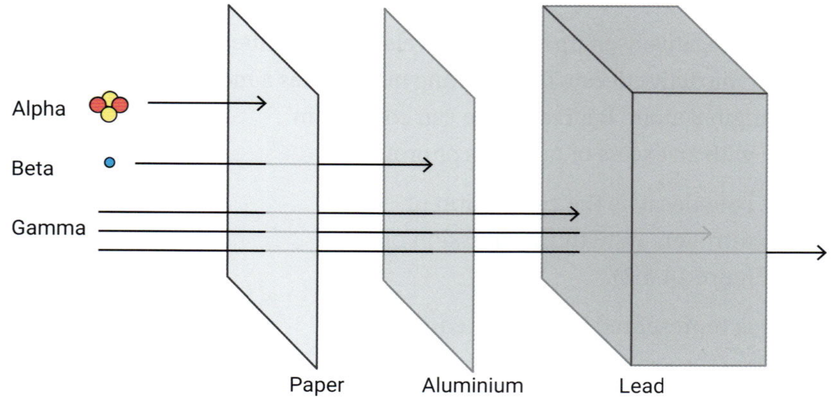

▲ **FIGURE 14.6.4** Alpha particles are stopped by paper and beta particles can be blocked by aluminium. Depending on the energy of the gamma radiation, it may be stopped by lead or concrete that is thick enough.

ionising power
the ability of radiation to change the structure of materials it passes through

Radiation also has the property of **ionising power**. Radiation can change the structure of materials it passes through by causing atoms to lose electrons and form positively charged particles. Particles like this with a positive or negative charge are called ions. Ionisation of atoms changes the way they function. When this happens to DNA or proteins in the tissue of living things, it can cause cancer, cell death or other cell damage. Alpha particles are the strongest ionising radiation because their mass and charge are more likely to affect the electrons in atoms. Beta particles, having lower mass and charge, are less likely to affect atoms, while gamma radiation usually causes damage due to energy transfer inside the material.

The true damage that radiation causes is a combination of both ionising power and penetrating power. Alpha particles cause a lot of damage but cannot pass through skin. Alpha radiation is only dangerous to humans if it is accidentally ingested (swallowed). Gamma radiation is the type of radiation most likely to pass into human tissues and cause internal damage, as seen in Figure 14.6.5.

9780170491785

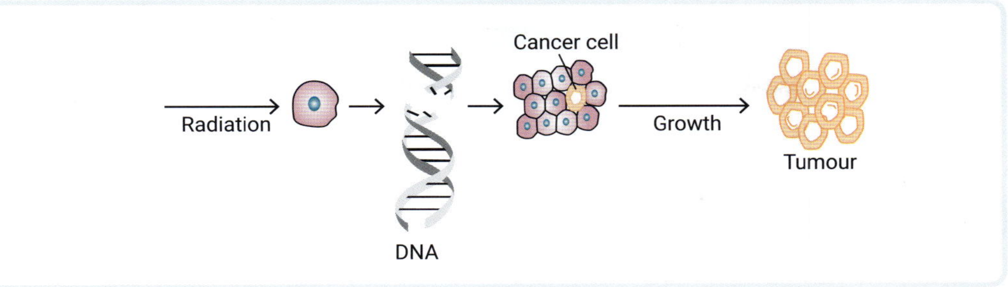

▲ FIGURE 14.6.5 DNA damage due to ionising radiation

14.6 LEARNING CHECK

1 **Describe** the composition of alpha particles, beta particles and gamma radiation in symbols and words, or with labelled diagrams.

2 **Create** a table to **summarise** and **compare** the relative penetrating power and ionising power of the different types of radiation. Use 'high', 'medium' and 'low' to rank the three types of radiation.

3 **Predict** the atomic number and mass number of the daughter nucleus that forms when uranium-238 ($^{238}_{92}U$) undergoes alpha decay. Write a nuclear equation to represent this decay.

4 **Explain** why a neutron would convert to a proton during beta decay by considering the stability of the final product.

5 A high-energy barium-137 isotope undergoes gamma decay. **Predict** and **explain** the mass number of the final daughter nuclei.

6 A scientist was trying to **determine** the thickness of lead that was needed to completely block the beta radiation from a sample of sodium-22. To do this, the scientist measured the radiation that passed through different thicknesses of lead. The results are shown in Table 14.6.2.

▼ TABLE 14.6.2 Radiation that passes through different thicknesses of lead

Lead thickness (mm)	Radiation count
0	35 500
2.5	23 500
5	15 000
7.5	10 000
10	6500
12.5	4000

a **Construct** a graph with radiation count on the vertical axis, and lead thickness on the horizontal axis. The horizontal axis should cover values from 0 to 30 mm.

b **Draw** a curve of best fit and use this to **estimate** the thickness of lead that would be needed to completely block the radiation from the sodium-22 radioisotope.

c Are you confident in the value you estimated? Why or why not?

d Suggest an improvement to the experiment that would increase your confidence in your estimated value.

14.7 Radioactive dating

BY THE END OF THIS MODULE, YOU WILL BE ABLE TO:

- ✓ explain how radioactive isotopes are used to date materials
- ✓ calculate the age of materials using known data.

Video activity
Radioactive dating

GET THINKING

Examine the photographs in Figure 14.7.1. What element do they have in common? How do you think radioactive decay could be used to find out how old an object is?

Dating materials

carbon-14 dating
the use of the carbon-14 decay process to determine the age of materials containing carbon

uranium–lead dating
the use of either the uranium-238 or uranium-235 decay process to determine the age of rocks

One use of radioactive isotopes is for dating materials in a process called radioactive or radiometric dating. **Carbon-14 dating** is used to estimate the age of materials that contain carbon. This includes anything that was once a living organism. **Uranium–lead dating** is used to estimate the age of rocks that are between 1 million and 4.5 billion years old. Both methods use the known half-lives of isotopes to determine how many half-lives have passed. This allows scientists to calculate the age of the material.

Carbon-14 dating

Carbon-14 dating is widely used by scientists to date plant or animal remains. Materials such as wooden buildings and ships, paper and parchments, charcoal, twigs and seeds can be carbon-dated. Animal remains including bones, shells, leather and hair can also be carbon-dated (Figure 14.7.1).

Mark Spowart/Alamy Stock Photo

Melnikov Dmitriy/Shutterstock.com

▲ **FIGURE 14.7.1** Carbon-dating can be used to estimate the age of wooden ships and human/animal remains such as those of woolly mammoths.

Carbon-14 dating uses the radioactive isotope carbon-14 (6 protons, 8 neutrons). The amount of carbon-14 in Earth's atmosphere is small but very consistent (one out of 1 000 000 000 000 atoms of carbon on Earth is carbon-14).

9780170491785

Plants take in carbon dioxide from the atmosphere when they photosynthesise. A small, consistent percentage of the carbon taken in is carbon-14. This carbon-14 moves through the food chain, so the same percentage of carbon-14 is present in all living organisms. While a plant or an animal is alive, any carbon-14 lost through radioactive decay is replaced by more carbon-14 as the plant photosynthesises or the animal eats. If you know the mass of the plant or animal when it was alive, you can accurately determine the mass of carbon-14 it contained.

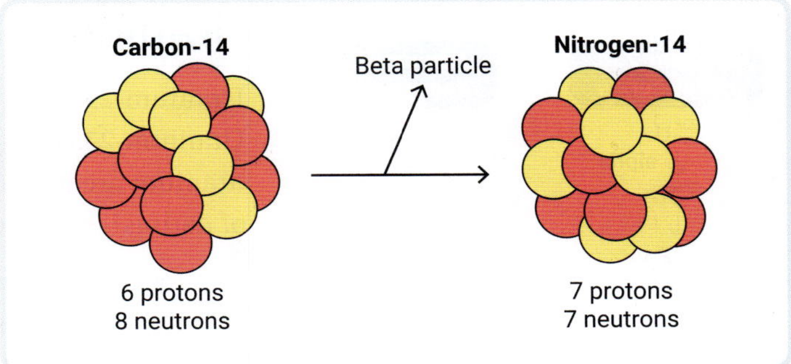

▲ **FIGURE 14.7.2** The radioactive decay of carbon-14 into nitrogen-14 with release of beta particles. The beta particles are lost from the carbon-14 isotopes.

Once a plant or an animal dies, it no longer takes in new carbon-14. The carbon-14 left behind decays at a consistent rate to nitrogen-14, as seen in Figure 14.7.2.

Scientists measure the mass of carbon-14 in the remains of the plant or animal, calculating how much must have been present originally, as shown in Figure 14.7.3. The missing mass is due to decay and can be used to determine the age as follows.

- The tested item was once alive, so a specific amount of carbon-14 was present in the object when it was a living organism.
- Scientists determine the remaining amount of carbon-14 in a sample, based on the rate of the production of beta particles as the carbon-14 decays.
- For example, if only 25% of the original carbon-14 remains, it indicates that two half-lives have passed: 100% → 50% (1 half-life) → 25% (2 half-lives).

Because the half-life of carbon-14 is known to be 5730 years, the object is approximately 11 460 years old.

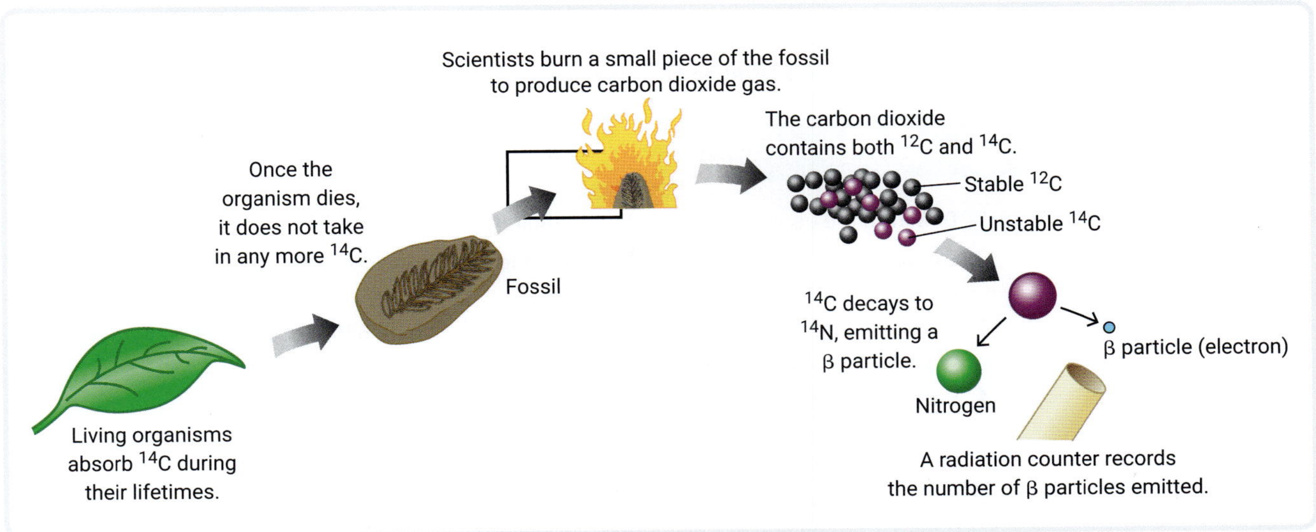

▲ **FIGURE 14.7.3** All living organisms contain carbon-14 at a fixed percentage. After the organism dies, the age of its remains can be determined by measuring how much carbon-14 has decayed.

Uranium-238 decay series	Uranium-235 decay series
$^{238}_{92}U$	$^{235}_{92}U$
↓ alpha	↓ alpha
$^{234}_{90}Th$	$^{231}_{90}Th$
↓ beta	↓ beta
$^{234}_{91}Pa$	$^{231}_{91}Pa$
↓ beta	↓ alpha
$^{234}_{92}U$	$^{227}_{89}Ac$
↓ alpha	↓ beta
$^{230}_{90}Th$	$^{227}_{90}Th$
↓ alpha	↓ alpha
$^{226}_{88}Ra$	$^{223}_{88}Ra$
↓ alpha	↓ alpha
$^{222}_{86}Rn$	$^{219}_{86}Rn$
↓ alpha	↓ alpha
$^{218}_{84}Po$	$^{215}_{84}Po$
↓ alpha	↓ alpha
$^{214}_{82}Pb$	$^{211}_{82}Pb$
↓ beta	↓ beta
$^{214}_{83}Bi$	$^{211}_{83}Bi$
↓ beta	↓ alpha
$^{214}_{84}Po$	$^{207}_{81}Ti$
↓ alpha	↓ beta
$^{210}_{82}Pb$	$^{207}_{82}Pb$
↓ beta	
$^{210}_{83}Bi$	
↓ beta	
$^{210}_{84}Po$	
↓ alpha	
$^{206}_{82}Pb$	

▲ **FIGURE 14.7.4** The uranium to lead decay series

Uranium–lead dating

Uranium-238 decays in a series of steps to become lead-206. Uranium-235 decays to lead-207 (Figure 14.7.4). Both isotopes of uranium are used to estimate the age of rocks (Figure 14.7.5). This type of dating is called uranium–lead dating.

As a rock forms, it may develop crystals. When rock crystals form, uranium atoms are trapped inside them. Over time, the uranium-238 atoms decay to lead-206, with a half-life of 4.47 billion years. The uranium-235 atoms decay to lead-207, with a half-life of 710 million years.

Both decay series can be used to date rocks by determining the ratio of uranium and lead isotopes present in the rock. In general terms, the more lead that is present, the older the rock is.

- If a rock contains a 1:1 ratio of uranium-238 to lead-206 this means half the uranium-238 has decayed (one half-life), and the rock is approximately 4.47 billion years old.
- If a rock contains a 1:3 ratio of uranium-235 to lead-207 this means the uranium-235 has been through two half-lives (100 per cent → 50 per cent → 25 per cent) and the rock is approximately 1420 million years old.

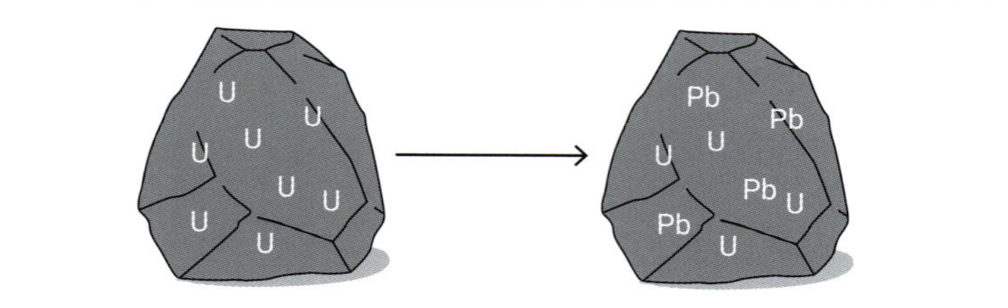

▲ **FIGURE 14.7.5** Uranium decays to lead in rocks.

14.7 LEARNING CHECK

1. **Describe** the uses of carbon-14 and uranium–lead dating.
2. **Explain** why carbon-14 could not be used to **determine** the age of wooden buildings in Australia from the 1900s.
3. A sample of preserved wood is determined as being 28 650 years old. What percentage of carbon-14 would remain in the preserved wood now?
4. Earth is 4.54 billion years old, and rock formation is a continuous process, with rocks constantly forming throughout Earth's history. **Explain** why both uranium-238 and uranium-235 are needed to **determine** the age of rocks on Earth.
5. Figure 14.7.1 shows two examples of discoveries that have been made using carbon-14 dating. **Choose** one of the images and **research** a notable discovery in this area. **Create** a two-slide presentation showing key features of the discovery and the role of carbon-14 dating.

9780170491785

WORKING SCIENTIFICALLY

14.8 Graphing data

SCIENCE SKILLS IN FOCUS

IN THIS MODULE, YOU WILL FOCUS ON LEARNING AND IMPROVING THIS SKILL:

▶ using provided data to form conclusions, identify trends and make predictions.

Graphing quantitative data

Graphs visually represent quantitative data and allow the visualisation of trends and patterns, if any exist. You have probably already created both column and line graphs. Column graphs are used to represent discrete data, where the data only has certain values or is organised into categories. Examples of discrete data include types of animals or colours. Line graphs are used for continuous data, where data can take an infinite number of values. Examples of continuous data include measurements such as length, mass, speed or time.

Graphing tips

Whether you are creating column or line graphs, there are several features you should always include in a graph:

- a title that specifically describes the data presented in the graph
- labels on the horizontal and vertical axis, including units if appropriate
- a scale on each axis that increases in regular increments
- a line of best fit for line graphs (this can be a straight line or a curved line)
- a key if multiple sets of data are being graphed on the same axis.

Video
Science skills in a minute: Representing data

DATA ANALYSIS – MAKING PREDICTIONS ABOUT RADIOACTIVITY

AIM

To use tables and graphs to solve problems and make predictions about radioactivity

PART A: FIND THE HALF-LIFE

Use Table 14.8.1 to determine the half-life of the element. Hint: You should use graph paper or Excel to draw a decay curve as your first step.

1 **Justify** the method you used to determine the half-life.

2 **Describe** two limitations you encountered while determining the half-life that reduce the reliability of the answer.

▼ **TABLE 14.8.1** Activity of a radioactive isotope

Time (s)	Activity (counts)
0	7112
84	6650
220	5964
346	5392
467	4894
600	4400

PART B: PREDICTING THE BEST RADIOACTIVE ISOTOPE TO USE

The radioactive isotopes in Table 14.8.2 include three that are used for medical purposes and two that are used in industry.

▼ **TABLE 14.8.2** Some isotopes, their half-lives and the radiation they produce

Isotope	Half-life	Type of radiation produced
Iodine-123	13 hours	Gamma
Cobalt-57	272 days	Gamma
Thallium-201	73 hours	Gamma
Americium-241	432 years	Alpha and gamma
Sodium-24	15 hours	Beta

1. a **Identify** two isotopes from the table that could be used for medical purposes.
   b **Justify** your answer by referring to two properties of isotopes commonly used in medicine.
2. Sodium-24 is used in industry for tracking water movement through household pipes. It has a short half-life. Suggest why an isotope with a very short half-life is used for this purpose.
3. **Justify** which of the five isotopes is not suitable for use in medicine.

CONCLUSION

Write a conclusion for your investigation.

9780170491785

14.9 Nuclear fission

BY THE END OF THIS MODULE, YOU WILL BE ABLE TO:

- ✓ describe the process of nuclear fission
- ✓ compare controlled and uncontrolled nuclear fission
- ✓ explain how fission is controlled in a nuclear reactor.

GET THINKING

Imagine you're a scientist trying to harness energy from atoms. You know that when certain atoms split, they release a huge amount of energy. However, this process is unpredictable, and if not controlled properly, it could cause massive destruction. What might be some of the dangers or challenges in controlling nuclear fission?

What is nuclear fission?

Nuclear fission is the splitting of a nucleus into smaller fragments. The nucleus of an atom is held together by very strong bonds. There is a lot of energy present in these bonds that is released during a nuclear fission reaction.

Most nuclear fission reactions start by firing neutrons at a large nucleus. We will use uranium as an example, because it is commonly used in nuclear power reactors. A neutron is fired at a uranium-235 nucleus and is captured, forming the isotope uranium-236. The mass number has increased by one because there is one extra particle in the nucleus. This new nucleus is very unstable and starts to **fission** into two components. As you can see in Figure 14.9.1, the two larger products formed are krypton-92 and barium-141. There are also three neutrons produced in this reaction. Figure 14.9.1 is an expansion of Figure 14.1.1 to show what is happening in more detail.

fission
to undergo the process of nuclear fission

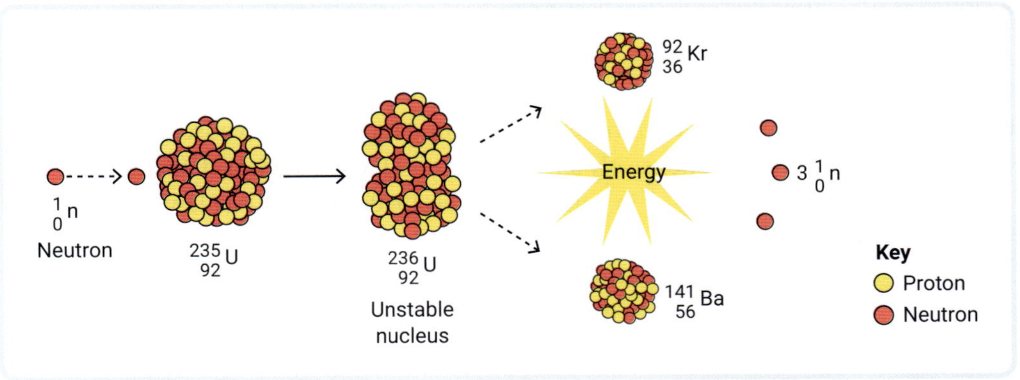

▲ **FIGURE 14.9.1** Uranium-235 fission results in two smaller nuclei and three neutrons being released.

Uncontrolled and controlled fission

The neutrons released during fission can go on and strike other uranium nuclei. This releases more neutrons, which strike more uranium atoms, with this process continuing until there is no more uranium to fission.

uncontrolled fission when the rate of fission is not managed and very large amounts of energy are created in a short period of time

This process happens very quickly. Figure 14.9.2 shows that fission of one uranium-235 atom can cause three more nuclei to fission. Then nine more neutrons are released, then 27 neutrons – the growth is very fast. This process is called **uncontrolled fission**.

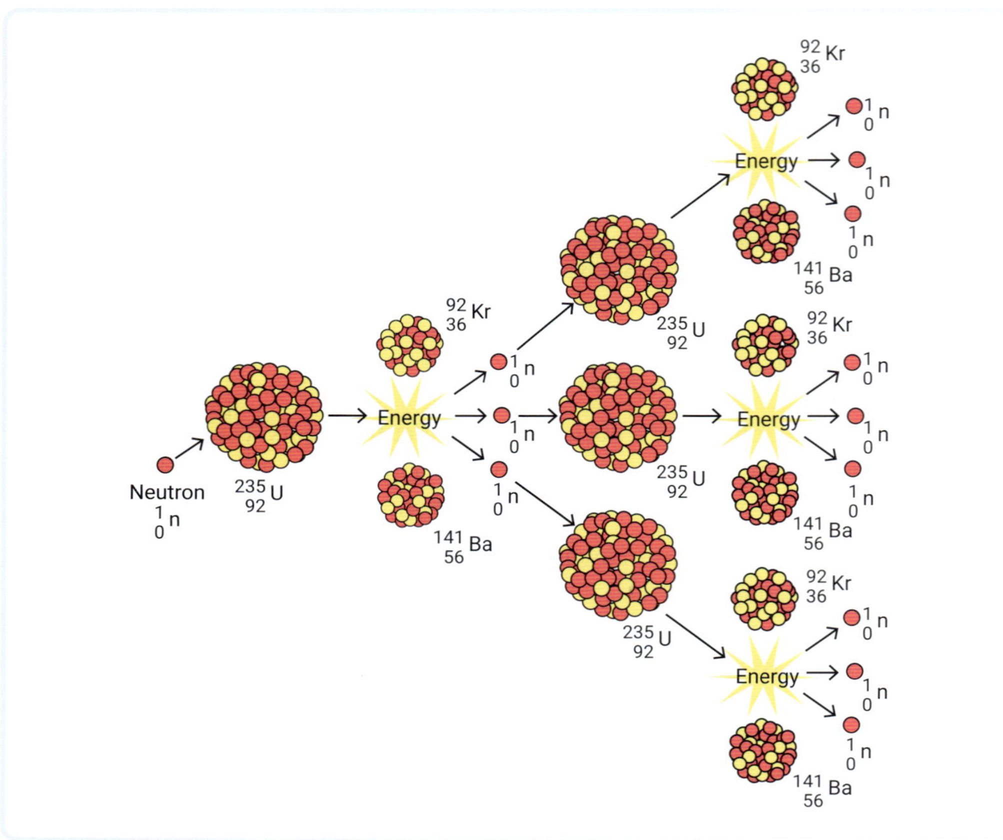

▲ **FIGURE 14.9.2** Uncontrolled nuclear fission can result in a large amount of energy being released in a short period of time.

controlled fission when the rate of fission is managed and a constant amount of energy is produced

Uncontrolled fission has been used in nuclear weapons. A large amount of energy is released in a very short space of time, and a lot of damage can be caused. Nuclear weapons have been tested extensively since the 1940s all over the world, including in Australia in the 1950s and 1960s (Figure 14.9.3). The United States dropped nuclear bombs on Hiroshima and Nagasaki in Japan during World War II, with devastating consequences for the Japanese people.

Galerie Bilderwelt/Getty Images

▲ **FIGURE 14.9.3** Nuclear testing at Maralinga, South Australia, in the 1950s, to determine the strength and side effects of these weapons

Controlled fission involves capturing some of the neutrons released from fission so that a smaller number of neutrons go on to create more fission reactions. This way the amount of energy produced can be controlled and adjusted.

Controlled fission is used in nuclear power reactors and research reactors. Nuclear power reactors are used across the world to generate electricity through the fission of uranium, thorium and plutonium. Other reactors use fission to create isotopes that are used in medicine, environmental monitoring, industrial processes and scientific research. Australia has only one operating nuclear reactor, which is a research reactor at Lucas Heights in Sydney. You will learn more about this reactor in Module 14.14.

14.9

Nuclear reactors and controlled fission

There are a lot of different types of nuclear reactors in use, but most of them have the same key components. Figure 14.9.4 shows the core of a nuclear reactor. The core is the heart of a nuclear power station, producing heat energy that heats water, creating steam to turn turbines and generate electricity.

The uranium fuel is contained in the **fuel rods**. These rods are fixed in place in the core. When neutrons strike the fuel rods, fission of uranium occurs, producing more neutrons that can go on to strike more fuel rods, creating a continuous reaction.

To prevent this reaction from becoming uncontrolled, movable **control rods** are inserted into the core. Control rods are made from materials that absorb neutrons, such as cadmium. If three neutrons are produced per fission, the control rods will absorb two of the three neutrons. This means only one of the neutrons produced will go on to cause another fission reaction.

When the ratio of uranium to neutrons is 1:1, the reaction is controlled and the amount of energy produced is constant. If more energy is required, some of the control rods are raised out of the core. As the 1:1 ratio increases, more fission occurs and more energy is produced. The rate of energy production is slowed by inserting more control rods into the core.

fuel rods
rods inside the nuclear reactor core that contain the uranium fuel

control rods
moveable rods that absorb neutrons and are lowered and raised to control the rate of fission in a nuclear reactor

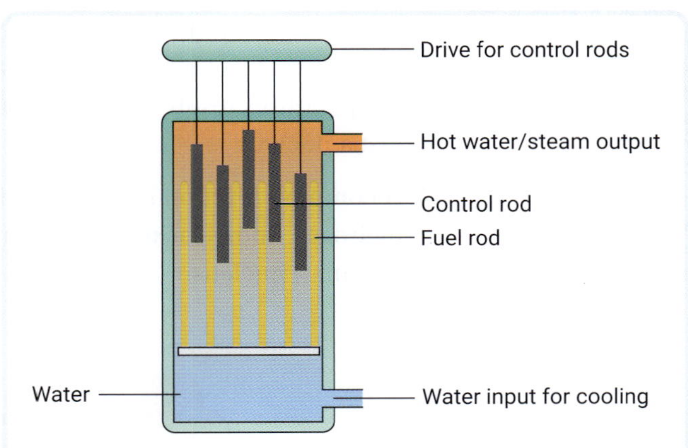

▲ **FIGURE 14.9.4** The core of a nuclear reactor has components to control the rate of fusion and energy production.

14.9 LEARNING CHECK

1. **Identify** the products of the fission of uranium-235.
2. **Describe** the process of nuclear fission.
3. **Describe** how control rods are used in nuclear power reactors.
4. **Explain** the difference between controlled and uncontrolled nuclear fission reactions.

14.10 Nuclear fusion

BY THE END OF THIS MODULE, YOU WILL BE ABLE TO:

- ✓ describe the process of nuclear fusion
- ✓ describe fusion in different types of stars.

GET THINKING

In Module 14.2 you learned that stars are where all natural elements form. Given that there was only hydrogen and helium at the start of the universe, how do you think all the other elements formed? Compare your ideas with the person next to you.

plasma
electrically charged gas abundant in stars; often called the fourth state of matter

deuterium
an isotope of hydrogen with one proton and one neutron

tritium
an isotope of hydrogen with one proton and two neutrons

What is nuclear fusion?

Nuclear fusion occurs when two smaller nuclei combine to form a larger nucleus while releasing very large amounts of energy. Fusion reactions happen in the cores of stars, where temperatures are extremely hot. The core of our Sun is around 15 million degrees Celsius. At this temperature, the atoms are in the state of matter called **plasma**. The high temperatures result in the electrons being ripped away from the atom, leaving behind just a nucleus.

A common reaction in the core of stars is shown in Figure 14.10.1. Two isotopes of hydrogen known as **deuterium** (hydrogen-2) and **tritium** (hydrogen-3) fuse to form a helium-4 nucleus.

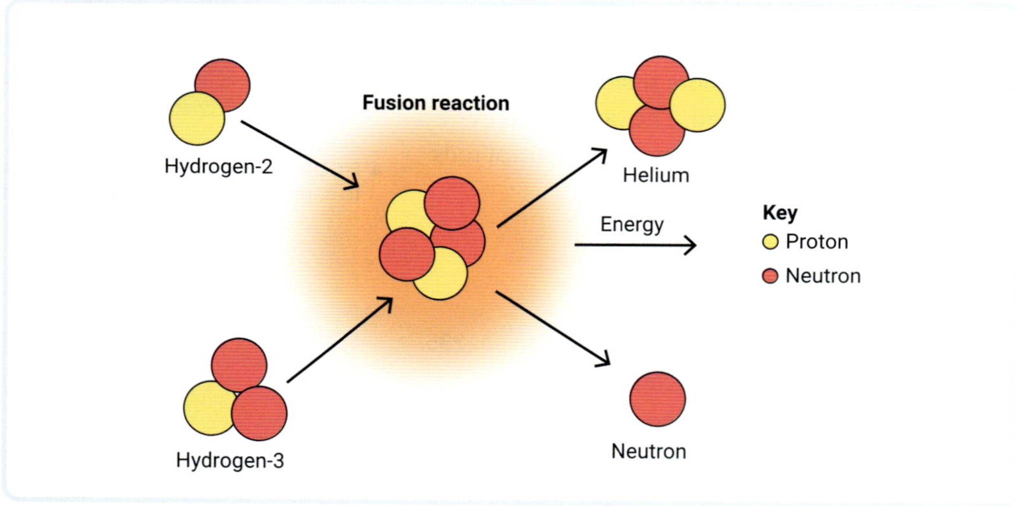

▲ **FIGURE 14.10.1** Two isotopes of hydrogen fuse to produce helium in the cores of stars.

Fusion in stars

As you learned in Module 14.2, stars are formed in areas of hydrogen gas and dust called nebulae. All elements, from hydrogen to uranium, are formed in stars through the process of nuclear fusion.

9780170491785

Main sequence stars, like our Sun, all fuse hydrogen to helium. This phase of a star's life results in the majority of the hydrogen in the core being fused. When the hydrogen levels drop too far, the star is unable to fuse any more hydrogen and moves into a new phase. The next phase is either a red giant or red supergiant, depending on the size of the original star.

In giants and supergiants, the core is hotter and denser than in a main sequence star, so it is able to fuse larger nuclei. As nuclei increase in size, they become more positively charged. This makes it harder for two nuclei to collide because they have stronger electrostatic forces repelling each other. The hot, dense conditions in a red supergiant allow these larger nuclei to overcome these forces, collide and fuse.

Eventually the core builds up a series of shells of the elements that it has fused, as seen in Figure 14.10.2.

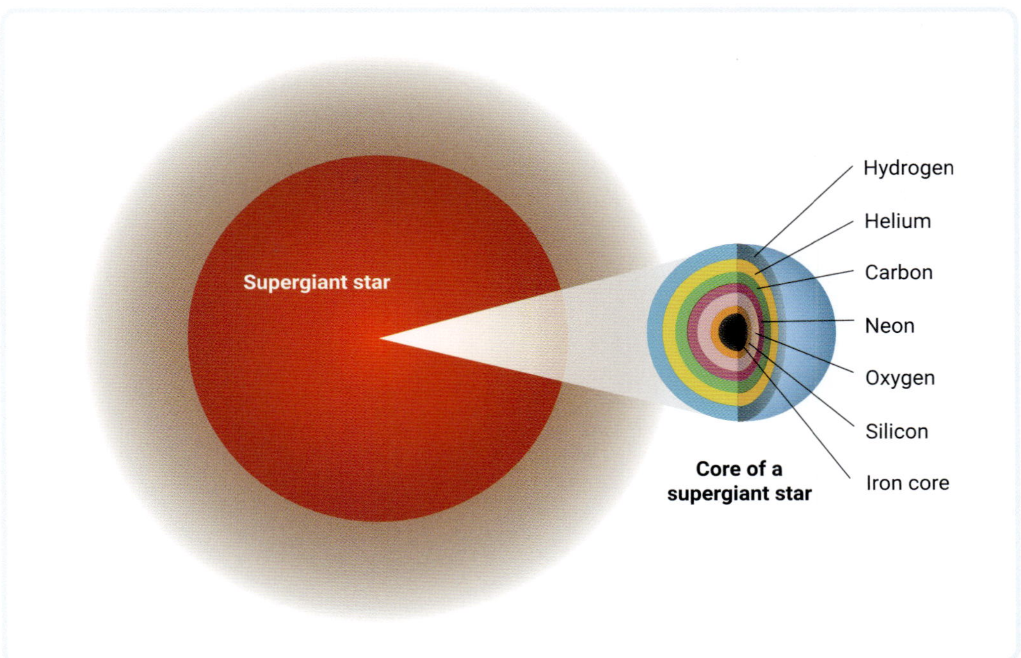

▲ **FIGURE 14.10.2** Stars form layers of elements as a result of nuclear fusion.

Fusion on Earth

Research scientists are trying to replicate fusion for electricity generation. If fusion could be used for this purpose, it would have a number of advantages over using nuclear fission or fossil fuels for power generation.

- The fuel for fusion is easier to obtain, because the hydrogen isotopes required can be generated from water.
- The energy gained from the same mass of fuel is around four times larger than fission, and several million times larger than fossil fuels.
- Fusion reactors would not generate the levels of nuclear waste that fission reactions do.

Currently, it is difficult to replicate the conditions necessary to have nuclear fusion reactions that could be used for energy generation. Stars have massive gravitational forces that help to drive fusion, which are not possible to produce on Earth. This means that we need to create temperatures of around 100 million degrees. This is a massive challenge for scientists.

14.10 LEARNING CHECK

1 **Describe** what happens during nuclear fusion.
2 **Explain** the difference in fusion in main sequence stars and in red supergiant stars.
3 **Describe** the challenges facing scientists trying to replicate nuclear fusion on Earth.
4 **Analyse** why the potential use of nuclear fusion for electricity generation is preferred over nuclear fission.

14.11 Uses of radiation

BY THE END OF THIS MODULE, YOU WILL BE ABLE TO:

✓ describe how radiation is used for different purposes
✓ explain why different types of radiation are used for different purposes
✓ predict the best type of radiation for a particular purpose.

GET THINKING

Skim over the examples of radioactive isotopes in this module. How many of them have you heard of? Make a list of two or three examples that interest you for further research later in the module.

Video
Science skills in a minute: Uses of radiation

Uses of radioactive isotopes

Radioactive isotopes are used for a range of applications such as medicine, plumbing, construction and scientific research. The choice of radioactive isotope depends on its properties. You may recall that radiation can have high, medium or low ionising power and penetrating power. Ionising power is the ability to damage atoms. Penetrating power is the ability to pass through materials.

Radioactivity in medicine

Radioactivity has many uses in medicine. Doctors use it to find out what illness a patient is suffering from and to treat illnesses such as cancer.

Nuclear medicine monitors what happens to certain chemicals as they pass through the body. It can be used to check if an organ is functioning properly. These chemicals, called tracers, are made with radioactive isotopes (Figure 14.11.1 and Table 14.11.1). This allows doctors to follow their path through the body by monitoring the radiation they emit.

Isotopes used in nuclear medicine have the following properties.

- Short half-life: radioactive substances should not stay in the human body longer than necessary. Most radioactive isotopes used in medicine have a half-life of only hours or days.
- Emission of detectable radiation: the radioactive isotope should emit radiation that has a high enough penetrating power to leave the human body and be detected, but preferably a low ionising power so that it does not damage cells.

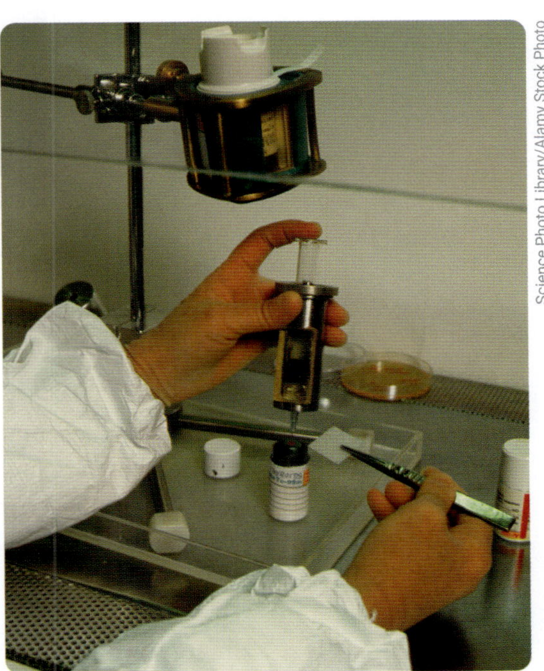

Science Photo Library/Alamy Stock Photo

▲ FIGURE 14.11.1 A chemical tracer containing a radioactive isotope

▼ TABLE 14.11.1 Medical radioactive isotopes

Isotope	Use
Technetium-99	Diagnosis of heart, lung, brain, thyroid and blood flow problems
Fluorine-18	Study of brain function (Figure 14.11.2) Diagnosis of epilepsy, heart disease and some types of cancer
Iodine-123	Diagnosis of thyroid disease and some types of cancer

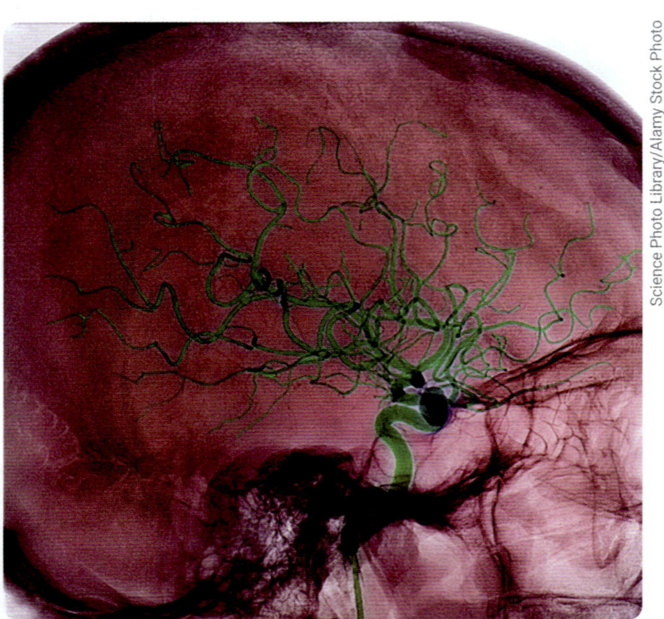
Science Photo Library/Alamy Stock Photo

▲ FIGURE 14.11.2 Arterial brain scans involve injecting a radioactive isotope such as fluorine-18 into a blood vessel and taking an image.

radiotherapy the treatment of cancer by radiation

The use of radiation in the treatment of cancer is called **radiotherapy**. Cancer cells actively divide and are more likely to be killed by radiation than healthy cells. Radiotherapists target the radiation at the cancer cells.

Radiotherapy machines often use cobalt-60 as a source of gamma radiation. Gamma radiation has very high energy and if targeted correctly will damage cancer cells. It does this by transferring this energy to the chemical bonds, which break and cause cell damage and death. Cobalt-60 has a half-life of 5.2 years. This longer half-life is not a problem for the patient because the isotope is in the machine and not in the body.

▼ TABLE 14.11.2 Medical treatment isotopes

Isotope	Half-life	Use
Iodine-131	8.03 days	Treatment of thyroid disease
Phosphorus-32	14.26 days	Treatment of excess red blood cells
Yttrium-90	64 hours	Liver cancer therapy
Iridium-132	73.83 days	Treatment of cancer in the body

Radioactivity in industry

Radioactive isotopes have industrial applications. Caesium-137 is used to determine the thickness of steel sheets, paper, aluminium foil and plastic film during production (Figure 14.11.3). The caesium-137 is placed on one side of the material, and a detector, which measures the amount of radiation, is placed on the other. Thicker materials absorb more radiation, so the amount of radiation detected means the material's thickness can be accurately measured.

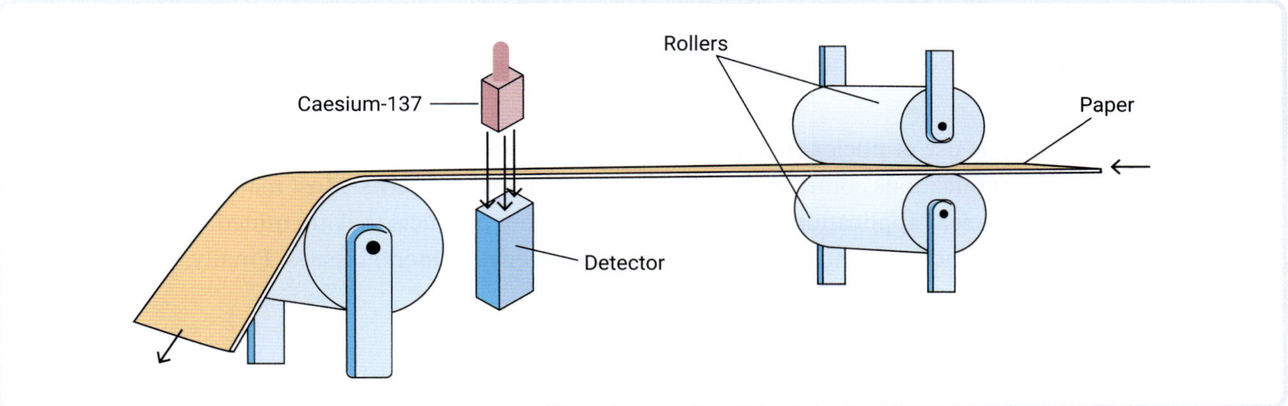

▲ **FIGURE 14.11.3** Radiation is used to measure the thickness of paper.

Gold-198 is used as a tracer to follow the movement of sewage and other wastes through waterways in a similar way to how radioactive tracers are used in medicine.

Leaks can be detected in underground water pipes or oil lines by adding a radioactive isotope, such as sodium-24, and following the passage of the tracer in the pipe with a detector. The detector will show any leaks as the radioactive isotope escapes the pipes through cracks.

14.11 LEARNING CHECK

1 **Describe** two uses of medical radioactive isotopes.
2 **Describe** how a radioactive tracer works.
3 **Explain** why medical isotopes generally have short half-lives of hours or days.
4 **Predict** the penetrating power of the sodium-24 isotope used in leak detection in pipes (high, medium or low). **Justify** your prediction.
5 Using one of the examples of radioactive isotopes in industry, **research** the properties of the isotope and **create** a one-slide presentation that links the properties of the isotope to its use.

14.12 Environmental impacts of nuclear reactions

BY THE END OF THIS MODULE, YOU WILL BE ABLE TO:

✓ describe environmental concerns around the use of nuclear reactions.

Quiz
Nuclear reactions and the environment

GET THINKING

Most human actions have environmental consequences. Our use of metals, fossil fuels and other materials involves mining, pollution and many other issues you have covered in your Science studies. Do you think there is anything we should stop using due to environmental concerns? Write down your reasons for answering yes or no to this question.

The use of nuclear reactions for electricity production, research, medicine and industrial applications comes with various environmental concerns. Most material used in nuclear applications is mined, or comes from products that are mined. The refining and production of nuclear material may cause environmental issues. Finally, the waste from nuclear processes must be dealt with.

Mining

About two-thirds of uranium used in nuclear reactions is mined in Australia, Canada and Kazakhstan. Australia currently has operating mines in the Northern Territory (Ranger) and South Australia (Olympic Dam and Beverley). Mining of other kinds has been covered in other parts of your Science studies, and the potential environmental issues are the same.

To build a mine usually requires clearing vegetation from a large area of land, causing habitat loss for animals. Mining can contaminate the soil and any surrounding waterways, poisoning animals and plants and causing long-term harm, even after the mine shuts down.

David Wall/Alamy Stock Photo

▲ **FIGURE 14.12.1** Tailings dams at uranium mines contain radioactive waste.

Production and use

Uranium processing begins at the mine site. The waste from this processing is usually placed into a tailings dam (Figure 14.12.1). These dams contain significant amounts of material that will remain radioactive for thousands of years. One of the isotopes commonly present is thorium-230, with a half-life of 75 380 years. Tailings dams must have solid barriers to contain waste and stop it from escaping into surrounding waterways. Once the mine shuts down, the area cannot be built over or used for another purpose. Sites of former mines are managed and monitored constantly for safety concerns.

9780170491785

14.12

Nuclear waste

Any use of a radioactive isotope will generate waste. All of the gloves, gowns, needles and containers that contact radioactive isotopes during medical treatments must be disposed of safely. The fuel rods and other materials from nuclear power reactors must be stored in a safe place for a very long time, as must any used isotopes from industrial applications.

Nuclear waste is a problem because it contains radioisotopes that will remain radioactive for many thousands of years. It often needs to be securely transported overland or overseas to reach approved waste storage sites. Figure 14.12.2 shows waste from a power reactor being transported for long-term storage.

The majority of nuclear waste is stored underground, in geologically stable areas. These areas are unlikely to experience earthquakes or other seismic events that could damage the stored containers. Figure 14.12.3 shows an underground storage facility for nuclear waste. Metal containers are used to store the waste, carefully sealed, then deposited into natural caves or purpose-built spaces. Often the containers are surrounded by steel, concrete and other materials. This prevents any radiation escaping, making the facility safe for people living or working nearby.

CHARLY TRIBALLEAU/Getty Images

▲ **FIGURE 14.12.2** Transport of nuclear materials requires planning, caution and specialised equipment.

dpa picture alliance archive/Alamy Stock Photo

▲ **FIGURE 14.12.3** Underground storage for nuclear waste is used across the world in geologically stable areas.

14.12 LEARNING CHECK

1 **Identify** two environmental concerns raised by the use of nuclear reactions.
2 **Outline** the purpose of a tailings dam.
3 **Explain** the conditions required for underground storage of nuclear waste.
4 **Discuss** the advantages, disadvantages and ethical considerations of using nuclear reactions in society.

ABORIGINAL & TORRES STRAIT ISLANDER SCIENCE CONTEXTS

14.13 Deep-time history of Aboriginal Peoples in Australia

IN THIS MODULE, YOU WILL:

✓ explore dating methods that provide scientific evidence of how long Aboriginal and Torres Strait Islander Peoples have lived on the Australian continent.

© Joe Sambono

▲ FIGURE 14.13.1 Examples of Aboriginal Peoples' ground-edge axes

Radiocarbon dating in Australia

Archaeologists have used radiocarbon dating to study the patterns of movement of humans on the Australian continent. The discovery of human remains (burials) in the remote Willandra Lakes region (New South Wales) has been one of the most important archaeological discoveries. The remains, known as Mungo Man and Mungo Lady, are of immense importance to the Ngiyampaa, the Mutthi Mutthi and the Paakantji Peoples. Radiocarbon dating of the remains estimated them to be at least 25 000 years old. Further analysis of this site using other dating techniques suggests that humans have been in this region for more than 40 000 years. Archaeologists often refer to this time span of human occupation as the deep-time history of Australia.

There have been many years of heated debate about the ownership and perceived cultural disrespect of these important human remains. This has included disagreement about whether the remains should or should not be returned to their respective ancestral descendants. The recent reburial of some of these remains has resulted in further debate and pain for some people in the community.

courtesy of Amy Roberts and the River Murray and Mallee Aboriginal Corporation

▲ FIGURE 14.13.2 Shellfish remains at a midden site. Radiocarbon dating of the shells can indicate how long humans have been in the area.

Radiocarbon dating of Mungo Man and Mungo Lady revealed human occupation of the Australian continent for many thousands of years. Findings of artefacts and dating of cave paintings suggest Aboriginal Peoples' occupation of Australia is even longer. Many Aboriginal Peoples believe that their ancestors originated in Australia, on their own clan countries and that Aboriginal Peoples have been on this continent since time immemorial.

The earliest ground-edge axe recovered from Bunuba Country in the south of the West Kimberley region in Western Australia has been dated to about 49 000 years ago. You can see an example of a ground-edge axe in Figure 14.13.1.

Radiocarbon dating of freshwater mussel shells at a midden site on Meru Country in the Pike River region of South Australia showed that they are about 29 000 years old. Shell midden sites are places along waterways where groups of people repeatedly gathered, often using the same site over long periods. The shells come from Aboriginal and Torres Strait Islander Peoples collecting, cooking and eating shellfish. The shells contain carbon in the form of calcium carbonate, which can be extracted for dating (Figure 14.13.2).

9780170491785

Other dating methods

Often, dating of paintings and artefacts relies on a combination of methods. In the northern Kimberly region (Western Australia), mud wasp nests from rock shelters were used to date rock paintings. The nests were built on top of the paintings and so provide some information about the age of the work. These wasp nests were dated using quartz sand grains within the mud. The nests are believed to be up to 18 000 years old, so the paintings underneath them are even older. Figure 14.13.3 shows wasp nests covering rock paintings in Guugu Yimithirr Country, northern Queensland.

Radiocarbon dating and sand grain analysis dated tools of the Mirarr People (Kakadu region of the Northern Territory). An excavation uncovered ground-edge hatchets, tools used for seed grinding and ochre crayons used to make pigments. These artefacts are thought to be at least 65 000 years old.

Important

Please remember to never touch or disturb Aboriginal and Torres Strait Islander Peoples' artefacts. Report any finds to your local cultural heritage authority.

© Joe Sambono

FIGURE 14.13.3 Remains of wasp nests over rock paintings in Guugu Yimithirr Country, northern Queensland. By dating the nests, scientists can estimate how old the rock paintings are.

ACTIVITY

Aboriginal and Torres Strait Islander Peoples' occupation of Australia

1 Use the information in this module to **create** a table that compares dating studies. Include information such as the:
   a Traditional Owners of the site or name of the Country/Place.
   b material that was used for dating.
   c dating method.
   d age of the material.

2 Why are these studies about Aboriginal and Torres Strait Islander Peoples' occupation of Australia important? What have you learned about Aboriginal and Torres Strait Islander Peoples and their respective cultures?

3 How can such studies be done in ways that are respectful to the Traditional Owners of the sites?

4 What are some of the limitations of these methods of dating materials?

14.14 Using radioactive isotopes to track contaminants and pollutants in the environment

BY THE END OF THIS MODULE, YOU WILL BE ABLE TO:

- ✓ explain how societal values influence scientific research
- ✓ explain how radioactivity is used to identify and track contaminants in the environment.

The Australian Nuclear Science and Technology Organisation (ANSTO) conducts research in a range of science disciplines, including medicine, environmental science and materials development.

ANSTO operates the Open Pool Australian Lightwater (OPAL) nuclear reactor at Lucas Heights in Sydney (Figure 14.14.1). It is a research reactor that produces radioactive isotopes. The OPAL reactor, along with other machines at the facility, can also help to analyse samples of water, soil and air to determine if specific radioactive isotopes are present.

Air, water and soil pollution is a problem in Australia and around the world (Figure 14.14.2). We need accurate information about the location and amount of pollution in the environment (Figure 14.14.3). Farmers need to know that the soil they are growing crops in is not polluted. City gardeners are encouraged to test soils if growing food in case the soil is contaminated by previous land use. People living in cities need to know if air pollution levels are high so they can protect their health. It is vital for everyone that waterways are clean and do not contain toxic contaminants that could get into drinking water or recreational areas.

Image supplied by ANSTO

▲ **FIGURE 14.14.1** The OPAL reactor at Lucas Heights in Sydney

PradeepGaurs/Shutterstock.com

▲ **FIGURE 14.14.2** Air pollution in some cities such as Delhi, India, can be a health hazard to people living there.

UBLIC HEALTH ENGLAND / SCIENCE PHOTO LIBRARY

▲ **FIGURE 14.14.3** A scientist measures levels of radioactive isotopes in the environment.

9780170491785

The scientific team at ANSTO conducts research and monitoring of the environment to:

- monitor changes to the composition of the atmosphere to ensure air quality
- analyse how the soil composition is changing
- analyse water samples to identify any contaminants.

Scientists use radioactive isotopes to:

- identify specific contaminants in soil, air and water so we know what hazards are present (Figure 14.14.4)
- trace the path of nutrients through soils, water through waterways and underground water sources by using radioactive isotopes as trackers
- monitor how pollutant levels change over time by measuring the amount of a known radioactive isotope in the atmosphere, water or soil.

Istock.com/iStock Signature

▲ **FIGURE 14.14.4** Pollution from industry can affect waterways.

14.14 LEARNING CHECK

1. How does having a nuclear research reactor in Australia help scientists monitor the environment?
2. **Describe** two ways that radioactive isotopes are used to help reduce pollution.
3. **Discuss** (give positives and negatives for) the use of radioactive isotopes to monitor the environment.

14 REVIEW

REMEMBERING

1 **Identify** one place where nuclear fission occurs and one place where nuclear fusion occurs.

2 **Identify** the number of protons and neutrons in the following elements.

a $^{27}_{13}Al$ b $^{65}_{30}Zn$

3 **Compare** the number of subatomic particles in the isotopes oxygen-16 and oxygen-18. The atomic number of oxygen is 8.

4 **Describe** the two main stages in the evolution of the universe.

5 **Compare** the structure of alpha particles, beta particles and gamma radiation.

6 **Describe** three different uses of radioisotopes. Include one medical and one industrial example in your response.

UNDERSTANDING

7 **Explain** the differences between chemical reactions and nuclear reactions.

8 The notation $^{A}_{Z}X$ is used to represent atomic structure. **Explain** why the notation of hydrogen is unique among all the elements.

9 Carbon-14 has a half-life of 5730 years. **Explain** what information you can **determine** from this statement.

10 **Describe** the forces that act in the nucleus of atoms and **explain** why some isotopes are unstable.

11 **Explain** the role of fuel rods and control rods in a nuclear reactor.

APPLYING

12 A student claims that all elements are created in stars like our Sun. **Assess** this claim and **explain** why the claim is true or not.

13 The diagram at the top of the next column shows three isotopes of a common element. Use the diagram to write the correct notation for each of the three isotopes in the format 'element-X' (e.g. oxygen-16).

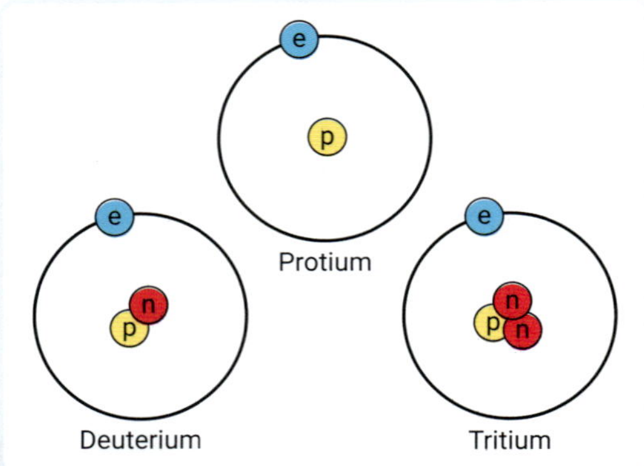

14 Scientists who work with alpha radiation wear only gloves to protect themselves. Scientists working with beta radiation often work with robotic arms with the radiation source contained within a metal-lined box. **Explain** why the precautions differ with different types of radiation sources.

ANALYSING

15 **Compare** the processes of nuclear fission and nuclear fusion.

16 The following graph shows the decay curve for a radioisotope. The measurement of radiation on the vertical axis is 'counts per minute', which is how much radiation is recorded on a detector.

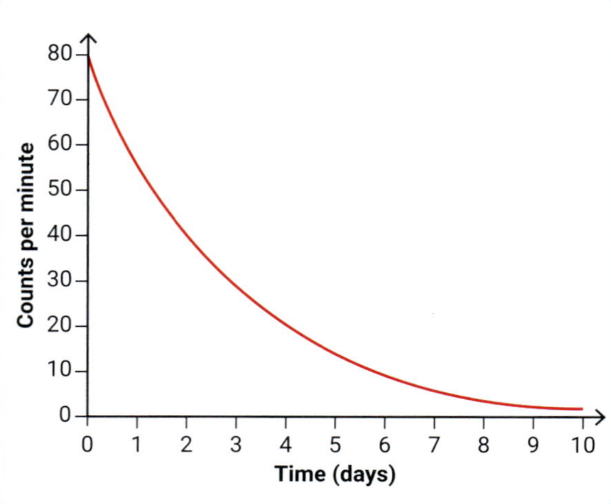

9780170491785

a Using the graph, **determine** the half-life of this radioisotope.

b **Determine** the counts per minute that would be recorded after three half-lives.

17 The radioisotope polonium-208 has 84 protons. By comparing the products of the beta decay and alpha decay of polonium-208, **explain** why this radioisotope undergoes alpha decay. Write nuclear equations for both the alpha and beta decay of polonium-208.

18 Carbon-14 dating is not considered accurate for objects less than 4000 years old or more than 60 000 years old. **Explain** why, by considering the half-life of carbon-14.

19 **Explain** how dating the remains of wasp nests can be used to **determine** the age of Aboriginal Peoples' rock paintings.

EVALUATING

20 **Sketch** a radioisotope decay curve and use it to **explain** why it is difficult to **predict** masses after more than five half-lives have passed.

21 Scientists have found fossilised skulls of hominids (human ancestors) embedded in rocks estimated to be 2–5 million years old. This means that the hominids died at the same time as the rocks formed, so they are the same age. **Evaluate** the pros and cons of using carbon-14 dating and uranium–lead dating to **determine** the age of the skulls.

22 A scientist determines that a rock has equal amounts of uranium and lead and thus states one half-life has passed and the rock is 710 million years old. **Assess** the accuracy of this statement. If there is information missing, rewrite the statement correctly and **justify** your answer.

CREATING

23 **Create** a flow chart to show how elements formed after the Big Bang.

24 **Create** a decay curve graph for carbon-14 that you could use to **estimate** the age of materials. Include an explanation of how you could use the decay curve. Ask a classmate to try it out with an example (make up one based on your decay curve). Get feedback on your explanation. Make improvements based on the feedback until you have an explanation a classmate can use.

SCIENCE IN DEPTH STUDY

1 Connect what you've learned

In this chapter, you have learned about the structure of the atom, and how it can change through radioactive decay.

Draw a mind map to show how the structure of the atom links to the different types of radiation. Use the key words from each module to prompt you.

2 Check your thinking

At the start of the chapter, you were asked 'What have media, books, movies or comic books told you about radiation?' and to reflect on how the image of the radioactive warning signs made you feel. How have your feelings about the warning signs changed? If they haven't, why do you think you still feel that way?

At the start of the chapter, you were asked if you knew any examples of how radiation is used. How has your understanding of the uses of radiation changed?

3 Get into action

What do you think the view of wider society on radiation is? How do you think you may be able to educate people on the science involved? Have a discussion with your parents about their knowledge of radiation, and the advantages and disadvantages of radioactivity.

4 Communicate

Make a poster or presentation showing why radioactivity is useful and how it affects your everyday life.

9780170491785

WAVES AND MOTION

SYLLABUS OUTCOMES

A STUDENT:

- describes the features and applications of different forms of waves SC5-WAM-01
- explains the motion of objects using Newton's laws of motion SC5-WAM-02
- follows a planned procedure to undertake safe, ethical, valid and reliable investigations SC5-WS-04
- selects and uses a range of tools to process and represent data SC5-WS-05

© 2023 NSW Education Standards Authority

CHAPTERS RELATED TO THIS FOCUS AREA ARE:

15 Waves

SCIENCE IN DEPTH

▲ **FIGURE 15.0.1** Sound barriers are often installed on busy roads.

Have you ever considered sound to be a type of pollution? Although it's invisible, noise pollution can pose a risk to living organisms, including humans. Traffic and construction noise can cause hearing loss and high blood pressure for humans. Noise pollution from ships can interfere with how dolphins and whales use sound to find their way through the ocean.

On major highways, engineered acoustic barriers are often installed to protect residents and nearby wildlife from the sounds of passing cars and trucks.

- **Do you think there are times when the noise *you* produce might be harmful to other people or animals?**
- **Can you think of some other examples of ways we can reduce the negative effects of sound?**

#15 DIVE INTO SCIENCE!

At the end of this chapter, you can complete Science in Depth Study #15. You can use the information you learn in this chapter to complete the project.

Assessments
- Prior knowledge quiz
- Chapter review questions
- End-of-chapter test
- Depth study: Research project and flyer

Videos
- Science skills in a minute: Assessing models **(15.2)**
- Video activities: What is sound? **(15.6)**; What is light? **(15.8)**

Science skills resources
- Science skills in practice: Assessing scientific models **(15.2)**
- Extra science investigations: Slinky extension **(15.2)**; Observing sound waves **(15.6)**; Travelling sound **(15.6)**; Mixing light **(15.8)**; Reflection and refraction **(15.9)**

Interactive and other resources
- Simulation: Waves introduction **(15.1)**; Bending light **(15.9)**
- Crossword: Types of waves **(15.1)**
- Label: Features of waves **(15.3)**
- Activity sheets: Visualising frequency **(15.3)**; Sound waves **(15.6)**
- Worksheets: Electromagnetic spectrum **(15.8)**

Nelson MindTap

To access resources above, visit **cengage.com.au/nelsonmindtap**

15.1 Types of waves

BY THE END OF THIS MODULE, YOU WILL BE ABLE TO:

- ✓ describe how mechanical and electromagnetic waves transfer energy
- ✓ compare transverse and longitudinal mechanical waves by identifying and describing their features.

Interactive resources
Simulation: Waves introduction

Crossword: Types of waves

GET THINKING

Imagine a Mexican wave in a stadium at a football match. If you can't imagine this, look up a video of it online. Why do you think it is called a 'wave'? What does it have in common with other waves you know about?

vacuum
a space where there is no matter (no particles)

mechanical wave
a wave that passes energy through matter by vibrations of particles

medium
the matter or substance through which a wave passes (made from particles)

Classifying waves

There are two main types of waves: mechanical and electromagnetic. They are classified according to whether they can or cannot transmit energy through a **vacuum**.

Mechanical waves transfer energy through matter by the vibrations of particles (Figure 15.1.1). The matter they move through is referred to as the **medium**. Mechanical waves cannot travel through a vacuum where there are no particles, such as in outer space. Mechanical waves include sound waves, ocean waves, seismic waves and the waves you make when you flick a piece of rope or a string.

WhiteJack/Shutterstock.com

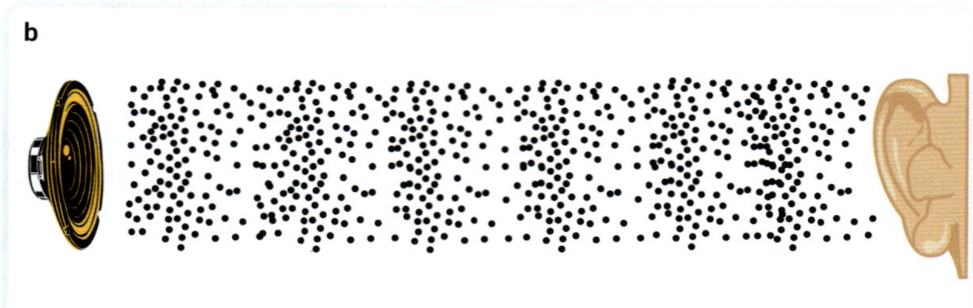

▲ **FIGURE 15.1.1** Some examples of mechanical waves: (a) a wave moving through water; (b) a sound wave

electromagnetic wave
a wave that transfers energy through space by electric and magnetic fields

Electromagnetic waves are waves that transfer energy by oscillation of electric and magnetic fields (Figure 15.1.2).

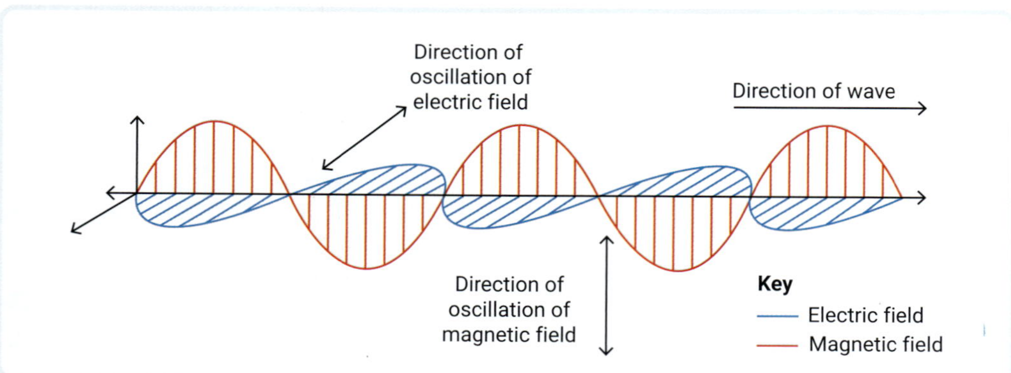

▲ **FIGURE 15.1.2** In an electromagnetic wave, electric and magnetic fields oscillate at 90° (perpendicular) to the direction of the wave.

9780170491785

Light, infrared (heat) radiation, X-rays and ultraviolet (UV) rays are all electromagnetic waves (Figure 15.1.3). Electromagnetic waves do not use particle movement to transfer energy, and can travel through a vacuum. This is why light and heat are able to travel through outer space from the Sun to Earth. Electromagnetic waves are covered in more detail in Module 15.8.

Waves can also be classified by the direction of **oscillation** relative to the direction of energy transfer. An oscillation is a repeated vibration back and forth or up and down.

Transverse waves transfer energy by oscillations that are in a direction that is perpendicular (at right angles) to the direction that energy is travelling, like a wave in the ocean. You can see a transverse wave in Figure 15.1.4. As you can see, the oscillations in the transverse wave are up and down as the wave passes from left to right. On a transverse wave, the highest points are **crests** and the lowest points are **troughs**.

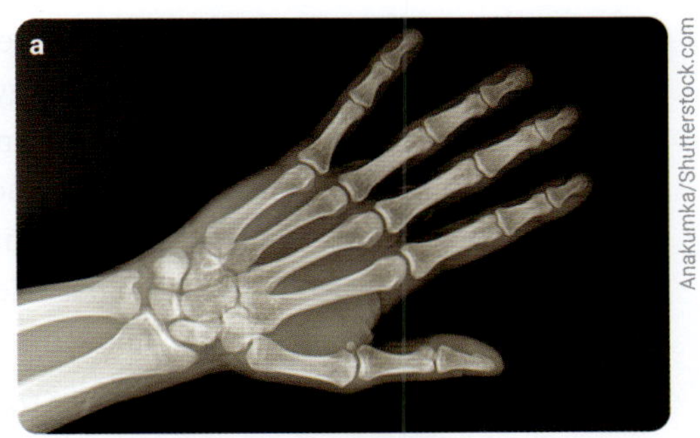

Anakumka/Shutterstock.com

ROBERT BANKS/Nine Publishing

FIGURE 15.1.3 Some examples of electromagnetic waves: (a) X-rays can pass through your body; (b) microwaves are used in speed detection radar guns.

oscillation
movement back and forth in a regular rhythm or pattern

transverse wave
a wave in which oscillations are at right angles to the direction in which the wave is travelling

crest
the highest point on a transverse wave

trough
the lowest point on a transverse wave

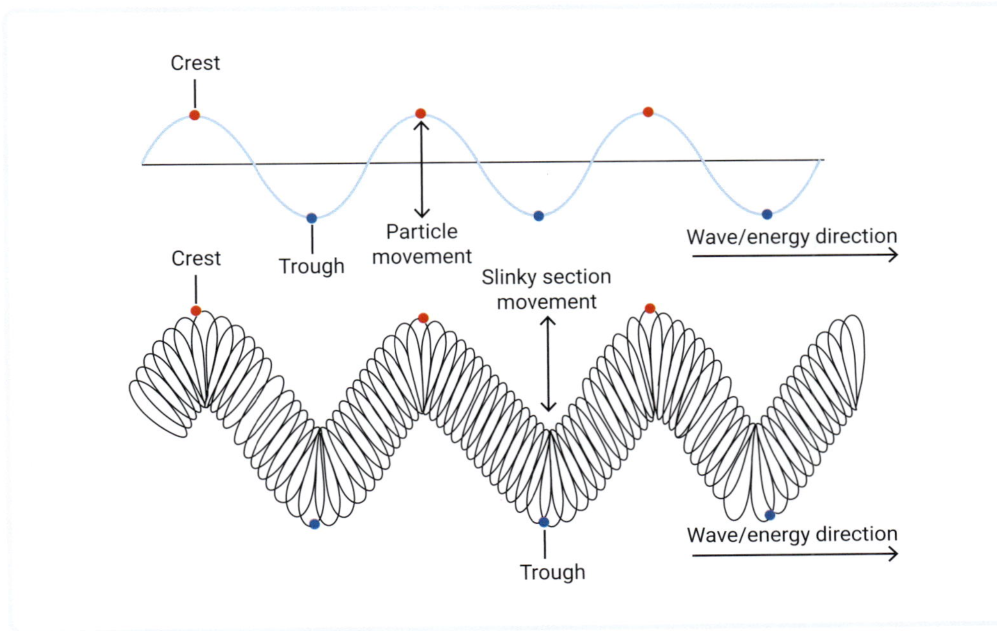

FIGURE 15.1.4 In a transverse wave, the particles oscillate at right angles to the direction in which the wave is moving. You can model this type of wave with a slinky.

longitudinal wave
a wave in which oscillations are parallel to the direction in which the wave is travelling

compression
a high-pressure region of a longitudinal mechanical wave where particles are pushed close together

rarefaction
a low-pressure region of a longitudinal mechanical wave where particles are spread apart

Longitudinal waves are waves that transfer energy by the oscillation of particles in a direction parallel to the direction the energy is travelling, like a sound wave. You can see a longitudinal wave in Figure 15.1.5. This means that if the wave is travelling left to right, the particles in the wave also vibrate left and right as the wave passes. The particles are compressed together and spread apart as the wave passes. On a longitudinal wave, there are zones of **compression** where the particles are closest together and zones of **rarefaction** where the particles are most spread out.

You will model waves using slinkies in Module 15.2.

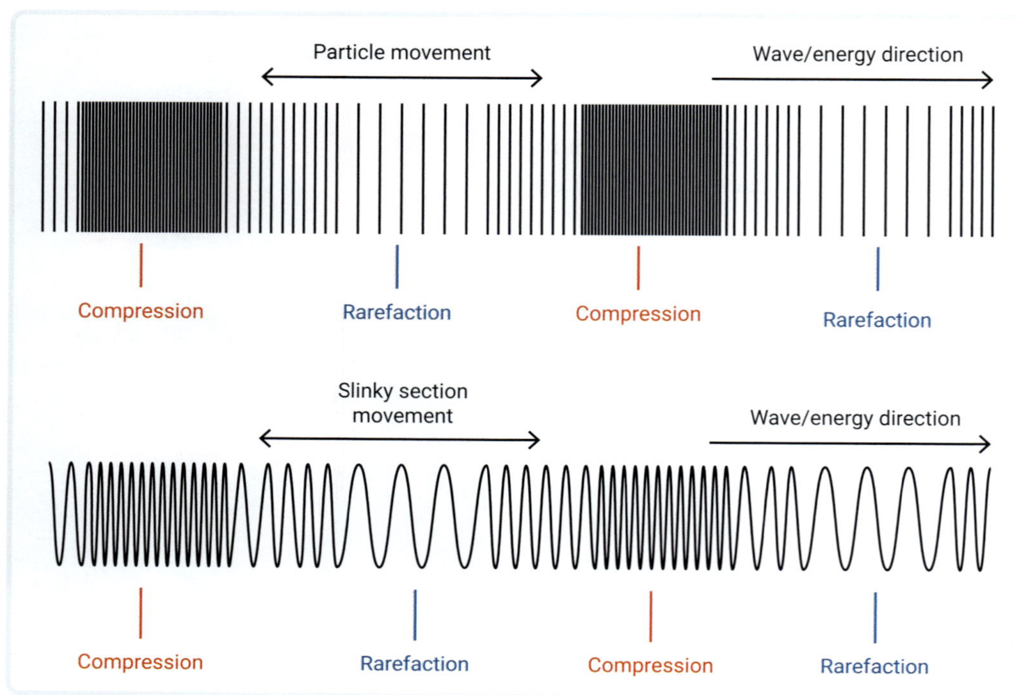

▲ **FIGURE 15.1.5** In a longitudinal wave, the particle oscillates in the same direction as the wave. You can model this type of wave with a slinky.

amplitude
the maximum oscillation of a particle in a wave

wavelength
the length of one full wave; the distance between two crests or two troughs; measured in metres (m)

frequency
the number of waves produced, or passing a point per unit of time; measured in hertz (Hz)

period
the time taken for one full wave to pass a particular point

speed
a measure of how fast a wave is travelling; a measure of the distance covered by the wave per unit of time; measured in metres per second (m/s)

Features of mechanical waves

Mechanical waves can differ from each other in some significant features: **amplitude**, **wavelength**, **frequency**, **period** and **speed** (Table 15.1.1).

In Module 15.3, you will see how these features can be measured and the formulas used to calculate them in some worked examples.

9780170491785

▼ TABLE 15.1.1 A summary of wave features

Feature	Definition	Symbol	Units
Amplitude	• A measure of the extent of oscillation • The greater the energy of a wave, the greater the oscillation size and the larger the amplitude	A	metres (m)
Wavelength	• A measure of how long each wave is • Can be measured from any point on a wave to the same point on the next wave	λ (Greek letter lambda)	metres (m)
Frequency	• A measure of how many full waves pass by a point each second	f	hertz (Hz) or per second (/s)
Period	• The time it takes for one full wave to pass a particular point	T	seconds
Speed	• How far a wave travels in a certain time • Depends on the medium that a wave is travelling through	v	m/s

15.1

15.1 LEARNING CHECK

1 **Distinguish** between mechanical waves and electromagnetic waves.

2 **Define:**
   a crest.
   b trough.
   c compression.
   d rarefaction.

3 Copy and complete the table to **summarise** wave features.

Feature	Definition	Symbol	Units
Wavelength			
Frequency			
Speed			
Amplitude			

4 a Copy and complete the following sentence about transverse waves.
   The ______ in a transverse wave are in a ______ direction relative to the direction of wave motion.
   b Write a similar sentence to the one in part **a** for longitudinal waves that describes the direction of movement.

5 **Compare** longitudinal waves and transverse waves.

WORKING SCIENTIFICALLY

15.2 Assessing scientific models

SCIENCE SKILLS IN FOCUS

IN THIS MODULE, YOU WILL FOCUS ON LEARNING AND IMPROVING THESE SKILLS:

- using a model to visualise and understand wave behaviour
- evaluating the usefulness of a model by identifying strengths and limitations.

Evaluating scientific models

Scientific models are not perfect. Sometimes they do a really great job of explaining or modelling a concept and have many strengths. However, there are often still limitations or issues that mean we can't use them to understand certain ideas. Discussing these strengths and weaknesses allows us to evaluate the quality of a model. Evaluating scientific models lets us determine how useful the model is and understand how and when it can be applied.

Strengths of scientific models

When identifying strengths of a scientific model, considerations include:

- Which ideas or observations about the concept being modelled does the model show well?
- In what ways does the model explain or help you understand observations about the phenomena being modelled?
- Does the model predict outcomes that can then be tested for the phenomena being modelled?
- Is the model easy and cheap to use?

Limitations of scientific models

When identifying limitations of a model, considerations include:

- Are there circumstances when the model behaves differently from the concept being modelled?
- What are the physical differences between the model and the idea?
- In what ways does the model make it harder to understand observations about the phenomena being modelled?

There may be improvements or extensions to the model that can be made to minimise the limitations. However, in most cases, limitations cannot be avoided and need to be considered when using a model. The model may only be applicable for certain ideas relating to the concept and not all. But it can still be very helpful to understand these specific ideas or behaviours.

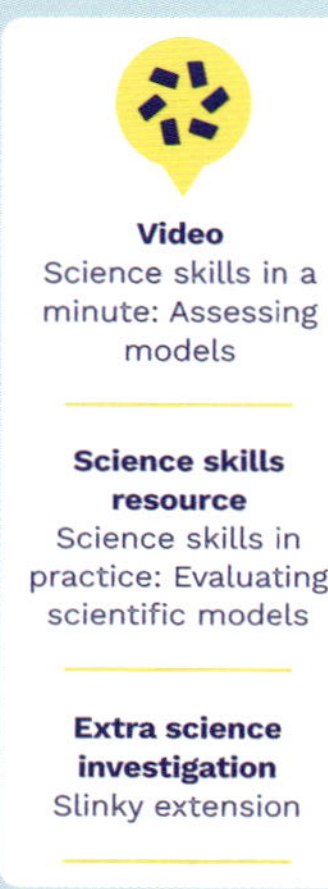

9780170491785

MODELLING WAVES WITH SLINKIES

AIM

To model energy transfer by waves (sound and light) using slinkies and evaluate the usefulness of the model

MATERIALS AND EQUIPMENT

- ☑ slinky that can be stretched approximately 5 m
- ☑ masking tape
- ☑ camera

PROCEDURE

1. Work in groups of three.
2. Fix a small bit of masking tape around a central part of the slinky.
3. Between two people, hold each end of the slinky firmly and stretch it out along the ground so that it extends 4–5 m in length.
4. One person should then move their hand from side to side in a repeating pattern to generate a transverse wave along the ground.
5. A third person should then take a photo or a video of the transverse wave.
6. Make some different transverse waves by changing the frequency of the hand movement. Make a wave with a long wavelength (slower hand movement) and one with a short wavelength (faster hand movement).
7. The second person holding the slinky should send a compressional or longitudinal wave along the slinky by pulsing their hand forwards and backwards towards the other person.
8. Take a photo or a video of the longitudinal wave.
9. Make some different longitudinal waves by changing the frequency of the hand movement. Make a wave with a long wavelength and one with a short wavelength.

RESULTS

Include a copy of the pictures or videos you have taken.

ANALYSIS

1. As the transverse wave moved from one person to the other, **describe** the motion of the point on the slinky with the masking tape.
2. As the longitudinal wave moved from one person to the other, **describe** the motion of the point on the slinky with the masking tape.
3. **Explain** where and how the energy of each wave was generated or transferred to the slinky.
4. Considering what you know about sound waves, **identify** some strengths and limitations of this model for understanding sound and hearing.
5. Considering what you know about light or other electromagnetic waves, **identify** some strengths and limitations of this model for understanding light and seeing.
6. Suggest any improvements that could enhance the usefulness of this model.

CONCLUSION

Overall, do you think the model was a good one or not? Give reasons for your answer.

15.3 Measuring the features of waves

BY THE END OF THIS MODULE, YOU WILL BE ABLE TO:

- ✓ use wave graphs to identify the amplitude, wavelength and period of a wave
- ✓ use the formula for frequency and speed to calculate these features of a wave.

Interactive resource
Label: Features of waves

Other resource
Activity sheet: Visualising frequency

GET THINKING

Can you think of examples of when we might need to measure the features of waves, such as a wave's speed, frequency or amplitude? Why do you think this information can be helpful?

Measuring amplitude

The amplitude of a transverse wave measures half the distance from a crest to the trough. It can be measured from the middle of the wave to a crest or to a trough. The amplitude for a longitudinal wave can be measured by how far particles have travelled when compressed together at a compression point or how far they have travelled apart at a rarefaction point. This is usually quite hard to measure.

Measuring wavelength

Wavelength can be measured from any point on a wave to the same point on the next wave. The easiest ways to measure a wavelength for each type of wave are:

- transverse wave: the distance from crest to crest or from trough to trough
- longitudinal wave: the distance from compression to compression or from rarefaction to rarefaction.

This can be done when observing waves in different mediums such as water (Figure 15.3.1) or using distance wave graphs or diagrams, as shown in Figure 15.3.2.

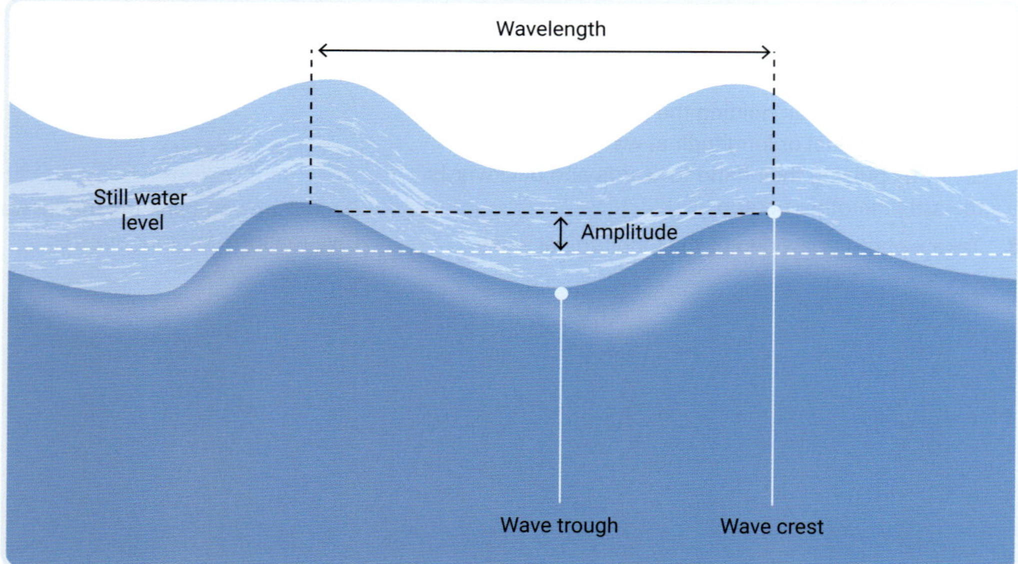

▲ **FIGURE 15.3.1** Amplitude and wavelength can be measured when observing waves moving through the water.

9780170491785

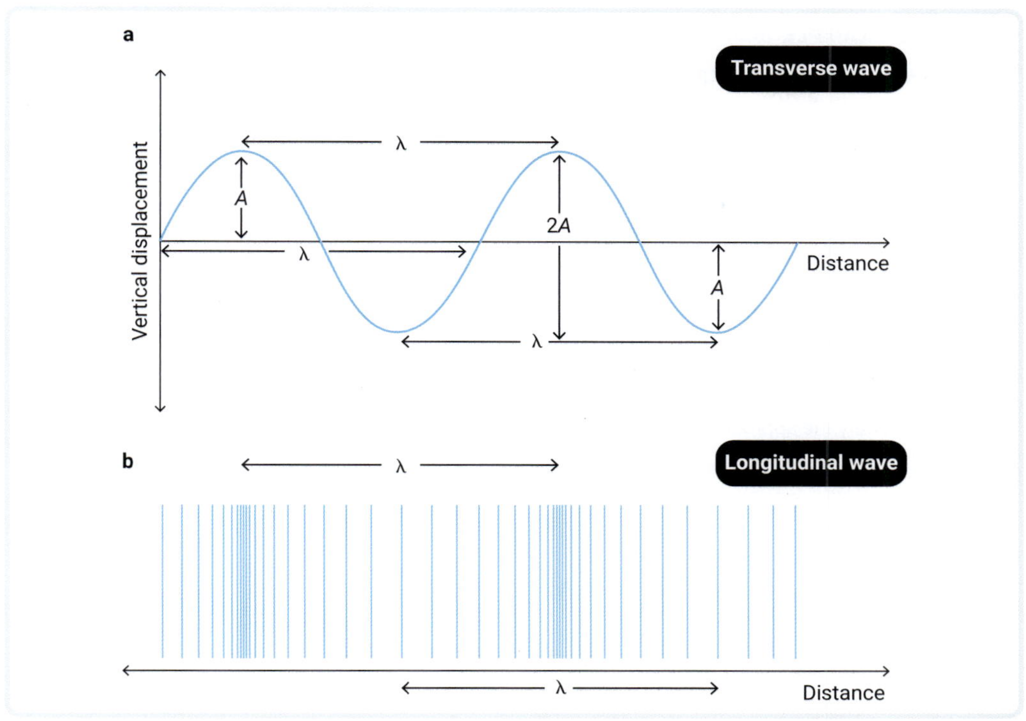

▲ **FIGURE 15.3.2** (a) Graph of vertical displacement versus distance for a horizontal transverse wave, showing amplitude (*A*) and wavelenth (λ). (b) Image of a longitudinal wave showing wavelenth (λ). For a longitudinal wave, amplitude (*A*) is the maximum distance a particle moves from its rest point.

Measuring frequency

The frequency (f) can be determined by counting how many full waves pass by a single point in a certain amount of time:

$$f = \frac{\text{number of full waves passed in time period}}{\text{length of time}}$$

where length of time is measured in seconds (s) and frequency is measured in hertz (Hz).

Ostranitsa Stanislav/Shutterstock.com

▲ **FIGURE 15.3.3** You can work out the frequency of waves by counting how many full waves pass by a particular point in a given time.

WORKED EXAMPLE 15.3.1

During a 4-minute period, 48 water waves pass a point in a river. Calculate the frequency of the waves.

THINKING PROCESS	WORKING
Step 1: Identify the known and unknown variables.	Length of time = 4 minutes Number of waves = 48 Frequency (f) = ?
Step 2: Ensure that all known variables are given in appropriate units. Convert if required.	Length of time = 4 minutes = 4 × 60 seconds = 240 s
Step 3: Identify the appropriate relationship.	$f = \frac{\text{number of full waves passed in time period}}{\text{length of time period}}$
Step 4: Substitute known values.	$f = \frac{48\text{ waves}}{240\text{ s}}$
Step 5: Rearrange, using algebra if necessary, and solve. State the answer.	$f = 0.2\text{ Hz}$ The frequency of the waves passing the point in the river is 0.2 Hz.

Measuring period

The period (T) is determined by measuring the time it takes for one wave to pass a point or by counting the number of full waves that pass by a single point in a certain amount of time:

$$T = \frac{\text{length of time period}}{\text{number of full waves passed in time period}}$$

where length of time and period are both measured in seconds (s).

WORKED EXAMPLE 15.3.2

During a 4-minute period, 48 water waves pass a point in a river. Calculate the period of each wave.

THINKING PROCESS	WORKING
Step 1: Identify the known and unknown variables.	Length of time = 4 minutes Number of waves = 48 Period (T) = ?
Step 2: Ensure that all known variables are given in appropriate units. Convert if required.	Length of time = 4 minutes = 4 × 60 seconds = 240 s
Step 3: Identify the appropriate relationship.	$T = \frac{\text{length of time period}}{\text{number of full waves passed in time period}}$
Step 4: Substitute known values.	$T = \frac{240\text{ s}}{48\text{ waves}}$
Step 5: Rearrange, using algebra if necessary, and solve. State the answer.	$T = 5\text{ s}$ The period of each wave is 5 s.

If a wave's motion is represented versus time (see Figure 15.3.4) instead of distance (as it was in Figure 15.3.2), then the period can be measured directly as the duration of one full wave.

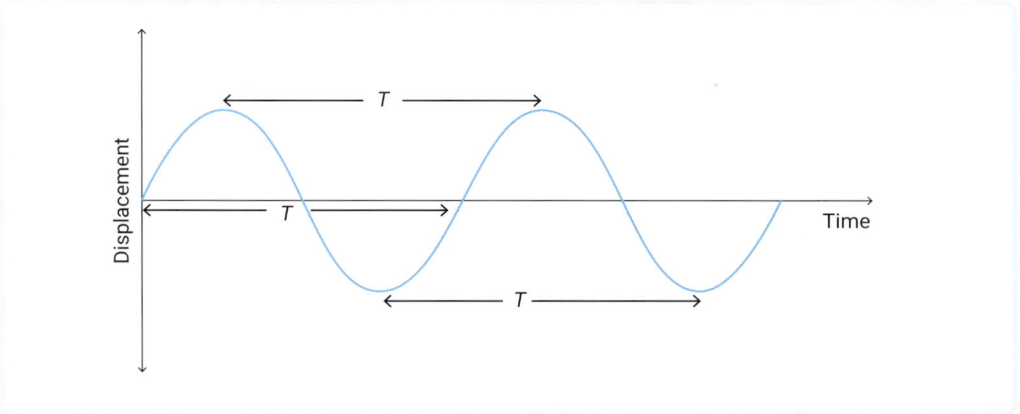

▲ **FIGURE 15.3.4** Graph of displacement versus time for a wave, showing period (T)

Measuring speed

Wave speed (v) can be calculated by measuring how far a wave travels in a certain time:

$$v = \frac{\text{distance travelled by wave}}{\text{length of time}}$$

where distance travelled is measured in metres (m), time is measured in seconds (s) and speed is measured in metres per second (m/s). Radar guns use this method (Figure 15.3.5).

Paul Kingsley/Alamy Stock Photo

▲ **FIGURE 15.3.5** Radar guns determine a car's speed by sending out radio waves and measuring their 'echo,' as they travel through air at a constant speed.

WORKED EXAMPLE 15.3.3

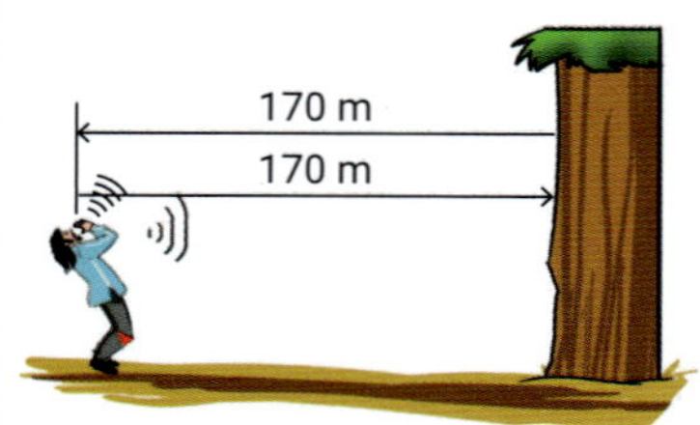

A person stands 170 m from a vertical cliff and shouts. After 1 s, the person hears their shout return to them (Figure 15.3.6). Calculate the speed of the sound wave.

FIGURE 15.3.6 It takes 1 s for the echo of the shout to return to the person.

THINKING PROCESS	WORKING
Step 1: Identify the known and unknown variables.	Distance travelled by wave = 170 m + 170 m = 340 m Length of time period = 1 s Speed (v) = ?
Step 2: Identify the appropriate relationship.	$v = \dfrac{\text{distance travelled by wave}}{\text{length of time period}}$
Step 3: Substitute known values.	$v = \dfrac{340\text{ m}}{1\text{ s}}$
Step 4: Rearrange, using algebra if necessary, and solve. State the answer.	v = 340 m/s The speed of the sound wave is 340 m/s.

15.3 LEARNING CHECK

1 Using Figure 15.3.7, **identify** the:
  a wavelength. b amplitude.
  c number of full wavelengths shown.

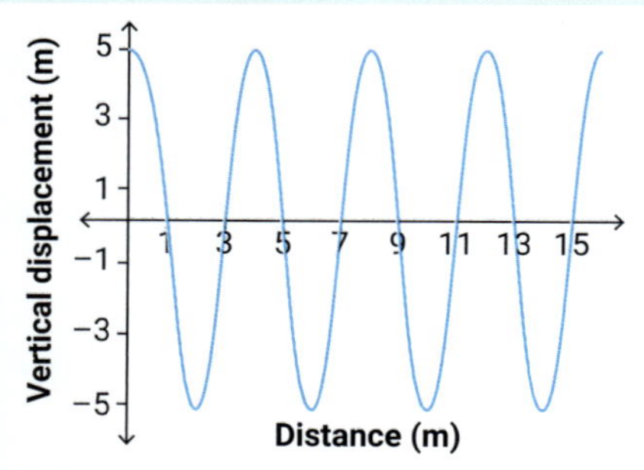

FIGURE 15.3.7 A graph of vertical displacement against distance for a transverse wave

2 Using Figure 15.3.8 (notice that the label on the horizontal axis is different from in Figure 15.3.7):
  a **identify** the period of the wave.
  b **calculate** the frequency of the wave.

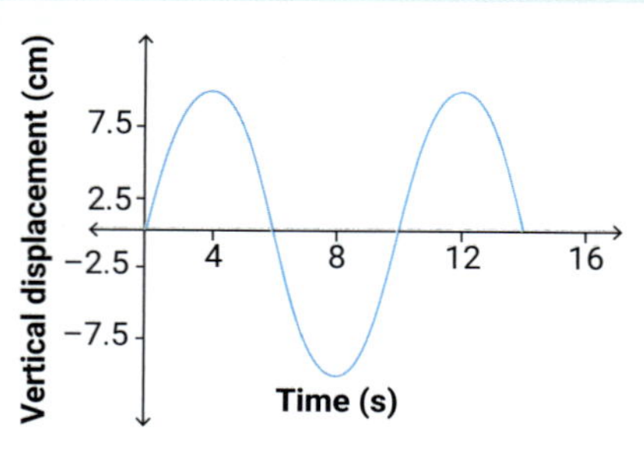

FIGURE 15.3.8 A graph of vertical displacement against time for a transverse wave

15.4 More wave calculations

BY THE END OF THIS MODULE, YOU WILL BE ABLE TO:

- ✓ calculate the frequency of a wave when given the period and vice versa
- ✓ calculate the frequency, wavelength and speed of a wave using the wave equation.

Frequency and period calculations

Period and frequency are reciprocals of one another. A high frequency means a low period and a low frequency means a high period. The period can be calculated from the frequency and vice versa.

$$\text{Period} = \frac{1}{\text{frequency}} \text{ and frequency} = \frac{1}{\text{period}}$$

$$T = \frac{1}{f} \text{ and } f = \frac{1}{T}$$

where period is measured in seconds (s) and frequency is measured in hertz (Hz).

WORKED EXAMPLE 15.4.1

A sound wave takes 0.02 s to pass a particular point. Calculate the frequency of the sound wave.

THINKING PROCESS	WORKING
Step 1: Identify the known and unknown variables.	$T = 0.02$ s $f = ?$
Step 2: Identify the appropriate relationship.	$f = \frac{1}{T}$
Step 3: Substitute known values.	$f = \frac{1}{0.02 \text{ s}}$
Step 4: Rearrange, using algebra if necessary, and solve. State the answer.	$f = 50$ Hz The frequency of the wave is 50 Hz.

WORKED EXAMPLE 15.4.2

A wave machine in a wave pool generates waves at a frequency of 0.5 Hz. Calculate the period of each wave.

THINKING PROCESS	WORKING
Step 1: Identify the known and unknown variables.	$f = 0.5$ Hz $T = ?$
Step 2: Identify the appropriate relationship.	$T = \frac{1}{f}$
Step 3: Substitute known values.	$T = \frac{1}{0.5 \text{ Hz}}$
Step 4: Rearrange, using algebra if necessary, and solve. State the answer.	$T = 2$ s The period of each wave is 2 s.

The wave equation

wave equation
a mathematical relationship between the speed, wavelength and frequency of a wave

Another important relationship that can be used to analyse waves is the **wave equation**. The wave equation is a mathematical relationship between the speed, wavelength and frequency of a wave:

$$\text{Wavelength} = \frac{\text{speed}}{\text{frequency}}$$

$$\lambda = \frac{v}{f}$$

where speed is measured in metres per second (m/s), wavelength is measured in metres (m) and frequency is measured in hertz (Hz).

WORKED EXAMPLE 15.4.3

A wave has a frequency of 50 Hz and a wavelength of 10 m. Calculate the speed of the wave.

THINKING PROCESS	WORKING
Step 1: Identify the known and unknown variables.	$f = 50$ Hz $\lambda = 10$ m $v = ?$
Step 2: Identify the appropriate relationship.	$\lambda = \frac{v}{f}$
Step 3: Substitute known values.	$10\text{ m} = \frac{v}{50\text{ Hz}}$
Step 4: Rearrange, using algebra if necessary, and solve. State the answer.	$v = 500$ m/s The speed of the wave is 500 m/s.

WORKED EXAMPLE 15.4.4

A wave rolling through a harbour is travelling at 3 m/s and has a wavelength of 0.5 m. Calculate the frequency of the wave.

THINKING PROCESS	WORKING
Step 1: Identify the known and unknown variables.	$v = 3$ m/s $\lambda = 0.5$ m $f = ?$
Step 2: Identify the appropriate relationship.	$\lambda = \frac{v}{f}$
Step 3: Substitute known values.	$0.5\text{ m} = \frac{(3\text{ m/s})}{f}$
Step 4: Rearrange, using algebra if necessary, and solve. State the answer.	$f = \frac{3\text{ m/s}}{0.5\text{ m}}$ $= 6$ Hz The frequency of the wave is 6 Hz.

15.4

WORKED EXAMPLE 15.4.5

A wave made within a skipping rope is travelling at 8 m/s and 16 waves are produced every second. Calculate the wavelength of the wave.

THINKING PROCESS	WORKING
Step 1: Identify the known and unknown variables.	$v = 8$ m/s $f = 16$ Hz $\lambda = ?$
Step 2: Identify the appropriate relationship.	$\lambda = \frac{v}{f}$
Step 3: Substitute values and solve. State the answer.	$\lambda = \frac{8 \text{ m/s}}{16 \text{ Hz}}$ $= 0.5$ m The wavelength of the wave is 0.5 m.

15.4 LEARNING CHECK

1 A tap is dripping into a bowl of water. Drops of water **produce** water waves from the point where the drops hit the surface of the water (Figure 15.4.1). If the frequency of drops from the tap is 5 Hz, **calculate** the period of each wave produced.

WhiteJack/Shutterstock.com

▲ **FIGURE 15.4.1** Dripping water making waves along the water's surface

2 At the beach, the distance between two consecutive wave crests is 4 m. If two waves reach the shoreline every 10 seconds, calculate the:
   a frequency of the waves.
   b speed of the waves.

3 Red light travels at a speed of 300 000 000 m/s and has a wavelength of 0.000 0007 m. **Calculate** the frequency of the waves of red light.

Abdul Razak Latif/Shutterstock.com

▲ **FIGURE 15.4.2** Red light travels at the speed of light but has a different frequency and wavelength from other colours of light. Red light has the longest wavelength of all the colours so it can be seen from the greatest distances.

WORKING SCIENTIFICALLY

15.5 Using simulation models

SCIENCE SKILLS IN FOCUS

IN THIS MODULE, YOU WILL FOCUS ON LEARNING AND IMPROVING THESE SKILLS:

- using a simulation by following a planned procedure
- selecting and using a range of tools to process and represent data.

MODELLING TRANSVERSE WAVES USING A SIMULATION

AIM

To use a two-dimensional wave simulation model to investigate transverse waves

MATERIALS

☑ laptop with internet access

☑ 'Wave on a string' simulation (MindTap) (Figure 15.5.1)

PROCEDURE

SETTING UP

1 Open the simulation and make the following adjustments to the settings in the simulation:
   - Change the wave generation mode to 'Oscillate'.
   - Change the end type to 'No end'.
   - Choose 'Slow motion'. You can change to 'Normal' at any time in this activity if preferred.
   - Turn damping to 'None'.
   - Turn on the 'Ruler' tool.

2 Press the pause button and practise using the ruler tool to measure the wavelength of one full wave. You can click and drag the ruler tool to where you need it.

3 Unpause the simulation so that it is running and ready for Part A.

Weblink
PhET: Wave on a string

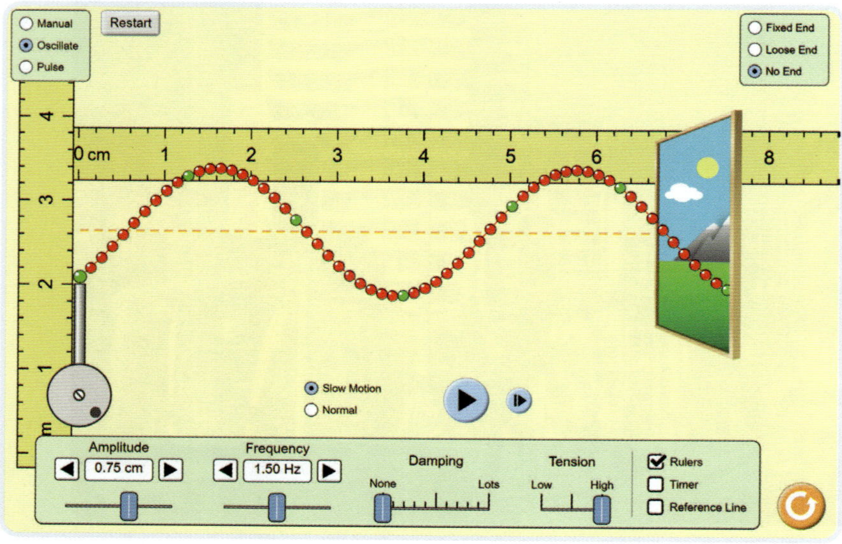

▲ FIGURE 15.5.1 Screenshot of simulation set-up

9780170491785

PART A: RELATIONSHIP BETWEEN AMPLITUDE AND WAVELENGTH

How does a change in amplitude affect the wavelength of a transverse wave when modelled on a string?

4 In your workbook or on your device, set up a results table for an investigation into how amplitude (independent variable) affects the wavelength of the wave (dependent variable). An example table has been provided in Table 15.5.1.

5 Complete the first row of the table for the starting amplitude and wavelength of the wave.

6 Increase the amplitude of the wave by increments of 0.10 cm to 0.85 cm.

7 Measure and record the wavelength of the wave in the results table. Do this as you have done in the 'Setting up' stage.

8 Repeat steps 6 and 7 four more times until a maximum amplitude of 1.25 cm is reached. Don't forget to unpause the simulation before making each change.

PART B: A SECOND RELATIONSHIP

9 Choose another variable (frequency or tension) to investigate how it might affect the wavelength.

10 Write a research question similar to that in Part A to reflect the direction of your investigation.

11 Set up a results table for this second research question.

12 Complete the first row of the table for the starting settings of your simulation.

13 Make four changes to your independent variable and record the resulting wavelengths.

14 It is best to sequence the rows of your results table so that the independent variable is increasing. This will make it easier to identify trends.

RESULTS

1 Complete the results table for Part A (Table 15.5.1).

▼ **TABLE 15.5.1** Part A results table: amplitude and wavelength data for a transverse wave

Amplitude (cm)	Wavelength (cm)

2 Include your results table for Part B here.

ANALYSIS

1 Complete the following sentence to **identify** a relationship in Part A.

As the amplitude of the wave increased, the wavelength ___________.

2 **Examine** the data collected in Part B and describe the relationship using a sentence similar to Question 1.

3 **Discuss** how this simulation has been helpful to understand relationships between variables for transverse waves.

4 Suggest how this simulation might need to change if it were to model longitudinal waves instead.

CONCLUSION

Summarise the three main things you have learned about waves and simulations after completing this inquiry activity.

15.6 Sound waves and hearing

BY THE END OF THIS MODULE, YOU WILL BE ABLE TO:

- ✓ describe how sound energy is transferred
- ✓ construct representations of sound waves and identify key features, including amplitude and frequency
- ✓ compare the volume and pitch of different sound waves
- ✓ describe the structure and function of the human ear and explain how we can hear.

Video activity
What is sound?

Extra science investigations
Observing sound waves
Travelling sound

Other resource
Activity sheet: Sound waves

GET THINKING

List two things you already know about sound. Write down two things about sound that you don't know how to explain but would like to learn more about.

Making and moving sound

Sounds are caused by the transfer of kinetic energy through a medium by longitudinal waves. When you clap your hands, the kinetic energy of your hands is transferred to the air particles around your hands. These vibrations pass through the air to your ears where the particle vibrations are turned into sounds by your ears and brain. A similar thing happens when you bang a drum and the vibration of the drum skin passes a repeating wave of compressions through the air, as shown in Figure 15.6.1.

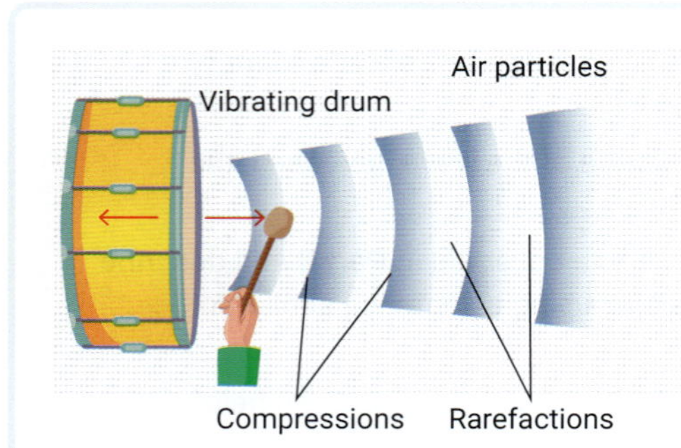

▲ **FIGURE 15.6.1** When a drum is struck, the vibrations of the skin cause the air to be bunched in compressions and stretched apart in rarefactions. The particles of air vibrate about a middle position, striking neighbouring particles. This results in sound energy being transferred as a longitudinal wave.

When you hear sounds under water, the kinetic energy is passed between water molecules to the air particles trapped in your ear canal and received by your ears in the same way. Because sound is transferred between particles, it requires a medium; both air and water provide these particles. Sound cannot travel through a vacuum because it contains no particles. Sound is a mechanical wave and transfers energy by vibration of particles.

Figure 15.6.2 shows the air particles in a silent room. A sound wave then travels through the air, and the air particles compress together and spread apart in a repeated pattern.

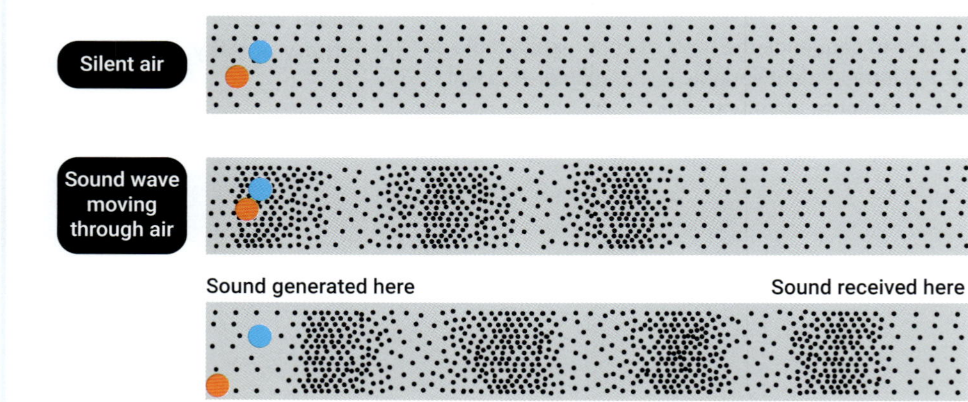

▲ **FIGURE 15.6.2** Air particles in a silent room and then when a sound wave moves through the air. The enlarged coloured dots are two air particles highlighted to show their movement as the wave passes.

The particles oscillate backwards and forwards as the sound waves move past them, forming a longitudinal mechanical wave. When this happens to the air particles inside your ears, you hear it as a sound.

When sound waves hit a hard boundary, the energy is not always absorbed by the material; instead, the sound waves may reflect off the surface and travel back in the opposite direction. This is what happens when you hear an echo (Figure 15.6.3).

▲ **FIGURE 15.6.3** Sound waves rebound from hard surfaces, creating echoes.

Features of sound waves

Sound waves can vary in amplitude, frequency and wavelength.

The amplitude of a sound wave describes how far the particles move either side of a middle position. This is referred to as the displacement of the particles. When more energy is being transferred, the particles are compressed and spread further and so the amplitude of the wave is greater.

Sound waves carrying more energy are louder than other waves. The amplitude of a sound wave is related to the **volume** of the sound wave. Figure 15.6.4 shows a low-volume and a high-volume sound wave. A low-volume sound wave could be if you whistled a high note very softly. A high-volume sound wave might be the wave when the same note is whistled very loudly.

volume
the measure of how loud a sound is; measured in decibels (dB)

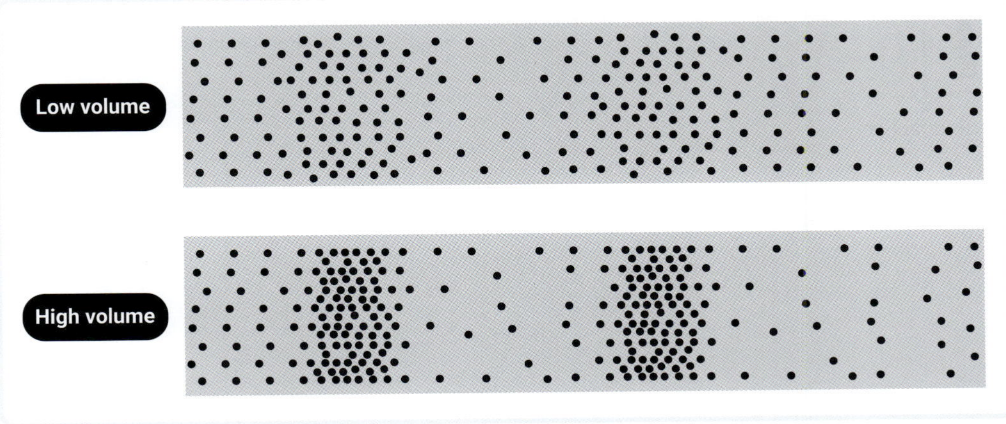

▲ **FIGURE 15.6.4** Two sound waves of the same pitch moving through air, one low volume and one high volume

pitch the degree of frequency (high or low) of a sound or note

The frequency of a sound wave matches the frequency of the vibrations that cause the sound wave to travel through the medium. A higher frequency sound is heard as a higher note or higher **pitch** (Figure 15.6.5).

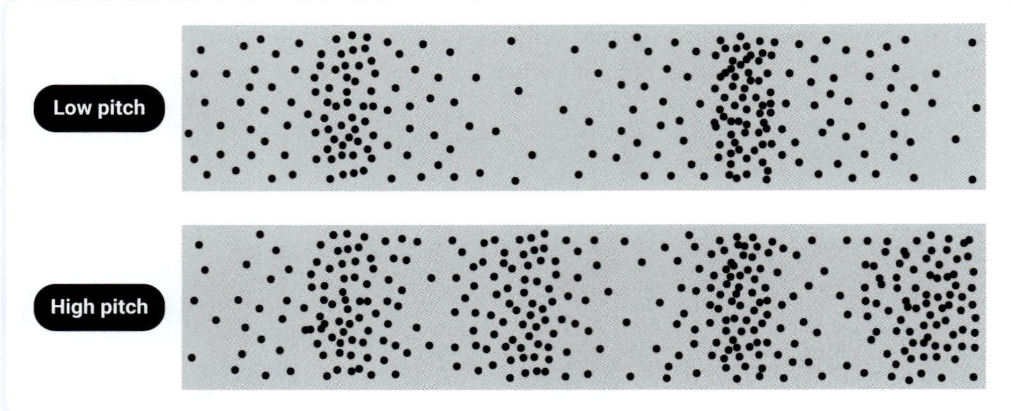

▲ **FIGURE 15.6.5** Two sound waves of the same volume moving through air, one low pitch and one high pitch

The wavelength of a sound wave can vary and will depend on how fast the wave is travelling and the frequency of the sound.

Visualising sound waves

Sound waves are often represented graphically; for example, as seen in the compression graphs in Figures 15.6.6 and 15.6.7.

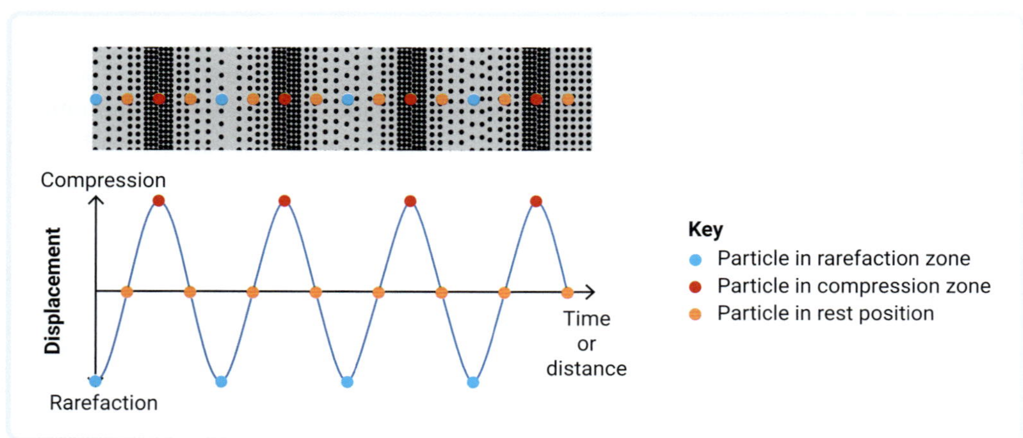

▲ **FIGURE 15.6.6** Converting a sound wave to a compression graph makes it easy to visualise and identify key features.

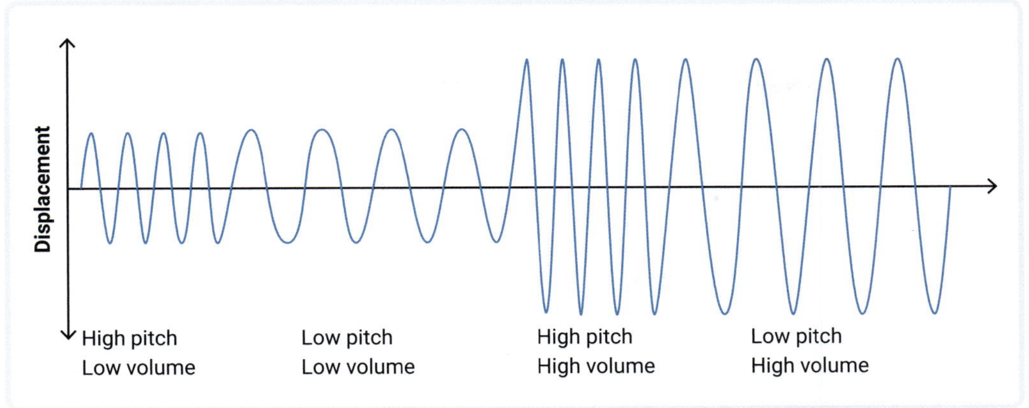

▲ **FIGURE 15.6.7** Sound waves of varying pitch (frequency) and volume (amplitude)

How we hear sound

The ear is made up of many smaller structures that allow humans to hear sounds. Figure 15.6.8 shows the various structures within the ear and how these structures allow hearing.

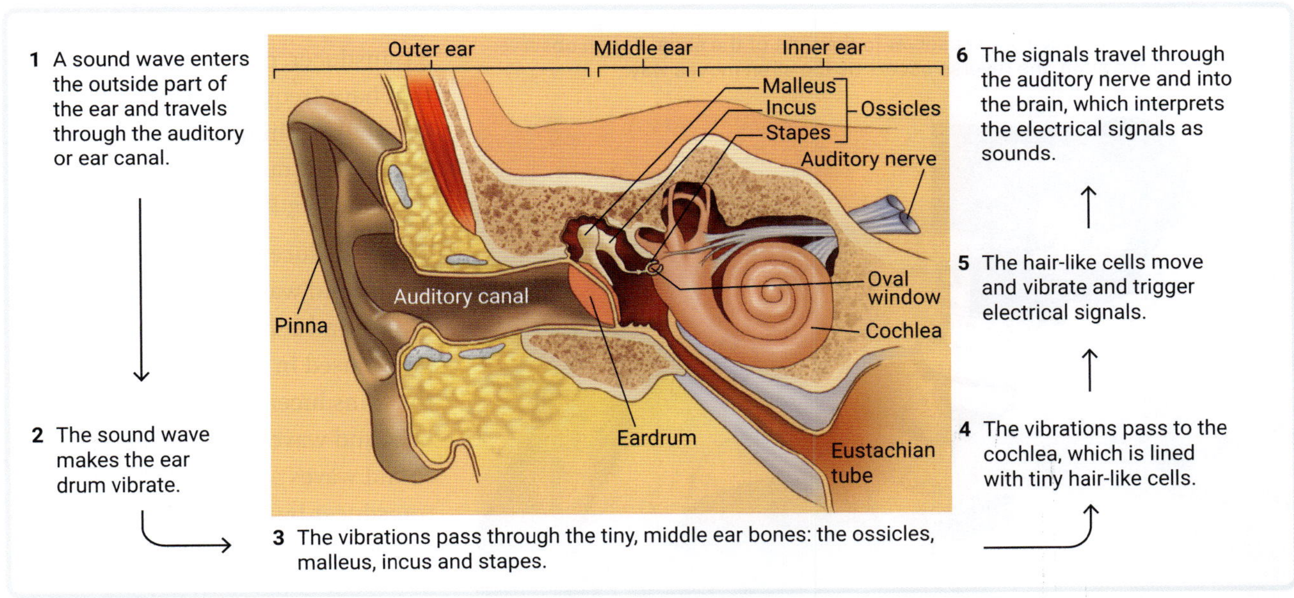

▲ **FIGURE 15.6.8** The structure of the human ear and the process of hearing

The Doppler effect

If an object that makes sound is moving, such as the siren on a speeding fire truck, the pitch of the sound you hear is affected. This is called the Doppler effect.

If the moving source of the sound is getting closer to you (such as an fire truck moving towards you), then the sound waves will be 'bunched up' or compressed. This means the wavelengths arriving at your ear are shortened, with a reduction in the time between compressions. As a result, you will hear a sound that is a higher pitch than the actual pitch of the sound being emitted.

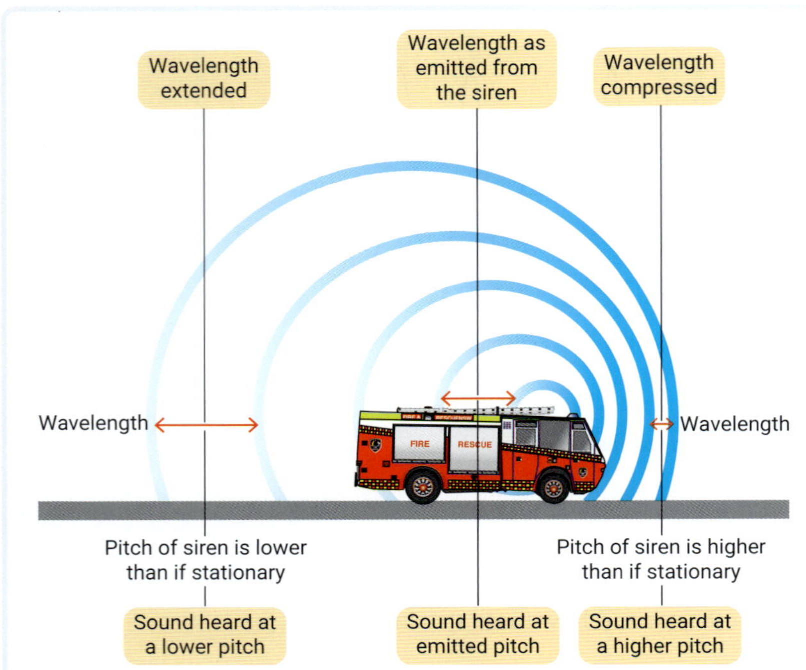

▲ **FIGURE 15.6.9** The Doppler effect for a siren moving from left to right. The siren 'chases' the waves it emits, causing a decrease in the wavelength of the sound waves measured by an observer to the right of the siren. For an observer to the left of the siren, the reverse occurs as the siren moves way.

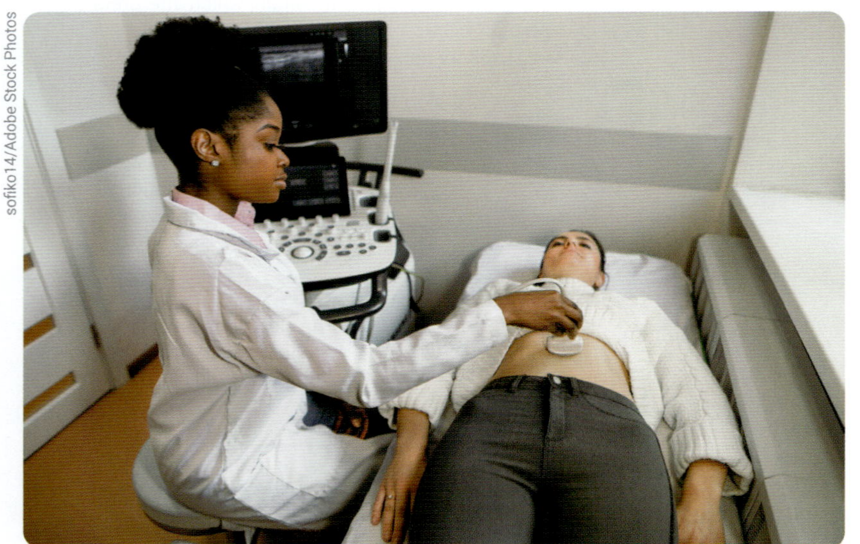

sofiko14/Adobe Stock Photos

▲ **FIGURE 15.6.10** An ultrasound can produce an image of internal organs and structures.

In contrast, if the source is moving away from you, the waves are 'stretched out' or extended. This means the wavelengths arriving at your ear are made longer, with more time between compressions. As a result, you will hear a sound that is a lower pitch than the actual pitch of the sound being emitted. Figure 15.6.9 shows an example of the Doppler effect.

Sound waves in medical diagnosis

Sound waves are used in medical diagnosis in a process called ultrasound. Diagnostic ultrasound is a non-invasive technique used to take images of the inside of the body. Ultrasound probes, known as a transducer, produce sound waves that have frequencies above the threshold of human hearing. Most ultrasound transducers are placed on the skin (Figure 15.6.10).

The transducer can both emit ultrasound waves and detect the ultrasound echoes reflected back. When used in an ultrasound scanner, the transducer sends out a beam of sound waves into the body. The sound waves are reflected back to the transducer by boundaries between tissues in the path of the wave, such as the boundary between soft tissue and bone. When the echoes hit the transducer, they generate electrical signals that are sent to the ultrasound scanner. The scanner calculates the distance from the transducer to the tissue boundary by measuring the speed of sound and the time it takes for each echo to return. These distances are used to generate two-dimensional (2D) images of tissues and organs. Ultrasound images are displayed in either 2D, 3D, or 4D (3D in motion) images.

The images produced by ultrasound help doctors by visualising changes or differences within the body's structures or organs. Doctors use ultrasound to diagnose injuries, study internal organs, detect diseases and observe the development of a foetus.

15.6

Observing sound

ACTIVITY

1 Put different amounts of water in three bottles.
2 Blow over the top of the bottles to produce a sound.
3 Write a conclusion about whether higher frequency sound waves are produced in fuller or emptier bottles. Give reasons for your answer.

15.6 LEARNING CHECK

1 a **Draw** a compression graph for a sound wave similar to the one in Figure 15.6.11, with an amplitude of 2 units and wavelength of 4 units.

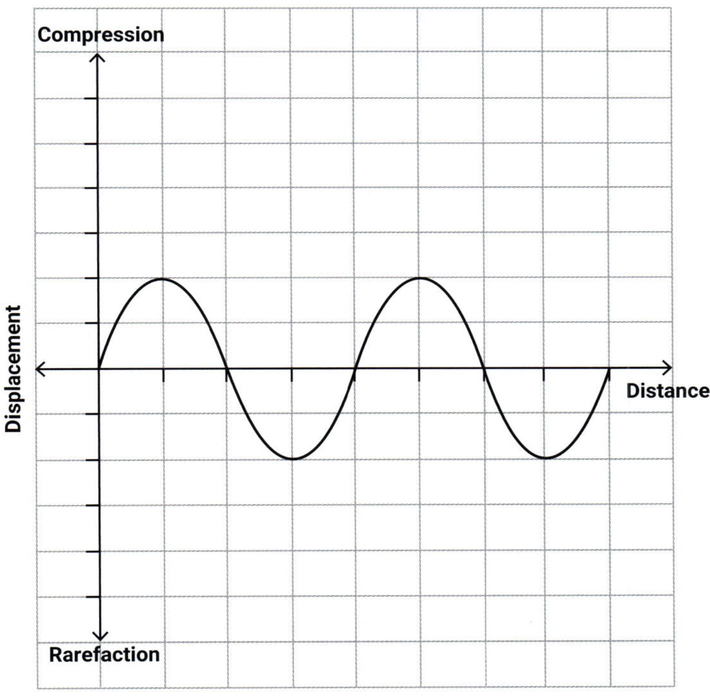

▲ **FIGURE 15.6.11** A sound wave with an amplitude of 2 units and wavelength of 4 units

b Using a second colour, **draw** a sound wave over the top that is twice as loud as the wave in part **a**.

c Using a third colour, **draw** a sound wave over the top that has a lower pitch than the wave in part **a**.

d Using a fourth colour, **draw** a sound wave that is half as loud and that has a higher pitch than the wave in part **a**.

2 As people get older, the tiny hair-like cells inside the cochlea can become damaged. **Infer** what this might mean for the hearing of people as they age.

WORKING SCIENTIFICALLY

15.7 Identifying random errors

SCIENCE SKILLS IN FOCUS

IN THIS MODULE, YOU WILL FOCUS ON LEARNING AND IMPROVING THESE SKILLS:

- using quantitative measurements and mathematical formula to calculate the speed of sound waves
- identifying sources of random errors.

What are random errors?

Random errors are errors that happen during an investigation that are normally due to chance. These types of errors are often caused by slight variations in how an instrument is used or due to fluctuations in the environment. We can classify random errors as observational or environmental.

Examples of random errors include changes in the air temperature, pressure and humidity inside a laboratory, measurements that fall between two scale markings on an analog scale or a digital scale that doesn't settle on a value, and slight variations in measurement equipment or technique.

How to identify random errors

Random errors can be hard to detect because they are not predictable. The best approach is to reduce the likelihood of these errors happening in the first place. You can do this by:

- identifying potential sources of random errors before you begin your investigation
- taking extra care when taking measurements
- double-checking your results table to avoid any 'typos'
- conducting multiple trials in your investigations and increasing the size of your datasets.

MEASURING THE SPEED OF SOUND

AIM

To use echoes to calculate the speed of sound

MATERIALS AND EQUIPMENT

- ☑ stopwatch
- ☑ large reflecting surface, preferably outdoors
- ☑ 2 wooden blocks
- ☑ tape measure or trundle wheel

PROCEDURE

1. Read the procedure in full before beginning and note any potential sources of random errors.
2. Work in pairs. Stand at a measured distance (s) of 50–100 m or more from a large reflecting surface (e.g. a wall or building).
3. Clap the wooden blocks together to make a loud sound – the louder the better! Listen for the echo from the reflecting surface.
4. Next, clap the wooden blocks regularly and at quite a fast rate. Adjust the clapping rate until the echo from the reflecting surface overlaps with the next clap.
5. Use the stopwatch to measure the time taken for 11 claps at this rate.

RESULTS

1. Record the distance to the reflecting surface.
2. Calculate the distance to the reflecting surface and back to the clapper ($2 \times s$).
3. Record the time taken for 11 claps at the clapping rate that has the claps coinciding with the echoes.

ANALYSIS

1. **Calculate** the time taken for one clap to travel to the reflective surface and back (t). Do this by dividing the time recorded in step 3 of the Results section by 10. (We divide by 10 because it takes 10-time intervals for 11 claps to coincide with the echoes.)

9780170491785

2 **Calculate** the speed of sound by dividing the total distance the sound travels by the time it takes to travel that distance.

3 **Calculate** the speed of sound, c:

speed of sound $= \frac{2 \times s}{t}$

4 What sources of random error did you **identify** for this investigation?

5 What changes could you make to this investigation to reduce the likelihood of random errors occurring?

CONCLUSION

1 Provide a realistic estimate of the speed of sound.

2 **Compare** your calculated speed of sound in air to the standard average value of 340 m/s.

EXTENSION

The speed of sound is different in different media (Table 15.7.1).

TABLE 15.7.1 The speed of sound in different media

Material	Speed of sound (m/s)
Rubber	60
Air at 20°C	343
Air at 40°C	355
Water	1481
Gold	3240
Glass	4540
Copper	4600
Aluminium	6320

1 Why do you think the speed of sound is different in different media?

2 Give one reason why sound travels faster in warmer air.

3 **Explain** why sound travels faster through:

   a water than air.

   b gold than water.

15.8 Light waves and the electromagnetic spectrum

BY THE END OF THIS MODULE, YOU WILL BE ABLE TO:

- ✓ recognise that visible light is a type of electromagnetic wave and each type has different wave features
- ✓ explain light phenomena and behaviour using the wave model, including how we see different colours.

Video activity
What is light?

Extra science investigation
Mixing light

Other resource
Worksheet: Electromagnetic spectrum

GET THINKING

You are probably already familiar with terms like gamma rays, X-rays, microwaves and radio waves, even if you didn't know they were all part of the same 'family' of waves as visible light. What features of waves can you use to explain the differences between these waves? For example, why are X-rays and gamma rays dangerous, but radio waves are not? Why do microwaves heat food but visible light doesn't?

The electromagnetic spectrum

electromagnetic radiation
energy that travels in electromagnetic waves

electromagnetic spectrum
the range of electromagnetic waves arranged in sequence from high-energy gamma rays to low-energy radio waves

Electromagnetic waves do not require particles to transfer energy and so can travel through a vacuum. Electromagnetic waves are sometimes referred to as **electromagnetic radiation**. Electromagnetic waves carry different amounts of energy, depending on their frequency and wavelength. Scientists have classified the types of radiation into seven main groups on the **electromagnetic spectrum**. Figure 15.8.1 shows the electromagnetic spectrum, including gamma radiation, X-ray radiation, ultraviolet (UV) radiation, visible light, infrared (IR) radiation, microwave radiation and radio wave radiation.

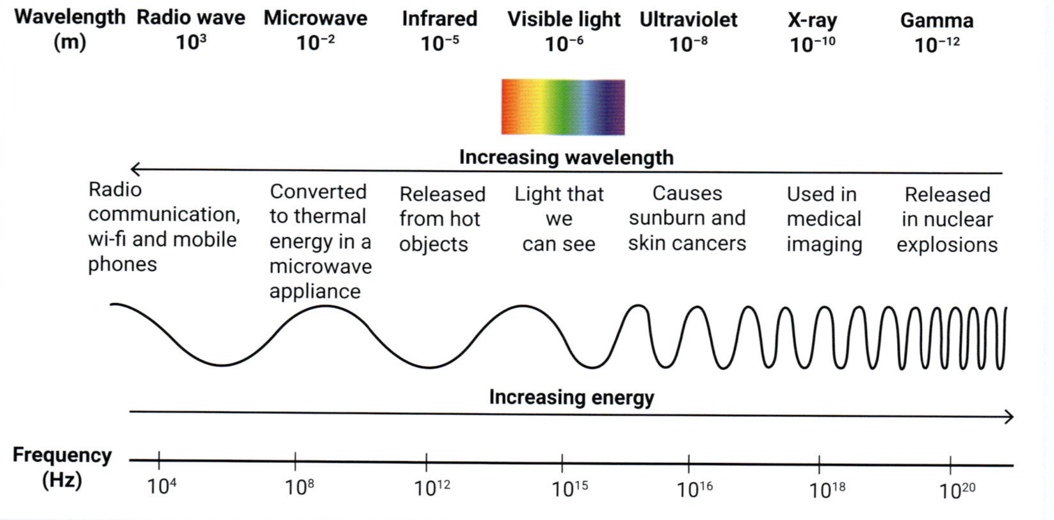

▶ **FIGURE 15.8.1** The different electromagnetic waves of the electromagnetic spectrum

The waves of the electromagnetic spectrum are an important part of our lives. We use light to see, infrared radiation from the Sun heats Earth, and we warm food with microwaves. Without electromagnetic waves, there would be no X-ray machines, wireless internet or mobile phones. Figure 15.8.1 provides some more information about each type of electromagnetic radiation.

9780170491785

A **spectrum** is a scale. The electromagnetic spectrum shows the specific frequency range and wavelength range of each type of radiation. This determines which group an electromagnetic wave belongs to. The spectrum also orders types of radiation by how much energy they carry. Radio waves have the lowest frequency and longest wavelength, and they carry the least amount of energy.

spectrum
a range or a scale; also, the different bands of colour that are visible when white light is refracted through a prism, as seen in rainbows

Gamma rays have the highest frequency and shortest wavelength, and they carry the most energy. The higher the frequency of electromagnetic radiation, the more energy carried by the wave.

Electromagnetic waves can have different frequencies and wavelengths, but they all have the same speed when travelling through the same medium. In a vacuum, the speed of electromagnetic radiation is 300 000 km/s or 3×10^8 m/s. This value is often referred to as the speed of light because light is one type of electromagnetic radiation.

Light: an electromagnetic wave

Visible light is the only part of the electromagnetic spectrum we can see. Humans cannot see radiation outside of the wavelength range of about 380–750 nm (nanometres) (3.8×10^{-7} m to 7.5×10^{-7} m). Other animals can detect different parts of the electromagnetic spectrum. For example, some insects and birds can see ultraviolet light.

Visible light can be many different colours depending on the wavelength. Scientists often refer to seven different colours of the visible light spectrum as ROYGBIV: red, orange, yellow, green, blue, indigo and violet. Violet light has the shortest wavelength and carries the most energy.

When our eyes receive all colours of the visible light spectrum at the same time, the light we see is white. The different colours in white light can be separated by passing it through a prism. Water droplets in the air can also act like a prism, splitting the white light from the Sun to form rainbows (Figure 15.8.2). Next time you see a rainbow, remember ROYGBIV and see if you can spot each colour!

Vilma.K/Shutterstock.com

▲ **FIGURE 15.8.2** A rainbow forms when white light passes through water droplets and is split into different colours.

Where does light come from?

Light is produced when electromagnetic energy is released from somewhere and has a wavelength in the visible light section of the spectrum. Sometimes when objects get extremely hot and the radiation has more energy than infrared heat radiation, light is released. Flames are an example. The Sun is constantly undergoing nuclear reactions that release large amount of radiation, including visible light, UV light and infrared radiation. Some living things can produce their own light through a process of bioluminescence (Figure 15.8.3).

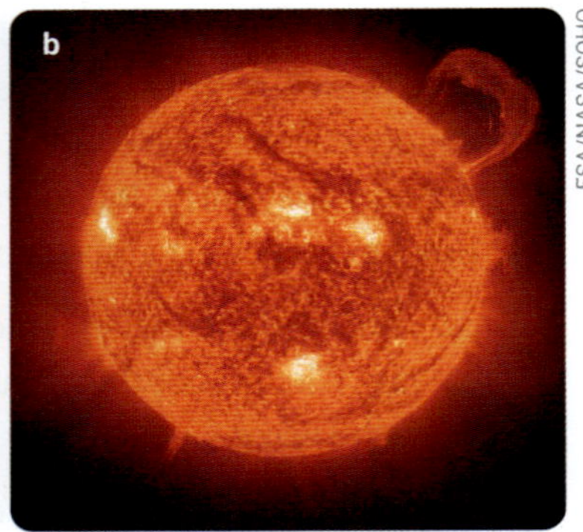

▲ **FIGURE 15.8.3** Light can be produced when energy is released such as (**a**) in a biological reaction (bioluminescence) or (**b**) in a nuclear reaction.

However, most light that we see does not come directly from its source. We can see objects because light reflects off them. When light (or any electromagnetic radiation) strikes a surface, several things can happen. The light can:

- be reflected off the surface
- be absorbed by the material of the surface and transformed into another energy form (usually thermal energy)
- be scattered by the surface
- pass into the material and be refracted.

These four behaviours of waves will be covered in more detail in Module 15.9.

reflection
the process of a wave bouncing off a surface and returning through the same medium

absorption
the process of taking in radiation and converting it to other types of energy

Even though light is a wave, it is often visualised as an arrow or a ray in diagrams to show the direction of motion of the light wave. The **reflection** and **absorption** of light is shown in Figure 15.8.4.

Mirrors are surfaces that reflect all wavelengths of light. Black surfaces absorb most wavelengths of light, and transparent objects transmit light through them. Other surfaces reflect and absorb different wavelengths of light. This is the reason we see different colours on different surfaces.

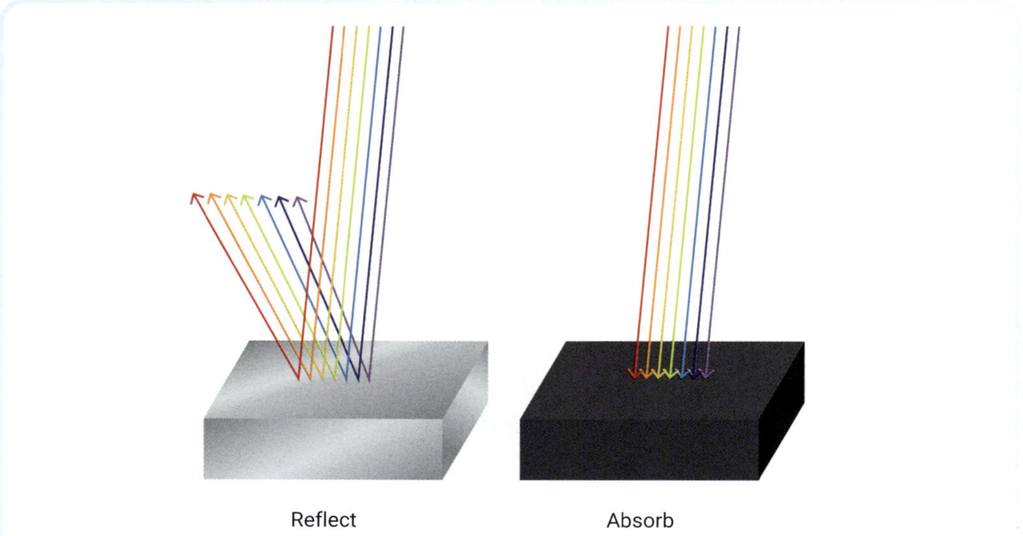

▲ **FIGURE 15.8.4** Reflection and absorption of light when it strikes a surface

Seeing colours

In a dark room where there is no light, everything is black. When a light is turned on or sunlight enters through a window, a red delicious apple in the room appears red to our eyes. White light containing all ROYGBIV colours shines onto the apple. Because of the chemicals in the apple skin, it absorbs orange, yellow, green, blue, indigo and violet (OYGBIV) light, but reflects red light off the surface (Figure 15.8.5). The reflected light travels to our eyes and our brain interprets the colour of the object. Black surfaces are 'colourless' because they absorb all colours of light.

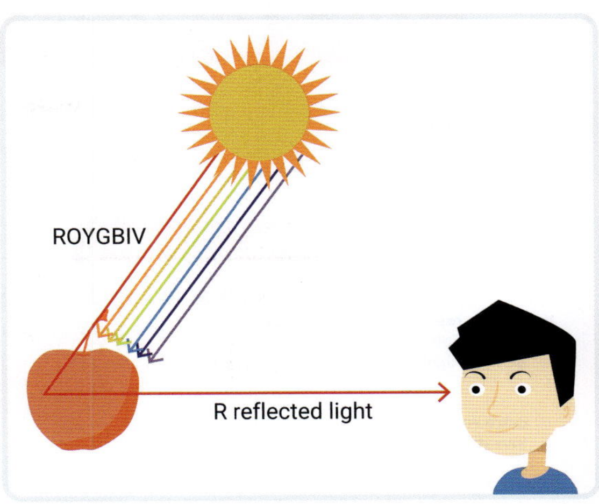

▲ **FIGURE 15.8.5** We see different colours when certain colours of light are reflected from surfaces to our eyes.

15.8 LEARNING CHECK

1 Copy and complete the following sentences about electromagnetic waves.
  - a Waves that have short wavelengths have ________ frequencies and carry ________ energy. An example of this is a ________ wave.
  - b Waves that have long wavelengths have ________ frequencies and carry ________ energy. An example of this is a ________ wave.
  - c Electromagnetic waves that are visible to the human eye are known as visible light. They have wavelengths longer than ________ light but shorter than ________ radiation.
  - d Compared to green light, blue light has a ________ frequency and ________ wavelength.

2 **Calculate** the frequency of ultraviolet light with a wavelength of 0.000 000 350 m.

3 Look at Figure 15.8.6 and answer the following questions.
  - a **Explain** why leaves appear green in the daylight.
  - b **Predict** what colour the leaves would appear if you shone a red light on them in a dark room. **Explain** your answer.

4 **Research** the types of electromagnetic waves and **summarise** the common uses of each in areas such as medicine, sanitation and communication.

IgorZh/Shutterstock.com

▲ **FIGURE 15.8.6** Why do leaves appear green?

15.9 The behaviour of light

BY THE END OF THIS MODULE, YOU WILL BE ABLE TO:

- ✓ describe how light waves reflect and refract
- ✓ describe the structure and function of the human eye and explain how we can see.

Interactive resource
Simulation: Bending light

Extra science investigation
Reflection and refraction

GET THINKING

Understanding the behaviour of light means that we can use light to help us do different things. Make a list of specific examples where we use light to help us. How many can you think of?

Reflection of light

When waves meet surfaces or boundaries between substances, they can reflect off the surface and travel back through the first medium. Reflection is the process of waves bouncing off a surface and returning. Water waves bounce off seawalls and sound waves reflect off surfaces as echoes. Light behaves like a wave and will reflect off certain surfaces, such as sunlight bouncing off a shiny surface, making it appear bright to your eyes. Scientists have used experiments to develop the law of reflection, which helps us to understand how reflection happens (Figure 15.9.1).

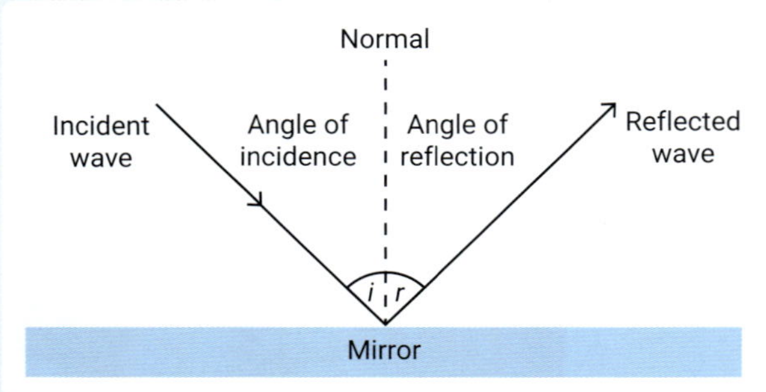

Law of reflection
When a wave of light reflects off a surface, the angle of incidence is equal to the angle of reflection

angle of incidence = angle of reflection
$i = r$

▲ **FIGURE 15.9.1** The law of reflection

incident wave
a wave of light that is approaching a surface

reflected wave
a wave of light that has been reflected off a surface

The **incident wave** is the wave that is approaching the surface. The **reflected wave** is the wave that has been reflected off the surface. The normal is an imaginary line that is perpendicular to the surface at the point where the incident wave meets the surface.

✩ ACTIVITY

Observing the law of reflection

Use a single-slit light box or a laser beam to **demonstrate** the law of reflection. Point it on an angle at a mirror and use a protractor to measure the incident angle and reflected angle. Do not look directly at a light or laser source.

Based on your measurements, **evaluate** whether the law of reflection has been demonstrated or not.

Refraction of light

When waves travel from one medium into another, their speed and direction can be changed. **Refraction** of light is when light rays bend as they pass into a new medium. In Figure 15.9.2, the light that is reflected off the part of the pencil submerged in the water follows a bent path to the eye compared to the light reflected from the dry part. This is because the light reflected from the lower part of the pencil changes direction as it moves from the water into the air.

refraction
the transmission of a wave from one medium into another with a resulting change of direction

Kuki Ladron de Guevara/Shutterstock.com

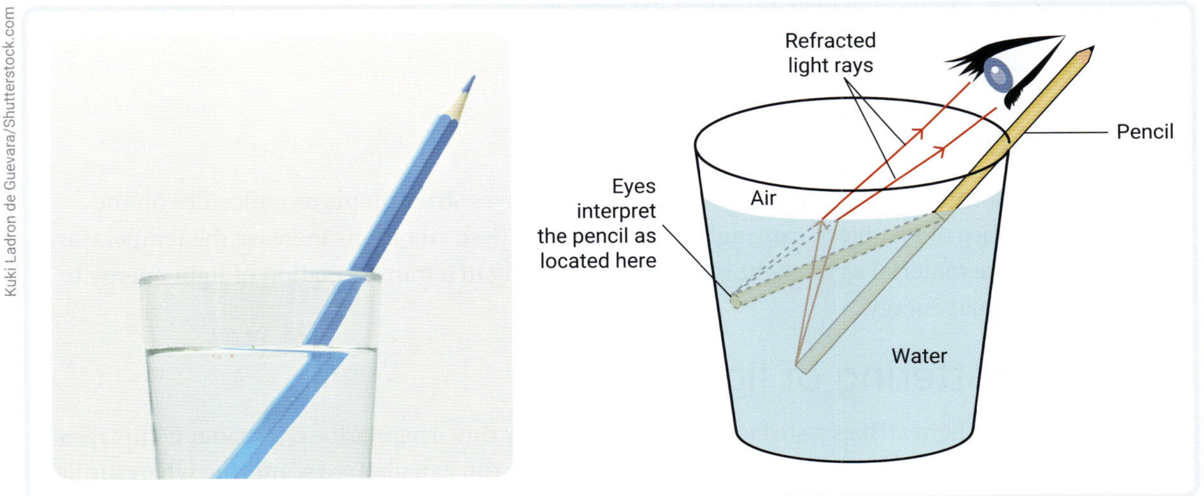

▲ **FIGURE 15.9.2** Light undergoes refraction and bends in new directions when passing from water to air.

The amount of refraction of light is determined by the change in optical density of the substance. When light travels from a less dense to a more dense medium, such as from air to water, the light bends towards the normal to the boundary. When light travels from a more dense to a less dense medium, the light bends away from the normal to the boundary. This phenomenon is easier to imagine by using the analogy of a car's wheels, as the car travels from the fast medium of the road to the slow medium of sand (Figure 15.9.3).

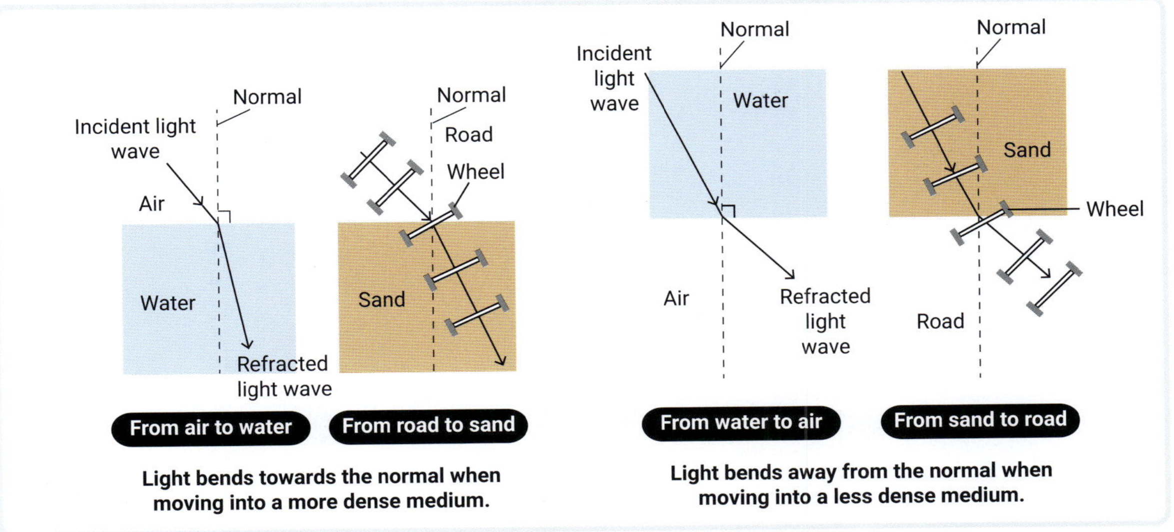

▲ **FIGURE 15.9.3** Light changes direction when entering a new medium on an angle in the same way that a car might change direction when entering sand.

☆ ACTIVITY

Observing refraction

1 Place a coin under a transparent container such as a jar, glass or beaker.
2 Observe the coin through the side of the glass.
3 Slowly pour water from a jug or bottle into the container while focusing on the coin.

Analysis

1 What did you **observe**?
2 How can you use your knowledge of refraction to **explain** your observations?
3 Review your response when you have reached the end of the chapter. Has your answer changed?

Absorption of light

Absorption of light happens when the light waves strike atoms or molecules on the surface of an object, causing them to vibrate. These vibrations increase the temperature of the material as the light is absorbed, resulting in a transformation of light energy to thermal energy.

Scattering of light

When light strikes a surface that features many tiny irregularities, it scatters. This results in the light travelling in a variety of directions. You can observe scattering when sunlight passes through clouds. When this happens, the light is scattered by tiny water droplets in the clouds, which reduces the light intensity that reaches you. Scattering is also responsible for making the sky appear blue during the day and red or orange at sunrise and sunset.

Light waves and seeing

We see an object because light is reflected from the object's surface and enters our eyes. The light travels through the eye and is received by cells that send signals to the brain where we interpret the information as images. Figure 15.9.4 shows how light travels through the eye.

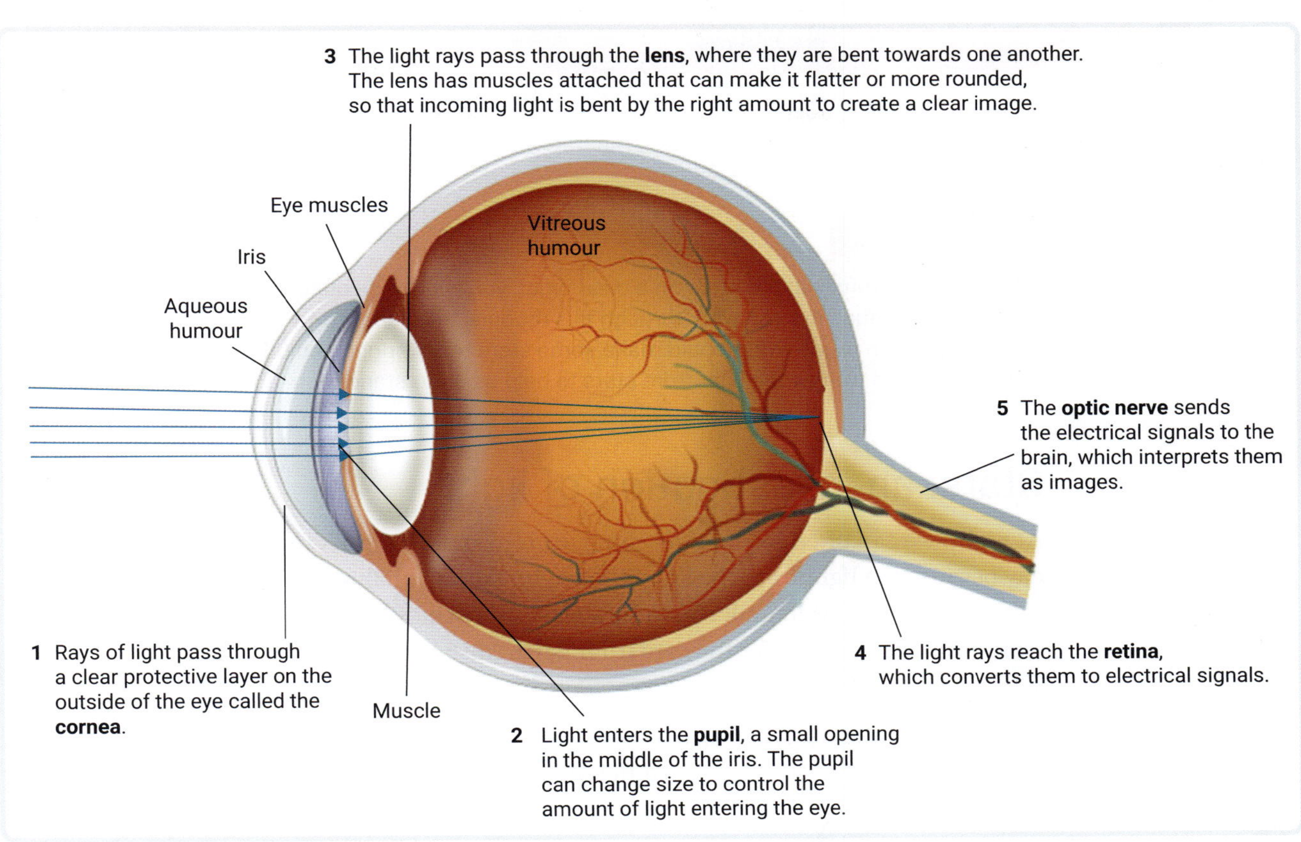

▲ **FIGURE 15.9.4** The structure of the human eye and the pathway of light through the eye

15.9 LEARNING CHECK

1 **Draw** a diagram showing the law of reflection. On your diagram, **identify** the:
   a incident wave.
   b reflected wave.
   c normal.
   d angle of incidence.
   e angle of reflection.

2 **Define** refraction.

3 **Construct** a diagram of the refraction that occurs when a wave of light travels out of water (more dense) and into air (less dense).

4 **Describe** the function of the retina in the eye.

15.10 Starlight

BY THE END OF THIS MODULE, YOU WILL BE ABLE TO:

✓ explain how scientists use light spectra and the brightness of a star to identify its composition, movement, surface temperature and distance from Earth.

GET THINKING

If you look carefully at the stars on a clear night, you will notice quite a bit of variety in the light they produce. As you work through this module, think about why the brightness of stars varies so much. What could make some stars really bright, and others appear quite dull? How can we use the light from stars to help us understand more about them?

Stars: an investigation of light

At night, we can often see light reaching us from stars. We can't see this light during the day because the light from the Sun is so bright. We can analyse the light that reaches us from stars to help us to understand how stars move (movement), what they are made of (composition) and how far they are from Earth (distance).

Analysing a star's light can involve examining the brightness of the light that comes from individual stars. It can also involve analysing the star's spectrum, which is all the different wavelengths that make up the light reaching us from the star. This is done using an advanced spectrometer. The light strikes a grating to break the light received into its component wavelengths, producing a spectrum on the detector (Figure 15.10.1).

Source
Mirrors
Grating
Detector

▲ **FIGURE 15.10.1** A spectrometer

Composition

Scientists can use a spectrometer on Earth or on a satellite to analyse the composition of stars. This is because every element (such as hydrogen, helium and lithium) produces (emits) a unique light spectrum, like a fingerprint. The emission spectrum of hydrogen is shown in Figure 15.10.2. You can see that when a full rainbow of coloured light (a continuous spectrum) is shone through hydrogen, the hydrogen absorbs light of exactly the same wavelengths. Using a very high-powered telescope, such as the Hubble Space Telescope, light emitted from a distant star is collected to enable scientists to compare the star's light

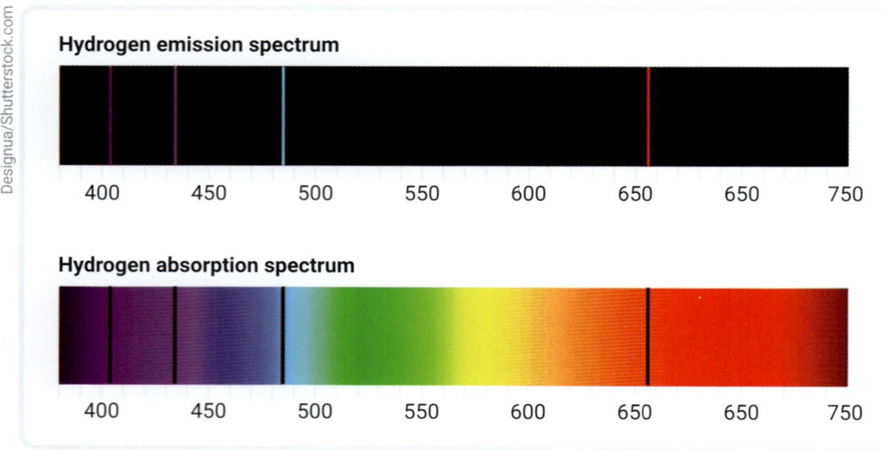

Designua/Shutterstock.com

▲ **FIGURE 15.10.2** The absorption and emission spectra of hydrogen

absorption spectrum to a known element's **absorption spectrum** to determine which elements are present in the star.

absorption spectrum the dark lines characteristic of an element or compound in a continuous spectrum

redshift the shift of a star's spectrum towards the red end as the star moves away from Earth

blueshift the shift of a star's spectrum towards the blue end as the star moves toward Earth

15.10

Movement

The light our telescopes receive is affected by the movement of stars. This is caused by the Doppler effect (explained in relation to sound in Module 15.6). The effect we observe depends on which direction a star is moving in relation to us here on Earth. Depending on whether a star is moving towards us or away from us, the wavelength of light we observe can shift slightly.

Visible light has wavelengths from 400 nm (violet light) up 700 nm (red light). If we are observing light from a star that is moving away from us, the waves will be 'stretched' to a longer wavelength. This results in the dark lines on the absorption spectrum moving towards the red end of the spectrum. This is called a **redshift**. If we are observing a star moving towards us, the waves will be 'shortened'. This will result in the dark lines on the absorption spectrum moving towards the blue end of the spectrum. This is called a **blueshift**. Scientists compare the amount that the lines move to a laboratory absorption spectrum (Figure 15.10.3). This indicates how fast the star is moving towards or away from us.

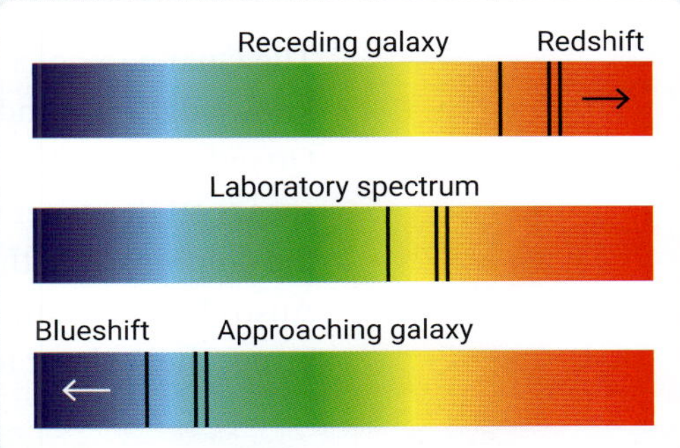

▲ **FIGURE 15.10.3** Comparing a redshift and blueshift. The arrows show the redshift of the absorption line spectrum from a light source moving away from Earth and the blueshift of a light source moving towards Earth. The central spectrum is an unshifted absorption line spectrum from a light source that is not moving.

Temperature

All hot objects, including stars, radiate energy. The amount of radiated energy at each wavelength can be used to determine the surface temperature of the object. To some extent, we can perceive the differences in radiated energy at each wavelength as colour. Therefore, the colour of a star gives an estimate of its surface temperature. More sophisticated equipment can give a more exact value. The hottest stars are blue, and stars range in temperature downwards through white, yellow and orange, and finally to red stars, which have the lowest surface temperature.

Distance

There are many ways to measure the distance between stars and Earth. The simplest way is to observe the star's brightness as seen from Earth and then compare this to a star's total energy output, which can be determined if the surface temperature and surface area are known.

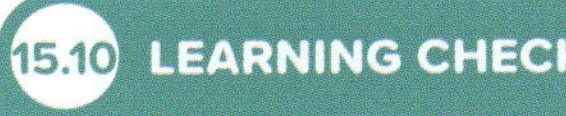

15.10 LEARNING CHECK

1 **State** what aspect of a star's appearance can be used to approximate its surface temperature.
2 **Outline** why the brightness of stars varies so much when we view them from Earth.
3 **Describe** how we use light to **determine** the composition of stars.
4 **Explain** how we can use the light from a star to **determine** how it is moving relative to Earth.

15.11 Traditional musical, hunting and communication instruments

ABORIGINAL & TORRES STRAIT ISLANDER SCIENCE CONTEXTS

IN THIS MODULE, YOU WILL:

✓ investigate the materials used to transfer sound energy in Aboriginal and Torres Strait Islander Peoples' instruments.

Aboriginal and Torres Strait Islander Peoples' instruments

Sound is a wave that is transmitted from one location to another through a medium, such as air or water. For many thousands of years, Aboriginal and Torres Strait Islander Peoples have used scientific understanding of the transfer of sound waves for hunting, communication and music.

The Quandamooka Peoples of Minjerribah (south-east Queensland) applied the knowledge of sound travelling in water to hunt fish in partnership with dolphins. Quandamooka Peoples called the dolphins by making a specific sound by slapping spears on the water surface and digging them into the sand. The sound waves alerted the dolphins, and the dolphins herded fish into waiting fishing nets. In other parts of Australia, Aboriginal Peoples clapped stones together under water. This startles fish and brings them closer to the surface for easier capture.

istock.com/iStock Signature

© Joe Sambono

▲ **FIGURE 15.11.1** (a) A bottlenose dolphin; (b) a dolphin place marker designed by artist Belinda Close in Pulan (Amity Point), Queensland, represents Quandamooka Peoples.

Aboriginal and Torres Strait Islander Peoples use many instruments to generate sound waves for communication and music. The sound generated by any instrument depends on the materials used in its construction.

Idiophones are instruments that generate sound waves by vibrating themselves, without the use of strings or membranes. Instruments that are struck on each other, such as boomerangs or clapsticks, are examples of Aboriginal and Torres Strait Islander Peoples idiophones. The sound produced depends on the type of wood used to construct the instrument, the length of the instrument and the position at which they are struck together.

Rattles and shakers are also idiophones. The sound waves are generated indirectly by the operator making two materials strike together. Shakers or rattles made from *kulaps* are produced by many of the Peoples of the Torres Strait Islands (Figure 15.11.2). *Kulaps* are the hard, brown shells of seeds of the matchbox bean vine. The shells are cut in half and attached to a length of twine to make the rattle.

Membranophones are instruments that generate sound waves through the vibration of a stretched membrane. Sound waves are produced when the stretched membrane is struck by hand or a drumstick. Aboriginal and Torres Strait Islander Peoples across the continent have used skin from animals, including goanna, kangaroo, possum and lizards, to make drums. For example, the Tjungundji and Yupangathi Peoples from the Cape York Peninsula in Queensland make a drum by stretching goanna skin across a hollow pandanus stem.

© Joe Sambono

▲ **FIGURE 15.11.2** A rattle made from *kulaps*, the seeds of a matchbox bean vine

Making a drum

☆ ACTIVITY 1

Aboriginal and Torres Strait Islander Peoples select particular materials to generate a desired sound. You will make a musical instrument and investigate how the materials used affect the sound produced.

Materials and equipment

- plastic bowl
- waxed paper
- soft cloth (e.g. dishwashing cloth or damp chamois)
- stretchy fabric (e.g. Lycra)
- rubber bands, tape or zip ties

Procedure

1. Cover the plastic bowl with waxed paper and fix it in place with rubber bands, tape or zip ties.
2. Tap the paper with an implement or your hand.
3. Record the sound wave produced using an app or a website such as the one at the weblink.
4. Change the material on your drum.
5. Record the sound waves again.
6. Repeat with a third material.

Weblink
Virtual oscilloscope

ACTIVITY 2

Making a shaker

Materials and equipment

- empty plastic bottle with a lid
- uncooked rice
- uncooked popcorn
- small pebbles

Procedure

1 Quarter fill the bottle with uncooked rice and seal it with the lid.
2 Shake the bottle to produce sound waves.
3 Record the sound wave produced using an app or the same weblink provided in activity 1.

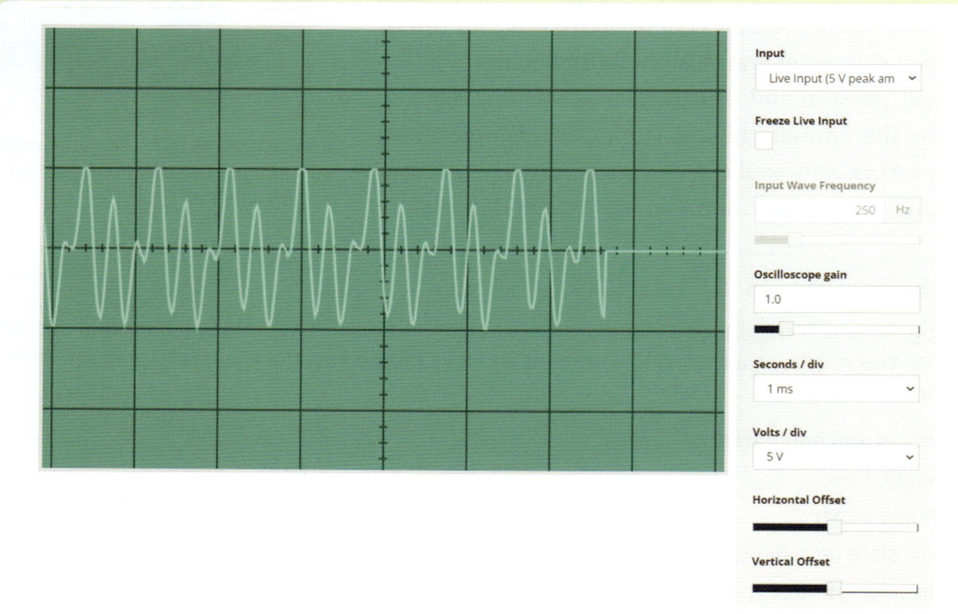

▲ **FIGURE 15.11.3** A sound wave recorded using a virtual oscilloscope

4 Change the material inside the bottle, making sure to use the same amount.
5 Shake the bottle to produce sound waves.
6 Record the sound waves again.
7 Repeat with a third material.

Analysis

1 How did the material used affect the sound waves produced?
2 Was this a fair test? How did you control the variables?
3 How would the sound waves be affected by the amount of material in the shaker?
4 How would the sound waves be affected by the thickness of the material used to make the drum in Activity 1?
5 How would the sound waves be affected by the tightness of the material stretched over the drum in Activity 1?

You can **test** these other variables to see how they affect the sound waves produced.

academo.org

9780170491785

SCIENCE IN CONTEXT

15.12 Optical fibres

BY THE END OF THIS MODULE, YOU WILL BE ABLE TO:

✓ explain how optical fibres can be used for communication.

Communication over a distance

Human eyes and ears can only see and hear over limited distances, so we rely on technology to communicate over long distances. Over time, telecommunication technology has evolved and become more effective. Early technologies transmitted signals through electrical wires or via radio waves. However, both of these methods had limitations:

- only small amounts of information could be transmitted at one time
- the range of communication was limited
- communication could be unclear due to interference from electrical sources such as thunderstorms
- signals could be weak.

Today, there are many different technologies, such as satellites and wireless 4G and 5G networks that allow us to communicate over long distances without these issues. Another important telecommunication technology is networks of optical fibres (or fibre optics), which rely on the reflection of light to work.

DATA SCIENCE 

Learn more about how data scientists rely on telecommunications for big datasets in **Module 2.5.**

Total internal reflection

As seen in Module 15.9, refraction occurs when light travelling in one transparent medium meets a boundary with another transparent medium. Recall that if the light moves from a denser medium (one with a higher refractive index, medium 1) to a less dense medium (one with a lower refractive index, medium 2), the light bends away from the normal, as shown in Figure 15.12.1a. We can express this as θ_2 (the angle of the reflected wave from the normal) being greater than θ_1 (the angle of the incident wave from the normal).

If an incident wave travels towards a less dense medium (medium 2) at a large enough angle, it won't pass through medium 2. Instead, it will travel along the boundary between the two mediums (Figure 15.12.1b). The angle of the incident wave from the normal at which this happens is called the critical angle or θ_c.

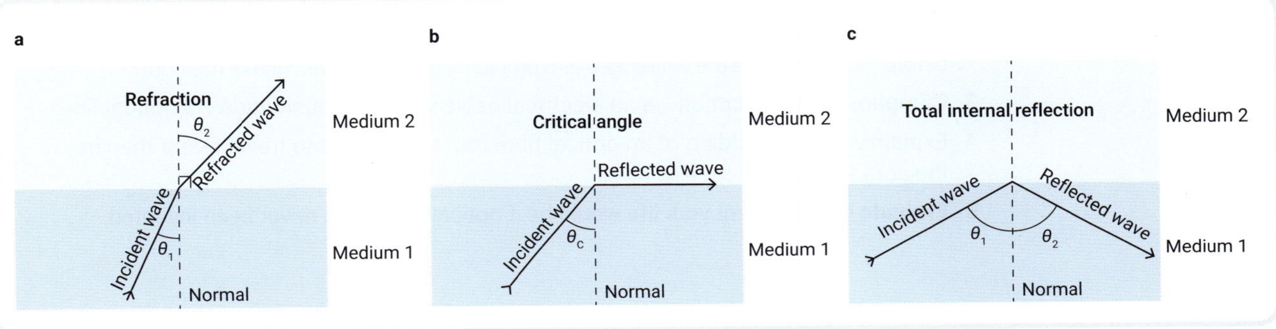

▲ **FIGURE 15.12.1** (a) Refraction causes this incident wave to bend away from the normal. (b) At the critical angle, the wave bends and travels along the boundary between the two mediums. (c) At angles greater than the critical angle, the wave goes through total internal reflection.

If the angle of the incident wave becomes greater than the critical angle, the light will reflect off the boundary back into medium 1 (Figure 15.12.1c). This is called total internal reflection.

How optical fibres work

One of the most significant applications of total internal reflection is the optical fibre. Optical fibres are made from a transparent material that causes light to travel slightly slower than a surrounding cladding material, as seen in Figure 15.12.2a.

When a ray of light passes into the fibre, it undergoes total internal reflection all the way to the other end of the fibre (Figure 15.12.2b). Australia is connected to places like New Zealand, Hawaii and Singapore by these fibres. The light can travel a very long way at very high speed. Light in optical fibres does not decrease in strength and does not suffer from interference. Also, vast quantities of information can be conveyed simultaneously along very thin fibres. As a result, optical fibre is a powerful communication technology that has completely revolutionised how we communicate.

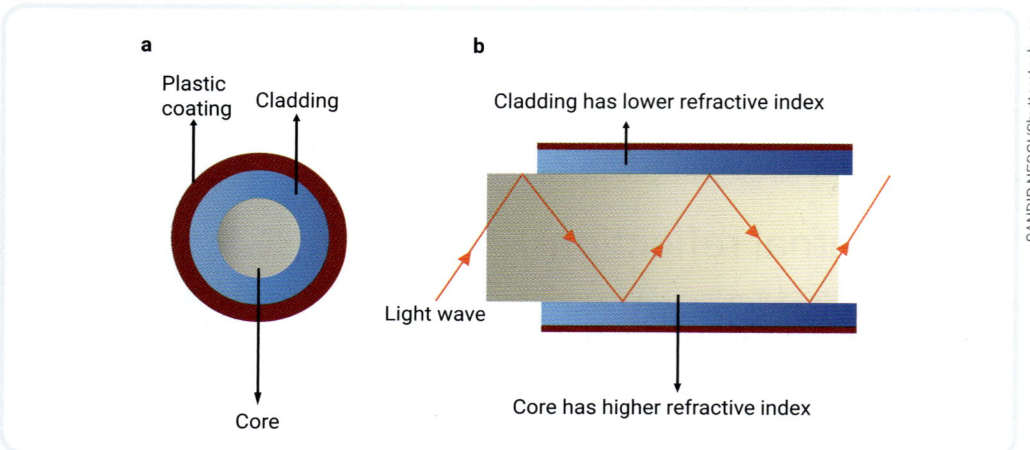

▲ **FIGURE 15.12.2** **(a) Optical fibres consist of a core of glass or plastic, surrounded by a plastic cladding and coating. (b) Light is internally reflected along the core of the optical fibre.**

15.12 LEARNING CHECK

1. **List** the ways that humans communicate with each other over distance. **Identify** all the communication techniques in your list that rely on light or sound waves.
2. **Research** the ways that important information (for example, a legal contract) was communicated before 1970. **Describe** the challenges and problems associated with these methods.
3. **Compare** communication via an electrical cable with communication via optical fibre.
4. **Explain** why the cladding of an optical fibre must allow light to travel faster than in the core.
5. **Evaluate** how different your life would be if optical fibres had never been invented.

9780170491785

15 REVIEW

REMEMBERING

1 Imagine this image is a photograph of a transverse wave (moving from left to right) at a specific moment in time. Copy the wave diagram and label its key parts.

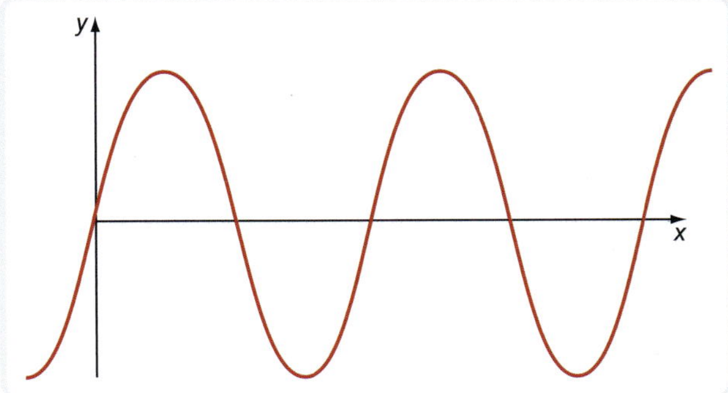

2 **Draw** a diagram of a longitudinal wave to show regions of rarefaction and compression.

3 **Describe** the transfer of sound energy through a medium.

4 Copy and **annotate** the following diagram of the human eye to show how light travels through the eye.

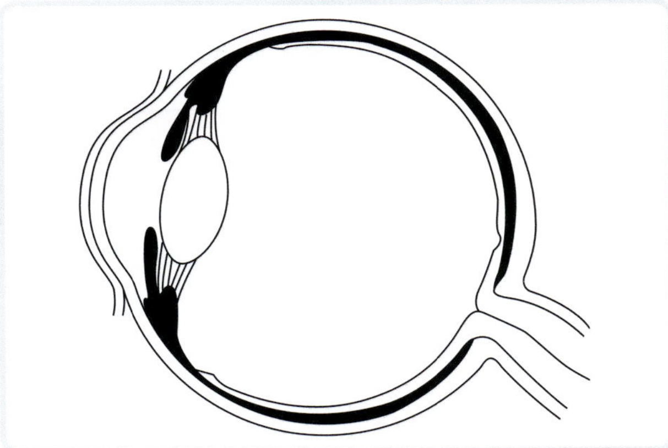

UNDERSTANDING

5 **Describe** what happens to sound energy when it meets the ear drum in the human ear.

6 **Research** and **summarise** in terms of waves how a microwave oven heats food.

7 **Describe** the path of light waves through the human eye by listing, in order, the structures that they travel through.

APPLYING

8 A longitudinal wave has a compression zone every 0.005 m, and 20 compression zones pass by a particular point every second. **Calculate** the speed of the longitudinal wave.

9 **Explain** in terms of closeness of particles why sound travels faster in water than it does in air.

10 During an accident, the malleus inside your ear is broken and no longer functions. **Predict** the possible consequences of this.

11 Complete this sentence: When light travels from air to water, it bends _______ the normal.

ANALYSING

12 Use the wave graph to answer the following questions.

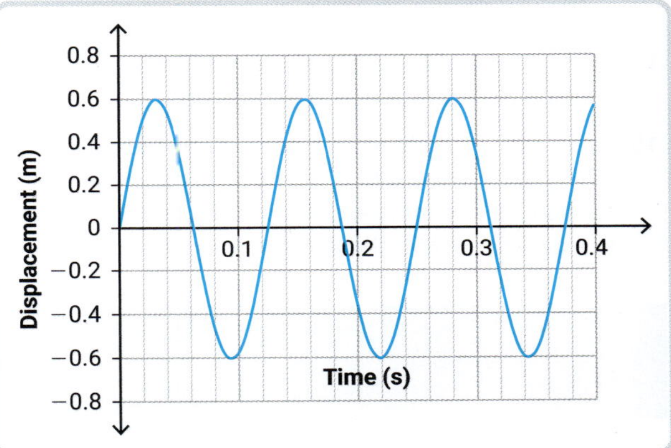

- **a** **Identify** the period of the wave.
- **b** **Calculate** the frequency of the wave.
- **c** **Explain** why the wavelength cannot be determined for this wave.

13 Infrared cameras are used by emergency rescue workers looking for lost bushwalkers or skiers. **Explain** why these cameras are most effective in winter and at night when looking for lost people.

14 **Explain** how the Quandamooka Peoples' traditional method of hunting fish relates to their knowledge of sound.

15 If you wanted to **model** a low-volume high-pitched sound wave using a slinky, what would you have to do to the slinky? **Justify** your reasoning.

EVALUATING

16 Cataracts occur when the eye lens becomes cloudy or opaque and this can be caused by excessive UV radiation. **Discuss** whether fashion sunglasses without UV protection should be sold in Australia.

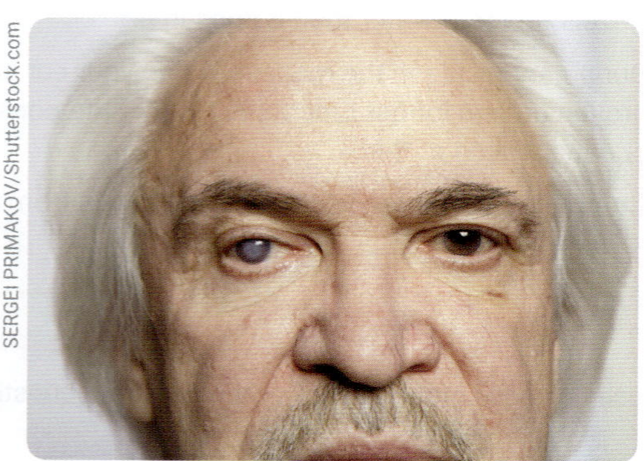
SERGEI PRIMAKOV/Shutterstock.com

17 **Discuss** why physicists use the light from stars when seeking to **determine** their properties.

18 Scientists use both the wave model and the particle model to **describe** light energy and its movement. **Assess** the strengths and limitations of using models to **explain** scientific phenomena, using this example.

CREATING

19 **Explain** why astronomers observing the spectra from stars can only make a best guess at the star's speed, composition, temperature and distance from Earth.

20 **Create** a generalised flow chart for the transfer of energy through the eyes and ears. **Annotate** the flow chart with specific examples within the eye and the ear.

21 **Draw** a picture of a cup of tea sitting on a bench beside an open refrigerator. Imagine your drawing is an infrared photo where the infrared radiation from each object has been converted to colours of the rainbow where the blue represents low-energy radiation and red represents high-energy radiation, as shown in the colour spectrum below.

Colour in your drawing as it would appear in an infrared photo. You can add other household or kitchen items if you wish. Some suggestions are:

- a freshly boiled kettle
- a cat lying on the floor
- bread sitting on the bench
- snow falling outside the window.

Low-energy radiation

High-energy radiation

9780170491785

SCIENCE IN DEPTH STUDY

1 Connect what you've learned

In this chapter, you've learned about scientific models for energy transfer, waves, sound, light and heat. Create a mind map to show how the main ideas you have learned about are connected.

2 Check your thinking

Consider the photo of acoustic highway barriers at the beginning of the chapter and the image of the person on this page. Considering what you now know about sound, hearing and the behaviour of waves, compare the two noise protection methods and explain how they are similar and how they are different.

3 Get into action

For each of the sounds you identified at the start of the chapter as having negative impacts on you, other people or animals:

- explain why the sound is harmful in this context
- suggest one way in which the negative impact could be reduced. It might be a strategy to reduce the noise produced or the effect of the noise
- describe the benefits of the strategy you are suggesting.

4 Communicate

Consider one environment where noise pollution might be a problem. Create a flyer that could be posted in this environment to encourage people to reduce the amount of noise created in this place. Include information about why the noise is a problem and how it could be reduced.

16 Motion

SCIENCE IN DEPTH

NASA/Joel Kowsky

▲ **FIGURE 16.0.1** The launch of the Mars 2020 spacecraft, with the Perseverance rover aboard, 30 July 2020

With a mission to search for extraterrestrial life and another habitable planet, humans have been launching rockets and rovers in an attempt to explore Mars since the 1960s. The first exploration device to reach Mars landed in 1971. As of 2021, there are two United States rovers and one Chinese rover operational on Mars, surveying and studying its surface and atmosphere. NASA's most recent landing was the rover Perseverance, which has one key objective (of many) to search for signs of ancient microbial life. But how did they all get there? What is the shortest distance between Earth and Mars? How long did it take to reach Mars? How fast did they travel? How were these speeds achieved? Motion in outer space can be explained by the same principles of motion that apply here on Earth.

- **What do you already know about rockets and how they travel?**
- **How does a rocket's motion change when it is close to Earth compared to when it is far away?**
- **How many similarities and differences can you think of between a real rocket and a toy rocket launched from the ground with a chemical reaction?**

 **DIVE INTO SCIENCE!**

At the end of this chapter, you can complete Science in Depth Study #16. You can use the information you learn in this chapter to complete the project.

Assessments
- Prior knowledge quiz
- Chapter review questions
- End-of-chapter test
- Depth study: Research project and webpage

Videos
- Science skills in a minute: Linear data **(16.5)**; Validating scientific work **(16.9)**
- Video activities: Newton's laws of motion **(16.6)**; Newton's second law of motion **(16.7)**; Science of golf **(16.12)**; Female crash test dummies **(16.14)**

Science skills resources
- Science skills in practice: Analysing linear data **(16.5)**; Validating scientific work **(16.9)**
- Extra science investigations: Using ticker timers **(16.4)**; Investigating Newton's second and third laws of motion **(16.7)**; Test Newton's second law **(16.7)**; Motion in traffic **(16.12)**; Stopping distance **(16.12)**; Crumple zones **(16.14)**

Interactive and other resources
- Simulation: Forces and motion **(16.6)**
- Label: Speed–time graphs **(16.3)**
- Quizzes: Distance **(16.1)**; Calculating speed **(16.2)**
- Activity sheets: Head-on smash at 60 km/h **(16.12)**
- Worksheets: Converting units **(16.1)**; Measuring distances **(16.1)**; Newton's second law **(16.7)**

Nelson MindTap

To access resources above, visit **cengage.com.au/nelsonmindtap**

16.1 Distance travelled

BY THE END OF THIS MODULE, YOU WILL BE ABLE TO:

✓ understand the measurement and calculation of distance travelled by moving bodies
✓ construct and interpret distance–time graphs to represent the motion of moving bodies.

Quiz
Distance

Other resources
Worksheets:
Converting units

Measuring distances

GET THINKING

Flip through this module and make a list of the headings. Add these headings to a mind map and draw lines that link terms where you think there is a connection. Annotate each of the links with a description of why you think these terms are connected.

SI units

In science, especially physics, there are many different types of units in which different quantities can be measured. This can be confusing at times, especially when you have to change units to make them compatible with one another in formulas. The International System of Units, also known as SI, is a globally accepted list of measurement units that set a standardised unit for all types of measurements. Table 16.1.1 shows the SI units for some commonly used measurements, many of which we will use in this chapter about motion.

▼ **TABLE 16.1.1** The SI units for some commonly used measurements

Measurement	SI unit name	SI unit symbol
Length or distance	metre	m
Time	second	s
Temperature	kelvin	K
Mass	kilogram	kg
Speed	metres per second	m s^{-1} or m/s
Acceleration	metres per second squared	ms^{-2} or m/s^2
Force	newton	N
Energy	joule	J

body
an object, person or thing that has mass

position
the location of a body

distance
the length of the pathway taken by a moving body

DPPI Media/Alamy Stock Photo

▲ **FIGURE 16.1.1** In most races, the person who completes the same distance in the shortest amount of time is the winner.

Distance as a quantity

In physics, we often refer to an object or thing as a **body**. A body can be anything made of matter and includes living things like humans and large objects like cars. When a body is moving or 'in motion', the **position** or location of the body changes. When a ball is thrown, the position changes from someone's hand to different points in the air to the landing site. The body will travel a certain path to change its position and the length of this path is known as the **distance** travelled by the body. Distance also features prominently in many fields of athletics (Figure 16.1.1).

9780170491785

The symbol used for distance in physics is usually a lowercase *d*. The SI unit for distance is metres (m) but distance can often be given in millimetres (mm) or centimetres (cm) for small distances, or kilometres (km) for bigger distances. Some conversions for distance measurements are shown in Figure 16.1.2.

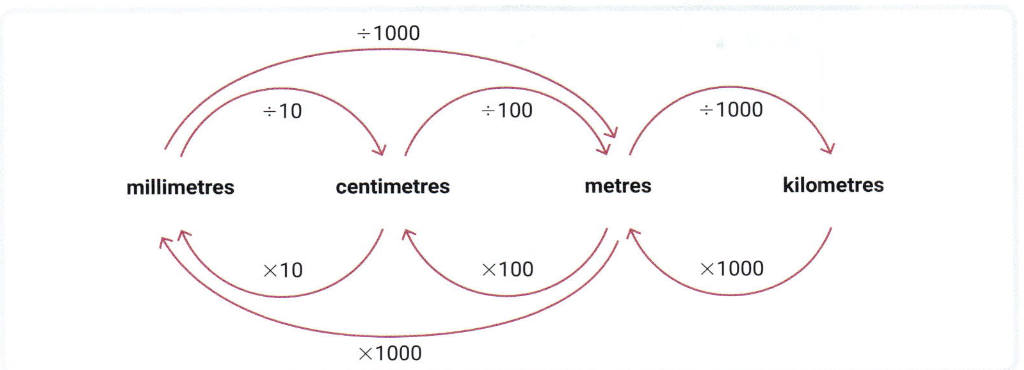

▲ **FIGURE 16.1.2** A summary of distance unit conversions

Measuring distance

Distances can be measured using a variety of tools. Shorter distances might be measured with a ruler or a tape measure, while longer distances could be measured with a measuring wheel (Figure 16.1.3), an odometer such as in a car, or using GPS software. Laser measuring devices allow you to point a laser from a start position to an end position and the device will determine the straight-line distance between them (Figure 16.1.4).

I AM NIKOM/Shutterstock.com

▲ **FIGURE 16.1.3** Measuring wheels can measure longer distances by multiplying how many circumferences of the circle have been traced along the ground.

Istock.com/iStock Essentials

▲ **FIGURE 16.1.4** Laser measuring devices measure straight-line distances by means of reflected light.

When a body's movement has been in a straight line, the distance travelled is simply the length of the straight path. When the movement has been on a path that changes direction, the distance travelled by a body is always the total length of the path the body has taken during that time. For example, if you were to walk from one corner of a rugby field to the opposite corner along the field's boundary lines, the distance travelled is half the perimeter of the whole field, not the shorter distance across the field between the end position and the start position (Figure 16.1.5).

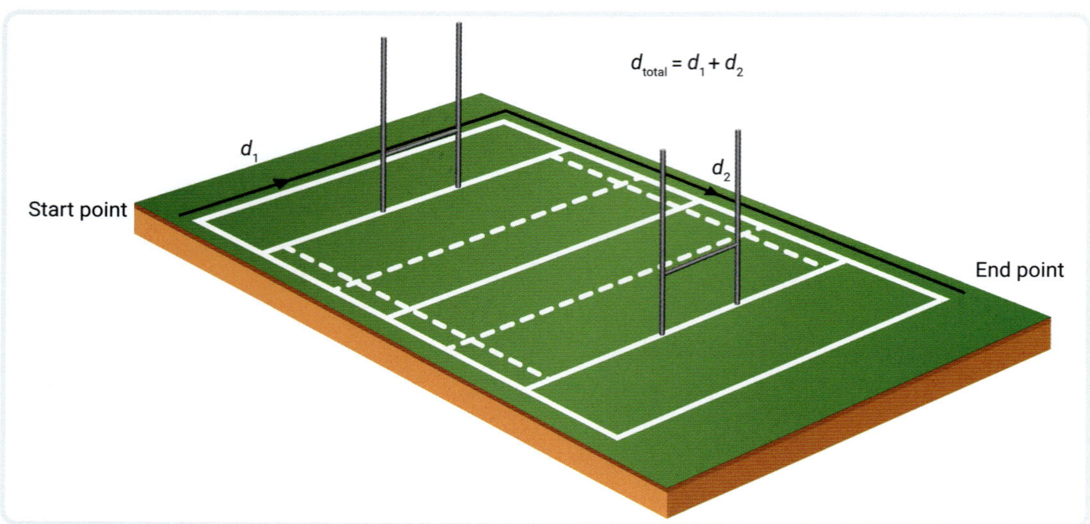

▲ **FIGURE 16.1.5** The distance travelled is the total length of the path taken.

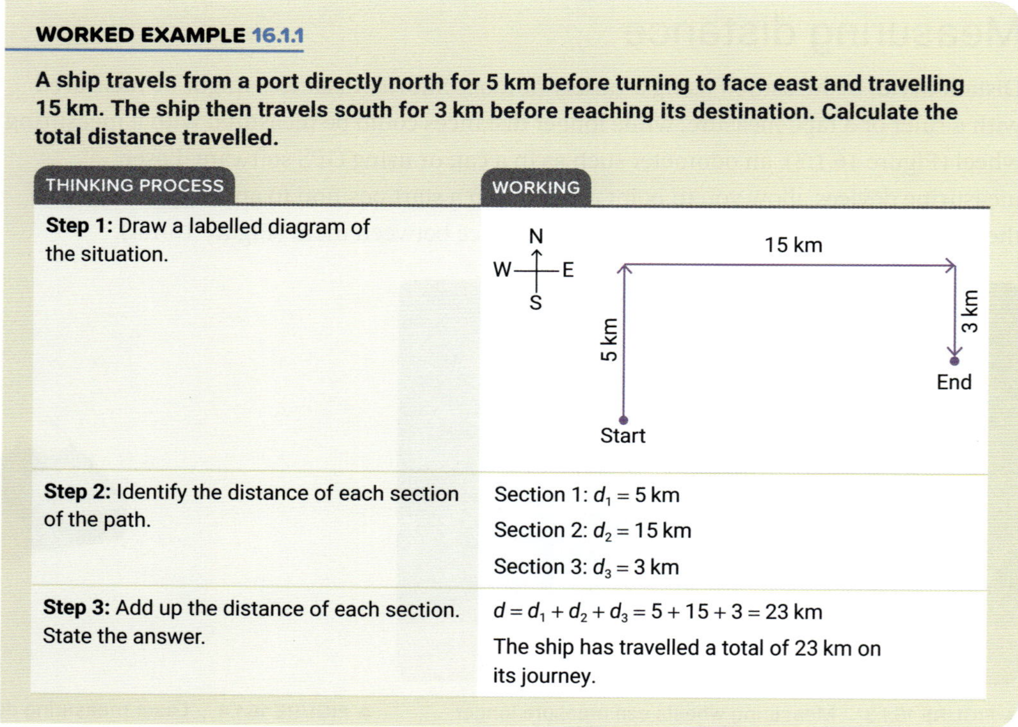

WORKED EXAMPLE 16.1.1

A ship travels from a port directly north for 5 km before turning to face east and travelling 15 km. The ship then travels south for 3 km before reaching its destination. Calculate the total distance travelled.

THINKING PROCESS	WORKING
Step 1: Draw a labelled diagram of the situation.	N, W, E, S 15 km 5 km 3 km Start End
Step 2: Identify the distance of each section of the path.	Section 1: $d_1 = 5$ km Section 2: $d_2 = 15$ km Section 3: $d_3 = 3$ km
Step 3: Add up the distance of each section. State the answer.	$d = d_1 + d_2 + d_3 = 5 + 15 + 3 = 23$ km The ship has travelled a total of 23 km on its journey.

Distance–time graphs

Visualising the journey of a moving body can be made easier by using different types of graphs. One graph that is particularly helpful is a distance–time graph (Figure 16.1.6), which plots the distance travelled by a moving body against time. As time increases on the x-axis, the y-axis plots the total distance the body has travelled at that time.

9780170491785

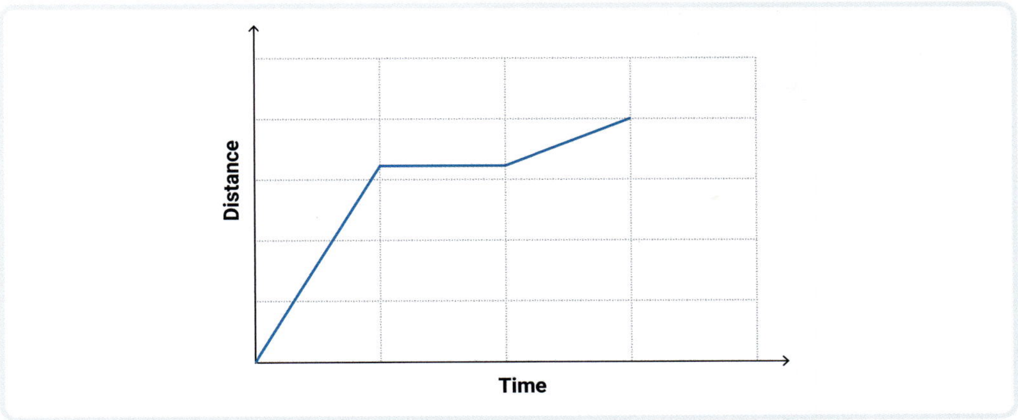

▲ **FIGURE 16.1.6** A distance–time graph for a body

Distance–time graphs can show the total distance travelled by a body and the amount of distance covered in different time periods. Sections where the body is not moving are shown as horizontal lines because no more distance is being covered in that time. When the position of a body is not changing, we refer to the body as being **stationary**, or at **rest**.

stationary
keeping a constant position or not moving in any direction; at rest

rest
the state of being stationary, having a speed of 0 m/s

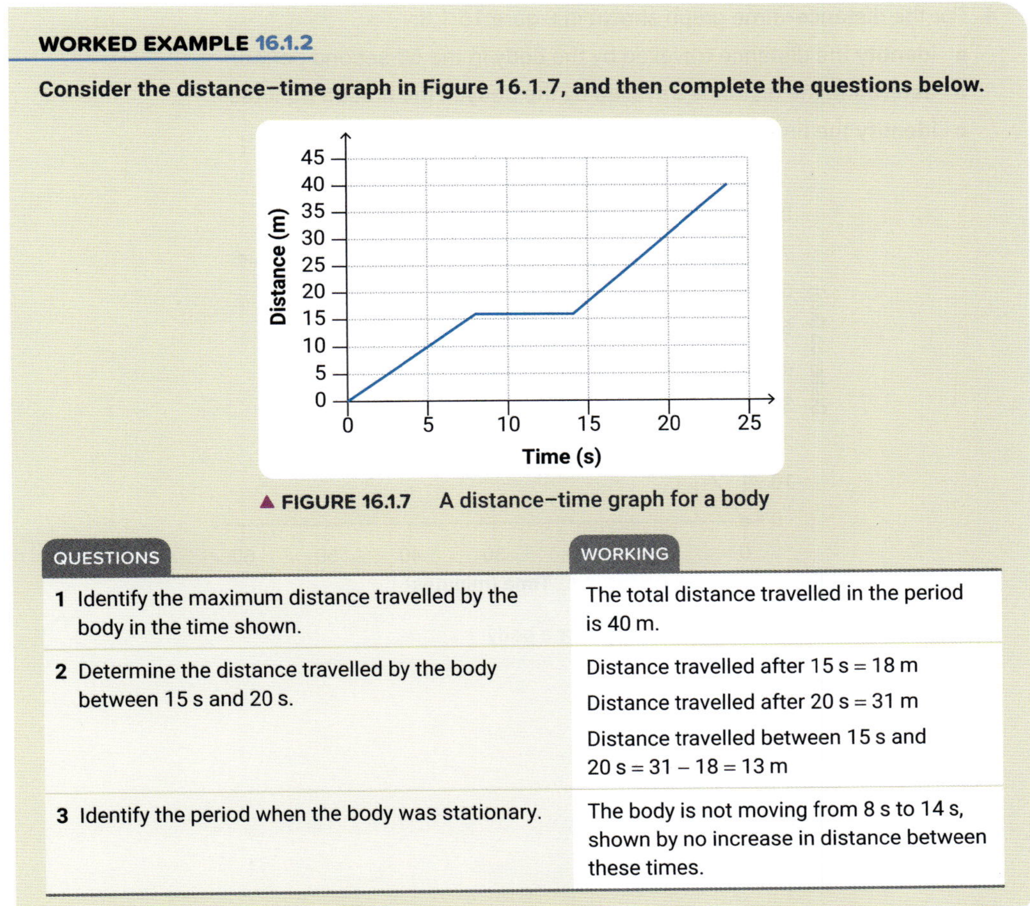

WORKED EXAMPLE 16.1.2

Consider the distance–time graph in Figure 16.1.7, and then complete the questions below.

▲ **FIGURE 16.1.7** A distance–time graph for a body

QUESTIONS	WORKING
1 Identify the maximum distance travelled by the body in the time shown.	The total distance travelled in the period is 40 m.
2 Determine the distance travelled by the body between 15 s and 20 s.	Distance travelled after 15 s = 18 m Distance travelled after 20 s = 31 m Distance travelled between 15 s and 20 s = 31 − 18 = 13 m
3 Identify the period when the body was stationary.	The body is not moving from 8 s to 14 s, shown by no increase in distance between these times.

16.1 LEARNING CHECK

1 Convert the following measurements into the units stated.

   a 0.987 km to m

   b 46 700 mm to m

   c 33.4 m to cm

   d 0.556 m to mm

   e 4.0 km to cm

   f 60 000 mm to km

2 **Determine** the total distance (in metres) travelled by each moving body.

   a From home, Sal ran 100 m down to the local running track and did one lap of the 500 m track before running home again.

   b A plane flew 125 km east before turning and travelling 42 000 m north.

   c A dog ran the perimeter of a square yard three times. The yard is 6 m wide.

   d Charlie walked 400 m to the corner shop to buy an ice cream. This took 4 minutes. He then stayed in the shop for 3 minutes. Charlie then walked another 320 m to the post office in 2 minutes before turning around and walking home in 6 minutes.

3 **Construct** a distance–time graph for Charlie's journey in Question 2**d**.

4 For the distance–time graph shown in Figure 16.1.8:

   a **identify** the distance travelled by the body in the 60-second period.

   b **determine** the distance travelled by the body in the first 25 minutes.

   c **identify** the time period when the body was stationary.

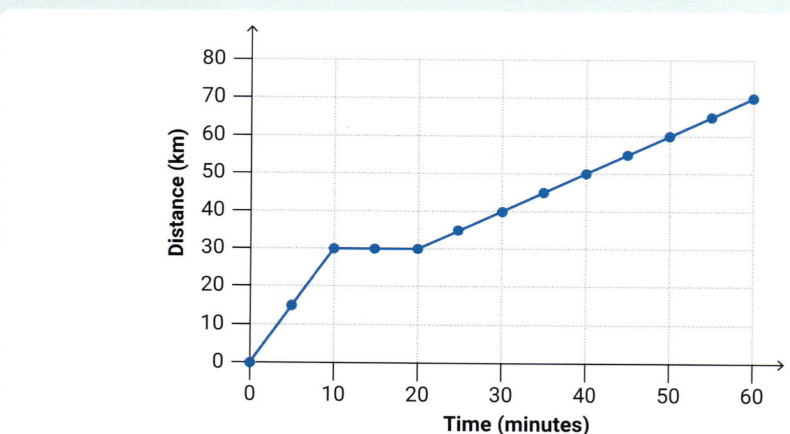

▲ **FIGURE 16.1.8** A distance–time graph for a body

9780170491785

16.2 Speed

16.2

BY THE END OF THIS MODULE, YOU WILL BE ABLE TO:

- ✓ understand the measurement and calculation of speed
- ✓ construct and interpret speed–time graphs to represent the motion of moving bodies
- ✓ interpret distance–time graphs and use them to calculate speed.

GET THINKING

Make a list of words that you associate with the term 'speed'. As you work through this module, tick them off if they come up.

speed
a measure of how much distance a body covers per unit of time; typically metres per second (m/s)

What is speed?

Speed is a measurement that describes how fast a body is travelling (Figure 16.2.1). Speed is a rate measurement that indicates how much distance is covered per unit of time. It is usually represented by a lowercase v. Speed can commonly be measured in kilometres per hour (km/h or km h^{-1}) but the SI unit for speed is metres per second (m/s or m s^{-1}). If a body is travelling at a speed of 2 m/s, it covers 2 m of distance every second. Bodies that are at rest have no speed or a speed of 0 m/s and are stationary.

Harry Collins Photography/Shutterstock.com

▲ **FIGURE 16.2.1** Peregrine falcons dive vertically at a speed of up to 400 km/h, which is about 111 m/s.

To convert between speed units, the following rules can be used:

$$\text{Speed in } \frac{\text{km}}{\text{h}} = \text{speed in } \frac{\text{m}}{\text{s}} \times 3.6$$

$$\text{Speed in } \frac{\text{m}}{\text{s}} = \text{speed in } \frac{\text{km}}{\text{h}} \div 3.6$$

The box below shows how these rules can be derived from smaller steps converting the distance and time measurements.

Converting units of speed

Convert metres to kilometres:

$$2\,\frac{\text{m}}{\text{s}} = 2 \div 1000\,\frac{\text{km}}{\text{s}} = 0.002\,\frac{\text{km}}{\text{s}}$$

Convert seconds to minutes:

$$0.002\,\frac{\text{km}}{\text{s}} = 0.002 \times 60\,\frac{\text{km}}{\text{min}} = 0.12\,\frac{\text{km}}{\text{min}}$$

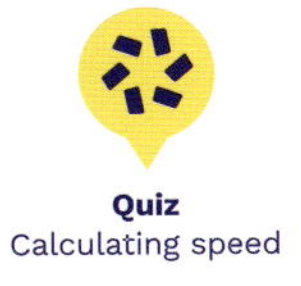

Quiz
Calculating speed

Convert minutes to hours:

$$0.12\frac{\text{km}}{\text{h}} = 0.12 \times 60\frac{\text{km}}{\text{h}} = 7.2\frac{\text{km}}{\text{h}}$$

Combining steps into one step:

$$2\frac{\text{m}}{\text{s}} \div 1000 \times 60 \times 60 = 7.2\frac{\text{km}}{\text{h}}$$

$$2\frac{\text{m}}{\text{s}} \times 3.6 = 7.2\frac{\text{km}}{\text{h}}$$

Rules

To convert from m/s to km/h, multiply by 3.6.

To convert from km/h to m/s, divide by 3.6.

Speed–time graphs

Some bodies travel at a constant speed, such as a car travelling down a highway at the speed limit of 100 km/h (27.8 m/s). Or bodies can travel with a varying speed. Speed–time graphs are helpful tools to analyse the speed of a body during a journey. Figure 16.2.2 shows a speed–time graph for a 45-second journey where a body starts with a speed of 0 m/s, increases speed to a speed of 60 m/s in the first 10 seconds before remaining at a constant speed for 20 seconds and then slowing down to stop again.

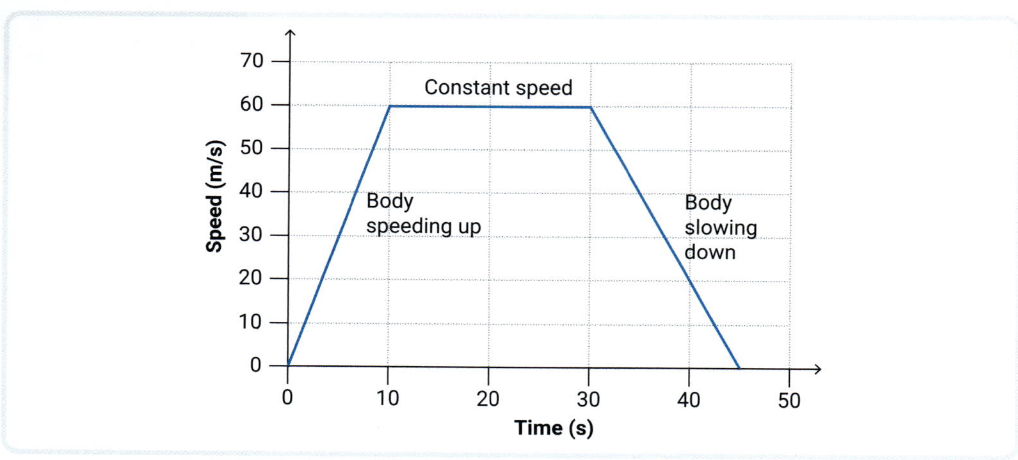

▲ **FIGURE 16.2.2** A speed–time graph for a moving body

Measuring and calculating speed

Speed is generally not measured directly as a single measurement. Speed can be calculated using the following formula:

$$\text{Speed} = \frac{\text{total distance travelled}}{\text{total time taken}}$$

$$v = \frac{d}{t}$$

where v is speed measured in metres per second (m/s), d is the total distance travelled measured in metres (m) and t is the total time taken measured in seconds (s).

16.2

WORKED EXAMPLE 16.2.1

Calculate the speed of a runner who runs 100 m in 25 s.

THINKING PROCESS	WORKING
Step 1: Identify known and unknown variables. Ensure known values are in SI units.	$v = ?$ $d = 100$ m $t = 25$ s
Step 2: Choose an appropriate formula.	$v = \frac{d}{t}$
Step 3: Substitute known variables into the formula.	$v = \frac{100}{25}$
Step 4: Rearrange using algebra (if necessary) and solve. State the answer.	$v = 4$ m/s The runner is running at a speed of 4 m/s.

This formula can also be used to calculate distance or time depending on the information given.

WORKED EXAMPLE 16.2.2

Calculate how long it would take a bee to fly 20 m if it were to fly at a speed of 0.25 m/s.

THINKING PROCESS	WORKING
Step 1: Identify known and unknown variables. Ensure known values are in SI units.	$t = ?$ $d = 20$ m $v = 0.25$ m/s
Step 2: Choose an appropriate formula.	$v = \frac{d}{t}$
Step 3: Substitute known values into the formula.	$0.25 = \frac{20}{t}$
Step 4: Rearrange using algebra (if necessary) and solve. State the answer.	$0.25 \times t = \frac{20}{t} \times t$ $0.25t = 20$ $t = \frac{20}{0.25}$ $t = 80$ s It would take the bee 80 s or 1 minute and 20 seconds to fly 20 m.

ACTIVITY

Measuring speed

For this activity, you will need:

- a stopwatch
- space outside to run.

Go outside and measure a distance of 20 m. Mark both ends of this distance with some objects.

Use a stopwatch to determine how long it takes you to run the 20 m.

a **Calculate** your average running speed.

b **Evaluate** whether you think you were running at a constant speed for the whole 20 m.

c **Explain** how you measured the 20 m distance and **discuss** whether you think this was an accurate measurement or not.

16.2 LEARNING CHECK

1 Convert the following speed measurements to the units stated.

a 23 km/h to m/s

b 10 m/s to km/h

2 **Calculate** the speed of a dog that runs at a constant speed and covers 100 m in 30 s.

3 **Calculate** the time it would take for a snail to crawl along the edge of a 50 m pool if its speed was 0.0072 km/h (Figure 16.2.3).

4 Speed cameras are used on roads to detect drivers who drive above the speed limit (Figure 16.2.4). The cameras take time-stamped photos of cars as they move over markings on the road that are a fixed distance apart. Based on what you know about measuring and calculating speed, **explain** how the speed camera is programmed to be able to determine how fast the driver is going.

▲ **FIGURE 16.2.3** Snails move very slowly.

PhotoIris2021/Shutterstock.com

▲ **FIGURE 16.2.4** Speed cameras take photos of cars moving across road markings.

Istock.com/iStock Signature

16.3 Using graphs to determine speed

BY THE END OF THIS MODULE, YOU WILL BE ABLE TO:

✓ interpret distance–time graphs and use them to calculate average speed.

Interpreting distance–time graphs

Information about the speed of a body can be gained by interpreting a distance–time graph. Figure 16.3.1 shows the distance–time graph and corresponding speed–time graph for a car's 1-minute journey.

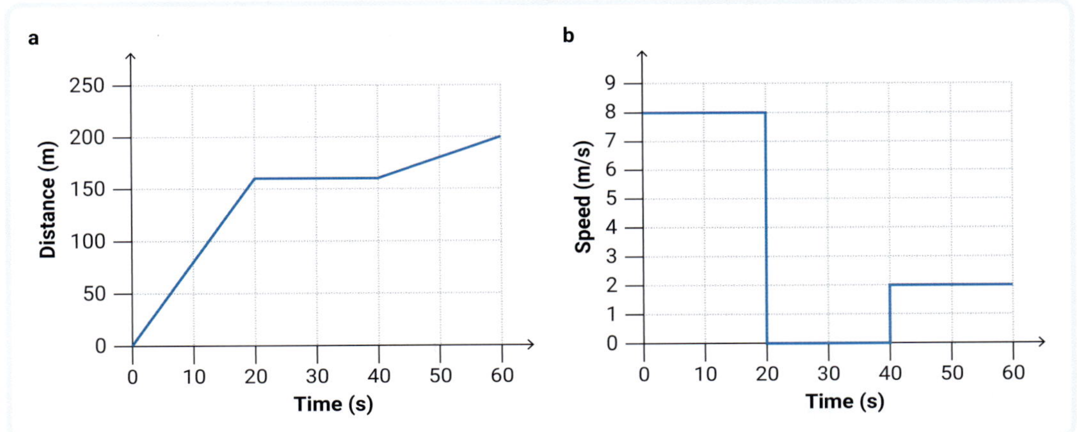

▲ FIGURE 16.3.1 (a) A distance–time graph and (b) a speed–time graph for a car's journey

The speed–time graph shows that the car travels at a constant speed of 8 m/s for the first 20 s of the journey. This can be seen on the distance–time graph (Figure 16.3.1a), as the distance is increasing at a constant rate from 0 s to 20 s. When the car is not moving, the distance–time graph shows a horizontal line from 20 s to 40 s. The same time period on the speed–time graph shows the car's speed of 0 m/s.

For the last 20 s of the journey, the car has a constant speed of 2 m/s, as shown on the speed–time graph. On the distance–time graph, the distance increases, but the slope is less steep than it was for the first 20 s. Since the car is travelling much more slowly, it covers less distance in the same period of time. This results in a less steep line and a smaller **gradient**. The gradient is the slope of a linear section of the distance–time graph and can be used to calculate the speed. Figure 16.3.2 shows how the gradient of a distance–time graph is equal to the speed of the body.

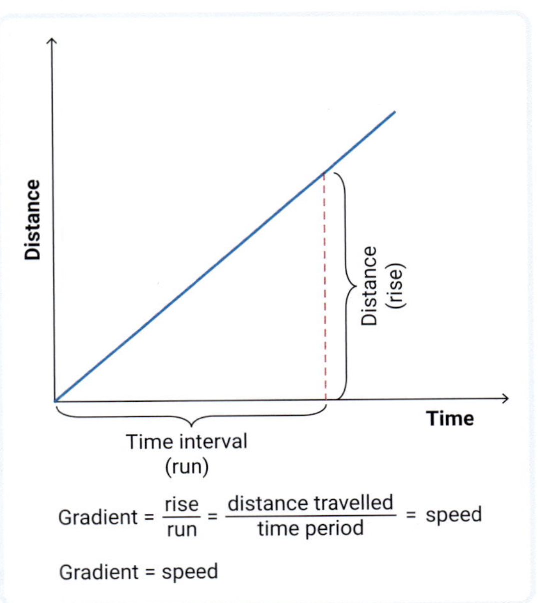

▲ FIGURE 16.3.2 The gradient of a distance–time graph is the speed.

gradient
a measure of the slope of a straight line on a graph

WORKED EXAMPLE 16.3.1

Use the distance–time graph Figure 16.3.1a to determine the speed of a car throughout its journey.

THINKING PROCESS	WORKING
Step 1: Determine the car's speed from (0, 0) to (20, 160).	$\text{Speed} = \text{gradient} = \frac{\text{rise}}{\text{run}}$ $= \frac{160-0}{20-0} = \frac{160}{20} = 8\text{ m/s}$
Step 2: Determine the car's speed from (20, 160) to (40, 160).	$\text{Speed} = \text{gradient} = \frac{\text{rise}}{\text{run}}$ $= \frac{160-160}{40-20} = \frac{0}{20} = 0\text{ m/s}$
Step 3: Determine the car's speed from (40, 160) to (60, 200). State the full answer.	$\text{Speed} = \text{gradient} = \frac{\text{rise}}{\text{run}}$ $= \frac{200-160}{60-40} = \frac{40}{20} = 2\text{ m/s}$ The car is travelling at a speed of 8 m/s for 20 s, is stationary for 20 s and then travels at a speed of 2 m/s for the final 20 s.

Average speed

Sometimes the speed of a moving body will be changing and the distance–time graph does not have a constant gradient. Figure 16.3.3 shows a distance–time graph for a body that is constantly getting faster across the period shown.

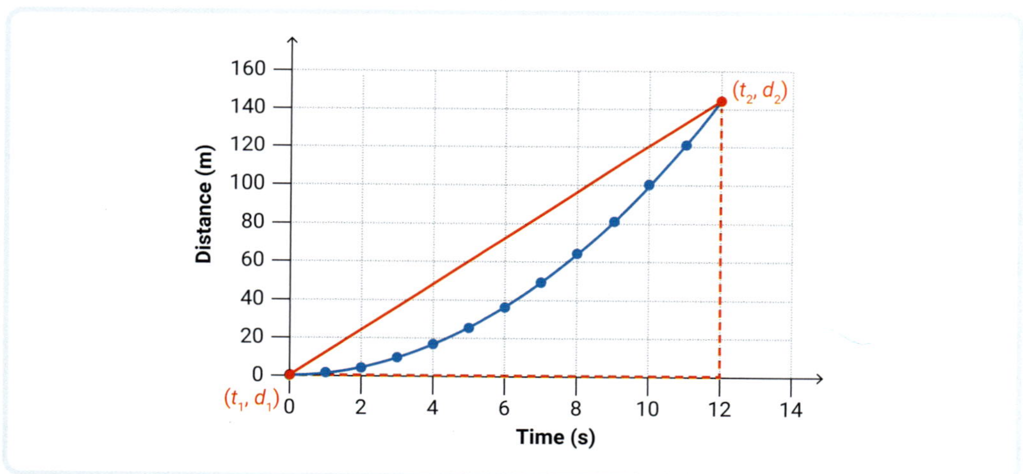

▲ **FIGURE 16.3.3** The average speed over a certain period can be determined for a body that is moving with changing speed.

The speed and, therefore, the gradient are increasing every second. We can determine the average speed of the body over the period by using the same formula for speed when it is constant.

9780170491785

$$\text{Average speed} = \frac{\text{total distance covered}}{\text{time period}}$$

$$v_{av} = \frac{d_2 - d_1}{t_2 - t_1}$$

where d_2 is the distance (m) covered at the end of the time period, d_1 is the distance (m) covered at the start of the time period, t_2 is the end time of the time period, and t_1 is the start time of the time period.

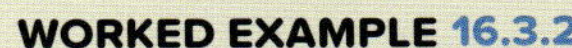

WORKED EXAMPLE 16.3.2

Consider Figure 16.3.4 and determine the average speed of the moving body in the third second (between $t = 2$ s and $t = 3$ s).

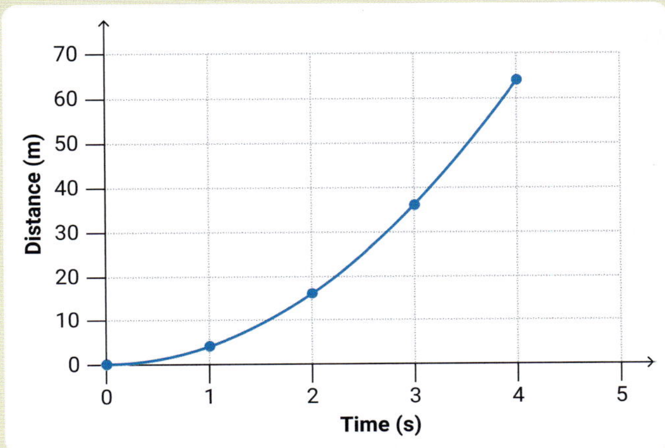

▲ **FIGURE 16.3.4** A distance–time graph for a body with changing speed (accelerating)

THINKING PROCESS	WORKING
Step 1: Identify known and unknown variables.	$v_{av} = ?$ $t_1 = 2$ s $t_2 = 3$ s $d_1 = 16$ m $d_2 = 36$ m

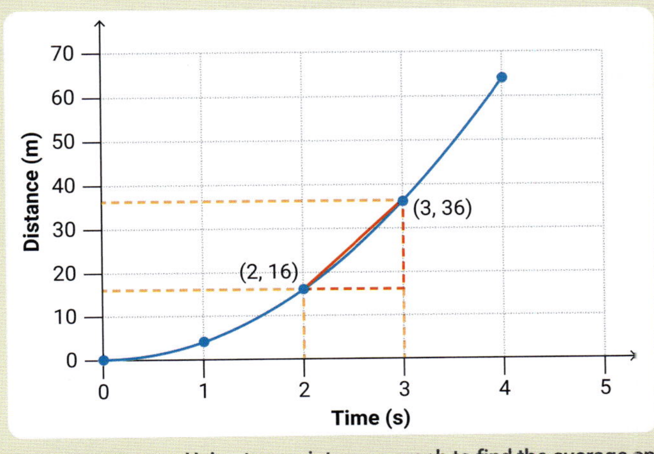

▲ **FIGURE 16.3.5** Using two points on a graph to find the average speed

Step 2: Identify appropriate formula.	$v_{av} = \frac{d_2 - d_1}{t_2 - t_1}$
Step 3: Substitute known variables into the formula.	$v_{av} = \frac{36 - 16}{3 - 2}$
Step 4: Rearrange using algebra (if necessary) and solve. State the answer.	$v_{av} = \frac{20}{1} = 20$ m/s The average speed of the body in the third second is 20 m/s.

16.3 LEARNING CHECK

Consider the distance–time graph shown in Figure 16.3.6.

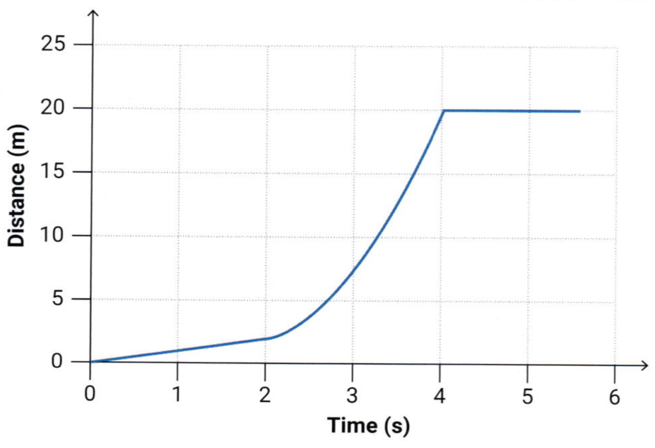

▲ **FIGURE 16.3.6** A distance–time graph for a moving body

1. **Identify** the total distance covered by the body.
2. **Identify** when the body was stationary.
3. **Identify** when the body was travelling with a constant speed.
4. At what time has the object travelled a distance of 7.5 m?
5. **Determine** the speed of the body in the first 2 seconds of its motion.
6. **Determine** the average speed of the body between 3 s and 4 s.
7. **Analyse** the shape of the graph between $t = 2$ s and $t = 3$ s and **explain** the meaning of this shape in regards to the motion of the object.

16.4 Acceleration

BY THE END OF THIS MODULE, YOU WILL BE ABLE TO:

- ✓ understand the measurement and calculation of acceleration
- ✓ construct and interpret speed–time graphs to represent the motion of moving bodies
- ✓ interpret distance–time graphs and speed–time graphs in terms of acceleration and use speed–time graphs to calculate acceleration.

GET THINKING

This module is all about acceleration. You might have already heard this word or similar ones such as accelerate or decelerate. Make a list of 10 words that you think have some connection to the term 'acceleration'.

Extra science investigation
Using ticker timers

Defining acceleration

When a parked car starts and then travels down the road, the speed doesn't change from 0 km/h to 60 km/h (16.7 m/s) instantaneously. The driver of the car presses their foot on the accelerator pedal to increase the speed of the car gradually up to the cruising speed. In this period of increasing speed, the car is accelerating or experiencing **acceleration** (Figure 16.4.1).

acceleration
the rate of change of speed

1000 Words/Shutterstock.com

▲ **FIGURE 16.4.1** Cars experience acceleration when their speed is changing.

Acceleration is a measure of how fast the speed of an object changes with time. It is represented with a lowercase *a*, and the units for acceleration are metres per second squared (m/s^2) or metres per second per second ($m\,s^{-1}\,s^{-1}$). The car, for example, might accelerate at a rate of $3\,m\,s^{-1}\,s^{-1}$ where the speed increases by 3 m/s every second.

When starting from rest, after 5 seconds the car will be travelling at 15 m/s. It will have increased its speed from 0 to 3 m/s, 3 m/s to 6 m/s, 6 m/s to 9 m/s, then to 12 m/s and finally to 15 m/s.

NASA/Joel Kowsky

▲ **FIGURE 16.4.2** A rocket taking off is an example of a body experiencing acceleration.

Acceleration also refers to motion where the speed is decreasing. If the car were to slow from 15 to 0 m/s over 5 seconds, the acceleration would be $-3\,m/s^2$ because the speed decreases by 3 m/s every second. Negative acceleration, where a body slows, can be referred to as **deceleration**.

deceleration
negative acceleration; when speed is decreasing over time

Another example of a body experiencing acceleration is a rocket. When a rocket takes off, it uses very strong **thrust** forces to accelerate it at a rate of $90\,m/s^2$ (Figure 16.4.2). After 3 seconds, the rocket is already travelling at a speed of 270 m/s (about 972 km/h).

thrust
a force that makes an object move in the opposite direction as a result of expelling mass or fuel

After finishing a race, a runner will decelerate from their top speed to stationary (Figure 16.4.3). A sprinter who has reached about 10 m/s will typically have an acceleration of $-2\,m/s^2$ when slowing from this speed to 0 m/s over a 5-second period after crossing the finish line.

PA Images/Alamy Stock Photo

▲ **FIGURE 16.4.3** Runners quickly decelerate at the end of a sprint.

Measuring and calculating acceleration

In the same way that speed can be calculated by the change in distance over time, acceleration can be calculated by the change in speed over time.

$$\text{Acceleration} = \frac{\text{final speed} - \text{initial speed}}{\text{time}}$$

$$a = \frac{v - u}{t}$$

WORKED EXAMPLE 16.4.1

Calculate the acceleration of a car that's speed changes from 14 m/s to 8 m/s over a 4 s period.

THINKING PROCESS	WORKING
Step 1: Identify known and unknown variables.	$a = ?$ $u = 14$ m/s $v = 8$ m/s $t = 4$ s
Step 2: Identify appropriate formula.	$a = \frac{v - u}{t}$
Step 3: Substitute known variables into the formula.	$a = \frac{8 - 14}{4}$
Step 4: Rearrange using algebra (if necessary) and solve. State the answer.	$a = \frac{-6}{4}$ $= -1.5$ m/s^2 The car's acceleration is -1.5 m/s^2 or is a deceleration of 1.5 m/s^2.

where a is acceleration (m/s^2), v is the final speed of the body (m/s), u is the initial speed of the body (m/s) and t is the time.

Speed–time graphs allow us to see how speed is changing with time and can help us calculate the acceleration of moving bodies. Figure 16.4.4 shows a speed–time graph for a body that's speed is increasing at a constant rate. The rate at which this speed is changing is the acceleration and can be found by the gradient of the speed–time graph.

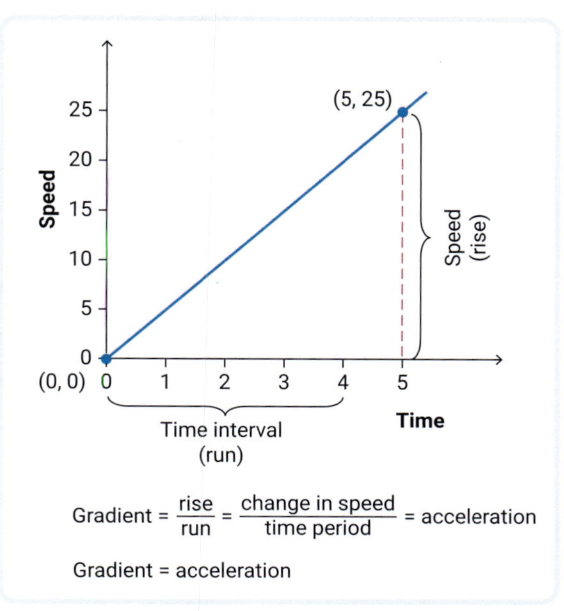

▲ **FIGURE 16.4.4** The gradient of a speed–time graph is the acceleration.

WORKED EXAMPLE 16.4.2

Use the speed–time graph in Figure 16.4.4 to determine the acceleration of the moving body.

THINKING PROCESS	WORKING
Step 1: Find the two points on the graph.	The points are (0, 0) and (5, 25). $t_1 = 0\ \text{s}$ $u = 0\ \text{m/s}$ $t_2 = 5\ \text{s}$ $v = 25\ \text{m/s}$
Step 2: Identify appropriate formula.	$\text{Acceleration} = \text{gradient} = \frac{\text{rise}}{\text{run}} = \frac{v-u}{t_2-t_1}$
Step 3: Substitute known variables into the formula. State the answer.	$a = \frac{25-0}{5-0} = \frac{25}{5} = 5\ \text{m/s}^2$ The body is accelerating at a rate of $5\ \text{m/s}^2$.

Distance–time graphs can indicate when a body is accelerating or decelerating based on how the speed (gradient) is changing. For example, Figure 16.4.5 shows a body that has a constant speed for the first 2 seconds before slowing for 2 seconds, evident by the slope gradually decreasing during the period of 2 s to 4 s.

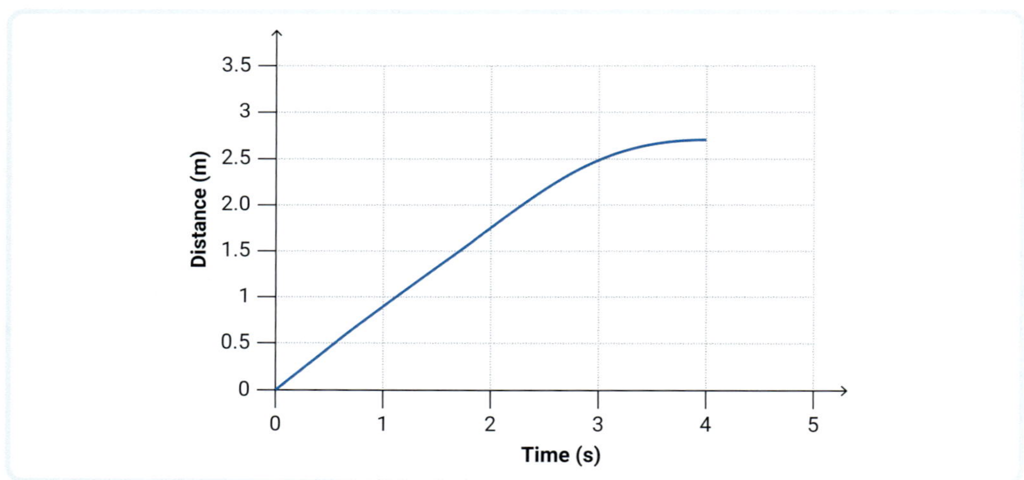

▲ **FIGURE 16.4.5** Distance–time graphs show acceleration by how the gradient is changing.

A summary of the general shape of distance–time graphs and speed–time graphs for different types of acceleration is shown in Table 16.4.1.

▼ **TABLE 16.4.1** The general shape of motion graphs for different accelerations

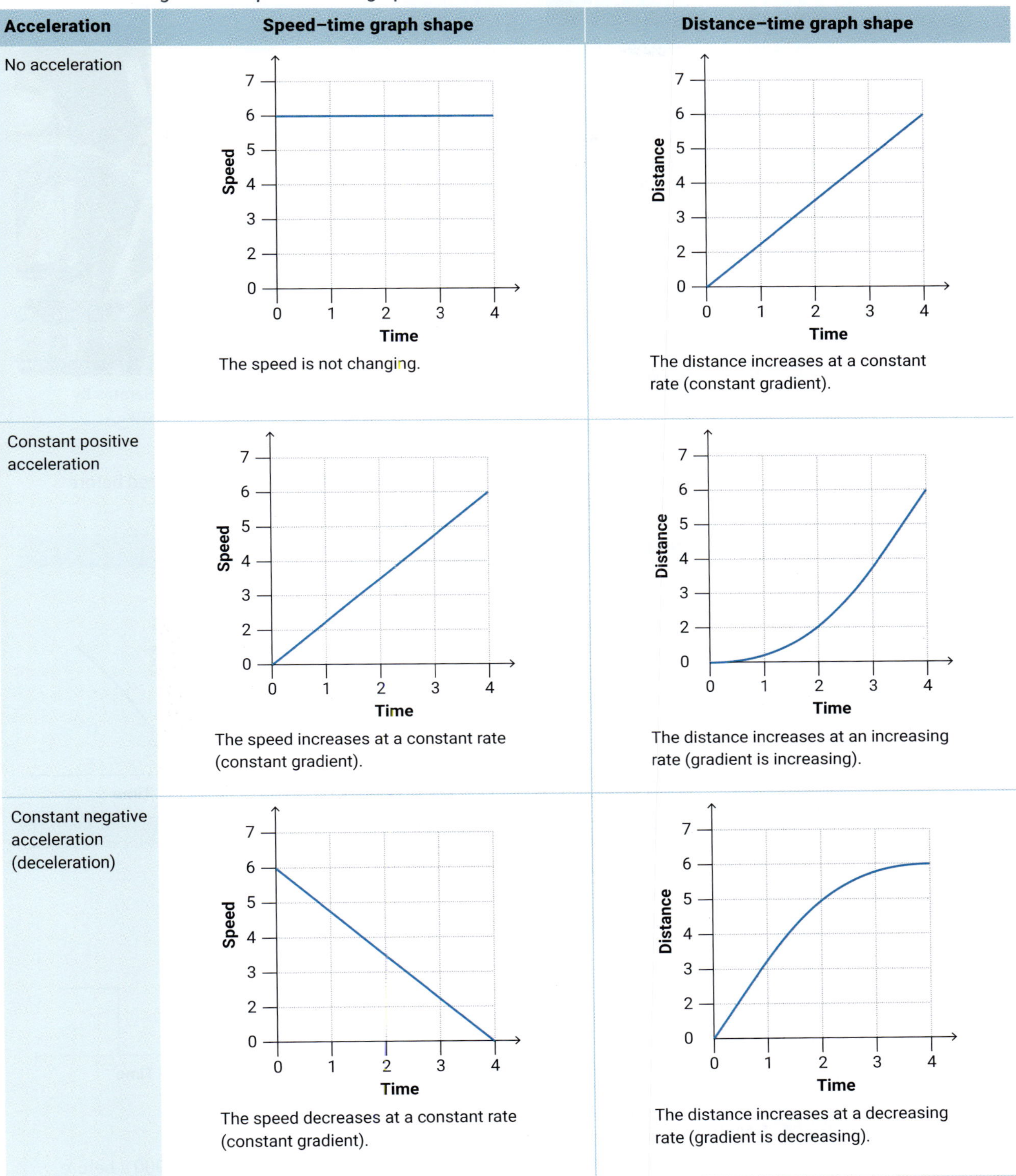

Acceleration	Speed–time graph shape	Distance–time graph shape
No acceleration	The speed is not changing.	The distance increases at a constant rate (constant gradient).
Constant positive acceleration	The speed increases at a constant rate (constant gradient).	The distance increases at an increasing rate (gradient is increasing).
Constant negative acceleration (deceleration)	The speed decreases at a constant rate (constant gradient).	The distance increases at a decreasing rate (gradient is decreasing).

16.4 LEARNING CHECK

1 **Calculate** the acceleration of a bike that starts from rest and is travelling at 9 m/s after 6 seconds (Figure 16.4.6).

2 **Calculate** the time it would take for a car to accelerate from 5 to 15 m/s with an acceleration of 2 m/s^2.

3 Match the following descriptions to one of the distance–time graphs A–C and one of the speed–time graphs D–E.

   a Pat the dog sat still for a few seconds and then raced off, accelerating at a constant rate.

   b Grandpa walked at a constant speed for a few seconds, stopped for a rest and then continued at a slower speed than before.

   c Jess rode a bike, starting from rest, and accelerated to a constant speed before cruising at this speed.

▲ **FIGURE 16.4.6** A bike accelerates by pedalling with an applied force.

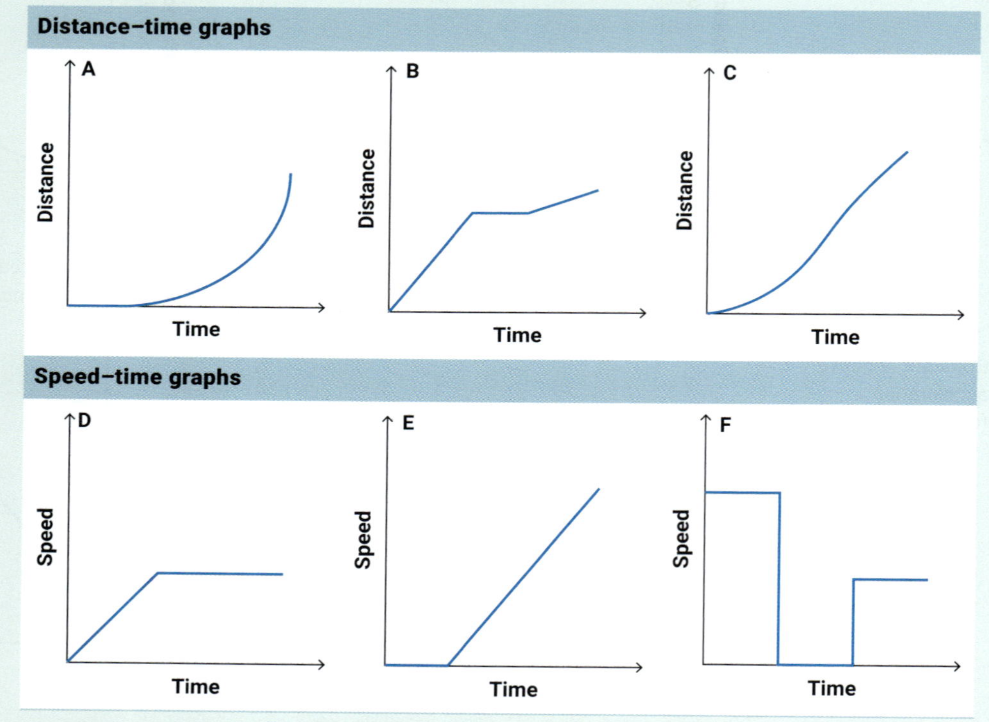

4 **Construct** a speed–time graph for the following situation.

An aeroplane starts from rest and accelerates at a rate of 0.25 m/s^2 for 900 s before cruising at this final speed for 500 s. The aeroplane then decelerates for 200 s until it is at a speed of 180 m/s.

5 Return to your response to the 'Get thinking' activity for this module. For each of the 10 words in your list, **describe** how this word is connected to the idea of acceleration. If you now think a word isn't connected to acceleration after completing the module, **explain** why.

9780170491785

WORKING SCIENTIFICALLY

16.5 Analysing linear data

SCIENCE SKILLS IN FOCUS

IN THIS MODULE, YOU WILL FOCUS ON LEARNING AND IMPROVING THESE SKILLS:

- analysing proportional data using two techniques to identify relationships and determine constant values
- selecting appropriate mathematical relationships and graphical analysis to organise and process data to identify trends and data anomalies or errors
- investigating the relationship between speed, time and acceleration by collecting and processing data in tables and graphs.

Using graphs to analyse linear data

Scientists often use collected experimental data to identify the relationship between variables and to determine or verify physical quantities. In this module, we will learn about how this can be done for variables that should be directly proportional to one another. Directly proportional means that, when one variable is multiplied, the other variable multiplies by the same amount, so there is a linear relationship between the two variables (Figure 16.5.1). For example, if you walk for three times as long at a constant speed, the distance travelled is three times as far.

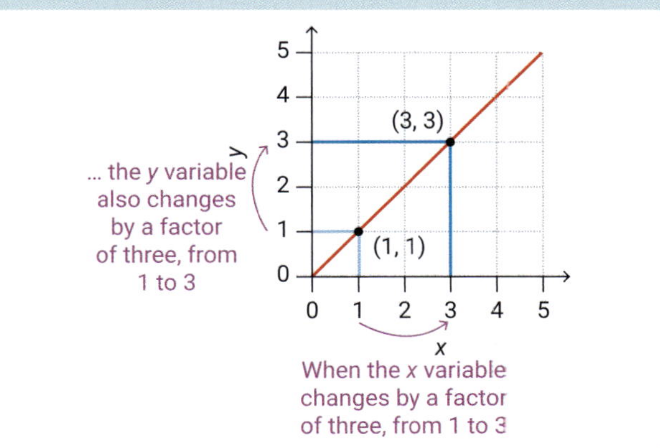

▲ FIGURE 16.5.1 Directly proportional data

To investigate this, we will use some data from an experiment that investigated Newton's second law. In the experiment, a ball of unknown mass was accelerated at four different rates and the force required to accelerate the ball was measured. The relationship should theoretically be linear because force is directly proportional to acceleration when mass is constant (double the acceleration requires double the force). We know this theoretical relationship as Newton's second law:

$$F = ma$$

The results of the experiment are presented in Table 16.5.1.

▼ TABLE 16.5.1 The experimental results for the force required to accelerate a ball at different rates

Acceleration (m/s^2), independent variable (x)	Force (N), dependent variable (y)
0	0
0.4	0.81
0.6	1.14
0.8	1.60
1.0	2.00

Non-graphical analysis methods

Using the data from Table 16.5.1, we can start to identify trends by describing how the dependent variable (force) changes as the independent variable (acceleration) increases. However, it is difficult to determine a mathematical relationship just by looking at the numbers. Since we know the theoretical relationship should be Newton's second law, we could apply this mathematical formula to determine the mass of the ball. To identify the trend accurately using the formula $F = ma$, we need to calculate the mass for each data point and then calculate the average mass.

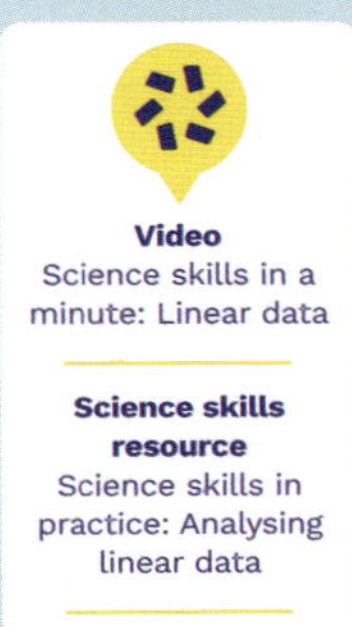

Video
Science skills in a minute: Linear data

Science skills resource
Science skills in practice: Analysing linear data

Table 16.5.2 demonstrates how to use all data to find an average with the formula $F = ma$.

▼ **TABLE 16.5.2** Non-graphical analysis methods to determine the mass of the ball

Acceleration (m/s^2), independent variable	Force (N), dependent variable	Mass (kg)
0	0	–
0.4	0.84	2.1
0.6	1.14	1.8
0.8	1.60	2.0
1.0	2.00	2.0

$$m_{av} = \frac{2.1 + 1.8 + 2.0 + 2.0}{4} = 1.98 \text{ kg}$$

This average mass value would be more reliable than if we only used one point to determine a trend, as the average is less likely to be affected by error. The more data points we have, the more we can minimise the impact of random errors. However, this method is still limited, as it considers all data points equally when determining the mass.

Graphical analysis method

If a graphical approach is used, a linear relationship can be easily identified for the experimental data by the shape of a graph. Figure 16.5.2 shows a scatter plot of the data with the dependent variable (force) on the *y*-axis and the independent variable (acceleration) on the *x*-axis.

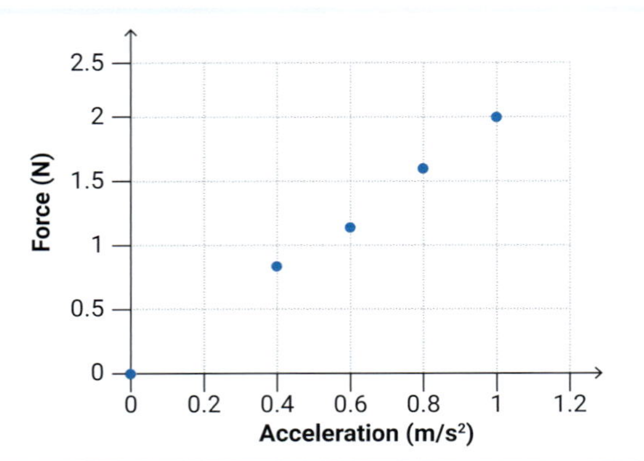

▲ **FIGURE 16.5.2** Graphing force against acceleration

Problem data points can be identified easily when they don't fit the general trend. This means we can disregard them when fitting a linear trendline. Fitting a trendline can be done by hand or using graphing software. The trendline shows us the experimental relationship between the data. If we know the theoretical relationship, we can align it with the trendline to interpret and give meaning to the gradient and *y*-intercept of the trendline. Figure 16.5.3 shows that a linear trendline has been fitted to the data so that it goes through the most points possible. The gradient and *y*-intercept for the straight line can be determined by hand or using graphical analysis software. We can then interpret the data's experimental relationship by comparing it to the theoretical relationship.

Using this method, the mass of the ball can be found since it is equal to the gradient of the straight line, which is 1.99 kg (Figure 16.5.3).

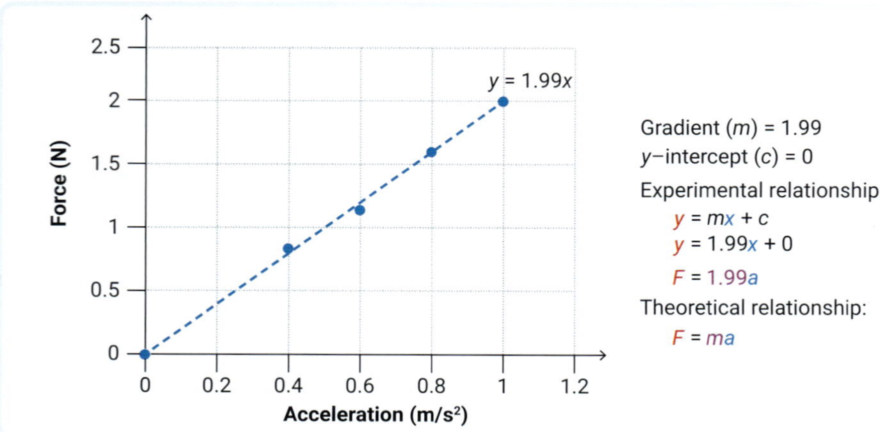

▲ **FIGURE 16.5.3** Using a graphical method to determine the mass of the ball

9780170491785

INVESTIGATING ACCELERATION, SPEED AND TIME RELATIONSHIPS USING MOTION SIMULATION

In this activity, you will apply the methods from the blue Science skills in focus box to investigate the acceleration, speed and time relationship:

$$a = \frac{v - u}{t}$$

During the investigation, a constant force will be applied to a body, and Newton's second law tells us that a constant force on a constant mass will result in constant acceleration. The body will start from rest with an initial speed of 0 m/s. Therefore, this relationship can be rearranged as:

$$v = at$$

This form of the equation shows us that it is, theoretically, a linear relationship, with the speed (v) being the dependent variable (y), acceleration (a) being the constant variable (m) and time (t) being the independent variable (x).

AIM

To investigate the relationship between time and speed for a body experiencing a constant net force, and apply mathematical and graphical analyses to interpret the raw data

MATERIALS AND EQUIPMENT

- ☑ PhET simulation 'Forces and Motion: Basics'

  https://phet.colorado.edu/en/simulations/forces-and
  -motion-basics
- ☑ spreadsheet software
- ☑ stopwatch

PROCEDURE

1. Construct an appropriate results table for the relationship you are investigating.
2. Open the simulation and choose the 'Motion' module from the start-up page.
3. Pause the simulation using the pause button to set up for the experiment.
4. Use the following settings for your experiment.
   a. Uncheck 'Force'.
   b. Check 'Values'.
   c. Leave 'Masses' unchecked.
   d. Check 'Speed'.
   e. Set the 'Applied Force' to 30 newtons.
   f. Use the box only as the mass on the skateboard.
5. Start the timer at the same time as you press the Play button on the simulation.
6. Record the speed of the box every 5 seconds for 30 seconds in your results table.

RESULTS

Present the results table appropriately, including headings and units.

ANALYSIS

1. **Apply** a graphical method to **determine** the acceleration of the body using a linear trendline. Use the spreadsheet software to generate the graph and add a trendline with an equation.
2. Based on the graph produced, **justify** why a linear trendline is appropriate for this data.
3. **Identify** if there were any anomalous data points based on the graph produced.
4. **Identify** possible limitations or random errors that may have occurred when collecting data in this investigation.
5. Use Newton's second law to **calculate** the mass of the block.
6. On the simulation, turn on 'Masses' to see the mass of the box you used. **Compare** this mass with your experimentally determined mass. **Discuss** reasons for the difference.

CONCLUSION

1. Write a conclusion about the relationship between speed and time for a body starting from rest and experiencing a constant net force.
2. Write a conclusion about the acceleration of the box used in your simulation when a net force of 30 N was applied.

16.6 Newton's laws of motion

BY THE END OF THIS MODULE, YOU WILL BE ABLE TO:

✓ recall Newton's laws of motion

✓ use Newton's laws of motion to explain the relationship between forces and motion and make predictions about the motion of bodies when experiencing different forces.

Video activity
Newton's laws of motion

Interactive resource
Simulation: Forces and motion

GET THINKING

Imagine this scenario: a box, a toy car and a ball are on a flat surface (like a table). Predict what will happen if:

- you push the box gently
- you push the box harder
- the box runs into the toy car
- you push the ball, and it bumps into the box.

Write your predictions for each situation and, as you progress in this chapter, double-check them.

Forces and free-body diagrams

As you might recall from Stage 4, pushes, pulls, twists and squeezes are all **forces**. Pushes, pulls, twists and squeezes are all applied by one object on another object. Forces can hold an object in place, such as those that are acting on a ladder leaning against a wall. A force applied to an object can result in a change in the object's shape or motion. Changes in motion include a change in:

- direction, such as when an object moves around a corner or when it bounces off a surface
- speed, such as when an object goes faster or slower.

Often, a change of direction and a change of speed both occur to an object as a result of the action of a force.

Stock liberi/Shutterstock.com

▲ **FIGURE 16.6.1** 1 N is approximately equal to the force required to hold an apple.

force
a push, pull, twist or squeeze

free-body diagram
a diagram that shows the forces acting on a single body as arrows

magnitude
size or extent

Force is represented by F and is measured in newtons (N). One newton is approximately equal to the force required to hold an apple (Figure 16.6.1).

Forces are often represented on **free-body diagrams**. A free-body diagram shows the forces acting on a single body. In a formal free-body diagram, the body is represented by a rectangle. However, sometimes the diagram uses a simplified drawing of the body being examined. Arrows represent forces and point in the direction that the force is acting. Force arrows are different lengths depending on the **magnitude** (size) of the force. The arrow always starts in the centre of the body, regardless of where on the body they are being applied (Figure 16.6.2).

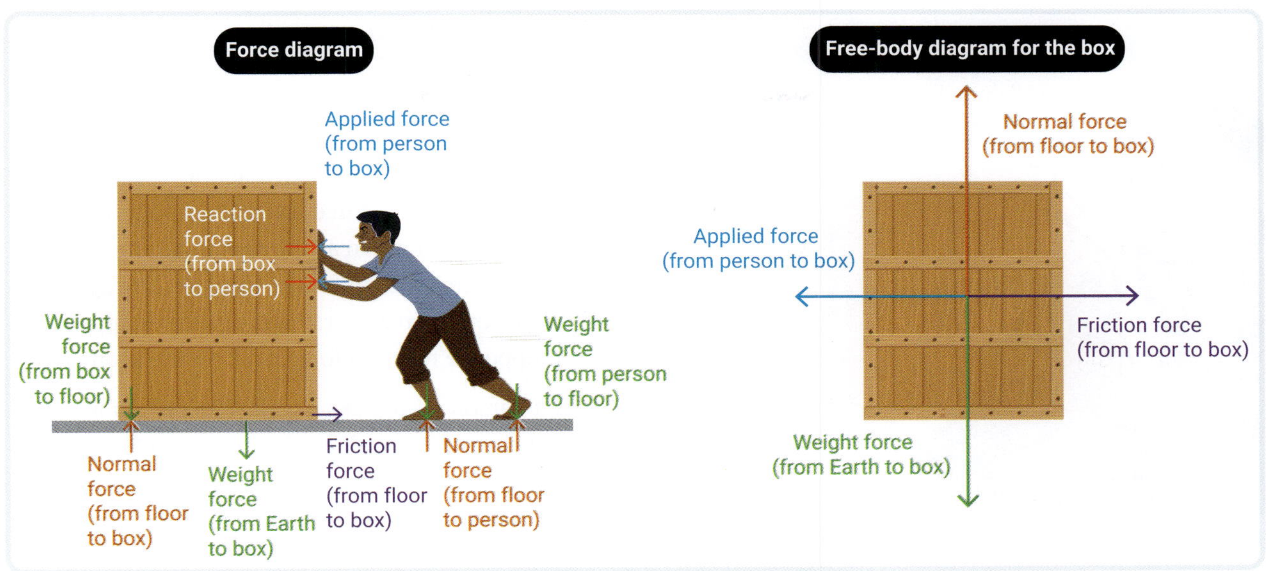

▲ **FIGURE 16.6.2** A free-body diagram for a box when it is being pushed only includes the forces acting on the box.

Newton's three laws of motion

Sir Isaac Newton (1643–1727) was an English physicist and is considered one of the world's most influential scientists. While studying motion and trying to make sense of how and why things move, he developed three physical rules or 'laws' by which all his observations could be explained (Figure 16.6.3). These laws are accepted universally as appropriate to explain and predict motion of most bodies on Earth. The three laws are:

bilha golan/Shutterstock.com

▲ **FIGURE 16.6.3** Stories often suggest that Isaac Newton had some of his big ideas about gravity and motion while sitting under an apple tree and watching the apples fall to the ground.

1 A body will remain at rest or at a constant speed unless acted on by unbalanced forces.
2 The acceleration of a body is directly proportional to the unbalanced force applied to it and inversely proportional to the body's mass.
3 For every force acting, there is an equal and opposite force reacting.

When learning about these laws, it is best to look at the third law first, followed by the first law and then the second law. Newton's second law was introduced in Module 16.5 and will be covered in more depth in the next module.

DATA SCIENCE

Learn more about how observations are used to form investigable questions in **Module 2.1**.

▲ **FIGURE 16.6.4** Reaction forces when kicking a wall

agent
a body applying a force on another body

receiver
the body receiving a force from another body

Newton's third law

Newton's third law states that for every acting force, there is an equal and opposite reaction force. This means that all forces come in pairs, often called action–reaction pairs. When one body applies a force on another body, the body applying the force is known as the **agent** and the body receiving the force is known as the **receiver**. The agent applies a force on the receiver and the receiver applies an equal and opposite reaction force back on the agent.

Consider a person who kicks a wall. The person applies a force to the wall. The wall applies an equal force back on the person's foot, away from the wall. The harder the person kicks the wall, the more force is pushed back on them (Figure 16.6.4).

Some other examples of Newton's third law in action are described in Table 16.6.1.

▼ **TABLE 16.6.1** Examples of Newton's third law

Example	Agent	Receiver	Outcome of forces
A foot kicking a ball	Foot	Ball	The foot applies a force forwards on the ball and the ball applies the same force back on the foot. The foot feels the force from the ball pushing back on it. The ball accelerates away from the foot.
A person diving from a diving block	Person	Block	The person applies a force back on the block. The block applies the same force forwards on the person, resulting in the person accelerating forwards.

Table 16.6.2 describes some common reaction forces that are harder to imagine or see in action.

▼ **TABLE 16.6.2** The types of reaction forces

Reaction force	Description	Example	
Tension	Tension is a reaction force that is exerted by a rope, string or wire on the bodies attached to either end when a force is applied to the rope, string or wire.	A crane holding a heavy mass	The weight applies a force down on the crane cable and the crane cable applies a tension force, pulling the weight back up.
Upthrust	Upthrust is a reaction force that a liquid or gas exerts on an object floating in it.	A small boat floating in water	The boat applies a force down on the water and the water applies a force up on the boat.
Normal	Normal force is a reaction force that a surface exerts on another body when the body applies a force to the surface.	A book sitting on a table	The book applies a force down on the table and the table applies a normal force up on the book.

Newton's first law

Newton's first law is often known as the law of **inertia**. Inertia is the concept that bodies will continue to stay stationary (at rest) or moving at a constant speed unless acted on by unbalanced forces. When a car is travelling at a constant speed and suddenly stops, objects and people in the car might continue moving at the same speed because of inertia. The force that made the car stop was applied to the car and not to the objects and occupants. If occupants are wearing a seatbelt or objects are restrained then there is a force applied that will prevent them continuing at a constant speed.

inertia
a property of a body that resists changes to its motion

For there to be changes in the speed of a body, the body must experience acceleration, which is caused by unbalanced forces acting on the body. Table 16.6.3 shows the different effects balanced and unbalanced forces have on motion.

▼ **TABLE 16.6.3** The different effects of balanced and unbalanced forces on bodies

Forces	Acceleration	Possible effects on motion
Balanced forces	No acceleration	• Body remains stationary • Body remains at constant velocity
Unbalanced forces	Acceleration	• Body starts moving • Body speeds up • Body stops • Body slows down • Body changes direction

Figure 16.6.5 shows the different scenarios for balanced and unbalanced horizontal forces acting on a car.

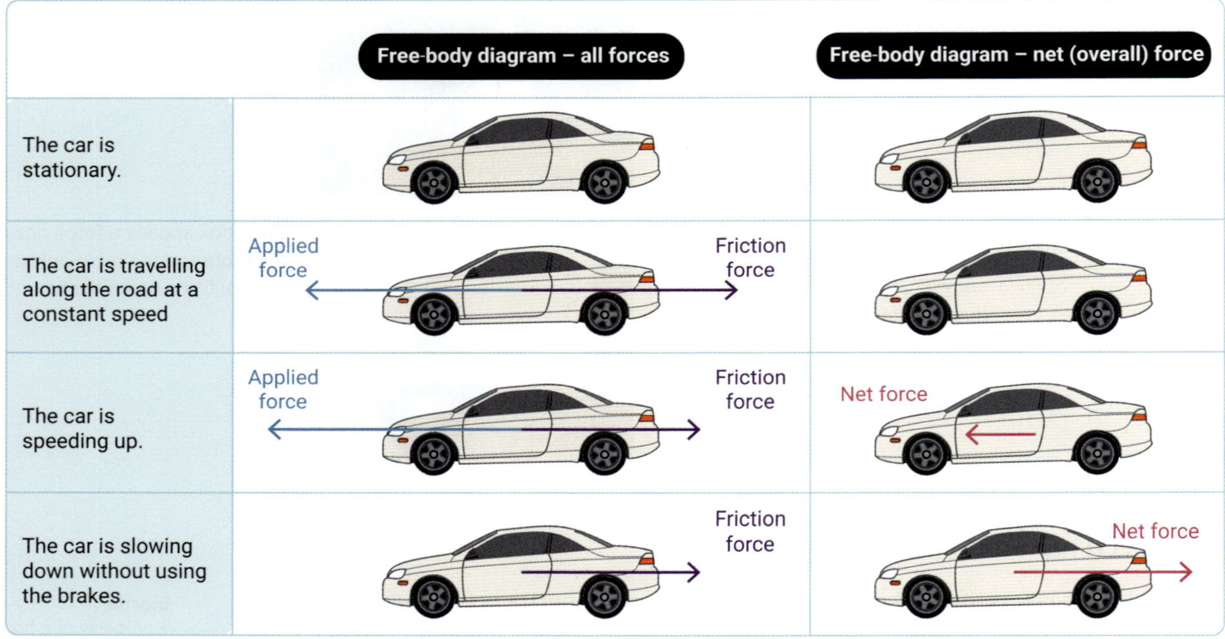

▲ **FIGURE 16.6.5** Different scenarios for balanced and unbalanced horizontal forces acting on a car

16.6 LEARNING CHECK

1 **Define**:
   a force.
   b newton.
   c receiver.
   d normal force.

2 **Construct** a free-body diagram for a bike that is speeding up while experiencing friction.

3 When there is only a small amount of tomato sauce left in a bottle, you can shake the bottle forwards very fast and stop quickly to make the sauce more accessible to the opening of the bottle (Figure 16.6.6). **Explain** how this is an example of inertia.

Istock.com/iStock Signature

▲ **FIGURE 16.6.6** Shaking a bottle to pour out sauce is an example of inertia.

4 **Classify** each of the free-body diagrams shown in Figure 16.6.7 as showing balanced or unbalanced forces.

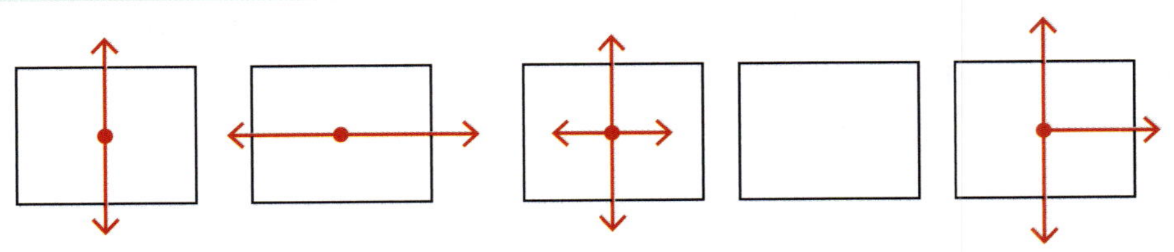

▲ **FIGURE 16.6.7** Examples of free-body diagrams

16.7 Newton's second law

BY THE END OF THIS MODULE, YOU WILL BE ABLE TO:

- ✓ recall Newton's second law in conceptual and quantitative contexts
- ✓ use the mathematical relationship for Newton's second law to calculate force, mass and acceleration for moving bodies.

Video activity
Newton's second law of motion

Extra science investigations
Investigating Newton's second and third laws of motion

Test Newton's second law

Other resource
Worksheet: Newton's second law

GET THINKING

In Module 16.6, we described Newton's second law as: 'The acceleration of a body is directly proportional to the unbalanced force applied to it and inversely proportional to the body's mass.' Talk to a partner or write what you think this means in general and what it means for two bodies of the same mass experiencing different forces.

Relating force, mass and acceleration

Newton's second law states that the acceleration of a body is directly proportional to the unbalanced force applied to it and inversely proportional to the body's mass. In mathematical terms, this law directly translates to the formula:

$$F_{net} = ma$$

where F_{net} is the overall force experienced by a body (N), m is the mass of the body (kg) and a is the acceleration of the body (m/s^2).

The equation shows us that when a greater net force is applied to the same body, the acceleration increases. It also shows us that greater masses require greater forces to accelerate them at the same rate. When there is no overall force acting on a body, there is no acceleration and the speed remains constant. This situation is an application of both Newton's first and second laws. This formula can be used in problems where we use mathematical equations to make predictions about the motion of bodies experiencing forces.

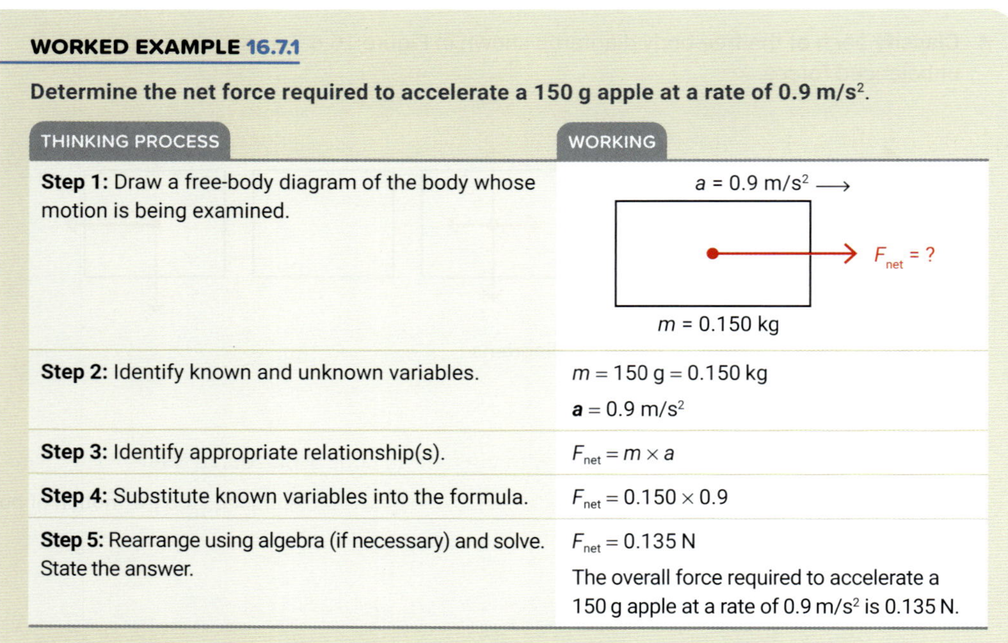

WORKED EXAMPLE 16.7.1

Determine the net force required to accelerate a 150 g apple at a rate of 0.9 m/s^2.

THINKING PROCESS	WORKING
Step 1: Draw a free-body diagram of the body whose motion is being examined.	a = 0.9 m/s^2 ⟶ F_{net} = ? m = 0.150 kg
Step 2: Identify known and unknown variables.	$m = 150\text{ g} = 0.150\text{ kg}$ $\mathbf{a} = 0.9\text{ m/s}^2$
Step 3: Identify appropriate relationship(s).	$F_{net} = m \times a$
Step 4: Substitute known variables into the formula.	$F_{net} = 0.150 \times 0.9$
Step 5: Rearrange using algebra (if necessary) and solve. State the answer.	$F_{net} = 0.135\text{ N}$ The overall force required to accelerate a 150 g apple at a rate of 0.9 m/s^2 is 0.135 N.

WORKED EXAMPLE 16.7.2

A 60 kg skier experiences a gravitational force of 164 N pulling them down a slope and a friction force in the opposite direction of 50 N. Determine the acceleration of the skier down the slope.

THINKING PROCESS	WORKING
Step 1: Draw a free-body diagram of the body whose motion is being examined.	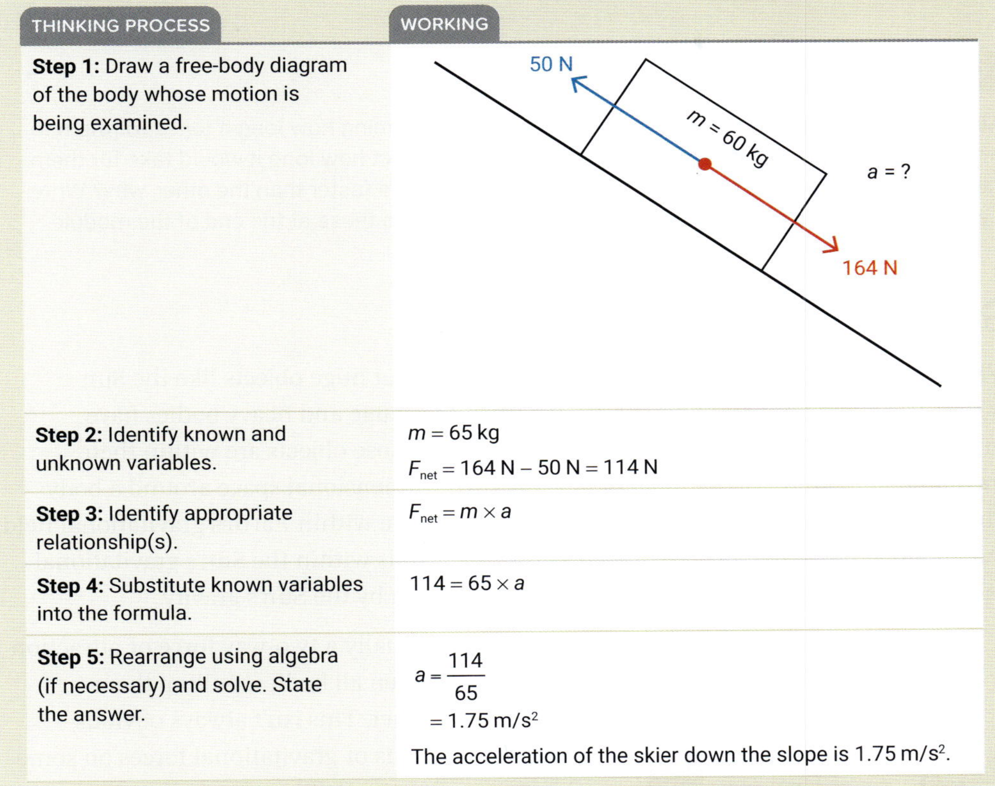
Step 2: Identify known and unknown variables.	$m = 65$ kg $F_{net} = 164\text{ N} - 50\text{ N} = 114\text{ N}$
Step 3: Identify appropriate relationship(s).	$F_{net} = m \times a$
Step 4: Substitute known variables into the formula.	$114 = 65 \times a$
Step 5: Rearrange using algebra (if necessary) and solve. State the answer.	$a = \frac{114}{65}$ $= 1.75\text{ m/s}^2$ The acceleration of the skier down the slope is 1.75 m/s^2.

16.7 LEARNING CHECK

1 **Recall** Newton's second law and **identify** what each symbol represents and the SI units for each.

2 **Calculate** the net force required from a bow to accelerate an arrow of mass 0.120 kg at a rate of 5 m/s^2.

3 **Determine** the mass of a block that accelerates at a rate of 0.12 m/s^2 when a net force of 16 N is applied to it.

4 A model rocket has a mass of 1.1 kg and experiences a net force of 2.5 N from a chemical reaction. **Determine** the acceleration of the rocket.

5 Two siblings are fighting over a remote control that has a mass of 250 g. Sibling A pulls with a force of 25 N on the remote while sibling B pulls in the opposite direction with a force of 10 N.

   a **Determine** which sibling will end up with the remote control closest to them.

   b **Determine** the acceleration of the remote control towards that sibling.

16.8 Acceleration due to gravity

BY THE END OF THIS MODULE, YOU WILL BE ABLE TO:

- ✓ recognise that acceleration due to gravity is constant on Earth's surface and recall its value
- ✓ solve problems for falling and rising objects using acceleration due to gravity.

GET THINKING

Imagine dropping a brick and a feather off a ledge and timing how long it takes for them to reach the ground. Why do the objects fall? Can you predict how long it would take for them to reach the ground? Which object falls faster? If one falls faster than the other, why? Write a brief answer to each question and we will come back to these at the end of the module.

Gravity – a two-way force

gravity
the force of attraction between Earth and objects within its gravitational field

gravitational field
the region where the pull of Earth's gravity is experienced

When we first learn about **gravity**, we usually talk about huge objects like the Sun or Earth as having gravitational forces. We learn that huge and heavy bodies have an attractive force that pulls objects towards them if those objects are within their **gravitational field**. A gravitational field is the three-dimensional space around a body where its gravitational force can be experienced. We are within Earth's gravitational field and are pulled towards Earth by Earth's gravity. Earth is within the Sun's gravitational field and so is pulled towards the Sun and held in orbit by the Sun's gravity.

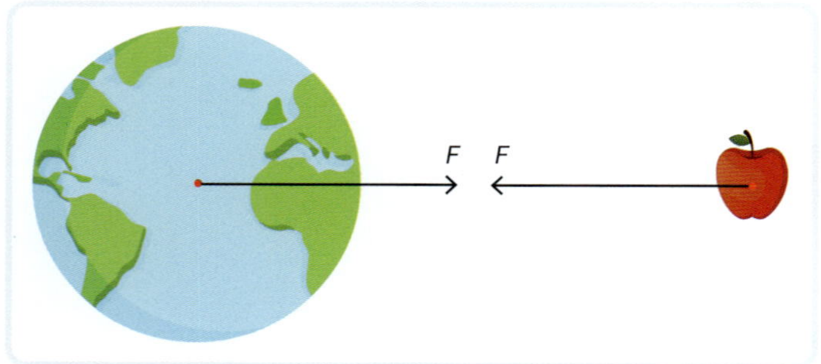

▲ **FIGURE 16.8.1** Gravity is a force that attracts two bodies towards one another. The bodies accelerate differently depending on their mass.

Yet gravity is actually a two-way force of attraction that exists between all bodies, pulling them towards each other. This isn't always obvious because the effects of gravitational forces on some objects are often negligible – so small they aren't worth considering. When an apple is dropped, there is an attractive force between the apple and Earth, pulling the apple towards Earth and pulling Earth towards the apple (Figure 16.8.1). The size of this force is big enough to accelerate the apple because it does not have much mass, but the force is so insignificant that Earth would only accelerate towards the apple by an immeasurably tiny amount.

The gravitational force pulling a body that is close to Earth's surface towards Earth is determined by the mass of Earth and the mass of the body. Heavier bodies are attracted more strongly towards Earth and lighter bodies experience weaker gravitational attraction towards Earth.

Acceleration due to gravity

Galileo Galilei was an Italian physicist in the 1600s who studied gravity and its effects on the motion of objects. In a famous experiment where Galileo supposedly dropped cannonballs from the top of the Leaning Tower of Pisa, he discovered that no matter the mass, objects always accelerated towards Earth's surface at the same rate

9780170491785

(Figure 16.8.2). In other words, the acceleration of bodies due to gravity is constant. However, this is not our everyday experience. We see that a leaf floats slowly to the ground from a tree while an apple drops very quickly. This difference is caused by other forces acting on the bodies as they fall. Forces such as wind or air resistance may push the leaf back upwards, while the apple doesn't experience the same effect of these resistance forces. When the only force acting on each object is Earth's gravity, both bodies will accelerate at the same rate towards Earth's surface. If you were to drop a feather and a hammer in a vacuum, you would see them fall at the same speed.

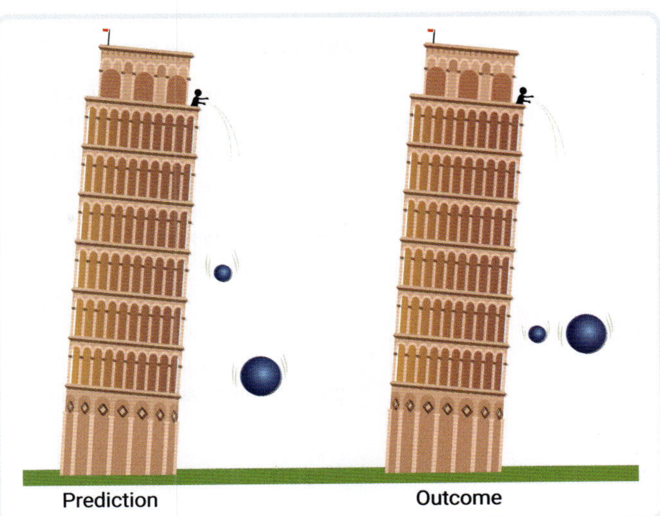

▲ **FIGURE 16.8.2** Galileo's famous Leaning Tower of Pisa experiment suggested that all objects fall at the same rate due to gravity's pull.

Experiments can be conducted to determine exactly what this rate of acceleration is. It can vary slightly, though the accepted value for acceleration due to gravity on Earth's surface is 9.8 m/s^2. This value is often symbolised as g.

With this value, predictions can be made about falling and rising objects. When bodies are falling and the only force acting is gravity's pull, the acceleration of the body is 9.8 m/s^2. When a body has been projected upwards from Earth's surface and the only force acting is gravity's pull, the body slows at the same rate and hence the acceleration is –9.8 m/s^2.

WORKED EXAMPLE 16.8.1

A coin is dropped from the top of a building and falls for 5 seconds. The only force acting on the coin is gravity. Determine the final speed of the coin.

THINKING PROCESS	WORKING
Step 1: Identify known and unknown variables.	$a = g = 9.8$ m/s^2 $u = 0$ m/s (We know this because the coin was dropped.) $t = 5$ s $v = ?$
Step 2: Identify appropriate relationship(s).	$a = \frac{v - u}{t}$
Step 3: Substitute known variables into the formula.	$9.8 = \frac{v - 0}{5}$
Step 4: Rearrange using algebra (if necessary) and solve. State the answer.	$9.8 \times 5 = \frac{v}{5} \times 5$ $v = 49$ m/s The coin would be travelling at a speed of 49 m/s after 5 seconds.

WORKED EXAMPLE 16.8.2

A netball is thrown directly upwards with a speed of 8 m/s. Determine how long it travels upwards until it stops (and the speed of the netball is 0 m/s) before falling back towards Earth. Assume that gravity is the only force acting on the netball.

THINKING PROCESS	WORKING
Step 1: Identify known and unknown variables.	$a = -g = -9.8\ \text{m/s}^2$ $u = 8\ \text{m/s}$ $v = 0\ \text{m/s}$ $t = ?$
Step 2: Identify appropriate relationship(s).	$a = \frac{v-u}{t}$
Step 3: Substitute known variables into the formula.	$-9.8 = \frac{0-8}{t}$
Step 4: Rearrange using algebra (if necessary) and solve. State the answer.	$-9.8 \times t = \frac{-8}{t} \times t$ $-9.8 \times t = -8$ $\frac{-9.8 \times t}{-9.8} = \frac{-8}{-9.8}$ $t = 0.82\ \text{s}$ The netball will stop moving and then start to fall back towards Earth after 0.82 seconds.

Weight force

Newton's second law of motion allows us to quantify the force of gravity that pulls a body towards Earth's surface. For a specific body, the force of gravity pulling it down is often known as the **weight force** or force of weight, as you learned in Stage 4. The weight force for any object can be calculated using a specialised case of Newton's second law:

weight force
the force of gravity pulling an object towards Earth

$$F_w = mg$$

where F_w is the weight force (N), m is the mass of the body (kg) and g is the acceleration due to gravity and equal to $9.8\ \text{m/s}^2$.

WORKED EXAMPLE 16.8.3

A 30 kg dog sits on a couch. Calculate the weight force that is exerted on the couch by the dog due to gravity's pull.

THINKING PROCESS	WORKING
Step 1: Identify known and unknown variables.	$m = 30$ kg $g = 9.8$ m/s^2 $F_w = ?$
Step 2: Identify appropriate relationship(s).	$F_w = mg$
Step 3: Substitute known variables into the formula.	$F_w = 30 \times 9.8$
Step 4: Rearrange using algebra (if necessary) and solve. State the answer.	$F_w = 294$ N The dog exerts a force of 294 N on the couch due to gravity's pull.

☆ ACTIVITY

Find two objects that have different masses and won't be damaged if dropped from a 2 m height. With a partner, time how long it takes for each mass to reach the ground when dropped from a 2 m height. **Discuss** similarities and differences in the time, using the ideas learned in this module about acceleration due to gravity.

16.8 LEARNING CHECK

1 A ball is thrown into the air at an initial speed of 20 m/s. **Calculate** how long it will travel upward before the speed has halved.

2 A scale that measures weight actually measures the weight force exerted on the scale and converts it to a mass value based on the acceleration due to gravity on Earth. If the weight force recorded was 539 N, **determine** the mass of the object on the scale.

3 If a 60 kg person used a scale from Earth on Mars, where the acceleration due to gravity is 3.72 m/s^2, **determine** the:

  a weight force recorded by the scale.

  b mass presented by the scale.

4 A 120 kg provisions package dropped from a plane to some trapped hikers takes 6 seconds to reach the ground.

  a **Calculate** the speed of the package when it hits the ground.

  b Use your answer to part **a** to **explain** why these kinds of packages require parachutes to aid their landing.

WORKING SCIENTIFICALLY

16.9 Validating the work of scientists

SCIENCE SKILLS IN FOCUS

IN THIS MODULE, YOU WILL FOCUS ON LEARNING AND IMPROVING THESE SKILLS:

- selecting and using appropriate mathematical relationships and graphical analysis to organise and process data
- analysing and identifying trends and data anomalies or errors
- recognising how the work of scientists is validated by other scientists.

Published science reports and validity

In the past, when scientists completed their research, they would usually communicate their findings to other scientists by publishing an article in a scientific journal, giving a presentation at a conference or writing a book. These days, scientists also publish on the internet to spread information more quickly and easily to large audiences.

Scientists publish their research so that others can critique what they have done and check whether their conclusions are accurate. Having their method repeated by others increases the reliability and validity of the researcher's conclusions. It is very important for science to be validated so that we can develop accurate theories and laws, like the theory of gravity and Newton's second law.

Making an investigation valid

Improving validity is achieved by:

1. using the same type of apparatus
2. following the same procedure
3. using the same measuring devices
4. using the same variables (controlled, independent and dependent)
5. statistically testing the new results to compare them with the original results.

Statistical testing

As you learned in Chapter 1, random errors can affect the accuracy of results. If a researcher carries out statistical testing of their results, it can reveal the effect of errors that may be outside their control. The more the research is replicated, and the more the replications produce similar statistical results, the more confident the researcher can be that their scientific findings are accurate.

Technology and scientific theories

As technology improves, historic research should also be replicated. For example, scientists on space stations conduct all kinds of experiments involving gravity, to explore investigable questions and solve scientific problems.

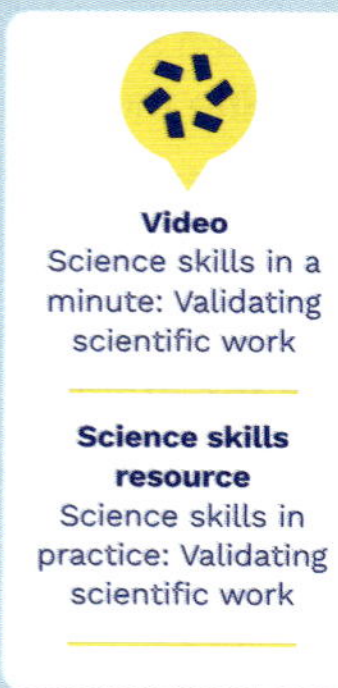

INVESTIGATING NEWTON'S SECOND LAW AND MEASURING ACCELERATION DUE TO GRAVITY

In this activity, you will apply the methods from the blue Science skills in focus box to investigate Newton's second law in the context of falling objects. Remember from Module 16.7 that Newton's second law is applied to the force produced by weight in the following way:

$$F = ma$$

$$F_w = mg$$

This equation shows a theoretically linear relationship with weight force (F_w) as the dependent variable (y), acceleration due to gravity (g) as constant (m as in gradient) and mass (m) as the independent variable (x).

AIM

To investigate the relationship between the mass of an object and its weight force

MATERIALS AND EQUIPMENT

- ☑ spring balance
- ☑ 5 different hanging masses
- ☑ spreadsheet software

PROCEDURE

1 Construct an appropriate results table for the relationship you are investigating.

2 Ensure the spring balance is calibrated and that when there is no mass hanging from the balance it reads 0 N.

3 Attach the lightest mass to the spring balance and record the force shown on the spring balance.

4 Increase the mass and record the force shown on the spring balance.

5 Repeat step 4 another three or four times so that there is a minimum of five data points recorded.

6 Repeat steps 3–5 one more time, so that two values are recorded for each mass used.

RESULTS

Present the results table appropriately, including headings, units and a column for average force value for each mass.

ANALYSIS

1 **Apply** a graphical method to **determine** the acceleration due to gravity using the average force value and a linear line of best fit. Use the spreadsheet software to generate the graph and add a trendline with an equation.

2 Based on the graph produced, **justify** why a linear trendline is appropriate for the data.

3 Based on the graph produced, **identify** if there were any anomalous data points.

4 **Identify** variations of the recorded force values above and below the average force value for each mass tested.

  a Is there random error in the measured values?

  b How could the effect of this random error be minimised, if the experiment was to be repeated?

  c Suggest some reasons that the random error may have occurred.

5 **Compare** the experimental value for g with the theoretical value of 9.8 m/s^2. You might include a percentage error calculation using the formula:

$$\text{Percentage error} = \left(\frac{|\text{theoretical value experimental value}|}{\text{theoretical value}}\right) \times 100\%$$

6 How has conducting this experiment helped validate Newton's second law and how it is used in relation to gravity?

7 What things could you do to improve the validity of this experiment?

CONCLUSION

1 Formulate a conclusion about the relationship between mass and weight force for a hanging body on Earth's surface.

2 Formulate a conclusion about the acceleration due to gravity.

16.10 Vector quantities and displacement

BY THE END OF THIS MODULE, YOU WILL BE ABLE TO:

- ✓ distinguish between scalar and vector quantities and give examples of each
- ✓ distinguish between distance and displacement
- ✓ perform calculations for displacement using vector addition.

GET THINKING

What does displacement mean to you? Where have you heard it before? Write some ideas about what you think displacement could be before you start this module.

Scalar and vector quantities

scalar quantity a measurement that has a magnitude only and no direction

vector quantity a measurement that has both a magnitude and a direction

A **scalar quantity** is a quantity that has a magnitude only. Scalar quantities include mass, time and temperature. A **vector quantity** is a quantity that has a magnitude and a direction. One vector quantity that you have learned about is force. When a force of 10 N is applied to a body, the direction of this force is important. Is the force acting upwards on the body? From the left? The direction of the vector quantity is just as important as the magnitude of the quantity.

When a vector quantity is provided, it is usually stated with a magnitude and a direction. For example, if you push a block to the left with a magnitude of 15 N, the force applied is 15 N to the left. Vector quantities are usually distinguished from scalar quantities by including a half-arrowhead above the symbol. Force as a vector is symbolised as $\overline{F}$, whereas F symbolises just the magnitude of the force – the scalar quantity. So, $F = 15$ N and $\overline{F} = 15$ N to the left.

Some scalar and vector quantities, their symbols and SI units are given in Table 16.10.1.

TABLE 16.10.1 Some scalar quantities and vector quantities and their SI units

Scalar quantity	Symbol	SI unit	Vector quantity	Symbol	SI unit
Distance	d	m	Displacement	$\overline{s}$	m
Speed	v or u	m/s	Velocity	$\overline{v}$ or $\overline{u}$	m/s
Time	t	s	Acceleration	$\overline{a}$	m/s^2
Mass	m	kg	Force	$\overline{F}$	N

Displacement versus distance

displacement the distance and direction that a body's position is relative to the starting point; measured in metres (m)

Distance is a scalar quantity because it does not require a direction to be measured. A jog around the neighbourhood could take many twists and turns, but the total distance could still be measured as 2 km. **Displacement** is a length measurement that also has a direction, making it a vector quantity. Displacement is defined as the position reached by a body relative to the starting point.

9780170491785

Figure 16.10.1 shows the jog distance to be 2 km, but the displacement is measured as the distance the person is at the end of the jog from where they started. Their displacement at the end of the jog is 850 m east.

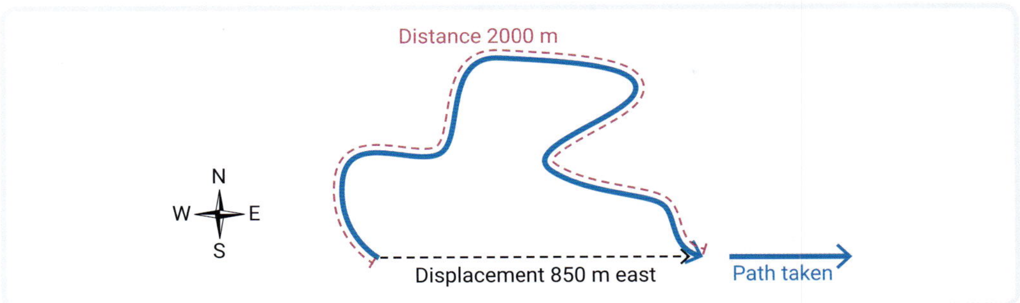

▲ **FIGURE 16.10.1** The displacement is not the length of the total path taken but the distance and direction from the starting point.

When displacements occur in one dimension (along a single straight line), we often use positive and negative signs to indicate the direction of the displacement and to help with calculations. The positive and negative might represent up and down or left and right. An **origin** or starting position can be defined and distances in one direction are considered positive displacements and distances in the opposite direction negative. In Figure 16.10.2, a body at the position marked A in the second diagram would have a displacement of +25 cm and a body at position B would have a displacement of −40 cm.

origin
a defined point that indicates the starting point or a displacement of zero

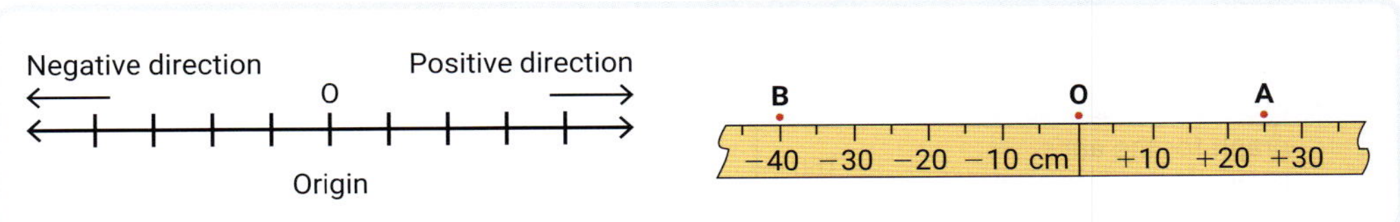

▲ **FIGURE 16.10.2** Defining an origin, a positive direction and a negative direction allows signs to be used for displacement directions.

Displacement–time graphs look similar to distance–time graphs, except that displacement can go below the *x*-axis. That's because negative displacements measure distance in the opposite direction to positive displacements. The other difference is that a displacement–time graph can have a negative gradient, which represents motion in a negative direction.

WORKED EXAMPLE 16.10.1

a Use the labelled ruler in Figure 16.10.3 to describe the motion of an ant, given the displacement–time graph for the ant in Figure 16.10.4.

b Determine the total distance travelled by the ant.

c Draw a distance–time graph to show the ant's journey.

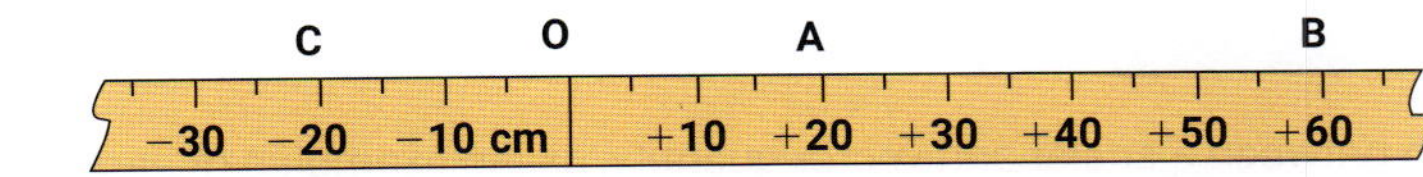

▲ **FIGURE 16.10.3** A defined origin and some labelled positions for an ant's journey

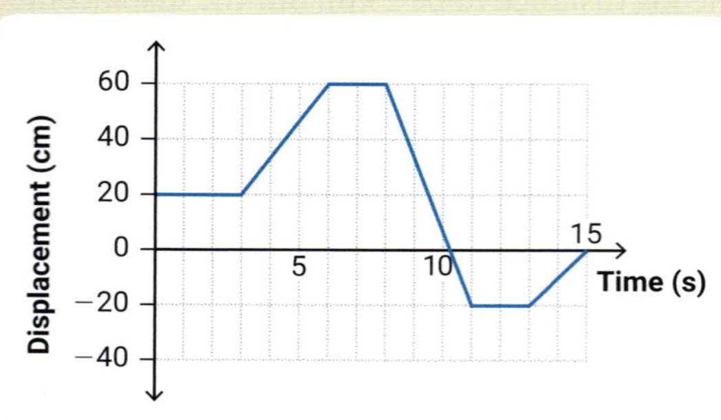

▲ **FIGURE 16.10.4** A displacement–time graph for an ant's journey

The gradient of the displacement–time graph is positive (going up) or negative (going down) depending on the direction the ant was travelling.

QUESTIONS	WORKING
a Use the labelled ruler in Figure 16.10.3 to describe the motion of an ant, given the displacement–time graph for the ant in Figure 16.10.4	The ant starts at position A and stays there for 3 seconds. Then the ant moves 40 cm to the right at a constant speed to position B over a period of 3 seconds before stopping for 2 seconds. The ant then turns around and moves 80 cm to the left at a constant speed, passes the origin and arrives at position C 3 seconds later. The ant stays there for 2 seconds and then moves right again for 20 cm at a constant speed and arrives at the origin 2 seconds later.
b Determine the total distance travelled by the ant.	$d = 40 + 80 + 20 = 140$ cm
c Draw a distance–time graph for the ant's journey.	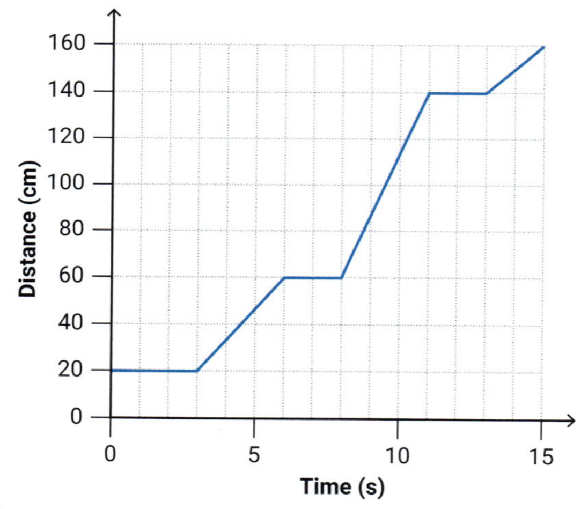

16.10 LEARNING CHECK

1 **Distinguish** between a scalar and vector quantity, giving examples of each.

2 **Evaluate** whether temperature is a scalar or vector quantity, giving reasons for your answer.

9780170491785

16.11 Velocity

BY THE END OF THIS MODULE, YOU WILL BE ABLE TO:

- ✓ distinguish between speed and velocity
- ✓ perform calculations for velocity in one dimension using vector addition
- ✓ recall the formula for calculating average velocity and use it in calculations.

GET THINKING

Write down two things you already know about speed and why you think speed and direction are related.

Velocity versus speed

velocity
the rate of change of displacement with time, a vector quantity for speed; measured in metres per second (m/s) and includes a direction

Velocity and speed both measure the rate at which a body moves over time. Speed is a scalar quantity that does not have a direction. **Velocity** is the vector quantity of a measure very similar to speed and includes a direction.

Figure 16.11.1 shows how speed and velocity differ. Travelling at 25 m/s south is different from travelling at 25 m/s north, even though the speed is the same.

By defining one of these directions as positive and the other as negative, it is easier to use a plus or minus sign in front of the number to give a quantitative value for velocity.

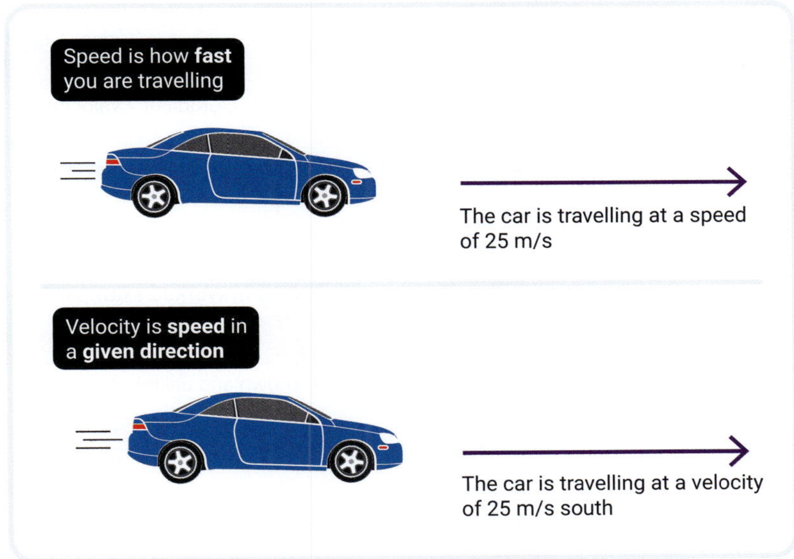

▲ **FIGURE 16.11.1** **Velocity is how fast something is travelling in a particular direction.**

We can see another difference between speed and velocity when we compare average speed and average velocity. We learned in Module 16.3 that the average speed for a journey could be calculated by dividing the total distance travelled by total time. Average velocity doesn't measure the rate at which distance changes over time. Instead, average velocity measures the rate at which *displacement* changes over time. Average velocity can be found using the formula:

$$\text{Average velocity} = \frac{\text{change in displacement}}{\text{time period}}$$

$$\vec{v}_{av} = \frac{\vec{s}_2 - \vec{s}_1}{t_2 - t_1}$$

where $\vec{s}_2$ is the displacement (m) of the body at the end of the time period, $\vec{s}_1$ is the starting point, t_2 is the end time of the time period and t_1 is the start time of the time period.

Velocity–time graphs are very similar to speed–time graphs except that they can go below the x-axis in the negative direction.

In the same way that speed is represented by the gradient on a distance–time graph, velocity is represented by the gradient on a displacement–time graph. A positive gradient on a displacement–time graph indicates a positive velocity, which is the speed in the *positive direction*. A negative gradient on a displacement–time graph indicates a negative velocity, which is the speed in the *negative direction*. This can be seen in Figure 16.11.2.

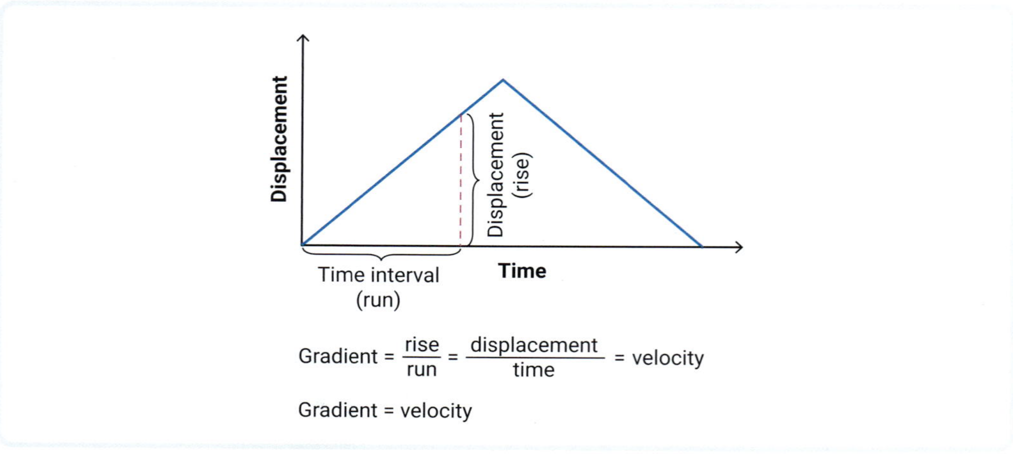

▲ **FIGURE 16.11.2** The gradient of a displacement–time graph is the velocity.

WORKED EXAMPLE 16.11.1

a For the ant's journey in Worked example 16.10.1, calculate the average speed of the ant.

THINKING PROCESS	WORKING
Step 1: Identify the known and unknown variables.	$d_1 = 0$ cm (at the start of the journey, the ant has not travelled any distance) $d_2 = 140$ cm $= 0.14$ m $t_1 = 0$ s $t_2 = 15$ s $v_{av} = ?$
Step 2: Identify appropriate relationship(s).	$v_{av} = \frac{d_2 - d_1}{t_2 - t_1}$
Step 3: Substitute known values.	$v_{av} = \frac{(0.14 - 0)}{15 - 0}$
Step 4: Rearrange algebraically (if necessary) and solve. State the answer.	$v_{av} = 0.093$ m/s Over the whole journey, the average speed is 0.093 m/s.

b For the ant's journey in Worked example 16.10.1, calculate the average velocity of the ant.

THINKING PROCESS	WORKING
Step 1: Identify known and unknown variables.	$\bar{s}_1 = +20$ cm $= 0.20$ m (at the start of the journey the ant is at position A, 20 cm to the right) $\bar{s}_2 = 0$ cm (at the end of the journey the ant is at the origin) $t_1 = 0$ s $t_2 = 15$ s $\bar{v}_{av} = ?$
Step 2: Identify appropriate relationship(s).	$v_{av} = \frac{s_2 - s_1}{t_2 - t_1}$
Step 3: Substitute known values.	$v_{av} = \frac{0 - 0.20}{15 - 0}$
Step 4: Rearrange algebraically (if necessary) and solve. State the answer.	$v_{av} = -0.013$ m/s Over the whole journey, the average velocity is 0.013 m/s to the left.

WORKED EXAMPLE 16.11.2

a Using the solutions from the ant's journey in Worked example 16.10.1, determine the speed of the ant throughout each section of the journey.

TIME PERIOD (s)	SPEED (cm/s)
0–3	0
3–6	$v = \frac{d}{t} = \frac{60 - 20}{3} = 13.3$
6–8	0
8–11	$v = \frac{d}{t} = \frac{60 + 20}{3} = 26.7$
11–13	0
13–15	$v = \frac{d}{t} = \frac{20}{2} = 10$

b Using the solutions from the ant's journey in Worked example 16.10.1, determine the velocity of the ant throughout each section of the journey.

Note: Based on the defined origin, the positive direction and the negative direction, right is positive and left is negative. Velocities can be found by adding a direction to the speed for each section.

TIME PERIOD (s)	SPEED (cm/s)	DIRECTION	VELOCITY (cm/s)
0–3	0	–	0
3–6	13.3	Right (+)	+13.3
6–8	0	–	
8–11	26.7	Left (–)	–26.7
11–13	0	–	
13–15	10	Right (+)	+10

The velocities could also be found by finding the gradient of the displacement–time graph.

TIME PERIOD (s)	GRADIENT = VELOCITY (cm/s)
0–3	$\bar{v} = \frac{\text{rise}}{\text{run}} = 0$
3–6	$\bar{v} = \frac{\text{rise}}{\text{run}} = \frac{s_2 - s_1}{t_2 - t_1} = \frac{60 - 20}{6 - 3} = \frac{40}{3} = +13.3\text{ cm/s}$
6–8	$\bar{v} = \frac{\text{rise}}{\text{run}} = 0$
8–11	$\bar{v} = \frac{\text{rise}}{\text{run}} = \frac{s_2 - s_1}{t_2 - t_1} = \frac{-20 - 60}{11 - 8} = \frac{-80}{3} = -26.7\text{ cm/s}$
11–13	$\bar{v} = \frac{\text{rise}}{\text{run}} = 0$
13–15	$v = \frac{\text{rise}}{\text{run}} = \frac{s_2 - s_1}{t_2 - t_1} = \frac{0 - -20}{15 - 30} = \frac{20}{2} = +10\text{ cm/s}$

We can construct a speed–time graph for the ant's journey, as shown in Figure 16.11.3.
We can also construct a velocity–time graph for the ant's journey, as shown in Figure 16.11.4.

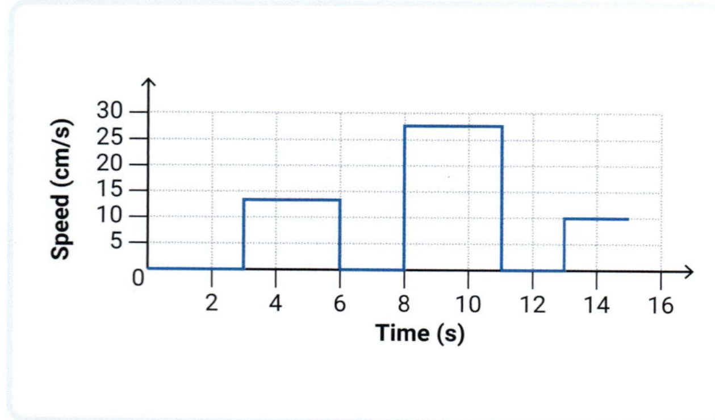

▲ **FIGURE 16.11.3** A speed–time graph for the ant's journey

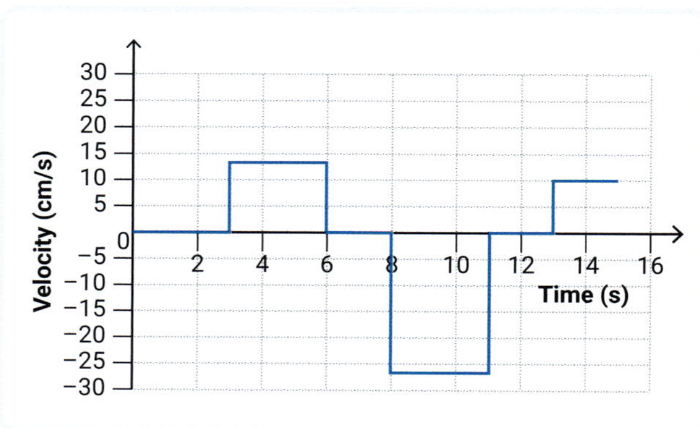

▲ **FIGURE 16.11.4** A velocity–time graph for the ant's journey

9780170491785

Acceleration – a vector quantity

Of course, speeds and velocities are not always constant – they change when a body is accelerating. Module 16.4 considered acceleration a vector quantity because it used positive accelerations and negative accelerations to compare speeding up with slowing down. However, since it is a vector quantity, acceleration should always have a direction. An example will help explain the idea of an object speeding up or slowing down while taking account of different directions.

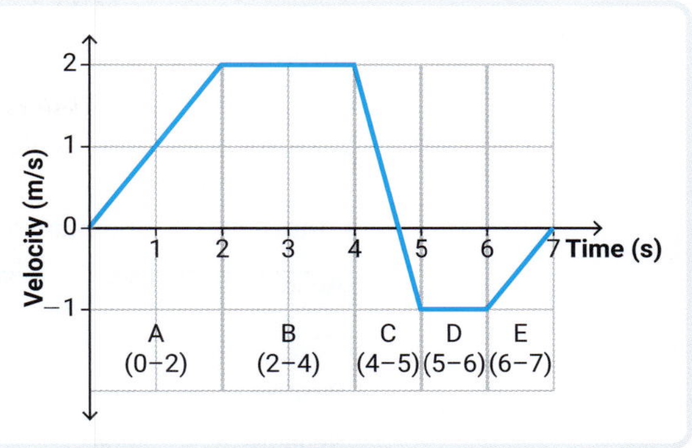

▲ **FIGURE 16.11.5** Acceleration can be found by calculating the gradient of a velocity–time graph.

Figure 16.11.5 shows the velocity on the journey of a body travelling in a forwards (+) and backwards (−) direction. The acceleration can be found by calculating the gradient of the velocity–time graph, similar to finding the velocity on a displacement–time graph. This has been broken down into five sections, and each section is summarised in Table 16.11.1.

▼ **TABLE 16.11.1** Investigating acceleration as a vector

Section	Time (s)	Velocity (m/s)	Direction	Acceleration (m/s²)	Acceleration meaning
A	0–2	0 to 2	Increasing positive velocity = forwards	$\vec{a} = \frac{\vec{v} - \vec{u}}{t_2 - t_1}$ $= \frac{2-0}{2-0} = +1$	The body is moving forwards and speeding up.
B	2–4	2	Constant positive velocity = forwards	$\vec{a} = \frac{\vec{v} - \vec{u}}{t_2 - t_1}$ $= \frac{2-2}{2-0} = 0$	The body is not changing speed or direction; acceleration is zero.
C	4–5	2 to −1	The body is moving forwards for 0.7 s (decreasing positive velocity) and then stops moving momentarily (velocity = 0) before moving backwards for 0.3 s (increasing negative velocity)	$\vec{a} = \frac{\vec{v} - \vec{u}}{t_2 - t_1}$ $= \frac{-1-2}{5-4} = -3$	The body is moving forwards and slowing down for 0.7 s and then speeds up in the backwards direction for 0.3 s.
D	5–6	−1	Constant negative velocity = backwards	$\vec{a} = \frac{\vec{v} - \vec{u}}{t_2 - t_1}$ $= \frac{-1--1}{6-5} = 0$	The body is not changing speed or direction; acceleration is zero.
E	6–7	−1 to 0	Decreasing negative velocity = backwards	$\vec{a} = \frac{\vec{v} - \vec{u}}{t_2 - t_1}$ $= \frac{0--1}{7-6} = +1$	The body is moving backwards and slowing down.

As seen in Table 16.11.1, positive and negative signs on an acceleration indicate different types of change in velocity, depending on the direction in which the body is travelling. A summary of how to interpret the value of acceleration as a vector quantity is given in Table 16.11.2.

▼ **TABLE 16.11.2** Summary of how to interpret acceleration values as a vector

Acceleration value	**Direction travelling**	
	Positive direction	**Negative direction**
Positive acceleration	Speeding up	Slowing down
Negative acceleration	Slowing down	Speeding up

16.11 LEARNING CHECK

Consider the displacement–time graph shown in Figure 16.11.6.

1 **Describe** the motion of the body for the journey.
2 **Determine** the total distance travelled.
3 **Calculate** the average speed.
4 **Calculate** the average velocity.
5 **Determine** the velocity throughout each section of the journey.
6 **Construct** a velocity–time graph for the journey.

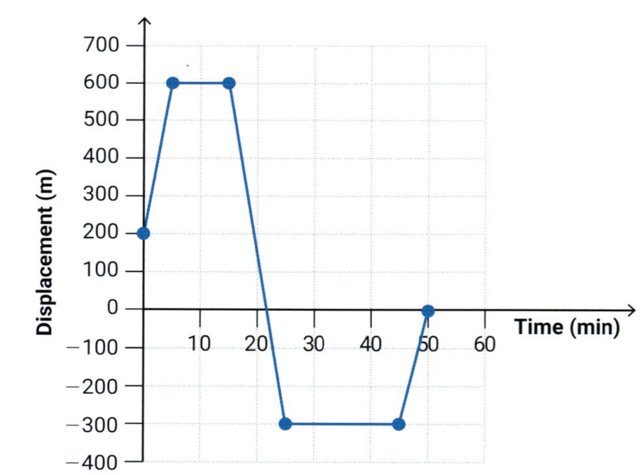

▲ **FIGURE 16.11.6** A displacement–time graph

16.12 Applying knowledge of motion

16.12

BY THE END OF THIS MODULE, YOU WILL BE ABLE TO:

✓ identify examples of how Newton's laws of motion are applied to improve the safety of cars and explain how they work

✓ identify examples of Newton's laws of motion in performance-enhancement strategies for sports and explain how they work.

GET THINKING

Newton's laws of motion apply to things we observe and do every day. Look at the subheadings in this module. How do you think Newton's laws could apply to car safety and sports performance?

Video activity
Science of golf

Extra science investigations
Motion in traffic
Stopping distance

Other resource
Activity sheet: Head-on smash at 60 km/h

Objects in motion are everywhere. Motion is a vital component of many parts of life, including transport, construction, sport and almost every industry. Newton's laws of motion are applied in all these settings to improve the safety and performance of many processes and operations. Some examples of their application in motor vehicle safety and sport performance are discussed in this module.

Motor vehicle safety

In 1925, there were 700 road deaths in Australia, which was about 230 deaths per 100 000 registered vehicles. In 2020, in Australia there were 1095 road deaths. However, factoring in the increased popularity of transport by car in the 21st century, this was equivalent to about five deaths per 100 000 registered vehicles. This dramatic reduction in road toll in the last 100 years has been a result of advances in car safety and design. Some of the most significant safety features in modern cars include seatbelts, airbags and headrests.

Seatbelts

When a car stops suddenly during a front-on collision, an unrestrained driver or passenger will continue moving. Newton's first law of motion explains that this is due to the inertia of the occupants. Without a seatbelt, the driver could hit the steering wheel or any of the car's occupants could pass through the windscreen. A seatbelt can apply an unbalanced force to occupants of the car, changing their velocity as the car's velocity changes and keeping them in their seats (Figure 16.12.1).

Monkey Business Images/Shutterstock.com

▲ **FIGURE 16.12.1** The operation of seatbelts is an example of the application of Newton's first law of motion to stop the motion of bodies during an accident.

Airbags

Despite the important role of seatbelts, parts of the upper body can sometimes continue to move forwards due to inertia. For passengers in the back seat, this is not as problematic because the seats in front of them are usually far away and are a softer barrier. In the front of a car, people's heads can collide with the dashboard, which can cause head injuries. Supplemental restraint systems in modern cars are activated when the car suffers an impact to release airbags automatically. Airbags apply a force to the upper body to decelerate the head and shoulders instead of allowing them to continue at the speed they were travelling before the collision (Figure 16.12.2).

Patti McConville/Alamy Stock Photo

▲ **FIGURE 16.12.2** Airbags provide an unbalanced force to the upper body of front seat car passengers to decelerate their bodies after a car stops suddenly.

Headrests

Different dangers exist when a car experiences a rear-end collision, since the car is being propelled forward quickly instead of suddenly stopping. The seat of the car applies a force onto most of a passenger's body, moving their body forward with the rest of the car. In old cars, there were often no headrests (Figure 16.12.3). This meant that when the body was propelled forward with the car, the passengers' heads remained at rest since they hadn't experienced the same applied force from the car seat. The fast movement of the body forward without the head moving with it can cause neck injuries known as whiplash, or spinal injuries. Headrests exist in modern cars to apply the same force to the head as the body, so that the passenger's body and head move forward as a whole.

ClassicStock/Alamy Stock Photo

▲ **FIGURE 16.12.3** Old cars did not have headrests, a safety feature to avoid whiplash injuries.

9780170491785

Using Newton's laws to analyse sport

The sporting world is another area where Newton's laws of motion are applied. Understanding how things move is the key to enhancing performance in most sports.

Golf

▲ **FIGURE 16.12.4** Swinging a golf club through the stroke, including the follow-through

A golfer swings a club to hit a small golf ball out onto the fairway. Part of a golfer's technique is to have a follow-through swing and to accelerate the ball to a high speed so that it can reach long distances. When the golf club makes contact with the ball, the club exerts a force on the golf ball and, in agreement with Newton's first law, the unbalanced forces cause the ball to accelerate. According to Newton's second law, the smaller the mass, the greater the acceleration of the ball for the same applied force. This is why golf balls are light (less than 46 g). Inertia also allows the golf club to follow through its swing, as the forces applied to the club are not great enough to change its motion.

According to Newton's third law, when the club applies a force on the ball, the ball applies an equal and opposite reaction force back on the club. If this force were to cause a significant acceleration for the club in the opposite direction to the swing, this would interfere with the technique of the swing and the club would not be able to follow through, as shown in Figure 16.12.4. Since a golf club is heavier than the ball, the effect of the force on the club is reduced. Newton's second law shows us that when the mass is greater, the acceleration effect of an applied force is much smaller. The same sized force that causes the ball to accelerate out onto the golf course has a deceleration effect of about 10 per cent on the club because of their difference in mass, and the club slows down only slightly during the swing.

net force
the overall force acting on a body; measured in newtons (N)

Cycling

Inertia plays a major role in the performance of a cyclist. A cyclist applies force through pedalling to reach a certain speed, and the inertia of the bicycle and cyclist allows them to move along with this speed until the friction between the tyres and the road slows them down. One of the other resistance forces that cause the cyclist to exert more force to reach higher speeds is air resistance. Air resistance lowers the **net force** forwards for the cyclist and reduces the acceleration. Each of the forces affecting the motion of a cyclist is demonstrated in Figure 16.12.5.

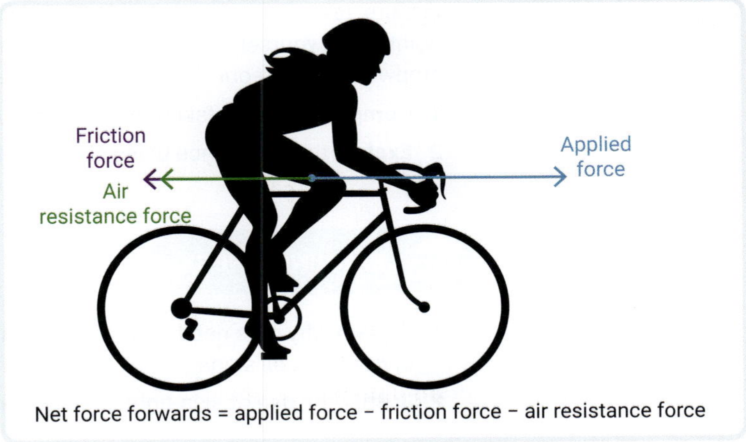

▲ **FIGURE 16.12.5** A summary of the forces affecting the motion of a cyclist

PA Images/Alamy Stock Photo

▲ **FIGURE 16.12.6** Cyclists use many strategies to reduce air resistance and enhance their speed and performance.

Professional cyclists use many strategies to decrease air resistance (Figure 16.12.6). Some of these strategies include:

- carefully engineered aerodynamic bikes that have a specialised shape to decrease air resistance
- aerodynamic helmets
- skin-tight elastane clothing
- hunched compact riding positions
- shaving their legs to minimise the air resistance caused by their leg hairs.

ACTIVITY

In pairs, take turns completing the following. Run at a jog and when you reach a certain point, bring yourself to a stop. Get your partner to time how long it took you to stop using a stopwatch. Now sprint to the same point and time how long it takes to stop.

1 **Compare** the time taken to slow to a stop from the two different speeds.
2 **Explain** this difference using Newton's laws of motion.

16.12 LEARNING CHECK

1 **Explain** why cars need both seatbelts and airbags in the front seat to protect occupants in front-on collisions.
2 **Justify** the use of skin-tight clothing for a racing cyclist rather than baggy clothing.

9780170491785

16.13 Spear-throwers

IN THIS MODULE, YOU WILL:

✓ investigate the effect of Aboriginal and Torres Strait Islander Peoples' spear-throwers on the speed and impact force of an object.

Spear construction

Aboriginal and Torres Strait Islander Peoples use throwing spears for many purposes, including hunting and fishing. Throwing spears are often made with several different materials. The shaft of a spear is usually made of wood for strength and flexibility. The tip is constructed from a material that can withstand the force of the impact when it strikes the intended target. Prior to colonisation, the Gadigal People of the Eora Nation (Sydney region, New South Wales) constructed fishing spears using stingray barbs as tips. Multiple stingray barbs were attached to the spear shaft in an orientation that ensured the tip could not be dislodged from the flesh of the fish on impact (Figure 16.13.1).

© Joe Sambono

▲ **FIGURE 16.13.1** Stingray barbs used as spear tips.

Use of spear-throwers

Spears can be launched by hand or by using a spear-thrower. A typical spear-thrower consists of a shaft with a small peg that holds the base of the spear, and a handgrip to help launch the spear (Figure 16.13.2).

© Joe Sambono

▲ **FIGURE 16.13.2** A peg on a spear-thrower, used to hold the base of a spear

The spear is placed onto the spear-thrower and is launched with a throwing motion (Figure 16.13.3).

© Joe Sambono

▲ **FIGURE 16.13.3** How a spear is placed on a spear-thrower

The development of spear-throwers demonstrates Aboriginal and Torres Strait Islander Peoples' knowledge of the relationship between mass, force and acceleration, just as Newton's second law. A spear-thrower is designed to increase the acceleration of the spear and therefore the speed of a spear. A spear-thrower ensures the spear reaches a higher speed, allowing the spear to travel further and be more effective on impact with the target.

Different combinations of spear and spear-thrower are related to their target species (Figure 16.13.4). Heavier spears launched with a spear-thrower have a lower velocity than lighter spears.

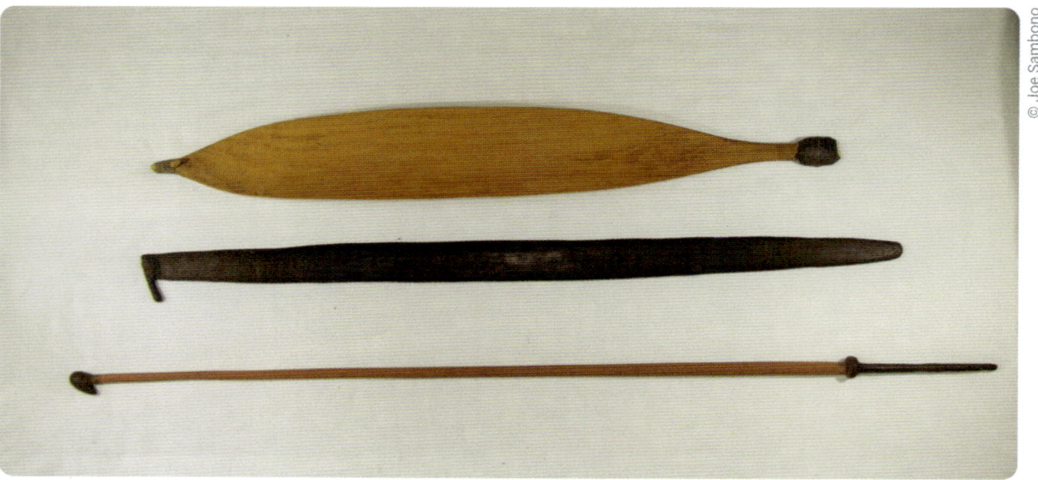

© Joe Sambono

▲ **FIGURE 16.13.4** The weight and shape of the spear-thrower and spear affect speed, energy and the force of impact.

Investigating the relationship between mass and the force of impact

ACTIVITY

It is not culturally respectful to use a traditional spear-thrower without consulting Elders or Knowledge Holders. It is also not safe to throw spears or sticks. In this activity, you will use a ball thrower to launch a projectile. Like a spear-thrower, a ball thrower acts to increase the launch speed of an object – in this case a ball.

Materials and equipment

- balls with different masses
- ball thrower
- velocity speed gun (or timer)
- measuring tape
- scales

Make balls with different masses by drilling a small hole in a tennis ball and filling it with different amounts of dry sand. Seal the hole with glue. Balls made this way will ensure the hold on the ball in the thrower and the friction through the air are controlled variables in the experiment. Alternatively, balls of different weights (e.g. plastic balls, tennis balls or cricket balls) can be used, although they may not fit in the ball thrower. They may also introduce additional variables to the investigation.

Procedure

1. Throw a tennis ball with your arm.
2. Use a speed sensor to record the speed of the ball. If a speed sensor is not available, estimate the speed by measuring the distance travelled and time from launch to landing. The speed can then be calculated using the ideas from Module 16.2.
3. Measure the distance the ball is thrown. You will need to do several trials. Record your data in a table.
4. Throw a tennis ball using the ball thrower. Try to ensure you are throwing it at the same angle.
5. Measure or estimate the speed of the ball. Record your data in a table.
6. Investigate the effect of the mass of the ball by repeating steps 1–4 with balls of different masses. Record your data in the table.

Analysis

1. What did you **observe**?
2. How does the lever (ball thrower) affect the speed of the ball?
3. What effect did the mass of the projectile have on the speed?
4. The speed or average speed of the ball measured or estimated in step 5 will be directly related to the acceleration of the ball caused by the force applied to it by the arm or by the thrower.

   We can reasonably assume that in each case the force applied by the arm stayed approximately the same throughout and that the force applied by the thrower also stayed approximately the same throughout.

   Using the speed or average speed of the ball measured or estimated in step 5 as a measure of acceleration, check if your observations are consistent with Newton's second law.
5. How did Aboriginal and Torres Strait Islander Peoples' development of spear-throwers overcome the limitations of being able to hit targets at long distances and with light projectiles?

SCIENCE IN CONTEXT

16.14 Crash test dummies for vehicle safety

BY THE END OF THIS MODULE, YOU WILL BE ABLE TO:

✓ evaluate the significance of using different car crash test dummies for vehicle safety.

Video activity
Female crash test dummies

Extra science investigation
Crumple zones

Crash testing motor vehicles

Each year in Australia and New Zealand, thousands of adults and children die or are seriously injured in vehicle collisions. Although this number is large, it would be significantly higher if it were not for the safety features that have been mandated for motor vehicles in these countries since the 1960s. Before then, most cars had no built-in protection for vehicle occupants and lacked today's standard safety features such as seatbelts, airbags and sophisticated anti-lock braking systems. These days, safety ratings for new cars are a major selling point. The Australasian New Car Assessment Program (ANCAP) performs safety tests and publishes independent safety ratings for new cars.

ANCAP safety tests include destructive crash tests in a lab to simulate common types of road incidents and to assess how these crashes would affect the driver and passengers within different cars. Crash tests use sophisticated crash test dummies that collect impact data for potential human injuries during collisions. The forces experienced by the dummies when travelling at different speeds and decelerating at different rates can be studied to make recommendations about car safety features.

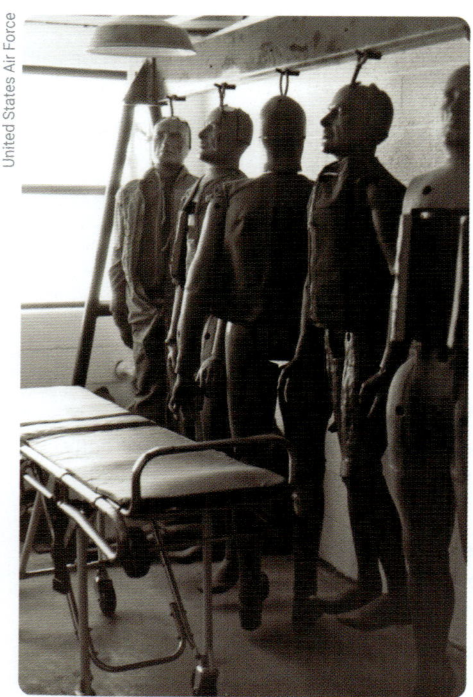

United States Air Force

▲ **FIGURE 16.14.1** The first crash test dummy, Sierra Sam, was developed in the 1940s for the United States Air Force.

The development of the crash test dummy

In the 1940s, researchers and automobile engineers used human cadavers in crash tests. Obviously, there were some ethical issues associated with this practice. This led to the United States Air Force creating Sierra Sam, the first crash test dummy (Figure 16.14.1). Sam was modelled on the average-sized American male.

A more advanced crash test dummy, Hybrid I, was developed by General Motors in the 1970s. Their third-generation dummy, Hybrid III, is the most commonly used dummy in the world today. These days, dummies are made in a range of sizes to reflect different sized bodies (Figure 16.14.2).

9780170491785

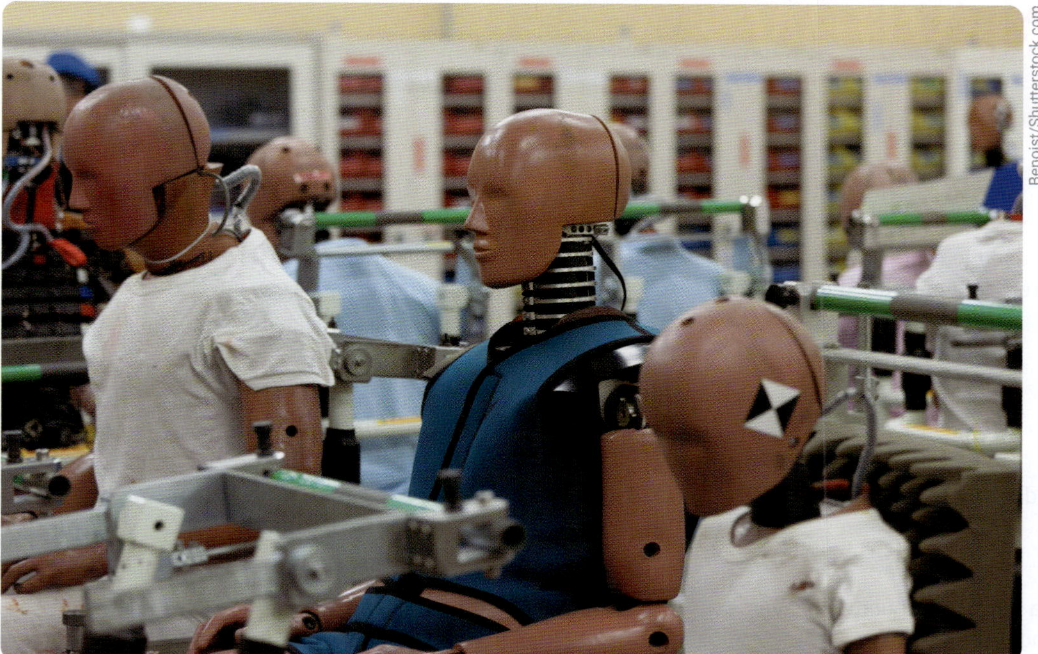

Benoist/Shutterstock.com

▲ **FIGURE 16.14.2** Crash test dummies come in different sizes to show how accidents may impact people with different body types.

Today's crash test dummies are extremely sophisticated, costing between $500 000 and $1 million because of the complex technology and sensors used within them. Unfortunately, in many countries it is not mandated that safety testing uses both male and female dummies, with many only using male dummies. There are many dangerous implications of this, such as seatbelts and airbags designed for men not offering the same protection for women and children, who tend to be smaller than the average man.

There are still questions posed about how well a handful of dummies can accurately predict the impacts of a crash on very diverse body sizes and shapes. Many research institutes are developing virtual testing systems, which many consider to be the future of crash testing. In these tests, computer-generated models of humans with varying size, skeletons and muscle composition can be used to make more accurate predictions about crash impacts for all drivers and passengers. This data can then be used to design and develop modern safety features for road vehicles.

16.14 LEARNING CHECK

1 **Explain** what a crash test dummy is and what it is used for.
2 **Suggest** why the first crash test dummy and those that are commonly used today are based on the male body.
3 **Evaluate** the importance of crash test dummies to society.
4 **Compare** the following scenarios for crash testing to **evaluate** the advantages and disadvantages of each.
   a Using only male crash test dummies
   b Using male and female Hybrid III crash test dummies
   c Using a virtual crash test system for varied body types and compositions

16 REVIEW

REMEMBERING

1 **State** the symbol used to represent each of the following measurements and the SI units they are measured in.
   a Distance
   b Speed
   c Acceleration
   d Force
   e Acceleration due to gravity
   f Mass
   g Time

2 **State** the formula used to calculate the average speed of a moving body.

3 **Define** acceleration in words and using a mathematical formula.

4 **Recall** Newton's three laws of motion.

5 Copy and complete the following sentence: According to Newton's second law, the net force applied to a body is proportional to its ____ and its _____.

6 **Recall** the form of Newton's second law that is specifically used for the force of gravity pulling bodies towards Earth's surface. **Identify** each variable and its SI units.

UNDERSTANDING

7 Imagine your friend had never heard the word 'distance' before. How would you **describe** to them what it means?

8 **Describe** the concept of inertia, giving an example in your response.

9 **Explain** what a negative acceleration value means for a car travelling on a road.

10 **Explain** why the normal force is considered a type of reaction force and is an example of Newton's third law.

11 Copy and complete the following sentence: According to Newton's second law, for a body experiencing a constant net force, if the mass increases the acceleration will _____.

12 **Explain** why the gravitational attraction force between Earth and an elephant is different from the gravitational attraction force between Earth and a puppy.

APPLYING

13 If it takes a snail 2 minutes to travel 10 cm, **determine** how many metres the snail can travel in 1 hour.

14 Provide an example of Newton's third law in action.

15 A dog sitting on a skateboard is pushed by a small child, who applies a force of 45.5 N. The frictional force that resists the forward motion of the skateboard is 12.6 N. If the dog weighs 15 kg and the skateboard is accelerating at a rate of 1.6 m/s^2, **determine** the mass of the skateboard.

16 For a 120 g apple that falls from a very tall tree, **determine** the:
   a weight force pulling the apple towards the ground.
   b speed of the apple when it hits the ground if it is falling for 2.4 seconds.

17 **Explain** the need for headrests in cars as a safety feature.

ANALYSING

18 What knowledge of motion do Aboriginal and Torres Strait Islander Peoples use when developing and using spear-throwers?

19 The distance–time graph below shows the journey of a teenager's afternoon jog.
   a **Identify** the distance of the entire jog.
   b **Identify** how long the teenager was out on their jog.
   c **Identify** when and where the teenager took a break.
   d **Determine** the average speed of the teenager across the whole journey.

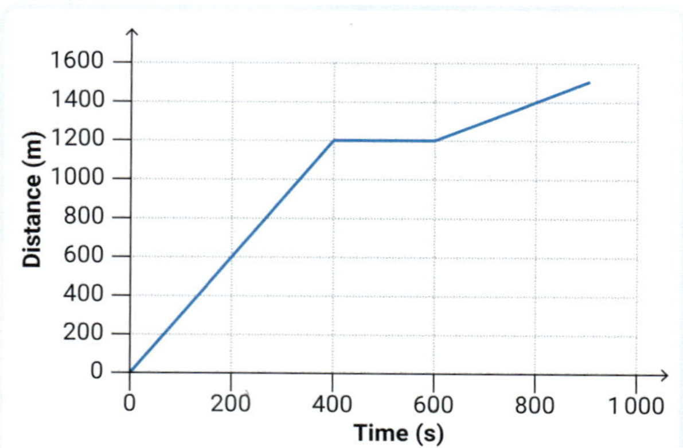

9780170491785

20 A cyclist is riding down a hill. Their speed is recorded on the speed–time graph shown below.

a **Identify** the maximum speed of the cyclist.

b **Identify** the period in which the cyclist is travelling at a constant speed.

c **Identify** the periods when the cyclist is accelerating.

d At what time did the cyclist most likely start using the brakes? **Justify** your answer.

e **Determine** the acceleration of the bicycle in the first 25 seconds.

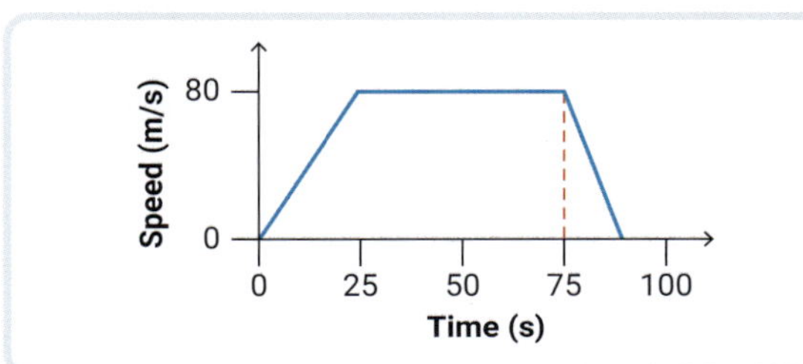

21 Two dogs, Spot and Max, were each sitting on a 2 kg skateboard being pushed by their owner, and their speed after 5 seconds was recorded. Use the information in the following table to **determine** the mass of each dog.

Dog	Force applied by owner (N)	Speed after 5 s (m/s)
Spot	20	5
Max	16	2.5

22 Copy and complete the following table for a body that has been thrown into the air, assuming there is no air resistance.

Time (t)	Speed (m/s)	Acceleration (m/s^2)
0	60	
2	40.4	
4		
	11	

EVALUATING

23 When a hazard such as a pedestrian appears on the road, the driver of a car should react by putting their foot on the brake to decelerate the car to a stop before they reach the point of the hazard (the pedestrian). **Evaluate** the importance of having a fast reaction time when driving a car on suburban roads.

24 When you stand on scales, the scales actually measure the weight force your body exerts on them and converts this force to a mass using Newton's second law and the acceleration due to gravity ($g = 9.8\ m/s^2$), before displaying a mass value on the screen. Imagine standing on a set of scales on the Moon or another planet, and **discuss** the implications this may have on your weight, as shown by the scales.

25 During a triathlon, a triathlete swims through water, runs along the road and cycles on the road. **List** five tips for a triathlete that they could employ before and during the race to help enhance their performance. **Justify** your recommendations using Newton's laws of motion.

CREATING

26 **Create** an infographic of a person driving a car; include annotations that **describe** different ways Newton's laws of motion are relevant to driving. Include at least one application of each law.

27 **Create** a promotional poster for a piece of cycling equipment that is marketed for professional cyclists with the selling point that it will improve their performance. Use what you have learned from this chapter, particularly Module 16.12, to complete this.

SCIENCE IN DEPTH STUDY

1 Connect what you've learned

In this chapter, you've learned about many ideas within the topic of motion and many applications for these concepts. Create a mind map to show how these main ideas are connected, annotating the connections with explanations of their relationship.

▲ The Perseverance rover landing on Mars

2 Check your thinking

When introducing the Mars rovers at the start of this chapter, we asked the following questions.

- How did they all get there?
- How far is it between Earth and Mars?
- How long did it take to fly there?
- How fast did they travel?
- How were these speeds achieved?

Perform some research to find answers to these questions. Write a short paragraph about what you think the biggest challenges were for NASA and the China National Space Administration's astrophysicists in designing the rockets and rovers and planning for their journey to Mars.

3 Get into action

Imagine you are working with the team at NASA that planned the landing of the Perseverance rover on Mars. Using the topics from this chapter, make a list of 10 key questions you and your team would need to ask and answer to ensure that the rover landed safely on the surface of Mars. Explain why each of your questions is important for this project.

4 Communicate

Create an infographic titled 'Planning for Perseverance's Landing' using a photo or a sketch similar to the one on this page. Use the information from your action plan as the labels and annotations.

Glossary

absolute dating determining the age of a fossil in years

absolute zero the lowest possible temperature (0 K or −273.15°C); the point at which there is no particle movement and no energy

absorption the process of taking in radiation and converting it to other types of energy

absorption spectrum the dark lines characteristic of an element or compound in a continuous spectrum

acceleration the rate of change of speed

accuracy how close a measurement is to the correct value

accuracy (instruments) how close the instrument measures to the true value

achievable capable of being completed successfully

acid a substance that can donate a hydrogen ion when in solution

acid mine water water from mines that has become acidic through a chemical reaction and contains dissolved minerals

acid rain rain that is acidic due to chemicals dissolved in water in the atmosphere

acrylic an opaque transparent polymer used in place of glass

activation energy the minimum energy particles need to have for a successful collision to occur

activity series a list of metals ranked by their chemical reactivity

adaptation a change in how something is done so that the impact of an issue is reduced

aerosol fine droplets of saliva or mucus containing pathogens

afferent direction from the sensory receptor towards the central nervous system

agent a body applying a force on another body

alcohol an organic molecule containing an -OH group, such as ethanol

alkane a hydrocarbon containing only carbon and hydrogen, connected by single bonds

alkene a hydrocarbon that contains one or more double bonds between carbon atoms

allele an alternative form of a gene

allele frequency the measure of how common an allele is in a population

allopatric speciation speciation due to a barrier

allotrope an alternative form of an element with different physical and chemical properties

alternative medicine therapy or practice unsupported by scientific evidence or medical experts that aims to improve the health of people

ammeter an instrument used to measure current (the rate of electricity flow in a circuit)

amplitude the maximum oscillation of a particle in a wave

analogous structures structures in different species that have similar function but are anatomically different

anaphase the third phase of cell division, when the chromosomes are pulled apart

aneuploidy having additional or missing chromosomes

anion an ion with a negative charge

antibody a Y-shaped protein that binds to one type of antigen

antigen a substance on a pathogen that stimulates the adaptive immune response

apoptosis programmed cell death

artificial immunity the protection obtained from the production of antibodies after receiving a vaccine or antibodies

artificial isotope an isotope of an element produced in a nuclear reactor or particle accelerator

artificial selection the process whereby humans breed organisms for desired traits

asexual reproduction a method of reproduction that involves one individual producing an identical copy of itself

atmosphere the gaseous layer surrounding Earth that is retained by Earth's gravity

atom the smallest part of an element that gives the element its chemical properties

atomic number (*Z*) the number of protons in the nucleus of an atom, which is the same for every atom of the same element

autosomal the inheritance of genes on autosomes

autosomes all of the chromosomes in a cell, except for the sex chromosomes

axon a long, thin fibre that carries electrical impulses from the cell body of a neuron towards the next neuron

B cell a type of white blood cell involved in the humoral (antibody)-mediated immune response

balanced chemical equation a chemical equation in which there are the same number of atoms of each element on each side

ball-and-stick model a three-dimensional representation of a molecule that uses balls to represent atoms (often coloured differently for each element) and sticks to represent the bonds between them

base (biology) a part of the structure of DNA that codes genetic information about an organism

base (chemistry) a substance that can produce hydroxide, oxide or carbonate ions in solution

batteries devices that transform chemical energy into electrical energy

beta particle a negatively charged particle identical to an electron, emitted when an unstable nucleus decays

bias a strong preference for one thing or idea over another

Big Bang theory the generally accepted theory for the formation of the universe

binary fission a method of asexual reproduction in bacteria in which a parent cell splits into two identical daughter cells

bioaccumulation the gradual accumulation of substances in an organism

biochemical evidence evidence of evolution based on the fact the same enzymes are found in the cells of most organisms

biodegradable a substance that can be broken down by bacteria or other living organisms

biodiversity the variety of living species on Earth, including plants, animals, bacteria and fungi

biofuels fuels that are made from biological sources that can replace fossil fuels

bioinformatics the science of collecting and analysing complex biological data

biomagnification the increase in the concentration of a substance in organisms as you move up a food chain

biopolymer a polymer made from plant or animal materials

biosphere all aspects of Earth and the atmosphere that support life, including all living things

biotechnology a branch of science that manipulates living organisms, systems or DNA to produce products that are useful to humans

bitumen a solid used in road surfaces

bivariate analysis an analysis involving two variables that identifies and explains trends, correlations or groupings

blood glucose the amount of glucose in blood

blueshift the shift of a star's spectrum towards the blue end as the star moves toward Earth

body an object, person or thing that has mass

booster an additional dose of a vaccine, designed to provide a higher level of protection when antibodies have decreased

box plot graphical representation of the distribution of data

brain a very complex organ; the coordinating centre of the CNS

brain stem consists of the midbrain, the pons and the medulla oblongata; controls automatic and involuntary activities such as breathing, heart rate, digestion and vomiting

budding a method of asexual reproduction in which a bud (a growth on the parent body) forms, grows and then separates or spreads

calibration the process of adjusting and verifying the accuracy of a measuring instrument or device

cancer uncontrolled cell division resulting in a growth or tumour

capsid the protein coating that protects the genetic material inside a virus

carbon footprint a measure of the amount of greenhouse gases released into the atmosphere by an individual or group

carbon sequestration the process of capturing and storing atmospheric carbon dioxide

carbon sink something that naturally stores large quantities of carbon dioxide

carbon-14 dating the use of the carbon-14 decay process to determine the age of materials containing carbon

carboxylic acid an organic molecule containing a -COOH group, such as ethanoic acid

carcinogen an agent that increases the likelihood of developing cancer

catalyst a substance that increases the rate of a chemical reaction by lowering the activation energy of the reaction without itself being changed

cation an ion with a positive charge

causation when a change in one variable causes a change in another variable

cause-and-effect relationship a relationship in which a change in one variable causes a change in another variable; also known as causation

cell body the part of a neuron that contains the cytoplasm, including the nucleus

cell cycle the series of events that takes place in a cell as it grows and divides

cell-mediated immunity the adaptive immune response prompted by T cells on infected host cells

central nervous system (CNS) the brain and spinal cord

centrioles organelles in the cell that produce spindle fibres

centromere the point on a chromosome where the two chromatids are joined

cerebellum a smaller section at the back of the brain; controls movement and balance

cerebrum the largest part of the brain; controls movement, learning, emotion, memory and perception

chemical bonding how atoms combine together to form larger structures; the joining of atoms via the transfer, loss or sharing of electrons to achieve a stable electron configuration

chemical equations word equations or balanced equations that are used to represent the substances in chemical reactions

chemical formula symbols and numbers used to represent the composition of a chemical

chemical properties the properties of a substance that determine how it reacts when combined with other substances

chemical reaction a process that occurs when a substance changes to produce a new substance

chiasma a point on the chromatids where crossing over occurs

chromatid one half of a duplicated chromosome

chromatin unpackaged DNA found within the nucleus of a non-dividing cell

chromosomal mutation a change in the number of chromosomes or parts of a chromosome

chromosome a thread-like structure found in the cell, composed of DNA

cilia the hair-like extensions of mucous membranes that pushes pathogens up and out of the body

circuit diagram a simple diagram of an electrical circuit that uses standard symbols

circular economy an economy using behaviours that reduce the rate of resource use and waste creation

cite to give reference or credit to a secondary source of data or information

claim a statement asserting something is true, real or a fact

climate the average weather conditions in a particular region over an extended period

climate change a change in global or regional climate patterns

coal a solid form of a fossil fuel

coefficient a number placed in front of a chemical formula to balance an equation

cogeneration the process of using waste thermal energy in power generation for another purpose

collision theory the theory that a reaction will only occur if particles come into contact with sufficient energy and at the correct orientation to cause a successful collision

combustion a chemical reaction where a fuel combines with oxygen to produce energy

common ancestor the ancestor that two or more descendants have in common

comparative anatomy the study of the body structures of species to show the adaptive changes made from a common ancestor

complete combustion combustion that occurs with plentiful oxygen; it produces carbon dioxide and high amounts of energy

complete loop an unbroken, continuous path for electricity

component any part of an electrical circuit

compound two or more different elements joined by a chemical bond

compression a high-pressure region of a longitudinal mechanical wave where particles are pushed close together

computer analysis a detailed examination of something using a computer

confirmation bias the intentional selection of data to support a case or opinion

confounding variable a variable that influences both the independent and dependent variable, producing a false relationship between them

consequence the result of a decision or action

continental climate a climate zone with large variation in temperature between seasons and irregular patterns of precipitation

continental ice sheet an extensive sheet of permanent ice covering a large area of the land surface

continuous data data from measurements that may include whole numbers and any value between them

control group a trial or group in an investigation that doesn't have the independent variable and is used as a comparison

control rods moveable rods that absorb neutrons and are lowered and raised to control the rate of fission in a nuclear reactor

controlled fission when the rate of fission is managed and a constant amount of energy is produced

controlled variable a factor that needs to be kept the same throughout a science investigation so that any changes in it do not influence the results

convection current a flow of materials (such as air, water or molten rock) and energy caused by differences in densities due to temperature differences

coordinating centre an organ or tissue that receives and processes information from receptor cells and coordinates a response

COP26 the 26th Conference of the Parties to the United Nations Framework Convention on Climate Change

Coriolis effect the apparent deflection of large masses of air and water due to Earth's rotation on its axis

corpus callosum a collection of axons that extend across the two hemispheres of the brain that compares and combines the sensory inputs from the left and right sides of the body

correlation a trend in data in which one variable changes consistently as the other variable changes

corrosion the breakdown of a metallic substance due to chemical reactions with substances in the environment, such as oxygen or water

covalent bond an electrostatic force of attraction between a shared pair of electrons and the nuclei of the atoms sharing the electrons

covalent molecule a distinct structure formed when two or more non-metal atoms join through covalent bonding

covalent network a structure formed when non-metal atoms join in a covalently bonded lattice

crest the highest point on a transverse wave

crossing over the exchange of genetic material between non-sister chromatids on homologous chromosomes

crude oil an unrefined liquid form of a fossil fuel

current the amount of charge that passes a point in an electrical circuit in one second

cyanobacteria colonies of blue-green algae capable of photosynthesis

cytokinesis the division of the cytoplasm after mitosis

data logger an electronic device that automatically records and stores data over time

daughter cells the cells that are produced as a result of cell division

deceleration negative acceleration; when speed is decreasing over time

decomposition reaction a reaction involving the breakdown of compounds into simpler elements or compounds

deep ocean currents the water movement occurring below a depth of 400 m caused by changes in water temperature and salinity

deforestation the removal of large areas of forest to enable the land to be used for other purposes

deletion the removal of one or more nucleotides from a DNA sequence

delocalised electrons electrons in a metallic lattice that do not belong to any particular atom

dendrite a branching network at the end of a neuron that receives information from other neurons

deoxyribonucleic acid, see DNA

deoxyribose sugar one of the components of a nucleotide in DNA

dependent variable the factor that may be affected by the independent variable; the factor that can be measured or counted

descriptive statistics values derived from a dataset, such as a mean or maximum value, that summarise the dataset

deuterium an isotope of hydrogen with one proton and one neutron

diploid the full complement of DNA, represented as $2n$

discrete data data where there is only a limited number of possible values

disease an abnormal condition or disorder that affects the normal functioning of the body, either wholly or partially, and is associated with specific causes and symptoms

disorder a group of symptoms that disrupts your normal body functions but does not have a known cause

dispersion the extent to which values in a dataset are above or below the mean (average)

displacement the distance and direction that a body's position is relative to the starting point; measured in metres (m)

displacement reaction a reaction involving the replacement of atoms in compounds

distance the length of the pathway taken by a moving body

distillation a process used to separate solutions that collects both the solute and the solvent

distillation column the tower used in crude oil distillation, which has a temperature gradient that separates components as they condense

DNA the molecule that makes up the genetic material inside the nucleus of cells and in some viruses

DNA profile an image used to determine an individual's DNA characteristics

dominant an inheritance that identifies the dominant allele in a genotype

dominant allele the allele that will be expressed in the phenotype

double-blind study research method where neither the participants in a study nor the researchers know which groups are receiving treatments

double displacement reaction a reaction in which atoms of the reactants displace each other from their compounds

double helix the shape of DNA, similar to a twisted ladder

dry climate a climate zone where seasonal evaporation exceeds seasonal precipitation

ductile able to be stretched into a wire

Earth system Earth as a complex whole, made up of the four interacting spheres: atmosphere, biosphere, hydrosphere and geosphere

ecological footprint the amount of the environment necessary to supply the goods and services that support a lifestyle

effector a muscle or gland that receives a message from the coordinating centre and carries out a response

efferent direction from the central nervous system to the effector (muscle or gland)

electrical circuit a complete loop that can conduct a continuous flow of electricity so that electrical potential energy can be transferred

electrically conductive material a substance that enables electricity to flow through it

electricity tariff the price a consumer pays for the energy they use

electromagnetic radiation energy that travels in electromagnetic waves such as infrared and visible light

electromagnetic spectrum the range of electromagnetic waves arranged in sequence from high-energy gamma rays to low-energy radio waves

electromagnetic wave a wave that transfers energy through space by electric and magnetic fields

electron a negatively charged particle that moves in space around the nucleus of an atom

electron configuration the arrangement of electrons in energy shells in an atom

electronegativity the ability of an atom to attract shared electrons in a bond

electrostatic bond the force of attraction between positive and negative particles

electrostatic force a force of attraction between oppositely charged particles

element a pure substance made up of only one type of atom; it cannot be broken down into a simpler substance

embryology the study of the early stages of development

empirical evidence evidence collected through scientific observation or experimentation

endocrine system a network of glands that secrete hormones into the bloodstream to be transported to target cells

energy efficiency a measure of how much input energy is converted to useful output energy, often stated as a percentage

energy flow diagram a visual representation of energy movement using arrows and boxes

energy shell a level around a nucleus containing electrons of the same energy

energy transformation the changing of one type of energy into another type of energy

enhanced greenhouse effect strengthening of the natural greenhouse effect due to an increase in greenhouse gases caused by human activities

ethene an alkane with two carbon atoms

eutrophication the accumulation of nutrients in a water body leading to the rapid increase in algae and bacteria

evaporative cooling the cooling effect that occurs when water evaporates

evidence verified information that supports or refutes a statement

evolution the gradual change in characteristics of a species over many generations resulting in a new and different species

exabyte a very large amount of information, equal to a billion gigabytes

excited the state of an atom or electron when it absorbs energy

exothermic describes a chemical reaction that releases energy

extrapolation the estimation of an unknown value by extending a trend beyond the known values

extreme weather event unexpected, unusual, severe or unseasonal weather

fact a piece of information that is supported by verified evidence

fair test an investigation that is conducted correctly to answer a scientific question

feedstock raw material that is used in an industrial process

fever a non-specific response to infection where the body temperature exceeds the normal 37°C

filial first set of offspring from a cross between parents

finite resource a limited resource

first filial generation the first set of offspring from a parent cross

fission to undergo the process of nuclear fission

flammability the ability of a substance to catch fire

force a push, pull, twist or squeeze

fossil the remains or traces of living organisms, commonly preserved in sedimentary rocks

fossil fuel a substance containing hydrocarbons that is used as an energy source; takes millions of years to form from the remains of dead plants and animals

fossil record the collection of all known classified fossils and the information they provide about past life

fractional distillation the process of separating a mixture into multiple components based on their boiling points

fragmentation a method of asexual reproduction in which a body part breaks off from a parent body and then develops into an offspring

framework a systematic structure used to plan or build a concept

free-body diagram a diagram that shows the forces acting on a single body as arrows

frequency the number of waves produced, or passing a point per unit of time; measured in hertz (Hz)

friction a resistance force that results when two surfaces rub against one another

frictional heat loss waste thermal energy produced as a result of the vibration of particles when moving past each other in a process where thermal energy production is not intended or useful

fuel rods rods inside the nuclear reactor core that contain the uranium fuel

fuse to undergo the process of nuclear fusion

G

gametes the sex cells of a sexually reproducing organism

gamma radiation high-energy electromagnetic energy emitted when an unstable nucleus decays

gel electrophoresis a technology used to separate DNA by size to produce a DNA profile

gene a section of DNA that codes for a protein or a certain trait

gene mutation a change in the DNA sequence of a gene

gene pool the total range of genetic material available in a population

gene therapy the introduction of functional genes into cells to replace defective or missing genes to treat genetic disorders

generation a group of organisms that start life and reproduce at around the same time; also, the time between one group and their offspring reproducing

generator a device that transforms kinetic energy into electrical energy

genetic code the sequence of nitrogen-rich bases in an organism's DNA

genetic disorder disease symptoms produced when a DNA sequence is different from normal

genetic drift the change in allele frequency seen in small populations due to chance events from one generation to the next

genetic engineering the deliberate modification of an organism's DNA

genetically modified organism (GMO) an organism whose genes have been altered in the laboratory to produce a desired trait

geneticists scientists who study genetics and inheritance

genome the complete set of genetic material present in an organism

genotype the combination of alleles for a specific gene

geosphere the rocks, minerals and landforms of the surface and interior of Earth

geothermal energy heat that is trapped in rocks close to heat sources deep within Earth's crust

germline cells the cells that form the ovum and the sperm

germline mutations mutations that occur in sperm or ova

gland a tissue that releases hormones

global convective cells three large atmospheric pressure cells that occur in both the northern and southern hemispheres

glucagon a hormone secreted by the pancreas; breaks down glycogen into glucose

glycogen a store of glucose in the liver and muscles

gonads the sex organs of an organism; where meiosis occurs

gradient a measure of the slope of a straight line on a graph

gravitational field the region where the pull of Earth's gravity is experienced

gravity the force of attraction between Earth and objects within its gravitational field

greenhouse effect a natural process that traps energy within the atmosphere, raising Earth's surface temperature

greenhouse gases gases in the atmosphere that can trap heat and affect global surface temperatures and other aspects of the climate

group a vertical column on the periodic table

H

half-life the time it takes for half of the nuclei in a sample of an isotope to decay

haploid having one copy of each chromosome, represented as n

hazard something that has the potential to harm

health a condition of total physical, mental, and social wellness, going beyond just the absence of illness or disability

heavy metal metallic elements with relatively high density and toxic properties such as lead, mercury, arsenic and cadmium

hemisphere one of the two symmetrical halves of the brain

herd immunity a level of protection gained when a certain percentage of a community is immune to an infectious disease, which breaks the chain of transmission

herd immunity threshold the level of population immunity needed to stop the spread of a specific disease

heterozygous having two different alleles on homologous chromosomes

homeopathy an alternative medicine based on unsupported scientific ideas

homeostasis the maintenance of a constant internal environment necessary for survival

homeostatic related to homeostasis

homologous carrying the same genes for characteristics at the same locations on a chromosome

homologous pairs maternal and paternal chromosomes with genes found at the same location

homologous structures body parts that can be found in a range of species that have similar structures but different functions

homozygous dominant having two of the same dominant alleles on homologous chromosomes

homozygous recessive having two of the same recessive alleles on homologous chromosomes

hormone a chemical messenger

host an organism infected with a pathogen

human germline engineering editing the genome of germ cells so that genetic changes can be passed on to the next generation

humoral immunity the adaptive immune response prompted by B cells, and antibodies specific to the antigen, on pathogens found outside of host cells

hydrocarbon organic molecules that contain only hydrogen and carbon

hydroelectricity electrical energy produced by transforming gravitational energy of falling water into kinetic energy to drive a turbine

hydrogen bond a type of attraction between atoms of different molecules

hydrosphere the parts of Earth containing all forms of water

hypothalamus an almond-sized structure of the brain; plays a major role in homeostasis

hypothesis a testable explanation for something based on existing knowledge; a testable statement of the predicted relationship between the independent and dependent variables

incidence the number of new disease cases appearing in a population over a period of time

incident wave a wave of light that is approaching a surface

incomplete combustion combustion that occurs in limited oxygen conditions, producing carbon monoxide, carbon and less energy than complete combustion

inconsistency in the data a data point that does not follow the pattern of other data from the same experiment or across repeated experiments

independent variable the factor that you choose to vary in your investigation

index fossil a fossil that can be used to compare the relative age of rock strata from different locations

indicator a chemical that changes colour in acidic and basic solutions

inertia a property of a body that resists changes to its motion

inflammation a response triggered by damaged cells or a pathogen in the body; characterised by redness, swelling, heat and pain

inflation the initial rapid expansion of space-time just after the Big Bang

inorganic substances that are not primarily carbon based, such as metals and ionic salts

insertion the addition of one or more nucleotides into a DNA sequence

insulating reducing the transfer/loss of heat

insulation material used to minimise heat transfer

insulin a hormone secreted by the pancreas; controls how much glucose is in the blood

intermolecular forces forces of attraction between covalent molecules

interneuron a short neuron that sends nerve impulses between sensory and motor neurons within the CNS

interphase the resting phase of the cell cycle

interpolation the estimation of an unknown value within the range of known data for a relationship

inversion mutation a chromosomal mutation in which part of a chromosome is reversed end-to-end

investigable question a question that can be answered using a fair test investigation

ion a charged particle formed when an atom loses or gains valence electrons

ionic formula the chemical representation of an ionic substance showing the number and type of atoms present

ionic salt a chemical containing a metal ion and a non-metal ion

ionisation energy the energy required to remove an electron from the valence shell of an atom

ionising power the ability of radiation to change the structure of materials it passes through

isolated system a system in which no energy or matter is exchanged with the surroundings

isolation a mechanism or barrier that separates breeding populations

isotopes atoms of an element with the same number of protons but different numbers of neutrons

iterative describes something that is repeated multiple times, often in a cycle or process

karyotype a picture of an organism's complete set of chromosomes arranged using size, genetic composition, and centromere positions

kinetic energy the energy an object has because of its motion

kJ/mol kilojoules per mol; the unit of the amount of energy a substance has

Köppen climate classification a system used since 1900 that classifies climate zones based on temperature, amount and type of precipitation, and vegetation

land clearing the removal of natural vegetation and habitats, such as forests

large dataset a dataset large enough to be statistically reliable and require computer analysis

law of conservation of energy when energy is transferred or transformed, the total amount of energy remains the same

law of conservation of mass the total mass of reactants and products in a chemical reaction is equal

likelihood the chance something will happen

line of best fit the line that best represents the trend of a set of data points

liquid petroleum gas (LPG) a mixture of propane and butane used as barbecue gas and as vehicle fuel

literature review a study of reports about previous related experiments and investigations

load any component in a circuit that transforms electrical energy

longitudinal wave a wave in which oscillations are parallel to the direction in which the wave is travelling

9780170491785

lustrous shiny when cut or polished

lymphocyte a type of white blood cell involved in adaptive immune responses

lysosome a cell organelle that contains digestive enzymes

lysozyme an enzyme that breaks down pathogens

magnitude size or extent

main sequence star a star that is fusing hydrogen to helium

malleable able to be beaten into different shapes

mass number (*A*) the total number of protons and neutrons in the nucleus of an atom

master mix a premixed solution used in PCR techniques to make copies of DNA

maximum value the largest value in a dataset

mean the calculated 'central' value of a set of numbers; an average

mechanical wave a wave that passes energy through matter by vibrations of particles

median the middle value in an ordered set of numbers

medium the matter or substance through which a wave passes (made from particles)

meiosis cell division producing cells that will specialise into gametes with half the number of chromosomes of the parent cell

memory cell a mature B cell or T cell that 'remembers' information about a specific antigen

metal hydroxide an ionic salt containing a metal ion and a hydroxide ion

metal oxide a chemical substance made up of a metal and oxygen

metallic bond the force of attraction between metal cations and delocalised electrons

metallic lattice the organised structure formed with rows of metal cations surrounded by delocalised electrons

metaphase the second phase of cell division, when chromosomes line up in the centre of the cell

method steps that were taken during an investigation, written in past tense

microplastics very small pieces of plastic debris found in the environment

mineral a naturally occurring inorganic solid with a neatly ordered crystal structure and characteristic composition

minimum value the smallest value in a dataset

mitigation reducing the severity of an impact or event

mitosis cell division for growth, replacement and repair of somatic cells

mode the value that occurs most frequently in a dataset

mode of inheritance the manner in which a genetic trait or disorder is passed from one generation to the next

molecule a group of atoms bonded together

monogenic trait a trait that is determined by a single gene

monohybrid cross a cross between two organisms with two alleles at one gene location

monomer a small molecule building block used to make polymers

motor neuron a neuron that sends electrical impulses from the CNS to effectors

mucus a thick, sticky liquid in our airways that captures pathogens

mutagen an agent that increases the likelihood of mutation

mutation a spontaneous and permanent change to a DNA sequence

myelin sheath a protective coat around an axon that increases the speed of nerve impulses

natural gas a gaseous form of a fossil fuel; usually methane, propane or butane

natural immunity the protection obtained from the production of antibodies after exposure to an infectious disease

natural isotope an isotope of an element found in nature

natural pollutants pollutants from natural processes

natural selection the process in which an environmental factor acts on a population, resulting in some individuals being more likely to survive and reproduce

negative correlation a correlation where one variable decreases as the other variable increases

negative feedback a mechanism in which a response reduces a condition or factor back to its optimal range

nerve a collection of fibres, surrounded by a protective coat, that transmit messages as nerve impulses to and from the CNS

nerve fibre the section of a nerve cell that carries nerve impulses away from the cell body

nerve impulse an electrical message that is transmitted along nerves to and from the CNS

net force the overall force acting on a body; measured in newtons (N)

net zero greenhouse emissions when emissions of greenhouse gas do not exceed the amount of gases absorbed or stored

neuron a specialised cell that can transmit nerve impulses; also known as a nerve cell

neurotransmitter a chemical signal that delivers a message across a synapse, to be converted back to an electrical impulse

neutron an uncharged particle in the nucleus of an atom

New Age a philosophy that seeks healing and self-development through practices such as meditation, astrology and alternative medicine

nitrogenous base a base that contains nitrogen: adenine (A); thymine (T); cytosine (C) and guanine (G)

nomenclature a system of names; for example, the naming system used for chemicals

non-disjunction the incorrect separation of chromosomes or chromatids at the centromere, resulting in gametes with an unusual chromosome number

non-investigable question a question not easily answered by scientific investigation

non-renewable describes resources that are either non-replaceable or replaceable at a slower rate than they are used

non-renewable energy source a source of energy that is finite in nature; that is, used at a faster rate than it can be produced

nuclear fission nuclear reactions where a large atom splits into two or more smaller atoms with a release of energy

nuclear fusion nuclear reactions where two or more smaller atoms combine to form a larger atom with a release of energy

nuclear reactions reactions involving a change in composition of the nucleus of an atom

nucleotide the building block of DNA

nucleus the dense centre of an atom; it contains protons and neutrons, so it is positively charged

null hypothesis the hypothesis that a relationship between variables being studied does not exist

nylon a synthetic fibre used in clothing

observation data collected through the senses (sight, smell, taste, touch or hearing) or with measuring equipment

occupational related to an occupation or profession

offspring a new organism produced by asexual or sexual reproduction

opaque cannot be seen through

open system a system that allows the transfer of matter and energy in and out of the system

ore a rock that contains one or more minerals containing valuable substances

organic covalent molecules that contain carbon

origin a defined point that indicates the starting point or a displacement of zero

oscillation movement back and forth in a regular rhythm or pattern

outlier a value that differs significantly from other values in a dataset

overharvesting harvesting a resource faster than the resource can recover

oxidation a chemical process in which oxygen is added to a substance

pandemic a disease that has spread rapidly across the world, having grown from an epidemic

parallel circuit an electrical circuit where the current can travel along two or more paths

parthenogenesis a method of asexual reproduction in which an unfertilised egg matures and develops into an offspring without fertilisation by sperm

particle accelerator a machine that enables high-speed, high-energy collisions between atoms to produce artificial isotopes and new elements

particulates very small particles of a substance, especially those produced by burning fuels

pathogen an organism or virus that causes disease in its host

PCR, see polymerase chain reaction

pedigree chart a diagram showing patterns of inheritance over generations; also called a family tree

penetrating power the ability of radiation to pass through materials

period (chemistry) a horizontal row on the periodic table

period (physics) the time taken for one full wave to pass a particular point

periodic table a method of arranging elements by increasing atomic number

peripheral nervous system (PNS) the network of nerves outside the brain and spinal cord

permafrost permanently frozen soil, sediment or rock

persist stay or remain

pesticide a poisonous substance used to control pests

pH a measure of how acidic or basic a substance is

phagocyte a specialised white blood cell that can engulf and destroy pathogens

phagocytosis a process in which blood cells called phagocytes engulf and destroy a pathogen

phenotype the observable characteristics of the genotype

phosphate group one of the components of a nucleotide in DNA

photon a form of elementary energy particle

photosynthetic able to photosynthesise, converting carbon dioxide and water into sugar and oxygen

phylogenetic tree a diagram representing lines of evolutionary descent from a common ancestor

physical properties the properties of a substance that can be observed or examined without changing its composition

pitch the degree of frequency (high or low) of a sound or note

pituitary gland a gland that produces and releases several hormones; is connected to, and is controlled by, the hypothalamus

placebo effect a positive effect due to a patient's belief in a treatment

plasma electrically charged gas abundant in stars; often called the fourth state of matter

plasmid a small circular fragment of DNA in the cytoplasm of bacteria

plastic a common name for a group of polymers made from crude oil components

polar climate a climate zone with ice, snow and temperatures too low to support most vegetation

pollutant a substance introduced into an environment that can be harmful

pollution the introduction of harmful substances into the environment

polyester a synthetic fibre used in clothing

polyethylene a polymer formed from ethene monomers

polymer a large molecule made up of many repeating units called monomers, joined together to form a long chain

polymerase chain reaction (PCR) a technology used to make many copies of DNA

population a group of individuals of the same species living in the same place at the same time

position the location of a body

positive correlation a correlation where one variable increases as the other variable increases

9780170491785

positive feedback a mechanism in which a response reinforces the condition or factor until an end point

power pack an adjustable power supply used in the science laboratory that plugs into a power point

precipitate a solid formed from certain combinations of positive and negative ions

precipitation liquid or solid water in the form of rain, snow, sleet or hail

precision how close repeated measurements of the same thing are to each other

precision (instruments) the level of variation between measurements by an instrument

precision cancer medicine a branch of medicine that uses a patient's genetic make-up and knowledge of their body and lifestyle to develop individualised cancer treatments

predisposition a likelihood of developing a particular disease or condition

procedure a set of instructions to follow; written in the present tense

products chemical substances that form in chemical reactions

progeny offspring

prognosis predictions made by health professional about how a disease will progress

prophase the first phase of cell division, when chromosomes duplicate and condense

proteins a group of large, complex molecules vital to cell replication, growth and repair

proton a positively charged particle in the nucleus of an atom

pseudoscience beliefs or practices that claim to be scientific but have not been supported by fair tests

purebred having the same alleles for a given gene; see homozygous dominant/ recessive

qualitative data non-numerical information that relates to a quality, type, choice or opinion

quantitative data numerical information that is counted or measured and expressed as quantities in numbers

quark a type of elementary particle that is the fundamental component of matter

radiation a stream of particles and/or energy from a radioactive source

radioactive decay the spontaneous disintegration of certain atomic nuclei accompanied by the emission of alpha particles, beta particles or gamma radiation

radioactive isotope (radioisotope) an isotope that is unstable and undergoes radioactive decay to become more stable

radiometric dating a dating method that measures the decay of radioactive isotopes to determine the age of fossils

radiotherapy the treatment of cancer by radiation

random assortment the way chromosomes line up during metaphase I of meiosis, resulting in random combinations of genes

random error a small, variable error caused by slight variations in an instrument or the environment

range a measure of spread calculated by subtracting the smallest number from the largest number in a dataset

rarefaction a low-pressure region of a longitudinal mechanical wave where particles are spread apart

rate of reaction a measurement of how fast a reaction is proceeding

rationale the grounds that explain the evidence that supports the claims of the investigation

reactants chemical substances that, when added together, react to form products in a chemical reaction

reasoning the process of drawing conclusions from facts

receiver the body receiving a force from another body

receptor a specialised cell that detects a stimulus; may be internal or external

recessive an inheritance that identifies the recessive allele in a genotype

recessive allele the allele that is masked by a dominant allele and is only expressed in the homozygote

recombinant DNA DNA that has been manipulated by combining DNA from other species

recycling the process of converting wastes into new materials or products

red giant a star that is fusing smaller elements larger than helium

red supergiant a star that is fusing elements up to iron

redshift the shift of a star's spectrum towards the red end as the start moves away from Earth

reflected wave a wave of light that has been reflected off a surface

reflection the process of a wave bouncing off a surface and returning through the same medium

refraction the transmission of a wave from one medium into another with a resulting change of direction

refutable able to be proven wrong

relative dating determining if a fossil or rock is older or younger than another

relative humidity the amount of moisture that air holds compared with the amount it could hold if saturated at a given temperature

reliability how similar the results of the same experiment are

reliable consistent; able to be trusted

remediation the process of cleaning up pollution or contaminants from the environment

renewable describes resources that can be replaced at a rate greater than they are used

renewable energy source a source of energy that can be produced at a faster rate than it can be used

reserve the amount of a resource available for mining or harvesting at a profit now and in the future

reservoir (biology) a source of infection such as a habitat or population of organisms where a pathogen can replicate and survive for long periods

reservoir (environmental science) a place where something is collected or stored

resistor a component that regulates the flow of electricity in an electrical circuit

resource something used to complete a task or achieve a goal

resource (non-renewable resource) an estimate of how much of a useful non-renewable substance exists

response the action a body takes due to detection of a stimulus, usually to return conditions inside the body to the normal range

rest the state of being stationary, having a speed of 0 m/s

restriction enzymes special enzymes that cut DNA at specific recognition sites; isolated from bacteria

ribonucleuc acid, see RNA

risk of harm the chance that damage or injury will occur

RNA ribonucleic acid, a molecule similar to DNA found in the cytoplasm of cells; the genetic material found in many viruses

S

salinity the concentration of dissolved salt

Sankey diagram a type of flow chart that uses arrows of various sizes to indicate the amount of energy transferring or transforming in a system

scalar quantity a measurement that has a magnitude only and no direction

scientific claim a statement based on evidence from a scientific investigation

second filial generation the set of offspring from the first filial parent cross

secondary source a publication, information or data that has been written or collected by another person

secrete produce and release a substance (often a hormone or enzyme)

selection bias selecting subjects that do not accurately represent the population they are drawn from

selection pressure the effect the selective agent has on the population

selective agent the environmental factor acting on the population

sensory neuron a neuron that sends electrical impulses from receptors to the CNS

series circuit an electrical circuit where wires and components are connected end-to-end

sex chromosomes a pair of chromosomes that determine the sex of an individual

sexual reproduction a method of reproduction that involves two parents producing offspring that are not identical to the parent or each other (except for identical twins or triplets)

silicon a common element, used in many electronic devices; the main component of solar cells

singularity a point where density and matter are infinite

smelting the process of extracting a metal from the ore that contains it

soil degradation physical, chemical and biological decline in soil quality

solar cell a device that transforms light energy into electrical energy

solar power electricity generated directly from sunlight, using solar cells

somatic relating to the cells that make up the body other than the reproductive cells

space-filling molecule a representation of molecules that approximates their physical appearance, where atoms are represented as spheres of different sizes

speciation a process in which two groups become so genetically different they can no longer breed with each other under natural conditions to produce fertile offspring

species a group of organisms capable of reproducing under natural conditions to produce fertile offspring

specific clearly identified or defined

spectrum a range or a scale; also, the different bands of colour that are visible when white light is refracted through a prism, as seen in rainbows

speed (of a body) a measure of how much distance a body covers per unit of time; typically metres per second (m/s)

speed (of a wave) a measure of how fast a wave is travelling; a measure of the distance covered by the wave per unit of time; measured in metres per second (m/s)

spindle fibres protein structures that separate the chromosomes during cell division

spurious false, or not what it appears to be

staggered cut a cut from a restriction enzyme that results in 'sticky ends', overhanging unpaired nucleotides

standard deviation a calculated measure of spread representing the spread of data around the mean in a dataset

standard of living a person's level of wealth and access to food, shelter and safety

stationary keeping a constant position or not moving in any direction; at rest

statistical reliability the ability to produce consistent results when repeatedly analysed using the same methods

stimulus a change in a factor above or below the optimal range

storm surge a brief increase in sea levels due to the combined effect of storms, low air pressure, strong winds and tides

straight-chain alkane a single, continuous alkane carbon chain with no branches

stratigraphy comparing strata or layers of rock to determine the relative age of fossils

strong nuclear force a force of attraction between particles in the nucleus of an atom

structural formula a two-dimensional representation of a molecule that shows bonds between atoms drawn as lines

struggle for existence the competition between individuals for required resources such as food, water or space

subatomic particle a particle inside an atom, such as a proton, a neutron or an electron

substitution the swapping of a nucleotide within a DNA sequence

successful collision a collision that results in products forming

supernova an explosion at the end of the life cycle of a star that produces heavy elements larger than iron

surface currents currents in the top 100 m of the ocean

surface temperature the measure of the relative hotness or coldness of Earth's surface

survival of the fittest the idea that individuals with the best suited characteristics will survive, reproduce and pass their traits on to the next generation

sustainability consuming resources in a way that meets our current needs but also allows for the needs of future generations

sustainable development development that meets our current needs without reducing the ability of future generations to meet their needs

sustainable use using resources in ways that do not harm the environment and leave the resource available for future generations

switch a component that can complete or disconnect the conducting path of electricity in an electrical circuit

sympatric speciation speciation due to reproductive isolation

symptom something a person experiences that may indicate the existence of disease

synapse the gap between two neurons

synthesis reaction a reaction in which two or more elements and compounds combine to form a more complex substance

systematic something done in an organised way

systematic error an error that occurs in the same way with each measurement, normally due to a mistake in the procedure

T cell a type of white blood cell involved in the cell-mediated immune response

Teflon a non-stick polymer coating for uses including cooking equipment, waterproofing fabric and coating medical devices

telophase the fourth and final phase of cell division, when the nucleus re-forms and the chromosomes unravel

temperate climate a climate zone with moderate temperature and precipitation that exists between the extremes of tropical climates and polar climates

thermal cycling the repeated process of heating and cooling required in PCR

thermal expansion an increase in the volume of materials as they get hotter

thermal waste hot or cold water or air, released into the environment by industry or energy production

thermohaline the circulation of dense, cold-water currents deep in Earth's oceans

thrust a force that makes an object move in the opposite direction as a result of expelling mass or fuel

tidal energy electricity generated from the ebb and flow of the tides

tolerance range the range of a particular condition inside the body that an organism can survive

toxic able to cause harmful health effects

toxicity a measure of how poisonous a substance is

transgenic organism an organism that contains DNA sequences from an unrelated organism that have been artificially introduced

translocation the result when part of a chromosome detaches and reattaches to a different chromosome

transmission the transfer of a pathogen from one organism or reservoir to a new host

transparent see-through

transverse wave a wave in which oscillations are at right angles to the direction in which the wave is travelling

treatment groups all the trials or groups in an investigation that contain variations of the independent variable

tritium an isotope of hydrogen with one proton and two neutrons

trough the lowest point on a transverse wave

ubiquitous proteins proteins that are found in nearly all organisms and carry out the same function

uncertainty range of possible values or outcomes that may be due to limitations in measurement, variability or incomplete data

uncontrolled fission when the rate of fission is not managed and very large amounts of energy are created in a short period of time

univariate analysis an analysis that involves only one variable that describes the data

upcycling the creation of a product of higher value from a used item

upwelling the movement of water from the depths of the ocean to the surface, bringing nutrients and carbon dioxide with it

uranium–lead dating the use of either the uranium-238 or uranium-235 decay process to determine the age of rocks

useful output energy the output energy of a process or action that is intended and useful

vaccination administration of a vaccine to provide protection from a specific disease

vaccine a substance that stimulates a response in the immune system to produce antibodies

vacuum a space where there is no matter (no particles)

valence electron an electron in the highest energy shell of an atom

valency the number of electrons an atom loses, gains or shares to form a stable electron configuration

valid describes the extent to which an investigation tests a hypothesis

validity the extent to which an investigation tests a hypothesis

variance a measurement of the amount of spread in the values in a dataset

variation a difference in characteristics due to different genes

vector an organism that transmits a pathogen from an animal or a plant to another animal or plant

vector quantity a measurement that has both a magnitude and a direction

velocity the rate of change of displacement with time, a vector quantity for speed; measured in metres per second (m/s) and includes a direction

verify to demonstrate an argument is true by providing evidence

vestigial organs organs that are retained in a species despite no longer being functional as they were in the ancestral species

viral envelope the outer layer of many viruses

viscosity a liquid's resistance to flowing

voltage a measure of the difference in electrical energy between two points in a circuit

voltmeter an instrument used to measure voltage (the energy transformed when electricity passes through a component in a circuit)

volume the measure of how loud a sound is; measured in decibels (dB)

waste output energy the output energy of a process or an action that is not intended and not useful for the main purpose of the process or action

wave equation a mathematical relationship between the speed, wavelength and frequency of a wave

wavelength the length of one full wave; the distance between two crests or two troughs; measured in metres (m)

weather the conditions in the lower part of the atmosphere at a given time or place

weight force the force of gravity pulling an object towards Earth

white blood cell a specialised cell in the blood that defends the body against disease

wind power electricity generated by harnessing the kinetic energy of wind to drive a turbine

zygote a diploid ($2n$) cell resulting from the joining of two haploid gametes

9780170491785

Index

A

9780170491785

D

9780170491785

F

G

9780170491785

H

M

9780170491785

P

Q

R

S

T

Additional credits

Chapter 1

- **Page 19: Table 1.6.2:** first column top to bottom: MARTYN F. CHILLMAID/Science Source, Natasa Re/Shutterstock.com, AjayTvm/Shutterstock.com, mllevphoto/Shutterstock.com.
- **Page 21: Figure 1.7.3:** top row left to right: Rabbitmindphoto/Shutterstock.com, Mehmet Cetin/Shutterstock.com; bottom row left to right: LightField Studios/Shutterstock.com, Rabbitmindphoto/Shutterstock.com.
- **Page 27: Figure 1.9.1:** left: Deemerwha studio/Shutterstock.com; right: paaruOK/Shutterstock.com.

Chapter 2

- **Page 55: Figure 2.2.6:** top row left to right: zhangyang13576997233/Shutterstock.com, SARIN KUNTHONG/Shutterstock.com; bottom row left to right: Wojmac/Shutterstock.com, Bjoern Wylezich/Shutterstock.com
- **Page 82: Unnumbered figure:** Phillip Wittke/Alamy Stock Photo.

Chapter 3

- **Page 93: Figure 3.2.5:** left to right: Rafael Ben-Ari/Getty Images, Nerthuz/Shutterstock, Creative Family/Shutterstock.
- **Page 97: Figure 3.3.3:** left to right: supachai sumrubsuk/Shutterstock.com, GSDesign/Shutterstock.com, DK samco/Shutterstock.com.
- **Page 97: 3.3 Learning Check:** left to right: AlexLMX/Shutterstock.com, Hodoimg/Shutterstock.com, Pixel-Shot/Shutterstock.com.
- **Page 114: Unnumbered figure:** left column: © Department of Climate Change, Energy, the Environment and Water 2023. Understand the Zoned Energy Rating Label, https://www.energyrating.gov.au/industry-information/understand-requirements/labelling/understand-zoned-energy-rating-label.
- **Page 115: Unnumbered figures:** top: ktsdesign/Shutterstock.com; bottom: Frame Stock Footage/Shutterstock.com.

Chapter 4

- **Page 122: Table 4.2.1:** first column top to bottom: k_samurkas/Shutterstock.com, cosma/Shutterstock.com, iStock/the-lightwrite.
- **Page 132: Figure 4.5.4:** PhET Interactive Simulations, University of Colorado Boulder, https://phet.colorado.edu.
- **Page 146: Figure 4.9.1:** National Energy Productivity Plan 2015–2030: Annual Report 2016', Commonwealth of Australia, 2016. Licensed under CC BY 4.0.
- **Page 148: Figure 4.9.3:** Australian Energy Update 2023, Commonwealth of Australia 2, licensed under CC BY 4.0. https://www.energy.gov.au/sites/default/files/Australian%20Energy%20Update%202023_0.pdf.
- **Page 150: Figure 4.9.5:** Australian Energy Update 2023, Commonwealth of Australia 2, licensed under CC BY 4.0. https://www.energy.gov.au/sites/default/files/Australian%20Energy%20Update%202023_0.pdf.
- **Page 153: Figure 4.10.4:** left: mastersky/Shutterstock.com; right: Jason Benz Bennee/Shutterstock.com.
- **Page 156: Unnumbered figure:** Freedom Life/Shutterstock.com.

Chapter 5

- **Page 189: Unnumbered figures:** left to right: iStock.com/Pheelings Media, Max4e Photo/Shutterstock.com, coka/Shutterstock.com.

Chapter 6

- **Page 200: Figure 6.3.4:** MacIntyre et al. (July 24, 2020). Which mask works best? We filmed people coughing and sneezing to find out, The Conversation. https://theconversation.com/which-mask-works-best-we-filmed-people-coughing-and-sneezing-to-find-out-143173. Licensed under CC BY-ND.
- **Page 211: Figure 6.7.6:** Courtesy: National Human Genome Research Institute, https://www.genome.gov/genetics-glossary/Lysosome.
- **Page 223: Figures 6.10.4 & 6.10.5:** Diabetes: Australian facts, Australian Institute of Health and Welfare, Licensed under CC BY 4.0. https://www.aihw.gov.au/reports/diabetes/diabetes/contents/how-common-is-diabetes/type-2-diabetes.
- **Page 231: Figure 6.13.1:** DVIDS/US Army Acquisition Support Center. The appearance of US Department of Defense (DoD) visual information does not imply or constitute DoD endorsement.
- **Page 234: Unnumbered figures:** top: Mario Tama/Getty Images; bottom: SA Health.

Chapter 7

- **Page 237: Figure 7.0.1:** left to right: Maxx-Studio/Shutterstock.com, Spok83/Shutterstock.com, Pixelwind/Shutterstock.com.
- **Page 239: Figure 7.1.2:** left to right: Pics by Bel/Alamy Stock Photo, Bloomberg/Getty Images, fotosr52/Adobe Stock.
- **Page 295: Unnumbered figure:** Alf Manciagli/Shutterstock.com.

Chapter 8

- **Page 306: Figure 8.4.1:** left to right: nevodka/Shutterstock.com, Mykhailo Baidala/Shutterstock.com, Olga_DigitalWork/Shutterstock.com.
- **Page 334: Unnumbered figure:** Greg Brave/Shutterstock.com.

Chapter 9

- **Page 344: Figure 9.2.3:** Wikimedia/LangeLeslie and Anna Wright CC BY-SA 4.0 (https://creativecommons.org/licenses/by-sa/4.0/deed.en).
- **Page 355: Figure 9.4.4:** Adapted from Indigenous Weather Knowledge, http://www.bom.gov.au/iwk/climate_zones/map_1.shtml, © Copyright Commonwealth of Australia 2025 Bureau of Meteorology. This map is licensed under the Creative Commons Attribution 4.0 International Licence, HYPERLINK "https://urldefense.com/v3/__https://creativecommons.org/licenses/by/4.0/__;!!MXVguWEtGgZw!Mhh8gorOVWrsMoZHBOdZuy39fRRwEOfSiJddYpaUNJdgcWS0l3xF5ttVq5GLqY6_0YmCC6tI4bo8kTGlwj3lYBh3OqvNuPs$"https://creativecommons.org/licenses/by/4.0/. Martyman, based on BoM data (CC BY-3.0 SA)/Bureau of Metrology.

- **Page 358: Figure 9.5.2:** NASA Earth Observatory images by Lauren Dauphin, using data from Curtis, P.G., et al. (2018), data from Goldman, Elizabeth, et al. (2020), and Landsat data from the U.S. Geological Survey.
- **Page 378: Figure 9.12.2:** R. Bliege Bird, N. Taylor, D. W. Bird, B. F. Codding, C. Taylor, F. Walsh.
- **Page 379: Figure 9.13.1:** NASA/Goddard Space Flight Center.
- **Page 379: Figure 9.13.2:** NASA Earth Observatory image by Lauren Dauphin, using VIIRS data from NASA EOSDIS/LANCE and GIBS/Worldview, and the Suomi National Polar-orbiting Partnership.
- **Page 383: Unnumbered figure:** Samantha Happe/Shutterstock.com.

Chapter 10

- **Page 405: Figure 10.6.2:** UNEP (2013) Metal Recycling: Opportunities, Limits, Infrastructure, A Report of the Working Group on the Global Metal Flows to the International Resource Panel. Reuter, M. A.; Hudson, C.; van Schaik, A.; Heiskanen, K.; Meskers, C.; Hagelüken, C.
- **Page 416: Unnumbered figure:** Alexandra Barbu/Dreamstime LLC.

Chapter 11

- **Page 420: Figure 11.1.1:** left: Donaldson Collection/Michael Ochs Archives/Getty Images; right: The Ava Helen and Linus Pauling Papers (MSS Pauling), Oregon State University Special Collections and Archives Research Center, Corvallis, Oregon.
- **Page 422: Figure 11.1.5:** Jerome Walker & Dennis Myts.
- **Page 430: Figure 11.5.1:** valda butterworth/Shutterstock.com, Hanneke Wetzer/Shutterstock.com.
- **Page 450: Figure 11.10.4:** Adapted from https://www.topperlearning.com/answer/mechanism-for-sexual-determination-in-honey-bees/gmp56lll.
- **Page 450: Figure 11.10.5:** Adapted from BOGOBiology, https://www.youtube.com/watch?v=sg7xe8C9DmQ.
- **Page 471: Figure 11.17.1:** Figure adapted from diagram of Gurruṯu skin names cycle, Chris Matthews, 'Indigenous perspectives in maths: Understanding Gurruṯu', *Teacher Magazine* April 27 2020 https://www.teachermagazine.com/au_en/articles/indigenous-perspectives-in-maths-understanding-gurruu.
- **Page 477: Unnumbered figure:** Istock.com/iStock Signature.

Chapter 12

- **Page 481: Figure 12.1.2:** Cornell, B. 2016. Random Assortment. [ONLINE] Available at: http://ib.bioninja.com.au. Acessed October 2022.
- **Page 482: Figure 12.2.1:** Adapted from https://quizlet.com/462177544/ex24-killing-by-uv-light-flash-cards/, © Pearson Education.
- **Page 483: Figure 12.2.2:** Adapted from https://le.ac.uk/vgec/topics/genemutations-and-cancer/school-and-colleges.
- **Page 485: Figure 12.2.4:** Adapted from https://www.pathwayz.org/Tree/Plain/CHROMOSOMAL+MUTATIONS.
- **Page 487: Figure 12.3.1:** Hug, L., Baker, B., Anantharaman, K. et al. A new view of the tree of life. Nat Microbiol 1, 16048 (2016). https://doi.org/10.1038/nmicrobiol.2016.48.
- **Page 488: Figure 12.3.4:** left: Hulton Archive/Getty Images; right: The Natural History Museum/Alamy Stock Photo.
- **Page 490: Figure 12.4.2:** Schwab Lukas/Prisma by Dukas Presseagentur GmbH/Alamy Stock Photo.
- **Page 496: Figure 12.6.1:** top row left to right:Taṣhh1601/Shutterstock.com, Photomaster/Shutterstock.com; second row: © Cengage; third row: Seksan wangjaisuk/Shutterstock.com
- **Page 506: Table 12.9.1:** first row: VPC Travel Photo/Alamy Stock Photo; second row: blickwinkel/Alamy Stock Photo; third row: olpo/Shutterstock.com; fourth row: © 2016, Masao et al. from eLife, New footprints from Laetoli (Tanzania) provide evidence for marked body size variation in early hominins (CC BY 4.0).
- **Page 514: Figure 12.11.6:** left: Oliver Thompson-Holmes/Alamy Stock Photo; right: Auscape International Pty Ltd/Alamy Stock Photo.
- **Page 524: Unnumbered figure:** TOON WELL/Shutterstock.com.

Chapter 13

- **Page 540: Figure 13.5.2:** Science Photo Library/Alamy Stock Photo.
- **Page 544: Figure 13.6.2:** first row left to right: Karol Kozlowski/Shutterstock.com, Vile_82/Shutterstock.com, vvoe/Shutterstock.com; second row left to right: Aksenenko Olga/Shutterstock.com, Lorena Fernandez/Shutterstock.com, Anastasia_Fisechko/Shutterstock.com.
- **Page 585: Unnumbered figure:** Pool/Getty Images.

Chapter 14

- **Page 628: Unnumbered figure:** Image Source Trading Ltd/Shutterstock.com.

Chapter 15

- **Page 646: Figure 15.5.1:** PhET Interactive Simulations, University of Colorado Boulder, https://phet.colorado.edu (CC BY 4.0)
- **Page 673: Unnumbered figure:** doomu/Shutterstock.com.

Chapter 16

- **Page 700: Table 16.6.1:** first row: alphaspirit.it/Shutterstock.com; second row: wavebreakmedia/Shutterstock.com.
- **Page 701: Table 16.6.2:** first row: demamiel62/Shutterstock.com; second row: Jim Vallee/Shutterstock.com; third row: Istock.com/iStock Essentials.
- **Page 732: Unnumbered figure:** NASA/JPL-Caltech.